T0270917

DISCRETE HARMONIC ANALYSIS

This self-contained book introduces readers to discrete harmonic analysis with an emphasis on the Discrete Fourier Transform and the Fast Fourier Transform on finite groups and finite fields, as well as their noncommutative versions. It also features applications to number theory, graph theory, and representation theory of finite groups. Beginning with elementary material on algebra and number theory, the book then delves into advanced topics from the frontiers of current research, including spectral analysis of the DFT, spectral graph theory and expanders, representation theory of finite groups and multiplicity-free triples, Tao's uncertainty principle for cyclic groups, harmonic analysis on $GL(2,F_q)$, and applications of the Heisenberg group to DFT and FFT. With numerous examples, figures, and more than 160 exercises to aid understanding, this book will be a valuable reference for graduate students and researchers in mathematics, engineering, and computer science.

Tullio Ceccherini-Silberstein is Professor of Mathematical Analysis at Università del Sannio, Benevento. He is also an editor of the EMS journal *Groups, Geometry, and Dynamics*. He has written more than 80 research articles on topics ranging from functional and harmonic analysis to group theory, ergodic theory and dynamical systems, and theoretical computer science. He has also coauthored four monographs and four proceedings volumes.

Fabio Scarabotti is Professor of Mathematical Analysis at Sapienza Università di Roma. He has written more than 40 research articles on subjects ranging from harmonic analysis to group theory, combinatorics, ergodic theory and dynamical systems, and theoretical computer science. He has also coauthored three monographs.

Filippo Tolli is Professor of Mathematical Analysis at Università Roma Tre. He has written more than 30 research articles ranging from harmonic analysis to group theory, combinatorics, Lie groups, and partial differential equations. He has also coauthored three monographs.

Already published
133 G. Malle & D. Testerman *Linear algebraic groups and finite groups of Lie type*
134 P. Li *Geometric analysis*
135 F. Maggi *Sets of finite perimeter and geometric variational problems*
136 M. Brodmann & R. Y. Sharp *Local cohomology (2nd Edition)*
137 C. Muscalu & W. Schlag *Classical and multilinear harmonic analysis, I*
138 C. Muscalu & W. Schlag *Classical and multilinear harmonic analysis, II*
139 B. Helffer *Spectral theory and its applications*
140 R. Pemantle & M. C. Wilson *Analytic combinatorics in several variables*
141 B. Branner & N. Fagella *Quasiconformal surgery in holomorphic dynamics*
142 R. M. Dudley *Uniform central limit theorems (2nd Edition)*
143 T. Leinster *Basic category theory*
144 I. Arzhantsev, U. Derenthal, J. Hausen & A. Laface *Cox rings*
145 M. Viana *Lectures on Lyapunov exponents*
146 J.-H. Evertse & K. Győry *Unit equations in Diophantine number theory*
147 A. Prasad *Representation theory*
148 S. R. Garcia, J. Mashreghi & W. T. Ross *Introduction to model spaces and their operators*
149 C. Godsil & K. Meagher *Erdős–Ko–Rado theorems: Algebraic approaches*
150 P. Mattila *Fourier analysis and Hausdorff dimension*
151 M. Viana & K. Oliveira *Foundations of ergodic theory*
152 V. I. Paulsen & M. Raghupathi *An introduction to the theory of reproducing kernel Hilbert spaces*
153 R. Beals & R. Wong *Special functions and orthogonal polynomials*
154 V. Jurdjevic *Optimal control and geometry: Integrable systems*
155 G. Pisier *Martingales in Banach spaces*
156 C. T. C. Wall *Differential topology*
157 J. C. Robinson, J. L. Rodrigo & W. Sadowski *The three-dimensional Navier–Stokes equations*
158 D. Huybrechts *Lectures on K3 surfaces*
159 H. Matsumoto & S. Taniguchi *Stochastic analysis*
160 A. Borodin & G. Olshanski *Representations of the infinite symmetric group*
161 P. Webb *Finite group representations for the pure mathematician*
162 C. J. Bishop & Y. Peres *Fractals in probability and analysis*
163 A. Bovier *Gaussian processes on trees*
164 P. Schneider *Galois representations and* (ϕ, Γ)-*modules*
165 P. Gille & T. Szamuely *Central simple algebras and Galois cohomology (2nd Edition)*
166 D. Li & H. Queffelec *Introduction to Banach spaces, I*
167 D. Li & H. Queffelec *Introduction to Banach spaces, II*
168 J. Carlson, S. Müller-Stach & C. Peters *Period mappings and period domains (2nd Edition)*
169 J. M. Landsberg *Geometry and complexity theory*
170 J. S. Milne *Algebraic groups*
171 J. Gough & J. Kupsch *Quantum fields and processes*
172 T. Ceccherini-Silberstein, F. Scarabotti & F. Tolli *Discrete harmonic analysis*

Discrete Harmonic Analysis
Representations, Number Theory, Expanders, and the Fourier Transform

TULLIO CECCHERINI-SILBERSTEIN
Università degli Studi del Sannio, Benevento, Italy

FABIO SCARABOTTI
Università degli Studi di Roma 'La Sapienza', Italy

FILIPPO TOLLI
Università degli Studi Roma Tre, Italy

CAMBRIDGE
UNIVERSITY PRESS

CAMBRIDGE
UNIVERSITY PRESS

University Printing House, Cambridge CB2 8BS, United Kingdom

One Liberty Plaza, 20th Floor, New York, NY 10006, USA

477 Williamstown Road, Port Melbourne, VIC 3207, Australia

314–321, 3rd Floor, Plot 3, Splendor Forum, Jasola District Centre, New Delhi - 110025, India

79 Anson Road, #06-04/06, Singapore 079906

Cambridge University Press is part of the University of Cambridge.

It furthers the University's mission by disseminating knowledge in the pursuit of
education, learning, and research at the highest international levels of excellence.

www.cambridge.org
Information on this title: www.cambridge.org/9781107182332
DOI: 10.1017/9781316856383

First published 2018

A catalogue record for this publication is available from the British Library.

Library of Congress Cataloging-in-Publication Data
Names: Ceccherini-Silberstein, Tullio, author. | Scarabotti, Fabio,
author. | Tolli, Filippo, 1968– author.
Title: Discrete harmonic analysis : representations, number theory,
expanders, and the fourier transform / Tullio Ceccherini-Silberstein
(Universitaa degli Studi del Sannio, Benevento, Italy), Fabio Scarabotti
(Universitaa degli Studi di Roma 'La Sapienza', Italy), Filippo Tolli
(Universitaa Roma Tre, Italy).
Other titles: Harmonic analysis
Description: Cambridge : Cambridge University Press, 2018. | Series: Cambridge studies
in advanced mathematics | Includes bibliographical references and index.
Identifiers: LCCN 2017057902 | ISBN 9781107182332 (hardback : alk. paper)
Subjects: LCSH: Harmonic analysis. | Fourier transformations. | Finite
groups. | Finite fields (Algebra)
Classification: LCC QA403 .C4285 2018 | DDC 515/.2433 – dc23 LC record available at
https://lccn.loc.gov/2017057902

ISBN 978-1-107-18233-2 Hardback

To the memory of my six (sic!) grandparents:
Tullio Levi-Civita, Padua 29.3.1873 – Rome 29.12.1941
Libera Trevisani Levi-Civita, Verona 17.5.1890 – Rome 11.12.1973
Riccardo Vittorio **Ceccherini** Stame, Rome 7.6.1903 – Rome 16.10.1991
Piera Paoletti Ceccherini, Florence 9.8.1896 – Rome 8.5.1973
Walter **Silberstein**, Vienna 14.7.1911 – Auschwitz 11.7.1944
Edith Hahn Silberstein, Vienna 7.2.1911 – Auschwitz 23.5.1944

To my parents, Cristina, Nadiya, and Virginia

To the memory of my father, and to my mother

Contents

Preface   *page* xi

Part I   Finite Abelian groups and the DFT

1   Finite Abelian groups   3
   1.1   Preliminaries in number theory   3
   1.2   Structure theory of finite Abelian groups: preliminary
       results   10
   1.3   Structure theory of finite Abelian groups: the theorems   18
   1.4   Generalities on endomorphisms and automorphisms of
       finite Abelian groups   26
   1.5   Endomorphisms and automorphisms of finite cyclic
       groups   30
   1.6   The endomorphism ring of a finite Abelian p-group   35
   1.7   The automorphisms of a finite Abelian p-group   40
   1.8   The cardinality of $\mathrm{Aut}(A)$ 42

2   The Fourier transform on finite Abelian groups   46
   2.1   Some notation   46
   2.2   Characters of finite cyclic groups   48
   2.3   Characters of finite Abelian groups   50
   2.4   The Fourier transform   53
   2.5   Poisson's formulas and the uncertainty principle   59
   2.6   Tao's uncertainty principle for cyclic groups   62

3   Dirichlet's theorem on primes in arithmetic progressions   74
   3.1   Analytic preliminaries   74
   3.2   Preliminaries on multiplicative characters   84
   3.3   Dirichlet L-functions   89

3.4	Euler's theorem	96
3.5	Dirichlet's theorem	99
4	**Spectral analysis of the DFT and number theory**	**101**
4.1	Preliminary results	101
4.2	The decomposition into eigenspaces	107
4.3	Applications: some classical results by Gauss and Schur	115
4.4	Quadratic reciprocity and Gauss sums	116
5	**The Fast Fourier Transform**	**129**
5.1	A preliminary example	129
5.2	Stride permutations	131
5.3	Permutation matrices and Kronecker products	139
5.4	The matrix form of the FFT	151
5.5	Algorithmic aspects of the FFT	161
	Part II Finite fields and their characters	
6	**Finite fields**	**167**
6.1	Preliminaries on ring theory	167
6.2	Finite algebraic extensions	171
6.3	The structure of finite fields	176
6.4	The Frobenius automorphism	177
6.5	Existence and uniqueness of Galois fields	178
6.6	Subfields and irreducible polynomials	183
6.7	Hilbert Satz 90	187
6.8	Quadratic extensions	192
7	**Character theory of finite fields**	**197**
7.1	Generalities on additive and multiplicative characters	197
7.2	Decomposable characters	201
7.3	Generalized Kloosterman sums	203
7.4	Gauss sums	210
7.5	The Hasse-Davenport identity	213
7.6	Jacobi sums	217
7.7	On the number of solutions of equations	222
7.8	The FFT over a finite field	227
	Part III Graphs and expanders	
8	**Graphs and their products**	**235**
8.1	Graphs and their adjacency matrix	235

8.2	Strongly regular graphs	241
8.3	Bipartite graphs	245
8.4	The complete graph	247
8.5	The hypercube	248
8.6	The discrete circle	250
8.7	Tensor products	252
8.8	Cartesian, tensor, and lexicographic products of graphs	258
8.9	Wreath product of finite graphs	265
8.10	Lamplighter graphs and their spectral analysis	268
8.11	The lamplighter on the complete graph	270
8.12	The replacement product	273
8.13	The zig-zag product	277
8.14	Cayley graphs, semidirect products, replacement products, and zig-zag products	279
9	**Expanders and Ramanujan graphs**	**283**
9.1	The Alon-Milman-Dodziuk theorem	284
9.2	The Alon-Boppana-Serre theorem	295
9.3	Nilli's proof of the Alon-Boppana-Serre theorem	300
9.4	Ramanujan graphs	307
9.5	Expander graphs	309
9.6	The Margulis example	311
9.7	The Alon-Schwartz-Shapira estimate	320
9.8	Estimates of the first nontrivial eigenvalue for the Zig-Zag product	327
9.9	Explicit construction of expanders via the Zig-Zag product	338
	Part IV Harmonic analysis on finite linear groups	
10	**Representation theory of finite groups**	**343**
10.1	Representations, irreducibility, and equivalence	343
10.2	Schur's lemma and the orthogonality relations	349
10.3	The group algebra and the Fourier transform	361
10.4	Group actions and permutation characters	372
10.5	Conjugate representations and tensor products	380
10.6	The commutant of a representation	390
10.7	A noncommutative FFT	397
11	**Induced representations and Mackey theory**	**399**
11.1	Induced representations	399
11.2	Frobenius reciprocity	409

11.3	Preliminaries on Mackey's theory	413
11.4	Mackey's formula for invariants	414
11.5	Mackey's lemma	419
11.6	The Mackey-Wigner little group method	421
11.7	Semidirect products with an Abelian group	424

12 Fourier analysis on finite affine groups and finite Heisenberg groups — 426

12.1	Representation theory of the affine group $\mathrm{Aff}(\mathbb{F}_q)$	426
12.2	Representation theory of the affine group $\mathrm{Aff}(\mathbb{Z}/n\mathbb{Z})$	432
12.3	Representation theory of the Heisenberg group $H_3(\mathbb{Z}/n\mathbb{Z})$	437
12.4	The DFT revisited	443
12.5	The FFT revisited	447
12.6	Representation theory of the Heisenberg group $H_3(\mathbb{F}_q)$	457

13 Hecke algebras and multiplicity-free triples — 460

13.1	Preliminaries and notation	460
13.2	Hecke algebras	462
13.3	Commutative Hecke algebras	466
13.4	Spherical functions: intrinsic theory	469
13.5	Harmonic analysis on the Hecke algebra $\mathcal{H}(G, K, \chi)$	474

14 Representation theory of $\mathrm{GL}(2, \mathbb{F}_q)$ — 482

14.1	Matrices associated with linear operators	482
14.2	Canonical forms for $\mathfrak{M}_2(\mathbb{F})$	484
14.3	The finite case	488
14.4	Representation theory of the Borel subgroup	492
14.5	Parabolic induction	494
14.6	Cuspidal representations	501
14.7	Whittaker models and Bessel functions	512
14.8	Gamma coefficients	522
14.9	Character theory of $\mathrm{GL}(2, \mathbb{F}_q)$	527
14.10	Induced representations from $\mathrm{GL}(2, \mathbb{F}_q)$ to $\mathrm{GL}(2, \mathbb{F}_{q^m})$	533
14.11	Decomposition of tensor products	540

Appendix Chebyshëv polynomials	543
Bibliography	555
Index	563

Preface

The aim of the present monograph is to introduce the reader to some central topics in discrete harmonic analysis, namely character theory of finite Abelian groups, (additive and multiplicative) character theory of finite fields, graphs and expanders, and representation theory of finite (possibly not Abelian) groups, including spherical functions, associated Fourier transforms, and spectral analysis of invariant operators. An important transversal topic, which is present in several sections of the book, is constituted by tensor products, which are developed for matrices, graphs, and representations.

We have written the book to be as self-contained as possible: it only requires some elementary notions in linear algebra (including the spectral theorem and its applications), abstract algebra (first rudiments in the theory of finite groups and rings), and elementary number theory.

First of all, we study in detail the structure of finite Abelian groups and their automorphisms. We then introduce the corresponding character theory leading to a complete analysis of the Fourier transform, focusing on the connections with number theory. For instance, we deduce Gauss law of quadratic reciprocity from the spectral analysis of the Discrete Fourier Transform. Actually, characters of finite Abelian groups will appear also, as a fundamental tool in the proof of several deep results, in subsequent chapters, constituting this way the central topic and common thread of the whole book.

We also present Dirichlet's theorem on primes in arithmetic progressions, which is based on the character theory of finite Abelian groups as well as Tao's uncertainty principle for (finite) cyclic groups [157].

Our treatment also includes an exposition of the Fast Fourier Transform, focusing on the theoretical aspects related to its expressions in terms of factorizations and tensor products. This part of the monograph is inspired, at least partially, by the important work of Auslander and Tolimieri [15] and the papers by Davio [50] and Rose [130]. The book by Stein and Shakarchi [150] has been a fundamental source for our treatment of Dirichlet's theorem as well as for the first section of the chapter on the Fast Fourier Transform.

The second part of the book constitutes a self-contained introduction to the basic algebraic theory of finite fields and their characters. This includes, on the one hand, a complete study of the automorphisms, norms, traces, and quadratic extensions of finite fields and, on the other hand, additive characters and multiplicative characters and several associated sums (trigonometric and Gaussian) and the Fast Fourier Transform over finite fields. One of the main goals is to present the generalized Kloosterman sums from Piatetski-Shapiro's monograph [123], which will play a fundamental role in Chapter 14 on the representation theory of $GL(2, \mathbb{F}_q)$. We also introduce the reader to the study, initiated by André Weil [165], of the number of solutions of equations over finite fields and present the Hasse-Davenport identity [48], which relates the Gauss sums over a finite field and those over a finite extension.

The third part is devoted to harmonic analysis on finite graphs and several constructions such as the replacement product and the zig-zag product. The central themes are expanders and Ramanujan graphs. We present the basic theorems of Alon-Milman and Dodziuk, and of Alon-Boppana-Serre, on the isoperimetric constant and the spectral gap of a (finite, undirected, connected) regular graph, and their connections. We discuss a few examples with explicit computations showing optimality of the bounds given by the above theorems. We then give the basic definitions of expanders and describe three fundamental constructions due to Margulis, to Alon, Schwartz, and Shapira (based on the replacement product), and to Reingold, Vadhan, and Wigderson (based on the zig-zag product). In these constructions, the harmonic analysis on finite Abelian groups and finite fields we developed in the previous parts plays a crucial role. The presentation is inspired by the monographs by Terras [159], Lubotzky [99], and by Davidoff-Sarnak-Valette [49], as well as by the papers by Hoory-Linial-Wigderson [74], Alon-Schwartz-Shapira [10], and Alon-Lubotzky-Wigderson [8].

The final part of the present monograph is devoted to the representation theory of finite groups with emphasis on induced representations and Mackey theory. This includes a complete description of the irreducible representations of the affine groups and Heisenberg groups with coefficients in both the finite field $\mathbb{F}_q$ and the ring $\mathbb{Z}/n\mathbb{Z}$. Moreover, both the Discrete Fourier Transform and the Fast Fourier Transform are revisited, following Auslander-Tolimieri [15] and Schulte [142], in terms of two different realizations of a particular representation of the Heisenberg group. In Chapter 13 we develop, with a complete and original treatment, the basic theory of multiplicity-free triples, their associated spherical functions, and (commutative) Hecke algebras. This is a subject that has not yet received the attention it deserves. As far as we know, this notion is just mentioned in some exercises in Macdonald's book [105]. The classical theory of finite Gelfand pairs, which constitutes a particular yet fundamental case, was essentially covered in our first monograph [29]. The exposition culminates with a complete treatment of the representation theory of GL(2, $\mathbb{F}_q$), along the lines developed by Piatetski-Shapiro [123]: our approach, via multiplicity-free triples, constitutes our original contribution to the theory.

All this said, one can use this monograph as a textbook for at least four different courses on:

(i) **Finite Abelian groups, the DFT, and the FFT** (the structure of finite Abelian groups, their character theory, and the Fourier transforms): Sections 1.1, 1.2, and 1.3, and Chapters 2, 4, and 5. The remaining sections in Chapter 1 as well as Chapter 3 are optional.

(ii) **Finite commutative harmonic analysis** (the structure of finite Abelian groups, their character theory, and the Fourier transforms; Dirichlet's theorem; finite fields and their characters): Sections 1.1, 1.2, and 1.3, and Chapters 2, 3, 4, 6, and 7.

(iii) **Graph theory** (a brief introduction to finite graphs, various notions of graph products, spectral theory, and expanders): Sections 1.1, 1.2, 1.3, 2.1, 2.2, 2.3, and 2.4, and Chapters 8 and 9 (omitting, if necessary, the parts involving character theory of finite fields).

(iv) **Finite harmonic analysis** (representation theory of finite groups: from the basics to GL(2, $\mathbb{F}_q$)): Sections 1.1, 1.2, and 1.3, Chapters 2, 4, and 6, Sections 7.1, 7.2, 7.3, and 7.4, and the whole of Part IV (Section 12.5, Chapter 13, and Sections 14.7 and 14.8 may be omitted).

We thank Alfredo Donno for interesting discussions as well as for helping us with some figures. We also express our deep gratitude to Sam Harrison, Kaitlin Leach, Clare Dennison, Adam Kratoska, and Mark Fox from Cambridge University Press as well as the project manager Vijay Kumar Bhatia and the copyeditor Sara Barnes, for their constant encouragement and most precious help at all stages of the editing process.

Roma, 31 July 2017   TCS, FS, and FT

Part I

Finite Abelian groups and the DFT

1

Finite Abelian groups

This chapter contains an elementary, self-contained, but quite complete exposition of the structure theory of finite Abelian groups, including a detailed account on their endomorphisms and automorphisms. We also provide all the necessary background in number theory (only basic prerequisites are assumed).

1.1 Preliminaries in number theory

In this section we review some basic facts on elementary number theory. Most of the proofs are elementary and often left as exercises. More details can be found in the monographs by Apostol [13], Davenport [47], Herstein [71], Ireland and Rosen [79], Mac Lane and Birkhoff [113], Nagell [117], and Nathanson [118].

We denote by $\mathbb{N} = \{0, 1, 2, \ldots\}$ the set of natural numbers, and we recall that, by Peano's axioms (see [113]), every non-empty subset $A \subseteq \mathbb{N}$ admits a (unique) minimal element.

Also, a basic tool in elementary number theory is the *division (Euclidean) algorithm* (long division): let $a, b \in \mathbb{Z}$ such that $b \geq 1$, then there exist unique $q, r \in \mathbb{Z}$ with $0 \leq r < b$ such that

$$a = bq + r. \tag{1.1}$$

If $r = 0$, one says that b *divides* a and we write $b|a$.

Theorem 1.1.1 (Definition of the greatest common divisor) *Let* $a, b \in \mathbb{Z}$ *with* $(a, b) \neq (0, 0)$. *Then there exists a unique positive integer d satisfying the following conditions:*

(i) *$d|a$ and $d|b$;*
(ii) *if $d'|a$ and $d'|b$, then $d'|d$.*

Moreover, there exist (not necessarily unique) $m_0, n_0 \in \mathbb{Z}$ such that (Bézout identity)

$$d = m_0 a + n_0 b. \tag{1.2}$$

Definition 1.1.2 The positive integer d as in the above statement is called the *greatest common divisor* of a and b and it is denoted by $\gcd(a, b)$.

Proof of Theorem 1.1.1 Suppose that d_1 and d_2 are two positive integers satisfying conditions (i) and (ii). Then, by (ii) we have $d_1 | d_2$ and $d_2 | d_1$. This forces $d_1 = \pm d_2$, and therefore $d_1 = d_2$ by positivity. This proves uniqueness. In order to show existence, consider the set

$$\mathcal{I} = \{ma + nb : m, n \in \mathbb{Z}\} \subseteq \mathbb{Z}.$$

Note that if $z, z' \in \mathcal{I}$ then $z + z' \in \mathcal{I}$ and $-z \in \mathcal{I}$. As a consequence, $\mathcal{I}_+ = \mathcal{I} \cap (\mathbb{N} \setminus \{0\})$ is a non-empty subset of $\mathbb{N}$. Let $d = m_0 a + n_0 b$ denote the minimal element of $\mathcal{I}_+$: we claim that $\mathcal{I} = \{hd : h \in \mathbb{Z}\}$. Indeed, the inclusion $\supseteq$ is obvious, while if $k \in \mathcal{I}$, by the division algorithm we can find $q, r \in \mathbb{Z}$ such that $k = qd + r$ with $0 \leq r < d$. Now, since $r = k - qd \in \mathcal{I}_+ \cup \{0\}$, by minimality of d we necessarily have $r = 0$, that is, $k \in \{hd : h \in \mathbb{Z}\}$. This shows the other inclusion and proves our claim. Since $a = a \cdot 1 + b \cdot 0$, $b = a \cdot 0 + b \cdot 1 \in \mathcal{I}$, there exist $h_1, h_2 \in \mathbb{Z}$ such that $a = h_1 d$ and $b = h_2 d$, so that $d | a$ and $d | b$. On the other hand, if $d' | a$ and $d' | b$, say $a = h'_1 d'$ and $b = h'_2 d'$, with $h'_1, h'_2 \in \mathbb{Z}$, then $d = m_0 a + n_0 b = m_0 h'_1 d' + n_0 h'_2 d' = (m_0 h'_1 + n_0 h'_2) d'$ so that $d' | d$. This shows that $d = \gcd(a, b)$. $\qquad\square$

Remark 1.1.3 The set $\mathcal{I}$ is an *ideal* in the ring $\mathbb{Z}$, and $\mathbb{Z}$ is a *principal ideal domain* (see Section 6.1).

From the proof of Theorem 1.1.1 we immediately deduce the following:

Corollary 1.1.4 *Given $a, b, c \in \mathbb{Z}$ with $(a, b) \neq (0, 0)$, the linear equation*

$$na + mb = c$$

has a solution $(n, m) \in \mathbb{Z}^2$ if and only if $\gcd(a, b)$ divides c.

(See also Proposition 1.2.13 below.)

Exercise 1.1.5 Let $a_1, a_2, \ldots, a_n \in \mathbb{Z}$ with $(a_1, a_2, \ldots, a_n) \neq (0, 0, \ldots, 0)$.

   (1) Show that there exists a unique positive integer d satisfying the following conditions:

      (i) $d | a_i$ for all $i = 1, 2, \ldots, n$;

      (ii) if $d' | a_i$ for all $i = 1, 2, \ldots, n$, then $d' | d$.

In particular, setting $d_2 = \gcd(a_1, a_2)$ and $d_i = \gcd(d_{i-1}, a_i)$ for $i \geq 3$, show that $d = d_n$;

(2) show that there exist $m_i \in \mathbb{Z}$, $i = 1, 2, \ldots, n$, such that (generalized Bézout identity) $d = m_1 a_1 + m_2 a_2 + \ldots + m_n a_n$.

Definition 1.1.6 Let $a_1, a_2, \ldots, a_n \in \mathbb{Z}$ with $(a_1, a_2, \ldots, a_n) \neq (0, 0, \ldots, 0)$. The number d in Exercise 1.1.5 (1) is called the *greatest common divisor* of the a_is and it is denoted by $\gcd(a_1, a_2, \ldots, a_n)$. One says that $a_1, a_2, \ldots, a_n \in \mathbb{Z}$ are *relatively prime* provided $\gcd(a_1, a_2, \ldots, a_n) = 1$.

An integer $p > 1$ is said to be *prime* if its positive divisors are exactly 1 and p.

Exercise 1.1.7 (Euclidean algorithm) Let $a, b \in \mathbb{N}$ and suppose that $b \geq 1$ and $b \nmid a$. Set $r_0 = a$, $r_1 = b$, and recursively define, by the division algorithm,

$$r_k = r_{k+1} q_{k+1} + r_{k+2}$$

where $0 \leq r_{k+2} < r_{k+1}$, for all $k \geq 0$. Show that $\gcd(a, b) = r_n$ where $n \in \mathbb{N}$ is the largest index for which $r_n > 0$ (so that $r_{n+1} = 0$).

Exercise 1.1.8 Let $a, b, c \in \mathbb{Z}$ and p a prime number.

(1) Prove that if $\gcd(a, b) = 1$ and $a | bc$ then $a | c$;
(2) deduce that if $p | bc$ then $p | b$ or $p | c$.

Exercise 1.1.9 (Fundamental theorem of arithmetic) Let $n \geq 2$ be an integer. Show that there exists a unique *prime factorization*

$$n = p_1^{m_1} p_2^{m_2} \cdots p_h^{m_h}$$

where $p_1 < p_2 < \cdots < p_h$ are prime numbers, $m_1, m_2, \ldots, m_h \geq 1$ are the *multiplicities*, and $h \geq 1$.

Hint. For uniqueness, use induction combined with Exercise 1.1.8.

Exercise 1.1.10 Let $a_1, a_2, \ldots, a_n \geq 2$ be integers. Suppose that

$$a_j = p_1^{m_{1j}} p_2^{m_{2j}} \cdots p_h^{m_{hj}}$$

with distinct primes p_i and multiplicities $m_{ij} \geq 0$, for all $i = 1, 2, \ldots, h$ and $j = 1, 2, \ldots, n$. Show that

$$\gcd(a_1, a_2, \ldots, a_n) = p_1^{m_1} p_2^{m_2} \cdots p_h^{m_h}$$

where $m_i = \min\{m_{ij} : j = 1, 2, \ldots, n\}$ for all $i = 1, 2, \ldots, h$.

Exercise 1.1.11 (Euclid's proof of the infinitude of primes)

(1) Let $p_1, p_2, \ldots, p_n$, $n \geq 1$, be distinct primes. Show that the number $p_1 p_2 \cdots p_n + 1$ is not divisible by p_i for all $i = 1, 2, \ldots, n$;
(2) deduce that the set of prime numbers is infinite.

There are many other proofs of the infinitude of primes. Six of them (including Euclid's proof) are in the book by Aigner and Ziegler [5]. A deep generalization of this fact will be presented in Chapter 3.

Definition 1.1.12 Let $n \geq 1$ and $a, b \in \mathbb{Z}$. One says that a is *congruent* to b *modulo* n, and one writes $a \equiv b \bmod n$, provided $n | (a - b)$.

Exercise 1.1.13 Let $n \geq 1$.

(1) Show that the congruence relation $\equiv \bmod n$ is an equivalence relation;
(2) suppose that $a = nq + r$, with $0 \leq r < n$. Show that $a \equiv r \bmod n$;
(3) deduce that there are exactly n equivalence classes and that a complete list of representatives is provided by $0, 1, \ldots, n - 1$.

For $n \geq 1$ and $a \in \mathbb{Z}$ we denote by

$$\bar{a} = \{a + hn : h \in \mathbb{Z}\} \tag{1.3}$$

the equivalence class containing a.

We denote by $\mathbb{Z}/n\mathbb{Z} = \{\bar{a} : a \in \mathbb{Z}\} = \{\bar{0}, \bar{1}, \ldots, \overline{n-1}\}$ the corresponding quotient set.

Exercise 1.1.14 Let $n \geq 1$ and $a, b \in \mathbb{Z}$. Set

$$\bar{a} + \bar{b} = \overline{a + b} \quad \text{and} \quad \bar{a} \cdot \bar{b} = \overline{ab}. \tag{1.4}$$

(1) Show that the operations $+$ and $\cdot$ in (1.4) are well defined;
(2) show that $(\mathbb{Z}/n\mathbb{Z}, +)$ is a cyclic group;
(3) show that $(\mathbb{Z}/n\mathbb{Z}, +, \cdot)$ is a unital commutative ring;
(4) show that $\bar{a}$ is invertible in $(\mathbb{Z}/n\mathbb{Z}, +, \cdot)$ if and only if $\gcd(a, n) = 1$;
(5) deduce that if p is a prime, then $(\mathbb{Z}/p\mathbb{Z}, +, \cdot)$ is a field.

For (5), see also Corollary 6.1.13.

Notation 1.1.15 Let $n \geq 1$. For $k, m \in \mathbb{Z}$ we write

$$k\bar{m} = \bar{m} + \bar{m} + \cdots + \bar{m} \ (k \text{ summands})$$

if $k \geq 0$, and $k\bar{m} = -(|k|\bar{m})$ if $k < 0$, where $\bar{m}$ is as in (1.3).

The notation above is consistent with the fact that $(\mathbb{Z}/n\mathbb{Z}, +)$, as any Abelian group, is a $\mathbb{Z}$-module; see the monographs by Herstein [71], Lang [93], and Knapp [87].

Lemma 1.1.16 *Let r and s be positive integers with $\gcd(r, s) = 1$. Then for every $0 \le k \le rs - 1$ there exist unique $0 \le u \le r - 1$ and $0 \le v \le s - 1$ such that*

$$k \equiv us + vr \quad \mod rs. \tag{1.5}$$

Proof. As u and v vary, with $0 \le u \le r - 1$ and $0 \le v \le s - 1$, the expression $us + vr$ yields (at most) rs integers; therefore it suffices to show that these are all distinct mod rs. Indeed, for $0 \le u, u' \le r - 1$ and $0 \le v, v' \le s - 1$ we have (keeping in mind that $\gcd(r, s) = 1$):

$$us + vr \equiv u's + v'r \quad \mod rs \implies (u - u')s + (v - v')r \equiv 0 \quad \mod rs$$

$$(\text{by Exercise } 1.1.8.(1)) \implies \begin{cases} u \equiv u' \mod r \\ v \equiv v' \mod s \end{cases}$$

$$\implies u = u' \text{ and } v = v'. \qquad \square$$

Notation 1.1.17 For $n \ge 1$ we denote by

- $\mathbb{Z}_n$ the *additive* group $(\mathbb{Z}/n\mathbb{Z}, +)$ *of integers* mod n;
- C_n the *multiplicative* cyclic group of order n;
- $\mathbb{Z}/n\mathbb{Z}$ the *ring* $(\mathbb{Z}/n\mathbb{Z}, +, \cdot)$ *of integers* mod n.

When $n = p$ is a prime, we shall denote by $\mathbb{F}_p$ the finite field $\mathbb{Z}/p\mathbb{Z}$ (cf. Exercise 1.1.14.(5)).

Note that if C_n is generated by the element $a \in C_n$, then the map $\bar{k} \mapsto a^k$, for all $k \in \mathbb{Z}$, is well defined and establishes a natural group isomorphism of $\mathbb{Z}_n$ onto C_n.

We shall examine the structure of all finite fields in Section 6.3.

Definition 1.1.18 The *Euler totient function* is the map φ defined by

$$\varphi(n) = |\{m \in \mathbb{N} : 1 \le m \le n, \gcd(m, n) = 1\}|$$

for all $n \ge 1$, where $|\cdot|$ denotes cardinality. In words, the value $\varphi(n)$ equals the number of positive integers less than or equal to n that are relatively prime to n.

Proposition 1.1.19 *Let n be a positive integer. Then in the cyclic group $\mathbb{Z}_n$ there are exactly $\varphi(n)$ distinct generators.*

Proof. Let $1 \le m \le n - 1$ and suppose that $\gcd(m, n) = 1$. By Bézout identity, we can find $a, b \in \mathbb{Z}$ such that $am + bn = 1$. Let $1 \le h \le n - 1$ be such that $\bar{h} = \bar{a}$. Then, in $\mathbb{Z}_n$ we have $\bar{m} + \bar{m} + \cdots + \bar{m} = h\bar{m} = \overline{am} = \bar{1}$. As $\bar{1}$ clearly generates $\mathbb{Z}_n$, this shows that $\bar{m}$ generates $\mathbb{Z}_n$ as well. On the other hand, if $\gcd(m, n) = q > 1$, then we can find $h, k \in \mathbb{N}$ such that $m = hq$ and $n = kq$. Note that $1 \le k < n$. Then we have $k\bar{m} = \overline{km} = \overline{khq} = h\bar{n} = \bar{0}$ so that the (cyclic) subgroup generated by $\bar{m}$ in $\mathbb{Z}_n$ has order $\le k$ and therefore cannot equal the whole $\mathbb{Z}_n$. This shows that $\bar{m}$ is not a generator of $\mathbb{Z}_n$.

The statement then follows from the definition of $\varphi(n)$.   □

Proposition 1.1.20 (Gauss) *Let n be a positive integer. Then we have*

$$\sum_{\substack{1 \le r \le n \\ r \mid n}} \varphi(r) = n.$$

Proof. For every positive divisor r of n let us set

$$A(r) := \{k \in \mathbb{N} : 1 \le k \le n, \gcd(k, n) = n/r\}. \tag{1.6}$$

For $1 \le k \le n$ we clearly have $k \in A(r)$ with $r = n/\gcd(k, n)$, and such an r is unique, so that

$$\{1, 2, \ldots, n\} = \coprod_{\substack{1 \le r \le n \\ r \mid n}} A(r). \tag{1.7}$$

Now, for every $k \in A(r)$ there exists a unique positive integer j such that $k = j\frac{n}{r}$. It follows that $1 \le j \le r$ and

$$\frac{n}{r} = \gcd(k, n) = \gcd\left(j\frac{n}{r}, r\frac{n}{r}\right) = \frac{n}{r}\gcd(j, r)$$

so that $\gcd(j, r) = 1$. Conversely, if $r \mid n$ and $\gcd(j, r) = 1$, then $\gcd(j\frac{n}{r}, n) = \gcd(j\frac{n}{r}, r\frac{n}{r}) = \frac{n}{r}$. As a consequence, $A(r) = \{j\frac{n}{r} : \gcd(j, r) = 1\}$ so that

$$|A(r)| = \varphi(r) \tag{1.8}$$

and therefore, from (1.7) we deduce

$$n = \sum_{\substack{1 \le r \le n \\ r \mid n}} |A(r)| = \sum_{\substack{1 \le r \le n \\ r \mid n}} \varphi(r).$$

                                                                                    □

Theorem 1.1.21 *Let p be a prime number. The (multiplicative) group $\mathbb{F}_p^*$ of invertible elements in the field $\mathbb{F}_p$ is cyclic (of order $p - 1$).*

Proof. We first observe that $|\mathbb{F}_p^*| = |\{\overline{1}, \overline{2}, \ldots, \overline{p-1}\}| = p - 1$.

For every positive divisor r of $p - 1$ let us set

$$B(r) := \{\alpha \in \mathbb{F}_p^* : \alpha \text{ is of order } r\}.$$

Thus, if $\alpha \in B(r)$, we have $\alpha^r = 1$ and α generates a cyclic group $\langle \alpha \rangle$ of order r consisting exactly of all the solutions in $\mathbb{F}_p$ of the equation $x^r = 1$. That is, $B(r) \subseteq \langle \alpha \rangle$ (recall also that over any field, an equation of degree m has at most m solutions). By virtue of Proposition 1.1.19, $\langle \alpha \rangle$ has $\varphi(r)$ generators, namely the powers α^h with $1 \leq h \leq r$ and $\gcd(h, r) = 1$. As a consequence, if $B(r) \neq \varnothing$ we have $|B(r)| = \varphi(r)$. Therefore

$$p - 1 = |F_p^*| = \sum_{r|(p-1)} |B(r)| \leq \sum_{r|(p-1)} \varphi(r) = p - 1,$$

where the last equality follows from Proposition 1.1.20. Since the above is indeed an equality, we deduce that $B(r) \neq \varnothing$ for every r which divides $p - 1$. In particular, every element $\alpha \in B(p - 1)$ is of order $p - 1$ and therefore $\langle \alpha \rangle = \mathbb{F}_p^*$. $\qquad\square$

Exercise 1.1.22 (Fermat's little theorem) Show that if p is a prime, then for all $n \in \mathbb{Z}$ we have $n^p \equiv n \bmod p$ so that, if in addition $p \nmid n$, then $n^{p-1} \equiv 1 \bmod p$.

We end this section with the following well-known results (see also Remark 5.2.15), which we deduce from Theorem 1.1.1.

Corollary 1.1.23 (Chinese remainder theorem I) *Let r, s be two positive integers such that $\gcd(r, s) = 1$. Then for all $(a, b) \in \mathbb{Z}$ there exists $x = x(a, b) \in \mathbb{Z}$ solution to the system*

$$\begin{cases} x \equiv a \quad \bmod r \\ x \equiv b \quad \bmod s. \end{cases} \tag{1.9}$$

Proof. By Bézout identity, we can find $u, v \in \mathbb{Z}$ such that $1 = ur + vs$. We leave it to the reader to check that the quantities $a + (b - a)ur$ and $b + (a - b)vs$ are equal and constitute a solution to (1.9). $\qquad\square$

Exercise 1.1.24 With the notation from Corollary 1.1.23, set $\delta_1 = x(1, 0)$ and $\delta_2 = x(0, 1)$. Show that $x(a, b) = a\delta_1 + b\delta_2$.

Exercise 1.1.25 (Chinese remainder theorem II) Let $r_1, r_2, \ldots, r_n$ be positive integers such that $\gcd(r_i, r_j) = 1$ for all $1 \leq i < j \leq n$.

(a) Show that for all $(a_1, a_2, \ldots, a_n) \in \mathbb{Z}^n$ there exists a solution $x = x(a_1, a_2, \ldots, a_n) \in \mathbb{Z}$ of the system

$$\begin{cases} x \equiv a_1 & \mod r_1 \\ x \equiv a_2 & \mod r_2 \\ \cdots & \cdots \\ x \equiv a_n & \mod r_n; \end{cases} \tag{1.10}$$

(b) set $R = r_1 r_2 \cdots r_n$. Show that $y \in \mathbb{Z}$ is another solution to (1.10) if and only if $x \equiv y \mod R$.

Hint. For every $i = 1, 2, \ldots, n$ denote by $\delta_i \in \mathbb{Z}$ a solution to (1.9) with $a = 1, b = 0, r = r_i$, and $s = R/r_i$. Show that δ_i is a solution to (1.9) with $a = 1, b = 0, r = r_i$, and $s = r_j$, for all $j \neq i$. Then show that $x(a_1, a_2, \ldots, a_n) = a_1 \delta_1 + a_2 \delta_2 + \cdots + a_n \delta_n$.

Proposition 1.1.26 *Let $n \geq 1$, $m \in \mathbb{Z}$, and set $d = \gcd(m, n)$. Then, in the cyclic group $\mathbb{Z}_n$ the element $\overline{m}$ has order $\frac{n}{d}$.*

Proof. For $k \in \mathbb{Z}$ we have

$$\begin{aligned} km \equiv 0 \quad \mod n \quad &\Leftrightarrow n \mid km \\ &\Leftrightarrow \frac{n}{d} \mid k\frac{m}{d} \\ &\Leftrightarrow \frac{n}{d} \mid k, \end{aligned}$$

since $\frac{n}{d}$ and $\frac{m}{d}$ are relatively prime. $\square$

Exercise 1.1.27 Deduce Proposition 1.1.19 from Proposition 1.1.26.

1.2 Structure theory of finite Abelian groups: preliminary results

In this section we review some basic facts on finite Abelian groups and their structure. Our exposition is based on the following monographs: by Machì [102], Zappa [170], Kurzweil and Stellmacher [90], Kurosh [89], Rotman [132], Herstein [71], Nathanson [118], and on the papers [18, 72, 120].

We use additive notation. In particular, for $a \in \mathbb{Z}_n$ and $r \in \mathbb{N}$ we set $ra = a + a + \cdots + a$ (r summands). Moreover, for an element a (respectively a subset B) of an Abelian group A, we denote by $\langle a \rangle = \{ra : r \in \mathbb{N}\}$ (respectively $\langle B \rangle$) the subgroup of A generated by a (respectively B) and by $o(a) = |\langle a \rangle| \in \mathbb{N} \cup \{\infty\}$ the *order* of a.

Let A be a finite Abelian group and let $A_1, A_2, \ldots, A_k \leq A$, $k \geq 1$ be subgroups of A.

Definition 1.2.1 The *sum* of the subgroups $A_1, A_2, \ldots, A_k$ is the subgroup

$$B = A_1 + A_2 + \cdots + A_k \tag{1.11}$$

formed by all elements $a \in A$ which can be expressed as

$$a = a_1 + a_2 + \cdots + a_k \tag{1.12}$$

with $a_j \in A_j$, $j = 1, 2, \ldots, k$.

One says that the subgroup B in (1.11) is an *(internal) direct sum*, and we write

$$B = A_1 \oplus A_2 \oplus \cdots \oplus A_k, \tag{1.13}$$

provided that the expression (1.12) is unique for every $a \in B$.

Proposition 1.2.2 *The following conditions are equivalent for $B = A_1 + A_2 + \cdots + A_k$:*

(i) *B is a direct sum;*
(ii) *if $a_1 + a_2 + \cdots + a_k = 0$ with $a_j \in A_j$, $j = 1, 2, \ldots, k$, then $a_1 = a_2 = \cdots = a_k = 0$;*
(iii) *$(A_1 + A_2 + \cdots + A_{j-1} + A_{j+1} + \cdots + A_k) \cap A_j = \{0\}$ for all $j = 1, 2, \ldots, k$;*
(iv) *$|B| = |A_1| \cdot |A_2| \cdot \ldots \cdot |A_k|$.*

Moreover, if one of the above conditions holds and

$$A_j = B_{j,1} \oplus B_{j,2} \oplus \cdots \oplus B_{j,h_j},$$

where the $B_{j,i}$s are subgroups and $h_j \geq 1$, for all $j = 1, 2, \ldots, k$, then

$$B = \bigoplus_{j=1}^{k} \bigoplus_{i=1}^{h_j} B_{j,i}.$$

Proof. We leave it as an easy exercise. $\qquad\square$

Let now $B_1, B_2, \ldots, B_k$ be Abelian groups.

Definition 1.2.3 The *(external) direct sum* of the groups $B_1, B_2, \ldots, B_k$, denoted

$$B_1 \oplus B_2 \oplus \cdots \oplus B_k, \tag{1.14}$$

is the Cartesian product $B_1 \times B_2 \times \cdots \times B_k$ endowed with the group operation

$$(b_1, b_2, \ldots, b_k) + (b_1', b_2', \ldots, b_k') = (b_1 + b_1', b_2 + b_2', \ldots, b_k + b_k')$$

for all $b_i, b_i' \in B_i$, $i = 1, 2, \ldots, k$.

Note that

$$|B_1 \oplus B_2 \oplus \cdots \oplus B_k| = |B_1| \cdot |B_2| \cdot \ldots \cdot |B_k|. \qquad (1.15)$$

The notions of internal and external direct sum are strictly correlated:

Proposition 1.2.4

(i) *Let $B = B_1 \oplus B_2 \oplus \cdots \oplus B_k$ be an external direct sum. For every $j = 1, 2, \ldots, k$, denote by A_j the subgroup, isomorphic to B_j, consisting of all elements of B of the form $(0, 0, \ldots, 0, a_j, 0, \ldots, 0)$ with $a_j \in B_j$ in the jth coordinate. Then,*

$$B = A_1 \oplus A_2 \oplus \cdots \oplus A_k$$

as an internal direct sum;

(ii) *the internal direct sum (1.11) is isomorphic to the external direct sum of the groups $A_1, A_2, \ldots, A_k$.*

Proof. We leave it as an easy exercise. □

As a consequence, in the sequel, if $B \cong B_1 \oplus B_2 \oplus \cdots \oplus B_k$, by abuse of language we shall regard the groups B_j, $j = 1, 2, \ldots, k$, as subgroups of the Abelian group B.

We now focus on some basic results on cyclic groups and their structure.

Proposition 1.2.5 *Let r, s be two positive integers satisfying $\gcd(r, s) = 1$. Then if $n = rs$ we have*

$$\mathbb{Z}_n \cong \mathbb{Z}_r \oplus \mathbb{Z}_s.$$

Proof. Let a be a generator of $\mathbb{Z}_n$ and set $b = ra$ and $c = sa$. Since $sb = sra = na = 0$ and $kb = kra \neq 0$ for $0 \leq k < s$, we have that $o(b) = s$ and, similarly, $o(c) = r$. Moreover,

$$\langle b \rangle \cap \langle c \rangle = 0.$$

Indeed, if $kb = hc$ with $0 \leq k < s$ and $0 \leq h < r$ then

$$kra = hsa$$

with $0 \leq kr, hs < n$, which implies that $kr = hs$. Since $\gcd(r, s) = 1$ we necessarily have $s|k$ and $r|h$ (see Exercise 1.1.8.(1)) and this forces $h = k = 0$. Finally, by Bézout identity (cf. Theorem 1.1.1), there exist $u, v \in \mathbb{Z}$ such that $ru + sv = 1$ so that

$$a = 1a = ura + vsa = ub + vc.$$

This implies that $\mathbb{Z}_n = \langle b \rangle \oplus \langle c \rangle \cong \mathbb{Z}_r \oplus \mathbb{Z}_s$. $\qquad\qquad\square$

Definition 1.2.6 An Abelian group is termed *indecomposable* if it cannot be written as a direct sum of two or more nontrivial subgroups.

A *p-primary cyclic* group is a cyclic group of order a nontrivial power of a prime p.

From Proposition 1.2.5 we deduce:

Corollary 1.2.7 (Chinese remainder theorem III) *Let* $n = p_1^{k_1} p_2^{k_2} \cdots p_t^{k_t}$ *be the prime factorization of an integer $n \geq 2$. Then*

$$\mathbb{Z}_n \cong \mathbb{Z}_{p_1^{k_1}} \oplus \mathbb{Z}_{p_2^{k_2}} \oplus \cdots \oplus \mathbb{Z}_{p_t^{k_t}}. \tag{1.16}$$

That is, every cyclic group may be written as a direct sum of p-primary cyclic groups corresponding to distinct primes p.

Exercise 1.2.8 Show that the Chinese remainder theorem III (Corollary 1.2.7) is equivalent to the Chinese remainder theorem II (Exercise 1.1.25).

Corollary 1.2.9 *Let m and n be two positive integers and suppose that m divides n. Then $\mathbb{Z}_n$ contains an element of order m.*

Proof. Let $n = p_1^{k_1} p_2^{k_2} \cdots p_t^{k_t}$ be the prime factorization of n. Then we can write $m = p_1^{h_1} p_2^{h_2} \cdots p_t^{h_t}$ with $0 \leq h_i \leq k_i, i = 1, 2, \ldots, t$. In the notation of Corollary 1.2.7, let $a_1, a_2, \ldots, a_t$ be the generators of the primary cyclic subgroups in (1.16). We claim that the element

$$z = p_1^{k_1 - h_1} a_1 + p_2^{k_2 - h_2} a_2 + \cdots + p_t^{k_t - h_t} a_t$$

has order m. Indeed,

$$mz = \frac{m}{p_1^{h_1}} p_1^{k_1} a_1 + \frac{m}{p_2^{h_2}} p_2^{k_2} a_2 + \cdots + \frac{m}{p_t^{h_t}} p_t^{k_t} a_t = 0$$

and if $m' | m$ and $m' < m$, say $m' = p_1^{h_1'} p_2^{h_2'} \cdots p_t^{h_t'}$ (with $0 \leq h_i' \leq h_i$, for all $i = 1, 2, \ldots, t$, and there exists $1 \leq j \leq t$ such that $h_j' < h_j$) then

$$m'z = \frac{m'}{p_1^{h_1'}} p_1^{k_1 - h_1 + h_1'} a_1 + \frac{m'}{p_2^{h_2'}} p_2^{k_2 - h_2 + h_2'} a_2 + \cdots + \frac{m'}{p_t^{h_t'}} p_t^{k_t - h_t + h_t'} a_t \neq 0$$

since

$$\frac{m'}{p_j^{h_j'}} p_j^{k_j - h_j + h_j'} a_j \neq 0.$$

This proves the claim and the corollary. $\qquad\qquad\square$

Proposition 1.2.10 *Let p be a prime number and let a be a generator of the p-primary cyclic group $\mathbb{Z}_{p^k}$. Then every nontrivial subgroup of $\mathbb{Z}_{p^k}$ contains the element $p^{k-1}a$. In particular, $\mathbb{Z}_{p^k}$ is indecomposable.*

Proof. Let $x \in \mathbb{Z}_{p^k}$ be any nontrivial element. Then we can find $0 < s < p^k$ such that $x = sa$. We may decompose s in the form $s = p^h r$, with $0 \le h < k$ and $r \in \mathbb{N}$ such that $\gcd(p, r) = 1$. Then we can find $u, v \in \mathbb{Z}$ such that $ru + pv = 1$ so that

$$
\begin{aligned}
(p^{k-h-1}u)x &= p^{k-h-1}usa \\
&= p^{k-1}ura \\
&= p^{k-1}(1 - pv)a \\
&= p^{k-1}a
\end{aligned}
$$

that is, $p^{k-1}a \in \langle x \rangle$. This shows that every nontrivial subgroup of $\mathbb{Z}_{p^k}$ contains $p^{k-1}a$.

The last statement then follows from Proposition 1.2.2.(iii). $\qquad\square$

Corollary 1.2.11 *For every $n \ge 2$, the cyclic group $\mathbb{Z}_n$ has a unique decomposition as a direct sum of p-primary cyclic groups and it is given by* (1.16).

Proposition 1.2.12 *Let $n \ge 1$, and let a be a generator of the cyclic group $\mathbb{Z}_n$. Then every subgroup A of $\mathbb{Z}_n$ is cyclic and $A = \langle \frac{n}{m}a \rangle$ where $m = o(A)$. Conversely, for every divisor m of n there exists a unique subgroup $A_m \le \mathbb{Z}_n$ of order m.*

Proof. Let A be a non trivial subgroup of $\mathbb{Z}_n$. Set

$$
h = \min\{k \in \mathbb{N} : ka \in A\}
$$

and let us show that $A = \langle ha \rangle$. Indeed, if $sa \in A$, then, by the division algorithm, there exist $q \in \mathbb{N}$ and $0 \le r < h$ such that $s = qh + r$ so that

$$
ra = sa - qha \in A
$$

forcing $r = 0$ and $sa = qha \in \langle ha \rangle$.

On the other hand, if m divides n, then $o(\frac{n}{m}a) = m$. Indeed, $m\frac{n}{m}a = na = 0$, while if $0 < r < m$ then $r\frac{n}{m} < n$ so that $(r\frac{n}{m})a = r(\frac{n}{m}a) \ne 0$. This shows that $A_m = \langle \frac{n}{m}a \rangle$ (uniqueness follows from the first part). $\qquad\square$

Proposition 1.2.13 *Let $n \ge 1$, $a, b \in \mathbb{Z}$, and set $d = \gcd(a, n)$. Then the linear congruence*

$$
ma \equiv b \quad \mathrm{mod}\ n \tag{1.17}
$$

has a solution $m \geq 1$ if and only if

$$b \equiv 0 \quad \bmod d.$$

If this is the case, (1.17) has d distinct pairwise non-congruent solutions.

Proof. We have $ma \equiv b \bmod n$ if and only if there exists $k \in \mathbb{Z}$ such that $ma = b + kn$, that is, $b = ma - kn$. By Corollary 1.1.4, this last equation admits a solution $(m, k) \in \mathbb{Z}^2$ if and only if d divides b. By Proposition 1.1.26 the linear congruence

$$ha \equiv 0 \quad \bmod n$$

has exactly d non-congruent solutions, namely $h = \frac{n}{d}, 2\frac{n}{d}, \ldots, (d-1)\frac{n}{d}, n$. If $b \equiv 0 \bmod d$ and m_0 is a fixed solution of (1.17), then a complete list of pairwise non-congruent solutions of (1.17) is given by

$$m = m_0, m_0 + \frac{n}{d}, m_0 + 2\frac{n}{d}, \ldots, m_0 + (d-1)\frac{n}{d}. \qquad \square$$

Remark 1.2.14 We write Proposition 1.2.13 in a more abstract form by using multiplicative notation. Let $n \geq 1$, recall that C_n denotes the multiplicative cyclic group, and let $x \in C_n$ be a generator. Let $a \in \mathbb{Z}$ and set $d = \gcd(a, n)$. Given $z \in C_n$ consider the equation (in the variable y in C_n)

$$y^a = z. \tag{1.18}$$

- If $z = u^d$ for some $u \in C_n$, then (1.18) has d solutions;
- otherwise, (1.18) has no solutions.

(Just set $z = x^b$ and $y = x^m$, and consider the exponents.)

We now examine arbitrary Abelian groups (not necessarily cyclic). We begin with a kind of converse to Proposition 1.2.5.

Proposition 1.2.15 *Let A be a finite Abelian group. Let $a, b \in A$ and suppose that $\gcd(o(a), o(b)) = 1$. Then $o(a + b) = o(a)o(b)$.*

Proof. Set $o(a) = r$, $o(b) = s$ and observe that $rs(a + b) = rsa + rsb = s(ra) + r(sb) = 0$. Suppose now that $m \in \mathbb{N}$ satisfies $m(a + b) = 0$. As a consequence, $ma = -mb$ so that $sma = -msb = 0$ and therefore r divides sm. Since r and s are coprime, we deduce that r divides m. Analogously, s divides m. Since $\gcd(r, s) = 1$ this implies that m is a multiple of rs. Therefore rs is the order of $a + b$. $\qquad \square$

Remark 1.2.16 In general, we do *not* have $o(a + b) = \frac{o(a)o(b)}{\gcd(o(a),o(b))} = $ lcm$(o(a), o(b))$, where lcm denotes the least common multiple. For instance, just consider the case $a = -b$.

Proposition 1.2.17 *Let p be a prime number and $\mu_1 \geq \mu_2 \geq \cdots \geq \mu_h$ positive integers. Then the Abelian group*

$$A = \mathbb{Z}_{p^{\mu_1}} \oplus \mathbb{Z}_{p^{\mu_2}} \oplus \cdots \oplus \mathbb{Z}_{p^{\mu_h}}$$

is not cyclic.

Proof. The elements in A of maximal order are of the form $a_1 + a_2 + \cdots + a_h$, where a_1 is a generator of $\mathbb{Z}_{p^{\mu_1}}$ and $a_i \in \mathbb{Z}_{p^{\mu_i}}$ for $i = 2, 3, \ldots, h$; their order is p^{μ_1}. $\square$

Exercise 1.2.18 Let A be a finite Abelian group and $a, b \in A$. Show that A contains an element of order lcm$(o(a), o(b))$.

The following is, probably, the most difficult exercise in Herstein's book [71] (it is Exercise 26 in Section 2.5). Its difficulty relies on the fact that the author asked for a proof based only on tools developed up to Section 2.5 of his book. A proof in this style was published by Robert Beals [18].

Exercise 1.2.19 Let A be a finite Abelian group and $B, C \leq A$ subgroups of A with $|B| = m$ and $|C| = n$. Show that A contains a subgroup of order lcm(m, n).

Exercises 1.2.18 and 1.2.19 are quite easy once the whole structure theory of finite Abelian groups will be fully developed (in the remaining part of this chapter).

Proposition 1.2.20 *Let A be a finite Abelian group and $a \in A$ an element of maximal order. Then for all $b \in A$ one has that $o(b)$ divides $o(a)$.*

Proof. Fix $b \in A$ and let p^k be a prime power in the factorization of $o(b)$. Suppose that $o(a) = p^h m$, where $h \geq 0$ and $\gcd(p, m) = 1$. By Corollary 1.2.9, there exist $c \in \langle a \rangle$ with $o(c) = m$ and $d \in \langle b \rangle$ with $o(d) = p^k$. Then, by Proposition 1.2.15, $o(c + d) = p^k m$ so that, by maximality of $o(a)$, we necessarily have $k \leq h$. This shows that every prime power in the factorization of $o(b)$ divides $o(a)$. It follows that $o(b)$ divides $o(a)$. $\square$

Lemma 1.2.21 *Let A be a finite Abelian group, $a \in A$ an element of maximal order, $b \in A$ an arbitrary element, and denote by m the order of $b + \langle a \rangle$ in the quotient group $A/\langle a \rangle$. Then there exists c in the coset $b + \langle a \rangle$ such that $o(c) = m$.*

Proof. First of all we observe that $mb + \langle a \rangle = m(b + \langle a \rangle) = \langle a \rangle$ so that $mb \in \langle a \rangle$ and we can find $n \in \mathbb{N}$ such that

$$mb = na. \tag{1.19}$$

Setting

$$h = \mathrm{o}(a) \text{ and } t = \gcd(n, h),$$

by Proposition 1.1.26 we have $\mathrm{o}(na) = \frac{h}{t}$.

We claim that

$$\mathrm{o}(b) = \frac{mh}{t}. \tag{1.20}$$

Indeed, setting $r = \mathrm{o}(b)$, by (1.19) we have $\frac{mh}{t}b = \frac{h}{t}mb = \frac{h}{t}na = 0$ and this implies

$$r \mid \frac{mh}{t}. \tag{1.21}$$

Conversely, since $r(b + \langle a \rangle) = (rb + \langle a \rangle) = \langle a \rangle$ and, by hypothesis, $\mathrm{o}(b + \langle a \rangle) = m$, we have that m divides r. Thus we can find $q \in \mathbb{N}$ such that $r = qm$. As a consequence, by (1.19) we have

$$0 = rb = qmb = qna.$$

Since $\mathrm{o}(na) = \frac{h}{t}$, we deduce that $\frac{h}{t}$ divides q, that is, there exists $s \in \mathbb{N}$ such that $q = s\frac{h}{t}$. It follows that

$$r = qm = sm\frac{h}{t} = s\frac{mh}{t}$$

so that $\frac{mh}{t}$ divides r and, by (1.21),

$$\mathrm{o}(b) = r = \frac{mh}{t}.$$

Thus, the claim (1.20) follows.

From Proposition 1.2.20 it follows that $r = \frac{mh}{t}$ divides h (the order of a, which is maximal) and therefore, $m \mid t$. Thus we can find $k \in \mathbb{N}$ such that $t = km$. Setting $v = \frac{n}{t}$ (this is an integer since $t = \gcd(n, h)$) and recalling (1.19), we have

$$mb = na = vta = mvka. \tag{1.22}$$

Setting

$$c = b - vka$$

we have $b + \langle a \rangle = c + \langle a \rangle$ and by (1.22)

$$mc = mb - mvka = 0.$$

This shows that $o(c)|m$. Since $m = o(b + \langle a \rangle) = o(c + \langle a \rangle) \leq o(c) \leq m$, we deduce that $o(c) = m$. $\square$

1.3 Structure theory of finite Abelian groups: the theorems

In this section we present the three structure theorems for finite Abelian groups.

Theorem 1.3.1 (Invariant factors decomposition) *Let A be a finite Abelian group. Then there exists a* unique *finite sequence* $r_1, r_2, \ldots, r_k$, $k \geq 1$, *of positive integers such that*

    (i) r_j *divides* r_{j-1} *for all* $j = 2, 3, \ldots, k$;
    (ii) $|A| = r_1 r_2 \cdots r_k$;
    (iii) $A \cong \mathbb{Z}_{r_1} \oplus \mathbb{Z}_{r_2} \oplus \cdots \oplus \mathbb{Z}_{r_k}$.

Proof. First of all we show, by induction on $n = |A|$, that such a sequence exists. The case $n = 1$ is trivial (take $k = 1 = r_1$). Let now $n \geq 2$ and suppose the statement holds for all finite Abelian groups of order $1 \leq h \leq n - 1$. Let then $a_1 \in A$ such that $r_1 = o(a_1)$ is maximal and consider the quotient group $A' = A/\langle a_1 \rangle$. We have $|A'| = |A|/o(a_1) < n$ so that, by the inductive hypothesis, we can find a finite sequence $r_2, r_3, \ldots, r_k$ of positive integers such that r_j divides r_{j-1} for all $j = 3, 4, \ldots, k$,

$$|A'| = r_2 r_3 \cdots r_k \tag{1.23}$$

and

$$A' \cong \mathbb{Z}_{r_2} \oplus \mathbb{Z}_{r_3} \oplus \cdots \oplus \mathbb{Z}_{r_k}. \tag{1.24}$$

By virtue of Lemma 1.2.21, we can find elements $a_2, a_3, \ldots, a_k \in A$ such that the summand $\mathbb{Z}_{r_j}$ is generated by $a_j + \langle a_1 \rangle$ and

$$o(a_j) = r_j \tag{1.25}$$

for all $j = 3, 4, \ldots, k$. Clearly,

$$A = \langle a_1 \rangle + \langle a_2 \rangle + \cdots + \langle a_k \rangle. \tag{1.26}$$

Indeed, if $b \in A$ then by virtue of (1.24) we can find integers $m_2, m_3, \ldots, m_k$ such that

$$
\begin{aligned}
b + \langle a_1 \rangle &= m_2(a_2 + \langle a_1 \rangle) + m_3(a_3 + \langle a_1 \rangle) + \cdots + m_k(a_k + \langle a_1 \rangle) \\
&= (m_2 a_2 + \langle a_1 \rangle) + (m_3 a_3 + \langle a_1 \rangle) + \cdots + (m_k a_k + \langle a_1 \rangle) \\
&= (m_2 a_2 + m_3 a_3 + \cdots + m_k a_k) + \langle a_1 \rangle
\end{aligned}
$$

so that $b - (m_2 a_2 + m_3 a_3 + \cdots + m_k a_k) \in \langle a_1 \rangle$, and therefore we can find $m_1 \in \mathbb{N}$ such that $b = m_1 a_1 + m_2 a_2 + m_3 a_3 + \cdots + m_k a_k$. This shows (1.26).

From (1.23) and $o(a_1) = r_1$ we deduce that $|A| = r_1 |A'| = r_1 r_2 \cdots r_k$ (namely, condition (ii)) so that, by virtue of Proposition 1.2.2, the sum (1.26) is indeed a direct sum, and (iii) follows as well. Moreover, by Proposition 1.2.20 we deduce that r_2 divides r_1 so that, by induction, also (i) is satisfied.

We now turn to uniqueness of the sequence $r_1, r_2 \ldots, r_k$. Suppose that $s_1, s_2, \ldots, s_h$, $h \in \mathbb{N}$, is also a sequence of integers satisfying (i), (ii), and (iii). For every $j = 1, 2, \ldots, h$, we denote by $b_j \in A$ a generator of the summand $\mathbb{Z}_{s_j}$ so that, for every $c \in A$, we can find $n_1, n_2 \ldots, n_h \in \mathbb{N}$ such that $c = n_1 b_1 + n_2 b_2 + \cdots + n_h b_h$. From (i) we deduce that $s_1 c = 0$ so that $s_1 = o(b_1)$ is the maximal order of the elements of A so that (cf. the first part of the proof)

$$s_1 = r_1.$$

Suppose then that we have, for some $2 \leq j \leq \min\{h, k\}$,

$$s_1 = r_1, \ s_2 = r_2, \ \ldots, \ s_{j-1} = r_{j-1} \ \text{and} \ s_j \neq r_j. \tag{1.27}$$

To fix ideas, suppose that $s_j < r_j$ and denote by

$$B = \{s_j c : c \in A\}$$

the set of s_j-multiples of the elements of A. Clearly, B is a subgroup of A. Moreover (cf. (1.26)), if $c \in A$ we can find $m_1, m_2 \ldots, m_k \in \mathbb{N}$ such that $c = m_1 a_1 + m_2 a_2 + \cdots + m_k a_k$. Thus

$$s_j c = m_1 (s_j a_1) + m_2 (s_j a_2) + \cdots + m_k (s_j a_k),$$

which implies that

$$B = B_1 \oplus \langle s_j a_j \rangle \oplus B_2 \tag{1.28}$$

where $B_1 = \langle s_j a_1 \rangle \oplus \langle s_j a_2 \rangle \oplus \cdots \oplus \langle s_j a_{j-1} \rangle$ and $B_2 = \langle s_j a_{j+1} \rangle \oplus \langle s_j a_{j+2} \rangle \oplus \cdots \oplus \langle s_j a_k \rangle$, and each summand in $B_1 \oplus \langle s_j a_j \rangle$ is nontrivial since $s_j < r_i = o(a_i)$ for all $i = 1, 2, \ldots, j$; in particular,

$$o(s_j a_j) = \frac{o(a_j)}{\gcd(s_j, r_j)} = \frac{r_j}{\gcd(s_j, r_j)} > 1. \tag{1.29}$$

Similarly, we have

$$B = \langle s_j b_1 \rangle \oplus \langle s_j b_2 \rangle \oplus \cdots \oplus \langle s_j b_{j-1} \rangle, \tag{1.30}$$

since $s_j b_\ell = 0$ for $\ell = j, j+1, \ldots, h$. Note that

$$o(s_j a_i) = \frac{o(a_i)}{s_j} = \frac{r_i}{s_j} = \frac{s_i}{s_j} = o(s_j b_i), \tag{1.31}$$

for $i = 1, 2, \ldots, j - 1$. From (1.30) and (1.31) we deduce that $B = B_1$ so that, in particular, $\langle s_j a_j \rangle$ is trivial, a contradiction with (1.29). This shows that $h = k$ and $s_1 = r_1, s_2 = r_2, \ldots, s_h = r_h$, and uniqueness follows.   □

Definition 1.3.2 The positive integers satisfying (i), (ii), and (iii) in Theorem 1.3.1 are called the *invariant factors* of A.

Corollary 1.3.3 (Cauchy's theorem for Abelian groups) *Let A be a finite Abelian group. Suppose that p is a prime divisor of the order of A. Then A contains an element of order p.*

Proof. Let $r_1, r_2, \ldots, r_k$ denote the invariant factors of A. Since p divides $|A| = r_1 r_2 \cdots r_k$, by virtue of Exercise 1.1.8.(2), we can find $1 \le j \le k$ such that $p | r_j$ (in fact, by Theorem 1.3.1.(i), we always have $p | r_1$). From Corollary 1.2.9 we deduce that the subgroup $\mathbb{Z}_{r_j}$, and therefore A, contains an element of order p.   □

Remark 1.3.4 The above is a quite unusual proof of Cauchy's theorem for Abelian groups. Indeed, any book on group theory or on undergraduate algebra contains a direct proof of the more general result, namely the Cauchy theorem for not necessarily Abelian groups. Often (e.g. Robinson [129]), one deduces Cauchy's theorem from the even more general Sylow theorem. In other books (e.g. Herstein [71], Lang [93], Mac Lane and Birkhoff [113], and Rotman [132]) the Abelian case is proved as a first step towards the general case. Finally, in Machì's monograph [102] there is an elementary direct proof of the general result based on the paper by McKay [106] (cf. Exercise 1.3.6 below). In the next exercise we outline a direct proof of Corollary 1.3.3 following [120].

Exercise 1.3.5 Let A be a finite Abelian group. Suppose that p is a prime divisor of the order of A and let B be a proper maximal subgroup of A.

(1) Show that the quotient group A/B is cyclic of prime order;
(2) show that if p does not divide $|B|$ then there exists $c \in A$ such that $\langle c \rangle + B = A$ and $|\langle c \rangle / (\langle c \rangle \cap B)| = p$;
(3) use (1) and (2) to give (another) inductive proof of Corollary 1.3.3.

As mentioned above, in the next exercise we outline a direct proof of the general Cauchy theorem. We use some elementary notions on group actions that will be further developed in Section 10.4.

Exercise 1.3.6 Let G be a finite (not necessarily Abelian) group: we use multiplicative notation. Suppose that p is a prime divisor of the order of G and set

$$X = \{(g_1, g_2, \ldots, g_p) \in G^p : g_1 g_2 \cdots g_p = 1_G\}.$$

(1) Show that $|X| = |G|^{p-1}$;
(2) show that $\mathbb{Z}_p$ acts on X by cyclic permutations, namely that if $x = (g_1, g_2, \ldots, g_p) \in X$ and t is a fixed generator of $\mathbb{Z}_p$ then $tx = (g_2, g_3, \ldots, g_p, g_1) \in X$;
(3) for $x \in X$ denote by $\mathrm{Stab}_x = \{s \subset \mathbb{Z}_p : sx = x\}$ the *stabilizer* of x: show that Stab_x is a subgroup of $\mathbb{Z}_p$ and, from Lagrange's theorem, deduce that it is either trivial or the whole $\mathbb{Z}_p$;
(4) denote by $\mathbb{Z}_p x = \{sx : s \in \mathbb{Z}_p\}$ the *orbit* of $x \in X$ and show that $|\mathbb{Z}_p x| = p/|\mathrm{Stab}_x|$ (orbit-stabilizer theorem);
(5) deduce that the only possible orbit sizes are 1 and p;
(6) show that $\mathbb{Z}_p x = \{x\}$ if and only if there exists $g \in G$ such that $x = (g, g, \ldots, g)$, so that, necessarily, $g^p = 1_G$;
(7) let m (respectively n) denote the number of orbits of size 1 (respectively p): from (5) and (6) deduce that $m + np = |G|^{p-1}$ and $m \geq 1$;
(8) from (7) deduce that $m \geq 2$ (in fact m is divisible by p) and therefore, by (6), there exists $g \in G$ of period p.

Theorem 1.3.7 (Primary decomposition) *Let A be a finite Abelian group. Let*

$$|A| = p_1^{k_1} p_2^{k_2} \cdots p_t^{k_t} \tag{1.32}$$

be the prime factorization of the order of A. Then

$$A_i = \{a \in A : o(a) \text{ is a power of } p_i\}$$

is a subgroup of A of order $p_i^{k_i}$, for $i = 1, 2, \ldots, t$, and

$$A = A_1 \oplus A_2 \oplus \cdots \oplus A_t. \tag{1.33}$$

Proof. We first remark that, by virtue of Corollary 1.3.3, $A_i \neq \{0\}$, and we leave it as an exercise to check that A_i is a subgroup for $i = 1, 2, \ldots, t$.

Let $a \in A$. Then, since $o(a)$ divides $|A|$, there exists a nonempty subset $\{i_1, i_2, \ldots, i_m\}$ of $\{1, 2, \ldots, t\}$ and integers $1 \leq h_j \leq k_{i_j}, j = 1, 2, \ldots, m$, such that

$$o(a) = p_{i_1}^{h_1} p_{i_2}^{h_2} \cdots p_{i_m}^{h_m}.$$

By the Chinese remainder theorem III (Corollary 1.2.7), we have

$$\langle a \rangle = \mathbb{Z}_{p_{i_1}^{h_1}} \oplus \mathbb{Z}_{p_{i_2}^{h_2}} \oplus \cdots \oplus \mathbb{Z}_{p_{i_m}^{h_m}} \subseteq A_{i_1} + A_{i_2} + \cdots + A_{i_m}.$$

This shows that

$$A = A_1 + A_2 + \cdots + A_t. \tag{1.34}$$

We claim that the above sum is direct. Suppose that $a_1 + a_2 + \cdots + a_t = 0$, where $a_i \in A_i$, $i = 1, 2, \ldots, t$. Let $1 \leq i \leq t$. Then, after multiplying by $q_i = \frac{|A|}{p_i^{k_i}}$, we get $q_i a_i = 0$ and, since the order of a_i does not divide q_i, we necessarily have $a_i = 0$. Thus $a_1 = a_2 = \cdots = a_t = 0$ and from Proposition 1.2.2 the claim follows. This establishes (1.33).

Let $1 \leq i \leq t$. Since A_i only contains elements of order a power of p_i, from Corollary 1.3.3 we deduce that $|A_i| = p_i^{r_i}$ for some integer $r_i \geq 1$. Moreover, since the sum (1.34) is direct, we have $|A| = |A_1| \cdot |A_2| \cdot \ldots \cdot |A_t| = p_1^{r_1} p_2^{r_2} \cdots p_t^{r_t}$ so that, by uniqueness of the prime factorization (1.32) of $|A|$, we necessarily have $r_i = k_i$ for all $i = 1, 2, \ldots, t$, completing the proof. $\square$

Definition 1.3.8 Let p be a prime number. A group G is termed a *p-group* provided that every element has order a power of p.

Sylow's first theorem (see for instance Herstein [71]) states that if G is a finite group and p a prime number such that $|G| = p^n m$, where $n, m \geq 1$ with $\gcd(p, m) = 1$ (thus n is the maximal power of p dividing the order of G), then G contains a p-subgroup of order p^n: this is called a *p-Sylow subgroup* of G.

Thus, from Theorem 1.3.7, an Abelian version of Sylow's first theorem follows.

Definition 1.3.9 Let p be a prime number. An Abelian p-group is called a *p-primary group* (cf. Definition 1.2.6). Moreover, for $i = 1, 2, \ldots, t$, the subgroup A_i in (1.33) is termed the *p_i-primary component* of A.

The following relates and refines the statements of Theorem 1.3.1 and Theorem 1.3.7: we use the notation therein.

Corollary 1.3.10 (Structure theorem for finite Abelian groups) *Let A be a finite Abelian group. Then there exist unique positive integers h_i and m_{ij}, $i = 1, 2, \ldots, t$ and $j = 1, 2, \ldots, h_i$, satisfying $h_i \leq k_i$ and*

$$m_{i1} \geq m_{i2} \geq \cdots \geq m_{ih_i} \tag{1.35}$$

for all $i = 1, 2, \ldots, t$, such that the following holds:

$$A \cong \bigoplus_{i=1}^{t} \bigoplus_{j=1}^{h_i} \mathbb{Z}_{p_i^{m_{ij}}} \tag{1.36}$$

$$A_i \cong \bigoplus_{j=1}^{h_i} \mathbb{Z}_{p_i^{m_{ij}}} \tag{1.37}$$

for $i = 1, 2, \ldots, t$, and

$$\mathbb{Z}_{r_j} \cong \bigoplus_{\substack{1 \le i \le t: \\ h_i \ge j}} \mathbb{Z}_{p_i^{m_{ij}}} \tag{1.38}$$

for $j = 1, 2, \ldots, k$. In particular, $\sum_{j=1}^{h_i} m_{ij} = k_i$ for $i = 1, 2, \ldots, t$ and

$$\prod_{\substack{1 \le i \le t: \\ h_i \ge j}} p_i^{m_{ij}} = r_j \tag{1.39}$$

for all $j = 1, 2, \ldots, k$.

Proof. We shall present two proofs of this fundamental result: we can exchange the order of the applications of Theorem 1.3.1 and Theorem 1.3.7.

First proof. We apply Theorem 1.3.1 to each p-primary component A_i in (1.33): thus we can find $1 \le h_i \le k_i$ and $m_{i1} \ge m_{i2} \ge \cdots \ge m_{ih_i}$ such that (1.37) and therefore (1.36) hold. Uniqueness follows from uniqueness in Theorem 1.3.1 and uniqueness of the prime factorization of $|A|$. Let now $1 \le j \le k$. Then (1.35) implies that $\prod_{\substack{1 \le i \le t: \\ h_i \ge j}} p_i^{m_{ij}}$ divides $\prod_{\substack{1 \le i \le t: \\ h_i \ge j-1}} p_i^{m_{i,j-1}}$ so that, by Proposition 1.2.5 and uniqueness in Theorem 1.3.1, we deduce (1.39) and (1.38).

Second proof. Consider the invariant factors r_j, $j = 1, 2, \ldots, t$, in Theorem 1.3.1.(iii). Let $r_1 = p_1^{m_{11}} p_2^{m_{21}} \cdots p_t^{m_{t1}}$ denote the prime factorization of r_1 (so that $m_{i1} > 0$ for $i = 1, 2, \ldots, t$). Let $1 \le j \le k$. Since $r_j | r_{j-1}, \ldots, r_2 | r_1$, we can write $r_j = p_1^{m_{1j}} p_2^{m_{2j}} \cdots p_t^{m_{tj}}$ with $m_{i,j-1} \ge m_{ij} \ge 0$ for $i = 1, 2, \ldots, t$. Let us denote by h_i the largest j such that $m_{ij} > 0$ (equivalently, $m_{ih_i} > 0$ and $m_{i,h_i+1} = 0$). This way, $r_j = \prod_{\substack{1 \le i \le t: \\ h_i \ge j}} p_i^{m_{ij}}$ is the prime factorization of r_j and (1.39) follows. Applying Theorem 1.3.7 to each $\mathbb{Z}_{r_j}$, $j = 1, 2, \ldots, t$, we deduce (1.38). Finally, from the direct sum decomposition in Theorem 1.3.1.(iii), we deduce (1.36) and, by definition of A_i, (1.37). $\qquad\square$

Corollary 1.3.11 *A finite Abelian group is indecomposable if and only if it is a p-primary cyclic group for some prime p.*

Proof. The "if" part is Proposition 1.2.10. Conversely, if A is indecomposable, then in (1.36) we must have $t = 1$ and $h_1 = 1$. $\qquad\square$

Definition 1.3.12 The positive integers $m_{ij}, i = 1, 2, \ldots, t, j = 1, 2, \ldots, h_i,$ in Corollary 1.3.10 are called the *elementary divisors* of A.

In Corollary 1.3.10 we have shown that the invariant factors determine uniquely the elementary divisors, and vice versa. More precisely, given the prime factorization (1.32), from (1.39) we have a correspondence

$$(r_j)_{j=1}^k \leftrightarrow \left((h_i)_{i=1}^t, (m_{ij})_{\substack{1 \le i \le t \\ 1 \le j \le h_i}} \right).$$

Our next task is to compute the number of nonisomorphic Abelian groups of a given order $n \in \mathbb{N}$. For this purpose we introduce the following definitions.

Definition 1.3.13 Let $n \in \mathbb{N}$. A *partition* of n is a sequence

$$\lambda = (\lambda_1, \lambda_2, \cdots, \lambda_h)$$

of positive integers such that

$$\lambda_1 \ge \lambda_2 \ge \cdots \ge \lambda_h \text{ and } \lambda_1 + \lambda_2 + \cdots + \lambda_h = n.$$

We then write $\lambda \vdash n$.

We denote by $p(n) = |\{\lambda : \lambda \vdash n\}|$ the number of partitions of n.

The map $p \colon \mathbb{N} \to \mathbb{N}$ is called the *partition function*.

Let now A and B be two finite Abelian groups. Then $A \cong B$ if and only if, denoting by $(r_j^A)_{j=1}^{k_A}$ and $(r_j^B)_{j=1}^{k_B}$ the corresponding invariant factors, then $k_A = k_B$ and $r_j^A = r_j^B$ for all $j = 1, 2, \ldots, k_A$: we express this last condition by saying, with a slight abuse of language, that A and B have the same invariant factors. Equivalently, A and B are isomorphic if and only if $|A| = |B|$ and, denoting by $(m_{ij}^A)_{\substack{1 \le i \le t \\ 1 \le j \le h_i^A}}$ and $(m_{ij}^B)_{\substack{1 \le i \le t \\ 1 \le j \le h_i^B}}$ the corresponding elementary divisors, we have $h_i^A = h_i^B$ and $m_{ij}^A = m_{ij}^B$ for all $i = 1, 2, \ldots, t$ and $j = 1, 2, \ldots, h_i^A$. Again, with a slight abuse of language, this last condition may be expressed by saying that A and B have the same elementary divisors.

Proposition 1.3.14 *Let $n \ge 2$ and denote by $n = p_1^{k_1} p_2^{k_2} \cdots p_t^{k_t}$ its prime factorization. Then the number of nonisomorphic Abelian groups of order n is*

$$p(k_1)p(k_2) \cdots p(k_t).$$

Proof. Let A be an Abelian group of order n and denote by $(m_{ij}^A)_{\substack{1 \le i \le t \\ 1 \le j \le h_i^A}}$ the corresponding elementary divisors. Then for each $i = 1, 2, \ldots, t$ we have the partition $\mu_i = (m_{i1}, m_{i2}, \ldots, m_{ih_i}) \vdash k_i$. Since, by the above observations, the elementary divisors uniquely determine A (of the given order n) up to isomorphism, this ends the proof. $\square$

Remark 1.3.15 Theorem 1.3.1, Theorem 1.3.7, and Corollary 1.3.10 provide three different decompositions of a finite Abelian group. In Theorem 1.3.1 and Corollary 1.3.10, the *structure* of the decompositions is *unique* (that is, the invariant factors and the elementary divisors, respectively, are uniquely determined). On the one hand, the associated subgroups (namely the $\mathbb{Z}_{r_j}$, $j = 1, 2, \ldots, k$, and the $\mathbb{Z}_{p_i^{m_{ij}}}$, $i = 1, 2, \ldots, t$, $j = 1, 2, \ldots, h_i$, respectively) are *not* uniquely determined. This aspect will be discussed in Section 1.8 (see Corollary 1.8.4). On the other hand, the *subgroups* in the decomposition in Theorem 1.3.7 are *uniquely determined*.

We now give a characterization of the decomposition (1.36) in Corollary 1.3.10. First recall that, by Proposition 1.2.10, every p-primary cyclic group $\mathbb{Z}_{p^{m_{ij}}}$ is indecomposable.

Proposition 1.3.16 *With the notation from Corollary 1.3.10, let* $A = \bigoplus_{\mu=1}^{q} B_\mu$ *be a decomposition of A as a direct sum of* indecomposable *subgroups. Then* $q = \sum_{i=1}^{t} h_i$ *and there exists a bijection*

$$\mu : \{(i, j) : 1 \le i \le t, 1 \le j \le h_i\} \longrightarrow \{1, 2, \ldots, q\}$$

such that

$$\mathbb{Z}_{p_i^{m_{ij}}} \cong B_{\mu(i,j)} \tag{1.40}$$

for $i = 1, 2, \ldots, t$ *and* $j = 1, 2, \ldots, h_i$.

Proof. By Corollary 1.3.11, each B_μ is a p-primary cyclic group. Let $1 \le i \le t$. Then, in the notation of Theorem 1.3.7, we can find distinct indices $1 \le \mu(i, 1), \mu(i, 2), \ldots, \mu(i, k_i) \le q$ such that

$$A_i = B_{\mu(i,1)} \oplus B_{\mu(i,2)} \oplus \cdots \oplus B_{\mu(i,k_i)}$$

and $B_{\mu(i,1)}, B_{\mu(i,2)}, \ldots, B_{\mu(i,k_i)}$ are all the p_i-groups among the B_μs. Up to permuting the indices, if necessary, we may assume that

$$|B_{\mu(i,1)}| \ge |B_{\mu(i,2)}| \ge \cdots \ge |B_{\mu(i,k_i)}|$$

so that, necessarily, $|B_{\mu(i,j-1)}|$ divides $|B_{\mu(i,j)}|$ for $j = 2, 3, \ldots, k_i$. By applying the uniqueness assertion in Theorem 1.3.1, we deduce (1.40) (in particular, $k_i = h_i$ for all $i = 1, 2, \ldots, t$). The remaining part of the statement is now clear. $\square$

Proposition 1.3.17 *Let A be a finite Abelian group. Then, in the notation of Theorem 1.3.7, the following conditions are equivalent:*

(a) *A is cyclic;*

(b) *A contains exactly one subgroup of order p_i for every $i = 1, 2, \ldots, t$;*

(c) A_i *is cyclic for every* $i = 1, 2, \ldots, t$.

Proof. The implication (a) $\Rightarrow$ (b) follows immediately from Proposition 1.2.12.

Suppose that there exists $1 \leq i \leq t$ such that A_i is not cyclic. Then, in (1.37) (and with the notation therein) we necessarily have $h_i \geq 2$ so that A_i contains a subgroup B isomorphic to $\mathbb{Z}_{p_i^{m_{i1}}} \oplus \mathbb{Z}_{p_i^{m_{i2}}}$. By virtue of Cauchy's theorem (Corollary 1.3.3) applied to each direct component, B and therefore A contain two distinct subgroups of order p_i. This shows the implication (b) $\Rightarrow$ (c).

Suppose (c). Let $a_i \in A_i$ be a generator of A_i for every $i = 1, 2, \ldots, t$. Then, by Proposition 1.2.15, the element $a = a_1 a_2 \cdots a_t$ has order $\mathrm{o}(a) = \mathrm{o}(a_1)\mathrm{o}(a_2) \cdots \mathrm{o}(a_t) = |A_1| \cdot |A_2| \cdot \ldots \cdot |A_t| = |A|$ (the last equality follows from (1.33) and (1.15)). This shows that $A = \langle a \rangle$ is cyclic, and the implication (c) $\Rightarrow$ (a) follows as well. $\qquad\qquad\qquad\qquad\qquad\qquad\square$

Remark 1.3.18 The decomposition of a finite Abelian group as a direct sum of cyclic groups presented in (1.36) is the finer, while the one in Theorem 1.3.1.(iii) is the coarser.

1.4 Generalities on endomorphisms and automorphisms of finite Abelian groups

In the next sections we present a complete description of the automorphisms of finite Abelian groups in order to:

- clarify the structure theorem (cf. Remark 1.3.15);
- show examples for potential applications of Theorem 11.7.1.

We start with some basic general results.

Let A be a finite Abelian group. A map $\alpha \colon A \to A$ such that

$$\alpha(a + b) = \alpha(a) + \alpha(b)$$

for all $a, b \in A$ is called an *endomorphism* of A. We denote by End(A) the set of all endomorphisms of A.

Note that if $\alpha \in \mathrm{End}(A)$ then $\alpha(0) = 0$ and $\alpha(-a) = -\alpha(a)$ for all $a \in A$. Moreover, End(A) is a unital *ring*: for $\alpha, \beta \in \mathrm{End}(A)$ we define their sum $\alpha + \beta$ and their product $\alpha\beta$ by setting

$$(\alpha + \beta)(a) = \alpha(a) + \beta(a)$$

and, respectively,

$$(\alpha\beta)(a) = \alpha(\beta(a))$$

for all $a \in A$; the zero endomorphism $0 = 0_{\text{End}(A)} \in \text{End}(A)$ and the identity map $1 = \text{Id}_A \in \text{End}(A)$ defined by

$$0(a) = 0_A$$

and

$$1(a) = a$$

for all $a \in A$, are the zero and unital element of $\text{End}(A)$, respectively.

Let $\alpha \in \text{End}(A)$. We denote by $\text{Ker}(\alpha) = \{a \in A : \alpha(a) = 0\}$ the *kernel* of α. It is immediate that $\text{Ker}(\alpha)$ is a subgroup of A and that $\text{Ker}(\alpha) = \{0\}$ if and only if α is a bijective map.

Suppose now that α is bijective. Then the inverse map α^{-1} is also an endomorphism: indeed, if $a, b \in A$

$$\alpha[\alpha^{-1}(a+b)] = a + b = \alpha[\alpha^{-1}(a)] + \alpha[\alpha^{-1}(b)] = \alpha[\alpha^{-1}(a) + \alpha^{-1}(b)]$$

so that, by bijectivity, we have $\alpha^{-1}(a+b) = \alpha^{-1}(a) + \alpha^{-1}(b)$.

A bijective endomorphism of A is called an *automorphism* of A. It follows from the previous observation that the set

$$\text{Aut}(A) = \{\alpha \in \text{End}(A) : \text{Ker}(\alpha) = \{0\}\}$$

of all automorphisms of A is the group of *units* of $\text{End}(A)$.

Lemma 1.4.1 *Let A be a finite Abelian group and $m \in \mathbb{N}$. Then the map $\alpha_m \colon A \to A$ defined by $\alpha_m(a) = ma$ for all $a \in A$, is an endomorphism of A. Moreover, α_m is an automorphism if and only if $\gcd(m, |A|) = 1$.*

Proof. The fact that $\alpha_m \in \text{End}(A)$ follows immediately from the fact that A is Abelian. Let now $d = \gcd(m, |A|)$. If $d > 1$ and p is a prime dividing d, by Cauchy's theorem (Corollary 1.3.3) we can find $a \in A$ such that $o(a) = p$. As a consequence, $\alpha_m(a) = ma = \frac{m}{p}(pa) = \frac{m}{p}0 = 0$ so that α_m cannot be injective, that is, $\alpha_m \notin \text{Aut}(A)$. Conversely, if $d = 1$, then by Lagrange's theorem, A does not contain elements of order q for every integer $q \geq 2$ dividing m. As a consequence $\alpha_m(a) = ma \neq 0$ for all $a \in A \setminus \{0\}$, equivalently, $\text{Ker}(\alpha) = \{0\}$, so that $\alpha_m \in \text{Aut}(A)$. $\qquad\square$

Let R_1 and R_2 be two unital rings. We equip their Cartesian product $R_1 \times R_2$ with a structure of a unital ring by setting

$$(r_1, r_2) + (r_1' + r_2') = (r_1 + r_1', r_2 + r_2') \quad \text{and} \quad (r_1, r_2)(r_1', r_2') = (r_1 r_1', r_2 r_2')$$

for all $r_1, r_1' \in R_1$ and $r_2, r_2' \in R_2$. It is clear that the elements $(0, 0)$ and $(1, 1)$ are the zero and unit elements of $R_1 \times R_2$. Moreover if $(r_1, r_2) \in R_1 \times R_2$ we have $-(r_1, r_2) = (-r_1, -r_2)$ and (r_1, r_2) is a unit if and only if both r_1 and r_2 are and, if this is the case, $(r_1, r_2)^{-1} = (r_1^{-1}, r_2^{-1})$. In other words, denoting by $\mathcal{U}(R)$ the *group of units* of any unital ring R, we have

$$\mathcal{U}(R_1 \times R_2) = \mathcal{U}(R_1) \times \mathcal{U}(R_2). \tag{1.41}$$

Theorem 1.4.2 ([72]) *Let A and B be two finite Abelian groups. Suppose that* $\gcd(|A|, |B|) = 1$. *Then the map* $\Phi\colon \mathrm{End}(A) \times \mathrm{End}(B) \to \mathrm{End}(A \oplus B)$ *defined by*

$$[\Phi(\alpha, \beta)](a + b) = \alpha(a) + \beta(b)$$

for all $\alpha \in \mathrm{End}(A)$, $\beta \in \mathrm{End}(B)$, $a \in A$, *and* $b \in B$, *is a unital ring isomorphism. In particular,*

$$\mathrm{Aut}(A \oplus B) \cong \mathrm{Aut}(A) \times \mathrm{Aut}(B). \tag{1.42}$$

Proof. It is easy to check that $\Phi(\alpha, \beta) \in \mathrm{End}(A \oplus B)$. Let us show that Φ is a ring homomorphism. For $\alpha_1, \alpha_2 \in \mathrm{End}(A)$, $\beta_1, \beta_2 \in \mathrm{End}(B)$, $a \in A$, and $b \in B$ we have

$$
\begin{aligned}
{[}\Phi(\alpha_1, \beta_1) + \Phi(\alpha_2, \beta_2)](a + b) &= [\Phi(\alpha_1, \beta_1)](a + b) + [\Phi(\alpha_2, \beta_2)](a + b) \\
&= (\alpha_1(a) + \beta_1(b)) + (\alpha_2(a) + \beta_2(b)) \\
&= (\alpha_1(a) + \alpha_2(a)) + (\beta_1(b) + \beta_2(b)) \\
&= [\alpha_1 + \alpha_2](a) + [\beta_1 + \beta_2](b) \\
&= [\Phi(\alpha_1 + \alpha_2, \beta_1 + \beta_2)](a + b) \\
&= [\Phi((\alpha_1, \beta_1) + (\alpha_2, \beta_2))](a + b)
\end{aligned}
$$

and

$$
\begin{aligned}
{[}\Phi(\alpha_1, \beta_1)\Phi(\alpha_2, \beta_2)](a + b) &= \Phi(\alpha_1, \beta_1)[(\Phi(\alpha_2, \beta_2))(a + b)] \\
&= \Phi(\alpha_1, \beta_1)(\alpha_2(a) + \beta_2(b)) \\
&= \alpha_1(\alpha_2(a)) + \beta_1(\beta_2(b)) \\
&= [\Phi(\alpha_1\alpha_2, \beta_1\beta_2)](a + b) \\
&= [\Phi((\alpha_1, \beta_1)(\alpha_2\beta_2))](a + b)
\end{aligned}
$$

so that $\Phi((\alpha_1, \beta_1) + (\alpha_2, \beta_2)) = \Phi(\alpha_1, \beta_1) + \Phi(\alpha_2, \beta_2)$ and $\Phi((\alpha_1, \beta_1) (\alpha_2\beta_2)) = \Phi(\alpha_1, \beta_1)\Phi(\alpha_2\beta_2)$.

Moreover, it is straightforward that

$$\Phi(1, 1) = \Phi(\mathrm{Id}_A, \mathrm{Id}_B) = \mathrm{Id}_{A \oplus B} = 1. \tag{1.43}$$

This shows that Φ is a unital ring homomorphism.

Let us now show that $\mathrm{Ker}(\Phi) = \{(0, 0)\}$. Indeed, if $\alpha \in \mathrm{End}(A)$ and $\beta \in \mathrm{End}(B)$ satisfy $\Phi(\alpha, \beta) = 0$, then $\alpha(a) = \alpha(a) + \beta(0) = \Phi(\alpha, \beta)(a, 0) = 0$ for all $a \in A$ (respectively $\beta(b) = \alpha(0) + \beta(b) = \Phi(\alpha, \beta)(0, b) = 0$ for all $b \in B$) so that, necessarily, $\alpha = 0$ (respectively $\beta = 0$). This shows injectivity of Φ.

Let us show that Φ is surjective. Let $\omega \in \mathrm{End}(A \oplus B)$. Denoting by $\pi_A \colon A \times B \to A$ and $\pi_B \colon A \times B \to B$ the canonical projections (these are clearly group homomorphisms), we define a homomorphism $\gamma \colon B \to A$ by setting

$$\gamma(b) = \pi_A(\omega(0, b))$$

for all $b \in B$. Now, if $n = |A|$ we have, for all $b \in B$,

$$0 = n\gamma(b) = \gamma(nb).$$

Since by hypothesis $\gcd(n, |B|) = 1$, the map $\beta_n \colon B \to B$, defined by $\beta_n(b) = nb$ for all $b \in N$, is an isomorphism by Lemma 1.4.1. We deduce that $\gamma = 0$, that is,

$$\pi_A(\omega(0, b)) = 0 \tag{1.44}$$

for all $b \in B$. Exchanging the roles of A and B, we have

$$\pi_B(\omega(a, 0)) = 0 \tag{1.45}$$

for all $a \in A$. Consider the endomorphisms $\alpha = \alpha_\omega \in \mathrm{End}(A)$ and $\beta = \beta_\omega \in \mathrm{End}(B)$ defined by

$$\begin{aligned} \alpha(a) &= \pi_A(\omega(a, 0)) \\ \beta(b) &= \pi_B(\omega(0, b)) \end{aligned} \tag{1.46}$$

for all $a \in A$ and $b \in B$. Then, since $\pi_A + \pi_B = \mathrm{Id}_{A \oplus B}$, we have, for all $a \in A$ and $b \in B$

$$\begin{aligned} \omega(a, b) &= \omega(a, 0) + \omega(0, b) \\ &= [\pi_A + \pi_B](\omega(a, 0)) + [\pi_A + \pi_B](\omega(0, b)) \\ &= \pi_A(\omega(a, 0)) + \pi_B(\omega(a, 0)) \\ &\quad + \pi_A(\omega(0, b)) + \pi_B(\omega(0, b)) \\ (\text{by } (1.45) \text{ and } (1.44)) \quad &= \pi_A(\omega(a, 0)) + \pi_B(\omega(0, b)) \\ (\text{by } (1.46)) \quad &= \alpha(a) + \beta(b) \\ &= [\Phi(\alpha, \beta)](a, b). \end{aligned}$$

In other words,

$$\omega = \Phi(\alpha, \beta)$$

and therefore Φ is surjective.

Since Φ is unital, it establishes a group isomorphism between the corresponding groups of units, so that, keeping in mind (1.41), equation (1.42) follows. $\square$

1.5 Endomorphisms and automorphisms of finite cyclic groups

We turn to the study of the endomorphisms of a finite cyclic group. We keep in mind Notation 1.1.17 and (1.3), and recall that $\mathcal{U}(\mathbb{Z}/n\mathbb{Z}) \subseteq \mathbb{Z}/n\mathbb{Z}$ denotes the (multiplicative) group of units of $\mathbb{Z}/n\mathbb{Z}$.

Lemma 1.5.1 *For $n \geq 1$ we have $\mathcal{U}(\mathbb{Z}/n\mathbb{Z}) = \{\overline{m} \in \mathbb{Z}/n\mathbb{Z} : \gcd(n, m) = 1\}$.*

Proof. Indeed let $\overline{m} \in \mathbb{Z}/n\mathbb{Z}$ and set $d = \gcd(n, m)$. If $d > 1$ then, setting $s = m/d \in \mathbb{N}$ and $t = n/d \in \mathbb{N}$, we have $\overline{t} \neq \overline{0}$ and

$$\overline{m} \cdot \overline{t} = \overline{mt} = \overline{mn/d} = \overline{ns} = \overline{0}$$

thus showing that $\overline{m}$ is a zero-divisor and therefore is not invertible. On the other hand, if $d = 1$ by virtue of the Bézout identity (cf. (1.2)), we can find $a, b \in \mathbb{Z}$ such that $an + bm = 1$ so that

$$\overline{b} \cdot \overline{m} = \overline{bm} = \overline{1 - an} = \overline{1} - \overline{0} = \overline{1}.$$

This shows that $\overline{m}$ is invertible (with inverse $\overline{m}^{-1} = \overline{b}$). $\square$

Proposition 1.5.2 *For $n \geq 1$ we have $\mathrm{End}(\mathbb{Z}_n) \cong \mathbb{Z}/n\mathbb{Z}$.*

Proof. For $\overline{m} \in \mathbb{Z}/n\mathbb{Z}$ define $\psi_{\overline{m}} \in \mathrm{End}(\mathbb{Z}_n)$ by setting $\psi_{\overline{m}}(\overline{k}) = \overline{km} = \overline{mk}$ for all $\overline{k} \in \mathbb{Z}_n$. We claim that the map $\Psi \colon \mathbb{Z}/n\mathbb{Z} \to \mathrm{End}(\mathbb{Z}_n)$ defined by $\Psi(\overline{m}) = \psi_{\overline{m}}$ is a unital ring isomorphism. Let $0 \leq k, m, m' \leq n - 1$.

We have $[\psi_{\overline{m}}\psi_{\overline{m'}}](\overline{k}) = \psi_{\overline{m}}(\overline{m'k}) = \overline{mm'k} = \psi_{\overline{mm'}}(\overline{k}) = \psi_{\overline{mm'}}(\overline{k})$ thus showing that $\Psi(\overline{mm'}) = \Psi(\overline{m})\Psi(\overline{m'})$. Moreover, it is clear that $\Psi(\overline{1}) = \psi_{\overline{1}} = \mathrm{Id}_{\mathbb{Z}_n} = 1$, so that Ψ is a unital ring homomorphism.

Suppose that $\Psi(\overline{m}) = \Psi(\overline{m'})$. Then $\overline{m} = \psi_{\overline{m}}(\overline{1}) = \Psi(\overline{m})(\overline{1}) = \Psi(\overline{m'})(\overline{1}) = \psi_{\overline{m'}}(\overline{1}) = \overline{m'}$, showing that Ψ is injective.

Finally, let $\psi \in \mathrm{End}(\mathbb{Z}_n)$ and set $\overline{m} = \psi(\overline{1})$. Then we have

$$\psi(\overline{k}) = \psi(\overline{k1}) = \psi(\underbrace{\overline{1 + 1 + \cdots + 1}}_{k \text{ times}}) = k\psi(\overline{1}) = \overline{km} = \overline{km} = \psi_{\overline{m}}(\overline{k}).$$

In other words, $\psi = \psi_{\overline{m}} = \Psi(\overline{m})$. This shows that Ψ is also surjective, completing the proof. $\square$

Corollary 1.5.3 *For $n \geq 1$ we have* $\mathrm{Aut}(\mathbb{Z}_n) \cong \mathcal{U}(\mathbb{Z}/n\mathbb{Z})$. *In particular,* $\mathrm{Aut}(\mathbb{Z}_n)$ *is Abelian and*

$$|\mathrm{Aut}(\mathbb{Z}_n)| = \varphi(n), \tag{1.47}$$

where φ is Euler's totient function (cf. Definition 1.1.18).

Proof. The first statement follows from the fact that the map Ψ in the proof of Proposition 1.5.2 is a unital ring isomorphism and therefore establishes a group isomorphism between the corresponding groups of units. Moreover, since the ring $\mathbb{Z}/n\mathbb{Z}$ is commutative, we have that $\mathcal{U}(\mathbb{Z}/n\mathbb{Z})$ is Abelian. Finally, (1.47) is an immediate consequence of Lemma 1.5.1. $\qquad\square$

Exercise 1.5.4 Let $m \geq 1$ and $n \geq 2$ such that $\gcd(m, n) = 1$ and let p be a prime number such that $p \nmid m$.

(1) Prove the following (*Euler's identity*)

$$m^{\varphi(n)} \equiv 1 \bmod n;$$

(2) deduce the following (*Fermat's identity*)

$$m^{p-1} \equiv 1 \bmod p.$$

Recall that Theorem 1.1.21 may be expressed in the form: if p is a prime then $\mathcal{U}(\mathbb{Z}/p\mathbb{Z})$ is cyclic of order $p - 1$.

Exercise 1.5.5 Deduce Fermat's identity in Exercise 1.5.4 directly from Theorem 1.1.21.

In the remaining part of this section, we analyze more closely the structure of the Abelian group $\mathcal{U}(\mathbb{Z}/n\mathbb{Z}) \cong \mathrm{Aut}(\mathbb{Z}_n)$ focusing on its decomposition as a direct sum of cyclic groups (cf. Section 1.3). Actually, as these are multiplicative groups, we use multiplicative notation (cf. Notation 1.1.17) and decompose into direct products.

Proposition 1.5.6 *Let $n = p_1^{k_1} p_2^{k_2} \cdots p_t^{k_t}$ be the prime factorization of an integer $n \geq 2$. Then*

$$\begin{aligned}
\mathcal{U}(\mathbb{Z}/n\mathbb{Z}) &\cong \mathrm{Aut}(\mathbb{Z}_n) \\
&\cong \mathrm{Aut}(\mathbb{Z}_{p_1^{k_1}}) \times \mathrm{Aut}(\mathbb{Z}_{p_2^{k_2}}) \times \cdots \times \mathrm{Aut}(\mathbb{Z}_{p_t^{k_t}}) \\
&\cong \mathcal{U}(\mathbb{Z}/p_1^{k_1}\mathbb{Z}) \times \mathcal{U}(\mathbb{Z}/p_2^{k_2}\mathbb{Z}) \times \cdots \times \mathcal{U}(\mathbb{Z}/p_t^{k_t}\mathbb{Z}).
\end{aligned}$$

Proof. The first isomorphism follows from Corollary 1.5.3. The second from (1.42) and the Chinese remainder theorem III (Theorem 1.2.7). The last one follows again from Corollary 1.5.3. $\qquad\square$

We now determine the structure of $\mathcal{U}(\mathbb{Z}/p^k\mathbb{Z}) \cong \operatorname{Aut}(\mathbb{Z}_{p^k})$ for p prime and $k \geq 1$. This requires some nontrivial calculations in number theory; our treatment is inspired by the monographs by Nathanson [118], Ireland and Rosen [79], and Rotman [132]. We first observe that

$$|\mathcal{U}(\mathbb{Z}/p^k\mathbb{Z})| = \varphi(p^k) = p^k - p^{k-1} = (p-1)p^{k-1}. \qquad (1.48)$$

Indeed, the first equality follows from Corollary 1.5.3, while the second is a consequence of the fact that an integer $1 \leq n \leq p^k$ is divisible by p if and only if there exists $1 \leq h \leq p^{k-1}$ such that $n = ph$.

Theorem 1.5.7 *We have:* $\mathcal{U}(\mathbb{Z}/2\mathbb{Z}) = \{\overline{1}\}$, $\mathcal{U}(\mathbb{Z}/4\mathbb{Z}) = \langle \overline{-1} \rangle \cong C_2$ *and, for* $k \geq 3$,

$$\mathcal{U}(\mathbb{Z}/2^k\mathbb{Z}) = \langle \overline{-1} \rangle \times \langle \overline{5} \rangle \cong C_2 \times C_{2^{k-2}}. \qquad (1.49)$$

Proof. The first two assertions are trivial. Suppose that $k \geq 3$. We observe that (1.48) now becomes

$$|\mathcal{U}(\mathbb{Z}/2^k\mathbb{Z})| = 2^k - 2^{k-1} = 2^{k-1}. \qquad (1.50)$$

In particular the order of $\overline{5}$, as an element of (the Abelian multiplicative group) $\mathcal{U}(\mathbb{Z}/2^k\mathbb{Z})$, is $o(\overline{5}) = 2^r$ for some $1 \leq r \leq k - 1$.

Claim 1: For $k \geq 3$ we have $5^{2^{k-3}} \equiv 1 + 2^{k-1} \bmod 2^k$.

We proceed by induction on k. For $k = 3$ this is easy: indeed we have $5^1 = 5 \equiv 1 + 4 \bmod 8$.

Assume the congruence holds for some $k \geq 3$ and let us prove it for $k + 1$. Observe that there exists $h \in \mathbb{Z}$ such that

$$5^{2^{k-3}} = 1 + 2^{k-1} + h2^k. \qquad (1.51)$$

We have

$$5^{2^{(k+1)-3}} = 5^{2^{k-2}}$$
$$= \left(5^{2^{k-3}}\right)^2$$
$$\text{(by (1.51))} = \left(1 + 2^{k-1} + h2^k\right)^2$$
$$= 1 + 2^k + h2^{k+1} + 2^{2k-2} + (h + h^2)2^{2k}$$
$$\equiv 1 + 2^k \bmod 2^{k+1},$$

where the last congruence follows from the fact that, recalling that $k \geq 3$, $h2^{k+1} + 2^{k-3}2^{k+1} + (h + h^2)2^{k-1}2^{k+1} \equiv 0 \bmod 2^{k+1}$. The proof of the claim is completed.

It follows from Claim 1 that $r \geq k - 2$ since $1 + 2^{k-1} \not\equiv 1 \bmod 2^k$.

Moreover, the order of $\overline{-1}$, as an element of (the multiplicative group) $\mathcal{U}(\mathbb{Z}/2^k\mathbb{Z})$, is clearly $o(\overline{-1}) = 2$.

<u>Claim 2</u>: $\langle \overline{5} \rangle \cap \langle \overline{-1} \rangle = \{\overline{1}\}$.

Indeed, suppose by contradiction that $\overline{-1} \in \langle \overline{5} \rangle$. Then we can find a positive integer s such that $\overline{-1} = \overline{5}^s$, equivalently, $5^s \equiv -1 \bmod 2^k$ and therefore, *a fortiori*, $5^s \equiv -1 \bmod 4$. But this is impossible, since $5 \equiv 1 \bmod 4$ yields $5^s \equiv 1 \bmod 4$. The claim follows.

Recalling (1.50), we have

$$2^{k-1} = |\mathcal{U}(\mathbb{Z}/2^k\mathbb{Z})| \geq |\langle \overline{-1} \rangle \times \langle \overline{5} \rangle| = |\langle \overline{-1} \rangle| \cdot |\langle \overline{5} \rangle| = 2 \cdot 2^r \geq 2 \cdot 2^{k-2} = 2^{k-1}$$

so that $r = k - 2$, that is, $\langle \overline{5} \rangle \cong C_{2^{k-2}}$, and (1.49) follows. $\qquad\square$

Theorem 1.5.8 *Let $p \neq 2$ be a prime and $k \geq 1$. Then we have*

$$\mathcal{U}(\mathbb{Z}/p^k\mathbb{Z}) \cong C_{p^k - p^{k-1}}. \tag{1.52}$$

Proof. First of all, we note that for $k = 1$ the statement reduces to that of Theorem 1.1.21. Thus, we may assume $k \geq 2$.

Let $p - 1 = p_1^{k_1} p_2^{k_2} \cdots p_t^{k_t}$ denote the prime factorization of $p - 1$ and observe that $p_i \neq p$ for all $i = 1, 2, \ldots, t$. Since $\mathcal{U}(\mathbb{Z}/p^k\mathbb{Z})$ is Abelian and $|\mathcal{U}(\mathbb{Z}/p^k\mathbb{Z})| = (p - 1)p^{k-1}$ (by (1.48)), we can apply Theorem 1.3.7 and write $\mathcal{U}(\mathbb{Z}/p^k\mathbb{Z}) = G_1 \times G_2$ where $|G_1| = p - 1$ and $|G_2| = p^{k-1}$.

<u>Claim 1</u>: $G_1 \cong C_{p-1}$.

Consider the map $\Phi \colon \mathbb{Z}/p^k\mathbb{Z} \to \mathbb{Z}/p\mathbb{Z}$ defined by setting $\Phi(\overline{m}) = \widetilde{m}$ where $\overline{m} = m + p^k\mathbb{Z}$ and $\widetilde{m} = m + p\mathbb{Z}$, $m \in \mathbb{Z}$. We remark that Φ is well defined because if $m \equiv n \bmod p^k$ then $m \equiv n \bmod p$, equivalently, $\widetilde{m} \supseteq \overline{m}$, for all $m, n \in \mathbb{Z}$, so that the partition of $\mathbb{Z}$ induced by the congruence mod p^k is finer than the one induced by the congruence mod p. In particular, Φ is surjective. Let $m, n \in \mathbb{Z}$. Then we have

$$\Phi(\overline{m} \cdot \overline{n}) = \Phi(\overline{mn}) = \widetilde{mn} = \widetilde{m} \cdot \widetilde{n} = \Phi(\overline{m})\Phi(\overline{n})$$

so that the restriction ϕ of Φ to $\mathcal{U}(\mathbb{Z}/p^k\mathbb{Z})$ yields a group homomorphism of $\mathcal{U}(\mathbb{Z}/p^k\mathbb{Z})$ onto $\mathcal{U}(\mathbb{Z}/p\mathbb{Z})$.

Now, by Theorem 1.1.21, $\mathcal{U}(\mathbb{Z}/p\mathbb{Z}) \cong C_{p-1}$, and $|G_2| = p^{k-1}$. Thus every element $g_2 \in G_2$ has order $o(g_2) = p^h$ for some $0 \leq h \leq k - 1$. Its image under Φ has order $o(\Phi(g_2)) = p^{h'}$ for some $0 \leq h' \leq h$ but since $\gcd(p, p - 1) = 1$, necessarily $h' = 0$, that is, $g_2 \in \mathrm{Ker}(\Phi)$. This shows that $G_2 \subseteq \mathrm{Ker}(\Phi)$. Since

$$p^{k-1}(p - 1) = |\mathcal{U}(\mathbb{Z}/p^k\mathbb{Z})| = |\mathrm{Ker}(\Phi)| \cdot |\mathcal{U}(\mathbb{Z}/p\mathbb{Z})| = |\mathrm{Ker}(\Phi)|(p - 1),$$

we have that $|\text{Ker}(\Phi)| = p^{k-1}$ and therefore $G_2 = \text{Ker}(\Phi)$. Then

$$G_1 \cong \frac{G_1 \times G_2}{G_2} \cong \frac{\mathcal{U}(\mathbb{Z}/p^k\mathbb{Z})}{\text{Ker}(\Phi)} \cong C_{p-1},$$

and the claim follows. Notice that we have also proved that $G_2 = \{\overline{m} \in \mathbb{Z}/p^k\mathbb{Z} : m \equiv 1 \bmod p\}$.

<u>Claim 2</u>: $G_2 \cong C_{p^{k-1}}$.

We first prove, by induction on $h \in \mathbb{N}$, the following identities

$$(1+p)^{p^h} \equiv 1 \bmod p^{h+1} \tag{1.53}$$

and

$$(1+p)^{p^h} \not\equiv 1 \bmod p^{h+2}. \tag{1.54}$$

For $h = 0$ this is clear: (1.53) becomes $1 + p \equiv 1 \bmod p$ and (1.54) becomes $1 + p \not\equiv 1 \bmod p^2$. Assume the result for some $h \geq 0$ and let us prove it for $h + 1$. Now, (1.53) implies that $(1 + p)^{p^h} = 1 + rp^{h+1}$ for some $r \in \mathbb{Z}$, while (1.54) implies that $p \nmid r$. Therefore

$$
\begin{aligned}
(1+p)^{p^{h+1}} &= \left[(1+p)^{p^h} \right]^p \\
&= \left[1 + rp^{h+1} \right]^p \\
&= \sum_{j=0}^{p} \binom{p}{j} r^j p^{jh+j} \\
&= 1 + \binom{p}{1} rp^{h+1} + \left(\binom{p}{2} r^2 p^{2h+2} + \sum_{j=3}^{p} \binom{p}{j} r^j p^{jh+j} \right) \\
&= 1 + rp^{h+2} + sp^{h+3}
\end{aligned}
$$

where $s = \sum_{j=2}^{p} \binom{p}{j} r^j p^{(j-1)h+j-3} \in \mathbb{N}$ since, for all $h \geq 0$, $p | \binom{p}{2}$, so that $p^{h+3} | \binom{p}{2} p^{2h+2}$, and $p^{h+3} | p^{jh+j}$ for all $j \geq 3$.

We deduce that $(1 + p)^{p^{h+1}} \equiv 1 \bmod p^{h+2}$ and, since $p \nmid r$ by (1.54), $(1 + p)^{p^{h+1}} \not\equiv 1 \bmod p^{h+3}$. This proves the induction.

Taking $h = k - 1$ in (1.53) and $h = k - 2$ in (1.54), we deduce that the element $\overline{1+p} \in \mathcal{U}(\mathbb{Z}/p^k\mathbb{Z})$ has multiplicative order $o(\overline{1+p}) = p^{k-1}$ and therefore it generates a cyclic group of order p^{k-1}. Thus, the second claim follows as well.

Finally, from the two claims it follows that $\mathcal{U}(\mathbb{Z}/p^k\mathbb{Z}) = G_1 \times G_2 \cong C_{p-1} \times C_{p^{k-1}}$ and it is cyclic (of order $p^k - p^{k-1}$) by Proposition 1.2.15 (or Proposition 1.2.5). $\qquad\square$

Corollary 1.5.9 (Gauss) *Let $n \geq 2$. Then $\mathcal{U}(\mathbb{Z}/n\mathbb{Z})$ is cyclic if and only if one of the following cases holds: (i) $n = 2$, (ii) $n = 4$, (iii) $n = p^k$, (iv) $n = 2p^k$, where, in (iii) and (iv), p is an odd prime and $k \geq 1$.*

Proof. Consider the factorization (1.16). Suppose first that $t = 1$. If $p_1 = 2$, then, by Theorem 1.5.7, $\mathcal{U}(\mathbb{Z}/n\mathbb{Z})$ is cyclic if and only if $k_1 = 1$ or $k_1 = 2$ (note that for the "only if" part we should also invoke Proposition 1.2.17). This covers cases (i) and (ii). On the other hand, if $p_1 > 2$, then (iii) follows immediately from Theorem 1.5.8.

Suppose now that n is not a power of a prime, so that $t \geq 2$. If there exist $1 \leq i < j \leq t$ such that p_i and p_j are both odd, then, from Theorem 1.5.8, we deduce that $\mathcal{U}(\mathbb{Z}/n\mathbb{Z})$ contains a subgroup isomorphic to $C_{p_i^{k_i} - p_i^{k_i-1}} \times C_{p_j^{k_j} - p_j^{k_j-1}}$, where both $p_i^{k_i} - p_i^{k_i-1}$ and $p_j^{k_j} - p_j^{k_j-1}$ are even. As a consequence, $\mathcal{U}(\mathbb{Z}/n\mathbb{Z})$ contains a subgroup isomorphic to $C_2 \oplus C_2$, which is not cyclic (cf. Proposition 1.2.17). Since a subgroup of a cyclic group is also cyclic, this prevents $\mathcal{U}(\mathbb{Z}/n\mathbb{Z})$ from being cyclic.

It only remains the case when n is even (so that $p_1 = 2$) and $t = 2$. If $k_1 > 1$, then, also keeping in mind Theorem 1.5.7, $\mathcal{U}(\mathbb{Z}/n\mathbb{Z})$ contains a subgroup isomorphic to $C_2 \oplus C_{p_2^{k_2} - p_2^{k_2-1}}$. Since $p_2^{k_2} - p_2^{k_2-1}$ is even, by the argument above we deduce that $\mathcal{U}(\mathbb{Z}/n\mathbb{Z})$ cannot be cyclic. Finally, if $k_1 = 1$, so that $n = 2p_2^{k_2}$, we have $\mathcal{U}(\mathbb{Z}/n\mathbb{Z}) \cong C_{p_2^{k_2} - p_2^{k_2-1}}$. This covers the case (iv) and completes our analysis. $\qquad\square$

In the case where $\mathcal{U}(\mathbb{Z}/n\mathbb{Z})$ is cyclic (cf. Corollary 1.5.9), a generator of $\mathcal{U}(\mathbb{Z}/n\mathbb{Z})$ is called a *primitive root* mod n.

1.6 The endomorphism ring of a finite Abelian *p*-group

We now examine the structure of the endomorphism ring of a finite (not necessarily cyclic) Abelian group A. Observe that, by virtue of Theorem 1.3.7 and Theorem 1.4.2, it suffices to reduce to the case when A is a p-group. We thus suppose that

$$A = \bigoplus_{j=1}^{h} \mathbb{Z}_{p^{m_j}} \tag{1.55}$$

where p is prime and

$$1 \leq m_1 \leq m_2 \leq \cdots \leq m_h \tag{1.56}$$

(note that, in contrast with (1.35), in (1.56) we have reversed the order of the m_js). We closely follow the arguments in [72].

We first introduce some specific notation. If R is a unital commutative ring, we denote by $\mathfrak{M}_h(R)$ the set of all $h \times h$ matrices with coefficients in R. We now recall some basic facts of matrix theory; we refer to the monographs by Horn and Johnson [75] and by Lancaster and Tismenetsky [91] as a general reference for further details (although these books treat complex matrices, the results that we use can be easily adapted for $\mathfrak{M}_h(R)$; see also the book by Malcev [114]). Let $B = (b_{i,j})_{i,j=1}^h \in \mathfrak{M}_h(R)$. We denote by adj$(B)$ the *adjugate* of B (in [91], following an older terminology, the term "adjoint" is used instead), that is, the matrix whose (i, j)-entry is equal to $(-1)^{i+j}B_{j,i}$, where $B_{j,i}$ is the (j, i)-th *minor* (of order $h - 1$) of B, that is, the determinant of the matrix obtained by deleting row j and column i from B. Since these determinants are expressed as polynomials in the coefficients, we have that adj$(B) \in \mathfrak{M}_h(R)$ for all $B \in \mathfrak{M}_h(R)$. Moreover, adj(B) satisfies the fundamental identity

$$B \cdot \text{adj}(B) = \text{adj}(B) \cdot B = I \cdot \det(B). \tag{1.57}$$

As a consequence, B is invertible in $\mathfrak{M}_h(R)$ if and only if $\det(B)$ is an invertible element in R and, if this is the case, one has

$$B^{-1} = \det(B)^{-1}\text{adj}(B).$$

In particular, if B is invertible, adj(B) is the unique matrix satisfying (1.57). Moreover, if R is a field, then B is invertible if and only if $\det(B) \neq 0$.

Continuing with our purpose of setting notation, an element of $\mathbb{Z}^h$ (respectively A) will be represented by a column vector $\mathbf{n} = (n_j)_{j=1}^h$ (respectively $\overline{\mathbf{n}} = (\overline{n_j})_{j=1}^h$), where $n_j \in \mathbb{Z}$ (respectively $\overline{n_j} \in \mathbb{Z}/p^{m_j}\mathbb{Z}$) for $j = 1, 2, \ldots, h$. Note that we use the same notation for the different congruence classes mod p^{m_j}, $j = 1, 2, \ldots, h$. Also, for $j = 1, 2, \ldots, h$, we set $\delta_j = (\delta_{i,j})_{i=1}^h \in \mathbb{Z}^h$ (respectively $\mathbf{a}_j = (\overline{\delta_{i,j}})_{i=1}^h \in A$, where $\overline{\delta_{i,j}} \in \mathbb{Z}_{p^{m_i}}$). This way, we have $\mathbf{n} = \sum_{j=1}^h n_j\delta_j$ and $\overline{\mathbf{n}} = \sum_{j=1}^h n_j\mathbf{a}_j$ for all $\mathbf{n} \in \mathbb{Z}^h$. Moreover,

$$A = \langle \mathbf{a}_1 \rangle \oplus \langle \mathbf{a}_2 \rangle \oplus \cdots \oplus \langle \mathbf{a}_h \rangle. \tag{1.58}$$

Given a matrix $B = (b_{i,j})_{i,j=1}^h \in \mathfrak{M}_h(\mathbb{Z})$ and $\mathbf{n} \in \mathbb{Z}^h$, the usual product $B\mathbf{n}$ is given by $B\mathbf{n} = \sum_{i,j=1}^h b_{i,j}n_j\delta_i$. In other words, setting $\mathbf{b}_j = (b_{i,j})_{i=1}^h = \sum_{i=1}^h b_{i,j}\delta_i \in \mathbb{Z}^h$, we have

$$B\delta_j = \mathbf{b}_j = \sum_{i=1}^h b_{i,j}\delta_i, \tag{1.59}$$

for all $j = 1, 2, \ldots, h$.

Moreover, for all $j = 1, 2, \ldots, h$, we denote by $\pi_j : \mathbb{Z} \to \mathbb{Z}/p^{m_j}\mathbb{Z}$ the standard quotient map, that is $\pi_j(n_j) = \overline{n_j}$ for all $n_j \in \mathbb{Z}$, and by $\pi : \mathbb{Z}^h \to A$ the map defined by

$$\pi(\mathbf{n}) = \pi\left(\sum_{j=1}^{h} n_j \delta_j\right) = \sum_{j=1}^{h} \overline{n_j} \mathbf{a}_j = \overline{\mathbf{n}},$$

for all $\mathbf{n} \in \mathbb{Z}^h$. Note that π is a group homomorphism.

We now introduce a subring of $\mathfrak{M}_h(\mathbb{Z})$ that plays a fundamental role in the description of End(A). We set

$$\mathcal{R} = \mathcal{R}(p; m_1, m_2, \ldots, m_h)$$
$$= \{B = (b_{i,j})_{i,j=1}^h \in \mathfrak{M}_h(\mathbb{Z}) : p^{m_i - m_j} | b_{i,j}, \text{ for all } 1 \le j < i \le h\}. \quad (1.60)$$

The fact that $\mathcal{R}$ is a subring of $\mathfrak{M}_h(\mathbb{Z})$ will be proved below.

For instance, if $h = 4$, $m_1 = 1$, $m_2 = 3$, $m_3 = 4$ and $m_4 = 7$ then

$$\mathcal{R}(p; 1, 3, 4, 7) = \left\{ \begin{pmatrix} c_{1,1} & c_{1,2} & c_{1,3} & c_{1,4} \\ p^2 c_{2,1} & c_{2,2} & c_{2,3} & c_{2,4} \\ p^3 c_{3,1} & p c_{3,2} & c_{3,3} & c_{3,4} \\ p^6 c_{4,1} & p^4 c_{4,2} & p^3 c_{4,3} & c_{4,4} \end{pmatrix} : c_{i,j} \in \mathbb{Z}, \ i, j = 1, 2, 3, 4 \right\}.$$

Consider the diagonal matrix

$$P = \begin{pmatrix} p^{m_1} & 0 & \cdots & 0 & 0 \\ 0 & p^{m_2} & \cdots & 0 & 0 \\ \vdots & \vdots & \ddots & \vdots & \vdots \\ 0 & 0 & \cdots & p^{m_{h-1}} & 0 \\ 0 & 0 & \cdots & 0 & p^{m_h} \end{pmatrix}.$$

Proposition 1.6.1

(i) *A matrix $B \in \mathfrak{M}_h(\mathbb{Z})$ belongs to $\mathcal{R}$ if and only if it can be represented in the form*

$$B = PCP^{-1} \quad (1.61)$$

for some $C \in \mathfrak{M}_h(\mathbb{Z})$;

(ii) *$\mathcal{R}$ is a unital ring;*

(iii) *adj(B) $\in \mathcal{R}$ for all invertible $B \in \mathcal{R}$.*

Proof. (i) Let $C = (c_{i,j})_{i,j=1}^h \in \mathfrak{M}_h(\mathbb{Z})$ then

$$PCP^{-1} = (p^{m_i - m_j} c_{i,j})_{i,j=1}^h \quad (1.62)$$

clearly belongs to $\mathcal{R}$. Conversely, suppose that $B = (b_{i,j})_{i,j=1}^h \in \mathcal{R}$ and consider the matrix $C = (c_{i,j})_{i,j=1}^h \in \mathfrak{M}_h(\mathbb{Z})$ defined by

$$c_{i,j} = p^{m_j-m_i} b_{i,j} = \begin{cases} b_{i,j}/p^{m_i-m_j} & \text{if } i > j \\ p^{m_j-m_i} b_{i,j} & \text{if } i \le j. \end{cases}$$

From the right hand side above (1.56) and (1.60), it follows that, indeed, $c_{i,j} \in \mathbb{Z}$ for all $1 \le i, j \le h$. Moreover, we deduce from (1.62) that C satisfies (1.61).

(ii) Let $B_1, B_2 \in \mathcal{R}$. Then, by (i), there exist $C_1, C_2 \in \mathfrak{M}_h(\mathbb{Z})$ such that $B_1 = PC_1P^{-1}$ and $B_2 = PC_2P^{-1}$. It follows that $B_1 + B_2 = P(C_1 + C_2)P^{-1} \in \mathcal{R}$ and $B_1B_2 = PC_1C_2P^{-1} \in \mathcal{R}$. Moreover, it is clear from the definitions that the identity matrix $I \in \mathcal{R}$.

(iii) Let $B \in \mathcal{R}$ be invertible. Then, by (i), there exists $C \in \mathfrak{M}_h(\mathbb{Z})$ such that (1.61) holds: we deduce that $\det(B) = \det(C) \ne 0$. Setting $\widetilde{B} = P\mathrm{adj}(C)P^{-1}$ we have

$$\widetilde{B}B = P\mathrm{adj}(C)CP^{-1} = \det(B)I = PC\mathrm{adj}(C)P^{-1} = B\widetilde{B}$$

and, by uniqueness of the adjugate satisfying (1.57) for invertible elements, we deduce that $\widetilde{B} = \mathrm{adj}(B)$. It follows from (i) that $\mathrm{adj}(B) \in \mathcal{R}$. $\square$

We are now in position to describe $\mathrm{End}(A)$ as a quotient of the ring $\mathcal{R}$.

Theorem 1.6.2 *The map* $\Psi\colon \mathcal{R} \to \mathrm{End}(A)$ *defined by setting*

$$\Psi(B)\overline{\mathbf{n}} = \pi(B\mathbf{n}) \tag{1.63}$$

for all $\mathbf{n} \in \mathbb{Z}^h$ *and* $B \in \mathcal{R}$, *is well defined and is a surjective unital ring homomorphism. Moreover,*

$$\mathrm{Ker}(\Psi) = \{(b_{i,j})_{i,j=1}^h \in \mathcal{R} : p^{m_i} | b_{i,j} \text{ for all } i, j = 1, 2, \dots, h\} \tag{1.64}$$

so that $\mathrm{End}(A) \cong \mathcal{R}/\mathrm{Ker}(\Psi)$.

Proof. Let $B \in \mathcal{R}$. First of all, we verify that $\Psi(B)$ is well defined. Suppose that $\mathbf{n}, \mathbf{n}' \in \mathbb{Z}^h$ satisfy $\overline{\mathbf{n}} = \overline{\mathbf{n}'}$, that is, $n_j \equiv n_j' \bmod p^{m_j}$, equivalently $p^{m_j} | (n_j - n_j')$, for all $j = 1, 2, \dots, h$. Let also $B = (b_{i,j})_{i,j=1}^h \in \mathcal{R}$. Then we have

$$\pi(B\mathbf{n}) - \pi(B\mathbf{n}') = \pi(B(\mathbf{n} - \mathbf{n}')) = \pi\Big(\sum_{i=1}^h \sum_{j=1}^h b_{i,j}(n_j - n_j')\delta_i\Big) = \overline{\mathbf{0}}$$

since, if $i > j$,

$$b_{i,j}(n_j - n_j') = \frac{b_{i,j}}{p^{m_i-m_j}} \cdot \frac{n_j - n_j'}{p^{m_j}} \cdot p^{m_i}$$

where $\frac{b_{i,j}}{p^{m_i-m_j}} \in \mathbb{Z}$ by (1.60), and $\frac{n_j-n'_j}{p^{m_j}} \in \mathbb{Z}$ by our assumptions, while, if $i \leq j$, then $b_{i,j}(n_j - n'_j)$ is divisible by p^{m_j} and therefore by p^{m_i}, since $m_i \leq m_j$. Thus $\Psi(B)$ is well defined.

The fact that $\Psi(B) \in \text{End}(A)$ follows easily from the linearity of the maps π and $\mathbf{n} \mapsto B\mathbf{n}$.

In order to show that Ψ is surjective, let $M \in \text{End}(A)$. Then we can find $B = (b_{i,j})_{i,j=1}^h \in \mathfrak{M}_h(\mathbb{Z})$ such that $M(\mathbf{a}_j) = \sum_{i=1}^h b_{i,j}\mathbf{a}_i$, $j = 1, 2, \ldots, h$. Since $M(\overline{\mathbf{0}}) = \overline{\mathbf{0}}$ and $p^{m_j}\mathbf{a}_j = 0$, we get (since M is a homomorphism)

$$\overline{\mathbf{0}} = M(p^{m_j}\mathbf{a}_j) = p^{m_j}M(\mathbf{a}_j) = p^{m_j}\sum_{i=1}^h b_{i,j}\mathbf{a}_i = \sum_{i=1}^h p^{m_j}b_{i,j}\mathbf{a}_i$$

which forces $p^{m_j}b_{i,j} \equiv 0 \bmod p^{m_i}$ for all $i, j = 1, 2, \ldots, h$ (cf. Proposition 1.2.2). In particular, $p^{m_i-m_j}|b_{i,j}$ for all $1 \leq j < i \leq h$, so that $B \in \mathcal{R}$.

As a consequence, given $\mathbf{n} \in \mathbb{Z}$ we have

$$M(\overline{\mathbf{n}}) = M(\sum_{j=1}^h n_j\mathbf{a}_j) = \sum_{j=1}^h n_jM(\mathbf{a}_j) = \sum_{i,j=1}^h n_jb_{i,j}\mathbf{a}_i$$

$$= \pi(\sum_{i,j=1}^h n_jb_{i,j}\delta_i) = \pi(B\mathbf{n}) = \Psi(B)(\overline{\mathbf{n}}).$$

In other words, $\Psi(B) = M$ and surjectivity follows.

We now show that Ψ is a unital ring homomorphism and determine its kernel.

It is clear that $\Psi(I) = \text{Id}_A$, the identity endomorphism of A and $\Psi(0) = 0_A$, the zero endomorphism of A.

Let now $B = (b_{i,j})_{i,j=1}^h$, $B_1, B_2 \in \mathcal{R}$, and $n_1, n_2, \ldots, n_h \in \mathbb{Z}$. Then, we have

$$\Psi(B_1 + B_2)\overline{\mathbf{n}} = \pi((B_1 + B_2)\mathbf{n}) = \pi(B_1\mathbf{n} + B_2\mathbf{n}) = \pi(B_1\mathbf{n}) + \pi(B_2\mathbf{n})$$
$$= \Psi(B_1)\overline{\mathbf{n}} + \Psi(B_2)\overline{\mathbf{n}},$$

showing that $\Psi(B_1 + B_2) = \Psi(B_1) + \Psi(B_2)$. Similarly,

$$\Psi(B_1)\Psi(B_2)\overline{\mathbf{n}} = \Psi(B_1)\pi(B_2\mathbf{n}) = \pi(B_1B_2\mathbf{n}) = \Psi(B_1B_2)\overline{\mathbf{n}},$$

showing that $\Psi(B_1B_2) = \Psi(B_1)\Psi(B_2)$.

Finally,

$$B \in \text{Ker}(\Psi) \Leftrightarrow \Psi(B)\mathbf{a}_j = \overline{\mathbf{0}} \text{ for all } j = 1, 2, \ldots, h$$
$$\Leftrightarrow \pi(B\delta_j) = \overline{\mathbf{0}} \text{ for all } j = 1, 2, \ldots, h$$
$$\Leftrightarrow \pi_i(b_{i,j}) = 0 \text{ for all } i, j = 1, 2, \ldots, h$$
$$\Leftrightarrow p_i^{m_i}|b_{i,j} \text{ for all } i, j = 1, 2, \ldots, h,$$

and (1.64) follows. $\qquad\square$

Corollary 1.6.3 *In* (1.58) *we have*

$$\Psi(B)\mathbf{a}_j \in \langle \mathbf{a}_j \rangle$$

for $j = 1, 2, \ldots, h$, if and only if $p^{m_i} | b_{i,j}$ for $i \neq j$. Moreover, if this is the case, then there exists a diagonal matrix $B' \in \mathcal{R}$ such that $\Psi(B') = \Psi(B)$.

Proof. We have

$$\Psi(B)\mathbf{a}_j = \pi(B\delta_j)$$

$$(\text{by } (1.59)) = \sum_{i=1}^{h} \pi(b_{i,j})\mathbf{a}_i$$

and therefore

$$\Psi(B)\mathbf{a}_j \in \langle \mathbf{a}_j \rangle \Leftrightarrow b_{i,j} \equiv 0 \mod p^{m_i} \text{ for } i \neq j$$
$$\Leftrightarrow p^{m_i} | b_{i,j} \text{ for } i \neq j.$$

The last statement follows from (1.64). $\qquad\square$

1.7 The automorphisms of a finite Abelian p-group

Let p be a prime number and $h \geq 1$ be an integer. Recall that we denote by $\mathbb{F}_p$ the finite field $\mathbb{Z}/p\mathbb{Z}$ and by $\bar{n} \in \mathbb{F}_p$ the congruence class of $n \in \mathbb{Z}$ mod p. We denote by $GL(h, \mathbb{F}_p)$ the group of all invertible matrices in $\mathfrak{M}_h(\mathbb{F}_p)$. We need to introduce this group in order to characterize the invertible elements in $\text{End}(A)$, where A is a p-group as in (1.55).

Let now $B = (b_{i,j})_{i,j=1}^{h} \in \mathfrak{M}_h(\mathbb{Z})$. We set

$$\bar{B} = (\overline{b_{i,j}})_{i,j=1}^{h} \in \mathfrak{M}_h(\mathbb{F}_p). \tag{1.65}$$

As we remarked above, $\bar{B}$ is invertible in $\mathfrak{M}_h(\mathbb{F}_p)$ if and only if $\det \bar{B} \neq \bar{0}$. Since $\det(\bar{B}) = \overline{\det(B)}$, we have that $\bar{B} \in GL(h, \mathbb{F}_p)$ if and only if $p \nmid \det(B)$. Moreover, if this is the case, B is also invertible in $\mathcal{R}$ (and in $\mathfrak{M}_h(\mathbb{Z})$).

With the same notation from the previous section we have:

Theorem 1.7.1 *Let $B \in \mathcal{R}$ and set $M = \Psi(B) \in \text{End}(A)$. Then M is invertible (i.e. $M \in \text{Aut}(A)$) if and only if $\bar{B} \in GL(h, \mathbb{F}_p)$.*

Proof. Suppose first that $\bar{B}$ is invertible, so that p does not divide $\det(B)$. Then we can find $q \in \mathbb{Z}$ such that

$$q \cdot \det(B) \equiv 1 \mod p^{m_j} \text{ for all } j = 1, 2, \ldots, h.$$

Indeed, $\gcd(\det(B), p^{m_h}) = 1$ so that $\det(B)$ has an inverse q mod p^{m_h}, which is also an inverse mod p^{m_j} for all other js (recall that $m_h \geq m_j$). Let us set

$$C = q \cdot \text{adj}(B).$$

By Proposition 1.6.1.(iii), we have $C \in \mathcal{R}$. Moreover, $\Psi(C)\Psi(B) = \Psi(CB) = \Psi((q \cdot \det(B))I) = \text{Id}_A \in \text{End}(A)$ and, similarly, $\Psi(B)\Psi(C) = \text{Id}_A$, so that $M = \psi(B)$ is invertible, with inverse $\Psi(C)$.

Conversely, suppose that $M = \psi(B)$ is invertible. Recalling that Ψ is surjective, we can find $C \in \mathcal{R}$ such that $\Psi(C)$ is the inverse of M. It follows that $\Psi(I) = \text{Id}_A = \Psi(B)\Psi(C) = \Psi(BC)$, equivalently, $\Psi(BC - I) = 0$ (the trivial endomorphism of A), so that, by (1.64), p divides all coefficients of $BC - I$, and therefore

$$\overline{B} \cdot \overline{C} = \overline{BC} = \overline{I} \in \mathfrak{M}_h(\mathbb{F}_p).$$

It follows that $\overline{B} \in \text{GL}(h, \mathbb{F}_p)$. $\qquad\square$

We now need some basic notions on group actions that will be recalled with more details in Section 10.4.

Denote by $\mathcal{V}$ the set of all h-tuples $(A_1, A_2, \ldots, A_h)$ such that

- $A_1, A_2, \ldots, A_h$ are subgroups of A
- $A_j \cong \mathbb{Z}_{p^{m_j}}, j = 1, 2, \ldots, h$
- $A = A_1 \oplus A_2 \oplus \cdots \oplus A_h$.

In other words, $\mathcal{V}$ is the set of all invariant factors decompositions of A (see Theorem 1.3.1 and (1.55)). Then the group $\text{Aut}(A)$ acts on $\mathcal{V}$ and this action is clearly transitive. We want to identify the stabilizer of a fixed decomposition.

Corollary 1.7.2 *The stabilizer of the decomposition (1.58) is given by the set of all $\Psi(B)$, where*

$$B = \begin{pmatrix} b_1 & 0 & 0 & 0 \\ 0 & b_2 & 0 & 0 \\ \vdots & & \ddots & 0 \\ 0 & 0 & \cdots & b_h \end{pmatrix}$$

is diagonal with $\overline{b_i} \in \mathcal{U}(\mathbb{Z}/p^{m_i}\mathbb{Z})$, $i = 1, 2, \ldots, h$. In particular, its cardinality is equal to

$$(p-1)^h \prod_{i=1}^{h} p^{m_i - 1}.$$

Proof. It is an immediate consequence of Corollary 1.6.3, Corollary 1.5.3, and (1.48). □

1.8 The cardinality of Aut(A)

In this section we determine the cardinality of Aut(A), where A is a p-group as in (1.55). To this end, keeping in mind (1.56), we introduce the following numbers:

$$t_j = \max\{j \leq t \leq h : m_t = m_j\}$$

and

$$s_i = \min\{1 \leq s \leq i : m_s = m_i\}$$

for all $i, j = 1, 2, \ldots, h$. Note that $t_j \geq j$ and $s_i \leq i$ for all $i, j = 1, 2, \ldots, h$; in particular, $t_h = h$ and $s_1 = 1$.

Lemma 1.8.1 *For all $i, j = 1, 2, \ldots, h$ we have*

$$m_i > m_j \Leftrightarrow i > t_j \Leftrightarrow j < s_i$$

and

$$m_i \leq m_j \Leftrightarrow i \leq t_j \Leftrightarrow j \geq s_i.$$

Proof. The proof is an immediate consequence of the fact that $m_1 \leq m_2 \leq \cdots \leq m_h$, and it is left as an exercise. □

Corollary 1.8.2 *Let $B = (b_{i,j})_{i,j=1}^{h} \in \mathcal{R}$ and $1 \leq i, j \leq h$. Suppose that $i > t_j$ (equivalently, $j < s_i$). Then, with the notation as in (1.65), $\overline{b_{i,j}} = \overline{0}$.*

Proof. If $i > t_j$ (equivalently, $j < s_i$), then $m_i > m_j$ and, as $B \in \mathcal{R}$, we have $p^{m_i - m_j} | b_{i,j}$. □

Theorem 1.8.3

$$|\text{Aut}(A)| = \prod_{k=1}^{h} (p^{t_k} - p^{k-1}) \prod_{j=1}^{s_h} p^{m_j(h-t_j)} \prod_{i=1}^{h} p^{(m_i-1)(h-s_i+1)}.$$

Proof. Let $B \in \mathcal{R}$ and suppose that $\Psi(B) \in \text{End}(A)$ is invertible (i.e. $\Psi(B) \in \text{Aut}(A)$). Then, by virtue of Theorem 1.7.1, $\overline{B} \in \text{GL}(h, \mathbb{F}_p)$ and, by Corollary

1.8.2, $\overline{B} = (\overline{b_{i,j}})_{i,j=1}^{h} = (c_{i,j})_{i,j=1}^{h}$ is given by

$$
\begin{pmatrix}
c_{1,1} & c_{1,2} & \cdots & c_{1,h} \\
c_{2,1} & c_{2,2} & \cdots & c_{2,h} \\
\vdots & \vdots & \vdots & \vdots \\
c_{t_1,1} & c_{t_1,2} & \cdots & c_{t_1,h} \\
0 & c_{t_1+1,2} & \cdots & c_{t_1+1,h} \\
\vdots & \vdots & \vdots & \vdots \\
0 & c_{t_2,2} & \cdots & c_{t_2,h} \\
0 & 0 & \cdots & c_{t_2+1,h} \\
\vdots & \vdots & \ddots & \vdots \\
0 & 0 & \cdots & c_{t_h,h}
\end{pmatrix}
=
\begin{pmatrix}
c_{1,s_1} & c_{1,2} & \cdots & \cdots & \cdots & \cdots & c_{1,h} \\
0 & \cdots & 0 & c_{2,s_2} & \cdots & \cdots & c_{2,h} \\
\vdots & \vdots & \vdots & \vdots & \vdots & \ddots & \vdots \\
0 & 0 & \cdots & 0 & c_{h,s_h} & \cdots & c_{h,h}
\end{pmatrix}.
$$

$$(1.66)$$

Note that the two matrices above have the same 0 entries: by Corollary 1.8.2 and Lemma 1.8.1, $c_{i,j} = 0$ if $i > t_j$, equivalently, if $j < s_i$.

Using the left hand side in the above equality, we have the following counting: the first column may be chosen in $p^{t_1} - 1$ distinct ways (the -1 because we have to discard the $\overline{0}$-column), the second one in $p^{t_2} - p$ ways (the $-p$ because we have to discard the p multiples of the first column, since the two have to be independent).

Continuing this way, setting

$$\mathcal{G} = \{C \in \mathrm{GL}(h, \mathbb{F}_p) : C = \overline{B}, B \in \mathcal{R}, \Psi(B) \in \mathrm{Aut}(A)\},$$

we have that

$$|\mathcal{G}| = \prod_{k=1}^{h}(p^{t_k} - p^{k-1}). \qquad (1.67)$$

Let us now fix $C = \overline{B} \in \mathcal{G}$ as in (1.66), and set

$$\mathcal{M}_C = \{\Psi(B) : B \in \mathcal{R}, \overline{B} = C\} \subset \mathrm{Aut}(A).$$

We claim that

$$|\mathcal{M}_C| = \prod_{j=1}^{s_h} p^{m_j(h-t_j)} \prod_{i=1}^{h} p^{(m_i-1)(h-s_i+1)} \qquad (1.68)$$

(in particular, $n = |\mathcal{M}_C|$ is independent of $C \in \mathcal{G}$).

For each $1 \leq j \leq h$ there are exactly $h - t_j$ zeroes below the entry $c_{t_j,j}$ (cf. the left hand side of (1.66)) and the ith one (corresponding to the (i, j)-entry: note that $i > t_j \geq j$, equivalently, $m_i > m_j$) gives p^{m_j} distinct possibilities for the (i, j)-th entry of $B \in \mathcal{R}$ (each yielding a different $\Psi(B)$): by (1.60) and (1.64) it must be an element of $p^{m_i - m_j} \mathbb{Z}/p^{m_i} \mathbb{Z} \cong \mathbb{Z}/p^{m_j} \mathbb{Z}$. The last isomorphism follows from the elementary congruence: for $x, y \in \mathbb{Z}$, $xp^{m_i - m_j} \equiv yp^{m_i - m_j} \mod p^{m_i}$ if and only if $x \equiv y \mod p^{m_j}$.

This yields the first factor in the right hand side of (1.68). Note also that $t_j = h \Leftrightarrow m_j = m_h \Leftrightarrow j \geq s_h$.

On the other hand, for each $1 \leq i \leq h$ there are exactly $h - s_i + 1$ terms on the right of and including c_{i, s_i} (cf. the right hand side of (1.66)) and the jth one (corresponding to the (i, j)-entry: note that $j \geq s_i$, equivalently, $m_i \leq m_j$) gives rise to $p^{m_i - 1}$ distinct possibilities for the (i, j)-th entry of $B \in \mathcal{R}$: it must be equal to $c_{i,j} +$ an element of $p\mathbb{Z}/p^{m_i}\mathbb{Z} \cong \mathbb{Z}/p^{m_i - 1}\mathbb{Z}$ (again by virtue of (1.64)). This yields the second factor in the right hand side of (1.68) proving the claim. Since

$$|\mathrm{Aut}(A)| = \sum_{C \in \mathcal{G}} |\mathcal{M}_C| = |\mathcal{G}| \cdot n,$$

the statement follows from (1.67) and (1.68). $\square$

We now count the number of invariant factors decompositions of A (recall the notation preceding Corollary 1.7.2).

Corollary 1.8.4

$$|\mathcal{V}| = p^{h(h-1)/2} \prod_{k=1}^{h} \left(\sum_{\ell=0}^{t_k - k} p^{\ell} \right) \cdot \prod_{j=1}^{s_h} p^{m_j(h - t_j)} \cdot \prod_{i=1}^{h} p^{(m_i - 1)(h - s_i)}.$$

Proof. Divide the cardinality of $\mathrm{Aut}(A)$ in Theorem 1.8.3 by the cardinality of the stabilizer in Corollary 1.7.2. $\square$

Example 1.8.5 Suppose that $m_1 = m_2 = \cdots = m_h = m$. Then $\mathrm{Aut}(A)$ is group-isomorphic to $\mathrm{GL}(h, \mathbb{Z}/\mathbb{Z}_{p^m})$ (here, according with our notation, $\mathbb{Z}/\mathbb{Z}_{p^m}$ is no more a field if $m \geq 2$, but just a ring). Indeed, in this case, $\mathcal{R} \equiv \mathfrak{M}_h(\mathbb{Z})$ and, by (1.64), we have $\mathrm{End}(A) \cong \mathfrak{M}_h(\mathbb{Z}/\mathbb{Z}_{p^m})$. Now $t_j = h$ for $j = 1, 2, \ldots, h$ and $s_i = 1$ for $i = 1, 2, \ldots, h$, so that, by Theorem 1.8.3, we have

$$|\mathrm{Aut}(\underbrace{\mathbb{Z}_{p^m} \oplus \mathbb{Z}_{p^m} \oplus \cdots \oplus \mathbb{Z}_{p^m}}_{h \text{ times}})| = p^{(m-1)h^2} \cdot \prod_{k=1}^{h} (p^h - p^{k-1}).$$

Two particular cases are relevant. For $h = 1$, we find

$$|\mathrm{Aut}(\mathbb{Z}_{p^m})| = p^{m-1}(p - 1) = p^m - p^{m-1}$$

and this agrees with the results in Theorem 1.5.7 and Theorem 1.5.8 (but this follows also from the fact that $\varphi(p^m) = p^m - p^{m-1}$, cf. Corollary 1.5.3). If, in addition, one has $m_1 = m_2 = \cdots = m_h = 1$ we get

$$\mathrm{Aut}(\underbrace{\mathbb{Z}_p \oplus \mathbb{Z}_p \oplus \cdots \oplus \mathbb{Z}_p}_{h \text{ times}}) \cong \mathrm{GL}(h, \mathbb{F}_p)$$

and

$$|\mathrm{Aut}(\underbrace{\mathbb{Z}_p \oplus \mathbb{Z}_p \oplus \cdots \oplus \mathbb{Z}_p}_{h \text{ times}})| = |\mathrm{GL}(h, \mathbb{F}_p)| = \prod_{k=1}^{h}(p^h - p^{k-1}),$$

which coincides with (1.67), since $t_k = h$ for all $k = 1, 2, \ldots, h$.

2

The Fourier transform on finite Abelian groups

This chapter is a fairly complete exposition of the basic character theory and the Fourier transform on finite Abelian groups. Our presentation is inspired by our monograph [29], and the books by Terras [159] and Nathanson [118]; Section 2.6 contains a recent result of Terence Tao [157]. The results established here will be used and generalized in almost every subsequent chapter.

2.1 Some notation

In this section, we fix some basic notation and results of "harmonic analysis" on finite sets. Further notation and results will be developed in Section 8.7. These two sections constitute the core of the preliminaries in finite harmonic analysis.

Let X be a finite set and denote by $L(X) = \{f : X \to \mathbb{C}\}$ the vector space of all complex-valued functions defined on X. Clearly, $\dim L(X) = |X|$, where $|\cdot|$ denotes cardinality.

For $x \in X$ we denote by δ_x the *Dirac function* centered at x, that is, the element $\delta_x \in L(X)$ defined by

$$\delta_x(y) = \begin{cases} 1 & \text{if } y = x \\ 0 & \text{if } y \neq x \end{cases}$$

for all $y \in X$.

The set $\{\delta_x : x \in X\}$ is a natural basis for $L(X)$ and if $f \in L(X)$ then $f = \sum_{x \in X} f(x)\delta_x$.

The space $L(X)$ is endowed with the *scalar product* defined by setting

$$\langle f_1, f_2 \rangle = \sum_{x \in X} f_1(x)\overline{f_2(x)}$$

for $f_1, f_2 \in L(X)$, and we denote by $\|f\| = \sqrt{\langle f, f \rangle}$ the *norm* of $f \in L(X)$. Note that the basis $\{\delta_x : x \in X\}$ is orthonormal with respect to $\langle \cdot, \cdot \rangle$. Sometimes we shall write $\langle \cdot, \cdot \rangle_{L(X)}$ (respectively $\| \cdot \|_{L(X)}$) to emphasize the space where the scalar product (the norm) is defined, if other spaces are also considered.

For a subset $Y \subseteq X$, we regard $L(Y)$ as a subspace of $L(X)$ and we denote by $1_Y = \sum_{y \in Y} \delta_y \in L(X)$ the *characteristic function* of Y. In particular, if $Y = X$ we simply write $\mathbf{1}$ (the constant function with value 1) instead of 1_X.

For $Y_1, Y_2, \ldots, Y_m \subseteq X$ we write $X = Y_1 \bigsqcup Y_2 \bigsqcup \cdots \bigsqcup Y_m$ to indicate that the Y_js constitute a *partition* of X, that is $X = Y_1 \cup Y_2 \cup \cdots \cup Y_m$ and $Y_i \cap Y_j = \varnothing$ whenever $i \neq j$. In other words, the symbol $\bigsqcup$ denotes a *disjoint union*. In particular, if we write $Y \bigsqcup Y'$ we implicitly assume that $Y \cap Y' = \varnothing$. Note that if $X = Y_1 \bigsqcup Y_2 \bigsqcup \cdots \bigsqcup Y_m$ then $L(X) \cong L(Y_1) \oplus L(Y_2) \oplus \cdots \oplus L(Y_m)$.

If $A \colon L(X) \to L(X)$ is a linear operator, setting

$$a(x, y) = [A\delta_y](x) \tag{2.1}$$

for all $x, y \in X$, we have that

$$[Af](x) = \sum_{y \in X} a(x, y) f(y) \tag{2.2}$$

for all $x \in X$ and $f \in L(X)$, and we say that the *matrix* $a = (a(x, y))_{x, y \in X}$, *indexed* by X, *represents* the operator A. We denote by $\mathrm{End}(L(X))$ the complex vector space of all linear operators $A \colon L(X) \to L(X)$.

With our notation, the identity operator $I \in \mathrm{End}(L(X))$ is represented by the identity matrix, which may be expressed as $I = (\delta_x(y))_{x, y \in X}$.

If $A_1, A_2 \in \mathrm{End}(L(X))$ are represented by the matrices a_1 and a_2, respectively, then the composition $A = A_1 \circ A_2 \in \mathrm{End}(L(X))$ is represented by the corresponding product of matrices $a = a_1 \cdot a_2$ that is

$$a(x, y) = \sum_{z \in X} a_1(x, z) a_2(z, y).$$

For $k \in \mathbb{N}$ we denote by $a^k = \left(a^{(k)}(x, y)\right)_{x, y \in X}$ the product of k copies of a, namely, $a^{(0)} = I$, the identity matrix, and, for $k \geq 1$,

$$a^{(k)}(x, y) = \sum_{z \in X} a^{(k-1)}(x, z) a(z, y).$$

We remark that (2.2) can also be interpreted as the product of the matrix a with the column vector $f = (f(x))_{x \in X}$.

Given a matrix a and a column (respectively a row) vector f, we denote by a^T and by f^T the transposed matrix (i.e. $a^T(x, y) = a(y, x)$ for all $x, y \in X$) and the row (respectively column) transposed vector. This way, we also denote by $f^T A$ the function given by

$$[f^T A](y) = \sum_{x \in X} f(x) a(x, y). \qquad (2.3)$$

If X is a set of cardinality $|X| = n$ and $k \leq n$, then a k-subset of X is a subset $A \subseteq X$ such that $|A| = k$.

If $v_1, v_2, \ldots, v_m$ are vectors in a vector space V, then $\langle v_1, v_2, \ldots, v_m \rangle$ will denote their linear span.

We end with the most elementary tool of finite harmonic analysis. It will be used and rediscovered many times (see Proposition 8.1.4, Theorem 9.1.7, and Example 10.4.3).

Proposition 2.1.1 *Let X be a finite set and set $W_0 = \{f \in L(X) : f$ is constant$\}$ and $W_1 = \{f \in L(X) : \sum_{x \in X} f(x) = 0\}$. Then we have the following orthogonal decomposition:*

$$L(X) = W_0 \oplus W_1. \qquad (2.4)$$

Proof. Let $f \in L(X)$. Setting $f_0(x) = \frac{1}{|X|} \sum_{y \in X} f(y)$ for all $x \in X$ we have $f_0 \in W_0$ and $f_1 = f - f_0 \in W_1$, so that $L(X) = W_0 + W_1$. Moreover, it is immediate to check that $W_0 \perp W_1$, so that (2.4) is an orthogonal direct sum. $\qquad \square$

2.2 Characters of finite cyclic groups

Let $n \geq 2$ and denote, as usual, by $\mathbb{Z}_n = \{\overline{0}, \overline{1}, \ldots, \overline{n-1}\}$ the cyclic group of order n, written additively.

Recall (cf. Section 2.1) that $L(\mathbb{Z}_n)$ denotes the complex vector space of all functions $f : \mathbb{Z}_n \to \mathbb{C}$. Note that if $f \in L(\mathbb{Z}_n)$, then the function $F : \mathbb{Z} \to \mathbb{C}$ defined by $F(x) = f(\overline{x})$ for all $x \in \mathbb{Z}$ is n-*periodic* (namely $F(x + n) = F(x)$ for all $x \in \mathbb{Z}$) and the map $f \mapsto F$ establishes a bijective correspondence between the elements in $L(\mathbb{Z}_n)$ and the n-periodic complex functions on $\mathbb{Z}$.

In the following, by abuse of language, we shall identify f and F and use the same notation for the corresponding arguments: in particular, for $x \in \mathbb{Z}$ the (a priori improperly defined) expressions $f(x)$ and $F(\overline{x})$ stand for $f(\overline{x}) = F(x)$. More generally, we shall use the same notation for an element $x \in \mathbb{Z}$ and its image in $\mathbb{Z}_n$ (in other words, we shall omit the bar-symbol "—" in the notation for $\overline{x} \in \mathbb{Z}_n$) and we shall use the bar-symbol to denote conjugation of complex numbers. In particular, we shall use the symbols $\sum_{y=0}^{n-1}$ to denote the sum $\sum_{y \in \mathbb{Z}_n}$ over all elements of $\mathbb{Z}_n$, and we regard the Dirac functions $\delta_x, x \in \mathbb{Z}_n$, as elements in $L(\mathbb{Z}_n)$.

Let us set

$$\omega = \exp\frac{2\pi i}{n} = \cos\frac{2\pi}{n} + i\sin\frac{2\pi}{n} \in \mathbb{C}.$$

We recall that ω is an n-th primitive root of 1 and that the n-th complex roots of the unit are ω^k, $k = 0, 1, \ldots, n-1$. Note that $\omega^z = \omega^{z+n}$ for all $z \in \mathbb{Z}$ so that (cf. the comments above) the map $z \mapsto \omega^z$ defines an element of $L(\mathbb{Z}_n)$. More generally, for $x \in \mathbb{Z}_n$, we denote by $\chi_x \in L(\mathbb{Z}_n)$ the function $z \mapsto \omega^{zx}$.

Definition 2.2.1 The functions $\chi_x \in L(\mathbb{Z}_n)$ are called the *characters* of $\mathbb{Z}_n$.

Note that $\chi_x(y) = \chi_y(x) \in \mathbb{T} = \{z \in \mathbb{C} : |z| = 1\}$, $\chi_y(-x) = \overline{\chi_y(x)}$ for all $x, y \in \mathbb{Z}_n$, and $\chi_0 = \mathbf{1}$, the constant function.

The basic identity for the characters is

$$\chi_z(x + y) = \chi_z(x)\chi_z(y)$$

for all $x, y, z \in \mathbb{Z}_n$ and, in the following lemma, we prove that, in fact, it is a "characteristic" property of characters.

Lemma 2.2.2 *If $\phi\colon \mathbb{Z}_n \to \mathbb{T}$ satisfies $\phi(x+y) = \phi(x)\phi(y)$ for all $x, y \in \mathbb{Z}_n$, then $\phi = \chi_z$ for some $z \in \mathbb{Z}_n$.*

Proof. First note that since $\phi(0) = \phi(0+0) = \phi(0)\phi(0)$, we necessarily have $\phi(0) = 1$. As a consequence, $1 = \phi(0) = \phi(\underbrace{1 + 1 + \cdots + 1}_{n \text{ times}}) = \phi(1)^n$ and we deduce that $\phi(1)$ is an n-th root of 1. Therefore there exists $z \in \mathbb{Z}_n$ such that $\phi(1) = \omega^z$. This gives $\phi(x) = \phi(1)^x = \omega^{zx} = \chi_z(x)$ for all $x \in \mathbb{Z}_n$. $\qquad\square$

Lemma 2.2.3 (Orthogonality relations for characters of $\mathbb{Z}_n$) *Let χ and ψ be two characters of $\mathbb{Z}_n$. Then*

$$\langle \chi, \psi \rangle = n\delta_{\chi,\psi}. \tag{2.5}$$

Proof. Let $x_1, x_2 \in \mathbb{Z}_n$ be such that $\chi = \chi_{x_1}$ and $\psi = \chi_{x_2}$. Let us set $z = \omega^{x_1 - x_2}$ and observe that $\chi_{x_1}(y)\overline{\chi_{x_2}(y)} = \omega^{y(x_1-x_2)} = z^y$ for all $y \in \mathbb{Z}_n$ so that

$$\langle \chi, \psi \rangle = \langle \chi_{x_1}, \chi_{x_2} \rangle = \sum_{y=0}^{n-1} \chi_{x_1}(y)\overline{\chi_{x_2}(y)} = \sum_{y=0}^{n-1} z^y. \tag{2.6}$$

Suppose first that $\chi \neq \psi$, i.e. $x_1 \neq x_2$. Then z is a nontrivial root of the unity (i.e. $z^n - 1 = 0$ and $z - 1 \neq 0$) and from the identity

$$z^n - 1 = (z-1)(1 + z + \cdots + z^{n-1}) = (z-1)\sum_{y=0}^{n-1} z^y$$

we deduce that $\sum_{y=0}^{n-1} z^y = 0$ and the quantity (2.6) vanishes. On the other hand, if $\chi = \psi$, that is $x_1 = x_2$, then $z = \omega^{x_1-x_2} = 1$ and the quantity (2.6) equals n. $\square$

Note that if $\chi = \chi_{x_1}$ and $\psi = \chi_{x_2}$, then (2.5) may be expressed as

$$\langle \chi_{x_1}, \chi_{x_2} \rangle = n\delta_{x_1,x_2} \equiv n\delta_0(x_1 - x_2). \tag{2.7}$$

From the lemma and the fact that $\chi_x(y) = \chi_y(x)$ for all $x, y \in \mathbb{Z}_n$ we immediately deduce the following *dual orthogonality relations for characters of* $\mathbb{Z}_n$:

$$\sum_{x \in \mathbb{Z}_n} \chi_x(y_1)\overline{\chi_x(y_2)} = n\delta_0(y_1 - y_2) \tag{2.8}$$

for all $y_1, y_2 \in \mathbb{Z}_n$.

2.3 Characters of finite Abelian groups

Let A be a finite Abelian group, written additively.

Definition 2.3.1 A *character* of A is a map $\chi : A \to \mathbb{T}$ such that

$$\chi(x + y) = \chi(x)\chi(y)$$

for all $x, y \in A$.

The set $\widehat{A}$ of all characters of A is an Abelian group with respect to the product $\widehat{A} \times \widehat{A} \ni (\chi, \psi) \mapsto \chi \cdot \psi \in \widehat{A}$ defined by $(\chi \cdot \psi)(x) = \chi(x)\psi(x)$, for all $x \in A$. It is called the *dual* of A.

Remark 2.3.2 Note that if $A = \mathbb{Z}_n$, then Definition 2.3.1 coincides with Definition 2.2.1 and $\widehat{\mathbb{Z}}_n = \{\chi_x : x \in \mathbb{Z}_n\}$ is isomorphic to $\mathbb{Z}_n$. Indeed, since for all $x \in \mathbb{Z}_n$ we have $(\chi_1)^x = \chi_x$, then $\widehat{\mathbb{Z}}_n$ is the cyclic group (necessarily of order n) generated by χ_1 (alternatively, as $\chi_{x+y}(z) = \chi_z(x + y) = \chi_z(x)\chi_z(y) = \chi_x(z)\chi_y(z)$ for all $x, y, z \in \mathbb{Z}_n$, the map $x \mapsto \chi_x$ yields a surjective (and therefore bijective) group homomorphism $\mathbb{Z}_n \to \widehat{\mathbb{Z}}_n$).

Proposition 2.3.3 *Let A be a finite Abelian group and let*

$$A = \mathbb{Z}_{m_1} \oplus \mathbb{Z}_{m_2} \oplus \cdots \oplus \mathbb{Z}_{m_k} \tag{2.9}$$

be a decomposition of A as direct sum of cyclic groups (see, for instance Theorem 1.3.1 or Corollary 1.3.10). Set $\omega_j = \exp\frac{2\pi i}{m_j}$, $j = 1, 2, \ldots, k$, and, for $y = (y_1, y_2, \ldots, y_k) \in A$, define $\chi_y : A \to \mathbb{T}$ by setting

$$\chi_y(x) = \omega_1^{x_1 y_1} \omega_2^{x_2 y_2} \cdots \omega_k^{x_k y_k} \tag{2.10}$$

for all $x = (x_1, x_2, \ldots, x_k) \in A$. *Then* χ_y *is a character of* A, *every character of* A *is of this form, and distinct* y *yield distinct characters. In particular,* $|\widehat{A}| = |A|$.

Proof. The first assertion, namely that (2.10) defines a character of A, is straightforward:

$$\chi_y(x + x') = \omega_1^{(x_1 + x_1')y_1} \omega_2^{(x_2 + x_2')y_2} \cdots \omega_k^{(x_k + x_k')y_k}$$
$$= \omega_1^{x_1 y_1} \omega_2^{x_2 y_2} \cdots \omega_k^{x_k y_k} \cdot \omega_1^{x_1' y_1} \omega_2^{x_2' y_2} \cdots \omega_k^{x_k' y_k}$$
$$= \chi_y(x) \chi_y(x')$$

for all $y = (y_1, y_2, \ldots, y_k)$, $x = (x_1, x_2, \ldots, x_k)$, and $x' = (x_1', x_2', \ldots, x_k') \in A$.

Let us show that every character of A is of the form (2.10). Let $\chi : A \to \mathbb{T}$ be a character of A. We first observe that, for all $j = 1, 2, \ldots, k$, the restriction $\chi|_{\mathbb{Z}_{m_j}}$ of χ to the subgroup $\mathbb{Z}_{m_j} \leq A$ is a character of $\mathbb{Z}_{m_j}$ so that, by Lemma 2.2.2, there exists $y_j \in \mathbb{Z}_{m_j}$ such that $\chi|_{\mathbb{Z}_{m_j}} = \chi_{y_j}$. As a consequence, setting $y = (y_1, y_2, \ldots, y_k) \in A$, we have

$$\chi(x) = \chi(x_1, x_2, \ldots, x_k)$$
$$= \chi|_{\mathbb{Z}_{m_1}}(x_1) \chi|_{\mathbb{Z}_{m_2}}(x_2) \cdots \chi|_{\mathbb{Z}_{m_k}}(x_k)$$
$$= \chi_{y_1}(x_1) \chi_{y_2}(x_2) \cdots \chi_{y_k}(x_k)$$
$$= \omega_1^{x_1 y_1} \omega_2^{x_2 y_2} \cdots \omega_k^{x_k y_k}$$
$$= \chi_y(x)$$

for all $x = (x_1, x_2, \ldots, x_k) \in A$. This shows that $\widehat{A} = \{\chi_y : y \in A\}$. $\qquad \square$

Note that with the notation above we may write

$$\chi_y(x) = \prod_{j=1}^{k} \chi_{y_j}(x_j) \tag{2.11}$$

for all $x = (x_1, x_2, \ldots, x_k)$ and $y = (y_1, y_2, \ldots, y_k) \in A$.

Corollary 2.3.4 *Let A be a finite Abelian group. Then the dual group $\widehat{A}$ is isomorphic to A.*

Proof. With the notation in Proposition 2.3.3, it is straightforward to check that $\chi_{y+y'} = \chi_y \cdot \chi_{y'}$ for all $y, y' \in A$ (cf. the particular case where $A = \mathbb{Z}_n$ in Remark 2.3.2) so that the map $y \mapsto \chi_y$ yields a surjective (and therefore bijective, since $|A| = |\widehat{A}|$) group homomorphism from A onto $\widehat{A}$. $\qquad \square$

Proposition 2.3.5 (Orthogonality relations for characters of A) *Let $\chi, \psi \in \widehat{A}$ and $x, y \in A$. Then we have the* orthogonality relations

$$\langle \chi, \psi \rangle = |A| \delta_{\chi, \psi} \tag{2.12}$$

and the dual orthogonality relations

$$\sum_{\chi \in \widehat{A}} \chi(x)\overline{\chi(y)} = |A|\delta_{x,y} \equiv |A|\delta_0(x-y). \tag{2.13}$$

Proof. By virtue of Proposition 2.3.3 and the notation therein, we can find $x = (x_1, x_2, \ldots, x_k)$ and $y = (y_1, y_2, \ldots, y_k) \in A$ such that $\chi = \chi_x$ and $\psi = \chi_y$. Using the notation in (2.11) we then have

$$\langle \chi, \psi \rangle = \langle \chi_x, \chi_y \rangle = \sum_{z \in A} \chi_x(z)\overline{\chi_y(z)}$$

$$= \sum_{z \in A} \prod_{j=1}^{k} \chi_{x_j}(z_j)\overline{\chi_{y_j}(z_j)}$$

$$= \prod_{j=1}^{k} \sum_{z_j \in \mathbb{Z}_{m_j}} \chi_{x_j}(z_j)\overline{\chi_{y_j}(z_j)}$$

$$= \prod_{j=1}^{k} \langle \chi_{x_j}, \chi_{y_j} \rangle$$

$$(\text{by Lemma 2.2.3}) = \prod_{j=1}^{k} m_j \delta_{x_j, y_j}$$

$$= |A|\delta_{x,y} = |A|\delta_{\chi, \psi}. \qquad \square$$

We remark that the isomorphism in Corollary 2.3.4 (given by (2.11)) depends on the choice of the decomposition of A and therefore on the generators for the corresponding cyclic subgroups, that is, it depends on the coordinates. There is, however, an intrinsic isomorphism between A and the dual of $\widehat{A}$, called the *bidual* of A and denoted by $\widehat{\widehat{A}}$, given by

$$A \ni a \mapsto \psi_a \in \widehat{\widehat{A}}, \tag{2.14}$$

where $\psi_a(\chi) = \chi(a)$ for all $\chi \in \widehat{A}$.

Exercise 2.3.6 Prove that the map (2.14) is a group isomorphism.

This duality is similar to the (possibly more familiar) one coming from linear algebra. Recall that if V is a finite dimensional vector space over a field $\mathbb{F}$, the dual of V is the vector space V^* consisting of all $\mathbb{F}$-linear maps $f : V \to \mathbb{F}$. Then if $\{v_1, v_2, \ldots, v_d\} \subset V$ ($d = \dim_{\mathbb{F}} V$) is a basis for V and $\{v_1^*, v_2^*, \ldots, v_d^*\} \subset V^*$ is the dual basis (defined by $v_i^*(v_j) = \delta_{i,j}$ for all $i, j = 1, 2, \ldots, d$), then the map $v_i \mapsto v_i^*$ linearly extends to a (unique) vector space isomorphism $\varphi : V \to V^*$. Note that φ depends on the choice of basis $\{v_1, v_2, \ldots, v_d\}$. However, denoting by $V^{**} = (V^*)^*$ the bidual of V, the map $V \ni v \mapsto \psi_v \in V^{**}$ defined by $\psi_v(v^*) = v^*(v)$ for all $v^* \in V^*$ yields a canonical vector space isomorphism between V and V^{**}.

Returning back to group theory, the isomorphism $A \to \widehat{\widehat{A}}$ extends to locally compact Abelian groups: this is called *Pontrjagin duality*. As an example, if $\mathbb{T} = \{z \in \mathbb{C} : |z| = 1\}$ denotes the unit circle, then $\widehat{\mathbb{T}} \cong \mathbb{Z}$ and $\widehat{\widehat{\mathbb{T}}} \cong \mathbb{T}$ (this is the setting of *classical Fourier series*, see, for instance, the monographs on abstract harmonic analysis by Rudin [134], Katznelson [85], and Loomis [98]).

2.4 The Fourier transform

Let A be a finite Abelian group. We recall (cf. Section 2.1) that $L(A)$, the complex vector space of all functions $f \colon A \to \mathbb{C}$, is equipped with an inner product $\langle \cdot, \cdot \rangle_{L(A)}$ (for short $\langle \cdot, \cdot \rangle$) defined by

$$\langle f_1, f_2 \rangle = \sum_{x \in A} f_1(x)\overline{f_2(x)}$$

for all $f_1, f_2 \in L(A)$. We also denote by $\| \cdot \|_{L(A)}$ (for short $\| \cdot \|$) the associated norm.

Note the $\dim(L(A)) = |A|$ and therefore, by virtue of the orthogonality relations for characters (Proposition 2.3.5), the set $\{\chi_x : x \in A\}$ is an orthogonal basis for $L(A)$.

Definition 2.4.1 The *Fourier transform* of a function $f \in L(A)$ is the function $\widehat{f} \in L(\widehat{A})$ defined by

$$\widehat{f}(\chi) = \langle f, \chi \rangle = \sum_{y \in A} f(y)\overline{\chi(y)} \tag{2.15}$$

for all $\chi \in \widehat{A}$. Then $\widehat{f}(\chi)$ is called the *Fourier coefficient* of f with respect to χ. Moreover, we shall denote by $\mathcal{F}f = \frac{1}{\sqrt{|A|}}\widehat{f}$ the *normalized* Fourier transform of $f \in L(A)$.

When $A = \mathbb{Z}_n$ (the cyclic group of order n), and $f \in L(\mathbb{Z}_n)$ we shall call $\frac{1}{n}\widehat{f}$ the *Discrete Fourier transform* (briefly, *DFT*) of f.

The following two theorems express, in a functional form, the fact that the χs constitute an orthogonal basis of the space $L(A)$.

Theorem 2.4.2 (Fourier inversion formula) *For every $f \in L(A)$ we have*

$$f = \frac{1}{|A|} \sum_{\chi \in \widehat{A}} \widehat{f}(\chi)\chi. \tag{2.16}$$

Proof. Let $f \in L(A)$ and $x \in A$. Then

$$\frac{1}{|A|} \sum_{\chi \in \widehat{A}} \widehat{f}(\chi)\chi(x) = \frac{1}{|A|} \sum_{\chi \in \widehat{A}} \sum_{y \in A} f(y)\overline{\chi(y)}\chi(x) =$$

$$= \frac{1}{|A|} \sum_{y \in A} f(y) \sum_{\chi \in \widehat{A}} \overline{\chi(y)}\chi(x) =$$

$$\text{(by (2.13))} \quad = \frac{1}{|A|} \sum_{y \in A} f(y)|A|\delta_0(y - x) = f(x). \qquad \square$$

Theorem 2.4.3 (Plancherel and Parseval formulas) *For* $f, g \in L(A)$ *we have (*Plancherel formula*)*

$$\|\widehat{f}\|_{L(\widehat{A})} = \sqrt{|A|}\|f\|_{L(A)}$$

*and (*Parseval formula*)*

$$\langle \widehat{f}, \widehat{g} \rangle_{L(\widehat{A})} = |A|\langle f, g \rangle_{L(A)}.$$

Proof. We first prove the Parseval formula:

$$\langle \widehat{f}, \widehat{g} \rangle_{L(\widehat{A})} = \sum_{\chi \in \widehat{A}} \widehat{f}(\chi)\overline{\widehat{g}(\chi)}$$

$$= \sum_{\chi \in \widehat{A}} \left(\sum_{y_1 \in A} f(y_1)\overline{\chi(y_1)} \right) \left(\sum_{y_2 \in A} \overline{g(y_2)}\chi(y_2) \right)$$

$$= \sum_{y_1 \in A} \sum_{y_2 \in A} f(y_1)\overline{g(y_2)} \sum_{\chi \in \widehat{A}} \overline{\chi(y_1)}\chi(y_2) =$$

$$\text{(by (2.13))} \quad = |A| \sum_{y \in A} f(y)\overline{g(y)} = |A|\langle f, g \rangle_{L(A)}.$$

The Plancherel formula is immediately deduced from the Parseval formula by taking $g = f$. $\qquad \square$

Exercise 2.4.4 Show that $\widehat{\delta_x}(\chi) = \overline{\chi(x)}$ for all $x \in A$ and $\chi \in \widehat{A}$.

For $f_1, f_2 \in L(A)$ we define their *convolution* as the function $f_1 * f_2 \in L(A)$ given by

$$(f_1 * f_2)(x) = \sum_{y \in A} f_1(x - y)f_2(y)$$

for all $x \in A$.

Definition 2.4.5 An *algebra* over a field $\mathbb{F}$ is a vector space $\mathcal{A}$ over $\mathbb{F}$ endowed with a product such that $\mathcal{A}$ is a ring with respect to the sum and the product and the following associative laws, for the product and multiplication by a scalar, hold:

$$\alpha(AB) = (\alpha A)B = A(\alpha B)$$

for all $\alpha \in \mathbb{F}$ and $A, B \in \mathcal{A}$.

An algebra $\mathcal{A}$ is *commutative* (or *Abelian*) if it is commutative as a ring, namely if $AB = BA$ for all $A, B \in \mathcal{A}$; it is *unital* if it has a *unit*, that is, there exists an element $I \in \mathcal{A}$ such that $AI = IA = A$ for all $A \in \mathcal{A}$.

Given two algebras $\mathcal{A}_1$ and $\mathcal{A}_2$ over the field $\mathbb{F}$, a bijective linear map $\Phi \colon \mathcal{A}_1 \to \mathcal{A}_2$ such that $\Phi(ab) = \Phi(a)\Phi(b)$ for all $a, b \in \mathcal{A}_1$ is called an *isomorphism*. If such an isomorphism Φ exists, one says that the algebras $\mathcal{A}_1$ and $\mathcal{A}_2$ are isomorphic, and we write $\mathcal{A}_1 \cong \mathcal{A}_2$.

In the following proposition we present the main properties of the convolution product in $L(A)$.

Proposition 2.4.6 *For all $f, f_1, f_2, f_3 \in L(A)$ one has*

(i) $f_1 * f_2 = f_2 * f_1$ *(commutativity)*
(ii) $(f_1 * f_2) * f_3 = f_1 * (f_2 * f_3)$ *(associativity)*
(iii) $(f_1 + f_2) * f_3 = f_1 * f_3 + f_2 * f_3$ *(distributivity)*
(iv) $\widehat{f_1 * f_2} = \widehat{f_1} \cdot \widehat{f_2}$
(v) $\delta_0 * f = f * \delta_0 = f$.

In particular, $L(A)$ is a commutative algebra over $\mathbb{C}$ with unit $I = \delta_0$.

Proof. We prove only (iv), namely that the Fourier transform of the convolution of two functions equals the pointwise product of their Fourier transforms. Let $f_1, f_2 \in L(A)$ and $\chi \in \widehat{A}$. Then we have

$$\widehat{f_1 * f_2}(\chi) = \sum_{x \in A}(f_1 * f_2)(x)\overline{\chi(x)}$$

$$= \sum_{x \in A}\sum_{t \in A} f_1(x-t)f_2(t)\overline{\chi(x-t)}\,\overline{\chi(t)}$$

$$= \widehat{f_1}(\chi)\widehat{f_2}(\chi).$$

The other identities are left as an exercise. $\qquad\square$

The *translation operator* $T_x \in \text{End}(L(A))$, $x \in A$, is defined by:

$$(T_x f)(y) = f(y - x)$$

for all $x, y \in A$ and $f \in L(A)$.

Exercise 2.4.7 Show that $T_x f = f * \delta_x$ and $\widehat{T_x f}(\chi) = \overline{\chi(x)}\widehat{f}(\chi)$ for all $f \in L(A)$, $x \in A$, and $\chi \in \widehat{A}$.

Let $R \in \mathrm{End}(L(A))$. We say that R is *A-invariant* if it commutes with all translations, namely

$$RT_x = T_x R$$

for all $x \in A$. Also we say that R is a *convolution operator* provided there exists $h \in L(A)$ such that $Rf = f * h$ for all $f \in L(A)$: the function h is then called the *(convolution) kernel* of R and we write $R = R_h$.

Exercise 2.4.8

(1) Show that every convolution operator is A-invariant.
(2) Show that
   - $R_{h_1} + R_{h_2} = R_{h_1 + h_2}$;
   - $R_{\alpha h} = \alpha R_h$;
   - $R_{h_1} R_{h_2} = R_{h_1 * h_2}$
   for all $h_1, h_2, h \in L(A)$ and $\alpha \in \mathbb{C}$.
(3) Deduce that $\mathcal{R} = \{R_h : h \in L(A)\}$ is a commutative algebra isomorphic to $L(A)$.

$\mathcal{R}$ is called the *algebra of convolution operators* on A.

Lemma 2.4.9 *The linear operator R associated with the matrix $(r(x, y))_{x,y \in A}$ is A-invariant if and only if*

$$r(x - z, y - z) = r(x, y) \tag{2.17}$$

for all $x, y, z \in A$.

Proof. The linear operator R is A-invariant if and only if, for all $x, z \in A$ and $f \in L(A)$ one has $[T_z(Rf)](x) = [R(T_z f)](x)$, that is,

$$\sum_{u \in A} r(x - z, u) f(u) = \sum_{u \in A} r(x, u) f(u - z),$$

equivalently,

$$\sum_{u \in A} r(x - z, u - z) f(u - z) = \sum_{u \in A} r(x, u) f(u - z).$$

Since the δ_t, $t \in A$, constitute a basis for $L(A)$, taking $f = \delta_{y-z}$ for all $y \in A$, the last equality is in turn equivalent to (2.17). $\qquad \square$

Theorem 2.4.10 *The following conditions are equivalent for $R \in \mathrm{End}(L(A))$:*

(a) *R is A-invariant;*

(b) *R is a convolution operator;*

(c) *every $\chi \in \widehat{A}$ is an eigenvector of R.*

Proof. (a) $\Rightarrow$ (b): by Lemma 2.4.9, A-invariance yields $r(x, y) = r(x - y, 0)$ for all $x, y \in A$, so that if we define $h \in L(A)$ by setting

$$h(x) = r(x, 0) \tag{2.18}$$

for all $x \in A$, we then have $r(x, y) = h(x - y)$ and therefore

$$(Rf)(x) = \sum_{y \in A} h(x - y)f(y) = (h * f)(x)$$

and $R = R_h$ is a convolution operator.

(b) $\Rightarrow$ (c): let $h \in L(A)$ and $\chi \in \widehat{A}$. Suppose that $R = R_h$. Then

$$[R\chi](y) = \sum_{t \in A} \chi(y - t)h(t) = \chi(y) \sum_{t \in A} \overline{\chi(t)}h(t) = \widehat{h}(\chi)\chi(y). \tag{2.19}$$

This shows that every $\chi \in \widehat{A}$ is an eigenvector of R with eigenvalue $\widehat{h}(\chi)$.

Suppose now that every $\chi \in \widehat{A}$ is an eigenvector of R with eigenvalue $\lambda(\chi) \in \mathbb{C}$. Observe that

$$[T_x\chi](y) = \chi(y - x) = \overline{\chi(x)}\chi(y) \tag{2.20}$$

for all $x, y \in A$ and $\chi \in \widehat{A}$. For $\chi \in \widehat{A}$ and $x \in A$ we have

$$\begin{aligned} [RT_x](\chi) &= R(\overline{\chi(x)}\chi) \text{ (by (2.20))} \\ &= \overline{\chi(x)}\lambda(\chi)\chi \\ \text{(by (2.20))} &= \lambda(\chi)T_x(\chi) \\ &= T_x(\lambda(\chi)\chi) \\ &= [T_xR](\chi). \end{aligned}$$

By linearity of R and T_x, and by the Fourier inversion theorem, this shows that $[RT_x](f) = [T_xR](f)$ for all $f \in L(A)$, and (c) $\Rightarrow$ (a) follows as well. $\square$

From the proof of the previous theorem (cf. equation (2.19)) we extract the following.

Corollary 2.4.11 *Let $h \in L(A)$. Then $R_h(\chi) = \widehat{h}(\chi)\chi$ for every $\chi \in \widehat{A}$. In particular, R_h is diagonalizable, its eigenvectors are the characters of A, and its spectrum is given by $\sigma(R_h) = \{\widehat{h}(\chi) : \chi \in \widehat{A}\}$.* $\square$

Corollary 2.4.12 (Trace formula) *Let $h \in L(A)$. Then*

$$\text{Tr}(R_h) = \sum_{\chi \in \widehat{A}} \widehat{h}(\chi) = |A|h(0).$$

Proof. The first equality follows from the previous corollary since $\mathrm{Tr}(R_h) = \sum_{\lambda \in \sigma(R_h)} \lambda$. The second equality follows from the Fourier inversion formula, keeping in mind that $\chi(0) = 1$ for all $\chi \in \widehat{A}$. $\qquad\square$

Exercise 2.4.13 Consider the normalized Fourier transform (cf. Definition 2.4.1), that is, the map $\mathcal{F} \colon L(A) \to L(A)$ defined by

$$[\mathcal{F}f](x) = \frac{1}{\sqrt{|A|}} \widehat{f}(\chi_x) = \frac{1}{\sqrt{|A|}} \sum_{y \in A} f(y) \overline{\chi_x(y)} \qquad (2.21)$$

for all $f \in L(A)$ and $x \in A$ (χ_x as in Proposition 2.3.3).

(1) Show that $\mathcal{F} \in \mathrm{End}(L(A))$ and that it is an isometric bijection.
(2) Show that $\mathcal{F}^{-1}$ is given by $[\mathcal{F}^{-1}f](x) = \frac{1}{\sqrt{|A|}} \widehat{f}(\chi_{-x})$ for all $f \in L(A)$ and $x \in A$.

Definition 2.4.14 Let $f \in L(A)$. We define $f^- \in L(A)$ by setting $f^-(a) = f(-a)$ for all $a \in A$. Then f is called *even* (respectively *odd*) if $f = f^-$ (respectively $f = -f^-$). Similarly, for $\varphi \in L(\widehat{A})$ we set $\varphi^-(\chi) = \varphi(\overline{\chi})$ and we say that φ is even if $\varphi = \varphi^-$.

Exercise 2.4.15 Let $h \in L(A)$.

(1) Show that $\widehat{h^-} = (\widehat{h})^-$. Deduce that h is even if and only if $\widehat{h}$ is even;
(2) show that $\overline{\widehat{h}} = \widehat{(\overline{h})^-}$;
(3) deduce that the following conditions are equivalent:
    (a) h is real valued and even;
    (b) $\widehat{h}$ is real valued and even;
(4) show that $\sigma(R_h) \subset \mathbb{R} \Leftrightarrow h = (\overline{h})^-$.

Exercise 2.4.16 Let $n \geq 1$. A matrix of the form

$$\begin{pmatrix} a_0 & a_1 & a_2 & \cdots & \cdots & a_{n-1} \\ a_{n-1} & a_0 & a_1 & \cdots & \cdots & a_{n-2} \\ a_{n-2} & a_{n-1} & a_0 & \cdots & \cdots & a_{n-3} \\ \vdots & \vdots & \vdots & \cdots & \cdots & \vdots \\ a_1 & a_2 & a_3 & \cdots & \cdots & a_0 \end{pmatrix}$$

with $a_0, a_1, \ldots, a_{n-1} \in \mathbb{C}$ is said to be *circulant*. Denote by $\mathcal{C}_n$ the set of all $n \times n$ circulant matrices.

(1) Let $R, S \in \mathcal{C}_n$ and $\alpha, \beta \in \mathbb{C}$. Show that $RS = SR$ and that $RS, (\alpha R + \beta S) \in \mathcal{C}_n$. Deduce that $\mathcal{C}_n$ is a commutative algebra with unit.
(2) Show that $R \in \mathcal{C}_n$ if and only if its adjoint $R^* \in \mathcal{C}_n$, so that $\mathcal{C}_n$ is closed under adjunction.

(3) Let $\mathcal{B} = \{\delta_0, \delta_1, \ldots, \delta_{n-1}\} \subset L(\mathbb{Z}_n)$ so that $f = \sum_{x=0}^{n-1} f(x)\delta_x$ for every $f \in L(\mathbb{Z}_n)$. Show that $R \in \mathrm{End}(L(\mathbb{Z}_n))$ is a convolution operator if and only if the matrix representing it is circulant.

Hint. If $h \in L(\mathbb{Z}_n)$ is the kernel of R, then $R = R_h$ is represented, with respect to $\mathcal{B}$, by the (circulant) matrix $(h(y - x))_{x,y \in \mathbb{Z}_n}$.

Deduce that $\mathcal{C}_n$ is isomorphic to $L(\mathbb{Z}_n)$ as algebras.

(4) Let $\omega = \exp(\frac{2i\pi}{n}) \in \mathbb{T}$ and set

$$F_n = \frac{1}{\sqrt{n}} \begin{pmatrix} 1 & 1 & \cdots & \cdots & 1 \\ 1 & \omega^{-1} & \omega^{-2} & \cdots & \omega^{-(n-1)} \\ 1 & \omega^{-2} & \omega^{-4} & \cdots & \omega^{-2(n-1)} \\ \vdots & \vdots & \vdots & \cdots & \vdots \\ 1 & \omega^{-(n-1)} & \omega^{-2(n-1)} & \cdots & \omega^{-(n-1)^2} \end{pmatrix}. \quad (2.22)$$

Observe that $F_n \in \mathfrak{M}_n(\mathbb{C})$ is symmetric so that its adjoint F_n^* is equal to its conjugate $\overline{F_n}$. Show also that the orthogonality relations in Lemma 2.2.3 are equivalent to saying that F_n is a unitary matrix.

(5) Prove that a matrix $R \in \mathfrak{M}_n(\mathbb{C})$ is in $\mathcal{C}_n$ if and only if $F_n R F_n^*$ is diagonal. The map $\mathcal{C}_n \ni R \mapsto F_n R F_n^* \in \Delta_n$, where $\Delta_n \subseteq \mathfrak{M}_n(\mathbb{C})$ denotes the subalgebra of all diagonal matrices, is called the *discrete Fourier transform*, briefly *DFT*, on $\mathcal{C}_n$.

2.5 Poisson's formulas and the uncertainty principle

In this section, following the monographs by Nathanson [118] and Terras [159], we treat the finite analogue of two basic properties of the classical Fourier transform.

Let A be a finite Abelian group, B a subgroup of A, and consider the quotient group A/B.

For $f \in L(A/B)$ we define $\widetilde{f} \in L(A)$ by setting $\widetilde{f}(a) = f(a + B)$, for all $a \in A$. In other words, $\widetilde{f} = f \circ \pi$, where $\pi : A \to A/B$ is the canonical quotient map. $\widetilde{f}$ is called the *inflation* of f to A.

Note that the correspondence $f \mapsto \widetilde{f}$ yields an algebra isomorphism between $L(A/B)$ and the subalgebra of $L(A)$ consisting of all functions that are constant on the B-cosets. Moreover, if $\psi \in \widehat{A/B}$, then $\widetilde{\psi} \in \widehat{A}$: indeed $\widetilde{\psi} = \psi \circ \pi$ is a composition of group homomorphisms.

Exercise 2.5.1 Let $\chi \in \widehat{A}$. Show that there exists $\psi \in \widehat{A/B}$ such that $\chi = \widetilde{\psi}$ if and only if $\chi|_B \equiv \mathbf{1}_B$.

Theorem 2.5.2 (Poisson summation formulas) *Let $f \in L(A)$ and let $\mathcal{S} \subseteq A$ be a system of representatives of the B-cosets in A. Then*

$$\frac{1}{|B|} \sum_{b \in B} f(b) = \frac{1}{|A|} \sum_{\psi \in \widehat{A/B}} \widehat{f}(\widetilde{\psi}) \tag{2.23}$$

and

$$\sum_{c \in \mathcal{S}} \left| \sum_{b \in B} f(c+b) \right|^2 = \frac{|B|}{|A|} \sum_{\psi \in \widehat{A/B}} |\widehat{f}(\widetilde{\psi})|^2. \tag{2.24}$$

Proof. Define $f^\sharp \in L(A)$ by setting

$$f^\sharp(a) = \sum_{b \in B} f(a+b)$$

for all $a \in A$. Clearly, $f^\sharp$ is constant on the B-cosets in A. Moreover, for each $\chi \in \widehat{A}$,

$$\widehat{f^\sharp}(\chi) = \sum_{a \in A} f^\sharp(a)\overline{\chi(a)}$$

$$= \sum_{a \in A} \sum_{b \in B} f(a+b)\overline{\chi(a)}$$

$$(\text{setting } c = a+b) \; = \sum_{c \in A} \sum_{b \in B} f(c)\overline{\chi(c-b)}$$

$$= \left[\sum_{b \in B} \chi(b) \right] \cdot \widehat{f}(\chi)$$

$$(\text{by (2.12) applied to } \chi|_B \in \widehat{B}) \; = \begin{cases} |B|\widehat{f}(\chi) & \text{if } \chi|_B = \mathbf{1}_B \\ 0 & \text{otherwise.} \end{cases}$$

As a consequence, taking into account Exercise 2.5.1, $\widehat{f^\sharp}(\chi)$ equals $|B|\widehat{f}(\widetilde{\psi})$ if $\chi = \widetilde{\psi}$ for some $\psi \in \widehat{A/B}$, and vanishes otherwise.

Then, the Fourier inversion formula (cf. Theorem 2.4.2) applied to $f^\sharp$ gives

$$f^\sharp = \frac{|B|}{|A|} \sum_{\psi \in \widehat{A/B}} \widehat{f}(\widetilde{\psi})\widetilde{\psi}$$

that is,

$$\frac{1}{|B|} \sum_{b \in B} f(b+a) = \frac{1}{|A|} \sum_{\psi \in \widehat{A/B}} \widehat{f}(\widetilde{\psi})\widetilde{\psi}(a)$$

for all $a \in A$. In particular, when $a = 0$ we get (2.23). Moreover, applying the Plancherel formula (cf. Theorem 2.4.3) to the function $f^\sharp$, we get

$$\|f^\sharp\|^2_{L(A)} = \frac{1}{|A|}\|\widehat{f^\sharp}\|^2_{L(\widehat{A})} = \frac{|B|^2}{|A|}\sum_{\psi \in \widehat{A/B}}|\widehat{f}(\widetilde{\psi})|^2.$$

Since

$$\|f^\sharp\|^2_{L(A)} = \sum_{a \in A}|f^\sharp(a)|^2$$

$$-\sum_{c \in S}\sum_{b \in B}|f^\sharp(c+b)|^2$$

(since $f^\sharp$ is constant on B-cosets) $= \sum_{c \in S}|B| \cdot |f^\sharp(c)|^2$

$$= |B|\sum_{c \in S}\left|\sum_{b \in B}f(c+b)\right|^2,$$

(2.24) follows. $\qquad\square$

For $f \in L(A)$ we set

$$\text{supp}(f) = \{a \in A : f(a) \neq 0\} \subseteq A,$$

$$\|f\|_\infty = \max\{|f(a)| : a \in A\}$$

and

$$\text{supp}(\widehat{f}) = \{\chi \in \widehat{A} : \widehat{f}(\chi) \neq 0\} \subseteq \widehat{A}.$$

Lemma 2.5.3 *Let $f \in L(A)$. Then*

$$\|f\|^2_{L(A)} \leq \|f\|^2_\infty \cdot |\text{supp}(f)|.$$

Proof. This is a straightforward calculation:

$$\|f\|^2_{L(A)} = \sum_{a \in A}|f(a)|^2 = \sum_{a \in \text{supp}(f)}|f(a)|^2$$

$$\leq \sum_{a \in \text{supp}(f)}\|f\|^2_\infty = \|f\|^2_\infty \cdot |\text{supp}(f)|. \qquad\square$$

Theorem 2.5.4 (Uncertainty principle) *Let $f \in L(A)$ and suppose that $f \neq 0$. Then*

$$|\text{supp}(f)| \cdot |\text{supp}(\widehat{f})| \geq |A|. \qquad (2.25)$$

Proof. From the Fourier inversion formula (Theorem 2.4.2) and the fact $\|\chi\|_\infty \leq 1$ for all $\chi \in \widehat{A}$, we deduce that for every $a \in A$,

$$|f(a)| = \frac{1}{|A|} \left| \sum_{\chi \in \widehat{A}} \widehat{f}(\chi)\chi(a) \right|$$

$$\leq \frac{1}{|A|} \sum_{\chi \in \widehat{A}} |\widehat{f}(\chi)|$$

$$= \frac{1}{|A|} \sum_{\chi \in \mathrm{supp}(\widehat{f})} |\widehat{f}(\chi)|.$$

Taking the max over $a \in A$ and squaring, we get

$$\|f\|_\infty^2 \leq \frac{1}{|A|^2} \left(\sum_{\chi \in \mathrm{supp}(\widehat{f})} |\widehat{f}(\chi)| \right)^2$$

$$= \frac{1}{|A|^2} \left(\sum_{\chi \in \widehat{A}} \mathbf{1}_{\mathrm{supp}(\widehat{f})}(\chi) \cdot |\widehat{f}(\chi)| \right)^2$$

$$(\text{by the Cauchy-Schwarz inequality}) \leq \frac{1}{|A|^2} |\mathrm{supp}(\widehat{f})| \cdot \sum_{\chi \in \widehat{A}} |\widehat{f}(\chi)|^2$$

$$= \frac{1}{|A|^2} |\mathrm{supp}(\widehat{f})| \cdot \|\widehat{f}\|_{L(\widehat{A})}^2$$

$$(\text{by the Plancherel formula}) = \frac{1}{|A|} |\mathrm{supp}(\widehat{f})| \cdot \|f\|_{L(A)}^2$$

$$(\text{by Lemma 2.5.3}) \leq \frac{1}{|A|} \|f\|_\infty^2 \cdot |\mathrm{supp}(f)| \cdot |\mathrm{supp}(\widehat{f})|.$$

Since $f \neq 0$ we have $\|f\|_\infty > 0$ and therefore, comparing the first and the last terms in the above formula, we get the desired inequality. □

Remark 2.5.5 If we take $f = \delta_0$ (the Dirac function at the identity element of A), then $|\mathrm{supp}(\delta_0)| = 1$, while $\widehat{\delta_0}(\chi) = \overline{\chi(0)} = 1$ for all $\chi \in \widehat{A}$ so that $|\mathrm{supp}(\widehat{\delta_0})| = |A|$. In this case, $|\mathrm{supp}(\widehat{\delta_0})| \cdot |\mathrm{supp}(\delta_0)| = |A|$ showing that the lower bound in (2.25) is optimal.

2.6 Tao's uncertainty principle for cyclic groups

In this section we prove an uncertainty principle, due to Tao [157], which improves on the inequality (2.25) when the finite Abelian group A is cyclic of prime order. We first present some general preliminary material on number theory together with some specific tools developed in [157]. Recall that $\mathbb{Z}[x]$ denotes the ring of polynomials with integer coefficients.

Proposition 2.6.1 (Eisenstein's criterion) *Let* $q(x) = a_0 + a_1 x + \cdots + a_n x^n$ $\in \mathbb{Z}[x]$. *Suppose that there exists a prime p such that*

    (i) *p divides $a_0, a_1, \ldots, a_{n-1}$;*
    (ii) *p does not divide a_n;*
    (iii) *p^2 does not divide a_0.*

Then the polynomial q is irreducible over $\mathbb{Z}$.

Proof. By contradiction, suppose that

$$q(x) = (b_0 + b_1 x + \cdots + b_{n-k} x^{n-k})(c_0 + c_1 x + \cdots + c_k x^k)$$

with $1 \leq k < n$ and $b_0, b_1, \cdots, b_{n-k}, c_0, c_1, \ldots, c_k \in \mathbb{Z}$. Then we have

$$
\begin{cases}
a_0 & = b_0 c_0 \\
a_1 & = b_0 c_1 + b_1 c_0 \\
a_2 & = b_0 c_2 + b_1 c_1 + b_2 c_0 \\
\cdots & \cdots \\
a_n & = b_{n-k} a_k.
\end{cases}
$$

Since a_0 is divisible by p but not by p^2, only one of the integers b_0, c_0 is divisible by p. Suppose that b_0 is divisible by p and c_0 is not. Since a_1 is divisible by p, this forces b_1 to be divisible by p. Continuing this way, we deduce that $b_2, b_3, \ldots$ are divisible by p until we arrive to

$$a_{n-k} = b_0 c_{n-k} + b_1 c_{n-k-1} + \cdots + b_{n-k-1} c_1 + b_{n-k} c_0,$$

which forces b_{n-k} to be divisible by p. But this contradicts the second assumption, because $a_n = b_{n-k} c_k$. □

Example 2.6.2 Let p be a prime number. Then, the polynomial $q(x) = 1 + x + x^2 + \cdots + x^{p-2} + x^{p-1}$ is irreducible over $\mathbb{Z}$. Indeed, we have

$$q(x+1) = \frac{(x+1)^p - 1}{(x+1) - 1} = \binom{p}{p-1} + \binom{p}{p-2} x + \cdots + \binom{p}{1} x^{p-2} + x^{p-1}.$$

Since $\binom{p}{k} = \dfrac{p!}{k!(p-k)!}$, $k = 1, 2, \ldots, p-1$, is an integer divisible by p and $\binom{p}{p-1} = p$ is not divisible by p^2, by virtue of Eisenstein's criterion we deduce that $q(x+1)$ (and therefore $q(x)$) is irreducible over $\mathbb{Z}$.

Definition 2.6.3 A polynomial $q(x) \in \mathbb{Z}[x]$ is called *primitive* if its coefficients are relatively prime and its leading coefficients is positive.

Clearly, any $q(x) \in \mathbb{Z}[x]$ may be represented in the form $q(x) = \pm cq_1(x)$, where $c \in \mathbb{N}$, called the *content* of $q(x)$, is the greatest common divisor of its coefficients and $q_1(x)$ is primitive. Also, any $f(x) \in \mathbb{Q}[x]$ may be represented in the form $f(x) = \frac{c}{d}q(x)$, where $q(x) \in \mathbb{Z}[x]$ is primitive and $c, d \in \mathbb{Z}$.

Proposition 2.6.4 (Gauss lemma) *The product of two primitive polynomials is primitive.*

Proof. By contradiction, suppose that $q_1(x) = a_0 + a_1 x + \cdots + a_{n-1}x^{n-1} + a_n x^n$ and $q_2(x) = b_0 + b_1 x + \cdots + b_{m-1}x^{m-1} + b_m x^m$ are primitive polynomials, but their product $q_1(x)q_2(x) = c_0 + c_1 x + \cdots + c_{n+m-1}x^{n+m-1} + c_{n+m}x^{n+m}$ is not. This means that there exists a prime p that divides all the coefficients $c_0, c_1, c_2, \ldots, c_{n+m-1}, c_{n+m}$. By the primitivity of $q_1(x)$ and $q_2(x)$, we can find i (respectively j) the minimal index such that a_i (respectively b_j) is not divisible by p. Then, in the expression

$$c_{i+j} = a_i b_j + (a_{i-1}b_{j+1} + \cdots + a_0 b_{j+i} + a_{i+1}b_{j-1} + \cdots + a_{i+j}b_0)$$

all the summands are divisible by p except $a_i b_j$. Thus p does not divide c_{i+j}, and this is a contradiction. $\qquad\square$

Corollary 2.6.5 *A polynomial $q(x) \in \mathbb{Z}[x]$ which is irreducible over $\mathbb{Z}$ is also irreducible over $\mathbb{Q}$.*

Proof. Let $q(x) \in \mathbb{Z}[x]$ and suppose that it is reducible over $\mathbb{Q}$, say $q(x) = f_1(x)f_2(x)$, where both $f_1(x)$ and $f_2(x)$ belong to $\mathbb{Q}[x]$ and are nontrivial ($\deg f_1, \deg f_2 < \deg q$). For $i = 1, 2$, we can write

$$f_i(x) = \frac{a_i}{b_i}q_i(x),$$

where $q_i(x)$ is a primitive polynomial and $a_i, b_i \in \mathbb{Z}$ are relatively prime. Then

$$q(x) = \frac{a_1 a_2}{b_1 b_2}[q_1(x)q_2(x)]. \tag{2.26}$$

Since both $q(x)$ and $q_1(x)q_2(x)$ are integer valued, $a_1 a_2[q_1(x)q_2(x)]$ must be divisible by $b_1 b_2$. Let $b_1 = p_1^{m_1} p_2^{m_2} \cdots p_t^{m_t}$ be the prime factorization of b_1. Consider the prime power $p_1^{m_1}$. It cannot divide all coefficients of $q_1(x)q_2(x)$ because, by Gauss lemma, this polynomial is primitive. Also, it cannot divide a_1 because this is relatively prime with b_1. Therefore it must divide a_2.

Repeating the same argument with the other prime factors of b_1 we deduce that b_1 divides a_2. Similarly, b_2 divides a_1. Thus, we can find $c_1, c_2 \in \mathbb{Z}$ such that

$$a_1 = c_1 b_2 \quad \text{and} \quad a_2 = c_2 b_1.$$

Then (2.26) becomes

$$q(x) = c_1 c_2 q_1(x) q_2(x).$$

This shows that $q(x)$ is (also) reducible over $\mathbb{Z}$. $\qquad\square$

Corollary 2.6.6 *Let $p(x), q(x) \in \mathbb{Z}[x]$ and suppose that $p(x)$ is primitive and divides $q(x)$ over $\mathbb{Q}$. Then $p(x)$ divides $q(x)$ over $\mathbb{Z}$.*

Proof. Let $f(x) \in \mathbb{Q}[x]$ such that $q(x) = p(x)f(x)$. Also write $f(x) = \frac{a}{b}r(x)$ with $r(x)$ a primitive polynomial and $a, b \in \mathbb{Z}$ relatively prime. Thus

$$q(x) = \frac{a}{b}p(x)r(x),$$

where the polynomials $q(x)$ and $p(x)r(x)$ both have integer coefficients. By Gauss lemma, $p(x)r(x)$ is primitive and this forces $b = \pm 1$, concluding the proof. $\qquad\square$

Definition 2.6.7 A complex number α is called *algebraic* provided it is a *root* of some polynomial $q(x) \in \mathbb{Z}[x]$, that is, $q(\alpha) = 0$. A *minimal polynomial* of an algebraic number α is a primitive polynomial of least degree $q(x) \in \mathbb{Z}[x]$ such that $q(\alpha) = 0$.

Clearly, a minimal polynomial is irreducible over $\mathbb{Z}$ (and therefore over $\mathbb{Q}$ by Corollary 2.6.5). In Proposition 2.6.8 we shall establish its uniqueness. For the next proposition, we need the notion of a principal ideal. Roughly speaking, a principal ideal in a commutative unital ring $\mathcal{R}$ is a subset of the form $\mathcal{I} = f\mathcal{R}$ for some $f \in \mathcal{R}$, called a generator of $\mathcal{I}$: we refer to Section 6.1 for a more comprehensive treatment of this and of other related notions.

Proposition 2.6.8 *Let $\alpha \in \mathbb{C}$ be an algebraic number and let $p(x) \in \mathbb{Z}[x]$ be a minimal polynomial of α. Consider the ideal $\mathcal{I} = \{q(x) \in \mathbb{Z}[x] : q(\alpha) = 0\}$. Then $\mathcal{I}$ is principal and generated by $p(x)$. In particular, $p(x)$ is the unique primitive irreducible polynomial in $\mathcal{I}$.*

Proof. Consider the ideal $\widetilde{\mathcal{I}} = \{f(x) \in \mathbb{Q}[x] : f(\alpha) = 0\}$ in $\mathbb{Q}[x]$. Since every ideal in $\mathbb{Q}[x]$ is principal (see Exercise 6.1.6), $\widetilde{\mathcal{I}}$ is generated by some element $f_0(x)$ of least degree. By eliminating the denominators and changing signs of all coefficients, if necessary, we may suppose that $f_0(x)$ belongs to $\mathbb{Z}[x]$ and is primitive. Let $q(x) \in \mathcal{I} \subseteq \tilde{\mathcal{I}}$. Then we can find $f(x) \in \mathbb{Q}[x]$ such that $q(x) = f(x)f_0(x)$. Since $f_0(x)$ is primitive, from Corollary 2.6.6 we deduce that $f_0(x)$ divides $q(x)$ in $\mathbb{Z}[x]$. Moreover, if $q(x) = p(x)$, we deduce that $f_0(x) = p(x)$, by minimality of the degree of $p(x)$. This shows that $\mathcal{I}$ is principal, generated by $p(x)$. $\square$

Example 2.6.9 Let p be a prime. Consider the algebraic number $\omega = \exp(\frac{2\pi i}{p})$ and the polynomial $q(x) = \frac{x^p - 1}{x - 1} = 1 + x + x^2 + \cdots + x^{p-1}$. Then $q(x)$ is irreducible (cf. Example 2.6.2) and $q(\omega) = 0$. Then, by Proposition 2.6.8, $q(x)$ is the minimal polynomial of ω and every $f(x) \in \mathbb{Z}[x]$ such that $f(\omega) = 0$ is a multiple of $q(x)$ in $\mathbb{Z}[x]$.

Proposition 2.6.10 *Let $P(x_1, x_2, \ldots, x_n)$ be a polynomial in the variables $x_1, x_2, \ldots, x_n$ with integer coefficients. Suppose that, for some $i \neq j$,*

$$P(x_1, x_2, \ldots, x_n)|_{x_i = x_j} \equiv 0.$$

Then there exists a polynomial $Q(x_1, x_2, \ldots, x_n)$ with integer coefficients such that $P(x_1, x_2, \ldots, x_n) = (x_i - x_j)Q(x_1, x_2, \ldots, x_n)$.

Proof. For the sake of simplicity, suppose that $i = 1$ and $j = 2$ so that $P(x_1, x_1, \ldots, x_n) \equiv 0$. Let us denote by $P_1(x_1, x_2, \ldots, x_n)$ (respectively $P_2(x_1, x_2, \ldots, x_n)$) the sum of the monomials of $P(x_1, x_2, \ldots, x_n)$ with positive (respectively negative) coefficients so that

$$P(x_1, x_2, \ldots, x_n) = P_1(x_1, x_2, \ldots, x_n) + P_2(x_1, x_2, \ldots, x_n).$$

Note that

$$P_1(x_1, x_1, \ldots, x_n) = -P_2(x_1, x_1, \ldots, x_n),$$

since $P(x_1, x_1, \ldots, x_n) \equiv 0$. This implies that there exists a bijection between the monomials in $P_1(x_1, x_1, \ldots, x_n)$ and those in $P_2(x_1, x_1, \ldots, x_n)$. More precisely, let us fix $m > 0$ and $k, k_3, \ldots, k_n \geq 0$; then the monomial $mx_1^k x_3^{k_3} \cdots x_n^{k_n}$ appears in $P_1(x_1, x_1, \ldots, x_n)$ if and only if $-mx_1^k x_3^{k_3} \cdots x_n^{k_n}$ appears in $P_2(x_1, x_1, \ldots, x_n)$. Suppose this is the case. Then we can find $m_0, m_1, \ldots, m_k$ and $n_0, n_1, \ldots, n_k$ non-negative integers such that the sum of the monomials of $P(x_1, x_2, \ldots, x_n)$ whose variables x_i have degree k_i for $i = 1, 2 \ldots, n$ and $k_1 + k_2 = k$ is

$$\sum_{\ell=0}^{k} m_\ell x_1^{k-\ell} x_2^{\ell} x_3^{k_3} \cdots x_n^{k_n} - \sum_{\ell=0}^{k} n_\ell x_1^{k-\ell} x_2^{\ell} x_3^{k_3} \cdots x_n^{k_n} \qquad (2.27)$$

and

$$m_0 + m_1 + \cdots + m_k = n_0 + n_1 + \cdots + n_k = m \tag{2.28}$$

but also such that

$$m_\ell \neq 0 \Rightarrow n_\ell = 0 \qquad n_\ell \neq 0 \Rightarrow m_\ell = 0$$

(because, otherwise, there would be a cancellation). By virtue of (2.28) with every monomial $x_1^{k-\ell} x_2^\ell x_3^{k_3} \cdots x_n^{k_n}$ such that $m_\ell \neq 0$ we can (arbitrarily but bijectively) associate a monomial $x_1^{k-h} x_2^h x_3^{k_3} \cdots x_n^{k_n}$ with $m_h \neq 0$ and $h \neq \ell$. Now, for $h > \ell$ we have the identity

$$x_1^{k-\ell} x_2^\ell - x_1^{k-h} x_2^h = x_2^\ell x_1^{k-h} (x_1^{h-\ell} - x_2^{h-\ell})$$
$$= x_2^\ell x_1^{k-h} (x_1 - x_2)(x_1^{h-\ell-1} + x_1^{h-\ell-2} x_2 + \cdots + x_2^{h-\ell-1}).$$

Exchanging h with ℓ we get the analogous identity for $h < \ell$. This shows that (2.27) is divisible by $x_1 - x_2$.

Repeating the argument for each monomial $m x_1^k x_3^{k_3} \cdots x_n^{k_n}$ (with $m > 0$ and $k, k_3, \ldots, k_n \geq 0$) appearing in $P_1(x_1, x_1, \ldots, x_n)$, we deduce that, in fact, $P(x_1, x_2, \ldots, x_n)$ is divisible by $x_1 - x_2$. $\qquad\square$

Example 2.6.11 Consider the polynomial $P(x_1, x_2) = x_1^2 + x_1 x_2 - 2x_2^2$. We have $P_1(x_1, x_1) = 2x_1^2$ and $P_2(x_1, x_1) = -2x_1^2$, and $m = 2$. Moreover, $m_0 = m_1 = 1$ and $m_2 = 0$, while $n_0 = n_1 = 0$ and $n_2 = 2$. We have $P(x_1, x_2) = (x_1^2 - x_2^2) + (x_1 x_2 - x_2^2) = (x_1 - x_2)(x_1 + x_2) + (x_1 - x_2)x_2 = (x_1 - x_2)(x_1 + 2x_2)$, so that $Q(x_1, x_2) = x_1 + 2x_2$.

Lemma 2.6.12 *Let p be a prime, n a positive integer, and $P(x_1, x_2, \ldots, x_n)$ a polynomial with integer coefficients. Suppose that $\omega_1, \omega_2, \ldots, \omega_n$ are (not necessarily distinct) pth roots of unity such that $P(\omega_1, \omega_2, \ldots, \omega_n) = 0$. Then $P(1, 1, \ldots, 1)$ is divisible by p.*

Proof. Setting $\omega = \exp(\frac{2\pi i}{p})$ we can find integers $0 \leq k_j \leq p - 1$ such that $\omega_j = \omega^{k_j}$, for $j = 1, 2, \ldots, n$.

Define the polynomials $q(x), r(x) \in \mathbb{Z}[x]$ by setting

$$P(x^{k_1}, x^{k_2}, \ldots, x^{k_n}) = (x^p - 1)q(x) + r(x)$$

where $\deg r < p$. Then $r(\omega) = 0$ and since $\deg r < p$ we deduce that $r(x)$ is a multiple of the minimal polynomial of ω, that is (cf. Example 2.6.9), $r(x) = m(1 + x + x^2 + \cdots + x^{p-1})$ for some $m \in \mathbb{Z}$. It follows that $P(1, 1, \ldots, 1) = r(1) = mp$. $\qquad\square$

Theorem 2.6.13 (Chebotarëv) *Let p be a prime and $1 \le n \le p$. Let $\eta_1, \eta_2, \ldots, \eta_n$ (respectively $\xi_1, \xi_2, \ldots, \xi_n$) be distinct elements in $\{0, 1, \ldots, p-1\}$. Then the matrix*

$$A = \left(\exp \frac{2\pi i \eta_h \xi_k}{p} \right)^n_{h,k=1}$$

is non-singular.

Proof. Set $\omega_h = \exp(\frac{2\pi i \eta_h}{p})$ for $h = 1, 2, \ldots, n$. Note that the ω_hs are distinct pth roots of unity and $A = \left(\omega_h^{\xi_k} \right)^n_{1=h,k}$. Define the polynomial $D(x_1, x_2, \ldots, x_n)$ (with integer coefficients) by setting

$$D(x_1, x_2, \ldots, x_n) = \det \left(x_h^{\xi_k} \right)^n_{h,k=1}.$$

As the determinant is an alternating form, we have $D(x_1, x_2, \ldots, x_n)|_{x_h = x_k} \equiv 0$ whenever $1 \le h \ne k \le n$, so that, by recursively applying Proposition 2.6.10, we can find a polynomial $Q(x_1, x_2, \ldots, x_n)$ with integer coefficients such that

$$D(x_1, x_2, \ldots, x_n) = Q(x_1, x_2, \ldots, x_n) \prod_{1 \le h < k \le n} (x_k - x_h). \qquad (2.29)$$

To prove the theorem, it is equivalent to show that $Q(\omega_1, \omega_2, \ldots, \omega_n) \ne 0$ (because the ω_hs are all distinct) so that, by virtue of Lemma 2.6.12, it suffices to show that p does not divide $Q(1, 1, \ldots, 1)$. For this, we need the next three lemmas. Let us first introduce some useful notation.

Given an n-tuple $\mathbf{k} = (k_1, k_2, \ldots, k_n)$ of non-negative integers, we say that the (monomial) differential operator

$$L = L_{\mathbf{k}} = \left(x_1 \frac{\partial}{\partial x_1} \right)^{k_1} \left(x_2 \frac{\partial}{\partial x_2} \right)^{k_2} \cdots \left(x_n \frac{\partial}{\partial x_n} \right)^{k_n} \qquad (2.30)$$

is of *type* $\mathbf{k}$ and *order* $o(\mathbf{k}) = k_1 + k_2 + \cdots + k_n$.

Lemma 2.6.14 *Let L be a differential operator of type $\mathbf{k}$ and $F(x_1, x_2, \ldots, x_n)$ and $G(x_1, x_2, \ldots, x_n)$ two polynomials. Then*

$$L(FG) = \sum_{(\mathbf{i}, \mathbf{j})} L_{\mathbf{i}}(F) \cdot L_{\mathbf{j}}(G) \qquad (2.31)$$

where the sum runs over all pairs $(\mathbf{i}, \mathbf{j})$ such that (componentwise) $\mathbf{i} + \mathbf{j} = \mathbf{k}$. (and therefore $o(\mathbf{i}) + o(\mathbf{j}) = k$).

Proof. We proceed by induction on the order k of L. If $k = 0$ then L is the identity and the statement is trivial. Suppose we have shown the statement for all differential operators of order $\le k$ and let L be a differential operator of order $k + 1$. Up to renaming the variables, we may suppose that $L = \left(x_1 \frac{\partial}{\partial x_1} \right) L'$, where L' has order k. By the Leibniz rule and the inductive hypothesis we then have

$$L(FG) = \left(x_1 \frac{\partial}{\partial x_1}\right) L'(FG)$$

$$= \left(x_1 \frac{\partial}{\partial x_1}\right) \sum_{(\mathbf{i},\mathbf{j})} L_{\mathbf{i}}(F) \cdot L_{\mathbf{j}}(G)$$

$$= \sum_{(\mathbf{i}',\mathbf{j})} L_{\mathbf{i}'}(F) \cdot L_{\mathbf{j}}(G) + \sum_{(\mathbf{i},\mathbf{j}')} L_{\mathbf{i}}(F) \cdot L_{\mathbf{j}'}(G)$$

where $\mathbf{i}' = (i_1 + 1, i_2, \ldots, i_n)$ and $\mathbf{j}' = (j_1 + 1, j_2, \ldots, j_n)$, and, clearly, $o(\mathbf{i}') + o(\mathbf{j}) = o(\mathbf{i}) + o(\mathbf{j}') = k + 1$. $\qquad \square$

Lemma 2.6.15 *For $1 \leq j \leq n$ and $1 \leq h \leq j - 1$ we have*

$$\left(x_j \frac{\partial}{\partial x_j}\right)^h (x_j - x_1)(x_j - x_2) \cdots (x_j - x_{j-1})$$

$$= \sum_{t=1}^{h} a_{h,t} x_j^t \sum_{\mathbf{i}_t} \prod_{\substack{1 \leq i \leq j-1 \\ i \neq i_1, i_2, \ldots, i_t}} (x_j - x_i) \quad (2.32)$$

where $\sum_{\mathbf{i}_t}$ runs over all $\mathbf{i}_t = (i_1, i_2, \ldots, i_t)$ with $1 \leq i_1 < i_2 < \ldots < i_t \leq j - 1$ and the $a_{h,t} = a_{h,t}(j)$s are non-negative integers such that $a_{h,h} = h!$. In particular,

$$\left(x_j \frac{\partial}{\partial x_j}\right)^{j-1} (x_j - x_1)(x_j - x_2) \cdots (x_j - x_{j-1})$$

$$= (j-1)! x_j^{j-1}$$

$$+ \text{ terms containing at least one factor } (x_j - x_i)$$

with $1 \leq i < j$.

Proof. We proceed by induction on $h = 1, 2, \ldots, j - 1$. For $h = 1$ we have

$$\left(x_j \frac{\partial}{\partial x_j}\right)(x_j - x_1)(x_j - x_2) \cdots (x_j - x_{j-1})$$

$$= x_j(x_j - x_2)(x_j - x_3) \cdots (x_j - x_{j-1})$$

$$+ (x_j - x_1)x_j(x_j - x_3) \cdots (x_j - x_{j-1})$$

$$+ \cdots$$

$$+ (x_j - x_1)(x_j - x_2) \cdots (x_j - x_{j-3})x_j(x_j - x_{j-1})$$

$$+ (x_j - x_1)(x_j - x_2) \cdots (x_j - x_{j-2})x_j$$

$$= x_j \sum_{k=1}^{j-1}(x_j - x_1)(x_j - x_2) \cdots \widehat{(x_j - x_k)} \cdots (x_j - x_{j-1}),$$

where the factor $\widehat{}$ is omitted. Since $\left(x_j \frac{\partial}{\partial x_j}\right) x_j = x_j$, keeping in mind the previous calculation, we have

$$
\left(x_j \frac{\partial}{\partial x_j}\right)^2 (x_j - x_1)(x_j - x_2) \cdots (x_j - x_{j-1})
$$

$$
= \left(x_j \frac{\partial}{\partial x_j}\right) x_j \sum_{k=1}^{j-1} (x_j - x_1) \cdots \widehat{(x_j - x_k)} \cdots (x_j - x_{j-1})
$$

$$
= x_j \sum_{k=1}^{j-1} (x_j - x_1) \cdots \widehat{(x_j - x_k)} \cdots (x_j - x_{j-1})
$$

$$
+ 2x_j^2 \sum_{1 \le k < k' \le j-1} (x_j - x_1) \cdots \widehat{(x_j - x_k)} \cdots \widehat{(x_j - x_{k'})} \cdots (x_j - x_{j-1}).
$$

Suppose we have proved the formula (2.32) for $h < j - 1$. Then

$$
\left(x_j \frac{\partial}{\partial x_j}\right)^{h+1} (x_j - x_1)(x_j - x_2) \cdots (x_j - x_{j-1})
$$

$$
= \left(x_j \frac{\partial}{\partial x_j}\right) \sum_{t=1}^{h} a_{h,t} x_j^t \sum_{i_t} \prod_{\substack{1 \le i \le j-1 \\ i \ne i_1, i_2, \ldots, i_t}} (x_j - x_i)
$$

$$
= \sum_{t=1}^{h} a_{h,t} t x_j^t \sum_{i_t} \prod_{\substack{1 \le i \le j-1 \\ i \ne i_1, i_2, \ldots, i_t}} (x_j - x_i)
$$

$$
+ \sum_{t=1}^{h} a_{h,t} x_j^{t+1} \sum_{i_{t+1}} (t+1) \prod_{\substack{1 \le i \le j-1 \\ i \ne i_1, i_2, \ldots, i_{t+1}}} (x_j - x_i)
$$

$$
= \sum_{t=1}^{h+1} a_{h+1,t} x_j^t \sum_{i_t} \prod_{\substack{1 \le i \le j-1 \\ i \ne i_1, i_2, \ldots, i_t}} (x_j - x_i),
$$

where

$$
a_{h+1,t} = \begin{cases} a_{h,t} t + a_{h,t-1} t & \text{for } t = 1, 2, \ldots, h \\ a_{h,h}(h+1) = (h+1)! & \text{for } t = h+1. \end{cases}
$$
$\qquad\square$

Lemma 2.6.16 *Let $L = L_{(0,1,\ldots,n-1)}$, that is,*

$$
L = \left(x_1 \frac{\partial}{\partial x_1}\right)^0 \left(x_2 \frac{\partial}{\partial x_2}\right)^1 \cdots \left(x_n \frac{\partial}{\partial x_n}\right)^{n-1}.
$$

Then if $D(x_1, x_2, \ldots, x_n)$ and $Q(x_1, x_2, \ldots, x_n)$ are as in (2.29), we have

$$[LD](1, 1, \ldots, 1) = \prod_{j=1}^{n} (j-1)! Q(1, 1, \ldots, 1). \qquad (2.33)$$

Proof. By virtue of Lemma 2.6.14 and Lemma 2.6.15 we have

$$[LD](x_1, x_2, \ldots, x_n) = \prod_{j=1}^{n} (j-1)! x_j^{j-1} Q(x_1, x_2, \ldots, x_n)$$

$$+ \text{ terms containing at least one factor } (x_j - x_i)$$

with $1 \leq i < j$. In particular, taking $x_i = 1$ for $i = 1, 2, \ldots, n$ we deduce (2.33).

$\square$

End of the proof of Theorem 2.6.13 For L as in Lemma 2.6.16 we have (where $\mathfrak{S}_n$ denotes the symmetric group of degree n)

$$[LD](x_1, x_2, \ldots, x_n) = L \sum_{\sigma \in \mathfrak{S}_n} \varepsilon(\sigma) x_1^{\xi_{\sigma(1)}} x_2^{\xi_{\sigma(2)}} \cdots x_n^{\xi_{\sigma(n)}}$$

$$= \sum_{\sigma \in \mathfrak{S}_n} \varepsilon(\sigma) \xi_{\sigma(1)}^0 x_1^{\xi_{\sigma(1)}} \xi_{\sigma(2)}^1 x_2^{\xi_{\sigma(2)}} \cdots \xi_{\sigma(n)}^{n-1} x_n^{\xi_{\sigma(n)}}$$

since

$$\left(x_j \frac{\partial}{\partial x_j}\right)^{j-1} x_j^{\xi_{\sigma(j)}} = \xi_{\sigma(j)}^{j-1} x_j^{\xi_{\sigma(j)}}$$

for all $j = 1, 2, \ldots, n$. Thus

$$[LD](1, 1, \ldots, 1) = \sum_{\sigma \in \mathfrak{S}_n} \varepsilon(\sigma) \xi_{\sigma(1)}^0 \xi_{\sigma(2)}^1 \cdots \xi_{\sigma(n)}^{n-1}$$

$$= \begin{vmatrix} 1 & 1 & \cdots & 1 \\ \xi_1 & \xi_2 & \cdots & \xi_n \\ \xi_1^2 & \xi_2^2 & \cdots & \xi_n^2 \\ \vdots & \vdots & \ddots & \vdots \\ \xi_1^{n-1} & \xi_2^{n-1} & \cdots & \xi_n^{n-1} \end{vmatrix} = \prod_{1 \leq i < j \leq n} (\xi_j - \xi_i)$$

is the Vandermonde determinant (see, e.g. [91]). Since $\xi_j \neq \xi_i$ for $1 \leq i < j \leq n$, we deduce that $[LD](1, 1, \ldots, 1)$ is not divisible by p. Since also $\prod_{j=1}^{n} (j-1)!$ is not divisible by p (because $n \leq p$), from (2.33) we deduce that $Q(1, 1, \ldots, 1)$ is not divisible by p either. By virtue of Lemma 2.6.12, this completes the proof of Theorem 2.6.13. $\square$

Given a non-empty subset $A \subseteq \mathbb{Z}_p$ and a function $f \in L(A)$, in the following we shall denote by $\overline{f}$ its extension $\overline{f} \colon \mathbb{Z}_p \to \mathbb{C}$ defined by setting $\overline{f}(z) = 0$ for all $z \in \mathbb{Z}_p \setminus A$. For simplicity, we regard the DFT as a map $L(\mathbb{Z}_p) \to L(\mathbb{Z}_p)$. In other words, for $f \in L(\mathbb{Z}_p)$ and $x \in \mathbb{Z}_p$,

$$\widehat{f}(x) = \frac{1}{p} \sum_{y \in \mathbb{Z}_p} f(y) \omega^{-xy};$$

see also Exercise 2.4.13.

Corollary 2.6.17 *Let p be a prime. Let $A, B \subseteq \mathbb{Z}_p$ such that $|A| = |B|$. Then the linear map $T = T_{A,B} \colon L(A) \to L(B)$ defined by $Tf = \widehat{\overline{f}}|_B$ is invertible.*

Proof. Set $A = \{\xi_1, \xi_2, \ldots, \xi_n\}$ and $B = \{\eta_1, \eta_2, \ldots, \eta_n\}$ and consider the basis of $L(A)$ (respectively, of $L(B)$) consisting of the Dirac functions δ_{ξ_j}, with $j = 1, 2, \ldots, n$ (respectively, δ_{η_k}, with $k = 1, 2, \ldots, n$), and let $\omega = \exp(2\pi i/p)$. Then we have

$$[T\delta_{\xi_k}](\eta_h) = \widehat{\delta_{\xi_k}}(\eta_h) = \sum_{x \in \mathbb{Z}_p} \delta_{\xi_k}(x) \omega^{-x\eta_h} = \omega^{-\eta_h \xi_k}.$$

By virtue of Theorem 2.6.13 we have $\det\left(\left[T\delta_{\xi_k}\right](\eta_h)\right)_{h,k=1}^n \neq 0$, showing that T is indeed invertible. $\qquad\square$

We are now in a position to state and prove the main result of this section.

Theorem 2.6.18 (Tao) *Let p be a prime number and $f \in L(\mathbb{Z}_p)$ non-zero. Then*

$$|\mathrm{supp}(f)| + |\mathrm{supp}(\widehat{f})| \geq p + 1.$$

Conversely, if $\varnothing \neq A, A' \subseteq \mathbb{Z}_p$ are two subsets such that $|A| + |A'| = p + 1$, then there exists $f \in L(\mathbb{Z}_p)$ such that $\mathrm{supp}(f) = A$ and $\mathrm{supp}(\widehat{f}) = A'$.

Proof. Suppose, by contradiction, that, setting $\mathrm{supp}(f) = A$ and $\mathrm{supp}(\widehat{f}) = C$, one has $|A| + |C| \leq p$. Then we can find a subset $B \subseteq \mathbb{Z}_p$ such that $|B| = |A|$ and $C \cap B = \varnothing$. We deduce that $Tf = \widehat{\overline{f}}|_B$ is identically zero. Since $f \not\equiv 0$, this contradicts injectivity of T (Corollary 2.6.17).

Conversely, let $\varnothing \neq A, A' \subseteq \mathbb{Z}_p$ be two subsets such that $|A| + |A'| = p + 1$. Let $B \subseteq \mathbb{Z}_p$ such that $|B| = |A|$ and $B \cap A'$ reduces to a single element, say ξ. Note that $(\mathbb{Z}_p \setminus B) \cup \{\xi\} \supseteq A'$ so that, by taking cardinalities, $|A'| = p + 1 - |A| = p - |B| + 1 = |(\mathbb{Z}_p \setminus B) \cup \{\xi\}| \geq |A'|$, which yields

$$(\mathbb{Z}_p \setminus B) \cup \{\xi\} = A'. \tag{2.34}$$

Consider the map $T = T_{A,B} \colon L(A) \to L(B)$. By Corollary 2.6.17, we can find $g \in L(A)$ such that $Tg = \delta_\xi|_B$ so that $\widehat{\overline{g}}$ vanishes on $B \setminus \{\xi\}$ but $\widehat{\overline{g}}(\xi) \neq 0$. Setting

$f = \bar{g} \in L(\mathbb{Z}_p)$ we clearly have $\mathrm{supp}(f) \subseteq A$ and $\mathrm{supp}(\widehat{f}) \subseteq (\mathbb{Z}_p \setminus B) \cup \{\xi\}$. Let us show that indeed $\mathrm{supp}(f) = A$ and, moreover, $\mathrm{supp}(\widehat{f}) = A'$. By the first part of the theorem we have

$$p + 1 \leq |\mathrm{supp}(f)| + |\mathrm{supp}(\widehat{f})| \leq |A| + |\mathbb{Z}_p \setminus B| + 1$$
$$= |A| + (p - |B|) + 1 = p + 1$$

so that all inequalities above are indeed equalities. In particular, $\mathrm{supp}(f) = A$ and $\mathrm{supp}(\widehat{f}) = (\mathbb{Z}_p \setminus B) \cup \{\xi\} = A'$, where the last equality follows from (2.34). □

3

Dirichlet's theorem on primes in arithmetic progressions

In this chapter, we give an exposition on the celebrated Dirichlet theorem on primes in arithmetic progressions. It states that, if r and m are relatively prime positive integers, then the arithmetic progression $r, r + m, r + 2m, \ldots, r + km, \ldots$ contains infinitely many primes. For instance, there are infinitely many prime numbers of the form $1 + 4k$, $k \in \mathbb{N}$. There are several proofs of this theorem: some of them are based on algebraic number theory (see the monograph by Weyl [166]), others on analytic number theory (see the monograph by Serre [144]), but also elementary proofs are available (see the paper by Selberg [143]). By an elementary proof we mean a proof that does not use sophisticated methods of complex variables, algebraic geometry, or cohomology theory, but it may be technically very difficult.

Here, the character theory of finite Abelian groups is an essential ingredient, in particular, in order to define Dirichlet L-functions, which constitute one of the central objects in number theory. We have chosen to follow the exposition in the beautiful book by Stein and Shakarchi [150]. The authors have managed to reduce the proof to the use of very elementary analysis. We have also taken some material from the book by Knapp [88]. Other proofs may be found in the monographs by Apostol [13], Ireland and Rosen [79], and Nathanson [118].

3.1 Analytic preliminaries

In this section, we establish some elementary results on real and complex series. As in our main source [150], we avoid the use of complex analysis: just elementary properties of real and complex series will be used (up to and including existence of the radius of convergence for real and complex power series, elementary properties of uniform convergence, and differentiability of real power series). In several points we closely follow the exposition in [88].

From the well known expansion $\log(1+t) = \sum_{k=1}^{\infty} \frac{(-1)^{k+1}}{k} t^k$ for $t \in (-1, 1]$ we deduce that

$$\log \frac{1}{1-t} = -\log(1-t) = \sum_{k=1}^{\infty} \frac{t^k}{k}$$

for $t \in [-1, 1)$. We then define

$$\log \frac{1}{1-z} = \sum_{k=1}^{\infty} \frac{z^k}{k} \tag{3.1}$$

for all $z \in \mathbb{C}$, $|z| < 1$. With exp we denote the usual complex exponential: $\exp(x + iy) = e^x e^{iy} = e^x(\cos y + i \sin y)$ for all $x, y \in \mathbb{R}$. Also, $\Re z$ denotes the real part of $z \in \mathbb{C}$.

Proposition 3.1.1

(i) $|z| < 1$ if and only if $\Re \frac{1}{1-z} > \frac{1}{2}$.

(ii) $\exp(\log \frac{1}{1-z}) = \frac{1}{1-z}$ for all $|z| < 1$.

(iii) $\log \frac{1}{1-z} = z + R(z)$ where the error term $R(z)$ satisfies $|R(z)| < |z|^2$ if $|z| < \frac{1}{2}$.

(iv) $|\log \frac{1}{1-z}| \le \frac{3}{2}|z|$ if $|z| < \frac{1}{2}$.

Proof. (i) Setting $w = \frac{1}{1-z}$ we have $z = \frac{w-1}{w}$ and

$$|z| < 1 \Leftrightarrow |w - 1| < |w| \Leftrightarrow \Re w > \frac{1}{2}.$$

(ii) Consider the polar expression of z given by $z = \rho e^{i\theta}$ with $\rho \ge 0$ and $\theta \in \mathbb{R}$. We then have to show that

$$(1 - \rho e^{i\theta}) \exp\left(\sum_{k=1}^{\infty} \frac{\rho^k e^{ik\theta}}{k}\right) = 1. \tag{3.2}$$

For $\rho = 0$ it is trivially satisfied. By differentiating with respect to the *real* variable ρ, we get

$$\frac{d}{d\rho}\left[(1 - \rho e^{i\theta}) \exp(\sum_{k=1}^{\infty} \frac{\rho^k e^{ik\theta}}{k})\right]$$

$$= \left[-e^{i\theta} + (1 - \rho e^{i\theta}) e^{i\theta} \sum_{k=1}^{\infty} (\rho e^{i\theta})^{k-1}\right] \exp\left(\log \frac{1}{1-z}\right)$$

which vanishes since $\sum_{k=0}^{\infty} (\rho e^{i\theta})^k = \frac{1}{1 - \rho e^{i\theta}}$. Therefore, the left hand side of (3.2) is constant along each line $\theta = cost$ and it is equal to its value for $\rho = 0$. Thus (3.2) follows.

(iii)

$$|R(z)| = \left| \log \frac{1}{1-z} - z \right| = \left| \sum_{k=2}^{\infty} \frac{z^k}{k} \right|$$

$$\leq \sum_{k=2}^{\infty} \frac{|z|^k}{k} \leq \frac{|z|^2}{2} \sum_{k=0}^{\infty} |z|^k$$

$$(\text{for } |z| < \tfrac{1}{2}) \quad < \frac{|z|^2}{2} \sum_{k=0}^{\infty} \frac{1}{2^k} = |z|^2.$$

(iv)

$$\left| \log \frac{1}{1-z} \right| \leq \sum_{k=1}^{\infty} \frac{|z|^k}{k}$$

$$\leq |z| \left[1 + \sum_{k=2}^{\infty} \frac{|z|^{k-1}}{2} \right]$$

$$(\text{for } |z| < \tfrac{1}{2}) \quad < |z| \left[1 + \sum_{k=2}^{\infty} \frac{1}{2^k} \right]$$

$$= \frac{3}{2} |z|. \qquad \qquad \square$$

Definition 3.1.2 Let $(z_n)_{n \in \mathbb{N}}$ be a sequence of complex numbers. The associated *infinite product*, denoted $\prod_{n=1}^{\infty} z_n$, is the limit of the partial products $z_1 z_2 \cdots z_n$ as n tends to infinity, in formulæ,

$$\prod_{n=1}^{\infty} z_n = \lim_{n \to +\infty} \prod_{k=1}^{n} z_k.$$

The product is said to *converge* when the limit exists and is not zero. Otherwise, the product is said to *diverge*.

The following is one of the basic results in the theory of infinite products.

Proposition 3.1.3 *Let $(z_n)_{n \in \mathbb{N}}$ be a sequence of complex numbers and suppose that $|z_n| < 1$ for all $n \in \mathbb{N}$. Then the infinite product $\prod_{n=1}^{\infty} \frac{1}{1-|z_n|}$ converges if and only if the series $\sum_{n=1}^{\infty} |z_n|$ converges. Moreover, if this is the case, the infinite product $\prod_{n=1}^{\infty} \frac{1}{1-z_n}$ also converges and one has*

$$\prod_{n=1}^{\infty} \frac{1}{1-z_n} = \exp \left(\sum_{n=1}^{\infty} \log \frac{1}{1-z_n} \right). \tag{3.3}$$

Proof. The only if part follows from the elementary inequalities

$$1 + \sum_{k=1}^{n} |z_k| \leq \prod_{k=1}^{n} (1 + |z_k|) \leq \prod_{k=1}^{n} \frac{1}{1 - |z_k|}.$$

Suppose now that $\sum_{n=1}^{\infty} |z_n| < +\infty$. Then $\lim_{n \to +\infty} |z_n| = 0$ and, without loss of generality, we may assume that $|z_n| < \frac{1}{2}$. From Proposition 3.1.1.(ii) we get

$$\prod_{k=1}^{n} \frac{1}{1 - |z_k|} = \prod_{k=1}^{n} \exp\left(\log \frac{1}{1 - |z_k|}\right)$$

$$= \exp\left(\sum_{k=1}^{n} \log \frac{1}{1 - |z_k|}\right)$$

and Proposition 3.1.1.(iv) yields

$$\left| \log \frac{1}{1 - |z_k|} \right| \leq \frac{3}{2} |z_k|$$

for all $k \in \mathbb{N}$. From our assumptions we then deduce that $\sum_{k=1}^{\infty} \log \frac{1}{1 - |z_k|}$ converges absolutely. We conclude by invoking the continuity of exp. The proof of the convergence of $\prod_{n=1}^{\infty} \frac{1}{1 - z_n}$ is analogous. Moreover, this limit is nonzero and equals $\lim_{n \to +\infty} \exp(\sum_{k=1}^{n} \log \frac{1}{1 - z_k})$. $\qquad \square$

In what follows, we will often use *Abel's formula of summation by parts*: if $(z_n)_{n \in \mathbb{N}}$ and $(w_n)_{n \in \mathbb{N}}$ are complex sequences, then setting $Z_0 = 0$ and $Z_k = \sum_{i=1}^{k} z_i$, for $k \geq 1$, one has

$$\sum_{k=m}^{n} z_k w_k = \sum_{k=m}^{n-1} Z_k (w_k - w_{k+1}) + Z_n w_n - Z_{m-1} w_m \qquad (3.4)$$

for all $1 \leq m \leq n$. The proof is just an easy exercise.

Definition 3.1.4 Let $(a_n)_{n \in \mathbb{N}}$ be a sequence of complex numbers. The associated *Dirichlet series* is the series given by

$$\sum_{n=1}^{\infty} \frac{a_n}{n^s}$$

where s is a complex variable and $n^s = \exp(s \log n)$ for all $n \geq 1$.

Let $A \subset \mathbb{C}$ and $(f_n)_{n \in \mathbb{N}}$ a sequence of complex functions. One says that the series $\sum_{n=1}^{\infty} f_n(z)$ is *M-test convergent* on A if there exists a sequence $(M_n)_{n \in \mathbb{N}}$ of positive real numbers such that

- $|f_n(z)| \leq M_n$ for all $z \in A$ and $n \geq 1$;
- $\sum_{n=1}^{\infty} M_n < +\infty$.

Clearly, M-test convergence on A implies both uniform and absolute convergence on A. In the following we regard a Dirichlet series as a series of complex functions.

Proposition 3.1.5 *Let $(a_n)_{n \in \mathbb{N}}$ be a sequence of complex numbers. If the Dirichlet series $\sum_{n=1}^{\infty} \frac{a_n}{n^s}$ is convergent for $s = s_0$ then it is uniformly convergent on each compact subset contained in $\{s \in \mathbb{C} : \Re s > \Re s_0\}$ and it is absolutely convergent at each $s \in \mathbb{C}$ such that $\Re s > \Re s_0 + 1$.*

Proof. According with the notation in (3.4), set

$$z_n = \frac{a_n}{n^{s_0}}, \quad Z_n = \sum_{k=1}^{n} z_k, \quad \text{and} \quad w_n(s) = \frac{1}{n^{s-s_0}}$$

for all $n \geq 1$. Then $\sum_{n=1}^{\infty} z_n w_n(s)$ coincides with the Dirichlet series. Moreover the following holds:

(i) The sequence $(Z_n)_{n \in \mathbb{N}}$ converges (by hypothesis); in particular, it is bounded: $\exists H > 0$ such that $|Z_n| \leq H$ for all $n \geq 1$.

(ii) $\lim_{n \to +\infty} w_n(s) = 0$ uniformly on each set $\{s \in \mathbb{C} : \Re s \geq \mu\}$ with $\mu > \Re s_0$. Indeed, for $\Re s \geq \mu > \Re s_0$ we have

$$|n^{s-s_0}| = n^{\Re(s-s_0)} \geq n^{\mu - \Re s_0}$$

so that $\left| \frac{1}{n^{s-s_0}} \right| \leq \frac{1}{n^{\mu - \Re s_0}}$ which tends to 0 as $n \to +\infty$.

(iii) The series $\sum_{n=1}^{\infty} |w_n(s) - w_{n+1}(s)|$ is M-test convergent on every compact set $A \subseteq \{s \in \mathbb{C} : \Re s > \Re s_0\}$. Indeed, if $|s - s_0| \leq \delta$ and $\Re s - \Re s_0 \geq \eta > 0$, we have

$$\left| \frac{1}{n^{s-s_0}} - \frac{1}{(n+1)^{s-s_0}} \right| = \left| \int_n^{n+1} \frac{s - s_0}{t^{s-s_0+1}} dt \right|$$

$$\leq \sup_{n \leq t \leq n+1} \left| \frac{s - s_0}{t^{s-s_0+1}} \right|$$

$$= \sup_{n \leq t \leq n+1} \frac{|s - s_0|}{t^{\Re(s-s_0)+1}}$$

$$= \frac{|s - s_0|}{n^{\Re(s-s_0)+1}}$$

$$\leq \frac{\delta}{n^{\eta+1}}.$$

Then we can apply Cauchy's criterion for uniform convergence:

$$\left| \sum_{k=m}^{n} \frac{a_k}{k^s} \right| = \left| \sum_{k=m}^{n} z_k w_k(s) \right|$$

$$\text{(by (3.4))} \quad \leq \sum_{k=m}^{n} |Z_k| \cdot |w_k(s) - w_{k+1}(s)|$$

$$+ |Z_n| \cdot |w_n(s)| + |Z_{m-1}| \cdot |w_m(s)|$$

$$\text{(by (i), (ii), and (iii))} \quad \leq \sum_{k=m}^{n} \frac{H\delta}{k^{\eta+1}} + \frac{H}{n^{\mu - \Re s_0}} + \frac{H}{m^{\mu - \Re s_0}}.$$

Thus, for each $\varepsilon > 0$ there exists $N \in \mathbb{N}$ such that $\left| \sum_{k=m}^{n} \frac{a_k}{k^s} \right| < \varepsilon$ for all $n \geq m \geq N$ and $s \in A$, and uniform convergence is proved.

Finally, if $\Re s > \Re s_0 + 1$ then, setting $\eta' = \Re s - \Re s_0 - 1 > 0$, we have

$$\left| \frac{a_n}{n^s} \right| = \left| \frac{a_n}{n^{s_0}} \right| \cdot \left| \frac{1}{n^{s - s_0}} \right| = \left| \frac{a_n}{n^{s_0}} \right| \cdot \frac{1}{n^{\Re s - \Re s_0}} = \left| \frac{a_n}{n^{s_0}} \right| \frac{1}{n^{1 + \eta'}}$$

for all $n \geq 1$, so that boundedness of the sequence $\left(\left| \frac{a_n}{n^{s_0}} \right| \right)_{n \geq 1}$ yields absolute convergence of the Dirichlet series. $\qquad \square$

Remark 3.1.6 By a celebrated theorem of Weierstass (see [3, 133, 115]) if a series of analytic functions converges uniformly on each compact subset of a set $A \subset \mathbb{C}$, then the sum is analytic on A. Then Proposition 3.1.5 ensures that if a Dirichlet series converges at $s_0 \in \mathbb{C}$ then it is analytic on the region $\{ s \in \mathbb{C} : \Re s > \Re s_0 \}$. We will not use this important fact.

Proposition 3.1.7 *Let $(a_n)_{n \in \mathbb{N}}$ be a sequence of complex numbers. If the Dirichlet series $\sum_{n=1}^{\infty} \frac{a_n}{n^s}$ is absolutely convergent at $s = s_0$, then it is M-test convergent on $\{ s \in \mathbb{C} : \Re s \geq \Re s_0 \}$.*

Proof. Just note that

$$\left| \frac{a_n}{n^s} \right| = \left| \frac{a_n}{n^{s_0}} \right| \cdot \left| \frac{1}{n^{s - s_0}} \right| \leq \left| \frac{a_n}{n^{s_0}} \right| \cdot \frac{1}{n^{\Re s - \Re s_0}} \leq \left| \frac{a_n}{n^{s_0}} \right|$$

for all $n \geq 1$. $\qquad \square$

A sequence $(a_n)_{n \in \mathbb{N}}$ of complex numbers is called *strictly multiplicative* if

$$a_1 = 1 \quad \text{and} \quad a_{nm} = a_n a_m \quad \text{for all } n, m \geq 1. \tag{3.5}$$

We are now in position to state and prove one of the central results of this chapter. We use analytic methods to prove a number theoretical result from the algebraic property (3.5). Its consequence, Euler product formula (3.11), is a landmark in number theory.

Theorem 3.1.8 *Let $(a_n)_{n \in \mathbb{N}}$ be a strictly multiplicative sequence of complex numbers. Suppose that the associated Dirichlet series converges at $s \in \mathbb{C}$ and that $|a_p| < p^{\Re s}$ for each prime p. Then, for such an s, the Dirichlet series has the product expansion*

$$\sum_{n=1}^{\infty} \frac{a_n}{n^s} = \prod_{p \text{ prime}} \frac{1}{1 - a_p p^{-s}}.$$

Proof. First of all, the infinite product in the right hand side converges by Proposition 3.1.3 applied to the sequence $(\frac{a_p}{p^s})_{p \text{ prime}}$. For $n, m \geq 1$ we set

$$P_n = \{p \text{ prime} : p \leq n\}, \quad S_n = \sum_{k=1}^{n} \frac{a_k}{k^s}, \quad S = \sum_{k=1}^{\infty} \frac{a_k}{k^s},$$

$$\Xi_{n,m} = \prod_{p \in P_n} \left(\sum_{h=0}^{m} \frac{a_{p^h}}{p^{hs}} \right) = \prod_{p \in P_n} \left(1 + \frac{a_p}{p^s} + \frac{a_{p^2}}{p^{2s}} + \cdots + \frac{a_{p^m}}{p^{ms}} \right),$$

$$\Xi_n = \prod_{p \in P_n} \frac{1}{1 - a_p p^{-s}}, \quad \text{and} \quad \Xi = \prod_{p \text{ prime}} \frac{1}{1 - a_p p^{-s}}.$$

Note that we have to prove that $S = \Xi$. Then, since $(a_p)^k = a_{p^k}$ (by strict multiplicativity), the formula for the sum of a geometric series and an easy combinatorial argument yield

$$\begin{aligned}
\Xi_n - \Xi_{n,m} &= \prod_{p \in P_n} \left(\sum_{h=0}^{\infty} \frac{a_{p^h}}{p^{hs}} \right) - \prod_{p \in P_n} \left(\sum_{h=0}^{m} \frac{a_{p^h}}{p^{hs}} \right) \\
&= \prod_{p \in P_n} \left(\sum_{h=0}^{m} \frac{a_{p^h}}{p^{hs}} + \sum_{h=m+1}^{\infty} \frac{a_{p^h}}{p^{hs}} \right) - \prod_{p \in P_n} \left(\sum_{h=0}^{m} \frac{a_{p^h}}{p^{hs}} \right) \qquad (3.6) \\
&= \sum_{\substack{A \subseteq P_n: \\ A \neq \varnothing}} \left[\prod_{p \in P_n \setminus A} \left(\sum_{h=0}^{m} \frac{a_{p^h}}{p^{hs}} \right) \cdot \prod_{p \in A} \left(\sum_{h=m+1}^{\infty} \frac{a_{p^h}}{p^{hs}} \right) \right].
\end{aligned}$$

For $n; m \geq 1$ we also set

$$Q_{n,m} = \{k = p_1^{h_1} p_2^{h_2} \cdots p_t^{h_t} : p_i \text{ prime}, p_1, p_2, \ldots, p_t \leq n; h_1, h_2, \ldots, h_t \leq m\}.$$

Clearly, $1 \in Q_{n,m}$. Since the sequence $(a_n)_{n\in\mathbb{N}}$ is strictly multiplicative, if $k = p_1^{h_1} p_2^{h_2} \cdots p_t^{h_t}$ then

$$a_k = (a_{p_1})^{h_1} (a_{p_2})^{h_2} \cdots (a_{p_t})^{h_t} \text{ and } \frac{a_k}{k^s} = \frac{a_{p_1}^{h_1}}{p_1^{h_1 s}} \frac{a_{p_2}^{h_2}}{p_2^{h_2 s}} \cdots \frac{a_{p_t}^{h_t}}{p_t^{h_t s}}.$$

Then

$$\Xi_{n,m} = \prod_{p\in P_n} \left(\sum_{h=0}^{m} \frac{a_{p^h}}{p^{hs}} \right) = \sum_{k\in Q_{n,m}} \frac{a_k}{k^s} \tag{3.7}$$

because in evaluating the product we get all possible factorizations of integers in $Q_{n,m}$.

Let $\varepsilon > 0$. By the convergence assumption, we can find an integer n_ε such that, for all $n > n_\varepsilon$,

$$|S_n - S| < \varepsilon \quad \text{and} \quad |\Xi_n - \Xi| < \varepsilon. \tag{3.8}$$

Fix $n > n_\varepsilon$. Then, by virtue of (3.6), for m sufficiently large we have

$$
\begin{aligned}
|\Xi_n - \Xi_{n,m}| &\leq \sum_{\substack{A\subseteq P_n: \\ A\neq\varnothing}} \left[\prod_{p\in P_n\setminus A} \left(\sum_{h=0}^{m} \frac{|a_{p^h}|}{|p^{hs}|} \right) \cdot \prod_{p\in A} \left(\sum_{h=m+1}^{\infty} \frac{|a_{p^h}|}{|p^{hs}|} \right) \right] \\
&\leq 2^{|P_n|} \left(\sum_{k=1}^{\infty} \frac{|a_k|}{|k^s|} \right)^{|P_n|} \sum_{k=m+1}^{\infty} \frac{|a_k|}{|k^s|} \\
&< \varepsilon
\end{aligned}
\tag{3.9}
$$

because n is fixed, $\sum_{k=1}^{\infty} \frac{|a_k|}{|k^s|}$ converges, $A \neq \varnothing$, and, for any $p \in A$,

$$\sum_{h=m+1}^{\infty} \frac{|a_{p^h}|}{|p^{hs}|} \leq \sum_{k=m+1}^{\infty} \frac{|a_k|}{|k^s|}$$

which tends to 0 as $m \to +\infty$ (for the last inequality, just note that certainly $p^{m+1} \geq m + 1$).

Moreover, if in addition $m \geq \log_2 n$, we clearly have $Q_{n,m} \supseteq \{1, 2, \ldots, n\}$. As a consequence, (3.7) and (3.8) imply that

$$|\Xi_{n,m} - S_n| \leq \left| \sum_{\substack{k\in Q_{n,m}: \\ k>n}} \frac{a_k}{k^s} \right| \leq \sum_{k=n+1}^{\infty} \frac{|a_k|}{|k^s|} \leq \varepsilon. \tag{3.10}$$

Finally, from (3.8), (3.9), and (3.10), we deduce that

$$|S - \Xi| \leq |S - S_n| + |S_n - \Xi_{n,m}| + |\Xi_{n,m} - \Xi_n| + |\Xi_n - \Xi| \leq 4\varepsilon.$$

As ε was arbitrary, this ends the proof. $\qquad\square$

If $a_n = 1$ for all $n \in \mathbb{N}$, then the sequence is strictly multiplicative and the associated Dirichlet series is the celebrated *Riemann zeta function*

$$\zeta(s) = \sum_{n=1}^{\infty} \frac{1}{n^s}.$$

From the equality $|\frac{1}{n^s}| = \frac{1}{n^{\Re s}}$ we deduce that this series converges absolutely at each $s \in \mathbb{C}$ with $\Re s > 1$. From Theorem 3.1.8 we deduce, as a particular case, the *Euler product formula*

$$\zeta(s) = \prod_{p \ prime} \frac{1}{1 - p^{-s}} \tag{3.11}$$

for all $s \in \mathbb{C}$ with $\Re s > 1$.

Remark 3.1.9

(i) Examining the proof of Theorem 3.1.8 in the case of the Riemann zeta function, that is, considering the expressions

$$\frac{1}{n^s} = \frac{1}{p_1^{sh_1} p_2^{sh_2} \cdots p_t^{sh_t}} \quad \text{and} \quad \frac{1}{1 - p^{-s}} = \sum_{h=0}^{\infty} \frac{1}{p^{sh}},$$

the identity

$$\sum_{n=1}^{\infty} \frac{1}{n^s} = \prod_{p \ prime} \frac{1}{1 - p^{-s}}$$

may be seen as an analytic formulation of the fundamental theorem of arithmetic (see Exercise 1.1.9).

(ii) Actually, the Riemann zeta function has a meromorphic continuation on the whole $\mathbb{C}$ with exactly one simple pole at $s = 1$ with residue 1. For this and other properties and applications of the Riemann zeta function we refer to [151].

We end this section by analyzing two remarkable asymptotic estimates for partial sums of particular values of the Riemann zeta function.

Proposition 3.1.10

(i) *There exists $\gamma > 0$ (the so-called Euler-Mascheroni constant) such that, for all $n \geq 1$,*

$$\sum_{k=1}^{n} \frac{1}{k} = \log n + \gamma + \mathcal{O}(\frac{1}{n}).$$

(ii) *There exists $\sigma \in \mathbb{R}$ such that, for all $n \geq 1$,*

$$\sum_{k=1}^{n} \frac{1}{\sqrt{k}} = 2\sqrt{n} + \sigma + \mathcal{O}(\frac{1}{\sqrt{n}}).$$

Proof. (i) Set

$$\gamma_k = \frac{1}{k} - \int_{k}^{k+1} \frac{1}{x} dx.$$

Since $\frac{1}{k+1} < \frac{1}{x} < \frac{1}{k}$ for $k < x < k+1$, we get

$$\frac{1}{k+1} < \int_{k}^{k+1} \frac{1}{x} dx < \frac{1}{k}$$

so that

$$0 < \gamma_k < \frac{1}{k} - \frac{1}{k+1}. \tag{3.12}$$

It follows that the series $\sum_{k=1}^{\infty} \gamma_k$ is convergent and has positive terms. Let us define γ as the sum of such a series.

Let $n \geq 1$. From (3.12) we get

$$\sum_{k=n+1}^{\infty} \gamma_k = \lim_{m \to \infty} \sum_{k=n+1}^{m} \gamma_k \leq \lim_{m \to \infty} \sum_{k=n+1}^{m} \left(\frac{1}{k} - \frac{1}{k+1} \right)$$

$$= \lim_{m \to \infty} \left(\frac{1}{n+1} - \frac{1}{m+1} \right) = \frac{1}{n+1} < \frac{1}{n}.$$

Finally, from

$$\gamma - \sum_{k=n+1}^{\infty} \gamma_k = \sum_{k=1}^{n} \frac{1}{k} - \sum_{k=1}^{n} \int_{k}^{k+1} \frac{1}{x} dx$$

$$= \sum_{k=1}^{n} \frac{1}{k} - \int_{1}^{n+1} \frac{1}{x} dx$$

$$= \sum_{k=1}^{n} \frac{1}{k} - \log(n+1)$$

we deduce (using $\frac{1}{n} \geq \log(1 + \frac{1}{n}) = \log(n+1) - \log n > 0$) that

$$|\sum_{k=1}^{n} \frac{1}{k} - \gamma - \log n| = |\log(1 + \frac{1}{n}) - \sum_{k=n+1}^{\infty} \gamma_k| \leq \frac{2}{n}.$$

(ii) We set

$$\eta_k = \frac{1}{\sqrt{k}} - \int_k^{k+1} \frac{1}{\sqrt{x}} dx.$$

Arguing as in the proof of (i), but replacing x and k by $\sqrt{x}$ and $\sqrt{k}$, respectively, we get

$$0 < \eta_k < \frac{1}{\sqrt{k}} - \frac{1}{\sqrt{k+1}}$$

which replaces (3.12). We deduce that the series $\sum_{k=1}^{\infty} \eta_k$ converges so that, denoting by η the sum of such a series, $\sum_{k=n+1}^{\infty} \eta_k \leq \frac{1}{\sqrt{n}}$ and

$$\eta - \sum_{k=n+1}^{\infty} \eta_k = \sum_{k=1}^{n} \frac{1}{\sqrt{k}} - 2\sqrt{n+1} + 2.$$

Finally, setting $\sigma = \eta - 2$, we get

$$\left| \sum_{k=1}^{n} \frac{1}{\sqrt{k}} - \sigma - 2\sqrt{n} \right| = \left| 2\left(\sqrt{n+1} - \sqrt{n}\right) - \sum_{k=n+1}^{\infty} \eta_k \right| \leq \frac{3}{\sqrt{n}},$$

where the last inequality follows from $\sqrt{n+1} - \sqrt{n} \leq \frac{1}{\sqrt{n}}$. □

We will also use the following elementary inequality: for $s > 1$

$$\zeta(s) \leq 1 + \sum_{n=2}^{\infty} \int_{n-1}^{n} \frac{1}{t^s} dt = 1 + \int_1^{+\infty} \frac{1}{t^s} ds = 1 + \frac{1}{s-1}, \qquad (3.13)$$

where the inequality follows from $\frac{1}{n^s} \leq \frac{1}{t^s}$, for $n - 1 \leq t \leq n$.

3.2 Preliminaries on multiplicative characters

In this section we consider the multiplicative characters of the ring $\mathbb{Z}/m\mathbb{Z}$, that is, the characters of the multiplicative Abelian group $\mathcal{U}(\widehat{\mathbb{Z}/m\mathbb{Z}})$ (see Section 1.4), where m is a positive integer. If $\psi \in \mathcal{U}(\widehat{\mathbb{Z}/m\mathbb{Z}})$ we extend it to the whole $\mathbb{Z}/m\mathbb{Z}$ by setting $\psi(x) = 0$ if $x \in \mathbb{Z}/m\mathbb{Z}$ is not invertible and then we think of it as an m-periodic function defined on $\mathbb{Z}$. More precisely, if $\psi \in \mathcal{U}(\widehat{\mathbb{Z}/m\mathbb{Z}})$, the associated *Dirichlet character* $\chi = \chi_\psi$ is the function $\chi : \mathbb{Z} \to \mathbb{T} \cup \{0\}$ defined by setting

$$\chi(n) = \begin{cases} \psi(\overline{n}) & \text{if } \gcd(n, m) = 1 \\ 0 & \text{otherwise,} \end{cases}$$

for all $n \in \mathbb{Z}$, where, as usual, $\bar{n} \in \mathbb{Z}/m\mathbb{Z}$ denotes the class $n + m\mathbb{Z}$. Clearly, $\chi(1) = \psi(\bar{1}) = 1$ and $\chi(nk) = \chi(n)\chi(k)$ for all $k, n \in \mathbb{Z}$; thus a Dirichlet character is strictly multiplicative (see (3.5)). The *principal Dirichlet character* mod m, denoted by χ_0, is the extension of the trivial character, that is,

$$\chi_0(n) = \begin{cases} 1 & \text{if } \gcd(n, m) = 1 \\ 0 & \text{otherwise,} \end{cases}$$

for all $n \in \mathbb{Z}$. We denote by $DC(m)$ the set of all Dirichlet characters mod m. From Corollary 1.5.3 and Corollary 2.3.4 we deduce that $|DC(m)| = \varphi(m)$. If $0 \leq n < m$ and $\gcd(n, m) = 1$, we define a variant Δ_n of the Dirac function, by setting,

$$\Delta_n(k) = \begin{cases} 1 & \text{if } k \equiv n \mod m \\ 0 & \text{otherwise,} \end{cases} \tag{3.14}$$

for all $k \in \mathbb{Z}$. In other words, Δ_n is the characteristic function of the class $\bar{n}$ mod m. Clearly, for the Abelian multiplicative group $\mathcal{U}(\mathbb{Z}/m\mathbb{Z})$ a Fourier analysis (as described in Section 2.4) is still valid: we may translate it in terms of the Dirichlet characters, as follows.

Proposition 3.2.1 *If* $\gcd(n, m) = 1$, *then, for all* $k \in \mathbb{Z}$,

$$\Delta_n(k) = \frac{1}{\varphi(m)} \sum_{\chi \in DC(m)} \overline{\chi(n)}\chi(k).$$

Proof. The Fourier transform of Δ_n (assuming $0 < n \leq m - 1$) yields

$$\widehat{\Delta_n}(\chi) = \sum_{h=0}^{m-1} \Delta_n(h)\chi(h) = \chi(n),$$

for all $\chi \in DC(m)$. Then we may apply the Fourier inversion formula (2.16). $\square$

We now describe some specific technical results on the Dirichlet characters. We begin with a cancellation property.

Lemma 3.2.2 *Let* $\chi \in DC(m)$. *If* $\chi \neq \chi_0$, *then*

$$\left| \sum_{k=1}^{n} \chi(k) \right| < m$$

for all $n \in \mathbb{N}$.

Proof. Indeed, the orthogonality relations for characters (Proposition 2.3.5) yield

$$\sum_{k=hm+1}^{(h+1)m} \chi(k) = \sum_{k=hm+1}^{(h+1)m} \chi(k)\overline{\chi_0(k)} = 0,$$

for all $h \in \mathbb{N}$. Therefore, if $n = qm + r$, with $0 \leq r < m$, we have

$$\sum_{k=1}^{n} \chi(k) = \sum_{h=0}^{q-1} \sum_{k=hm+1}^{(h+1)m} \chi(k) + \sum_{k=qm+1}^{qm+r} \chi(k) = \sum_{k=1}^{r} \chi(k)$$

so that

$$\left| \sum_{k=1}^{n} \chi(k) \right| \leq \sum_{k=1}^{r} |\chi(k)| \leq r < m. \qquad \square$$

Lemma 3.2.3 *For all $\chi \in DC(m)$, $\chi \neq \chi_0$, and for all positive integers $h < n$, we have the following asymptotic estimates:*

$$\sum_{k=h}^{n} \frac{\chi(k)}{\sqrt{k}} = \mathcal{O}(\frac{1}{\sqrt{h}}); \tag{3.15}$$

$$\sum_{k=h}^{n} \frac{\chi(k)}{k} = \mathcal{O}(\frac{1}{h}). \tag{3.16}$$

Proof. First of all, by applying the mean value theorem to the function $f(x) = \frac{1}{\sqrt{x}}$ we get

$$\frac{1}{\sqrt{k+1}} - \frac{1}{\sqrt{k}} = [(k+1) - k]f'(\xi) = -\frac{1}{2\xi^{3/2}}$$

for some $\xi \in [k, k+1]$ so that

$$0 \leq \frac{1}{\sqrt{k}} - \frac{1}{\sqrt{k+1}} \leq \frac{1}{2k\sqrt{k}}. \tag{3.17}$$

Using (3.4) with $z_n = \chi(n)$, $w_n = \frac{1}{\sqrt{n}}$, and $Z_n = \sum_{k=1}^{n} \chi(k)$, we have

$$\sum_{k=h}^{n} \frac{\chi(k)}{\sqrt{k}} = \sum_{k=h}^{n-1} Z_k \left(\frac{1}{\sqrt{k}} - \frac{1}{\sqrt{k+1}} \right) + \frac{Z_n}{\sqrt{n}} - \frac{Z_{h-1}}{\sqrt{h}}.$$

But, by Lemma 3.2.2, $|Z_k| \leq m$, so that (3.17) yields

$$\left| \sum_{k=h}^{n-1} Z_k \left(\frac{1}{\sqrt{k}} - \frac{1}{\sqrt{k+1}} \right) \right| \leq \frac{m}{2} \sum_{k=h}^{\infty} \frac{1}{k^{3/2}} \leq \frac{m}{2} \int_{h-1}^{+\infty} \frac{1}{x^{3/2}} dx = \frac{m}{\sqrt{h-1}} \leq \frac{2m}{\sqrt{h}}$$

for $h \geq 2$, which, together with the trivial estimate $\left| \frac{Z_n}{\sqrt{n}} - \frac{Z_{h-1}}{\sqrt{h}} \right| \leq \frac{2m}{\sqrt{h}}$, proves (3.15). The proof of (3.16) is similar, but now one uses the inequality (for $h \geq 2$)

$$\sum_{k=h}^{n-1} \left(\frac{1}{k} - \frac{1}{k+1} \right) \leq \sum_{k=h}^{\infty} \frac{1}{k^2} \leq \int_{h-1}^{+\infty} \frac{1}{x^2} dx = \frac{1}{h-1} \leq \frac{2}{h}. \qquad \square$$

Definition 3.2.4 A Dirichlet character $\chi \in DC(m)$ is called *real* if $\chi(n) \in \mathbb{R}$ (so that $\chi(n) \in \{-1, 0, 1\}$) for all $n \in \mathbb{Z}$.

Lemma 3.2.5 *If* $\chi \in DC(m)$ *is real, then, for all* $n \in \mathbb{N}$, *we have*

$$\sum_{\substack{k \in \mathbb{N}: \\ k|n}} \chi(k) \geq \begin{cases} 0 & \text{for all } n \in \mathbb{N} \\ 1 & \text{if } n \text{ is a square.} \end{cases}$$

Proof. If $n = p^h$, p prime, then the divisors of n are $1, p, \ldots, p^{h-1}, p^h$ so that

$$\sum_{\substack{k \in \mathbb{N}: \\ k|n}} \chi(k) = \chi(1) + \chi(p) + \cdots + \chi(p^{h-1}) + \chi(p^h)$$

$$= \chi(1) + \chi(p) + \cdots + \chi(p)^{h-1} + \chi(p)^h$$

$$= \begin{cases} h+1 & \text{if } \chi(p) = 1 \\ 1 & \text{if } \chi(p) = -1 \text{ and } h \text{ is even} \\ 0 & \text{if } \chi(p) = -1 \text{ and } h \text{ is odd} \\ 1 & \text{if } \chi(p) = 0. \end{cases}$$

Note also that $\chi(p) = 0$ if and only if $p|m$. If $n = p_1^{h_1} p_2^{h_2} \cdots p_t^{h_t}$ is the prime factorization of n as the product of distinct primes, then

$$\sum_{\substack{k \in \mathbb{N}: \\ k|n}} \chi(k) = \prod_{j=1}^{t} \left[\chi(1) + \chi(p_j) + \chi(p_j)^2 + \cdots + \chi(p_j)^{h_j} \right]$$

so that the sum in the left hand side vanishes if and only if $\chi(p_j) = -1$ and h_j is <u>odd</u> for at least one $j \in \{1, 2, \ldots, t\}$, otherwise the sum is ≥ 1. $\qquad \square$

For the last result of this section, we make use of a simple technique developed by Dirichlet (but for another problem in number theory, the so-called *divisor problem*; see [150]). For $f \colon \mathbb{N} \times \mathbb{N} \to \mathbb{C}$ and $h \in \mathbb{N}$ we set

$$S_h = \sum_{\substack{n,k \in \mathbb{N}: \\ nk \leq h}} f(n, k).$$

We can write this sum in the following useful ways:

$$S_h = \sum_{\ell=1}^{h} \sum_{\substack{n,k\in\mathbb{N}: \\ nk=\ell}} f(n,k) \quad \text{(summation along hyperbolas)}$$

$$= \sum_{n=1}^{h} \sum_{k=1}^{h/n} f(n,k) \quad \text{(vertical summation)}$$

$$= \sum_{k=1}^{h} \sum_{n=1}^{h/k} f(n,k) \quad \text{(horizontal summation)}.$$

Proposition 3.2.6 *Let* $\chi \in DC(m)$, $\chi \neq \chi_0$, *and suppose that* χ *is real. Set*

$$f(n,k) = \frac{\chi(k)}{\sqrt{nk}}$$

for all $n, k \geq 1$ *and*

$$S_h = \sum_{\substack{n,k\in\mathbb{N}: \\ nk\leq h}} f(n,k)$$

for all $h \geq 1$. *Then there exists a constant* $c > 0$ *such that, for all* $h \geq 1$,

$$S_h \geq c \log h.$$

Proof. Using summation along hyperbolas, we get

$$S_h = \sum_{\ell=1}^{h} \sum_{\substack{n,k\in\mathbb{N}: \\ nk=\ell}} \frac{\chi(k)}{\sqrt{nk}}$$

$$= \sum_{\ell=1}^{h} \frac{1}{\sqrt{\ell}} \sum_{\substack{k\in\mathbb{N}: \\ k|\ell}} \chi(k)$$

$$\text{(by Lemma 3.2.5 and } \ell = t^2) \geq \sum_{t=1}^{\sqrt{h}} \frac{1}{t}$$

$$\text{(by Proposition 3.1.10.(i))} \geq c \log h,$$

for some $c > 0$ sufficiently small. $\square$

3.3 Dirichlet *L*-functions

Definition 3.3.1 Let $m \in \mathbb{N}$ and $\chi \in DC(m)$. The associated *Dirichlet L-function* is the complex function $L(\cdot, \chi)$ defined by setting

$$L(s, \chi) = \sum_{n=1}^{\infty} \frac{\chi(n)}{n^s}$$

for all $s \in \mathbb{C}$ where the series converges.

Since $|\chi(n)| \leq 1$ for all $n \in \mathbb{N}$, the function $L(s, \chi)$ is defined for all $s \in \mathbb{C}$ with $\Re s > 1$, because for these values the series is absolutely convergent:

$$\left| \frac{\chi(n)}{n^s} \right| \leq \frac{1}{n^{\Re s}}.$$

We limit ourselves to give the most elementary properties of *L*-functions, following again our main reference [150]. More extensive treatments may be found in [13, 81]. For instance, $L(s, \chi)$ may be extended to an analytic (respectively, meromorphic with just a simple pole at $s = 1$) to the whole $\mathbb{C}$, if $\chi \neq \chi_0$ (respectively, $\chi = \chi_0$).

From Theorem 3.1.8, since any $\chi \in DC(m)$ is strictly multiplicative, we deduce that

$$L(s, \chi) = \prod_{p \text{ prime}} \frac{1}{1 - \chi(p)p^{-s}} \qquad \text{(Dirichlet formula)}$$

for all $s \in \mathbb{C}$ with $\Re s > 1$. In the case $\chi = \mathbf{1}$, Dirichlet formula reduces to Euler product formula (see (3.11)).

Proposition 3.3.2 *Let $m = p_1^{h_1} p_2^{h_2} \cdots p_t^{h_t}$ be the factorization of m into powers of distinct primes, then*

$$L(s, \chi_0) = \prod_{j=1}^{t} (1 - p_j^{-s}) \cdot \zeta(s),$$

for all $s \in \mathbb{C}$ with $\Re s > 1$.

Proof. Indeed, by Dirichlet formula,

$$L(s, \chi_0) = \prod_{\substack{p \text{ prime:} \\ p \nmid m}} \frac{1}{1 - p^{-s}}$$

since

$$\chi_0(p) = \begin{cases} 1 & \text{if } p \nmid m \\ 0 & \text{if } p \mid m. \end{cases}$$

$\square$

Following [150], we now focus our study to the case $s \in \mathbb{R}$, that is, we analyze $L(\cdot, \chi)$ mainly as a function of a real variable. This leads to a more elementary and simpler proof and more specific statements. However, note that, in general, $L(s, \chi) \in \mathbb{C}$, even if $s \in \mathbb{R}$.

Proposition 3.3.3 *Let* $\chi \in DC(m)$, $\chi \neq \chi_0$. *Then*

(i) $L(s, \chi)$ *converges for* $s > 0$ *and the convergence is uniform on each compact subset of* $(0, +\infty)$;
(ii) *the map* $s \mapsto L(s, \chi)$ *is* $C^1(0, +\infty)$;
(iii) *for* $s \to +\infty$

$$L(s, \chi) = 1 + \mathcal{O}(2^{-s}) \quad and \quad L'(s, \chi) = \mathcal{O}(2^{-s}).$$

Proof.   (i) Set $z_k = \chi(k)$ and $w_k = \frac{1}{k^s}$ in the summation by parts formula (3.4). Then, Lemma 3.2.2 yields $|Z_n| \leq m$ for all $n \in \mathbb{N}$ and therefore, by (3.4), for $0 < h \leq n$ and $s > 0$,

$$\left| \sum_{k=h}^{n} \frac{\chi(k)}{k^s} \right| \leq \sum_{k=h}^{n-1} m \left[\frac{1}{k^s} - \frac{1}{(k+1)^s} \right] + \frac{m}{h^s} + \frac{m}{n^s} = \frac{2m}{h^s}$$

which tends to 0 as $h \to +\infty$. Then, by the Cauchy criterion, the series defining $L(s, \chi)$ converges at all $s > 0$ and, moreover, it converges uniformly on each compact set in $(0, +\infty)$, by Proposition 3.1.5.

(ii) First of all, note that if we set $g(x) = x^{-s} \log x$ for $x > 0$, then $g'(x) = x^{-s-1}(1 - s \log x)$ and, for $x > 1$,

$$\begin{aligned} |g'(x)| &\leq x^{-s-1}(1 + s \log x) \\ &= x^{-s-1} + x^{-s-1} \log x^s \\ &\leq 3x^{-1-s/2} \end{aligned}$$

since $x^{-s} \leq x^{-s/2}$ and $\log x^s = 2 \log x^{s/2} \leq 2x^{s/2}$, for $x > 1$ and $s > 0$. By the mean value theorem, it follows that, for $k \in \mathbb{N}$,

$$\left| \frac{\log k}{k^s} - \frac{\log(k+1)}{(k+1)^s} \right| \leq \max_{[k, k+1]} g'(x) \leq \frac{3}{k^{1+s/2}}. \tag{3.18}$$

Then, by differentiating the series defining $L(s, \chi)$ we get

$$L'(s, \chi) = \sum_{n=2}^{\infty} -\frac{\log n}{n^s} \chi(n).$$

Setting $z_k = \chi(k)$ and $w_k = \frac{\log k}{k^s}$ in (3.4) and using $|Z_k| \leq m$ as in (i), we get

$$\left| \sum_{k=h}^{n} -\frac{\log k}{k^s} \chi(k) \right| \leq \sum_{k=h}^{n-1} m \left| \frac{\log(k+1)}{(k+1)^s} - \frac{\log k}{k^s} \right| + m \frac{\log h}{h^s} + m \frac{\log n}{n^s}$$

$$\text{(by (3.18))} \leq 3m \sum_{k=h}^{n-1} \frac{1}{k^{1+s/2}} + m \frac{\log h}{h^s} + m \frac{\log n}{n^s}$$

which tends to 0 uniformly in $s \in [\delta, +\infty)$, $\delta > 0$, as $h < n$ tend to $+\infty$. In other words, uniform convergence of $\sum_{k=1}^{\infty} \frac{1}{k^{1+s/2}}$ in $[\delta, +\infty)$, $\delta > 0$, together with the Cauchy criterion, ensures uniform convergence of the series of $L'(s, \chi)$.

(iii) Fix $s_0 > 1$ and set $C = \sum_{n=2}^{\infty} \frac{1}{n^{s_0}}$. Then for $s \geq s_0$ we have

$$|L(s, \chi) - 1| \leq \sum_{n=2}^{\infty} \frac{1}{n^s}$$

$$= 2^{-s} \sum_{n=2}^{\infty} \frac{1}{(n/2)^s}$$

$$\leq 2^{-s} \sum_{n=2}^{\infty} \frac{1}{(n/2)^{s_0}}$$

$$= 2^{s_0} C 2^{-s} = \mathcal{O}(2^{-s}).$$

Similarly,

$$|L'(s, \chi)| \leq \sum_{n=2}^{\infty} \frac{\log n}{n^s} = 2^{-s} \sum_{n=2}^{\infty} \frac{\log n}{(n/2)^s} = \mathcal{O}(2^{-s}). \qquad \square$$

Remark 3.3.4 Actually, from Proposition 3.3.3.(i) and elementary complex analysis, a stronger result than Proposition 3.3.3.(ii) follows, namely, that $L(s, \chi)$ is analytic on $\{s \in \mathbb{C} : \Re s > 0\}$; see [88]. But, as mentioned at the beginning of this section, this is not the strongest result: $L(s, \chi)$ has an analytic continuation on the whole $\mathbb{C}$, if $\chi \neq \chi_0$.

Corollary 3.3.5 *For $\chi \in DC(m)$, $\chi \neq \chi_0$, the integral*

$$\int_s^{+\infty} \frac{L'(t, \chi)}{L(t, \chi)} dt$$

is convergent for all $s > 1$.

Proof. From Proposition 3.3.3.(iii) it follows that

$$\frac{L'(t, \chi)}{L(t, \chi)} = \mathcal{O}(2^{-t})$$

as $t \to +\infty$. Note also that $L(t, \chi) \neq 0$ for $t > 1$, by Proposition 3.1.3 and Dirichlet product formula. $\qquad\square$

Proposition 3.3.6 *For $s > 1$ and $\chi \neq \chi_0$, <u>define</u> the logarithm of $L(s, \chi)$ by setting*

$$\log L(s, \chi) = -\int_s^{+\infty} \frac{L'(t, \chi)}{L(t, \chi)} dt.$$

Then, for $s > 1$, we have

$$\exp[\log L(s, \chi)] = L(s, \chi), \tag{3.19}$$

$$\log L(s, \chi) = \sum_{p \text{ prime}} \log \frac{1}{1 - \chi(p)p^{-s}} \tag{3.20}$$

where the logarithm in the right hand side is defined by means of (3.1), and

$$\prod_{\chi \in DC(m)} L(s, \chi) = \exp\left[\varphi(m) \sum_{p \text{ prime}} \sum_{k=1}^{\infty} \frac{\Delta_1(p^k)}{k p^{ks}} \right], \tag{3.21}$$

where Δ_1 is as in (3.14).

Proof. We have

$$\frac{d}{ds}\{L(s, \chi)\exp[-\log L(s, \chi)]\} = L'(s, \chi)\exp[-\log L(s, \chi)]$$

$$- L(s, \chi) \cdot \frac{L'(s, \chi)}{L(s, \chi)} \exp[-\log L(s, \chi)]$$

$$= 0$$

and by Proposition 3.3.3.(iii),

$$\lim_{s \to +\infty} L(s, \chi)\exp[-\log L(s, \chi)] = 1.$$

Since the argument of the above limit is constant, (3.19) follows.

We now prove (3.20). First of all, we note that by Proposition 3.1.1.(iv) and Proposition 3.1.3, the series at the right hand side is uniformly convergent on each interval $[\delta, +\infty)$, $\delta > 1$, so that it is continuous in $(1, +\infty)$. Moreover, for $s > 1$, the exponential of both sides of (3.20) is equal to $L(s, \chi)$. Indeed, for the left hand side this follows from (3.19), while, for the right hand side,

$$\exp\left[\sum_{p \text{ prime}} \log \frac{1}{1 - \chi(p)p^{-s}}\right] = \prod_{p \text{ prime}} \frac{1}{1 - \chi(p)p^{-s}} = L(s, \chi),$$

where the first equality follows from (3.3) and the second from Dirichlet product formula. Since exp has imaginary period equal to 2π, it follows that there exists an <u>integer valued</u> function h such that

$$\log L(s, \chi) = \sum_{p \text{ prime}} \log \frac{1}{1 - \chi(p)p^{-s}} + 2\pi i h(s).$$

But h is continuous, because both sides of (3.20) are continuous, and therefore it is constant. Since both sides of (3.20) tend to zero for $s \to +\infty$, this constant is equal to zero, and (3.20) is proved.

We now turn to the proof of (3.21). By (3.20) we have:

$$\prod_{\chi \in DC(m)} L(s, \chi) = \exp\left[\sum_{\chi \in DC(m)} \sum_{p \text{ prime}} \log \frac{1}{1 - \chi(p)p^{-s}}\right]$$

$$= \exp\left[\sum_{p \text{ prime}} \sum_{\chi \in DC(m)} \log \frac{1}{1 - \chi(p)p^{-s}}\right]$$

$$(\text{by (3.1)}) = \exp\left[\sum_{p \text{ prime}} \sum_{\chi \in DC(m)} \sum_{k=1}^{\infty} \frac{\chi(p^k)}{kp^{ks}}\right]$$

$$= \exp\left[\sum_{p \text{ prime}} \sum_{k=1}^{\infty} \frac{1}{kp^{ks}} \sum_{\chi \in DC(m)} \chi(p^k)\right]$$

$$(\text{by Proposition 3.2.1}) = \exp\left[\varphi(m) \sum_{p \text{ prime}} \sum_{k=1}^{\infty} \frac{\Delta_1(p^k)}{kp^{ks}}\right].$$

$\square$

Corollary 3.3.7 *For $s > 1$ the product in the left hand side of (3.21) is real and satisfies*

$$\prod_{\chi \in DC(m)} L(s, \chi) \geq 1. \tag{3.22}$$

Proof. The argument of the exponential in the right hand side of (3.21) is real and non-negative. $\square$

Lemma 3.3.8 *With the assumptions and notation as in Proposition 3.2.6 we have:*

$$S_h = 2\sqrt{h}L(1, \chi) + \mathcal{O}(1).$$

Proof. We partition $A_h = \{(n, k) \in \mathbb{N} \times \mathbb{N} : nk \leq h\}$, the summation region in the definition of S_h, into the regions

$$A_h^{(1)} = \left\{(n, k) \in \mathbb{N} \times \mathbb{N} : 1 \leq n \leq \sqrt{h}, \sqrt{h} < k \leq \frac{h}{n}\right\}$$

and

$$A_h^{(2)} = \left\{(n, k) \in \mathbb{N} \times \mathbb{N} : 1 \leq k \leq \sqrt{h}, 1 \leq n \leq \frac{h}{k}\right\}.$$

Correspondingly, $S_h = S_h^{(1)} + S_h^{(2)}$, where

$$S_h^{(1)} = \sum_{(n,k) \in A_h^{(1)}} \frac{\chi(k)}{\sqrt{nk}} = \sum_{n \leq \sqrt{h}} \frac{1}{\sqrt{n}} \left(\sum_{\sqrt{h} < k \leq \frac{h}{n}} \frac{\chi(k)}{\sqrt{k}}\right)$$

(the last equality follows from vertical summation) and

$$S_h^{(2)} = \sum_{(n,k) \in A_h^{(2)}} \frac{\chi(k)}{\sqrt{nk}} = \sum_{1 \leq k \leq \sqrt{h}} \frac{\chi(k)}{\sqrt{k}} \left(\sum_{n \leq \frac{h}{k}} \frac{1}{\sqrt{n}}\right)$$

(the last equality follows from horizontal summation). Then

$$\left|S_h^{(1)}\right| \leq \sum_{n \leq \sqrt{h}} \frac{1}{\sqrt{n}} \left|\sum_{\sqrt{h} < k \leq \frac{h}{n}} \frac{\chi(k)}{\sqrt{k}}\right|$$

$$\text{(by (3.15))} = \sum_{n \leq \sqrt{h}} \frac{1}{\sqrt{n}} \mathcal{O}\left(\frac{1}{\sqrt[4]{h}}\right) \qquad (3.23)$$

$$\text{(by Proposition 3.1.10.(ii))} = \mathcal{O}(1),$$

and, by Proposition 3.1.10.(ii),

$$S_h^{(2)} = \sum_{1 \leq k \leq \sqrt{h}} \frac{\chi(k)}{\sqrt{k}} \left[2\sqrt{\frac{h}{k}} + \sigma + \mathcal{O}\left(\sqrt{\frac{k}{h}}\right)\right] \qquad (3.24)$$

$$= 2\sqrt{h}L(1, \chi) + \mathcal{O}(1)$$

where in the last equality we have used the following estimates:

$$2\sqrt{h} \sum_{1 \leq k \leq \sqrt{h}} \frac{\chi(k)}{k} = 2\sqrt{h}L(1, \chi) - 2\sqrt{h} \sum_{k > \sqrt{h}} \frac{\chi(k)}{k}$$

$$\text{(by (3.16))} = 2\sqrt{h}L(1, \chi) + 2\sqrt{h} \, \mathcal{O}\left(\frac{1}{\sqrt{h}}\right)$$

$$= 2\sqrt{h}L(1, \chi) + \mathcal{O}(1),$$

by (3.15)

$$\sigma \sum_{1 \leq k \leq \sqrt{h}} \frac{\chi(k)}{\sqrt{k}} = \mathcal{O}(1),$$

and, finally, for some constant $C > 0$,

$$\left| \sum_{1 \leq k \leq \sqrt{h}} \frac{\chi(k)}{\sqrt{k}} \mathcal{O}\left(\sqrt{\frac{k}{h}}\right) \right| \leq \frac{C}{\sqrt{h}} \left| \sum_{1 \leq k \leq \sqrt{h}} \chi(k) \right| = \mathcal{O}(1).$$

From (3.23) and (3.24) the proof immediately follows. □

We are now in a position to state and prove the main technical result in the proof of the Dirichlet Theorem. Most of the preliminary results will be used, directly or indirectly, in its proof.

Theorem 3.3.9 (Dirichlet) *Let* $\chi \in DC(m)$ *and suppose that* $\chi \neq \chi_0$. *Then*

$$L(1, \chi) \neq 0.$$

Proof. First of all, we establish two simple inequalities. If $L(1, \chi) = 0$ then there exists $C_1 > 0$ such that

$$|L(s, \chi)| \leq C_1|s - 1| \tag{3.25}$$

for $1 \leq s \leq 2$ (this follows from the mean value theorem; recall also Proposition 3.3.3.(ii)), and there exists $C_2 > 0$ such that

$$|L(s, \chi_0)| \leq \frac{C_2}{|s - 1|} \tag{3.26}$$

for $1 < s \le 2$. Indeed, by Proposition 3.3.2 we have

$$|L(s, \chi_0)| \le \prod_{j=1}^{t} |1 - p_j^{-s}| \cdot |\zeta(s)|$$

$$(\text{by } (3.13)) \le C \left(1 + \frac{1}{s-1}\right)$$

$$\le \frac{C_2}{s-1},$$

where $C = \max_{1 \le s \le 2} \prod_{j=1}^{t} |1 - p_j^{-s}|$ and $C_2 = 2C$. The rest of the proof is divided into two cases.

<u>First case</u>: χ is *complex*, that is $\chi(n) \in \mathbb{C} \setminus \mathbb{R}$ for some $n \in \mathbb{Z}$. Therefore, $\chi \ne \overline{\chi}$. By contradiction, assume $L(1, \chi) = 0$. Then also $L(1, \overline{\chi}) = \overline{L(1, \chi)} = 0$. But then, taking into account (3.22), (3.25), (3.26), and the notation therein, we have, for $1 < s \le 2$,

$$1 \le \prod_{\chi' \in DC(m)} L(s, \chi') = L(s, \chi)L(s, \overline{\chi})L(s, \chi_0) \cdot \prod_{\substack{\chi' \in DC(m): \\ \chi' \ne \chi, \overline{\chi}, \chi_0}} L(s, \chi')$$

$$\le C_1^2 |s - 1|^2 \cdot \frac{C_2}{|s-1|} \cdot C_3 = C_1 C_2 C_3 |s - 1|,$$

where $C_3 > 0$ is a constant (cf. Proposition 3.3.3), a contradiction.

<u>Second case</u> $\chi \ne \chi_0$ is *real valued*, that is, $\chi(n) \in \{-1, 0, 1\}$ for all $n \in \mathbb{Z}$. On the one hand, by Proposition 3.2.6 and the notation therein, we have

$$S_h \ge c \log h$$

while, on the other hand, by Lemma 3.3.8, we have

$$S_h = \sqrt{h} L(1, \chi) + \mathcal{O}(1).$$

This clearly leads to a contradiction if $L(1, \chi) = 0$. $\qquad\square$

3.4 Euler's theorem

In this section we present a celebrated theorem of Euler. We begin with a further technical result, which is a consequence of Theorem 3.3.9.

Theorem 3.4.1 *Let $\chi \in DC(m)$. If $\chi \ne \chi_0$ then*

$$\sum_{p \text{ prime}} \frac{\chi(p)}{p^s} = \mathcal{O}(1)$$

for $s \to 1^+$.

Proof. By virtue of (3.20), for $s \to 1^+$ we have

$$\log L(s, \chi) = \sum_{p \text{ prime}} \log \frac{1}{1 - \chi(p)p^{-s}}$$

$$\text{(by Proposition 3.1.1.(iii))} = \sum_{p \text{ prime}} \frac{\chi(p)}{p^s} + \mathcal{O}\left(\sum_{p \text{ prime}} \frac{1}{p^{2s}}\right)$$

$$= \sum_{p \text{ prime}} \frac{\chi(p)}{p^s} + \mathcal{O}(1).$$

On the other hand, since $L'(t, \chi)$ and $L(t, \chi)$ are continuous in $(0, +\infty)$ (Proposition 3.3.3) and $L(1, \chi) \neq 0$ (Theorem 3.3.9), by Corollary 3.3.5 and Proposition 3.3.6 we have

$$\log L(s, \chi) = -\int_s^{+\infty} \frac{L'(t, \chi)}{L(t, \chi)} dt = \mathcal{O}(1)$$

for $s \to 1^+$. $\qquad\qquad\qquad\qquad\qquad\qquad\qquad\qquad\qquad\qquad \square$

We are now in a position to state and prove Euler's theorem. We give two proofs: the first one is Euler's original proof and follows from some of the results in the preceding sections; the second proof is due to Erdős and it is more elementary but based on a clever trick ([60]; see also [5]).

Theorem 3.4.2 (Euler)

$$\sum_{p \text{ prime}} \frac{1}{p} = +\infty.$$

Euler's proof For $s > 1$ the zeta function $\zeta(s)$ is real valued and, by virtue of Euler product formula (3.11), we have (here log is the usual real function)

$$\log \zeta(s) = \sum_{p \text{ prime}} \log \frac{1}{1 - p^{-s}}$$

$$\text{(by Proposition 3.1.1.(iii))} = \sum_{p \text{ prime}} \left[\frac{1}{p^s} + R\left(\frac{1}{p^s}\right)\right].$$

Moreover, again from Proposition 3.1.1.(iii) we deduce that

$$\left|\sum_{p \text{ prime}} R\left(\frac{1}{p^s}\right)\right| \leq \sum_{p \text{ prime}} \frac{1}{p^{2s}} \leq \sum_{n=1}^{\infty} \frac{1}{n^2} = \frac{\pi^2}{6}.$$

Therefore,

$$\sum_{p \text{ prime}} \frac{1}{p^s} \geq \log \zeta(s) - \frac{\pi^2}{6}$$

which tends to $+\infty$ for $s \to 1^+$, since $\zeta(s) = \sum_{n=1}^{\infty} \frac{1}{n^s}$ tends to $+\infty$ for $s \to 1^+$. $\square$

Erdős' proof By contradiction, assume that

$$\sum_{p \text{ prime}} \frac{1}{p} < +\infty.$$

Then there exists a partition $P \coprod Q$ of the set of all primes such that P is finite and

$$\sum_{p \in Q} \frac{1}{p} < \frac{1}{2}. \tag{3.27}$$

For $n \in \mathbb{N}$, set

$$A_n = \{k \in \mathbb{N} : k \leq n, k \text{ is divisible by at least one prime in } Q\}$$

$$B_n = \{k \in \mathbb{N} : k \leq n, k \text{ is divisible only by primes in } P\}.$$

Clearly,

$$\{1, 2, \ldots, n\} = A_n \coprod B_n. \tag{3.28}$$

From (3.27) we get

$$|A_n| \leq \sum_{p \in Q} \frac{n}{p} < \frac{n}{2} \tag{3.29}$$

because if $p \in Q$, then the multiples of p less than or equal to n are at most n/p. We now estimate the cardinality of B_n.

We uniquely write each $k \in B_n$ as the product of a square and a square-free integer

$$k = s_k^2 r_k,$$

in other words s_k is the largest divisor of k such that s_k^2 divides k. We first note that there are at most $2^{|P|}$ possible choices for r_k (this is a product of all primes in P each with exponent 0 or 1). Moreover, it is clear that $s_k \leq \sqrt{k} \leq \sqrt{n}$ so that, altogether

$$|B_n| \leq 2^{|P|} \sqrt{n}. \tag{3.30}$$

Then for

$$n = 2^{2|P|+4}$$

we have $2^{|P|} = \frac{\sqrt{n}}{4}$ and therefore, by virtue of (3.28),

$$
\begin{aligned}
n &= |A_n| + |B_n| \\
\text{(by (3.29) and (3.30))} \quad &\leq \frac{n}{2} + 2^{|P|}\sqrt{n} \\
&= \frac{n}{2} + \frac{n}{4} - \frac{3}{4}n,
\end{aligned}
$$

a contradiction. $\qquad\qquad\qquad\qquad\qquad\qquad\qquad\qquad\qquad\qquad\qquad\square$

3.5 Dirichlet's theorem

Theorem 3.5.1 (Dirichlet's theorem on primes in arithmetic progressions)
Let $m, r \in \mathbb{N}$ and suppose that $\gcd(m, r) = 1$. Then the arithmetic progression

$$r, r + m, r + 2m, r + 3m, \ldots, r + km, \ldots$$

contains infinitely many primes.

Proof. We show that

$$\lim_{s \to 1^+} \sum_{\substack{p \text{ prime:} \\ p \equiv r \bmod m}} \frac{1}{p^s} = +\infty, \tag{3.31}$$

from which it immediately follows that the set $\{p \text{ prime}: p \equiv r \mod m\}$ is infinite. (3.31) is clearly a generalization of Theorem 3.4.2, but it requires a lot more work. The first step is the use of the discrete Fourier inversion formula in Proposition 3.2.1 (with $n = r$ and $k = p$): for $s > 1$ we have

$$
\begin{aligned}
\sum_{\substack{p \text{ prime:} \\ p \equiv r \bmod m}} \frac{1}{p^s} &= \sum_{p \text{ prime}} \frac{\Delta_r(p)}{p^s} \\
&= \frac{1}{\varphi(m)} \sum_{\chi \in DC(m)} \overline{\chi(r)} \sum_{p \text{ prime}} \frac{\chi(p)}{p^s} \\
\text{(since } \chi_0(r) = 1\text{)} \quad &= \frac{1}{\varphi(m)} \sum_{p \text{ prime}} \frac{\chi_0(p)}{p^s} + \frac{1}{\varphi(m)} \sum_{\substack{\chi \in DC(m) \\ \chi \neq \chi_0}} \overline{\chi(r)} \sum_{p \text{ prime}} \frac{\chi(p)}{p^s}.
\end{aligned}
$$

Now, on the one hand, by Euler's theorem (Theorem 3.4.2) and the fact that there are only finitely many primes p dividing m,

$$\sum_{p \text{ prime}} \frac{\chi_0(p)}{p^s} = \sum_{p \nmid m} \frac{1}{p^s} \to +\infty$$

for $s \to 1^+$. On the other hand, for $\chi \neq \chi_0$ Theorem 3.4.1 ensures that the quantity $\sum_{p \text{ prime}} \frac{\chi(p)}{p^s}$ is bounded for $s \to 1^+$. $\square$

Remark 3.5.2 One of the most important and difficult results in number theory proved in recent years is the celebrated *Green-Tao theorem* [67], which states that the set of prime numbers contains arbitrarily long arithmetic progressions. This may be considered as a kind of "reciprocal" of Dirichlet's theorem, which ensures that certain arithmetic progressions contain infinitely many primes. The Green-Tao theorem, also, is a particular case of a celebrated conjecture, due to Erdős, on arithmetic progressions, which states that if A is an infinite subset of $\mathbb{N}$ such that $\sum_{n \in A} 1/n = +\infty$, then A contains arbitrarily long arithmetic progressions. Other particular cases of Erdős' conjecture are the celebrated theorems of Roth [131] and Szemerédi [155, 156], which we do not state here but for which we refer to the expository paper by Tao [158]. We only mention that Erdős' conjecture is still open and that a prize of 3000 USD is offered for its proof or disproof.

4

Spectral analysis of the DFT and number theory

In this chapter, following [104] and the exposition in [15], we present the spectral analysis of the normalized Fourier transform on $\mathbb{Z}_n$ (cf. Exercise 2.4.13). In the last two sections, as an application, we recover some classical results in number theory due to Gauss and Schur, including the celebrated law of quadratic reciprocity.

4.1 Preliminary results

We will use the notation and convention as in the beginning of Section 2.2.

This way, the normalized Fourier transform $\mathcal{F}\colon L(\mathbb{Z}_n) \to L(\mathbb{Z}_n)$ is given by

$$[\mathcal{F}f](m) = \frac{1}{\sqrt{n}} \sum_{k=0}^{n-1} f(k)\omega^{-km}$$

for all $f \in L(\mathbb{Z}_n)$ and $m \in \mathbb{Z}_n$; see Definition 2.4.1.

Similarly, the corresponding inverse Fourier transform $\mathcal{F}^{-1}\colon L(\mathbb{Z}_n) \to L(\mathbb{Z}_n)$ is given by

$$[\mathcal{F}^{-1}f](m) = \frac{1}{\sqrt{n}} \sum_{k=0}^{n-1} f(k)\omega^{km}$$

for all $f \in L(\mathbb{Z}_n)$ and $m \in \mathbb{Z}_n$. Note also that now Proposition 2.4.6.(iv) becomes

$$\mathcal{F}(f_1 * f_2) = \sqrt{n}\, \mathcal{F}(f_1)\mathcal{F}(f_2).$$

Recall (cf. Definition 2.4.14) that for $f \in L(\mathbb{Z}_n)$ we denote by $f^- \in L(\mathbb{Z}_n)$ the function defined by $f^-(x) = f(-x)$ for all $x \in \mathbb{Z}_n$.

Lemma 4.1.1   (i) $\mathcal{F}^{-1}\mathcal{F} = \mathcal{F}\mathcal{F}^{-1} = \mathrm{id}_{L(\mathbb{Z}_n)}$.

(ii) $\mathcal{F}$ and $\mathcal{F}^{-1}$ are unitary operators.

(iii) $\mathcal{F}^2 f = f^-$ for all $f \in L(\mathbb{Z}_n)$.

(iv) $\mathcal{F}\chi_m = \sqrt{n}\delta_m$ for all $m \in \mathbb{Z}_n$.

(v) $\mathcal{F}\delta_m = \frac{1}{\sqrt{n}}\chi_{-m} = \frac{1}{\sqrt{n}}\chi_{n-m}$.

Proof. (i) and (ii) are just a reformulation of the Fourier inversion formula (Theorem 2.4.2) and the Plancherel formula (Theorem 2.4.3), respectively; they can also be immediately deduced from the orthogonality relations (Proposition 2.3.5).

(iii) Let $f \in L(\mathbb{Z}_n)$ and $m \in \mathbb{Z}_n$. Then

$$[\mathcal{F}^2 f](m) = \frac{1}{n}\sum_{h=0}^{n-1}\left(\sum_{k=0}^{n-1} f(k)\omega^{-kh}\right)\omega^{-hm}$$

$$= \sum_{k=0}^{n-1} f(k)\frac{1}{n}\sum_{h=0}^{n-1}\chi_{-k}(h)\overline{\chi_m(h)}$$

$$(\text{by } (2.7)) = \sum_{k=0}^{n-1} f(k)\delta_0(-k-m)$$

$$= f(-m).$$

(iv) Let $m, h \in \mathbb{Z}_n$. Then

$$[\mathcal{F}\chi_m](h) = \frac{1}{\sqrt{n}}\sum_{k=0}^{n-1}\chi_m(k)\overline{\chi_h(k)}$$

$$(\text{by } (2.7)) = \frac{1}{\sqrt{n}}n\delta_0(m-h)$$

$$= \sqrt{n}\delta_m(h).$$

(v) Let $m, h \in \mathbb{Z}_n$. Then

$$[\mathcal{F}\delta_m](h) = \frac{1}{\sqrt{n}}\sum_{k=0}^{n-1}\delta_m(k)\omega^{-hk}$$

$$= \frac{1}{\sqrt{n}}\omega^{-mh}$$

$$= \frac{1}{\sqrt{n}}\chi_{-m}(h). \qquad \square$$

Proposition 4.1.2 *Let* $m \in \mathbb{Z}_n$.

(i) $\mathcal{F}^4 = \mathrm{id}_{L(\mathbb{Z}_n)}$;
(ii) $\mathcal{F}^2 \delta_m = \delta_{-m} \equiv \delta_{n-m}$;
(iii) $\mathcal{F}^2 \chi_m = \chi_{-m} \equiv \chi_{n-m}$.

Proof. (i), (ii), and (iii) follow immediately from Lemma 4.1.1 after observing that $(f^-)^- = f$ for all $f \in L(\mathbb{Z}_n)$, $(\chi_m)^- = \chi_{-m}$, and $(\delta_m)^- = \delta_{-m}$. $\qquad\square$

Theorem 4.1.3 *The characteristic polynomial* $p(\lambda) \in \mathbb{C}[\lambda]$ *of* $\mathcal{F}^2$ *is given by*

$$p(\lambda) = \begin{cases} (\lambda - 1)^{\frac{n+1}{2}} (\lambda + 1)^{\frac{n-1}{2}} & \textit{if } n \textit{ is odd} \\ (\lambda - 1)^{\frac{n+2}{2}} (\lambda + 1)^{\frac{n-2}{2}} & \textit{if } n \textit{ is even.} \end{cases}$$

Proof. By virtue of Proposition 4.1.2.(ii), the matrix $A_n \in \mathfrak{M}_n(\mathbb{C})$ representing $\mathcal{F}^2$ in the basis $\{\delta_0, \delta_1, \ldots, \delta_{n-1}\}$ is given by

$$A_n = \begin{pmatrix} 1 & 0 & \cdots & 0 & 0 \\ 0 & 0 & \cdots & 0 & 1 \\ 0 & 0 & \cdots & 1 & 0 \\ \vdots & \vdots & \ddots & \vdots & \vdots \\ 0 & 1 & \cdots & 0 & 0 \end{pmatrix}.$$

For $1 \leq k \leq n - 1$ define $B_k \in \mathfrak{M}_k(\mathbb{C})$ by setting

$$B_k = \begin{pmatrix} 0 & 0 & \cdots & 0 & 1 \\ 0 & 0 & \cdots & 1 & 0 \\ \vdots & \vdots & \ddots & \vdots & \vdots \\ 0 & 1 & 0 & \cdots & 0 \\ 1 & 0 & 0 & \cdots & 0 \end{pmatrix}.$$

Then

$$\det(\lambda I_n - A_n) = (\lambda - 1) \det(\lambda I_{n-1} - B_{n-1}) \tag{4.1}$$

and

$$\det(\lambda I_{n-1} - B_{n-1}) = \begin{pmatrix} \lambda & 0 & \cdots & 0 & -1 \\ 0 & \lambda & \cdots & -1 & 0 \\ \vdots & \vdots & \ddots & \vdots & \vdots \\ 0 & -1 & \cdots & \lambda & 0 \\ -1 & 0 & \cdots & 0 & \lambda \end{pmatrix}$$

$$= \lambda \begin{pmatrix} \lambda & \cdots & -1 & 0 \\ \vdots & \ddots & \vdots & \vdots \\ -1 & \cdots & \lambda & 0 \\ 0 & \cdots & 0 & \lambda \end{pmatrix} + (-1)^{n-2} \begin{pmatrix} 0 & \lambda & \cdots & -1 \\ \vdots & \vdots & \vdots & \vdots \\ 0 & -1 & \cdots & \lambda \\ -1 & 0 & \cdots & 0 \end{pmatrix}$$

$$= \lambda^2 \det(\lambda I_{n-3} - B_{n-3}) + (-1)^{2n-5} \det(\lambda I_{n-3} - B_{n-3})$$

$$= (\lambda^2 - 1) \det(\lambda I_{n-3} - B_{n-3})$$

so that, keeping in mind (4.1),

$$\det(\lambda I_n - A_n) = (\lambda^2 - 1)(\lambda - 1) \det(\lambda I_{n-3} - B_{n-3})$$

$$= (\lambda^2 - 1) \det(\lambda I_{n-2} - A_{n-2}).$$

Since

$$\det(\lambda I_3 - A_3) = \begin{vmatrix} \lambda - 1 & 0 & 0 \\ 0 & \lambda & -1 \\ 0 & -1 & \lambda \end{vmatrix} = (\lambda - 1)(\lambda^2 - 1) = (\lambda - 1)^2(\lambda + 1)$$

and

$$\det(\lambda I_2 - A_2) = \begin{vmatrix} \lambda - 1 & 0 \\ 0 & \lambda - 1 \end{vmatrix} = (\lambda - 1)^2,$$

the statement follows by induction. □

By virtue of Proposition 4.1.2.(i), the minimal polynomial of $\mathcal{F}$ divides $\lambda^4 - 1$, and therefore its eigenvalues are among $\pm 1, \pm i$; see [91] for the relations among eigenvalues and the minimal polynomial. Let us show that from the trace $\mathrm{Tr}\mathcal{F}$ of $\mathcal{F}$ we can recover the geometric/algebraic multiplicity of these eigenvalues.

Proposition 4.1.4 *Suppose that* $\mathrm{Tr}(\mathcal{F}) = \alpha + i\beta$. *Denote by* m_1 *(respectively* m_2, m_3, m_4*) the multiplicity of* 1 *(respectively* $-1, i, -i$*). If* n *is odd (respectively* *even), then the m_i's constitute the unique solution of the linear system*

$$\begin{cases} m_1 - m_2 & = \alpha \\ m_3 - m_4 & = \beta \\ m_1 + m_2 & = \frac{n+1}{2} \ (\text{respectively } \frac{n+2}{2}) \\ m_3 + m_4 & = \frac{n-1}{2} \ (\text{respectively } \frac{n-2}{2}). \end{cases}$$

Proof. By definition of the trace, we immediately have $\text{Tr}(\mathcal{F}) = m_1 - m_2 + i(m_3 - m_4)$: this explains the first two equations. Moreover, $m_1 + m_2$ (respectively $m_3 + m_4$) is the multiplicity of 1 (respectively -1) as an eigenvalue of $\mathcal{F}^2$. Thus the last two equations follow from Theorem 4.1.3. $\square$

In what follows, for $x \in \mathbb{R}$, we denote by $[x] \in \mathbb{Z}$ the greatest integer less than or equal to x. Setting $\nu = [n/2] + 1$ we consider the functions

$$\delta_0 \text{ and } \delta_j + \delta_{n-j} \text{ for } j = 1, 2, \ldots, \nu - 1 \tag{4.2}$$

and

$$\delta_k - \delta_{n-k} \text{ for } k = 1, 2, \ldots, n - \nu. \tag{4.3}$$

For example, if $n = 4$ then $\nu = 3$ and the functions in (4.2) are $\delta_0, \delta_1 + \delta_3 \equiv \delta_1 + \delta_{-1}$, and $2\delta_2 \equiv \delta_2 + \delta_{-2}$ (note that these are even functions), while there is only one in (4.3), namely $\delta_1 - \delta_3 \equiv \delta_1 - \delta_{-1}$ (note that this is, in turn, an odd function).

If $n = 5$, then $\nu = 3$ and the functions in (4.2) are $\delta_0, \delta_1 + \delta_4 \equiv \delta_1 - \delta_{-1}$, and $\delta_2 + \delta_3 \equiv \delta_2 + \delta_{-2}$ (note that these are even functions), while those in (4.3) are $\delta_1 - \delta_4 \equiv \delta_1 - \delta_{-1}$, and $\delta_2 - \delta_3 \equiv \delta_2 - \delta_{-2}$ (note that these are, in turn, odd functions).

Note that, more generally, if $n = 2h$ is even, then $\nu = h + 1$ and $\delta_{\nu-1} + \delta_{n-\nu+1} = \delta_h + \delta_{-h} = 2\delta_h$.

Moreover, we observe that $\nu - 1 = [n/2] \leq n/2$, and $j \leq n - j \Leftrightarrow j \leq n/2$ (respectively $n - \nu = n - 1 - [n/2] < n/2$, and $k < n - k \Leftrightarrow k < n/2$). It follows that the n functions in (4.2) and (4.3) are all distinct and nontrivial.

Let $L_+(\mathbb{Z}_n) \subseteq L(\mathbb{Z}_n)$ (respectively $L_-(\mathbb{Z}_n) \subseteq L(\mathbb{Z}_n)$) denote the subspace of complex valued even (respectively odd) functions on $\mathbb{Z}_n$.

Proposition 4.1.5 *The functions in (4.2) are even, i.e. belong to $L_+(\mathbb{Z}_n)$, while those in (4.3) are odd, i.e. belong to $L_-(\mathbb{Z}_n)$. Moreover, the functions in (4.2) and (4.3) altogether form an orthogonal basis of the whole $L(\mathbb{Z}_n)$. In particular, we have the orthogonal decomposition*

$$L(\mathbb{Z}_n) = L_+(\mathbb{Z}_n) \oplus L_-(\mathbb{Z}_n) \tag{4.4}$$

and $\dim L_+(\mathbb{Z}_n) = \nu$ *and* $\dim L_-(\mathbb{Z}_n) = n - \nu$. *Moreover,* (4.4) *is the spectral decomposition of* $\mathcal{F}^2$: $L_+(\mathbb{Z}_n)$ *is the eigenspace corresponding to* 1 *and* $L_-(\mathbb{Z}_n)$ *is the eigenspace corresponding to* -1.

Proof. Since $\delta_s(-t) = \delta_{-s}(t) = \delta_{n-s}(t)$ for all $s, t \in \mathbb{Z}_n$, it is clear that the functions in (4.2) (respectively (4.3)) are even (respectively odd). The mutual orthogonality of functions in (4.2) (respectively (4.3)) is obvious since their supports are disjoint. On the other hand, any function in (4.2) is orthogonal to any function in (4.3) since either their supports are disjoint, or they have the same support, say $\{s, t\}$, and then $\langle \delta_s + \delta_t, \delta_s - \delta_t \rangle = \langle \delta_s, \delta_s \rangle - \langle \delta_t, \delta_t \rangle = 0$. Finally, it is clear that n orthogonal functions constitute a basis of $L(\mathbb{Z}_n)$. The remaining statements are now clear; in particular, the last statement follows from Lemma 4.1.1.(iii) or from Proposition 4.1.2.(ii). $\qquad\square$

Lemma 4.1.6 *Let* $f \in L(\mathbb{Z}_n)$ *be an eigenvector of* $\mathcal{F}$. *Then either* f *is even and its associated eigenvalue is* 1 *or* -1, *or* f *is odd and its associated eigenvalue is* i *or* $-i$.

Proof. Let λ denote the eigenvalue associated with f, that is, $\mathcal{F}f = \lambda f$. Then $\mathcal{F}^2 f = \lambda^2 f$. We now express f in the basis in Proposition 4.1.5, that is,

$$f = a_0 \delta_0 + \sum_{j=1}^{\nu-1} a_j (\delta_j + \delta_{n-j}) + \sum_{k=1}^{n-\nu} b_k (\delta_k - \delta_{n-k})$$

with $a_0, a_1, \ldots, a_{\nu-1}, b_1, b_2, \ldots, b_{n-\nu} \in \mathbb{C}$. Then, by Proposition 4.1.2.(ii) we have

$$\mathcal{F}^2 f = a_0 \delta_0 + \sum_{j=1}^{\nu-1} a_j (\delta_j + \delta_{n-j}) - \sum_{k=1}^{n-\nu} b_k (\delta_k - \delta_{n-k})$$

so that the condition $\mathcal{F}^2 f = \lambda^2 f$ yields

$$a_0 \delta_0 + \sum_{j=1}^{\nu-1} a_j (\delta_{n-j} + \delta_j) - \sum_{k=1}^{n-\nu} b_k (\delta_k - \delta_{n-k})$$

$$= \lambda^2 a_0 \delta_0 + \sum_{j=1}^{\nu-1} \lambda^2 a_j (\delta_j + \delta_{n-j}) + \sum_{k=1}^{n-\nu} \lambda^2 b_k (\delta_k - \delta_{n-k}) \quad (4.5)$$

that is,

$$(\lambda^2 - 1)a_j = 0 \text{ for } j = 0, 1, \ldots, \nu - 1$$
$$(\lambda^2 + 1)b_k = 0 \text{ for } k = 1, 2 \ldots, n - \nu.$$

It follows that if $\lambda = \pm i$ then $a_j = 0$ for $j = 1, 2, \ldots, \nu - 1$, and therefore f is odd, while if $\lambda = \pm 1$ then $b_k = 0$ for $k = 1, 2, \ldots, n - \nu$, and therefore f is even. $\qquad\square$

Exercise 4.1.7 Let $\nu = [n/2] + 1$ as above. Let $f \in L(\mathbb{Z}_n)$.

Show that if f is *even*, then

$$\mathcal{F}f(m) = \frac{1}{\sqrt{n}}f(0) + \frac{2}{\sqrt{n}}\sum_{k=1}^{\nu-2} f(k) \cos \frac{2km\pi}{n}$$
$$+ \begin{cases} \frac{2}{\sqrt{n}}f(\nu - 1)\cos \frac{2(\nu-1)m\pi}{n} & \text{if } n \text{ is odd} \\ \frac{1}{\sqrt{n}}f(\nu - 1)(-1)^m & \text{if } n \text{ is even} \end{cases} \tag{4.6}$$

for all $m \in \mathbb{Z}_n$, and $\mathcal{F}f = \mathcal{F}^{-1}f$.

Show that if f is *odd*, then

$$\mathcal{F}f(m) = \frac{-2i}{\sqrt{n}}\sum_{k=1}^{n-\nu} f(k) \sin \frac{2km\pi}{n}$$

for all $m \in \mathbb{Z}_n$, and $\mathcal{F}^{-1}f = -\mathcal{F}f$.

Exercise 4.1.8 (cf. [55])

(1) Suppose that $F \in L(\mathbb{Z}_n)$ is *even* and define $T \in \text{End}(L(\mathbb{Z}_n))$ by setting

$$[Tf](x) = [f * F](x) + \sqrt{n}[\mathcal{F}F](x)f(x)$$

for all $f \in L(\mathbb{Z}_n)$ and $x \in \mathbb{Z}_n$. Show that

$$T\mathcal{F} = \mathcal{F}T.$$

(2) Deduce from (1) that the matrix

$$\begin{pmatrix} 2 & 1 & 0 & 0 & \cdots & 0 & 0 & 1 \\ 1 & 2\cos\frac{2\pi}{n} & 1 & 0 & \cdots & 0 & 0 & 0 \\ 0 & 1 & 2\cos\frac{4\pi}{n} & 1 & \cdots & 0 & 0 & 0 \\ \vdots & \vdots & & & \vdots & \vdots & & \vdots \\ 0 & 0 & 0 & 0 & \cdots & 1 & 2\cos\frac{2(n-2)\pi}{n} & 1 \\ 1 & 0 & 0 & 0 & \cdots & 0 & 1 & 2\cos\frac{2(n-1)\pi}{n} \end{pmatrix}$$

commutes with the matrix (2.22) of the Fourier transform.

4.2 The decomposition into eigenspaces

This section and the next one are among the most important sections of the book. We achieve a complete spectral theory of the DFT on $\mathbb{Z}_n$ by showing a decomposition into eigenspaces together with a careful computation of their dimensions.

Let us now set

$$W_1 = \{\mathcal{F}g + g : g \in L_+(\mathbb{Z}_n)\}$$
$$W_2 = \{\mathcal{F}g - g : g \in L_+(\mathbb{Z}_n)\}$$
$$W_3 = \{i\mathcal{F}g - g : g \in L_-(\mathbb{Z}_n)\}$$
$$W_4 = \{i\mathcal{F}g + g : g \in L_-(\mathbb{Z}_n)\}.$$

Theorem 4.2.1 *For the Fourier transform $\mathcal{F}$ the following holds:*

- W_1 *is the eigenspace corresponding to* 1
- W_2 *is the eigenspace corresponding to* -1
- W_3 *is the eigenspace corresponding to* i
- W_4 *is the eigenspace corresponding to* $-i$

so that

$$L_+(\mathbb{Z}_n) = W_1 \oplus W_2 \text{ and } L_-(\mathbb{Z}_n) = W_3 \oplus W_4$$

and therefore

$$L(\mathbb{Z}_n) = W_1 \oplus W_2 \oplus W_3 \oplus W_4$$

is the decomposition of $L(\mathbb{Z}_n)$ into the eigenspaces of $\mathcal{F}$.

Proof. First of all, we show that each W_j, $j = 1, 2, 3, 4$, is an eigenspace. Indeed, if $g \in L_+(\mathbb{Z}_n)$ then, by virtue of Lemma 4.1.1.(iii), $\mathcal{F}^2g = g$, and therefore the functions $f_+ = \mathcal{F}g + g \in W_1$ and $f_- = \mathcal{F}g - g \in W_2$ satisfy:

$$\mathcal{F}f_\pm = \mathcal{F}(\mathcal{F}g \pm g)$$
$$= g \pm \mathcal{F}g$$
$$= \pm(\mathcal{F}g \pm g)$$
$$= \pm f_\pm.$$

Similarly, if $g \in L_-(\mathbb{Z}_n)$ then, again by virtue of Lemma 4.1.1.(iii), $\mathcal{F}^2g = -g$, so that the functions $f_i = i\mathcal{F}g - g \in W_3$ and $f_{-i} = i\mathcal{F}g + g \in W_4$ satisfy:

$$\mathcal{F}f_{\pm i} = \mathcal{F}(i\mathcal{F}g \mp g)$$
$$= -ig \mp \mathcal{F}g$$
$$= \pm i(i\mathcal{F}g \mp g)$$
$$= \pm if_{\pm i}.$$

For the converse, we use repeatedly Lemma 4.1.6. Thus, if $\mathcal{F}f = f$, then f is even and $f = \mathcal{F}g + g$, with $g = \frac{1}{2}f \in L_+(\mathbb{Z}_n)$; if $\mathcal{F}f = -f$, then f is still even and $f = \mathcal{F}g - g$ with $g = -\frac{1}{2}f \in L_+(\mathbb{Z}_n)$; if $\mathcal{F}f = if$, then f is odd and $f = i\mathcal{F}g - g$ with $g = -\frac{1}{2}f \in L_-(\mathbb{Z}_n)$; finally, if $\mathcal{F}f = -if$, then f is odd and $f = i\mathcal{F}g + g$ with $g = \frac{1}{2}f \in L_-(\mathbb{Z}_n)$.

Since $\mathcal{F}$ is unitary, $L(\mathbb{Z}_n)$ can be expressed as the direct orthogonal sum of its eigenspaces and the remaining statements are trivial. $\qquad\square$

Exercise 4.2.2 Let W be a finite dimensional Hermitian space and $T: W \to W$ a unitary operator. Suppose that $T^4 = I_W$. Show that the eigenspaces of T^2 may be used to construct the eigenspaces of T as in Theorem 4.2.1.

Exercise 4.2.3 Let W be a finite dimensional Hermitian space and $T: W \to W$ a unitary operator. Suppose that $T^n = I_W$ for some positive integer n and let ω be an n-th root of unity.

(1) Show that a vector $w \in W$ satisfies $Tw = \omega w$ if and only if there exists $v \in W$ such that

$$w = T^{n-1}v + \omega T^{n-2}v + \cdots + \omega^{n-1}v.$$

(2) Suppose that $n = hk$ with $1 < h, k < n$ and set $S = T^h$ (so that $S^k = I$). Show that $w \in W$ satisfies $Tw = \omega w$ if and only if $w = T^{h-1}v + \omega T^{h-2}v + \cdots + \omega^{h-1}v$ for some $v \in W$ such that $Sv = \omega^h v$.

We are now in a position to exhibit suitable bases for the spaces $W_1, W_2, W_3,$ and W_4 in Theorem 4.2.1. One of the main tools is the notion of a Chebyshëv set: we refer to the Appendix for the corresponding definition and related properties. Moreover, we work separately on each of the spaces $W_1, W_2, W_3,$ and W_4, and we summarize the results in Theorem 4.3.1. In particular, for each space we consider four different cases, corresponding to the congruence modulo 4 of n.

Theorem 4.2.4 *Let* $n = 4m + r$, *with* $r \in \{0, 1, 2, 3\}$. *Then the functions* $u_0, u_1, \ldots, u_m \in W_1$ *defined by setting*

$$u_0 = \sqrt{n}(\mathcal{F}\delta_0 + \delta_0),$$

$$u_j = \frac{\sqrt{n}}{2}[\mathcal{F}(\delta_j + \delta_{-j}) + \delta_j + \delta_{-j}]$$

for $j = 1, 2, \ldots, m - 1$, *and*

$$u_m = \begin{cases} \sqrt{n}(\mathcal{F}\delta_{2m} + \delta_{2m}) & \text{if } n = 4m \\ \frac{\sqrt{n}}{2}[\mathcal{F}(\delta_{2m} + \delta_{-2m}) + \delta_{2m} + \delta_{-2m}] & \text{if } n = 4m + 1 \\ \frac{\sqrt{n}}{2}[\mathcal{F}(\delta_m + \delta_{-m}) + \delta_m + \delta_{-m}] & \text{if } n = 4m + 2, 4m + 3 \end{cases}$$

are linearly independent.

Proof. We divide the proof into the four cases corresponding to the possible values of r.

$n = 4m$. It suffices to show that the restrictions of $u_0, u_1, \ldots, u_m$ to the set $\{m, m+1, \ldots, 2m\} \subseteq \mathbb{Z}_n$ are linearly independent. Therefore, we consider the $(m+1)$-dimensional vectors:

$$\mathbf{z}_j = (u_j(m), u_j(m+1), \ldots, u_j(2m)) \tag{4.7}$$

for $j = 0, 1, \ldots, m$. By virtue of Lemma 4.1.1.(v) we have:

- $u_0 = \chi_0 + \sqrt{n}\delta_0$ and therefore $\mathbf{z}_0 = (1, 1, \ldots, 1)$;
- $u_j = \frac{1}{2}(\chi_j + \chi_{-j}) + \frac{\sqrt{n}}{2}(\delta_j + \delta_{-j})$ and therefore, since $\frac{1}{2}(\chi_j + \chi_{-j})(m+k) = \cos\frac{\pi j(m+k)}{2m}$,

$$\mathbf{z}_j = \left(\cos\frac{\pi}{2}j, \cos\frac{\pi(m+1)}{2m}j, \ldots, \cos\frac{\pi(m+k)}{2m}j, \ldots, \cos(\pi j) \right)$$

for $j = 1, 2, \ldots, m-1$;

- $u_m = \chi_{2m} + \sqrt{n}\delta_{2m}$ and, since $\chi_{2m}(m+k) = \cos\pi(m+k) + i\sin\pi(m+k) = (-1)^{m+k}$,

$$\mathbf{z}_m = ((-1)^m, (-1)^{m+1}, \ldots, (-1)^{2m-1}, 1 + \sqrt{n}).$$

We conclude by using Proposition A.2.(ii) applied to the Chebyshëv set $\{1, \cos\theta, \ldots, \cos(m-1)\theta\}$ (cf. Proposition A.3) with $t_k = \frac{\pi(m+k)}{2m}$, for $k = 0, 1, \ldots, m$.

$n = 4m + 1$. Following the previous case, we consider again the vectors (4.7):

- $\mathbf{z}_0 = (1, 1, \ldots, 1)$;
- since $\frac{1}{2}(\chi_j + \chi_{-j})(m+k) = \cos\frac{2\pi(m+k)}{4m+1}j$,

$$\mathbf{z}_j = \left(\cos\frac{2m\pi}{4m+1}j, \cos\frac{2\pi(m+1)}{4m+1}j, \ldots, \cos\frac{2\pi(m+k)}{4m+1}j, \ldots, \cos\frac{4m\pi}{4m+1}j \right)$$

for $j = 1, 2, \ldots, m-1$;

- since $\frac{1}{2}(\chi_{2m} + \chi_{-2m})(k+m) = \cos\frac{4m(m+k)\pi}{4m+1}$,

$$\mathbf{z}_m = \left(\cos\frac{4m^2\pi}{4m+1}, \cos\frac{4m(m+1)\pi}{4m+1}, \ldots, \cos\frac{4m(2m-1)\pi}{4m+1}, \cos\frac{8m^2\pi}{4m+1} + \frac{\sqrt{n}}{2} \right).$$

Thus we can conclude as in the previous case by taking $t_k = \frac{2\pi(m+k)}{4m+1}$ and $s_k = \cos\frac{4m(m+k)\pi}{4m+1}$ for $k = 0, 1, \ldots, m-1$, and $s_m = \cos\frac{8m^2\pi}{4m+1} + \frac{\sqrt{n}}{2}$. Just note that

$$\cos\frac{4m(m+k)\pi}{4m+1} = \cos\left[(m+k)\pi - \frac{m+k}{4m+1}\pi \right] = (-1)^{m+k}\cos\frac{(m+k)\pi}{4m+1}$$

and $\frac{(m+k)\pi}{4m+1} < \frac{\pi}{2}$ for $k = 0, 1, \ldots, m$ so that the s_ks alternate in sign, and, for $k = m - 1$ one has $(-1)^{2m-1} = -1$ so that $s_{m-1} < 0$, while $s_m = \cos \frac{2m\pi}{4m+1} + \frac{\sqrt{n}}{2} > 0$.

$n = 4m + 2$. We proceed as in the previous cases, now appealing to Proposition A.2.(i) and replacing (4.7) by

$$\mathbf{z}_j = (u_j(m+1), u_j(m+2), \ldots, u_j(2m+1)).$$

From the equality

$$\frac{1}{2}(\chi_j + \chi_{-j})(m+k) = \cos \frac{2\pi(m+k)j}{4m+2} = \cos \frac{\pi(m+k)j}{2m+1}$$

we get the $(m+1)$-dimensional vectors

$$\mathbf{z}_j = \left(\cos \frac{(m+1)\pi}{2m+1} j, \cos \frac{(m+2)\pi}{2m+1} j, \ldots, \cos \frac{(m+k)\pi}{2m+1} j, \ldots, \cos \pi j \right)$$

for $j = 0, 1, \ldots, m$. The Chebyshëv set is again $\{1, \cos\theta, \ldots, \cos m\theta\}$ and $t_k = \frac{\pi(m+k)}{2m+1}$, for $k = 1, 2, \ldots, m+1$.

$n = 4m + 3$. Now $\frac{1}{2}(\chi_j + \chi_{-j})(m+k) = \cos \frac{2\pi(m+k)j}{4m+3}$ so that, as in the preceding case,

$$\mathbf{z}_j = \left(\cos \frac{2\pi(m+1)}{4m+3} j, \cos \frac{2\pi(m+2)}{4m+3} j, \ldots, \cos \frac{2\pi(2m+1)}{4m+3} j \right)$$

for $j = 0, 1, \ldots, m$, and we may apply Proposition A.2.(i) with the same Chebyshëv set as in the previous case and $t_k = \frac{2\pi(m+k)}{4m+3}$, for $k = 1, 2, \ldots, m+1$. □

Theorem 4.2.5 *Let $n = 4m + r$, with $r \in \{0, 1, 2, 3\}$. Consider the functions $v_0, v_1, \ldots, v_m \in W_2$ defined by*

$$v_0 = \sqrt{n}(\mathcal{F}\delta_0 - \delta_0)$$

and

$$v_j = \frac{\sqrt{n}}{2}[\mathcal{F}(\delta_j + \delta_{-j}) - (\delta_j + \delta_{-j})]$$

for $j = 1, 2, \ldots, m$. Then the following holds:

- *if $n = 4m, 4m + 1$, then the functions $v_0, v_1, \ldots, v_{m-1}$ are linearly independent;*
- *if $n = 4m + 2, 4m + 3$, then the functions $v_0, v_1, \ldots, v_m$ are linearly independent.*

Proof. As for the proof of Theorem 4.2.4, we divide the proof into the four cases corresponding to the possible values of r.

$\underline{n = 4m}$. Arguing as in the cases $n = 4m + 2$ and $n = 4m + 3$ in the proof of Theorem 4.2.4, and evaluating the functions at the points $\{m + k : k = 1, 2, \ldots, m\}$ we get the vectors

$$\mathbf{z}_j = \left(\cos \frac{\pi(m+1)}{2m} j, \cos \frac{\pi(m+2)}{2m} j, \ldots, \cos \frac{\pi(m+k)}{2m} j, \ldots, \cos \pi j \right)$$

for $j = 0, 1, \ldots, m - 1$, and we may apply Proposition A.2.(i) to the Chebyshëv set $\{1, \cos\theta, \ldots, \cos(m-1)\theta\}$ with $t_k = \frac{\pi(m+k)}{2m}$, for $k = 1, 2, \ldots, m$.

$\underline{n = 4m + 1}$. This is very similar to the previous case: now

$$\mathbf{z}_j = \left(\cos\frac{2\pi(m+1)}{4m+1} j, \cos\frac{2\pi(m+2)}{4m+1} j, \ldots, \cos\frac{2\pi(m+k)}{4m+1} j, \ldots, \cos\frac{4\pi m}{4m+1} j \right)$$

for $j = 0, 1, \ldots, m - 1$, and we may apply Proposition A.2 to the same Chebyshëv set as above and $t_k = \frac{2\pi(m+k)}{4m+1}$, for $k = 1, 2, \ldots, m$.

$\underline{n = 4m + 2}$. This leads exactly to the same vectors as in case $n = 4m + 2$ of Theorem 4.2.4, evaluating the functions at the points $\{m + k : k = 1, 2, \ldots, m + 1\}$.

$\underline{n = 4m + 3}$. This leads exactly to the same vectors as in case $n = 4m + 3$ of Theorem 4.2.4, evaluating the functions at the points $\{m + k : k = 1, 2, \ldots, m + 1\}$. $\qquad\square$

Theorem 4.2.6 *Let again $n = 4m + r$, with $r \in \{0, 1, 2, 3\}$. Consider the functions*

$$w_j = \frac{\sqrt{n}}{2} [i\mathcal{F}(\delta_j - \delta_{-j}) - (\delta_j - \delta_{-j})] \in W_3$$

for $j = 1, 2, \ldots, m$. Then the following holds:

- *if $n = 4m$ then the functions $w_1, w_2, \ldots, w_{m-1}$ are linearly independent;*
- *if $n = 4m + 1, 4m + 2, 4m + 3$ then the functions $w_1, w_2, \ldots, w_m$ are linearly independent.*

Proof. Here, we divide the proof into two cases.

$\underline{n = 4m}$. For $k \geq 1$ and $j \geq 1$, by virtue of Lemma 4.1.1.(v)

$$w_j(m+k) = \frac{i}{2}(\chi_{-j} - \chi_j)(m+k) = \sin\frac{\pi j(m+k)}{2m}.$$

Therefore, if we restrict to the set $\{m + k : k = 1, 2, \ldots, m - 1\}$ we get the $(m-1)$-dimensional vectors

$$\mathbf{z}_j = \left(\sin\frac{\pi(m+1)}{2m}j, \sin\frac{\pi(m+2)}{2m}j, \ldots, \sin\frac{\pi(m+k)}{2m}j, \ldots, \sin\frac{\pi(2m-1)}{2m}j\right)$$

for $j = 1, 2, \ldots, m - 1$, and we can apply Proposition A.2 to the Chebyshëv set $\{\sin\theta, \sin 2\theta, \ldots, \sin(m-1)\theta\}$ (cf. Proposition A.3) with $t_k = \frac{\pi(m+k)}{2m}$ for $k = 1, 2, \ldots, m - 1$.

$\underline{n = 4m + r, r = 1, 2, 3}$. Now we restrict to the set $\{m + k : k = 1, 2, \ldots, m\}$ obtaining the m-dimensional vectors

$$\mathbf{z}_j = \left(\sin\frac{2\pi(m+1)}{4m+r}j, \sin\frac{2\pi(m+2)}{4m+r}j, \ldots, \sin\frac{2\pi(m+k)}{4m+r}j, \ldots, \sin\frac{4\pi m}{4m+r}j\right)$$

for $j = 1, 2, \ldots, m$. Using the Chebyshëv set $\{\sin\theta, \sin 2\theta, \ldots, \sin m\theta\}$ (cf. Proposition A.3) with $t_k = \frac{2\pi(m+k)}{4m+r}$, for $k = 1, 2, \ldots, m$, we conclude the proof. $\square$

Theorem 4.2.7 *Let again $n = 4m + r$, with $r \in \{0, 1, 2, 3\}$. Consider the functions*

$$z_j = \frac{\sqrt{n}}{2}\left[i\mathcal{F}(\delta_j - \delta_{-j}) + \delta_j - \delta_{-j}\right]$$

for $j = 1, 2, \ldots, m - 1$,

$$z_m = \frac{\sqrt{n}}{2}\begin{cases} i\mathcal{F}(\delta_{2m-1} - \delta_{-2m+1}) + \delta_{2m-1} - \delta_{-2m+1} & \text{if } r = 0 \\ i\mathcal{F}(\delta_m - \delta_{-m}) + \delta_m - \delta_{-m} & \text{if } r = 1, 2, 3 \end{cases}$$

and, only for $r = 3$,

$$z_{m+1} = \frac{\sqrt{n}}{2}\left[i\mathcal{F}(\delta_{2m+1} - \delta_{-2m-1}) + \delta_{2m+1} - \delta_{-2m-1}\right].$$

Then, all these functions belong to W_4 (cf. Theorem 4.2.1) and the following holds:

- *if $r = 0, 1, 2$ then the functions $z_1, z_2, \ldots, z_m$ are linearly independent;*
- *if $r = 3$ then the functions $z_1, z_2, \ldots, z_m, z_{m+1}$ are linearly independent.*

Proof. We divide the proof into three cases.

<u>$n = 4m$</u>. We restrict the functions to the set $\{m + k : k = 0, 1, \ldots, m - 1\}$ obtaining the m-dimensional vectors

$$\mathbf{z}_j = \left(\sin \frac{\pi}{2} j, \sin \frac{\pi(m+1)}{2m} j, \ldots, \sin \frac{\pi(m+k)}{2m} j, \ldots, \sin \frac{\pi(2m-1)}{2m} j \right)$$

for $j = 1, 2, \ldots, m - 1$ and, since

$$\sin \frac{\pi(m+k)(2m-1)}{2m} = \sin \left[\pi(m+k) - \frac{\pi(m+k)}{2m} \right]$$

$$= (-1)^{m+k+1} \sin \frac{\pi(m+k)}{2m},$$

with $\sin \frac{\pi(m+k)}{2m} > 0$ (because $0 < \frac{\pi(m+k)}{2m} < \frac{\pi}{2}$), for $k = 0, 1, \ldots, m - 1$, and $z_m(2m - 1) = \sin \frac{(2m-1)\pi}{2m} + \frac{\sqrt{n}}{2} > 0$, we have

$$\mathbf{z}_m = \left((-1)^{m+1} \sin \frac{\pi}{2}, (-1)^{m+2} \sin \frac{\pi(m+1)}{2m}, \ldots \right.$$

$$\left. \ldots (-1)^{m+k+1} \sin \frac{\pi(m+k)}{2m}, \ldots, \sin \frac{\pi(2m-1)}{2m} + \frac{\sqrt{n}}{2} \right).$$

By Proposition A.2.(ii) with the Chebyshëv set $\{\sin \theta, \sin 2\theta, \ldots, \sin(m - 1)\theta\}$ with $t_k = \frac{\pi(m+k)}{2m}$, for $k = 0, 1, \ldots, m - 1$ and $s_k = (-1)^{m+k+1} \sin \frac{\pi(m+k)}{2m}$, for $k = 0, 1, \ldots, m - 2$, and $s_{m-1} = \sin \frac{\pi(2m-1)}{2m} + \frac{\sqrt{n}}{2}$, this completes the first case.

<u>$n = 4m + 1, 4m + 2$</u>. These cases lead to the same vectors in the corresponding cases in Theorem 4.2.6.

<u>$n = 4m + 3$</u>. We restrict the functions to the set $\{m + k : k = 1, 2, \ldots, m + 1\}$ obtaining the m-dimensional vectors

$$\mathbf{z}_j = \left(\sin \frac{2\pi(m+1)}{4m+3} j, \sin \frac{2\pi(m+2)}{4m+3} j, \ldots, \sin \frac{2\pi(2m+1)}{4m+3} j \right)$$

for $j = 1, 2, \ldots, m$.
Since,

$$\sin \frac{\pi(m+k)(4m+2)}{4m+3} = \sin \left[\pi(m+k) - \frac{\pi(m+k)}{4m+3} \right]$$

$$= (-1)^{m+k+1} \sin \frac{\pi(m+k)}{4m+3}$$

with $\sin \frac{\pi(m+k)}{4m+3} > 0$, for $k = 1, 2, \ldots, m$, and

$$z_{m+1}(2m+1) = \sin \frac{\pi(2m+1)}{4m+3} + \frac{\sqrt{n}}{2} > 0,$$

we conclude by using the Chebyshёv set $\{\sin \theta, \sin 2\theta, \ldots, \sin m\theta\}$ with $t_k = \frac{2\pi(m+k)}{4m+3}$, for $k = 1, 2, \ldots, m+1$, and $s_k = (-1)^{m+k+1} \sin \frac{\pi(m+k)}{4m+3}$, for $k = 1, 2, \ldots, m$, and $s_{m+1} = \sin \frac{\pi(2m+1)}{4m+3} + \frac{\sqrt{n}}{2}$. □

4.3 Applications: some classical results by Gauss and Schur

Theorem 4.3.1 (Schur) *With the notation in Theorem 4.2.1, the multiplicities of the eigenvalues of the DFT are given in Table 4.1 (recall, cf. Proposition 4.1.4, that $m_j = \dim W_j$, for $j = 1, 2, 3, 4$).*

Table 4.1. *The multiplicities of the eigenvalues of the DFT.*

n	m_1	m_2	m_3	m_4
$4m$	$m+1$	m	$m-1$	m
$4m+1$	$m+1$	m	m	m
$4m+2$	$m+1$	$m+1$	m	m
$4m+3$	$m+1$	$m+1$	m	$m+1$

Proof. Consider first the case $n = 4m$. Then the following holds:

- Theorem 4.2.4 implies $m_1 = \dim W_1 \geq m+1$;
- Theorem 4.2.5 implies $m_2 = \dim W_2 \geq m$;
- Theorem 4.2.6 implies $m_3 = \dim W_3 \geq m-1$;
- Theorem 4.2.7 implies $m_4 = \dim W_4 \geq m$.

Since $m_1 + m_2 + m_3 + m_4 = 4m$, all the inequalities above are indeed equalities.

The other cases can be handled similarly. □

Remark 4.3.2 In the previous theorems we have given the spectral analysis of the matrix (2.22) of the DFT, namely of $F_n = \frac{1}{\sqrt{n}}(\omega^{-jk})_{j,k=0}^{n-1}$. Other authors (for instance Auslander and Tolimieri [15] and Terras [159]) consider, instead, the matrix $\frac{1}{\sqrt{n}}(\omega^{jk})_{j,k=0}^{n-1}$ (the kth column is switched with the $(n-k)$th column).

Corollary 4.3.3 (Gauss, Schur) *The trace of $\mathcal{F}$ is given by*

$$\mathrm{Tr}(\mathcal{F}) = \begin{cases} 1 - i & \text{if } n \equiv 0 \mod 4 \\ 1 & \text{if } n \equiv 1 \mod 4 \\ 0 & \text{if } n \equiv 2 \mod 4 \\ -i & \text{if } n \equiv 3 \mod 4 \end{cases}$$

and its characteristic polynomial $p(\lambda) \in \mathbb{C}[\lambda]$ is

$$p(\lambda) = \begin{cases} (\lambda - 1)^2(\lambda + 1)(\lambda + i)(\lambda^4 - 1)^{(n-4)/4} & \text{if } n \equiv 0 \mod 4 \\ (\lambda - 1)(\lambda^4 - 1)^{(n-1)/4} & \text{if } n \equiv 1 \mod 4 \\ (\lambda^2 - 1)(\lambda^4 - 1)^{(n-2)/4} & \text{if } n \equiv 2 \mod 4 \\ (\lambda^2 - 1)(\lambda + i)(\lambda^4 - 1)^{(n-3)/4} & \text{if } n \equiv 3 \mod 4. \end{cases}$$

Corollary 4.3.4 (Gauss)

$$\sum_{k=0}^{n-1} \exp\left(\frac{2\pi i k^2}{n}\right) = \begin{cases} (1 + i)\sqrt{n} & \text{if } n \equiv 0 \mod 4 \\ \sqrt{n} & \text{if } n \equiv 1 \mod 4 \\ 0 & \text{if } n \equiv 2 \mod 4 \\ i\sqrt{n} & \text{if } n \equiv 3 \mod 4. \end{cases}$$

Proof.

$$\mathrm{Tr}(\mathcal{F}) = \sum_{k=0}^{n-1} \langle \mathcal{F}\delta_k, \delta_k \rangle = \sum_{k=0}^{n-1} \frac{1}{\sqrt{n}} \chi_{-k}(k) = \frac{1}{\sqrt{n}} \sum_{k=0}^{n-1} \exp\left(-\frac{2\pi i k^2}{n}\right), \quad (4.8)$$

where the second equality follows from Lemma 4.1.1.(v). The statement then follows from Corollary 4.3.3 by conjugating both sides of (4.8). $\qquad\square$

The case $n \equiv 2 \mod 4$ is trivial, as it is shown in the following exercise.

Exercise 4.3.5 Suppose $n \equiv 2 \mod 4$. Prove the identity

$$\exp\left[\frac{2\pi i}{n}\left(k + \frac{n}{2}\right)^2\right] = -\exp\frac{2\pi i k^2}{n}$$

and deduce the case $n \equiv 2 \mod 4$ in Corollary 4.3.4.

4.4 Quadratic reciprocity and Gauss sums

This section is based on the monographs by Nathanson [118], Ireland and Rosen [79], Apostol [13], Terras [159], Nagell [117], and the paper [15] by Auslander and Tolimieri.

Definition 4.4.1 Let $n, m \in \mathbb{Z}$ with $\gcd(n, m) = 1$. We say that m is a *quadratic residue* mod n if the congruence

$$x^2 \equiv m \mod n \tag{4.9}$$

has a solution x in $\mathbb{Z}$; otherwise, we say that m is a *quadratic nonresidue* mod n.

This section is devoted to the study of the solvability of (4.9). It culminates with the celebrated Gauss law of quadratic reciprocity (Theorem 4.4.18).

Remark 4.4.2

(1) It is clear that $m = 1 + kn$ is a quadratic residue mod n for all $n \in \mathbb{Z} \setminus \{0\}$ and $k \in \mathbb{Z}$. Indeed, the congruence (4.9) has solution $x = 1$.

(2) Let $n, m \in \mathbb{Z}$ with $\gcd(n, m) = 1$, so that $\overline{m} \in \mathcal{U}(\mathbb{Z}/n\mathbb{Z})$ (cf. Lemma 1.5.1). Then m is a quadratic residue mod n if and only if $\overline{m}$ is a square in $\mathcal{U}(\mathbb{Z}/n\mathbb{Z})$ (that is, there exists $\overline{x} \in \mathcal{U}(\mathbb{Z}/n\mathbb{Z})$ such that $\overline{x}^2 = \overline{m}$).

(3) Let $n_1, n_2, m \in \mathbb{Z}$ with $\gcd(n_2, m) = 1$ and $n_1 | n_2$, and suppose that m is a quadratic residue mod n_2. Set $q = n_2/n_1 \in \mathbb{Z}$ and suppose that x is a solution of the congruence $x^2 \equiv m \mod n_2$. Then there exists $k \in \mathbb{Z}$ such that $x^2 = n_2 k + m = n_1(qk) + m$. This shows, in particular, that m is a quadratic residue mod n_1.

Proposition 4.4.3 *Let $n, m \in \mathbb{Z}$ with $\gcd(n, m) = 1$. Suppose that $n = n_1 n_2$ with $\gcd(n_1, n_2) = 1$. Then m is a quadratic residue mod n if and only if it is a quadratic residue mod n_i for $i = 1, 2$.*

Proof. The "only if" part is obvious. Conversely, suppose that there exist $x_i \in \mathbb{Z}$ such that $m \equiv x_i^2 \mod n_i$, $i = 1, 2$. By the Chinese reminder theorem I (Corollary 1.1.23), there exists $x \in \mathbb{Z}$ such that $x \equiv x_i \mod n_i$, $i = 1, 2$. Then, $x^2 \equiv x_i^2 \equiv m \mod n_i$, $i = 1, 2$, and $\gcd(n_1, n_2) = 1$ implies $x^2 \equiv m \mod n_1 n_2$. $\square$

Lemma 4.4.4 *Let $1 \leq \mu \leq 3$ and suppose that $m \in \mathbb{Z}$ is odd. Then the following conditions are equivalent:*

(a) *m is a quadratic residue* mod 2^μ;

(b) *$m \equiv 1$ mod 2^μ.*

Proof. Suppose that m is a quadratic residue mod 2^μ. Then we can find $x \in \mathbb{Z}$ such that $x^2 \equiv m \mod 2^\mu$. Note that x cannot be even (otherwise m itself would be even, contradicting the assumptions). Thus there exists $h \in \mathbb{Z}$ such that $x = 2h + 1$ and therefore $m \equiv x^2 = (2h + 1)^2 = 4h(h + 1) + 1 \equiv 1 \mod 2^\mu$, since $h(h + 1) \in 2\mathbb{Z}$. This shows the implication (a) $\Rightarrow$ (b).

Conversely, suppose that $m \equiv 1 \bmod 2^\mu$. Thus we can find $k \in \mathbb{Z}$ such that $m = 1 + 2^\mu k$ and it follows from Remark 4.4.2.(1) that $m = 1 + 2^\mu k$ is a quadratic residue mod 2^μ. $\qquad\square$

The following two theorems reduce the problem to the case n is an odd prime. To simplify notation, we denote by

$$|n| = 2^\mu p_1^{\mu_1} p_2^{\mu_2} \cdots p_k^{\mu_k} \tag{4.10}$$

the prime factorization of $|n|$ with the convention that if n is odd, then $\mu = 0$ and the factor 2^μ is, in fact, missing.

Theorem 4.4.5 *Let p be an odd prime. Then $m \in \mathbb{Z}$ is a quadratic residue mod p if and only if $m^{\frac{p-1}{2}} \equiv 1 \bmod p$.*

Proof. The multiplicative group $\mathbb{F}_p^*$ is cyclic of order $p - 1$ (cf. Theorem 1.1.21). Thus, we can find $1 \le y \le p - 1$ such that $\bar{y}$ generates $\mathbb{F}_p^*$. For $x \in \mathbb{Z}$ (respectively $m \in \mathbb{Z}$) such that $p \nmid x$ (respectively $p \nmid m$) we choose $1 \le s = s(x) \le p - 1$ (respectively $1 \le t = t(m) \le p - 1$) such that

$$\bar{y}^s = \bar{x} \ (\text{resp. } \bar{y}^t = \bar{m}), \text{ equivalently, } y^s \equiv x \ (\text{resp. } y^t \equiv m) \bmod p.$$

Then, $m \in \mathbb{Z}$ (with $\gcd(m, p) = 1$) is a quadratic residue mod p if and only if the equation $x^2 \equiv m \bmod p$ has a solution $x \in \mathbb{Z}$ and, with the above notation, this holds if and only if the equation $\bar{y}^{2s} = \bar{y}^t$, which in turn is equivalent to the congruence $2s \equiv t \bmod p - 1$, has a solution s (with $1 \le s \le p - 1$). But this is the case if and only if t is even (just take $s = t/2$). Now

$$t \text{ is even} \ \Leftrightarrow \ t\frac{p-1}{2} \equiv 0 \bmod p - 1 \ \Leftrightarrow \ (\bar{m})^{\frac{p-1}{2}} = (\bar{y})^{t\frac{p-1}{2}} = \bar{1},$$

where the last equality follows from $\bar{y}$ having order $p - 1$. $\qquad\square$

Theorem 4.4.6 *Let $n, m \in \mathbb{Z}$ with $\gcd(n, m) = 1$. Let (4.10) be the prime factorization of $|n|$. Then, m is a quadratic residue mod n if and only if the following conditions are satisfied:*

(i) $m^{\frac{p_j-1}{2}} \equiv 1 \bmod p_j$ *for $j = 1, 2, \ldots, k$;*

(ii) *and, (only) if n is even,*
- $m \equiv 1 \bmod 2^\mu$ *if $\mu = 1, 2$;*
- $m \equiv 1 \bmod 8$ *if $\mu \ge 3$.*

Proof. It follows from Proposition 4.4.3 that (4.9) has a solution (that is, m is a quadratic residue mod n) if and only if all the equations $x^2 \equiv m \bmod p_j^{\mu_j}$ for all $j = 1, 2, \ldots, k$ and, (only) if n is even, $x^2 \equiv m \bmod 2^\mu$, have a solution.

Claim 1. $m \in \mathbb{Z}$ *is a quadratic residue mod* 2^μ *if and only if*

- $m \equiv 1 \mod 2^\mu$ *if* $\mu = 1, 2$;
- $m \equiv 1 \mod 8$ *if* $\mu \geq 3$.

If $1 \leq \mu \leq 3$, the claim is equivalent to Lemma 4.4.4.

Suppose that $\mu > 3$ and that m is a quadratic residue mod 2^μ. Then, it follows from Remark 4.4.2.(3) with $n_1 = 8$ and $n_2 = 2^\mu$ that m is a quadratic residue mod 8. From Lemma 4.4.4 we deduce that $m \equiv 1 \mod 8$.

For the converse, suppose that $m \equiv 1 \mod 8$. We show, by induction on $t \geq 3$, that the congruence $x^2 \equiv m \mod 2^t$ has a solution in $\mathbb{Z}$. For $t = 3$, the statement follows from Lemma 4.4.4. Suppose now that for $t \geq 3$ there exists $x \in \mathbb{Z}$ such that $x^2 \equiv m \mod 2^t$ and let us show that there exists $y \in \mathbb{Z}$ such that $y^2 \equiv m \mod 2^{t+1}$. Let $q \in \mathbb{Z}$ be such that

$$x^2 - m = q 2^t \tag{4.11}$$

and observe that if q is even then we are done: just take $y = x$. Therefore, we suppose that q is odd. Set $y = x + 2^{t-1}$. Then we have

$$
\begin{aligned}
y^2 - m &= (x + 2^{t-1})^2 - m \\
&= x^2 - m + 2^t x + 2^{2t-2} \\
\text{(by (4.11))} \quad &= 2^t(q + x) + 2^{t+1} 2^{t-3} \\
&\equiv 0 \mod 2^{t+1},
\end{aligned}
$$

where the last equality follows from the fact that $q + x$ is even because x is odd (since m is odd). This completes the proof of the claim.

Claim 2. *Let p be an odd prime and $\mu \geq 1$. Then $m \in \mathbb{Z}$ is a quadratic residue mod p^μ if and only if m is a quadratic residue* mod p.

As in the previous claim, the "only if" part is obvious.

Conversely, we again proceed by induction. The basis is trivial. Suppose that $x^2 \equiv m \mod p^t$ with $t \geq 1$ and let us show that we can find $y \in \mathbb{Z}$ such that $y^2 \equiv m \mod p^{t+1}$. By the inductive hypothesis, we can find $q \in \mathbb{Z}$ such that

$$x^2 - m = q p^t \tag{4.12}$$

and observe that if q is a multiple of p, then we are done: just take $y = x$. Therefore we suppose that $p \nmid q$. By our assumption we also have $p \nmid x$ and therefore, since p is odd, $\gcd(2x, p) = 1$. By virtue of Bézout identity, we can find $a, b \in \mathbb{Z}$ such that $ap + 2bx = -q$, equivalently,

$$q + 2bx = -ap. \tag{4.13}$$

Set $y = x + p^t b$. Then we have

$$
\begin{aligned}
y^2 - m &= (x + p^t b)^2 - m \\
&= x^2 - m + 2bxp^t + p^{2t}b^2 \\
\text{(by (4.12))} \quad &= p^t(q + 2bx) + p^{t+1}p^{t-1}b^2 \\
\text{(by (4.13))} \quad &= p^{t+1}(p^{t-1}b^2 - a) \\
&\equiv 0 \mod p^{t+1}.
\end{aligned}
$$

This completes the proof of the claim.

The statement then follows from Theorem 4.4.5. $\qquad\square$

From now on, p is a fixed <u>odd</u> prime and we study quadratic residues mod p.

Definition 4.4.7 The *Legendre symbol* $\left(\dfrac{n}{p}\right)$ is defined by setting

$$
\left(\frac{n}{p}\right) = \begin{cases} 1 & \text{if } \gcd(n, p) = 1 \text{ and } n \text{ is a quadratic residue mod } p \\ -1 & \text{if } \gcd(n, p) = 1 \text{ and } n \text{ is a quadratic nonresidue mod } p \\ 0 & \text{if } p|n \end{cases}
$$

for every $n \in \mathbb{Z}$.

We now collect some basic properties of the Legendre symbol.

Proposition 4.4.8

(i) *The map* $n \mapsto \left(\dfrac{n}{p}\right)$ *is constant on the congruence classes mod p, and therefore it may be seen as a function defined on $\mathbb{F}_p$;*

(ii) $n^{\frac{p-1}{2}} \equiv \left(\dfrac{n}{p}\right)$ mod p *for all* $n \in \mathbb{Z}$;

(iii) $\left(\dfrac{mn}{p}\right) = \left(\dfrac{m}{p}\right)\left(\dfrac{n}{p}\right)$ *for all* $m, n \in \mathbb{Z}$;

(iv) $\left(\dfrac{-1}{p}\right) = (-1)^{\frac{p-1}{2}} = \begin{cases} 1 & \text{if } p \equiv 1 \mod 4 \\ -1 & \text{if } p \equiv -1 \mod 4. \end{cases}$

Proof. (i) This follows immediately from the definition of the Legendre symbol.

(ii) If $p|n$ this is trivial; otherwise, from the fact that the multiplicative group $\mathbb{F}_p^*$ has order $p - 1$, we have $n^{p-1} \equiv 1 \mod p$ (cf. Fermat's little theorem [Exercise 1.1.22]), which implies

$$
(n^{\frac{p-1}{2}} - 1) \cdot (n^{\frac{p-1}{2}} + 1) = n^{p-1} - 1 \equiv 0 \mod p,
$$

that is, $n^{\frac{p-1}{2}} \equiv \pm 1 \mod p$. By Theorem 4.4.5, $n^{\frac{p-1}{2}} \equiv 1 \mod p$ if and only if n is a quadratic residue mod p and therefore $n^{\frac{p-1}{2}} \equiv -1 \mod p$ if and only if n is a quadratic nonresidue. In both cases, the statement follows from the definition of the Legendre symbol.

(iii) Again, this is obvious if $p|n$ or if $p|m$, so that we may assume $p \nmid n$ and $p \nmid m$ (and therefore $p \nmid nm$). By (ii) we have

$$\left(\frac{nm}{p}\right) \equiv (nm)^{\frac{p-1}{2}} \mod p$$

$$\equiv n^{\frac{p-1}{2}} m^{\frac{p-1}{2}} \mod p$$

$$\equiv \left(\frac{n}{p}\right)\left(\frac{m}{p}\right) \mod p.$$

Since p is odd, $1 \not\equiv -1 \mod p$ and we deduce that $\left(\frac{nm}{p}\right) = \left(\frac{n}{p}\right)\left(\frac{m}{p}\right).$

(iv) This follows from (ii), after taking $n = -1$ therein. $\square$

Corollary 4.4.9 *Let $Q \subseteq \mathbb{Z}$ (respectively $P \subseteq \mathbb{Z}$) denote the set of quadratic residues (respectively nonresidues) mod p and denote by $\overline{Q}$ (respectively $\overline{P}$) its image in $\mathbb{F}_p$. Then $P \cdot P \subseteq Q = Q \cdot Q$ and $P \cdot Q = P$ (respectively $\overline{P} \cdot \overline{P} = \overline{Q} = \overline{Q} \cdot \overline{Q}$ and $\overline{P} \cdot \overline{Q} = \overline{P}$). Moreover,*

$$|\overline{Q}| = |\overline{P}| = \frac{p-1}{2}. \tag{4.14}$$

Proof. The inclusions $Q \cdot Q, P \cdot P \subseteq Q$, and $P \cdot Q \subseteq P$ follow immediately from Proposition 4.4.8.(iii). Since $1 \in Q$, the equalities $Q \cdot Q = Q$ and $P \cdot Q = P$ follow. Projecting onto $\mathbb{F}_p$ we have $\overline{P} \cdot \overline{P} \subseteq \overline{Q} = \overline{Q} \cdot \overline{Q}$ and $\overline{P} \cdot \overline{Q} = \overline{P}$. In order to show the equality $\overline{P} \cdot \overline{P} = \overline{Q}$ and determine the cardinalities of $\overline{Q}$ and $\overline{P}$, let us fix an element $\overline{n} \in \overline{P}$. We first observe that, since $Q, P \subseteq \mathbb{Z} \setminus p\mathbb{Z}$,

$$\overline{Q} \coprod \overline{P} = \mathbb{F}_p^*. \tag{4.15}$$

Since multiplication by $\overline{n}$ yields a bijection of $\mathbb{F}_p^*$, from (4.15) we deduce that

$$\overline{n}\overline{Q} \coprod \overline{n}\overline{P} = \mathbb{F}_p^*$$

so that, since $\overline{n}\overline{Q} \subseteq \overline{P}$ and $\overline{n}\overline{P} \subseteq \overline{Q}$, we necessarily have that the above inclusions are indeed equalities. In particular, $\overline{P} \cdot \overline{P} = \overline{Q}$ and (4.14) holds. $\square$

Exercise 4.4.10

(1) Deduce Corollary 4.4.9 from the proof of Theorem 4.4.5.
(2) Deduce Proposition 4.4.8.(iii) from Corollary 4.4.9 (which has been proved independently in (1)).

Definition 4.4.11 A finite subset $S \subseteq \mathbb{Z}$ of cardinality $|S| = \frac{p-1}{2}$ is called a *Gaussian set modulo p* if, for all $n \in \mathbb{Z}$ with $\gcd(n, p) = 1$, there exist $t_n \in S$ and $\varepsilon_n \in \{1, -1\}$ such that

$$n \equiv \varepsilon_n t_n \quad \text{mod } p. \tag{4.16}$$

Exercise 4.4.12

(1) Show that if S is a Gaussian set, then $r \not\equiv \pm s \mod p$ for all distinct $r, s \in S$.
(2) Show that the sets $S_1 = \{1, 2, \ldots, \frac{p-1}{2}\}$ and $S_2 = \{2, 4, \ldots, p-1\}$ are Gaussian sets modulo p.

Lemma 4.4.13 (Gauss' lemma) *Let S be a Gaussian set modulo p.*
Then, for every $n \in \mathbb{Z}$ with $\gcd(n, p) = 1$ we have

$$\left(\frac{n}{p}\right) = \prod_{s \in S} \varepsilon_{ns} = (-1)^k,$$

where $k = |\{s \in S : \varepsilon_{ns} = -1\}|$.

Proof. First of all, we show that for all $s, r \in S$

$$t_{ns} = t_{nr} \Leftrightarrow s = r.$$

Indeed, if $t_{ns} = t_{nr}$ then

$$
\begin{aligned}
nr &\equiv \varepsilon_{nr} t_{nr} \quad \text{mod } p \\
&\equiv \varepsilon_{nr} t_{ns} \quad \text{mod } p \\
&\equiv \pm \varepsilon_{ns} t_{ns} \quad \text{mod } p \\
&\equiv \pm ns \quad \text{mod } p
\end{aligned}
$$

that, after simplifying, yields $r \equiv \pm s \mod p$. By virtue of Exercise 4.4.12.(1), we deduce that $r = s$. In other words, the map $s \mapsto t_{ns}$ is a permutation of S so that

$$\prod_{s\in S} s \cdot \prod_{s\in S} \varepsilon_{ns} = \prod_{s\in S} t_{ns} \cdot \prod_{s\in S} \varepsilon_{ns}$$

$$= \prod_{s\in S} t_{ns}\varepsilon_{ns}$$

$$\text{(by (4.16))} \equiv \prod_{s\in S} sn \mod p$$

$$\text{(since } |S| = \tfrac{p-1}{2}) \equiv n^{\frac{p-1}{2}} \prod_{s\in S} s \mod p$$

$$\text{(by Proposition 4.4.8.(ii))} \equiv \left(\frac{n}{p}\right) \prod_{s\in S} s \mod p.$$

Simplifying by $\prod_{s\in S} s$, and taking into account that both $\prod_{s\in S} \varepsilon_{ns}$ and $\left(\frac{n}{p}\right)$ are equal to either 1 or -1 (and these are different mod p), the lemma follows. $\square$

Corollary 4.4.14

$$\left(\frac{2}{p}\right) = (-1)^{\frac{p^2-1}{8}} = \begin{cases} 1 & \text{if } p \equiv \pm 1 \mod 8 \\ -1 & \text{if } p \not\equiv \pm 1 \mod 8. \end{cases}$$

Proof. Take $S = \{1, 2, \ldots, \frac{p-1}{2}\}$ and $n = 2$. Then, by Gauss' lemma, we have $\left(\frac{2}{p}\right) = (-1)^k$, where k is the number of $s \in S$ such that $\varepsilon_{2s} = -1$. For every $s \in S$, we clearly have $2 \le 2s \le p - 1$. Since

$$2 \le 2s \le \frac{p-1}{2} \Rightarrow 2s \in S \Rightarrow \varepsilon_{2s} = 1$$

while, setting $t = p - 2s$,

$$\frac{p+1}{2} \le 2s \le p-1 \Rightarrow 1 \le p - 2s \le \frac{p-1}{2} \Rightarrow t \in S$$

$$\Rightarrow 2s = p - t \equiv -t \mod p \Rightarrow \varepsilon_{2s} = -1,$$

we deduce that k is equal to the number of $s \in S$ such that

$$\frac{p+1}{4} \le s \le \frac{p-1}{2}. \tag{4.17}$$

Now if, on the one hand, $p \equiv \pm 1 \mod 8$, then we can find $h \in \mathbb{Z}$ such that $p = 8h \pm 1$ and (4.17) becomes

$$2h + \frac{1}{4} \pm \frac{1}{4} \le s \le 4h - \frac{1}{2} \pm \frac{1}{2}$$

so that, in both cases, $k = 2h$ and $\left(\dfrac{2}{p}\right) = (-1)^{2h} = 1$.

If, on the other hand, $p \equiv \pm 3 \mod 8$, then we can find $h \in \mathbb{Z}$ such that $p = 8h \pm 3$ and (4.17) becomes

$$2h + \frac{1}{4} \pm \frac{3}{4} \le s \le 4h - \frac{1}{2} \pm \frac{3}{2}$$

so that $k = 2h \pm 1$ and, in both cases, $\left(\dfrac{2}{p}\right) = (-1)^{2h\pm1} = -1$. $\quad\square$

Now, following the monograph by Nathanson [118], we study the Legendre symbol as a character of the multiplicative group $\mathbb{F}_p^*$. We recall (cf. Section 2.2) that for $n, k \in \mathbb{Z} \setminus p\mathbb{Z}$ we have defined $\chi_n(k) = \exp\left(\frac{2\pi i n k}{p}\right)$.

For all $n \in \mathbb{Z}$ we set

$$\tau(p, n) = \sum_{k=1}^{p-1} \left(\frac{k}{p}\right) \chi_n(k). \tag{4.18}$$

Note that setting $\ell_p(n) = \left(\dfrac{n}{p}\right)$ for all $n \in \mathbb{Z}$ then, in the notation in Section 2.4, we have $\tau(p, n) = \widehat{\ell_p}(-n)$. Clearly, $\left(\dfrac{k}{p}\right)$ is a *multiplicative character* (cf. Proposition 4.4.8.(iii)), while χ_n is an *additive character*. Note also that

$$\sum_{k=1}^{p-1} \left(\frac{k}{p}\right) = 0. \tag{4.19}$$

Indeed, the left hand side in (4.19) may be seen as the scalar product of the nontrivial multiplicative character ℓ_p with the trivial multiplicative character, so that we may use Proposition 2.3.5 (for multiplicative characters of $\mathbb{F}_p^*$).

Theorem 4.4.15 (Gauss) *Let $n \in \mathbb{Z}$. Then the following holds:*

(i) $\tau(p, n) = \left(\dfrac{n}{p}\right) \tau(p, 1)$.

(ii) *If $\gcd(n, p) = 1$ then*

$$\tau(p, n) = \sum_{h=0}^{p-1} \exp\left(\frac{2\pi i h^2 n}{p}\right);$$

in particular,

$$\tau(p, 1) = \sum_{h=0}^{p-1} \exp\left(\frac{2\pi i h^2}{p}\right).$$

(iii)

$$\tau(p, 1) = \begin{cases} \sqrt{p} & \text{if } p \equiv 1 \mod 4 \\ i\sqrt{p} & \text{if } p \equiv 3 \mod 4 \end{cases} = i^{\frac{(p-1)^2}{4}} \sqrt{p}.$$

Proof. We first recall that $\chi_n(k) = \chi_1(nk)$. Assume $\gcd(n, p) = 1$ so that $\left(\dfrac{n}{p}\right) = \pm 1$ and, for $1 \le k \le p - 1$,

$$\left(\frac{k}{p}\right) = \left(\frac{k}{p}\right)\left(\frac{n}{p}\right)^2 = \left(\frac{nk}{p}\right)\left(\frac{n}{p}\right), \tag{4.20}$$

where the last equality follows from Proposition 4.4.8.(iii). Then

$$\tau(p, n) = \sum_{k=1}^{p-1} \left(\frac{k}{p}\right) \chi_n(k)$$

$$(\text{by } (4.20)) = \left(\frac{n}{p}\right) \sum_{k=1}^{p-1} \left(\frac{kn}{p}\right) \chi_n(k)$$

$$= \left(\frac{n}{p}\right) \sum_{k=1}^{p-1} \left(\frac{kn}{p}\right) \chi_1(kn)$$

$$= \left(\frac{n}{p}\right) \widehat{\ell_p}(-1)$$

$$= \left(\frac{n}{p}\right) \tau(p, 1).$$

It is easy to check, by means of (4.19), that if $p|n$ then $\tau(p, n) = 0$, and this ends the proof of (i).

(ii) Let P (respectively Q) be as in Corollary 4.4.9 and set $P' = P \cap \{1, 2, \ldots, p - 1\}$ (respectively $Q' = Q \cap \{1, 2, \ldots, p - 1\}$).

Let $k \in Q'$ and $h \in \{1, 2, \ldots, p - 1\}$ such that $h^2 \equiv k \mod p$. Then also $(p - h)^2 \equiv h^2 \equiv k \mod p$ and $p - h \not\equiv h \mod p$. Therefore

$$\sum_{h=1}^{p-1} \chi_1(nh^2) = 2 \sum_{k \in Q'} \chi_1(nk) \tag{4.21}$$

and

$$\tau(p, n) = \sum_{k=1}^{p-1} \left(\frac{k}{p}\right) \chi_1(nk)$$

$$= \sum_{k \in Q'} \chi_1(nk) - \sum_{k \in P'} \chi_1(nk)$$

$$= 1 + 2 \sum_{k \in Q'} \chi_1(nk) - \sum_{k=0}^{p-1} \chi_1(nk)$$

$$\text{(by (4.21) and (2.5))} = 1 + \sum_{h=1}^{p-1} \chi_1(nh^2)$$

$$= \sum_{h=0}^{p-1} \exp\left(\frac{2\pi i n h^2}{p}\right).$$

(iii) This follows from (ii) and Corollary 4.3.4. Moreover, it is immediate to check that

$$i^{\frac{(p-1)^2}{4}} = \begin{cases} 1 & \text{if } p \equiv 1 \mod 4 \\ i & \text{if } p \equiv 3 \mod 4. \end{cases}$$

$\square$

Definition 4.4.16 Given $m, n \in \mathbb{Z}, n \neq 0$, we define the *Gauss sum* $G(m, n)$ by setting

$$G(m, n) = \sum_{k=0}^{n-1} \exp\left(\frac{2\pi i m k^2}{n}\right)$$

(see also Definition 7.4.1 for Gauss sums over finite fields).

Observe that by virtue of Theorem 4.4.15.(ii), if $\gcd(p, n) = 1$ then

$$\tau(p, n) = G(n, p) \tag{4.22}$$

and that Corollary 4.3.4 may be reformulated in the form

$$G(1, n) = \begin{cases} (1+i)\sqrt{n} & \text{if } n \equiv 0 \mod 4 \\ \sqrt{n} & \text{if } n \equiv 1 \mod 4 \\ 0 & \text{if } n \equiv 2 \mod 4 \\ i\sqrt{n} & \text{if } n \equiv 3 \mod 4. \end{cases} \tag{4.23}$$

Proposition 4.4.17 *Let $m, r, s \in \mathbb{Z}, r, s \neq 0$, and suppose that $\gcd(r, s) = 1$. Then*

$$G(mr, s)G(ms, r) = G(m, sr).$$

Proof.

$$G(mr, s)G(ms, r) = \sum_{v=0}^{s-1} \exp\left(\frac{2\pi\, imrv^2}{s}\right) \cdot \sum_{u=0}^{r-1} \exp\left(\frac{2\pi\, imsu^2}{r}\right)$$

$$= \sum_{v=0}^{s-1} \sum_{u=0}^{r-1} \exp\left(2\pi\, im\frac{r^2v^2 + s^2u^2}{sr}\right)$$

$$(\text{since } \exp\left(2\pi\, im\tfrac{2uvsr}{sr}\right) = 1) \quad = \sum_{v=0}^{s-1} \sum_{u=0}^{r-1} \exp\left(2\pi\, im\frac{(rv + su)^2}{sr}\right)$$

$$(\text{by Lemma 1.1.16}) \quad = \sum_{k=0}^{sr-1} \exp\left(2\pi\, i\frac{mk^2}{sr}\right)$$

$$= G(m, sr).\qquad\qquad \square$$

We are now in a position to prove the main result of this section.

Theorem 4.4.18 (Gauss law of quadratic reciprocity) *Let* p, q *be distinct odd primes. Then*

$$\left(\frac{p}{q}\right)\left(\frac{q}{p}\right) = (-1)^{\frac{p-1}{2}\cdot\frac{q-1}{2}}.$$

Proof. By virtue of Theorem 4.4.15 we have

$$\tau(p, q) = \left(\frac{q}{p}\right)\tau(p, 1) = \left(\frac{q}{p}\right)i^{\frac{(p-1)^2}{4}}\sqrt{p}$$

and, exchanging p and q,

$$\tau(q, p) = \left(\frac{p}{q}\right)\tau(q, 1) = \left(\frac{p}{q}\right)i^{\frac{(q-1)^2}{4}}\sqrt{q}.$$

Moreover, from Proposition 4.4.17 (with $r = q$, $s = p$, and $m = 1$) and (4.22) we deduce that

$$\tau(p, q)\tau(q, p) = G(q, p)G(p, q)$$
$$= G(1, pq)$$
$$(\text{by (4.23)}) = i^{\frac{(pq-1)^2}{4}}\sqrt{pq}.$$

Then the equality

$$\left(\frac{p}{q}\right)\left(\frac{q}{p}\right)i^{\frac{(p-1)^2}{4} + \frac{(q-1)^2}{4}}\sqrt{pq} = i^{\frac{(pq-1)^2}{4}}\sqrt{pq}$$

yields the quadratic reciprocity law because

$$\frac{1}{4}\left[(pq-1)^2 - (p-1)^2 - (q-1)^2\right]$$
$$= \frac{1}{4}\left[-2(p-1)(q-1) + (p^2-1)(q^2-1)\right]$$

and

$$(4m+3)^2 \equiv (4m+1)^2 \equiv 1 \mod 4 \;\Rightarrow\; p^2 - 1 \equiv q^2 - 1 \equiv 0 \mod 4,$$

so that

$$i^{\frac{(p^2-1)(q^2-1)}{4}} = 1,$$

while

$$i^{\frac{-2(p-1)(q-1)}{4}} = (-1)^{\frac{p-1}{2} \cdot \frac{q-1}{2}}. \qquad \square$$

Exercise 4.4.19 From Theorem 4.4.18 deduce that

(1) if $p \equiv 1 \mod 4$ or $q \equiv 1 \mod 4$ then p is a quadratic residue mod q if and only if q is a quadratic residue mod p;

(2) if $p \equiv q \equiv 3 \mod 4$ then p is a quadratic residue mod q if and only if q is a quadratic nonresidue mod p.

For instance, using the congruences

$$179 \equiv 59 \equiv 3 \mod 4, \quad 179 \equiv 2 \mod 59, \quad \text{and} \quad 59 \equiv 3 \mod 8,$$

we get

$$\left(\frac{59}{179}\right) = -\left(\frac{179}{59}\right) = -\left(\frac{2}{59}\right) = 1,$$

where the last equality follows from Corollary 4.4.14.

Exercise 4.4.20 Deduce the following identities from Proposition 4.4.8 and Theorem 4.4.15: if $\gcd(n, p) = 1$ and p is an odd prime, then

$$\tau(p, n)^2 = \left(\frac{-1}{p}\right)p = (-1)^{\frac{p-1}{2}} p;$$

if q is another distinct odd prime

$$\tau(p, n)^{q-1} \equiv (-1)^{\frac{p-1}{2} \cdot \frac{q-1}{2}} \left(\frac{p}{q}\right) \mod q.$$

Another, more elementary proof of the Gauss law of quadratic reciprocity will be sketched in Exercise 6.5.7: it avoids Corollary 4.3.4 and, therefore, all the machinery on the spectral analysis of the DFT.

5

The Fast Fourier Transform

The *Fast Fourier Transform* (for brevity, *FFT*) is a numerical algorithm for the computation of the Discrete Fourier Transform. It is one of the most important algorithms, because it applies to an extremely wide class of numerical problems. It was discovered by Gauss who applied it to astronomical computations. It was rediscovered several times, and the most celebrated paper devoted to it is the seminal one by Cooley and Tukey [41] (one then often refers to this algorithm as the *Cooley-Tukey algorithm*).

However, as indicated in [15], this algorithm also has interesting theoretical interpretations. We will discuss this approach in Section 12.5.

In the present chapter, following the books by Tolimieri, An, and Lu [160] and by Van Loan [163], as well as the papers [50, 130, 168], we present a matrix theoretic approach to the FFT. Actually, [130] will constitute our main source, [50] is a fundamental inspiration for our treatment of stride permutations, and [160] has given us the general framework and the treatment of Rader's algorithm. Recent developments can be found in [46].

Before embarking on the formalism of Kronecker products and shuffle permutations, following the exposition in [150], we present the simplest example of the FFT.

5.1 A preliminary example

As in Section 2.2, set $\omega_n = \exp \frac{2\pi i}{n}$ (note that we have added the subscript n to ω). Then, the (unnormalized) Discrete Fourier Transform of $f \in L(\mathbb{Z}_n)$ (cf. Definition 2.4.1) is given by

$$\widehat{f}^n(m) = \frac{1}{n} \sum_{k=0}^{n-1} f(k)\omega_n^{-km}. \tag{5.1}$$

129

We have used the symbol $\widehat{}^n$ to emphasize the fact that we are computing the DFT of a function $f \in L(\mathbb{Z}_n)$. Then the computation of the Fourier coefficients of f requires:

- $n - 2$ multiplications to compute the numbers $\omega_n^2, \omega_n^3, \ldots, \omega_n^{n-1}$ (note that in (5.1) these numbers may occur with repetitions and do all appear in the expression of some of these coefficients);
- each coefficient $\widehat{f}^n(m)$ requires n multiplications (to compute $f(k)\omega_n^{-km}$), $n - 1$ sums, plus a final multiplication by $\frac{1}{n}$.

Therefore, to compute all Fourier coefficients, one needs (at most)

$$(n - 2) + n(n + (n - 1) + 1) = 2n^2 + n - 2 \leq 2n^2 + n = \mathcal{O}(n^2) \quad (5.2)$$

elementary operations. We denote by $\sharp n$ the *minimum* number of operations that are needed to compute all the Fourier coefficients of any function in $L(\mathbb{Z}_n)$.

Remark 5.1.1 Note that in the definition of $\sharp n$, the minimum is over all possible algorithms: we are not necessarily using the expression of the Fourier coefficients provided by their definition (i.e. by (5.1)).

We begin with a preliminary lemma.

Lemma 5.1.2

$$\sharp(2n) \leq 2\sharp n + 8n.$$

Proof. As above, we may compute the numbers $\omega_{2n}^k, k = 0, 1, \ldots, 2n - 1$, with $2n - 2$ multiplications. Note also that

$$\omega_{2n}^{2r} = \omega_n^r \quad \text{and} \quad \omega_{2n}^{2s+1} = \omega_{2n}\omega_n^s. \quad (5.3)$$

Then, for $f \in L(\mathbb{Z}_{2n})$, we define $f_0, f_1 \in L(\mathbb{Z}_n)$ by setting

$$f_0(k) = f(2k)$$
$$f_1(k) = f(2k + 1)$$

for all $k = 0, 1, \ldots, n - 1$. Then

$$\widehat{f}^{2n}(m) = \frac{1}{2n} \sum_{k=0}^{2n-1} f(k)\omega_{2n}^{-km}$$

$$(\text{by } (5.3)) \quad = \frac{1}{2}\left[\frac{1}{n}\sum_{r=0}^{n-1} f_0(r)\omega_n^{-rm} + \frac{1}{n}\sum_{s=0}^{n-1} f_1(s)\omega_{2n}^{-m}\omega_n^{-sm}\right] \quad (5.4)$$

$$= \frac{1}{2}\left[\widehat{f_0}^n(m) + \omega_{2n}^{-m}\widehat{f_1}^n(m)\right].$$

As an application of this formula, in order to compute the coefficients of f we need (at most):

- $2\sharp n$ operations to compute the coefficients of both f_0 and f_1,
- $2n - 2$ operations to compute the numbers ω_{2n}^k, $k = 0, 1, \ldots, 2n - 1$,
- $6n$ operations ($4n$ multiplications and $2n$ additions),

so that

$$\sharp(2n) \leq 2\sharp n + 8n - 2 \leq 2\sharp n + 8n. \qquad \square$$

Theorem 5.1.3 *Let $n = 2^h$. Then the Fourier coefficients of a function $f \in L(\mathbb{Z}_n)$ may be computed with at most $2^{h+2}h = 4n\log_2 n = \mathcal{O}(n\log n)$ operations.*

Proof. We proceed by induction on h. If $h = 1$ then $n = 2$ and the Fourier coefficients are

$$\widehat{f}^2(0) = \frac{1}{2}[f(0) + f(1)]$$

$$\widehat{f}^2(1) = \frac{1}{2}[f(0) + (-1)f(1)].$$

These computations require $5 < 8 = 2^{1+2} \cdot 1$ operations. Assume the statement for $n = 2^h$, so that $\sharp n \leq 2^{h+2}h$. By Lemma 5.1.2, for $2n = 2^{h+1}$ we have

$$\begin{aligned}
\sharp(2n) &\leq 2\sharp n + 8n \\
&\leq 2(2^{h+2}h) + 8 \cdot 2^h \\
&= 2^{h+3}(h + 1).
\end{aligned} \qquad \square$$

As the above result shows, a factorization of n yields an improvement on the computation of the DFT. We will explore this after the introduction of a couple of basic theoretical tools.

5.2 Stride permutations

Let n, m be two positive integers. By means of the Euclidean algorithm, any integer $0 \leq i \leq nm - 1$ may be (uniquely) represented in the following forms:

$$i = sm + r \quad 0 \leq s \leq n - 1, \quad 0 \leq r \leq m - 1 \tag{5.5}$$

$$i = \tilde{r}n + \tilde{s} \quad 0 \leq \tilde{s} \leq n - 1, \quad 0 \leq \tilde{r} \leq m - 1. \tag{5.6}$$

The expressions (5.5) and (5.6) are called the *(m, n)-representation* and the *(n, m)-representation* of i, respectively.

Definition 5.2.1 The *stride* (or *shuffle*) *permutation* is the bijection

$$\sigma(m, n): \{0, 1, \ldots, nm - 1\} \to \{0, 1, \ldots, nm - 1\}$$

defined by setting

$$\sigma(m, n)i \equiv \sigma(m, n)(sm + r) = rn + s$$

for every $0 \le i \le nm - 1$ represented in the form (5.5).

We now present an alternative description of $\sigma(m, n)$. Divide the <u>ordered</u> sequence $(0, 1, 2, \ldots, nm - 1)$ into n consecutive blocks (see Table 5.1), that is,

$$(0, 1, \ldots, nm - 1) = (\mathcal{B}_0, \mathcal{B}_1, \ldots, \mathcal{B}_{n-1})$$

where $\mathcal{B}_0 = (0, 1, \ldots, m - 1)$, $\mathcal{B}_1 = (m, m+1, \ldots, 2m - 1),\ldots, \mathcal{B}_s = (sm, sm + 1, \ldots, sm + r, \ldots, (s + 1)m - 1), \ldots,$ and $\mathcal{B}_{n-1} = ((n - 1)m, (n - 1)m + 1, \ldots, nm - 1)$. Then

$$(\sigma(m, n)0, \sigma(m, n)1, \ldots, \sigma(m, n)(nm - 1)) = (\mathcal{C}_0, \mathcal{C}_1, \ldots, \mathcal{C}_{n-1})$$

where the blocks $\mathcal{C}_0, \mathcal{C}_1, \ldots, \mathcal{C}_{n-1}$ are the ordered sequences defined by setting $\mathcal{C}_s = (s, s + n, \ldots, s + rn, \ldots, s + (m - 1)n)$ for all $s = 0, 1, \ldots, n - 1$.

Table 5.1. *The action of the stride permutation* $\sigma(m, n)$: *in the first array, the rows are the blocks* $\mathcal{B}s$, *while, in the second array, the rows are the blocks* $\mathcal{C}s$.

0	1	$\cdots$	$m-1$		0	n	$\cdots$	$(m-1)n$
m	$m+1$	$\cdots$	$2m-1$	$\xrightarrow{\sigma(m,n)}$	1	$n+1$	$\cdots$	$(m-1)n+1$
$\vdots$	$\vdots$	$\ddots$	$\vdots$		$\vdots$	$\vdots$	$\ddots$	$\vdots$
$(n-1)m$	$(n-1)m+1$	$\cdots$	$nm-1$		$n-1$	$2n-1$	$\cdots$	$mn-1$

For instance,

$$\sigma(3, 2)0 = 0 \ \sigma(3, 2)1 = 2 \ \sigma(3, 2)2 = 4$$
$$\sigma(3, 2)3 = 1 \ \sigma(3, 2)4 = 3 \ \sigma(3, 2)5 = 5.$$

Clearly, $\sigma(m, 1)$ and $\sigma(1, n)$ are the identity permutation and

$$\sigma(m, n)^{-1} = \sigma(n, m). \tag{5.7}$$

Let now m, n, k be positive integers. Then for any integer $0 \le i \le mnk - 1$ two applications of the Euclidean algorithm yield firstly $i = tmn + s_1$, with $0 \le t \le k - 1$ and $0 \le s_1 \le mn - 1$, and then $s_1 = sm + r$, with $0 \le s \le n - 1$

and $0 \le r \le m - 1$, so that we may write

$$i = tmn + sm + r. \tag{5.8}$$

We refer to (5.8) as to the (m, n, k)-*representation* of i. Moreover the positive integers t, s, r (or, to emphasize their ordering, the triple (t, s, r)) are called the coefficients of this representation.

Lemma 5.2.2 *Let $0 \le i < mnk - 1$ with (m, n, k)-representation as in (5.8). Then*

(i)

$$\sigma(mn, k)i = smk + rk + t,$$

that is, the $\sigma(mn, k)$-image of i is the number whose coefficients in the (k, m, n)-representation are (s, r, t); we then write (symbolically):

$$[(m, n, k); (t, s, r)] \stackrel{\sigma(mn,k)}{\to} [(k, m, n); (s, r, t)];$$

(ii)

$$\sigma(m, nk)i = rnk + tn + s,$$

that is, the $\sigma(m, nk)$-image of i is the number whose coefficients in the (n, k, m)-representation are (r, t, s) and we again write (symbolically):

$$[(m, n, k); (t, s, r)] \stackrel{\sigma(m,nk)}{\to} [(n, k, m); (r, t, s)].$$

Proof. We have

$$\sigma(mn, k)(tmn + sm + r) = \sigma(mn, k)[tmn + (sm + r)]$$
$$\text{(by Definition 5.2.1)} = (sm + r)k + t$$
$$= smk + rk + t$$

and this gives (i); moreover

$$\sigma(m, nk)(tmn + sm + r) = \sigma(m, nk)[(tn + s)m + r)]$$
$$\text{(by Definition 5.2.1)} = rnk + tn + s$$

and (ii) follows as well. □

Theorem 5.2.3 (Basic product identities) *Let m, n, k be positive integers. Then*

$$\sigma(mk, n)\sigma(mn, k) = \sigma(m, nk) \tag{5.9}$$

$$\sigma(n, mk)\sigma(m, nk) = \sigma(mn, k). \tag{5.10}$$

Proof. By two applications of Lemma 5.2.2.(i) we get

$$[(m, n, k); (t, s, r)] \overset{\sigma(mn,k)}{\to} [(k, m, n); (s, r, t)] \overset{\sigma(km,n)}{\to} [(n, k, m); (r, t, s)]$$

which coincides with $\sigma(m, kn)$ by Lemma 5.2.2.(ii). This proves (5.9).

By two applications of Lemma 5.2.2.(ii) we get

$$[(m, n, k); (t, s, r)] \overset{\sigma(m,nk)}{\to} [(n, k, m); (r, t, s)] \overset{\sigma(n,mk)}{\to} [(k, m, n); (s, r, t)]$$

which coincides with $\sigma(mn, k)$ by Lemma 5.2.2.(i). This proves (5.10). $\quad\square$

Definition 5.2.4 Let m, n, k be positive integers. We define the *partial stride permutations* $\iota(m, n, k)$ and $\tau(m, n, k)$ by setting

$$\iota(m, n, k)i = skm + tm + r$$

and

$$\tau(m, n, k)i = tmn + rn + s$$

for all $i = tmn + sm + r$ as in (5.8).

Note that in the definition of $\iota(m, n, k)$ we have $skm + tm + r = (sk + t)m + r$, that is, in $i = tmn + sm + r = (tn + s)m + r$ we replace $tn + s$ by $sk + t$. Moreover, we have the following (symbolic) representation

$$[(m, n, k); (t, s, r)] \overset{\iota(m,n,k)}{\to} [(m, k, n); (s, t, r)].$$

Analogously, in the definition of $\tau(m, n, k)$ we have $sm + r$ replaced by $rn + s$, and the corresponding (symbolic) representation is:

$$[(m, n, k); (t, s, r)] \overset{\tau(m,n,k)}{\to} [(n, m, k); (t, r, s)].$$

Theorem 5.2.5 (Product identities for partial strides) *We have*

$$\iota(n, m, k)\tau(m, n, k) = \sigma(m, nk) \tag{5.11}$$

and

$$\tau(m, k, n)\iota(m, n, k) = \sigma(mn, k). \tag{5.12}$$

Proof. We have

$$[(m, n, k); (t, s, r)] \overset{\tau(m,n,k)}{\to} [(n, m, k); (t, r, s)] \overset{\iota(n,m,k)}{\to} [(n, k, m); (r, t, s)]$$

which coincides with $\sigma(m, nk)$, proving (5.11). Similarly,

$$[(m, n, k); (t, s, r)] \overset{\iota(m,n,k)}{\to} [(m, k, n); (s, t, r)] \overset{\tau(m,k,n)}{\to} [(k, m, n); (s, r, t)]$$

which coincides with $\sigma(mn, k)$, proving (5.12). $\quad\square$

Theorem 5.2.6 (Mixed products identities)

$$\tau(k, m, n)\sigma(mn, k) = \iota(m, n, k) \tag{5.13}$$

$$\iota(n, k, m)\sigma(m, nk) = \tau(m, n, k) \tag{5.14}$$

$$\sigma(mk, n)\iota(m, n, k) = \tau(m, n, k) \tag{5.15}$$

$$\sigma(n, mk)\tau(m, n, k) = \iota(m, n, k).$$

Proof. The proofs are easy and left as exercises. $\qquad\square$

Corollary 5.2.7 (Similarity identity)

$$\sigma(mn, k)\tau(m, n, k)\sigma(k, mn) = \iota(k, m, n).$$

Proof. Starting by using (5.14) we have

$$\sigma(mn, k)\tau(m, n, k)\sigma(k, mn) = \sigma(mn, k)\iota(n, k, m)\sigma(m, nk)\sigma(k, mn)$$
$$\text{(by (5.15) and (5.10))} = \tau(n, k, m)\sigma(mk, n)$$
$$\text{(by (5.13))} = \iota(k, m, n). \qquad\square$$

Exercise 5.2.8 Give a direct proof of the similarity identity.

Notation 5.2.9 From now on, given integers $0 \le k < n$ and a map $f\colon \{0, 1, \ldots, n-1\} \to \{0, 1, \ldots, n-1\}$, we write "$f(k) = j \bmod n$" to indicate that, if $j \notin \{0, 1, \ldots, n-1\}$, then the value $f(k)$ equals the unique element $j' \in \{0, 1, \ldots, n-1\}$ such that $j' \equiv j \bmod n$. In other words, we regard $\{0, 1, \ldots, n-1\}$, the domain and codomain of f, as the additive group $\mathbb{Z}_n$.

Definition 5.2.10 Let $0 \le k \le m-1$ and suppose that $\gcd(k, m) = 1$. Then the *elementary congruence permutation* $\gamma(m, k)$ of $\{0, 1, \ldots, m-1\}$ is defined by setting

$$\gamma(m, k)j = kj \bmod m$$

for all $j = 0, 1, \ldots, m-1$ (recall Lemma 1.5.1).

Let also $0 \le h \le m-1$ and suppose that $\gcd(h, m) = 1$. Then the *product congruence permutation* $\gamma(m, k; n, h)$ of $\{0, 1, \ldots, nm-1\}$ is defined by setting

$$\gamma(m, k; n, h)i = s'm + r'$$

for every $i = sm + r$ as in (5.5) and $s' = hs \bmod n$ and $r' = kr \bmod m$.

The proof of the following proposition is trivial.

Proposition 5.2.11 *Let* $0 \le h, k \le m - 1$.

(i) *If* $\gcd(h, m) = \gcd(k, m) = 1$ *then* $\gamma(m, k)\gamma(m, h) = \gamma(m, hk \bmod m)$
$= \gamma(m, h)\gamma(m, k)$;

(ii) *if* $\gcd(k, m) = 1$ *then* $\gamma(m, k)^{-1} = \gamma(m, k^*)$, *where* k^* *denotes the inverse of* k *mod* m. $\qquad\qquad\square$

Definition 5.2.12 Suppose that $\gcd(n, m) = 1$. We define one more permutation of $\{0, 1, \ldots mn - 1\}$, denoted $\beta(m, n)$, by setting

$$\beta(m, n)i = s_1 m + r \qquad (5.16)$$

for all $i = sm + r$ as in (5.5), where $s_1 = s - m^*r \bmod n$ (here m^* denotes the inverse of m mod n).

Note that $\beta(m, n)$ defined above is indeed a permutation: for, with the notation as in Definition 5.2.12, if $0 \le s_0 \le n - 1$ and $0 \le r_0 \le m - 1$, we have that $\beta(m, n)i = s_0 m + r_0$ if and only if $s_1 = s_0$ and $r = r_0$, so that also $s = m^*r + s_0$ mod n.

Definition 5.2.13 Suppose that $\gcd(m, n) = 1$, $\gcd(k, m) = 1$, and $\gcd(h, n) = 1$. Let n^* be the inverse of $n \bmod m$. Then the *composite bijection permutation* $\pi(m, k; n, h)$ of $\{0, 1, \ldots, nm - 1\}$ is defined by setting

$$\pi(m, k; n, h)i = hsm + kn^*nr \quad \bmod nm$$

for all $i = sm + r$ as in (5.5).

Theorem 5.2.14 *In the notation of Definition 5.2.13,* $\pi(m, k; n, h)$ *is indeed a permutation and*

$$\beta(m, n)\gamma(m, k; n, h) = \pi(m, k; n, h). \qquad (5.17)$$

Moreover, its inverse is given by the map

$$j \mapsto sm + r \quad 0 \le j \le nm - 1,$$

where, denoting by k^* *(respectively* h^**) the inverse of* k *(respectively* h*) mod* m *(respectively mod* n*),*

$$\begin{cases} s = h^*m^*j & \bmod n \\ r = k^*j & \bmod m. \end{cases} \qquad (5.18)$$

Proof. It suffices to prove (5.17), since its left hand side is a permutation. We claim that if $0 \le n^* \le m - 1$ is the inverse of $n \bmod m$ and $0 \le m^* \le n - 1$ is the inverse of m mod n, then

$$mm^* + nn^* = 1 \qquad \text{mod } nm. \tag{5.19}$$

Indeed, recalling that $\gcd(m, n) = 1$, by virtue of Bézout identity (1.2), there exist $a, b \in \mathbb{Z}$ such that $an + bm = 1$. Clearly, this last identity implies that a (respectively b) is the inverse of n (respectively m) mod m (respectively mod n). If $a = \alpha m + a_1$, with $0 \le a_1 \le m - 1$, and $b = \beta n + b_1$, with $0 \le b_1 \le n - 1$, then

$$a_1 n + b_1 m + (\alpha + \beta)nm = 1$$

and we can take $n^* = a_1$ and $m^* = b_1$, proving the claim.

Now suppose $0 \le s \le n - 1$ and $0 \le r \le m - 1$. Then

$$\beta(m, n)\gamma(m, k; n, h)(sm + r) = \beta(m, n)(s'm + r') = s_1 m + r',$$

where (cf. Definition 5.2.10 and Definition 5.2.12)

$$kr = am + r' \text{ and } 0 \le r' \le m - 1$$
$$hs = bn + s' \text{ and } 0 \le s' \le n - 1,$$

for suitable $a, b \in \mathbb{Z}$, and

$$s' - m^* r' = cn + s_1 \text{ and } 0 \le s_1 \le n - 1,$$

for a suitable $c \in \mathbb{Z}$, and m^* as in (5.19). It follows that

$$s_1 = s' - m^* r' - cn = hs - bn - m^* kr + am^* m - cn.$$

Therefore

$$\begin{aligned} s_1 m + r' &= hsm - bnm - m^* mkr + am^* m^2 - cnm + kr - am \\ &= hsm + (1 - m^* m)kr - am(1 - m^* m) \text{ mod } nm \\ \text{(by (5.19))} \quad &= hsm + nn^* kr \text{ mod } nm, \end{aligned}$$

proving (5.17).

Finally, we prove the last assertion. Suppose that $0 \le j \le nm - 1$ and $\pi(m, k; n, h)(sm + r) = j$. Then

$$j = hsm + kn^* nr \qquad \text{mod } nm.$$

Multiplying by k^*, we get

$$k^* j = k^* hsm + k^* kn^* nr = r \qquad \text{mod } m,$$

while, multiplying by $h^* m^*$, we get

$$h^* m^* j = sh^* hm^* m + h^* m^* knn^* r = s \qquad \text{mod } n,$$

showing that conditions (5.18) are satisfied. $\qquad\qquad\qquad\qquad\qquad\square$

Remark 5.2.15 Two special cases of $\pi(m, k; n, h)$ are worth mentioning.

For $k = 1$ and $h = m^*$, we define the *Chinese remainder mapping* $c(m, n) = \pi(m, 1; n, m^*)$. We have

$$c(m, n)(ms + r) = mm^*s + nn^*r \mod nm.$$

Note that (cf. (5.18)), $j = mm^*s + nn^*r$ is a solution of the system

$$\begin{cases} j \equiv s \mod n \\ j \equiv r \mod m \end{cases}$$

(this explains the name of the map $c(m, n)$, cf. Corollary 1.1.23).

For $k = n$ and $h = 1$ we define the *Ruritanian map* $r(m, n) = \pi(m, n; n, 1)$. We have

$$r(m, n)(ms + r) = sm + n^2n^*r \mod nm$$

$$= sm + nr \mod nm$$

since $nn^* = 1 \mod m$ implies that

$$n^2n^* = n \mod nm. \tag{5.20}$$

Theorem 5.2.16 (Permutational Reverse Radix Identity) *If* $\gcd(m, n) = \gcd(k, m) = \gcd(h, n) = 1$, *then*

$$\pi(m, k; n, h)\gamma(m, n; n, m^*) = \pi(n, h; m, k)\sigma(m, n),$$

where, as usual, m^* denotes the inverse of m mod n.

Proof. For $0 \leq s \leq n - 1$ and $0 \leq r \leq m - 1$, by applying the definitions of γ and π, and setting

$$s' = sm^* \mod n \quad \text{and} \quad r' = rn \mod m, \tag{5.21}$$

we have

$$\pi(m, k; n, h)\gamma(m, n; n, m^*)(ms + r) = \pi(m, k; n, h)(s'm + r')$$

$$= hs'm + knn^*r' \mod nm$$

$$\text{(by (5.21))} = hsm^*m + kn^2n^*r \mod nm$$

$$\text{(by (5.20))} = hsm^*m + knr \mod nm.$$

On the other hand, applying the definition of $\sigma(m, n)$, we get

$$\pi(n, h; m, k)\sigma(m, n)(ms + r) = \pi(n, h; m, k)(rn + s)$$

$$= krn + hsmm^* \mod nm. \qquad \square$$

The Permutational Reverse Radix Identity in the cases discussed in Remark 5.2.15 may be expressed as follows.

Proposition 5.2.17

$$c(m, n) = c(n, m)\sigma(m, n) \quad and \quad r(m, n) = r(n, m)\sigma(m, n).$$

Proof. For $0 \le s \le n - 1$ and $0 \le r \le m - 1$ we have

$$c(n, m)\sigma(m, n)(ms + r) = c(n, m)(rn + s)$$
$$= rnn^* + mm^*s \quad \text{mod } nm$$
$$= c(m, n)(ms + r)$$

(note that $c(n, m) = \pi(n, 1; m, n^*)$) and

$$r(n, m)\sigma(m, n)(ms + r) = r(n, m)(rn + s)$$
$$= rn + sm \quad \text{mod } nm$$
$$= r(m, n)(ms + r)$$

(and now $r(n, m) = \pi(n, m; m, 1)$). $\qquad\qquad\qquad\qquad\qquad\square$

5.3 Permutation matrices and Kronecker products

We begin with some elementary but useful remarks on the product of matrices. Let $A = (a_{i,j})_{\substack{1 \le i \le n \\ 1 \le j \le m}}$ be an $n \times m$ matrix with complex coefficients.

Note that often we will actually use $\{0, 1, \ldots, n - 1\}$ (respectively $\{0, 1, \ldots, m - 1\}$) in place of $\{1, 2, \ldots, n\}$ (respectively $\{1, 2, \ldots, m\}$) as index sets.

We denote by A_{*j} its j-th column and by A_{i*} its i-th row, that is,

$$A_{*j} = \begin{bmatrix} a_{1,j} \\ a_{2,j} \\ \vdots \\ a_{n,j} \end{bmatrix} \quad \text{and} \quad A_{i*} = \begin{bmatrix} a_{i,1}, a_{i,2}, \cdots, a_{i,m} \end{bmatrix}$$

for $j = 1, 2, \ldots, m$ and $i = 1, 2, \ldots, n$. This way, we may decompose A as

$$A = [A_{*1}A_{*2} \cdots A_{*m}] = \begin{bmatrix} A_{1*} \\ A_{2*} \\ \vdots \\ A_{n*} \end{bmatrix}.$$

Let $B = (b_{j,k})_{\substack{1 \le j \le m \\ 1 \le k \le h}}$ be an $m \times h$ matrix. Then the product AB may be written in the following two forms. The first is:

$$AB = [(AB)_{*1}(AB)_{*2} \cdots (AB)_{*h}]$$

where, for $k = 1, 2, \ldots, h,$

$$(AB)_{*k} = \sum_{j=1}^{m} A_{*j} b_{j,k} = A(B_{*k}). \tag{5.22}$$

In other words, the k-th column of AB is the linear combination of the columns of A with coefficients $b_{1,k}, b_{2,k}, \ldots, b_{m,k}$ (the k-th column of B). The second one is:

$$AB = \begin{bmatrix} (AB)_{1*} \\ (AB)_{2*} \\ \vdots \\ (AB)_{n*} \end{bmatrix}$$

where, for $i = 1, 2, \ldots, n,$

$$(AB)_{i*} = \sum_{j=1}^{m} a_{i,j} B_{j*} = A_{i*}B. \tag{5.23}$$

That is, the i-th row of AB is the linear combinations of the rows of B with coefficients $a_{i,1}, a_{i,2}, \ldots, a_{i,m}$ (the i-th row of A).

With a permutation π of $\{1, 2, \ldots, n\}$ we associate the $n \times n$ *permutation matrix*

$$P_\pi = (\delta_{\pi(i),j})_{i,j=1}^{n}. \tag{5.24}$$

That is, the (i, j)-coefficient of P_π is equal to 1 if $j = \pi(i)$, and 0 otherwise. In other words, the i-th row of P_π is

$$(P_\pi)_{i*} = [0 \cdots 0\ 1\ 0 \cdots 0]$$

where the unique 1 is in the $\pi(i)$-th position (column). Noting that

$$\delta_{\pi(i),j} = \delta_{i,\pi^{-1}(j)}, \tag{5.25}$$

we can also conclude that the j-th column of P_π is

$$(P_\pi)_{*j} = \begin{bmatrix} 0 \\ \vdots \\ 0 \\ 1 \\ 0 \\ \vdots \\ 0 \end{bmatrix},$$

where the unique 1 is in the $\pi^{-1}(j)$-th position (row).

Lemma 5.3.1 (Product rules)

(i) *Let π, σ be permutations of $\{1, 2, \ldots, n\}$. Then*

$$P_\pi P_\sigma = P_{\sigma\pi}.$$

Moreover,

$$(P_\pi)^{-1} = P_{\pi^{-1}} = (P_\pi)^T. \tag{5.26}$$

(ii) *Let A (respectively B) be an $m \times n$ (respectively $n \times m$) matrix. Then*

$$AP_\pi = [A_{*1}A_{*2}\cdots A_{*n}]\,P_\pi = \left[A_{*\pi^{-1}(1)}A_{*\pi^{-1}(2)}\cdots A_{*\pi^{-1}(n)}\right],$$

while

$$P_\pi B = P_\pi \begin{bmatrix} B_{1*} \\ B_{2*} \\ \vdots \\ B_{n*} \end{bmatrix} = \begin{bmatrix} B_{\pi(1)*} \\ B_{\pi(2)*} \\ \vdots \\ B_{\pi(n)*} \end{bmatrix}.$$

Proof.

(i) The (i, j)-coefficient of the product $P_\pi P_\sigma$ is:

$$\sum_{k=1}^{n} \delta_{\pi(i),k}\delta_{\sigma(k),j} = \sum_{k=1}^{n} \delta_{\pi(i),k}\delta_{k,\sigma^{-1}(j)}$$

$$= \begin{cases} 1 & \text{if } \pi(i) = \sigma^{-1}(j) \\ 0 & \text{otherwise} \end{cases}$$

$$= \begin{cases} 1 & \text{if } j = \sigma(\pi(i)) \\ 0 & \text{otherwise} \end{cases}$$

$$= \delta_{\sigma(\pi(i)),j}.$$

Moreover, (5.26) follows from (5.25).

(ii) Taking into account (5.22) we have, for $j = 1, 2, \ldots, n$,

$$(AP_\pi)_{*j} = \sum_{k=1}^{n} A_{*k}\delta_{\pi(k),j}$$

$$= \sum_{k=1}^{n} A_{*k}\delta_{k,\pi^{-1}(j)}$$

$$= A_{*\pi^{-1}(j)}.$$

Similarly, by (5.23), for $i = 1, 2, \ldots, n$ we have

$$(P_\pi B)_{i*} = \sum_{k=1}^{n} \delta_{\pi(i),k}B_{k*} = B_{\pi(i)*}. \qquad \square$$

Corollary 5.3.2 *Let* $A = (a_{i,j})_{i,j=1}^n$ *be an* $n \times n$*-matrix. Then*

$$P_\pi A P_\pi^T = (a_{\pi(i),\pi(j)})_{i,j=1}^n.$$

In other words, multiplication on the left by P_π is equivalent to a permutation of the rows (in the i-th position we find the $\pi^{-1}(i)$-th row). Multiplication on the right by P_π is equivalent to a permutation of the columns (in the j-th position we find the $\pi(j)$-th column). Note also that if we set $Q_\pi = P_\pi^T$ then $Q_\pi Q_\sigma = Q_{\pi\sigma}$.

Definition 5.3.3 Let $A = (a_{i,j})_{i,j=1}^n$ and $B = (b_{i,j})_{i,j=1}^m$ be an $n \times n$ matrix and an $m \times m$ matrix, respectively. Then the *Kronecker product* of A and B is the $nm \times nm$ matrix $A \otimes B$ given in block form by

$$A \otimes B = \begin{pmatrix} a_{1,1}B & a_{1,2}B & \cdots & a_{1,n}B \\ a_{2,1}B & a_{2,2}B & \cdots & a_{2,n}B \\ \vdots & \vdots & \vdots & \vdots \\ a_{n,1}B & a_{n,2}B & \cdots & a_{n,n}B \end{pmatrix}.$$

This notion will be used in Section 8.7 and Section 10.5.

Example 5.3.4 Denote by I_n the $n \times n$ identity matrix. Then

$$I_n \otimes B = \begin{pmatrix} B & & & \\ & B & & \\ & & \ddots & \\ & & & B \end{pmatrix} \tag{5.27}$$

and

$$A \otimes I_m = \begin{pmatrix} a_{1,1}I_m & a_{1,2}I_m & \cdots & a_{1,n}I_m \\ a_{2,1}I_m & a_{2,2}I_m & \cdots & a_{2,n}I_m \\ \vdots & \vdots & \vdots & \vdots \\ a_{n,1}I_m & a_{n,2}I_m & \cdots & a_{n,n}I_m \end{pmatrix}.$$

In particular,

$$I_n \otimes I_m = I_{nm}. \tag{5.28}$$

Note that, in general, $A \otimes B$ is different from $B \otimes A$ (but we will show that they are similar).

Proposition 5.3.5 *The Kronecker product satisfies the following properties.*

(i) *Bilinearity:*

$$(\alpha_1 A_1 + \alpha_2 A_2) \otimes B = \alpha_1(A_1 \otimes B) + \alpha_2(A_2 \otimes B)$$

and

$$A \otimes (\beta_1 B_1 + \beta_2 B_2) = \beta_1 (A \otimes B) + \beta_2 (A \otimes B_2);$$

(ii) *associativity:*

$$(A \otimes B) \otimes E = A \otimes (B \otimes E);$$

(iii) *product rule:*

$$(A \otimes B)(C \otimes D) = (AC) \otimes (BD);$$

(iv)

$$A \otimes B = (A \otimes I_m)(I_n \otimes B) = (I_m \otimes A)(B \otimes I_n);$$

(v) *if both A, B are invertible then $A \otimes B$ is invertible and*

$$(A \otimes B)^{-1} = A^{-1} \otimes B^{-1};$$

(vi)

$$(A \otimes B)^T = A^T \otimes B^T,$$

for all $n \times n$ matrices A, A_1, A_2, C; $m \times m$ matrices B, B_1, B_2, D; $h \times h$ matrices E; and $\alpha_1, \alpha_2, \beta_1, \beta_2 \in \mathbb{C}$.

Proof. (i) and (ii) are easy exercises left to the reader.
(iii) If $C = (c_{i,j})_{i,j=1}^n$ then $(A \otimes B)(C \otimes D)$ equals

$$\begin{pmatrix} a_{1,1}B & a_{1,2}B & \cdots & a_{1,n}B \\ a_{2,1}B & a_{2,2}B & \cdots & a_{2,n}B \\ \vdots & \vdots & \vdots & \vdots \\ a_{n,1}B & a_{n,2}B & \cdots & a_{n,n}B \end{pmatrix} \begin{pmatrix} c_{1,1}D & c_{1,2}D & \cdots & c_{1,n}D \\ c_{2,1}D & c_{2,2}B & \cdots & c_{2,n}D \\ \vdots & \vdots & \vdots & \vdots \\ c_{n,1}D & c_{n,2}D & \cdots & c_{n,n}D \end{pmatrix}$$

$$= \begin{pmatrix} \left(\sum_{j=1}^n a_{1,j}c_{j,1}\right)BD & \left(\sum_{j=1}^n a_{1,j}c_{j,2}\right)BD & \cdots & \left(\sum_{j=1}^n a_{1,j}c_{j,n}\right)BD \\ \left(\sum_{j=1}^n a_{2,j}c_{j,1}\right)BD & \left(\sum_{j=1}^n a_{2,j}c_{j,2}\right)BD & \cdots & \left(\sum_{j=1}^n a_{2,j}c_{j,n}\right)BD \\ \vdots & \vdots & \vdots & \vdots \\ \left(\sum_{j=1}^n a_{n,j}c_{j,1}\right)BD & \left(\sum_{j=1}^n a_{n,j}c_{j,2}\right)BD & \cdots & \left(\sum_{j=1}^n a_{n,j}c_{j,n}\right)BD \end{pmatrix},$$

and this is exactly $(AC) \otimes (BD)$.
(iv) and (v) are easy consequences of (iii). Finally, (vi) is an easy exercise. $\square$

We now adopt the notation in [130]. We set

$$P_n^m = P_{\sigma(m,n)} \tag{5.29}$$

that is, P_n^m is the permutation matrix associated with the stride permutation $\sigma(m, n)$ (see Definition 5.2.1). Note that, by (5.26) and (5.7), we have

$$(P_n^m)^{-1} = (P_n^m)^T = P_m^n. \tag{5.30}$$

The following important result connects stride permutations and Kronecker products.

Proposition 5.3.6 (Similarity of tensor products by stride permutations)
Let $A = (a_{i,j})_{i,j=0}^{n-1}$ and $B = (b_{i,j})_{i,j=0}^{m-1}$. Then

$$P_m^n(A \otimes B)P_n^m = B \otimes A.$$

Proof. Denote by $(A \otimes B)_{i,i'}$ $(0 \leq, i, i' \leq nm - 1)$ the (i, i')-coefficient of $A \otimes B$. Then, in the notation of (5.5) and (5.6), the matrix $A \otimes B$ may be expressed as follows: if $i = sm + r$ and $i' = s'm + r'$, with $0 \leq r, r' \leq m - 1$ and $0 \leq s, s' \leq n - 1$, then

$$(A \otimes B)_{i,i'} = a_{s,s'}b_{r,r'}. \tag{5.31}$$

Moreover, if $j = rn + s$ and $j' = r'n + s$, with, as above, $0 \leq r, r' \leq m - 1$ and $0 \leq s, s' \leq n - 1$, then

$$(B \otimes A)_{j,j'} = b_{r,r'}a_{s,s'} \tag{5.32}$$

and

$$\begin{aligned} j = \sigma(m, n)i \ j' = \sigma(m, n)i' \\ i = \sigma(n, m)j \ i' = \sigma(n, m)j'. \end{aligned} \tag{5.33}$$

Therefore, taking into account Corollary 5.3.2 and (5.7), we have

$$\begin{aligned} \left[P_m^n(A \otimes B)P_n^m\right]_{j,j'} &= (A \otimes B)_{\sigma(n,m)j,\sigma(n,m)j'} \\ \text{(by (5.33))} &= (A \otimes B)_{i,i'} \\ \text{(by (5.31))} &= a_{s,s'}b_{r,r'} \\ \text{(by (5.32))} &= (B \otimes A)_{j,j'}. \quad \square \end{aligned}$$

We now examine the partial stride permutations introduced in Definition 5.2.4: we keep the same notation.

Proposition 5.3.7 *We have*

$$P_{\tau(m,n,k)} = I_k \otimes P_n^m$$

and

$$P_{\iota(m,n,k)} = P_k^n \otimes I_m.$$

Proof. Note that

$$(P_n^m)_{i,i'} = \delta_{\sigma(m,n)i,i'} = \delta_{r,r'}\delta_{s,s'} \tag{5.34}$$

if $i = sm + r$ and $i' = r'n + s'$, with $0 \le s, s' \le n - 1$ and $0 \le r, r' \le m - 1$. Therefore, if $i = tmn + sm + r$, with $0 \le t \le k - 1$, $0 \le s \le n - 1$, and $0 \le r \le m - 1$, and $i' = t'mn + r'n + s'$, with $0 \le t' \le k - 1$, $0 \le r' \le m - 1$, and $0 \le s' \le n - 1$, then (cf. Definition 5.2.4)

$$\tau(m, n, k)i = i' \leftrightarrow t = t', s = s', r = r'$$

so that

$$(P_{\tau(m,n,k)})_{i,i'} = \delta_{\tau(m,n,k)i,i'} = \delta_{t,t'}\delta_{r,r'}\delta_{s,s'}. \tag{5.35}$$

Similarly, by virtue of (5.31) (with n replaced by k and m replaced by nm), we have

$$\begin{aligned}
(I_k \otimes P_n^m)_{i,i'} &= \delta_{t,t'}(P_n^m)_{sm+r,r'n+s'} \\
\text{(by (5.34))} &= \delta_{t,t'}\delta_{r,r'}\delta_{s,s'}.
\end{aligned} \tag{5.36}$$

Comparing (5.35) and (5.36), we deduce the first identity.

Now suppose that $i' = s'km + t'm + r'$ with $0 \le t' \le k - 1$, $0 \le r' \le m - 1$, and $0 \le s' \le n - 1$, while i is as above. Then (cf. Definition 5.2.4)

$$\iota(m, n, k)i = i' \leftrightarrow t = t', s = s', r = r'$$

so that

$$(P_{\iota(m,n,k)})_{i,i'} = \delta_{\iota(m,n,k)i,i'} = \delta_{t,t'}\delta_{s,s'}\delta_{r,r'}, \tag{5.37}$$

while, writing i, i' in the forms $i' = (s'k + t')m + r'$ and $i = (tn + s)m + r$, we have

$$\begin{aligned}
(P_k^n \otimes I_m)_{i,i'} &= (P_k^n)_{tn+s,s'k+t'}\delta_{r,r'} \\
&= \delta_{\sigma(n,k)(tn+s),s'k+t'}\delta_{r,r'} \\
&= \delta_{s,s'}\delta_{r,r'}\delta_{t,t'},
\end{aligned} \tag{5.38}$$

where the first equality follows from (5.31). Comparing (5.37) and (5.38) we deduce the second identity. $\qquad\square$

By means of Lemma 5.3.1.(i) and of Proposition 5.3.7, all the identities in Theorem 5.2.3, Theorem 5.2.5, Theorem 5.2.6, and Corollary 5.2.7 may be translated into identities for permutation matrices. We list then in the following proposition.

Proposition 5.3.8 Basic product identities:

$$P_k^{mn} P_n^{mk} = P_{nk}^m$$
$$P_{nk}^m P_{mk}^n = P_k^{mn}. \tag{5.39}$$

Product identities for partial strides:

$$(I_k \otimes P_n^m)(P_k^m \otimes I_n) = P_{nk}^m$$
$$(P_k^n \otimes I_m)(I_n \otimes P_k^m) = P_k^{mn}.$$

Mixed product identities:

$$P_k^{mn}(I_n \otimes P_m^k) = P_k^n \otimes I_m$$
$$P_{nk}^m(P_m^k \otimes I_n) = I_k \otimes P_n^m$$
$$(P_k^n \otimes I_m)P_n^{mk} = I_k \otimes P_n^m$$
$$(I_k \otimes P_n^m)P_{mk}^n = P_k^n \otimes I_m.$$

Similarity identity:

$$P_{mn}^k(I_k \otimes P_n^m)P_k^{mn} = P_n^m \otimes I_k.$$

Proof. The proof is immediate and is left to the reader. We just note that, using the matrix formalism, the second identity follows from the first one by means of an application of (5.30). The same observation holds true for the other group of identities. Note also that the similarity identity is just a particular case of Proposition 5.3.6. $\square$

With the notation in Definition 5.2.10 we set

$$B_m^k = P_{\gamma(m,k)}. \tag{5.40}$$

Proposition 5.3.9

$$P_{\gamma(m,k;n,h)} = B_n^h \otimes B_m^k.$$

Proof. First note that, for $0 \le r, r' \le m - 1$,

$$(B_m^k)_{r,r'} = \delta_{\gamma(m,k)r,r'} = \begin{cases} 1 & \text{if } r' \equiv kr \mod m \\ 0 & \text{otherwise.} \end{cases} \tag{5.41}$$

Therefore, for $i = ms + r$ and $i' = ms' + r'$, with $0 \le s, s' \le n - 1$, and $0 \le r, r' \le m - 1$, by virtue of (5.31) we have

$$(B_n^h \otimes B_m^k)_{i,i'} = (B_n^h)_{s,s'} (B_m^k)_{r,r'}$$

$$(\text{by } (5.41)) = \begin{cases} 1 & \text{if } s' = hs \mod n \text{ and } r' = kr \mod m \\ 0 & \text{otherwise} \end{cases}$$

$$= \delta_{\gamma(m,k;n,h)i,i'}$$

$$= (P_{\gamma(m,k;n,h)})_{i,i'}. \qquad \square$$

Note also that, if $\gcd(k, m) = \gcd(h, m) = 1$, from Proposition 5.2.11 we get:

$$B_m^k B_m^h = B_m^{kh} = B_m^h B_m^k$$

and

$$(B_m^k)^{-1} \equiv (B_m^k)^T = B_m^{k*}, \qquad (5.42)$$

where, as usual, $k^*k = 1 \mod m$.

In order to describe the matrix formulations corresponding to $\beta(m, n)$ in (5.16), we introduce a few more definitions and notation. The *elementary circulant permutation matrix* of order n is the matrix

$$C_n = \begin{pmatrix} 0 & 0 & \cdots & & 0 & 1 \\ 1 & 0 & \cdots & & 0 & 0 \\ 0 & 1 & \cdots & & 0 & 0 \\ & & \ddots & & & \\ 0 & 0 & \cdots & 1 & 0 & 0 \\ 0 & 0 & \cdots & 0 & 1 & 0 \end{pmatrix};$$

(cf. Exercise 2.4.16). In other words, denoting by $\varepsilon = \varepsilon_n$ the permutation of $\{0, 1, \ldots, n - 1\}$ defined by setting $\varepsilon(i) = i - 1 \mod n$, then

$$(C_n)_{i,j} = \delta_{\varepsilon(i),j} \quad 0 \le i, j \le n - 1,$$

equivalently (cf. (5.24)),

$$C_n = P_\varepsilon. \qquad (5.43)$$

Clearly, $C_n^k = P_{\varepsilon^k}$ and therefore

$$(C_n^k)_{i,j} = \begin{cases} 1 & \text{if } i - k \equiv j \mod n \\ 0 & \text{otherwise.} \end{cases} \qquad (5.44)$$

We also define the *m-th block diagonal power* of an $n \times n$ matrix W, as the $mn \times mn$ matrix $D_m(W)$ defined by setting

$$D_m(W) = \begin{pmatrix} W^0 & & & & \\ & W^1 & & & \\ & & W^2 & & \\ & & & \ddots & \\ & & & & W^{m-1} \end{pmatrix} \tag{5.45}$$

where $W^0 = I_n$ and $W^i = WW^{i-1}$ for $i = 1, 2, \ldots, m - 1$. Note that, for $j = rn + s$ and $j' = r'n + s'$, with $0 \le r, r' \le m - 1$ and $0 \le s, s' \le n - 1$, we have

$$[D_m(W)]_{j, j'} = \delta_{r, r'} \cdot (W^r)_{s, s'}. \tag{5.46}$$

In what follows, for $0 \le k \le n - 1$, we set

$$Q_m^n(k) = P_n^m D_m(C_n^k) P_m^n. \tag{5.47}$$

Then, with the notation in Definition 5.2.12 we have

Proposition 5.3.10

$$P_{\beta(m,n)} = Q_m^n(m^*).$$

Proof. Let $i = sm + r$ and $i' = s'm + r'$, with $0 \le s, s' \le n - 1$ and $0 \le r, r' \le m - 1$. Then, setting $j = \sigma(m, n)i = rn + s$ and $j' = \sigma(m, n)i' = r'n + s'$, by virtue of Corollary 5.3.2 and (5.7), we have

$$[P_n^m D_m(C_n^{m^*}) P_m^n]_{i, i'} = [D_m(C_n^{m^*})]_{\sigma(m,n)i, \sigma(m,n)i'}$$
$$= [D_m(C_n^{m^*})]_{j, j'}$$
$$\text{(by (5.46))} = \delta_{r, r'}(C_n^{m^* r})_{s, s'}$$
$$\text{(by (5.44))} = \begin{cases} 1 & \text{if } r' = r \text{ and } s' = s - m^* r \mod n \\ 0 & \text{otherwise} \end{cases}$$
$$\text{(by (5.16))} = \delta_{\beta(m,n)i, i'}. \qquad \square$$

Finally, we define the permutation matrix corresponding to the composite bijection permutation by setting, with the same notation as in Definition 5.2.13,

$$\Xi_m^n(h, k) = P_{\pi(m,k;n,h)}. \tag{5.48}$$

Therefore, for $0 \le i, i' \le mn - 1$ with $i = sm + r$ as in (5.5), we have

$$[\Xi_m^n(h, k)]_{i, i'} = \begin{cases} 1 & \text{if } i' = hsm + knn^* r \mod nm \\ 0 & \text{otherwise.} \end{cases}$$

By means of Lemma 5.3.1.(i) we immediately get the matrix version of Theorem 5.2.14 and Theorem 5.2.16.

Theorem 5.3.11 *Suppose* $\gcd(n, m) = \gcd(k, m) = \gcd(h, n) = 1$, $mm^* = 1$ mod n and $nn^* = 1$ mod m. *Then we have:*

(i) Matrix Factorization of Composite Bijection Permutations

$$\Xi_m^n(h, k) = \left(B_n^h \otimes B_m^k\right) Q_m^n(m^*). \tag{5.49}$$

(ii) Reverse Radix Identity

$$\left(B_n^{m^*} \otimes B_m^n\right) \Xi_m^n(h, k) = P_n^m \Xi_n^m(k, h).$$

Denote by

$$C_m^n = P_{c(m,n)} \ (= \Xi_m^n(m^*, 1)) \text{ and } \mathcal{R}_m^n = P_{r(m,n)} \ (= \Xi_m^n(1, n)) \tag{5.50}$$

the permutation matrices associated with the Chinese remainder mapping and with the Ruritanian map (cf. Remark 5.2.15), respectively. Then from Proposition 5.2.17 we deduce the following symmetry relations.

Proposition 5.3.12

$$C_m^n = P_n^m C_n^m \ \text{ and } \ \mathcal{R}_m^n = P_n^m \mathcal{R}_n^m.$$

We need a generalization of (5.49).

Let n, m, h, k, ℓ be positive integers such that $\gcd(n, h) = \gcd(m, k) = 1$. We set

$$\Xi_m^n(h, k, \ell) = \left(B_n^h \otimes B_m^k\right) Q_m^n(\ell). \tag{5.51}$$

Therefore, by (5.49), if $\gcd(n, m) = 1$ then we have

$$\Xi_m^n(h, k) = \Xi_m^n(h, k, m^*), \tag{5.52}$$

where $mm^* = 1 \mod n$.

Before embarking on the study of the matrix formulation of the FFT, we show how to apply the machinery of stride and partial stride permutations to get some useful factorizations of tensor products.

Proposition 5.3.13 *For k, m, n positive integers and A an $n \times n$ matrix we have:*

$$I_k \otimes A \otimes I_m = P_{kn}^m (I_{km} \otimes A) P_m^{kn}$$

and

$$I_k \otimes A \otimes I_m = (I_k \otimes P_n^m)(I_{km} \otimes A)(I_k \otimes P_m^n)$$

(recall, cf. Proposition 5.3.7, that $I_k \otimes P_n^m = P_{\tau(m,n,k)}$).

Proof. First observe that $I_k \otimes A$ is a $kn \times kn$ matrix, so that

$$P_m^{kn}(I_k \otimes A \otimes I_m)P_{kn}^m = P_m^{kn}[(I_k \otimes A) \otimes I_m]P_{kn}^m$$
$$\text{(by Proposition 5.3.6)} = I_m \otimes (I_k \otimes A)$$
$$\text{(by Proposition 5.3.5.(ii) and (5.28))} = I_{mk} \otimes A.$$

Recalling that $(P_m^{kn})^{-1} = P_{kn}^m$ (cf. (5.30)) we get the first identity by conjugating with P_{kn}^m. Similarly,

$$(I_k \otimes P_m^n)(I_k \otimes A \otimes I_m)(I_k \otimes P_n^m) = (I_k \otimes P_m^n)[I_k \otimes (A \otimes I_m)](I_k \otimes P_n^m)$$
$$\text{(by Proposition 5.3.5.(iii))} = I_k \otimes [P_m^n(A \otimes I_m)P_n^m]$$
$$= I_k \otimes I_m \otimes A$$
$$= I_{km} \otimes A,$$

and the second identity follows as well. $\square$

We now introduce some further notation. Suppose that $n_1, n_2, \ldots, n_h$ are positive integers, $h \geq 3$, and A_j is an $n_j \times n_j$ matrix, for $j = 1, 2, \ldots, h$. Set

$$k_1 = 1 \quad \text{and} \quad k_j = n_1 n_2 \cdots n_{j-1} \text{ for } j = 2, 3, \ldots, h;$$
$$m_j = n_{j+1} n_{j+2} \cdots n_h \text{ for } j = 1, 2, \ldots h - 1, \quad \text{and} \quad m_h = 1$$

and, for $j = 1, 2, \ldots, h$,

$$X_j = I_{n_1} \otimes I_{n_2} \otimes \cdots I_{n_{j-1}} \otimes A_j \otimes I_{n_{j+1}} \otimes \cdots \otimes I_{n_h} = I_{k_j} \otimes A_j \otimes I_{m_j},$$

$$Y_j = I_{k_j m_j} \otimes A_j.$$

Finally, we set

$$Q_j = P_{\tau(m_{j+1}, n_{j+1}, k_{j+1}) \tau(n_j, m_j, k_j)}$$
$$= (I_{k_j} \otimes P_{m_j}^{n_j})(I_{k_{j+1}} \otimes P_{n_{j+1}}^{m_{j+1}}), \tag{5.53}$$

where the second equality follows from Proposition 5.3.7 and Lemma 5.3.1.(i).

Theorem 5.3.14 *With the above notation, the following factorization identities hold.*

(i) Fundamental factorization:

$$A_1 \otimes A_2 \otimes \cdots \otimes A_h = X_1 X_2 \cdots X_h.$$

(ii) Parallel tensor product factorization I:

$$A_1 \otimes A_2 \otimes \cdots \otimes A_h = P_{n_1}^{m_1} Y_1 P_{n_2}^{k_2 m_2} Y_2 P_{n_3}^{k_3 m_3} \cdots P_{n_{h-1}}^{k_{h-1} m_{h-1}} Y_{h-1} P_{n_h}^{k_h} Y_h$$

(iii) Parallel tensor product factorization II:

$$A_1 \otimes A_2 \otimes \cdots \otimes A_h = P_{n_1}^{m_1} Y_1 Q_1 Y_2 Q_2 \cdots Q_{h-2} Y_{h-1} Q_{h-1} Y_h.$$

Proof. The first identity is just an iterated form of Proposition 5.3.5.(ii)–(iv). For the second identity, first observe that Proposition 5.3.13 yields

$$X_j = P_{k_j n_j}^{m_j} Y_j P_{m_j}^{k_j n_j} \quad j = 1, 2, \ldots, h.$$

Moreover, since $k_{j+1} = k_j n_j$ and $m_j = n_{j+1} m_{j+1}$,

$$P_{m_j}^{k_j n_j} P_{k_{j+1} n_{j+1}}^{m_{j+1}} = P_{n_{j+1} m_{j+1}}^{k_{j+1}} P_{k_{j+1} n_{j+1}}^{m_{j+1}} = P_{n_{j+1}}^{k_{j+1} m_{j+1}},$$

where the last equality follows from (5.39) in Proposition 5.3.8. Therefore,

$$X_1 X_2 \cdots X_h = P_{n_1}^{m_1} Y_1 P_{m_1}^{n_1} P_{k_2 n_2}^{m_2} Y_2 P_{m_2}^{k_2 n_2} \cdots Y_j P_{m_j}^{k_j n_j} P_{k_{j+1} n_{j+1}}^{m_{j+1}} Y_{j+1} \cdots Y_h$$

$$= P_{n_1}^{m_1} Y_1 P_{n_2}^{k_2 m_2} Y_2 \cdots Y_j P_{n_{j+1}}^{k_{j+1} m_{j+1}} Y_{j+1} \cdots P_{n_{h-1}}^{k_{h-1} m_{h-1}} Y_{h-1} P_{n_h}^{k_h} Y_h.$$

Finally, from Proposition 5.3.13 we also deduce

$$X_j = (I_{k_j} \otimes P_{n_j}^{m_j})(I_{k_j m_j} \otimes A_j)(I_{k_j} \otimes P_{m_j}^{n_j})$$

which, by virtue of (5.53), immediately implies the last equality in the statement. $\square$

5.4 The matrix form of the FFT

This is the central section of the present chapter. It is devoted to the matrix form of several algorithms that reduce the matrix of the DFT to a tensor product of smaller matrices, when the size of the DFT is factorizable.

Let $\omega \in \mathbb{C}$ be an arbitrary n-th root of 1, that is, $\omega^n = 1$. Following [130], we define the $n \times n$ matrix $A_n(\omega)$ by setting

$$A_n(\omega) = (\omega^{ij})_{i,j=0}^{n-1}. \tag{5.54}$$

Clearly, $A_n(\omega)$ is symmetric. In the notation of Exercise 2.4.16.(4), we have

$$\frac{1}{\sqrt{n}}A_n(e^{-\frac{2\pi i}{n}}) = F_n.$$

Note also that if ω is a *primitive* n-th root of 1, then $A_n(\omega)^{-1}$ exists and

$$A_n(\omega)^{-1} = \frac{1}{n}A_n(\overline{\omega}).$$

The proof is similar to that one of Lemma 2.2.3. In general, if $\omega^r = 1$ and $\omega^h \neq 1$ for $0 \leq h \leq r - 1$, for some $r \geq 1$ (note that r necessarily divides n), then $\mathrm{rk}A_n(\omega) = r$.

Recall that C_n denotes the elementary circulant matrix (see (5.43)) and $D_n(\cdot)$ is the n-th diagonal power matrix (see (5.45)).

Proposition 5.4.1 (Eigenidentities) *Let n be a positive integer, $k \geq 0$, and ω an n-th root of 1. Then we have*

$$A_n(\omega)C_n^k = D_n(\omega^k)A_n(\omega) \quad and \quad C_n^k A_n(\omega) = A_n(\omega)D_n(\omega^{-k}).$$

Proof. From (5.43) we get, for $0 \leq i, j \leq n - 1$,

$$\begin{aligned}
[A_n(\omega)C_n^k]_{i,j} &= \sum_{h=0}^{n-1} \omega^{ih}\delta_{\varepsilon^k(h),j} \\
&= \sum_{h=0}^{n-1} \omega^{ih}\delta_{h,\varepsilon^{-k}(j)} \\
&= \omega^{i\varepsilon^{-k}(j)} \\
&= \omega^{i(j+k)} \\
&= \omega^{ik}\omega^{ij} \\
&= [D_n(\omega^k)A_n(\omega)]_{i,j},
\end{aligned}$$

proving the first equality.

The second equality follows from the first one, by transposing: observe that

$$(C_n^k)^T = (P_{\varepsilon^k})^T = P_{\varepsilon^{-k}} = C_n^{-k}$$

so that we must replace k with $-k$. $\qquad\square$

We also need the following transformation formula.

Proposition 5.4.2 *Suppose that $\gcd(h, n) = 1$ and $h^*h = 1 \bmod n$. Then*

$$A_n(\omega)B_n^h = A_n(\omega^{h^*}) \quad and \quad B_n^h A_n(\omega) = A_n(\omega^h).$$

Proof. For $0 \leq i, j \leq n - 1$ we have

$$[A_n(\omega)B_n^h]_{i,j} =_* [A_n(\omega)]_{i,\gamma(n,h)^{-1}j}$$
$$\text{(by Proposition 5.2.11.(ii))} = [A_n(\omega)]_{i,h^*j}$$
$$= \omega^{ijh^*}$$
$$= [A_n(\omega^{h^*})]_{i,j}$$

where $=_*$ follows from Lemma 5.3.1.(ii) and (5.40). This proves the first equality. The proof of the second one is similar and left to the reader. $\square$

Using the notation in (5.45), we define the *diagonal matrix of twiddle factors* by setting

$$T_m^n(\omega) = D_m(D_n(\omega)),$$

where now ω is an nm-th root of 1.

Note that, by virtue of (5.46), for $0 \leq r, r' \leq m - 1$ and $0 \leq s, s' \leq n - 1$, we have

$$[T_m^n(\omega)]_{rn+s, r'n+s'} = \delta_{r,r'}[D_n(\omega^r)]_{s,s'} = \delta_{r,r'}\delta_{s,s'}\omega^{rs}. \tag{5.55}$$

Proposition 5.4.3 *With the above notation we have*

$$P_n^m T_m^n(\omega)P_m^n = T_n^m(\omega).$$

Moreover, for integers k and h,

$$T_m^n(\omega^k)T_m^n(\omega^h) = T_m^n(\omega^{k+h}).$$

Proof. By virtue of Corollary 5.3.2 we have

$$[P_n^m T_m^n(\omega)P_m^n]_{sm+r,s'm+r'} = [T_m^n(\omega)]_{\sigma(m,n)(sm+r),\sigma(m,n)(s'm+r')}$$
$$= [T_m^n(\omega)]_{rn+s, r'n+s'}$$
$$\text{(by (5.55))} = \delta_{r,r'}\delta_{s,s'}\omega^{rs}$$
$$\text{(again by (5.55))} = [T_n^m(\omega)]_{sm+r,s'm+r'}.$$

The second identity is trivial. $\square$

Proposition 5.4.4 (Tensor form of the eigenidentities) *For n, m positive integers, ω an nm-th root of 1, and an integer k, we have*

$$D_m(C_n^k)[I_m \otimes A_n(\omega^m)] = [I_m \otimes A_n(\omega^m)]T_m^n(\omega^{-km})$$
$$[I_m \otimes A_n(\omega^m)]D_m(C_n^k) = T_m^n(\omega^{km})[I_m \otimes A_n(\omega^m)].$$

Proof. We only prove the first identity: the proof of the second one is similar and left to the reader.

$$D_m(C_n^k)[I_m \otimes A_n(\omega^m)]$$

$$= \begin{pmatrix} I_n & & & \\ & C_n^k & & \\ & & \ddots & \\ & & & C_n^{k(m-1)} \end{pmatrix} \begin{pmatrix} A_n(\omega^m) & & & \\ & A_n(\omega^m) & & \\ & & \ddots & \\ & & & A_n(\omega^m) \end{pmatrix}$$

$$= \begin{pmatrix} A_n(\omega^m) & & & \\ & C_n^k A_n(\omega^m) & & \\ & & \ddots & \\ & & & C_n^{k(m-1)} A_n(\omega^m) \end{pmatrix}$$

and, by Proposition 5.4.1, this equals

$$= \begin{pmatrix} A_n(\omega^m) & & & \\ & A_n(\omega^m)D_n(\omega^{-km}) & & \\ & & \ddots & \\ & & & A_n(\omega^m)D_n(\omega^{-km(m-1)}) \end{pmatrix}$$

$$= \begin{pmatrix} A_n(\omega^m) & & & \\ & A_n(\omega^m) & & \\ & & \ddots & \\ & & & A_n(\omega^m) \end{pmatrix}$$

$$\cdot \begin{pmatrix} I_n & & & \\ & D_n(\omega^{-km}) & & \\ & & \ddots & \\ & & & D_n(\omega^{-km(m-1)}) \end{pmatrix}$$

$$= [I_m \otimes A_n(\omega^m)]T_m^n(\omega^{-km}),$$

where the last identity follows from the definition of T_m^n and the identity $D_n(\omega^{-kmh}) = [D_n(\omega^{-km})]^h$. $\square$

We are now in position to prove the basic tensor product form of the FFT and to derive all its consequences.

Theorem 5.4.5 (General Radix Identity) *Let $n, m > 1$ be two positive integers and ω an nm-th root of 1. Then*

$$A_{nm}(\omega)P_m^n = [A_n(\omega^m) \otimes I_m]T_n^m(\omega)[I_n \otimes A_m(\omega^n)]. \tag{5.56}$$

Proof. Let $i = sm + r$, $i' = s'm + r'$, $j = \alpha m + \beta$, and $j' = \alpha'm + \beta'$, with $0 \le s, s', \alpha, \alpha' \le n - 1$ and $0 \le r, r', \beta, \beta' \le m - 1$. Then, on the one hand,

$$\{[A_n(\omega^m) \otimes I_m]T_n^m(\omega)[I_n \otimes A_m(\omega^n)]\}_{i,i'} = \sum_{j,j'=0}^{nm-1} [A_n(\omega^m) \otimes I_m]_{i,j}[T_n^m(\omega)]_{j,j'}$$

$$\cdot [I_n \otimes A_m(\omega^n)]_{j',i'}$$

$$\text{(by (5.31) and (5.55))} = \sum_{\alpha,\alpha'=0}^{n-1} \sum_{\beta,\beta'=0}^{m-1} [A_n(\omega^m)]_{s,\alpha}\delta_{r,\beta}$$

$$\cdot \delta_{\alpha,\alpha'}\delta_{\beta,\beta'}\omega^{\alpha\beta}\delta_{\alpha',s'}[A_m(\omega^n)]_{\beta',r'}$$

$$(\alpha = \alpha' = s' \text{ and } r = \beta = \beta') = [A_n(\omega^m)]_{s,s'}\omega^{s'r}[A_m(\omega^n)]_{r,r'}$$

$$\text{(by (5.54))} = \omega^{mss'+s'r+nrr'}.$$

On the other hand, by Lemma 5.3.1.(ii), (5.7), and (5.29),

$$[A_{nm}(\omega)P_m^n]_{i,i'} = [A_{nm}(\omega)]_{i,\sigma(m,n)i'}$$

$$= [A_{nm}(\omega)]_{sm+r,r'n+s'}$$

$$= \omega^{(sm+r)(r'n+s')}$$

$$(\omega^{nm} = 1) = \omega^{mss'+s'r+nrr'}. \qquad \square$$

We now show how, multiplying on the left and on the right the left hand side of the General Radix Identity (5.56) by suitable permutations, changes the diagonal matrix of twiddle factors in the right hand side (of (5.56)).

Theorem 5.4.6 (Twiddle Identity) *With the notation of Theorem 5.4.5, for arbitrary $k_1, k_2 \in \mathbb{Z}$ we have:*

$$Q_m^n(k_1)A_{nm}(\omega)[P_n^m Q_n^m(k_2)]^T$$

$$= [A_n(\omega^m) \otimes I_m]T_n^m(\omega^{1-k_1m-k_2n})[I_n \otimes A_m(\omega^n)].$$

Proof. First of all, note that $(C_m^{k_2})^T = C_m^{-k_2}$ (compare with (5.26) and (5.43)) and therefore, from (5.45) and (5.47) it follows that

$$[Q_n^m(k_2)]^T = Q_n^m(-k_2).$$

Therefore, taking into account Theorem 5.4.5,

$$Q_m^n(k_1)A_{nm}(\omega)[P_n^m Q_n^m(k_2)]^T = Q_m^n(k_1)[A_n(\omega^m) \otimes I_m]T_n^m(\omega)$$

$$\cdot [I_n \otimes A_m(\omega^n)]P_n^m Q_n^m(-k_2)P_m^n$$

$$\text{(by (5.47))} \quad = P_n^m D_m(C_n^{k_1})P_m^n[A_n(\omega^m) \otimes I_m]$$

$$\cdot T_n^m(\omega)[I_n \otimes A_m(\omega^n)]D_n(C_m^{-k_2})$$

$$\text{(by Proposition 5.3.6)} \quad = P_n^m D_m(C_n^{k_1})[I_m \otimes A_n(\omega^m)]P_n^m$$

$$\cdot T_n^m(\omega)[I_n \otimes A_m(\omega^n)]D_n(C_m^{-k_2})$$

$$\text{(by Proposition 5.4.4)} \quad = P_n^m[I_m \otimes A_n(\omega^m)]T_m^n(\omega^{-k_1 m})P_m^n T_n^m(\omega)$$

$$\cdot T_n^m(\omega^{-k_2 n})[I_n \otimes A_m(\omega^n)]$$

$$\text{(by Propositions 5.3.6 and 5.4.3)} \quad = [A_n(\omega^m) \otimes I_m]T_n^m(\omega^{1-k_1 m - k_2 n}).$$

$$\cdot [I_n \otimes A_m(\omega^n)]. \qquad \square$$

Corollary 5.4.7 *With the notation of (5.51) and supposing* $\gcd(k_i, m) = \gcd(h_i, n) = 1$, *for* $i = 1, 2$, *we have*

$$\Xi_m^n(h_1, k_1, \ell_1)A_{nm}(\omega)[P_n^m \Xi_n^m(k_2, h_2, \ell_2)]^T$$

$$= [A_n(\omega^{h_1 m}) \otimes B_m^{k_1}]T_n^m(\omega^{1-\ell_1 m - \ell_2 n})[B_n^{h_2^*} \otimes A_m(\omega^{k_2 n})], \quad (5.57)$$

where $h_2 h_2^* = 1 \mod n$.

Proof. This follows immediately from Proposition 5.3.5.(iii), Theorem 5.4.6, (5.51) and Proposition 5.4.2. Just note that, if $k_2 k_2^* = 1 \mod m$,

$$[P_n^m \Xi_n^m(k_2, h_2, \ell_2)]^T = [P_n^m(B_m^{k_2} \otimes B_n^{h_2})Q_n^m(\ell_2)]^T$$

$$\text{(by Proposition 5.3.6)} \quad = \{(B_n^{h_2} \otimes B_m^{k_2})[P_n^m Q_n^m(\ell_2)]\}^T$$

$$=_* [P_n^m Q_n^m(\ell_2)]^T (B_n^{h_2^*} \otimes B_m^{k_2^*}),$$

where, in $=_*$ we used the equality $[B_n^{h_2} \otimes B_m^{k_2}]^T = B_n^{h_2^*} \otimes B_m^{k_2^*}$, which follows from Proposition 5.3.5.(vi) and (5.42). $\qquad \square$

Remark 5.4.8 Note that Theorem 5.4.5 and Theorem 5.4.6 are particular cases of Corollary 5.4.7. Indeed, for $h_i = k_i = 1$, $i = 1, 2$, Corollary 5.4.7 reduces to Theorem 5.4.6, by virtue of (5.51). If, in addition, $\ell_1 = \ell_2 = 0$, then it reduces to Theorem 5.4.5.

Until now, we have determined algorithms for tensor product factorizations of the matrix $A_N(\omega)$, where $N = mn$ is an arbitrary factorization. In what follows, we examine the case when $\gcd(m, n) = 1$.

Theorem 5.4.9 (Twiddle Free Identity) *Suppose that* $\gcd(m, n) = 1$. *Then, with the notation and hypotheses of Corollary 5.4.7, we have*

$$\Xi_m^n(h_1, k_1)A_{nm}(\omega)[P_n^m \Xi_n^m(k_2, h_2)]^T = A_n(\omega^{h_1 h_2 m}) \otimes A_m(\omega^{k_2 k_1 n}).$$

Proof. In (5.57) choose $\ell_1 = m^*$ and $\ell_2 = n^*$ (where, as usual, $mm^* = 1$ mod n and $nn^* = 1$ mod m) and recall (5.52). Then, by (5.19), we have

$$1 - \ell_1 m - \ell_2 n = 1 - mm^* - nn^* = 0 \quad \text{mod } nm$$

so that the twiddle factor disappears and, by Proposition 5.3.5.(iii) and Proposition 5.4.2, the right hand side in (5.57) becomes

$$[A_n(\omega^{h_1 m}) \otimes B_m^{k_1}][B_n^{h_2^*} \otimes A_m(\omega^{k_2 n})] = A_n(\omega^{h_1 h_2 m}) \otimes A_m(\omega^{k_2 k_1 n}).$$

$\square$

A special case of Theorem 5.4.9, where only elementary circulant matrices and stride permutations are used, is of particular interest.

Corollary 5.4.10 *Suppose* $\gcd(m, n) = 1$ *and let* ω *be an nm-th root of* 1. *Then*

$$Q_m^n(m^*)A_{nm}(\omega)[P_n^m Q_n^m(n^*)]^T = A_n(\omega^m) \otimes A_m(\omega^n).$$

Proof. Set $h_1 = h_2 = k_1 = k_2 = 1$ in Theorem 5.4.9, and recall Theorem 5.3.11.(i). $\square$

Theorem 5.4.11 (Generalized Winograd's Method) *With the same notation and assumption of Theorem 5.4.9, we have*

$$\Xi_m^n(h_1, k_1)A_{nm}(\omega)[\Xi_m^n(h_2, k_2)]^T = A_n(\omega^{\alpha m}) \otimes A_m(\omega^{\beta n}),$$

where $\alpha = h_1 h_2 m \mod n$ *and* $\beta = k_1 k_2 n^* \mod m$.

Proof. Using the Reverse Radix Identity (Theorem 5.3.11.(ii)), the identity in Theorem 5.4.9 becomes

$$\Xi_m^n(h_1, k_1)A_{nm}(\omega)[\Xi_m^n(h_2, k_2)]^T (B_n^m \otimes B_m^{n^*}) = A_n(\omega^{h_1 h_2 m}) \otimes A_m(\omega^{k_1 k_2 n}).$$

Multiplying both sides on the right by $(B_n^m \otimes B_m^{n^*})^{-1} = B_n^{m^*} \otimes B_m^n$ and taking into account Proposition 5.4.2, the statement follows. $\square$

From the generalized Winograd's method we deduce the following four particular cases.

Corollary 5.4.12 (Winograd's Method [168]) *Suppose* $h_1 h_2 m = 1 \mod n$ *and* $\ell_1 \ell_2 n = 1 \mod m$. *Then*

$$\Xi_m^n(h_1, \ell_1 n)A_{nm}(\omega)[\Xi_m^n(h_2, \ell_2 n)]^T = A_n(\omega^m) \otimes A_m(\omega^n).$$

Proof. Just note that now $k_1 = \ell_1 n$, $k_2 = \ell_2 n$, and $\ell_1 \ell_2 n = 1 \bmod m$, which imply that $k_1 k_2 n^* = \ell_1 \ell_2 n^2 n^* = 1 \bmod m$.   □

Corollary 5.4.13 (Good's Method [66]) *With the notation in (5.50) we have*

$$C_m^n A_{nm}(\omega)[\mathcal{R}_m^n]^T = A_n(\omega^m) \otimes A_m(\omega^n).$$

Proof. Just set $h_1 = m^*$, $k_1 = 1$, $h_2 = 1$, and $k_2 = n$ in Theorem 5.4.11, so that $\alpha = mm^* = 1 \bmod n$ and $\beta = nn^* = 1 \bmod m$.   □

Corollary 5.4.14 (Similarity Identity) *Suppose   that $\gcd(k, m) = \gcd(h, n) = 1$. Then*

$$\Xi_m^n(h, k)A_{nm}(\omega)[\Xi_m^n(h, k)]^T = A_n(\omega^{\alpha m}) \otimes A_m(\omega^{\beta n})$$

where $\alpha = h^2 m \bmod n$ and $\beta = k^2 n^* \bmod m$.

Proof. Just set $h_1 = h_2 = h$ and $k_1 = k_2 = k$ in Theorem 5.4.11.   □

A special case of Corollary 5.4.14:

Corollary 5.4.15 (Winograd's Similarity)

$$C_m^n A_{nm}(\omega)[C_m^n]^T = A_n(\omega^{mm^*}) \otimes A_m(\omega^{nn^*}).$$

Proof. Set $h = m^*$ and $k = 1$ in Corollary 5.4.14.   □

For instance, for $n = 4$, $m = 3$, and $\omega = e^{i\pi/6}$, we have $m^* = 3$, $n^* = 1$, and

$$C_3^4 A_{12}(\omega)[C_3^4]^T = A_4(\omega^9) \otimes A_3(\omega^4).$$

We end this section with a brief description of the matrix form of the so-called *Rader-Winograd algorithm.* It was developed in [125]; see also [14] and, for the computational aspects, [15, 160, 163]. We consider first the case $n = p$, a prime number. By Theorem 1.1.21, $\mathbb{F}_p^*$ is cyclic of order $p - 1$. Let $\alpha \in \mathbb{F}_p^*$ be a generator and define the permutation ξ_p of $\{0, 1, \ldots, p - 1\}$ by setting $\xi_p(0) = 0$ and $\xi_p(k) = \alpha^{k-1} \bmod p$, for $k = 1, 2, \ldots, p - 1$. Then $Q_p = Q_p(\alpha) = P_{\xi_p}$ denotes the corresponding permutation matrix, as in (5.24). If ω is a nontrivial p-th root of 1, then, by Corollary 5.3.2,

$$Q_p A_p(\omega)Q_p^T = \left(\omega^{\xi_p(i)\xi_p(j)}\right)_{i,j=0}^{p-1},$$

that is,

$$Q_p A_p(\omega)Q_p^T = \begin{pmatrix} 1 & 1 & \cdots & 1 \\ 1 & & & \\ \vdots & & C_{p-1} & \\ 1 & & & \end{pmatrix}, \tag{5.58}$$

where

$$C_{p-1} = \left(\omega^{\alpha^{i+j}}\right)_{i,j=0}^{p-2},$$

is called the *core matrix*. Note that C_{p-1} is a symmetric $(p-1) \times (p-1)$ matrix, its (i,j)-entry only depends on the sum $i+j \mod p-1$ (i.e., it is a *Hankel matrix*: each ascending (from left to right) skew-diagonal is constant, see Example 5.4.16) and its first row is $(\omega, \omega^{\alpha}, \omega^{\alpha^2}, \cdots, \omega^{\alpha^{p-2}})$. The *Rader algorithm* consists in the use of (5.58) to compute the DFT on $\mathbb{Z}_p$. Explicitly, for $Y = (y_0, y_1, \dots, y_{p-1})^T$ we set $X = (x_0, x_1, \dots, x_{p-1})^T = Q_p Y$ so that

$$A_p(\omega)Y = Q_p^T \left[Q_p A_p(\omega)Q_p^T X\right] \tag{5.59}$$

and we have

$$\left[Q_p A_p(\omega)Q_p^T X\right]_0 = \sum_{k=0}^{p-1} x_k$$
$$\left[Q_p A_p(\omega)Q_p^T X\right]_j = x_0 + \sum_{k=1}^{p-1} \omega^{\alpha^{k+j-2}} x_k \text{ for } j = 1, 2, \dots, p-1.$$

In some papers, matrix (5.58) is replaced by

$$Q_p(\alpha)A_p(\omega)Q_p(-\alpha)^T = \begin{pmatrix} 1 & 1 & \cdots & 1 \\ 1 & & & \\ \vdots & & D_p & \\ 1 & & & \end{pmatrix}$$

with $D_p = \left(\omega^{\alpha^{i-j}}\right)_{i,j=0}^{p-2}$. Then, $\left[Q_p(\alpha)A_p(\omega)Q_p(-\alpha)^T X\right]_0 = \sum_{k=0}^{p-1} x_k$ and, for $j = 1, 2, \dots, p-1$,

$$\left[Q_p(\alpha)A_p(\omega)Q_p(-\alpha)^T X\right]_j = x_0 + \sum_{k=1}^{p-1} \omega^{\alpha^{j-k}} x_k,$$

which has a convolutional form.

Example 5.4.16 (Winograd) For $p = 7$ and $\alpha = 3$ we get

$$Q_7(3)A_7(\omega)Q_7(3)^T = \begin{pmatrix} 1 & 1 & 1 & 1 & 1 & 1 & 1 \\ 1 & \omega & \omega^3 & \omega^2 & \omega^6 & \omega^4 & \omega^5 \\ 1 & \omega^3 & \omega^2 & \omega^6 & \omega^4 & \omega^5 & \omega \\ 1 & \omega^2 & \omega^6 & \omega^4 & \omega^5 & \omega & \omega^3 \\ 1 & \omega^6 & \omega^4 & \omega^5 & \omega & \omega^3 & \omega^2 \\ 1 & \omega^4 & \omega^5 & \omega & \omega^3 & \omega^2 & \omega^6 \\ 1 & \omega^5 & \omega & \omega^3 & \omega^2 & \omega^6 & \omega^4 \end{pmatrix}.$$

Exercise 5.4.17 Fill in the details in Example 5.4.16.

For $n = p^h$, with p prime and $h \geq 2$, Winograd developed a variation of the Rader algorithm. We describe it only for the case $p \geq 3$. Recall that $\mathcal{U}(\mathbb{Z}/p^h\mathbb{Z})$ is a cyclic group of order $(p - 1)p^{h-1} = p^h - p^{h-1}$ (see Theorem 1.5.8). Then we deduce the following decomposition

$$\mathbb{Z}/p^h\mathbb{Z} = \{0\} \coprod \coprod_{j=1}^{h} p^{h-j}\mathcal{U}(\mathbb{Z}/p^j\mathbb{Z}). \tag{5.60}$$

Indeed, for $j = 1, 2, \ldots, h$ we have that $x \in \mathbb{Z}/p^j\mathbb{Z}$ is not invertible if and only if it is divisible by p, so that we have

$$\mathbb{Z}/p^j\mathbb{Z} = p(\mathbb{Z}/p^{j-1}\mathbb{Z}) \coprod \mathcal{U}(\mathbb{Z}/p^j\mathbb{Z}).$$

By iterating this relation we get (5.60). Fix a generator α_j of $\mathcal{U}(\mathbb{Z}/p^j\mathbb{Z})$, for $j = 1, 2, \ldots, h$. Using (5.60), we define a permutation ξ_{p^h} of $\{0, 1, \ldots, p^h - 1\}$ by setting $\xi_{p^h}(0) = 0$ and

$$\xi_{p^h}(k) = \alpha_j^{k-p^{j-1}} p^{h-j} \mod p^h$$

for $p^{j-1} \leq k \leq p^j - 1$ and $j = 1, 2, \ldots, h$. In other words, ξ_{p^h} maps the set $\{p^{j-1}, p^{j-1} + 1, \ldots, p^j - 1\}$ bijectively onto $p^{h-j}\mathcal{U}(\mathbb{Z}/p^j\mathbb{Z})$ for all $j = 1, 2, \ldots, h$. We then set

$$Q_{p^h} = P_{\xi_{p^h}}.$$

The matrix form of Winograd's generalization of the Rader algorithm is obtained as in (5.59) by applying

$$Q_{p^h} A_{p^h}(\omega) Q_{p^h}^T$$

with ω a p^h-th root of 1. The above matrix is symmetric, but no longer Hankel (though it is made up of blocks consisting of Hankel matrices; see Example (5.4.18) below).

Example 5.4.18 (Winograd) For $p = 3$, $h = 2$, $\alpha_1 = 2$, and $\alpha_2 = 2$ we get

$$\mathbb{Z}/9\mathbb{Z} = \{0\} \coprod 3\mathcal{U}(\mathbb{Z}/3\mathbb{Z}) \coprod \mathcal{U}(\mathbb{Z}/9\mathbb{Z}) = \{0\} \coprod \{3, 6\} \coprod \{1, 2, 4, 5, 7, 8\}$$

so that

$$\xi_9(0) = 0 \; \xi_9(1) = 3 \; \xi_9(2) = 6 \; \xi_9(3) = 1 \; \xi_9(4) = 2$$
$$\xi_9(5) = 4 \; \xi_9(6) = 8 \; \xi_9(7) = 7 \; \xi_9(8) = 5$$

and

$$Q_9 A_9(\omega) Q_9^T = \begin{pmatrix} 1 & 1 & 1 & 1 & 1 & 1 & 1 & 1 & 1 \\ 1 & 1 & 1 & \omega^3 & \omega^6 & \omega^3 & \omega^6 & \omega^3 & \omega^6 \\ 1 & 1 & 1 & \omega^6 & \omega^3 & \omega^6 & \omega^3 & \omega^6 & \omega^3 \\ 1 & \omega^3 & \omega^6 & \omega & \omega^2 & \omega^4 & \omega^8 & \omega^7 & \omega^5 \\ 1 & \omega^6 & \omega^3 & \omega^2 & \omega^4 & \omega^8 & \omega^7 & \omega^5 & \omega \\ 1 & \omega^3 & \omega^6 & \omega^4 & \omega^8 & \omega^7 & \omega^5 & \omega & \omega^2 \\ 1 & \omega^6 & \omega^3 & \omega^8 & \omega^7 & \omega^5 & \omega & \omega^2 & \omega^4 \\ 1 & \omega^3 & \omega^6 & \omega^7 & \omega^5 & \omega & \omega^2 & \omega^4 & \omega^8 \\ 1 & \omega^6 & \omega^3 & \omega^5 & \omega & \omega^2 & \omega^4 & \omega^8 & \omega^7 \end{pmatrix}.$$

Exercise 5.4.19 Fill in the details of the above example and show that the matrix is made up of the multiplication tables of the following three groups (written multiplicatively): the trivial group, $\mathcal{U}(\mathbb{Z}/3\mathbb{Z})$, and $\mathcal{U}(\mathbb{Z}/9\mathbb{Z})$.

Extensions of Rader's algorithm will be discussed in Section 7.8.

5.5 Algorithmic aspects of the FFT

In this section we examine some of the algorithmic aspects of the formulas obtained in Section 5.4. For a more complete discussion we refer to [160, 163].

First of all, we want to derive the general form of (5.4), which is also the basic nonmatrix form of the Cooley-Tukey algorithm. We consider the action of $A_{nm}(\omega)$ to a column vector $X = (x_0, x_1, \ldots, x_{nm-1})^T$. The General Radix Identity (Theorem 5.4.5) yields

$$A_{nm}(\omega) = [A_n(\omega^m) \otimes I_m] T_n^m(\omega) [I_n \otimes A_m(\omega^n)] P_n^m. \tag{5.61}$$

Therefore, arguing as in the proof of Theorem 5.4.5, and using the formulas established therein, from (5.61), for $j = sm + r$ and $j' = r'n + s'$, with $0 \leq s, s' \leq n - 1$ and $0 \leq r, r' \leq m - 1$, we get (by Lemma 5.3.1.(ii) and (5.29))

$$[A_{nm}(\omega)X]_j = \sum_{j'=0}^{nm-1} \{[A_n(\omega^m) \otimes I_m] T_n^m(\omega)[I_n \otimes A_m(\omega^n)]\}_{j,\sigma(n,m)j'} x_{j'}$$

$$= \sum_{r'=0}^{m-1} \sum_{s'=0}^{n-1} \{[A_n(\omega^m) \otimes I_m]$$

$$\cdot T_n^m(\omega)[I_n \otimes A_m(\omega^n)]\}_{sm+r,s'm+r'} x_{r'n+s'}$$

$$=_* \sum_{r'=0}^{m-1} \sum_{s'=0}^{n-1} \omega^{mss'+s'r+nrr'} x_{r'n+s'}$$

(where $=_*$ follows from the last equality in the first part of the proof of Theorem 5.4.5), that is,

$$[A_{nm}(\omega)X]_{sm+r} = \sum_{s'=0}^{n-1} \omega^{mss'} \omega^{s'r} \sum_{r'=0}^{m-1} \omega^{nrr'} x_{r'n+s'}. \tag{5.62}$$

The above is the nonmatrix form of the General Radix Identity and constitutes one of the basic formulations of the Cooley-Tukey algorithm.

Exercise 5.5.1 (5.61) is also called the *Decimation in time form of the Cooley-Tukey algorithm*. Prove the following equivalent formulas:

- (Decimation in Frequency)

$$A_{nm}(\omega) = P_m^n[I_n \otimes A_m(\omega^n)]T_n^m(\omega)[A_n(\omega^m) \otimes I_m];$$

- (Parallel Form)

$$A_{nm}(\omega) = P_n^m[I_m \otimes A_n(\omega^m)]P_m^nT_n^m(\omega)[I_n \otimes A_m(\omega^n)]P_n^m;$$

- (Vector Form)

$$A_{nm}(\omega) = [A_n(\omega^m) \otimes I_m]T_n^m(\omega)P_n^m[A_m(\omega^n) \otimes I_n].$$

Now, following [130], we examine the number of operations needed to compute the DFT by means of the General Radix Identity in Theorem 5.4.5 or, equivalently, in terms of (5.62). This way, we generalize the computation in Section 5.1. For the sake of clarity, we shall denote by $X^{(n)}$ (respectively $X^{(nm)}$) the vector $(x_0, x_1, \ldots, x_{n-1})^T$ (respectively $(x_0, x_1, \ldots, x_{nm-1})^T$). First of all, arguing as in the derivation of (5.2), we deduce that the n entries of the column matrix $A_n(\omega)X^{(n)}$ may be computed by means of at most

$$T_1(n) = [n + (n-1)]n + n - 2 = 2n^2 - 2 = \mathcal{O}(n^2) \tag{5.63}$$

operations.

Proposition 5.5.2 *Suppose we have an algorithm that computes $A_n(\omega)X^{(n)}$ (ω an n-th root of 1) by means of at most $T(n)$ operations. Then we can compute $A_{nm}(\omega)X^{(nm)}$ (ω an nm-th root of 1) by means of at most*

$$T(nm) \leq nT(m) + mT(n) + (m-1)(n-1)$$

operations.

Proof. Indeed, if we use (5.62), we need to compute

$$\sum_{r'=0}^{m-1} \omega^{nrr'} x_{nr'+s'} \quad \text{for } 0 \le r \le m-1 \text{ and } 0 \le s' \le n-1$$

and these may be seen as n DFT's with $A_m(\omega^n)$, namely,

$$A_m(\omega^n) X_{s'}^{(m)}$$

with

$$X_{s'}^{(m)} = (x_{s'}, x_{n+s'}, x_{2n+s'}, \dots, x_{(m-1)n+s'})^T$$

and $s' = 0, 1, \dots, n-1$. Then we must multiply these results by the numbers $\omega^{s'r}$ (note that, in general, only $(n-1)(m-1)$ of them are different from 1). Finally, we need to compute the external sum in (5.62) for $0 \le s \le n-1$ and $0 \le r \le m-1$, which, as before, may be seen as m DFTs with $A_n(\omega^m)$. $\quad\square$

For instance, from Proposition 5.5.2 and using (5.63), we get

$$T(nm) \le m \cdot 2(n^2 - 1) + n \cdot 2(m^2 - 1) + (n-1)(m-1)$$
$$= 2nm(n+m) + nm - 3(n+m) + 1.$$

This is a great improvement: if $n = m$ then $T_1(n^2) \sim 2n^4$ while $T(n^2) \sim 4n^3$.

Theorem 5.5.3 *Let M be a positive integer and let $M = m_1 m_2 \cdots m_k$ be a non-trivial factorization. Suppose that $T(m_j)$ operation are needed to compute the DFT with A_{m_j}. Then one can compute the DFT with A_M by means of at most*

$$T(M) \le \left\{ M \sum_{j=1}^{k} \frac{1}{m_j} [T(m_j) + m_j - 1] \right\} - M + 1$$

operations. Moreover, $T(M)$ does not depend on the order of the factors used in the factorization.

Proof. We deduce this from the General Radix Identity as in Proposition 5.5.2 by using induction on k. For $k = 2$ the theorem reduces to Proposition 5.5.2. Assume the result for $2 \le h \le k-1$ and let us set $m = m_1 m_2 \cdots m_\ell$ and

$n = m_{\ell+1} \cdots m_k$ for some $2 \le \ell \le k - 2$. Then $M = nm$ and

$$T(M) = T(nm)$$

(by Proposition 5.5.2) $\le nT(m) + mT(n) + (mn - n - m + 1)$

(by inductive hypothesis) $\le n \left(m \sum_{j=1}^{\ell} \frac{1}{m_j} [T(m_j) + m_j - 1] - m + 1 \right)$

$$+ m \left(n \sum_{j=\ell+1}^{k} \frac{1}{m_j} [T(m_j) + m_j - 1] - n + 1 \right)$$

$$+ mn - n - m + 1$$

$$= M \left\{ \sum_{j=1}^{k} \frac{1}{m_j} [T(m_j) + m_j - 1] \right\} - M + 1.$$

□

Some special cases of Theorem 5.5.3 are worth examining.

Corollary 5.5.4

$$T(M) \le M \sum_{j=1}^{k} (2m_j + 1) - M + 1.$$

Proof. This follows from Theorem 5.5.3 by using (5.63) and the elementary inequality $\frac{(2m+3)(m-1)}{m} \le 2m + 1$.   □

If $m_1 = m_2 = \cdots = m_k = m$, that is, $M = m^k$, we get the following generalization of Theorem 5.1.3.

Corollary 5.5.5

$$T(m^k) \le (2m + 1)m^k k - m^k + 1.$$

In particular, for m fixed and $k \to +\infty$, one gets

$$T(m^k) = \mathcal{O}(km^k),$$

equivalently, $T(M) = \mathcal{O}(M \log M)$.

Part II
Finite fields and their characters

6

Finite fields

This chapter is a self-contained introduction to the basic algebraic theory of finite fields. This includes a complete study of the automorphisms, norms, traces, and quadratic extensions of finite fields. Our treatment is inspired by a course given by Giuseppe Tallini in 1991 at the Istituto Nazionale di Alta Matematica "Francesco Severi" (INdAM) in Rome (cf. [141]). An alternative approach is in the monograph by Lidl and Niederreiter [96]. We also refer to the impressive volumes by Knapp [87, 88] for a very complete treatment at both a basic and an advanced level.

6.1 Preliminaries on ring theory

We start by recalling some basic notions and results in ring theory. Most of the proofs are elementary and left as exercises: we refer to the monographs by Herstein [71] and Lang [93] for more details. We also assume the most elementary facts on polynomials over a field: a good reference is the book by Kurosh [89].

Let $\mathcal{A}$ be a *commutative unital ring*. We denote by 0 the *zero* and by 1 the *(multiplicative) identity element* of $\mathcal{A}$.

$\mathcal{A}$ is said to be an *integral domain* if it contains no *zero divisors*, that is, if $a, b \in \mathcal{A}$ satisfy $ab = 0$ then $a = 0$ or $b = 0$.

An *ideal* of $\mathcal{A}$ is a subring $\mathcal{I} \subseteq \mathcal{A}$ such that $ai \in \mathcal{I}$ for all $a \in \mathcal{A}$ and $i \in \mathcal{I}$. Viewing $\mathcal{I}$ as a subgroup of the additive group $\mathcal{A}$, we can form the quotient group $\mathcal{A}/\mathcal{I} = \{(a + \mathcal{I}) : a \in A\}$ and then equip it with the multiplication defined by $(a + \mathcal{I})(b + \mathcal{I}) = (ab + \mathcal{I})$ for all $a, b \in A$. It is easy to check that this multiplication is well defined and that $\mathcal{A}/\mathcal{I}$ is a commutative unital ring, called the *quotient ring*: its zero is $(0 + \mathcal{I}) = \mathcal{I}$ and its unit element is $(1 + \mathcal{I})$.

An element $u \in \mathcal{A}$ is called *invertible*, or a *unit*, provided there exists an element $v \in \mathcal{A}$, necessarily unique, called the *inverse* of u, such that $uv = 1$.

A *field* is a commutative unital ring such that every nonzero element is invertible. In the sequel, we shall denote a field by the letters $\mathbb{F}$ and $\mathbb{E}$.

Exercise 6.1.1 Show that every finite integral domain is a field.
Hint. Use the pigeon-hole principle.

We denote by $\mathcal{A}[x]$ the commutative unital ring consisting of all polynomials

$$p(x) = a_n x^n + a_{n-1} x^{n-1} + \cdots + a_1 x + a_0 \tag{6.1}$$

with coefficients $a_0, a_1, \ldots, a_n$ in $\mathcal{A}$ in the indeterminate x. In (6.1) we implicitly assume that $a_n \neq 0$ and then denote by $\deg p = n$ the *degree* of the polynomial $p(x)$. If $a_n = 1$ one says that the polynomial $p(x)$ is *monic*.

Clearly, if $\mathcal{A}$ is an integral domain, so is $\mathcal{A}[x]$.

An ideal $\mathcal{I}$ in $\mathcal{A}$ is called *principal* provided there exists $a \in A$ such that $\mathcal{I} = a\mathcal{A} = \{ab : b \in \mathcal{A}\}$ and one then says that $\mathcal{I}$ is *generated* by a. A *principal ideal domain* is an integral domain in which every ideal is principal.

Exercise 6.1.2 Let $\mathcal{A}$ be an integral domain and let $a, b \in \mathcal{A}$. Suppose that the ideal $\mathcal{I} = \{xa + yb : x, y \in \mathcal{A}\}$ is principal. Show that every generator of $\mathcal{I}$ is a $\gcd(a, b)$ (the definition of a gcd in $\mathcal{A}$ is the same as in Theorem 1.1.1).

Exercise 6.1.3 Show that in a principal ideal domain any nondecreasing chain of ideals $\mathcal{I}_1 \subseteq \mathcal{I}_2 \subseteq \cdots \subseteq \mathcal{I}_n \subseteq \cdots$ must stabilize, that is, there exists $n_0 \in \mathbb{N}$ such that $I_n = I_{n_0}$ for all $n \geq n_0$.

Example 6.1.4 The ring $\mathbb{Z}$ of integers is a principal ideal domain. Let us show that if $\mathcal{I} \subseteq \mathbb{Z}$ is an ideal, then the *minimal primitive element* $a = \min\{i \in \mathcal{I} : i > 0\}$ generates $\mathcal{I}$. Indeed, given $m \in \mathcal{I}$, by Euclidean division we can find (unique) $q \in \mathbb{Z}$ and $r \in \mathbb{Z}$ such that $0 \leq r < a$ and $m = aq + r$. Since $r = m - aq \in \mathcal{I}$, by minimality of a we deduce that $r = 0$, showing that $m = aq$. Thus $\mathcal{I} = a\mathcal{A}$.

Exercise 6.1.5 Show that the integral domain $\mathbb{Z}[x]$ is *not* a principal ideal domain.
Hint. Show that the ideal generated by 2 and x cannot be generated by a single polynomial.

We recall that in the ring $\mathbb{F}[x]$ of all polynomials over a field $\mathbb{F}$ an analogue of (1.1) holds. This is the *Euclidean division of polynomials*: for $p, s \in \mathbb{F}[x]$ there exist unique $q, r \in \mathbb{F}[x]$ such that $p(x) = q(x)s(x) + r(x)$ and $0 \leq \deg r < \deg s$.

Exercise 6.1.6 Let $\mathbb{F}$ be a field. Show that $\mathbb{F}[x]$ is a principal ideal domain.
Hint. Use Euclidean division of polynomials.

Suppose that $\mathcal{A}$ is an integral domain. A nonzero noninvertible element $p \in \mathcal{A}$ is said to be *irreducible* if it cannot be expressed as a product $p = ab$ with $a, b \in \mathcal{A}$ noninvertible.

Exercise 6.1.7 Let $\mathcal{A}$ be a principal ideal domain and let $a, b, p \in \mathcal{A}$. Show that if p is irreducible and $p|ab$, then $p|a$ or $p|b$.
Hint. Use Exercise 6.1.2.

Example 6.1.8

(1) In the ring of integers, an element $p \in \mathbb{Z}$ is irreducible if and only if its absolute value $|p| \in \mathbb{N}$ is a prime number.

(2) If $\mathbb{F}$ is a field, then a polynomial $p(x) \in \mathbb{F}[x]$ is irreducible if and only if it is irreducible over $\mathbb{F}$ (in the usual sense of elementary algebra).

One then says that an integral domain $\mathcal{A}$ is a *unique factorization domain* (briefly, *UFD*) provided that every nonzero non-unit $a \in A$ can be written as a product $a = up_1 p_2 \cdots p_k$ of a unit $u \in \mathcal{A}$ and irreducible elements $p_1, p_2, \ldots, p_k \in \mathcal{A}$, and this factorization is *unique* in the following sense: if $a = vq_1 q_2 \cdots q_h$ is another factorization, with v a unit and $q_1, q_2, \ldots, q_h$ irreducible, then $h = k$ and, up to reordering the factors, $q_j = w_j p_j$, with w_j a unit, for all $j = 1, 2, \ldots, k$ (and therefore $v = u(w_1 w_2 \cdots w_k)^{-1}$).

Exercise 6.1.9 Show that every principal ideal domain is UFD.
Hint. For the existence of a factorization, consider the set B of all ideals of $\mathcal{A}$, whose generators do not admit factorization and use Exercise 6.1.3. For the uniqueness use Exercise 6.1.7.

Example 6.1.10

(1) $\mathbb{Z}$ is a UFD: every $n \in \mathbb{Z}$ can be written (uniquely) as a product

$$n = \varepsilon p_1^{\alpha_1} p_2^{\alpha_2} \cdots p_k^{\alpha_k}$$

where $\varepsilon \in \{1, -1\}$ and $p_1, p_2, \ldots, p_k \in \mathbb{N}$ are distinct prime numbers (the positive integers α_i's are the corresponding multiplicities).

(2) If $\mathbb{F}$ is a field, then $\mathbb{F}[x]$ is a UFD: every polynomial $p(x) \in \mathbb{F}[x]$ can be written (uniquely) as a product

$$p(x) = u p_1(x)^{\alpha_1} p_2(x)^{\alpha_2} \cdots p_k(x)^{\alpha_k}$$

where $u \in \mathbb{F}$ and $p_1(x), p_2(x), \ldots, p_k(x) \in \mathbb{F}[x]$ are distinct, monic, irreducible polynomials (the positive integers α_i's are the corresponding multiplicities).

A proper ideal $\mathcal{I} \subset \mathcal{A}$ is *maximal* if the following holds: whenever $\mathcal{I} \subseteq \mathcal{J} \subseteq \mathcal{A}$, where $\mathcal{J}$ is also an ideal, we necessarily have either $\mathcal{I} = \mathcal{J}$ or $\mathcal{J} = \mathcal{A}$.

Proposition 6.1.11 *Let $\mathcal{A}$ be a unital ring and $\mathcal{I} \subset \mathcal{A}$ an ideal. Then the quotient ring $\mathcal{A}/\mathcal{I}$ is a field if and only if $\mathcal{I}$ is maximal.*

Proof. Suppose that $\mathcal{I}$ is maximal. Let $a \in \mathcal{A} \setminus \mathcal{I}$ and let us show that the nonzero element $(a + \mathcal{I})$ of $\mathcal{A}/\mathcal{I}$ is a unit. Denote by $\mathcal{H} \subset \mathcal{A}/\mathcal{I}$ the ideal generated by $(a + \mathcal{I})$. Then if we denote by $\pi : \mathcal{A} \to \mathcal{A}/\mathcal{I}$ the canonical quotient homomorphism, we have that $\mathcal{J} = \pi^{-1}(\mathcal{H})$ is an ideal in $\mathcal{A}$, which contains $\mathcal{I}$ and a, so that $\mathcal{I} \subsetneq \mathcal{J}$. By maximality of $\mathcal{I}$ we have $\pi^{-1}(\mathcal{H}) = \mathcal{J} = \mathcal{A}$. Since $\mathcal{H}$ is generated by $(a + \mathcal{I})$, we can find $b \in \mathcal{A}$ such that $(1 + \mathcal{I}) = (a + \mathcal{I})(b + \mathcal{I})$ in $\mathcal{H}$. Thus $(b + \mathcal{I})$ is the inverse of $(a + \mathcal{I})$ in $\mathcal{A}/\mathcal{I}$. This shows that $\mathcal{A}/\mathcal{I}$ is a field.

Conversely, suppose that $\mathcal{A}/\mathcal{I}$ is a field. Let $\mathcal{J}$ be an ideal of $\mathcal{A}$ such that $\mathcal{I} \subsetneq \mathcal{J} \subseteq \mathcal{A}$. Let us show that $\mathcal{J} = \mathcal{A}$. Let $b \in \mathcal{J} \setminus \mathcal{I}$. Then $(b + \mathcal{I})$ is a nonzero element in $\mathcal{A}/\mathcal{I}$ and therefore we can find $a \in \mathcal{A}$ such that $(a + \mathcal{I})(b + \mathcal{I}) = (1 + \mathcal{I})$. It follows that

$$1 \in (ab + \mathcal{I}) \subseteq a\mathcal{J} + \mathcal{J} = \mathcal{J},$$

so that $\mathcal{J} = \mathcal{A}$. This shows that $\mathcal{I}$ is maximal. □

Proposition 6.1.12 *Let $\mathcal{A}$ be a principal ideal domain. If $a \in \mathcal{A}$ is a nonzero element, then the (principal) ideal $a\mathcal{A}$ generated by a is maximal if and only if a is irreducible.*

Proof. Suppose that a is not irreducible. Then we can find noninvertible elements $b, c \in \mathcal{A}$ such that $a = bc$. Let us show that $a\mathcal{A} \subsetneq b\mathcal{A} \subsetneq \mathcal{A}$. Indeed, if we had $b\mathcal{A} = \mathcal{A}$ we could find an element $b' \in \mathcal{A}$ such that $bb' = 1$, contradicting the fact that b is not invertible. On the other hand, if $a\mathcal{A} = b\mathcal{A}$ then $b \in a\mathcal{A}$ and we would find $d \in \mathcal{A}$ such that $b = ad$. As a consequence, $a = bc = adc$ yielding $a(1 - dc) = 0$. Since $\mathcal{A}$ is an integral domain and $a \neq 0$, we necessarily have $1 - dc = 0$, equivalently $dc = 1$, contradicting the fact that c is not invertible. This shows that the proper ideal $a\mathcal{A}$ is not maximal.

Conversely, suppose that a is irreducible and let us show that $a\mathcal{A}$ is a maximal ideal. Thus suppose that $\mathcal{J}$ is an ideal such that $a\mathcal{A} \subseteq \mathcal{J} \subseteq \mathcal{A}$. Since $\mathcal{A}$ is a principal ideal domain, we can find $b \in \mathcal{A}$ such that $\mathcal{J} = b\mathcal{A}$. Since $a \in b\mathcal{A}$ we can then find $c \in \mathcal{A}$ such that $a = bc$. By irreducibility of a, one of the two elements $b, c \in \mathcal{A}$ must be invertible. If b is invertible then $1 \in \mathcal{J}$ so that $\mathcal{J} = \mathcal{A}$. If c is invertible, then $b = ac^{-1} \in a\mathcal{A}$ so that $\mathcal{J} = b\mathcal{A} = a\mathcal{A}$. It follows that $a\mathcal{A}$ is a maximal ideal. □

Corollary 6.1.13 *Let* $n \in \mathbb{N}$. *Then the quotient ring* $\mathbb{Z}/n\mathbb{Z}$ *is a field if and only if* n *is a prime number.*

Recall that, for $p \in \mathbb{N}$ a prime number, we denote by $\mathbb{F}_p$ the field $\mathbb{Z}/p\mathbb{Z}$ (see Notation 1.1.17).

Corollary 6.1.14 *Let* $\mathbb{F}$ *be a field and* $p(x) \in \mathbb{F}[x]$. *Then the quotient ring* $\mathbb{F}[x]/p(x)\mathbb{F}[x]$ *is a field if and only if* $p(x)$ *is irreducible (over* $\mathbb{F}$).

Let $\mathbb{F}$ be a field. Consider the cyclic additive subgroup C generated by the identity element $1 \in \mathbb{F}$. The *characteristic* of $\mathbb{F}$, denoted char($\mathbb{F}$), is defined to be 0 if C is infinite (and therefore isomorphic to $\mathbb{Z}$) and equal to the cardinality of C otherwise. Let us show that in this last case char($\mathbb{F}$) is a prime number. Consider the map $\Phi \colon \mathbb{Z} \to \mathbb{F}$ defined by

$$\Phi(\pm n) = \pm(\underbrace{1 + 1 + \cdots + 1}_{n \text{ terms}}) \tag{6.2}$$

for all $n \in \mathbb{N}$. Then it is straightforward to see that Φ is a unital ring homomorphism, so that $\mathbb{Z}/\mathrm{Ker}(\Phi) \cong \Phi(\mathbb{Z}) = C$. If $\mathrm{Ker}(\Phi) = \{0\}$ then char($\mathbb{F}$) $= 0$. Otherwise, $\Phi(\mathbb{Z}) \subseteq \mathbb{F}$, being a finite integral domain is a field (cf. Exercise 6.1.1) and therefore, by Corollary 6.1.13, $\mathrm{Ker}(\Phi) = p\mathbb{Z}$ for some prime number p, so that char($\mathbb{F}$) $= p$.

6.2 Finite algebraic extensions

We now give a basic introduction to field extensions. More complete treatments can be found in the aforementioned monographs by Herstein [71], Lang [93], and Knapp [87, 88].

Let $\mathbb{F}$ and $\mathbb{E}$ be two fields and suppose that $\mathbb{F} \subseteq \mathbb{E}$. We say that $\mathbb{F}$ is a *subfield* of $\mathbb{E}$ or, equivalently, that $\mathbb{E}$ is an *extension* of $\mathbb{F}$.

Exercise 6.2.1 Show that $\mathbb{E}$ is a vector space over $\mathbb{F}$.

We denote by $[\mathbb{E} : \mathbb{F}]$ the corresponding dimension $\dim_{\mathbb{F}} \mathbb{E}$ (the cardinality of one (=any) vector basis of $\mathbb{E}$ over $\mathbb{F}$): it is called the *degree* of the extension. We say that $\mathbb{E}$ is a *finite* (resp. *infinite*) extension of $\mathbb{F}$ provided that $[\mathbb{E} : \mathbb{F}] < \infty$ (resp. $[\mathbb{E} : \mathbb{F}]$ is infinite).

An element $\alpha \in \mathbb{E}$ is called *algebraic* over $\mathbb{F}$ (or $\mathbb{F}$-*algebraic*) if there exists $p(x) \in \mathbb{F}[x]$ such that $p(\alpha) = 0$.

Let $\alpha \in \mathbb{E}$ be an $\mathbb{F}$-algebraic element. Then it is straightforward to check that the set $\mathcal{I}_\alpha = \{p \in \mathbb{F}[x] : p(\alpha) = 0\}$ is an ideal in $\mathbb{F}[x]$. It follows from

Exercise 6.1.6 that there exists a monic polynomial $q \in \mathbb{F}[x]$ such that $\mathcal{I}_\alpha$ is generated by q, i.e. $\mathcal{I}_\alpha = q(x)\mathbb{F}[x]$.

Exercise 6.2.2 Show that the monic polynomial $q \in \mathbb{F}[x]$ is unique and irreducible.

The polynomial q is called the *minimal polynomial* of α (over $\mathbb{F}$). It follows from Corollary 6.1.14 that $\mathbb{F}[x]/q(x)\mathbb{F}[x]$ is a field. On the other hand, consider the map

$$\Phi\colon\ \mathbb{F}[x] \to\ \mathbb{E}$$
$$p \mapsto p(\alpha).$$

We clearly have $\mathrm{Ker}(\Phi) = \mathcal{I}_\alpha = q(x)\mathbb{F}[x]$ and therefore $\mathbb{F}[x]/q(x)\mathbb{F}[x] = \mathbb{F}[x]/\mathrm{Ker}(\Phi)$ is isomorphic to the image $\mathrm{Im}(\Phi)$, which is a subfield of $\mathbb{E}$ containing α, denoted $\mathbb{F}[\alpha]$. We say that $\mathbb{F}[\alpha]$ is the subfield of $\mathbb{E}$ obtained by *adjoining* α to $\mathbb{F}$.

Exercise 6.2.3 Show that $\mathbb{F}[\alpha]$ is the subfield of $\mathbb{E}$ generated by $\mathbb{F}$ and α (that is, $\mathbb{F}[\alpha]$ is the intersection of all subfields of $\mathbb{E}$ containing $\mathbb{F}$ and α).

Proposition 6.2.4 *Let $\mathbb{E}$ be an extension of $\mathbb{F}$. Suppose $[\mathbb{E} : \mathbb{F}] < \infty$. Then every $\alpha \in \mathbb{E}$ is algebraic over $\mathbb{F}$.*

Proof. Let $\alpha \in \mathbb{E}$ and set $n = [\mathbb{E} : \mathbb{F}] = \dim_\mathbb{F}\mathbb{E}$. Then the $n + 1$ elements $1, \alpha, \alpha^2, \ldots, \alpha^n$ are linearly dependent over $\mathbb{F}$. It follows that there exists $a_0, a_1, \ldots, a_n \in \mathbb{F}$ such that $(a_0, a_1, \ldots, a_n) \neq (0, 0, \ldots, 0)$ and $a_0 + a_1\alpha + \cdots + a_n\alpha^n = 0$. Then the polynomial $q(x) = a_nx^n + \cdots + a_1x + a_0 \in \mathbb{F}[x]$ satisfies $q(\alpha) = 0$. This shows that α is algebraic over $\mathbb{F}$. $\qquad\square$

Proposition 6.2.5 *Let $\mathbb{E}$ be an extension of $\mathbb{F}$ and $\alpha \in \mathbb{E}$. Suppose that α is algebraic over $\mathbb{F}$ and denote by $q(x) \in \mathbb{F}[x]$ its minimal polynomial. Then setting $n = \deg(q)$ the following holds:*

   (i) *$\{1, \alpha, \alpha^2, \ldots, \alpha^{n-1}\}$ is a basis of $\mathbb{F}[\alpha]$ over $\mathbb{F}$;*
   (ii) *$\dim_\mathbb{F}\mathbb{F}[\alpha] = n$;*
   (iii) *$\mathbb{F}[\alpha] \cong \mathbb{F}[x]/q(x)\mathbb{F}[x]$.*

Moreover, let $\beta \in \mathbb{E}$ and suppose that $q(\beta) = 0$. Then the following holds:

   (iv) *β is algebraic over $\mathbb{F}$ and $q(x)$ is the minimal polynomial of β;*
   (v) *$\dim_\mathbb{F}\mathbb{F}[\beta] = n$;*
   (vi) *$\mathbb{F}[\alpha] \cong \mathbb{F}[\beta]$;*
   (vii) *if $\beta \in \mathbb{F}[\alpha]$ then $\mathbb{F}[\alpha] = \mathbb{F}[\beta]$.*

Proof. Let $q(x) = x^n + a_{n-1}x^{n-1} + \cdots + a_1 x + a_0$ and observe that $a_0 \neq 0$ by irreducibility (cf. Exercise 6.2.2). Since $q(\alpha) = 0$, we deduce that $\alpha^n = -(a_{n-1}\alpha^{n-1} + \cdots + a_1\alpha + a_0)$. After multiplying both sides by α^{m-n} we deduce that, more generally,

$$\alpha^m = -(a_{n-1}\alpha^{m-1} + \cdots + a_1\alpha^{m-n+1} + a_0\alpha^{m-n}) \tag{6.3}$$

for all $m \geq n$. Similarly, after multiplying the equation $q(\alpha) = 0$ by α^{-1}, we deduce that $\alpha^{-1} = -\frac{1}{a_0}(\alpha^{n-1} + a_{n-1}\alpha^{n-2} + \cdots + a_2\alpha + a_1)$ and, more generally,

$$\alpha^{-m} = -\frac{1}{a_0}(\alpha^{n-m} + a_{n-1}\alpha^{n-m-1} + \cdots + a_2\alpha^{2-m} + a_1\alpha^{1-m}) \tag{6.4}$$

for all $m \geq 1$. This shows that the n elements $1, \alpha, \alpha^2, \ldots, \alpha^{n-1}$ span $\mathbb{F}[\alpha]$ (recall Exercise (6.2.3)). Since $n = \deg(q)$ and q is the minimal polynomial of α, the above elements are also linearly independent and therefore constitute a basis for $\mathbb{F}[\alpha]$ over $\mathbb{F}$. This shows (i), and (ii) follows immediately thereafter. (iii) was observed when defining $\mathbb{F}[\alpha]$. (iv) follows from the obvious fact that every irreducible polynomial is the minimal polynomial of any of its roots. From this we deduce that the same relations (6.3) and (6.4) hold with α replaced by β, thus proving (v), while (vi) follows from (iii). Finally, suppose that $\beta \in \mathbb{F}[\alpha]$. Then $\mathbb{F}[\beta] = \{p(\beta) : p \in \mathbb{F}[x]\}$ is a subfield of $\mathbb{F}[\alpha]$ and, from (ii) and (v), we immediately deduce (vii). $\qquad\square$

Remark 6.2.6 With the above notation, one can also say that $\mathbb{F}[\alpha]$ is obtained from $\mathbb{F}$ by *adjoining a root* of (the irreducible polynomial) q. In a more abstract fashion, if q is any irreducible polynomial in $\mathbb{F}[x]$, then the field $\mathbb{F}[x]/q(x)\mathbb{F}[x]$ contains a subfield isomorphic to $\mathbb{F}$ (that we shall still denote by $\mathbb{F}$), namely the set of all elements of the form $a_0 + q(x)\mathbb{F}[x]$, where $a_0 \in \mathbb{F}$ is viewed as a polynomial of degree 0. Then the element $\alpha = x + q(x)\mathbb{F}[x] \in \mathbb{F}[x]/q(x)\mathbb{F}[x]$ is algebraic over $\mathbb{F}$: indeed, $q(\alpha) = q(x + q(x)\mathbb{F}[x]) = q(x) + q(x)\mathbb{F}[x] = 0 + q(x)\mathbb{F}[x] = 0$. As a consequence, $\mathbb{F}[x]/q(x)\mathbb{F}[x]$ is the *algebraic extension* of $\mathbb{F}$ by means of the (irreducible) polynomial $q(x)$.

When $\deg(q) = 2$ we call it a *quadratic extension*.

Example 6.2.7 The field $\mathbb{C} = \{a + ib : a, b \in \mathbb{R}\}$ of complex numbers is a quadratic extension of the field $\mathbb{R}$ of real numbers. The corresponding irreducible polynomial is $q(x) = x^2 + 1$.

Definition 6.2.8 Let $p(x) \in \mathbb{F}[x]$, say of degree $\deg(p) = n$. Then the smallest (= of minimal degree) field extension $\mathbb{E}$ of $\mathbb{F}$ containing elements $\alpha_1, \alpha_2, \ldots, \alpha_n$

such that $p(x) = (x - \alpha_1)(x - \alpha_2) \cdots (x - \alpha_n)$ is called a *splitting field* for the polynomial $p(x)$ *over* $\mathbb{F}$.

Exercise 6.2.9 (Existence and uniqueness of splitting fields)

(1) Prove that, in the above definition, the field $\mathbb{E}$ exists and is unique up to isomorphism.

Hint: existence is obtained by a repeated application of the constructions that have led to Proposition 6.2.5. Uniqueness is more difficult (we refer to the aforementioned references).

(2) Prove that, if p is irreducible (over $\mathbb{F}$), then $[\mathbb{E} : \mathbb{F}]$ divides $n!$, where $n = \deg(p)$.

Remark 6.2.10 Let $\mathbb{F} \subseteq \mathbb{G} \subseteq \mathbb{E}$ be fields and let $p(x) \in \mathbb{F}[x]$ (so that $p(x) \in \mathbb{G}[x]$). Then $\mathbb{E}$ is the splitting field of $p(x)$ over $\mathbb{F}$ if and only if it is the splitting field of $p(x)$ over $\mathbb{G}$.

Definition 6.2.11 Let $\mathbb{E}$ be an extension of $\mathbb{F}$. The *Galois group* $\mathbb{E}$ over $\mathbb{F}$, denoted $\mathrm{Gal}(\mathbb{E}/\mathbb{F})$, is the group of all automorphisms of $\mathbb{E}$ that fix $\mathbb{F}$ pointwise, in symbols:

$$\mathrm{Gal}(\mathbb{E}/\mathbb{F}) = \{\xi \in \mathrm{Aut}(\mathbb{E}) : \xi(\alpha) = \alpha \text{ for all } \alpha \in \mathbb{F}\}.$$

If we consider $\mathbb{E}$ as a vector space over $\mathbb{F}$, then every automorphism $\xi \in \mathrm{Gal}(\mathbb{E}/\mathbb{F})$ is $\mathbb{F}$-*linear*:

$$\xi(\alpha_1 \beta_1 + \alpha_2 \beta_2) = \alpha_1 \xi(\beta_1) + \alpha_2 \xi(\beta_2)$$

for all $\alpha_1, \alpha_2 \in \mathbb{F}$ and $\beta_1, \beta_2 \in \mathbb{E}$.

Proposition 6.2.12 $\mathrm{Gal}(\mathbb{E}/\mathbb{F})$ *is* $\mathbb{F}$-*linearly independent (as a subset of* $\mathrm{End}_{\mathbb{F}}(\mathbb{E})$, *the algebra of all* $\mathbb{F}$-*linear maps* $T : \mathbb{E} \to \mathbb{E}$*).*

Proof. Suppose, by contradiction, that there exist $\xi_1, \xi_2, \ldots, \xi_n \in \mathrm{Gal}(\mathbb{E}/\mathbb{F})$, all distinct, and $(\alpha_1, \alpha_2, \ldots, \alpha_n) \neq (0, 0, \ldots, 0)$ in $\mathbb{F}^n$ such that

$$\alpha_1 \xi_1 + \alpha_2 \xi_2 + \cdots + \alpha_n \xi_n = 0. \tag{6.5}$$

Up to reducing n if necessary, we may suppose that the length $n \geq 2$ of the non-trivial linear combination in the left hand side of (6.5) is minimal (in particular, $\alpha_i \neq 0$ for all $i = 1, 2, \ldots, n$).

Choose $\beta \in \mathbb{E}$ such that $\xi_1(\beta) \neq \xi_2(\beta)$. Then from (6.5) we deduce that

$$\sum_{k=1}^{n} \alpha_k \xi_k(\beta) \xi_k(\gamma) = \sum_{k=1}^{n} \alpha_k \xi_k(\beta \gamma) = 0$$

for all $\gamma \in \mathbb{E}$. It follows that

$$\alpha_1 \xi_1(\beta)\xi_1 + \alpha_2 \xi_2(\beta)\xi_2 + \cdots + \alpha_n \xi_n(\beta)\xi_n = 0 \qquad (6.6)$$

is another vanishing nontrivial linear combination of length n. But then, multiplying (6.5) by $\xi_1(\beta)$ and subtracting (6.6), we obtain

$$\alpha_2 \left(\xi_1(\beta) - \xi_2(\beta)\right)\xi_2 + \alpha_3 \left(\xi_1(\beta) - \xi_3(\beta)\right)\xi_3 + \cdots$$
$$+ \alpha_n \left(\xi_1(\beta) - \xi_n(\beta)\right)\xi_n = 0,$$

where the left hand side is nontrivial (because $\alpha_2 \left(\xi_1(\beta) - \xi_2(\beta)\right) \neq 0$) and of length at most $n - 1$, contradicting the minimality of n. This shows that the elements in $\mathrm{Gal}(\mathbb{E}/\mathbb{F})$ are $\mathbb{F}$-linearly independent. □

Theorem 6.2.13 *Let $\mathbb{E}$ be a finite extension of $\mathbb{F}$. Then $|\mathrm{Gal}(\mathbb{E}/\mathbb{F})| \leq [\mathbb{E} : \mathbb{F}]$.*

Proof. Let us set $n = [\mathbb{E} : \mathbb{F}]$ and let $\beta_1, \beta_2, \ldots, \beta_n \in \mathbb{E}$ constitute a basis of $\mathbb{E}$ as a vector space over $\mathbb{F}$. Suppose that $\xi_1, \xi_2, \ldots, \xi_m$ are distinct elements in $\mathrm{Gal}(\mathbb{E}/\mathbb{F})$. Consider the homogeneous linear system of n equations

$$\begin{cases} \alpha_1 \xi_1(\beta_1) + \alpha_2 \xi_2(\beta_1) + \cdots + \alpha_m \xi_m(\beta_1) = 0 \\ \alpha_1 \xi_1(\beta_2) + \alpha_2 \xi_2(\beta_2) + \cdots + \alpha_m \xi_m(\beta_2) = 0 \\ \cdots\cdots\cdots \quad \cdots \\ \alpha_1 \xi_1(\beta_n) + \alpha_2 \xi_2(\beta_n) + \cdots + \alpha_m \xi_m(\beta_n) = 0 \end{cases}$$

in the m variables $\alpha_1, \alpha_2, \ldots, \alpha_m$. It is a standard fact of linear algebra (over any arbitrary field) that if $m > n$ (i.e. the number of variables is greater than the number of equations) the above system has a nontrivial solution $(\overline{\alpha}_1, \overline{\alpha}_2, \ldots, \overline{\alpha}_m) \in \mathbb{E}^m$. Since the ξ_is are $\mathbb{F}$-linear and $\beta_1, \beta_2, \ldots, \beta_n$ constitute a basis for $\mathbb{E}$, we deduce that

$$\overline{\alpha}_1 \xi_1(\beta) + \overline{\alpha}_2 \xi_2(\beta) + \cdots + \overline{\alpha}_m \xi_m(\beta) = 0$$

for every $\beta \in \mathbb{E}$, that is, $\overline{\alpha}\xi_1 + \overline{\alpha}\xi_2 + \cdots + \overline{\alpha}_m \xi_m = 0$, contradicting Proposition 6.2.12. This shows that $m \leq n$ and therefore $|\mathrm{Gal}(\mathbb{E}/\mathbb{F})| \leq [\mathbb{E} : \mathbb{F}]$. □

Let $f(x) \in \mathbb{F}[x]$, say $f(x) = a_n x^n + a_{n-1} x^{n-1} + \cdots a_1 x + a_0$. Then the *derivative* of $f(x)$ is the polynomial $f'(x) \in \mathbb{F}[x]$ defined by setting

$$f'(x) := n a_n x^{n-1} + (n-1)a_{n-1} x^{n-2} + \cdots 2a_2 x + a_1.$$

Exercise 6.2.14 Show that the map $D \colon \mathbb{F}[x] \to \mathbb{F}[x]$ given by $D(f) = f'$ is $\mathbb{F}$-linear.

Note that if $\mathrm{char}(\mathbb{F}) = p > 0$, then $Dx^{kp} = kpx^{kp-1} = 0$ for all $k \geq 1$.

6.3 The structure of finite fields

Theorem 6.3.1 *Let $\mathbb{F}$ be a finite field. Then the following holds:*

(i) *There exists a prime number $p \in \mathbb{N}$ such that $char(\mathbb{F}) = p$;*

(ii) *$\mathbb{F}$ contains a subfield isomorphic to $\mathbb{F}_p$;*

(iii) *the additive group $(\mathbb{F}, +)$ is isomorphic to $\oplus_{i=1}^n \mathbb{F}_p$ for some $n \geq 1$;*

(iv) *there exists $n \geq 1$ such that $|\mathbb{F}| = p^n$.*

Proof. Consider the unital homomorphism $\Phi \colon \mathbb{Z} \to \mathbb{F}$ defined by (6.2). As we already observed at the end of Section 6.1, we have $\text{Ker}(\Phi) = p\mathbb{Z}$ with p a prime number. Moreover, $\text{Im}(\Phi) \cong \mathbb{Z}/\text{Ker}(\Phi) = \mathbb{Z}/p\mathbb{Z} = \mathbb{F}_p$ and this proves (i) and (ii). Let $n = [\mathbb{F} : \text{Im}(\Phi)]$; then $\mathbb{F}$ is a vector space of dimension n over $\text{Im}(\Phi) \cong \mathbb{F}_p$ and (iii) follows. Taking cardinalities, from (iii) we immediately deduce (iv). $\qquad\square$

In the sequel, with the notation from the above theorem, we shall denote by $q = p^n$ the cardinality of $\mathbb{F}$ and denote this field by $\mathbb{F}_q$.

Corollary 6.3.2 *Let $\mathbb{F}_q$ be a finite field of order $q = p^n$ and let $\mathbb{F}_r \subset \mathbb{F}_q$ be a subfield. Then there exists a divisor h of n such that $r = p^h$.*

Proof. Since $1 \in \mathbb{F}_r$, we clearly have $\text{char}(\mathbb{F}_r) = \text{char}(\mathbb{F}_q) = p$. Thus there exists an integer $h \geq 1$ such that $r = p^h$. Setting $s = [\mathbb{F}_q : \mathbb{F}_r]$, by Exercise 6.2.1 we have $p^n = q = r^s = (p^h)^s = p^{hs}$, so that $n = hs$. $\qquad\square$

In analogy with the particular case $q = p$ (cf. Theorem 1.1.21) we have the following:

Theorem 6.3.3 *The (multiplicative) group $\mathbb{F}_q^*$ of invertible elements in $\mathbb{F}_q$ is cyclic of order $q - 1$.*

Proof. The proof is identical to that of Theorem 1.1.21. $\qquad\square$

Definition 6.3.4 A generator of the cyclic group $\mathbb{F}_q^*$ is called a *primitive element* of $\mathbb{F}_q$.

Corollary 6.3.5 *$\mathbb{F}_q$ is the splitting field of the polynomial $x^q - x$ over $\mathbb{F}_p$ and consists exactly of the roots of this polynomial.*

Proof. First observe that $x^q - x \in \mathbb{F}_p[x]$. By Theorem 6.3.3, the multiplicative group $\mathbb{F}_q^*$ is cyclic of order $q - 1$. Therefore, every $\beta \in \mathbb{F}_q^*$ satisfies the equation $x^{q-1} = 1$, i.e. it is a root of the polynomial $x^q - x$. Since, clearly, 0 is also a root of this polynomial, it follows that $\mathbb{F}_q$ consists exactly of all the q roots of $x^q - x$. This shows that $\mathbb{F}_q$ is the splitting field of $x^q - x$ over $\mathbb{F}_p$. $\qquad\square$

Recalling that φ denotes Euler's totient function (cf. Definition 1.1.18), we have:

Corollary 6.3.6 *Let r be a divisor of $q - 1$. Then $\mathbb{F}_q^*$ contains $\varphi(r)$ elements of order r. In particular, there are $\varphi(q - 1)$ primitive elements of $\mathbb{F}_q$.* $\square$

6.4 The Frobenius automorphism

Let $\mathbb{F}_q$ be a finite field, where $q = p^n$. Then the map $\sigma : \mathbb{F}_q \to \mathbb{F}_q$ defined by

$$\sigma(\alpha) = \alpha^p$$

for all $\alpha \in \mathbb{F}_q$, is an automorphism. Indeed, for $\alpha, \beta \in \mathbb{F}_q$ we have

$$\sigma(\alpha + \beta) = (\alpha + \beta)^p$$
$$= \sum_{k=0}^{p} \binom{p}{k} \alpha^k \beta^{p-k}$$
$$= \alpha^p + \beta^p$$
$$= \sigma(\alpha) + \sigma(\beta),$$

because the integer $\binom{p}{k} = p \frac{(p-1)(p-2)\cdots(p-k+1)}{k!}$ is a multiple of p (since p is prime), and therefore $\binom{p}{k} \equiv 0 \bmod p$, for all $1 \le k \le p - 1$, and

$$\sigma(\alpha\beta) = (\alpha\beta)^p = (\alpha)^p(\beta)^p = \sigma(\alpha)\sigma(\beta).$$

One calls σ the *Frobenius automorphism* of $\mathbb{F}_q$.

Recall (cf. Theorem 6.3.1) that for $q = p^n$ the field $\mathbb{F}_q$ contains the subfield $\mathbb{F}_p$ and that $[\mathbb{F}_q : \mathbb{F}_p] = n$.

Theorem 6.4.1 *Let $q = p^n$. Then the following hold:*

(i) $\mathrm{Gal}(\mathbb{F}_q/\mathbb{F}_p)$ *is a cyclic group of order n;*
(ii) $\mathrm{Gal}(\mathbb{F}_q/\mathbb{F}_p)$ *is generated by the Frobenius automorphism σ;*
(iii) $\mathrm{Gal}(\mathbb{F}_q/\mathbb{F}_p) = \mathrm{Aut}(\mathbb{F}_q)$.

Proof. Let us first show that σ has order n. Clearly, $\sigma^k(\alpha) = \alpha^{p^k}$ for all $\alpha \in \mathbb{F}_q$ and $k \ge 1$. Since (in any field) the equation $x^{p^k} - x = 0$ has at most p^k solutions, there exists no $1 \le k < n$ such that $\sigma^k(\alpha) \equiv \alpha^{p^k} = \alpha$ for all $\alpha \in \mathbb{F}_q$.

On the other hand, it follows from Corollary 6.3.5 that $\sigma^n(\alpha) \equiv \alpha^q = \alpha$, for all $\alpha \in \mathbb{F}_q$. In other words, $\sigma^n = \mathrm{id}_{\mathbb{F}_q}$. This shows that the Frobenius automorphism σ has order n. Moreover, applying Corollary 6.3.5 to $\mathbb{F}_p^*$, we deduce that $\sigma(\alpha) \equiv \alpha^p = \alpha$ for all $\alpha \in \mathbb{F}_p$. This shows that σ fixes pointwise all elements in $\alpha \in \mathbb{F}_p$, that is, $\sigma \in \mathrm{Gal}(\mathbb{F}_q/\mathbb{F}_p)$. Since, by Theorem 6.2.13, $|\mathrm{Gal}(\mathbb{F}_q/\mathbb{F}_p)| \le [\mathbb{F}_q : \mathbb{F}_p] = n$, we deduce (i) and (ii).

Finally, let $\xi \in \mathrm{Aut}(\mathbb{F}_q)$. Then we have $\xi(0) = 0$, $\xi(1) = 1$, $\xi(2) = \xi(1 + 1) = \xi(1) + \xi(1) = 1 + 1 = 2$, and recursively, $\xi(k) = k$ for all

$k = 2, 3, \ldots, p - 1$ (but $\xi(p) = p\xi(1) = 0$). Thus ξ fixes $\mathbb{F}_p = \{0, 1, 2, \ldots, p - 1\}$ pointwise. This shows (iii). □

Corollary 6.4.2 *Every* $\alpha \in \mathbb{F}_q$ *has exactly one* p^k-*th root in* $\mathbb{F}_q$ *for* $k = 1, 2, \ldots, n$. □

Corollary 6.4.3 *The field* $\mathbb{F}_q$ *admits an involutory automorphism if and only if* n *is even. If this is the case, then it is given by* $\sigma^{n/2}$. □

A nontrivial *square* in a field $\mathbb{F}$ is an element $\alpha \in \mathbb{F}^*$ such that $\alpha \neq 1$ and $\alpha = \beta^2$ for some $\beta \in \mathbb{F}$.

Proposition 6.4.4 *If* $p = 2$ *then every element in* $\mathbb{F}_q^*$ *is a square. If* $p > 2$ *then there are* $\frac{q-1}{2}$ *squares in* $\mathbb{F}_q^*$.

Proof. The result for $p = 2$ follows immediately from Corollary 6.4.2 (with $k = 1$). Suppose $p > 2$ and denote by $\phi \colon \mathbb{F}_q^* \to \mathbb{F}_q^*$ the square map defined by $\phi(\beta) = \beta^2$ for all $\beta \in \mathbb{F}_q^*$. Note that for $\beta_1, \beta_2 \in \mathbb{F}_q$ one has $\phi(\beta_1) = \phi(\beta_2)$ if and only if $\beta_1 = \pm\beta_2$. This shows that ϕ is two-to-one. As a consequence, the number of squares in $\mathbb{F}_q^*$ equals $|\phi(\mathbb{F}_q^*)| = |\mathbb{F}_q^*|/2 = (q - 1)/2$. □

6.5 Existence and uniqueness of Galois fields

Definition 6.5.1 Let $f(x) \in \mathbb{F}_p[x]$ be an irreducible polynomial of degree n and denote by $f(x)\mathbb{F}_p[x]$ the ideal generated by $f(x)$. Then the field

$$\mathbb{F}_p[x]/f(x)\mathbb{F}_p[x]$$

is called a *Galois field* of order p^n (cf. Proposition 6.2.5 and Remark 6.2.6).

We shall not introduce a specific notation for Galois fields since for every prime number p and integer $n \geq 1$ all Galois fields of order $q = p^n$ are isomorphic (cf. Theorem 6.5.6), and we shall use the notation $\mathbb{F}_q$. In this section, we prove their existence and uniqueness. As usual, we denote by $\sigma \in \mathrm{Aut}(\mathbb{F}_q)$ the Frobenius automorphism.

Proposition 6.5.2 *Let* $f(x) = a_0 + a_1 x + \cdots + a_n x^n \in \mathbb{F}_p[x]$ *be an irreducible polynomial of degree* n *and let* $\mathbb{F}_q = \mathbb{F}_p[x]/f(x)\mathbb{F}_p[x]$ *be the associated Galois field. Let also* $\alpha \in \mathbb{F}_q$ *be a root of* f *(cf. Remark 6.2.6). Then the elements* $\alpha^{p^k} = \sigma^k(\alpha)$, $k = 0, 1, \ldots, n - 1$, *are all distinct and are the roots of* f. *In particular,* $\mathbb{F}_q$ *is the splitting field of* $f(x)$ *over* $\mathbb{F}_p$ *(cf. Definition 6.2.8) and*

$$f(x) = a_n(x - \alpha)(x - \alpha^p)(x - \alpha^{p^2}) \cdots (x - \alpha^{p^{n-1}}).$$

Proof. Since σ^k is an automorphism that fixes $\mathbb{F}_p$ pointwise, we have that $f(\sigma^k(\alpha)) = \sigma^k(f(\alpha)) = \sigma(0) = 0$, that is, $\sigma^k(\alpha)$ is a root of f, for all $k = 0, 1, \ldots, n-1$. Let us show that these elements are all distinct. Suppose that $\sigma^k(\alpha) = \sigma^j(\alpha)$, that is, $\alpha^{p^k} = \alpha^{p^j}$ for some $1 \le k < j \le n-1$. Set $\beta = \sigma^k(\alpha) = \alpha^{p^k}$ and $r = j - k \in \mathbb{N}$. We have

$$\sigma^r(\beta) = \beta^{p^r} = \beta^{p^{j-k}} = (\alpha^{p^k})^{p^{j-k}} = \alpha^{p^j} = \alpha^{p^k} = \beta. \tag{6.7}$$

Since $f(\beta) = 0$, from Proposition 6.2.5 we deduce that the elements

$$1, \beta, \beta^2, \ldots, \beta^{n-1}$$

constitute a vector space basis of $\mathbb{F}_q$ over $\mathbb{F}_p$. As a consequence, for every $\delta \in \mathbb{F}_q$ there exist $\eta_1, \eta_2, \ldots, \eta_n \in \mathbb{F}_p$ such that

$$\delta = \eta_1 + \eta_2 \beta + \cdots + \eta_n \beta^{n-1}.$$

Since $(\eta_i)^p = \eta_i$ for $i = 1, 2, \ldots, n$ and, by (6.7), $\beta^{p^r} = \beta$, we get

$$\begin{aligned}
\delta^{p^r} &= \sigma^r\left(\eta_1 + \eta_2 \beta + \cdots + \eta_n \beta^{n-1}\right) \\
&= \eta_1 + \eta_2 \beta^{p^r} + \eta_3 (\beta^{p^r})^2 + \cdots + \eta_n (\beta^{p^r})^{n-1} \\
&= \eta_1 + \eta_2 \beta + \cdots + \eta_n \beta^{n-1} \\
&= \delta.
\end{aligned}$$

Since δ was arbitrary, this contradicts Theorem 6.3.3, because $r < n$. $\qquad\square$

Proposition 6.5.3 *Let $f(x) \in \mathbb{F}_q[x]$ be an irreducible polynomial of degree m, and let $k \ge 1$. Then $f(x)$ divides $x^{q^k} - x$ if and only if m divides k.*

Proof. By Proposition 6.2.5 and Theorem 6.3.1, $\mathbb{F}_q[x]/f(x)\mathbb{F}_q[x]$ has q^m elements so that

$$\alpha^{q^m} = \alpha \quad \text{for all } \alpha \in \mathbb{F}_q[x]/f(x)\mathbb{F}_q[x] \tag{6.8}$$

(cf. Corollary 6.3.5). Taking $\alpha = x + f(x)\mathbb{F}_q[x]$, this yields

$$x^{q^m} - x \in f(x)\mathbb{F}_q[x]. \tag{6.9}$$

Let us show that for $s = 0, 1, 2, \ldots$ we have

$$x^{q^{sm}} - x \in f(x)\mathbb{F}_q[x]. \tag{6.10}$$

We proceed by induction. For $s = 0$, this is trivial and for $s = 1$ equation (6.10) reduces to (6.9). Let us prove the inductive step:

$$x^{q^{(s+1)m}} - x = \left(x^{q^{sm}}\right)^{q^m} - x$$

$$\text{(by (6.10))} \in \left(x + f(x)\mathbb{F}_q[x]\right)^{q^m} - x$$

$$\subseteq x^{q^m} - x + f(x)\mathbb{F}_q[x]$$

$$\text{(by (6.9))} = f(x)\mathbb{F}_q[x].$$

In particular, if m divides k then $f(x)$ divides $x^{q^k} - x$.

Let us prove the converse implication. Suppose that $f(x)$ divides $x^{q^k} - x$. Applying the Euclidean algorithm, we can find two non-negative integers s, r, with $0 \leq r \leq m - 1$, such that $k = sm + r$. We need to show that $r = 0$. By virtue of (6.9) we have

$$x^{q^{sm}} \in x + f(x)\mathbb{F}_q[x]$$

and therefore

$$x^{q^k} = x^{q^{sm+r}} = \left(x^{q^{sm}}\right)^{q^r} \in x^{q^r} + f(x)\mathbb{F}_q[x]. \tag{6.11}$$

Since $f(x)$ divides $x^{q^k} - x$, from (6.11) we deduce $x^{q^r} - x \in f(x)\mathbb{F}_q[x]$, equivalently, $f(x)$ also divides $x^{q^r} - x$. As a consequence, in the field $\mathbb{F}_q[x]/f(x)\mathbb{F}_q[x]$ every element α satisfies the identity

$$\alpha^{q^r} = \alpha$$

that contradicts (6.8), since $r < m$, unless $r = 0$. This shows that m divides k. $\qquad\square$

Proposition 6.5.4 *Let p and m be two primes and $q = p^h$ for some integer $h \geq 1$. Then in $\mathbb{F}_q[x]$ there exist exactly*

$$\frac{q^m - q}{m} > 0$$

distinct irreducible monic polynomials of degree m.

Proof. From the identity $\alpha^q = \alpha$ in $\mathbb{F}_q$, we deduce that $\alpha^{q^2} = \alpha^q = \alpha$ and, similarly, $\alpha^{q^3} = \alpha, \ldots, \alpha^{q^m} = \alpha$, for all $\alpha \in \mathbb{F}_q$. Therefore the polynomial $x^{q^m} - x$ is divisible by $x - \alpha$ for every $\alpha \in \mathbb{F}_q$ and therefore may be factorized as follows

$$x^{q^m} - x = f_1(x)f_2(x)\cdots f_r(x) \prod_{\alpha \in \mathbb{F}_q} (x - \alpha) \tag{6.12}$$

where $f_1, f_2, \ldots, f_r \in \mathbb{F}_q[x]$ are monic and irreducible. We claim that in the factorization (6.12) there cannot be two equal factors (it is square free), that is, one cannot have

$$x^{q^m} - x = f(x)^2 g(x),$$

where $f \in \mathbb{F}_q[x]$ has degree ≥ 1. Otherwise, by taking the derivative of both sides we would have that $q^m x^{q^m-1} - 1 = -1$ should equal $2f(x)f'(x)g(x) + f(x)^2 g'(x)$, that is,

$$-1 = f(x)\left(2f'(x)g(x) + f(x)g'(x)\right)$$

which is impossible since $\deg(f) \geq 1$. This proves our claim. In particular, in (6.12) for $j = 1, 2, \ldots, r$ we must have $\deg(f_j) \geq 2$ and therefore, by Proposition 6.5.3 and primality of m, $\deg(f_j) = m$.

In conclusion, $f_1, f_2, \ldots, f_r$ are distinct irreducible polynomials of degree m. Moreover, again by virtue of Proposition 6.5.3, they constitute the complete list of all irreducible monic polynomials of degree m. It follows that the degree of the right hand side of (6.12) is $mr + q$ and must equal q^m. This yields

$$r = \frac{q^m - q}{m},$$

completing the proof. □

Remark 6.5.5 The fact that the number $\frac{q^m-q}{m}$ is an integer is a particular case of Fermat's little theorem (cf. Exercise 1.1.22).

We are now in position to state and prove the main theorem of the theory of finite fields.

Theorem 6.5.6 (Main theorem: existence and uniqueness of Galois fields)
For every prime number p and integer $h \geq 1$ there exists a unique (up to isomorphism) finite field $\mathbb{F}_q$ of order $q = p^h$. It is the Galois field

$$\mathbb{F}_p[x]/\ell(x)\mathbb{F}_p[x],$$

where $\ell(x) = (x - \alpha)(x - \alpha^p)(x - \alpha^{p^2}) \cdots (x - \alpha^{p^{h-1}})$ and α is any generator of the cyclic group $\mathbb{F}_q^*$.

Proof. First of all, let us prove that a field with $q = p^h$ elements exists. Let

$$h = m_1 m_2 \cdots m_r \tag{6.13}$$

be a factorization of h into primes (repetitions are allowed). By Proposition 6.5.4, there exists an irreducible polynomial $f_1 \in \mathbb{F}_p[x]$ of degree m_1. Consider the field $\mathbb{F}_{p^{m_1}} = \mathbb{F}_p[x]/f_1(x)\mathbb{F}_p[x]$ and recall that it has p^{m_1} elements.

Now, again by Proposition 6.5.4, in $\mathbb{F}_{p^{m_1}}[x]$ there exists an irreducible polynomial f_2 of degree m_2, and so on. Eventually, we obtain a field $\mathbb{F}_q$ with $(p^{m_1 m_2 \cdots m_{r-1}})^{m_r} = p^{m_1 m_2 \cdots m_r} = p^h = q$ elements.

By Theorem 6.3.3, the group $\mathbb{F}_q^*$ is cyclic of order $q - 1$, and let α be a generator of $\mathbb{F}_q^*$. Then α is *algebraic* over $\mathbb{F}_p$, since it is a root of the polynomial $x^{q-1} - 1$, and, clearly,

$$\mathbb{F}_q = \mathbb{F}_p[\alpha].$$

Then, by Proposition 6.2.5, $\mathbb{F}_q$ is isomorphic to $\mathbb{F}_p[x]/\ell(x)\mathbb{F}_p[x]$, where $\ell(x) \in \mathbb{F}_p[x]$ is the minimal polynomial of α. It follows that $\mathbb{F}_q$ is a Galois field. Moreover, by Proposition 6.5.2, we have

$$\ell(x) = (x - \alpha)(x - \alpha^p)(x - \alpha^{p^2}) \cdots (x - \alpha^{p^{h-1}})$$

and

$$x^q - x = \ell(x)g(x) \tag{6.14}$$

with $g(x) \in \mathbb{F}_p[x]$, because α is a root of $x^q - x$, and $\ell(x)$ is its minimal polynomial, and the principal ideal $\mathcal{I}_\alpha = \{f \in \mathbb{F}_p[x] : f(\alpha) = 0\}$ is generated by $\ell(x)$.

Suppose now that $\mathbb{K}_q$ is another field with q elements. Let $\overline{\alpha} \in \mathbb{K}_q$ be a generator of the cyclic group $\mathbb{K}_q^*$. From the arguments above, we have that $\mathbb{F}_p[\overline{\alpha}] = \mathbb{K}_q$. Finally, it is straightforward that the map $\mathbb{F}_q = \mathbb{F}_p[\alpha] \to \mathbb{F}_p[\overline{\alpha}] = \mathbb{K}_q$, given by $f(\alpha) \mapsto f(\overline{\alpha})$ for all $f \in \mathbb{F}_p[x]$, is an isomorphism. $\qquad\square$

We now present, as an exercise, an elementary proof of Gauss law of quadratic reciprocity from [5]. This proof uses some facts on finite fields that we have already established. Let p and q be distinct odd primes and consider the field $\mathbb{F}_{q^{p-1}}$ and the cyclic group $\mathbb{F}_{q^{p-1}}^*$. By Fermat's little theorem (see Exercise 1.1.22), p divides $q^{p-1} - 1 = |\mathbb{F}_{q^{p-1}}^*|$, so that, by Corollary 1.2.9, $\mathbb{F}_{q^{p-1}}^*$ contains an element ζ of order p. We consider the *Gauss sum*

$$G_\zeta = \sum_{k=1}^{p-1} \left(\frac{k}{p}\right)\zeta^k,$$

where $\left(\dfrac{k}{p}\right)$ is the Legendre symbol (cf. Definition 4.4.7). Clearly, $G_\zeta \in \mathbb{F}_{q^{p-1}}$.

Exercise 6.5.7

(1) Prove that

$$G_\zeta^q = \left(\frac{q}{p}\right)G_\zeta. \tag{6.15}$$

Hint: Use the identities $(a + b)^q = a^q + b^q$ in $\mathbb{F}_{q^{p-1}}$ and

$$\left(\frac{k}{p}\right) = \left(\frac{kq^2}{p}\right) = \left(\frac{kq}{p}\right)\left(\frac{q}{p}\right),$$

where the last equality follows from Proposition 4.4.8.(iii).

(2) Suppose that $p \nmid h$ and show that

$$\sum_{j=1}^{p-1} \zeta^{jh} = -1.$$

(3) Show that

$$\sum_{h=1}^{p-2} \left(\frac{h}{p}\right) = -\left(\frac{-1}{p}\right).$$

(*Hint*: use Corollary 4.4.9), and deduce that

$$\sum_{h=1}^{p-2} \left(\frac{h}{p}\right) \sum_{j=1}^{p-1} \zeta^{(1+h)j} = \left(\frac{-1}{p}\right).$$

(4) From (2) and (3) deduce that

$$G_\zeta^2 = \left(\frac{-1}{p}\right)p.$$

(*Hint*: first prove that $G_\zeta^2 = \sum_{h=1}^{p-1} \left(\frac{h}{p}\right) \sum_{j=1}^{p-1} \zeta^{(1+h)j}$), so that, by Proposition 4.4.8.(iv),

$$G_\zeta^2 = p(-1)^{(p-1)/2}. \tag{6.16}$$

(5) From (6.15) and (6.16) deduce the Gauss law of quadratic reciprocity (Theorem 4.4.18).
Hint: start with the elementary identity $G_\zeta^q = G_\zeta (G_\zeta^2)^{(q-1)/2}$; use Proposition 4.4.8.(ii).

6.6 Subfields and irreducible polynomials

Proposition 6.6.1 *Let* $q = p^h$. *Then, for every divisor* m *of* h *there exists a unique subfield of* $\mathbb{F}_q$ *isomorphic to* $\mathbb{F}_{p^m}$. *Moreover all subfields are of this kind.*

Proof. Let $\mathbb{K}$ be a subfield of $\mathbb{F}_q$. Then $\mathbb{F}_q$ is a vector space over $\mathbb{K}$ and therefore the cardinality of $\mathbb{K}$ divides the cardinality of $\mathbb{F}_q$. By the uniqueness of Galois fields (Theorem 6.5.6), it follows that there exists an integer $m \leq h$ such that $\mathbb{K} = \mathbb{F}_{p^m} = \mathbb{F}_p/\ell(x)\mathbb{F}_p[x]$, where $\ell \in \mathbb{F}_p[x]$ is an irreducible polynomial of degree m. Since the equation $x^{p^h} - x = 0$ is satisfied by all elements in $\mathbb{F}_q \supseteq \mathbb{K}$ we deduce that $\ell(x)$ divides $x^{p^h} - x$ in $\mathbb{F}_p[x]$ (compare with (6.14)). Therefore, by virtue of Proposition 6.5.3, we have $m = \deg(\ell)$ must divide h (cf. Corollary 6.3.2).

In order to show that, conversely, if m divides h, then $\mathbb{F}_q = \mathbb{F}_{p^h}$ contains a subfield isomorphic to $\mathbb{F}_{p^m}$, we use the recursive construction of $\mathbb{F}_q$ in the proof of Theorem 6.5.6. Indeed, if we arrange the primes in the decomposition (6.13) of h in such a way that $m = m_1 m_2 \cdots m_i$ for some $1 \leq i \leq r$, then $\mathbb{F}_{p^m}$ appears, in the construction we alluded to above, as one of the intermediate fields between $\mathbb{F}_p$ and $\mathbb{F}_{p^h} = \mathbb{F}_q$. Uniqueness of the subfield $\mathbb{F}_{p^m}$ follows from the fact that its elements are precisely the roots of the polynomial $x^{p^m} - x \in \mathbb{F}_p[x]$. $\qquad\square$

Exercise 6.6.2 Show that the lattice of all subfields of $\mathbb{F}_q$ is isomorphic to the lattice of all divisors of m.

In the following, $\sigma \in \mathrm{Aut}(\mathbb{F}_q)$ denotes the Frobenius automorphism (cf. Section 6.4).

Proposition 6.6.3 *Let p be a prime number, $h \geq 1$ an integer, and $q = p^h$. Let also $1 \leq r \leq h - 1$. Then*

$$\mathbb{K} = \{\beta \in \mathbb{F}_q : \sigma^r(\beta) = \beta\} \tag{6.17}$$

coincides with the subfield of $\mathbb{F}_q$ isomorphic to $\mathbb{F}_{p^m}$, where $m = \gcd(h, r)$.
  On the other hand, if m divides h then

$$\mathrm{Gal}(\mathbb{F}_q/\mathbb{F}_{p^m}) \equiv \{\xi \in \mathrm{Aut}(\mathbb{F}_q) : \xi(\beta) = \beta \text{ for all } \beta \in \mathbb{F}_{p^m}\} = \langle \sigma^m \rangle.$$

Proof. First of all we observe that $\mathbb{K}$ is a subfield of $\mathbb{F}_q$. Therefore, by Proposition 6.6.1, there exists an integer m that divides h such that $\mathbb{K} = \mathbb{F}_{p^m}$.

Let us set $\tilde{\sigma} = \sigma|_{\mathbb{F}_{p^m}} \in \mathrm{Aut}(\mathbb{F}_{p^m})$. This is the Frobenius automorphism of $\mathbb{F}_{p^m}$ so that, by Theorem 6.4.1, $\mathrm{Aut}(\mathbb{F}_{p^m}) = \langle \tilde{\sigma} \rangle$. Now, for an integer $n \geq 0$ one has

$$\sigma^n(\beta) = \beta \text{ (i.e. } \tilde{\sigma}^n(\beta) = \beta) \ \forall \beta \in \mathbb{F}_{p^m} \Leftrightarrow m | n. \tag{6.18}$$

We deduce that m divides r and therefore also divides $\gcd(h, r)$. On the other hand, setting $m' = \gcd(h, r)$ and $\hat{\sigma} = \sigma|_{\mathbb{F}_{p^{m'}}} \in \mathrm{Aut}(\mathbb{F}_{p^{m'}})$, arguing as above, we have $\sigma^n(\beta') = \beta'$ (i.e. $\hat{\sigma}^n(\beta') = \beta'$) for all $\beta' \in \mathbb{F}_{p^{m'}}$ if and only if m' divides n. Thus, taking $n = r$ we have $\sigma^r(\beta') = \beta'$ for all $\beta' \in \mathbb{F}_{p^{m'}}$. Since $\mathbb{K} = \mathbb{F}_{p^m} \subseteq \mathbb{F}_{p^{m'}}$, this shows that $m = m' = \gcd(h, r)$.

Finally, $\mathrm{Gal}(\mathbb{F}_q/\mathbb{F}_{p^m})$, being a subgroup of the cyclic group $\mathrm{Gal}(\mathbb{F}_q/\mathbb{F}_p)$, is itself cyclic (cf. Proposition 1.2.12). By the above arguments, we have $\sigma^m \in \mathrm{Gal}(\mathbb{F}_q/\mathbb{F}_{p^m})$ and, by (6.18), we indeed have $\mathrm{Gal}(\mathbb{F}_q/\mathbb{F}_{p^m}) = \langle \sigma^m \rangle$. $\qquad\square$

The following is a generalization of Proposition 6.5.2.

Corollary 6.6.4 *Let $f \in \mathbb{F}_q[x]$ be an irreducible polynomial of degree n. Then $\mathbb{F}_{q^n}$ is the splitting field of f over $\mathbb{F}_q$. Moreover, if $\alpha \in \mathbb{F}_{q^n}$ is a root of f then $\alpha, \alpha^q, \ldots, \alpha^{q^{n-1}}$ are the roots of f and they are also distinct.*

Proof. Let $\mathbb{F}$ denote the splitting field of f over $\mathbb{F}_q$. Then we can find a positive integer $h \geq n$ such that $\mathbb{F} = \mathbb{F}_{q^h}$: indeed, denoting by $\alpha_1, \alpha_2, \ldots, \alpha_n \in \mathbb{F}$ the roots of f, by Proposition 6.2.5 and Theorem 6.6.1 we have $\mathbb{F}_{q^n} \cong \mathbb{F}_q[\alpha_1] \subseteq \mathbb{F}_q[\alpha_1, \alpha_2, \ldots, \alpha_n] = \mathbb{F} = \mathbb{F}_{q^h}$.

Let σ be the generator of $\mathrm{Gal}(\mathbb{F}_{q^n}, \mathbb{F}_q)$ given by $\sigma(\beta) = \beta^q$ for all $\beta \in \mathbb{F}_{q^n}$. Observe that σ is not the Frobenius automorphism, although we use the same symbol. Arguing as in the proof of Proposition 6.5.2, we deduce that $\alpha, \alpha^q, \ldots, \alpha^{q^{n-1}}$ are distinct roots of f and therefore exhaust all the roots of f. Then $\mathbb{F}_{q^n}$ contains all the roots of f, and therefore $n = h$, i.e. $\mathbb{F} = \mathbb{F}_{q^n}$. $\square$

Corollary 6.6.5 *With the notation from the previous corollary, if α is a root of f in $\mathbb{F}_{q^n}$, then f is a scalar multiple of the minimal polynomial of α over $\mathbb{F}_q$, and $\mathbb{F}_{q^n} = \mathbb{F}_q[\alpha]$.*

Notation 6.6.6 Let $\mathbb{F}$ be a finite field. We denote by $\mathbb{F}^{\mathrm{mon}}[x]$ (resp. $\mathbb{F}^{\mathrm{mon,irr}}[x]$) the set of monic (resp. monic irreducible) polynomials in $\mathbb{F}[x]$ and by $\mathbb{F}^{\mathrm{mon},k}[x]$ (resp. $\mathbb{F}^{\mathrm{mon,irr},k}[x]$) the set of monic (resp. monic irreducible) polynomials in $\mathbb{F}[x]$ of degree k.

In the proof of the following proposition, we need the most elementary facts on group actions (see the beginning of Section 10.4).

Proposition 6.6.7 *Let $f \in \mathbb{F}_q^{\mathrm{mon,irr}}[x]$ and $h \geq 1$. Choose $\widetilde{f} \in \mathbb{F}_{q^h}^{\mathrm{mon,irr}}[x]$ that divides f and set $d = d(\widetilde{f}) = \min\{1 \leq \ell \leq h : \sigma^\ell(\widetilde{f}) = \widetilde{f}\}$, where $\sigma(x) = x^q$ for all $x \in \mathbb{F}_{q^h}$. Then d divides h and*

$$f = \prod_{\ell=0}^{d-1} \sigma^\ell(\widetilde{f}) \tag{6.19}$$

is the (unique up to reordering the factors) factorization of f into $\mathbb{F}_{q^h}$-irreducible monic polynomials. Moreover, all factors are distinct, $\deg \sigma^\ell(\widetilde{f}) = \frac{\deg f}{d}$, for all $\ell = 0, 1, \ldots, d - 1$, and

$$d = d(\widetilde{f}) = \gcd(h, \deg f). \tag{6.20}$$

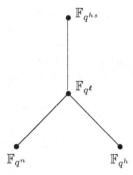

Figure 6.1. The inclusions of the fields $\mathbb{F}_{q^t}$, $t = n, h, \ell, hs$.

As a consequence we have, for all $k \geq 1$,

$$\mathbb{F}_{q^h}^{mon,irr,k}[x] = \coprod_{d \mid h} \coprod_{\substack{f \in \mathbb{F}_q^{mon,irr,dk}[x]: \\ \gcd(h/d,k)=1}} \{\tilde{f}, \sigma(\tilde{f}), \ldots, \sigma^{d-1}(\tilde{f})\}.$$

In other words, given $\tilde{f} \in \mathbb{F}_{q^h}^{mon,irr,k}[x]$ there exists a unique $f \in \mathbb{F}_q^{mon,irr}[x]$ such that $\tilde{f}$ divides f (clearly, $\deg f = d(\tilde{f}) \deg \tilde{f}$).

Proof. Every $\sigma^\ell(\tilde{f})$, for $\ell = 0, 1, \ldots, h-1$, is an $\mathbb{F}_{q^h}$-irreducible monic polynomial and divides f, since $\sigma(f) = f$. In other words, the Galois group $\text{Gal}(\mathbb{F}_{q^h}/\mathbb{F}_q)$ acts on the space of monic $\mathbb{F}_{q^h}$-irreducible divisors of f. We have that $d(\tilde{f})$ divides h because $\text{Gal}(\mathbb{F}_{q^h}/\mathbb{F}_q)$ is cyclic of order h and generated by σ (cf. Proposition 6.6.3), and the stabilizer of $\tilde{f}$ coincides with the set $\{\sigma^{dk} : k = 0, 1, \ldots, \frac{h}{d}\}$. Thus, the polynomial

$$\tilde{f}\sigma(\tilde{f}) \cdots \sigma^{d-1}(\tilde{f}), \tag{6.21}$$

a product of <u>distinct</u> $\mathbb{F}_{q^h}$-irreducible monic divisors of f, divides f. But (6.21) is also σ-invariant and monic, so that it belongs to $\mathbb{F}_q[x]$ and therefore must be equal to f (since f is irreducible over $\mathbb{F}_q$). This proves that the action described above is <u>transitive</u>. Moreover, since $\mathbb{F}_{q^d} = \{\alpha \in \mathbb{F}_{q^h} : \sigma^d(\alpha) = \alpha\}$ (by virtue of Proposition 6.6.3), we have $\tilde{f} \in \mathbb{F}_{q^d}[x]$.

Set $s = \deg \tilde{f}$ and $n = \deg f$. It follows from Corollary 6.6.4 that the splitting field of $\tilde{f}$ over $\mathbb{F}_{q^h}$ is $\mathbb{F}_{q^{hs}}$. Similarly, the splitting field of f over $\mathbb{F}_q$ is $\mathbb{F}_{q^n}$, so that, in particular, f, and therefore its factor $\tilde{f}$, split into linear factors over $\mathbb{F}_{q^n}$. Observe that, since $d \mid h$, say $h = ad$, and $n = sd$ (this follows from the fact that the polynomial in (6.21) coincides with f), we have $hs = ads = an$, so that $n \mid hs$.

Setting $\ell = \text{lcm}(h, n)$, we have the inclusion diagram as in Figure 6.1.

Since $\mathbb{F}_{q^n} \subseteq \mathbb{F}_{q^\ell}$, it follows that $\tilde{f}$ splits into linear factors over $\mathbb{F}_{q^\ell}$. Thus, since $\mathbb{F}_{q^h} \subseteq \mathbb{F}_{q^\ell} \subseteq \mathbb{F}_{q^{hs}}$, we deduce that $\mathbb{F}_{q^\ell} = \mathbb{F}_{q^{hs}}$, this being the splitting field of $\tilde{f}$ over $\mathbb{F}_{q^h}$. In particular, $hs = \mathrm{lcm}(h, n)$.

Setting $r = \gcd(h, n)$, we have

$$hs = \mathrm{lcm}(n, h) = \frac{hn}{\gcd(h, n)} = \frac{hsd}{r} \Rightarrow d = r,$$

and (6.20) follows. $\qquad\square$

Corollary 6.6.8 *Let $f \in \mathbb{F}_q[x]$ be irreducible and let $h \geq 2$. Then f is irreducible over $\mathbb{F}_{q^h}$ if and only if $\gcd(\deg f, h) = 1$.*

6.7 Hilbert Satz 90

We now specialize, to the case of finite fields, the theory of the norm and the trace for extensions of fields. A more general treatment may be found in [93]. Fix a prime number p, two integers $n \geq 1$ and $h > 1$, and set $q = p^n$. Let $\mathbb{E} = \mathbb{F}_{q^h} = \mathbb{F}_{p^{hn}}$ be the field with q^h elements and $\mathbb{F} = \mathbb{F}_q$ the unique subfield of $\mathbb{E}$ with q elements (cf. Proposition 6.6.1). By Proposition 6.6.3, the Galois group $\mathrm{Gal}(\mathbb{E}/\mathbb{F})$ is a cyclic group of order h: we denote by σ a generator of $\mathrm{Gal}(\mathbb{E}/\mathbb{F})$. We remark that here the notation is different from that in Proposition 6.6.3: for instance, σ is not the Frobenius automorphism of $\mathbb{E}$ but it can be chosen as its n-th power so that $\sigma(\alpha) = \alpha^{p^n} = \alpha^q$ for all $\alpha \in \mathbb{E}$ (see Corollary 6.6.4 and Proposition 6.6.7). We define the *trace* and the *norm* as the maps $\mathrm{Tr}_{\mathbb{E}/\mathbb{F}} : \mathbb{E} \to \mathbb{F}$ and $\mathrm{N}_{\mathbb{E}/\mathbb{F}} : \mathbb{E} \to \mathbb{F}$ given by

$$\mathrm{Tr}_{\mathbb{E}/\mathbb{F}}(\alpha) = \sum_{k=1}^{h} \sigma^k(\alpha) \tag{6.22}$$

and

$$\mathrm{N}_{\mathbb{E}/\mathbb{F}}(\alpha) = \prod_{k=1}^{h} \sigma^k(\alpha) \tag{6.23}$$

for all $\alpha \in \mathbb{E}$. Note that $\mathrm{Tr}_{\mathbb{E}/\mathbb{F}}(\alpha)$ (resp. $\mathrm{N}_{\mathbb{E}/\mathbb{F}}(\alpha)$) is indeed in $\mathbb{F}$:

$$\sigma\left(\mathrm{Tr}_{\mathbb{E}/\mathbb{F}}(\alpha)\right) = \sum_{k=1}^{h} \sigma^{k+1}(\alpha) = \sum_{k=2}^{h+1} \sigma^k(\alpha) = \mathrm{Tr}_{\mathbb{E}/\mathbb{F}}(\alpha) \tag{6.24}$$

(resp. $\sigma\left(\mathrm{N}_{\mathbb{E}/\mathbb{F}}(\alpha)\right) = \mathrm{N}_{\mathbb{E}/\mathbb{F}}(\alpha)$) because σ has order h. Moreover, it is clear that $\mathrm{Tr}_{\mathbb{E}/\mathbb{F}}(\alpha)$ (resp. $\mathrm{N}_{\mathbb{E}/\mathbb{F}}(\alpha)$) is independent of the choice of the generator σ in (6.22) (resp. (6.23)).

Proposition 6.7.1 (Transitivity of the trace and the norm) *Let* $\mathbb{E}, \mathbb{F}, \mathbb{G}$ *be finite fields such that* $\mathbb{F} \subseteq \mathbb{E} \subseteq \mathbb{G}$. *Then*

(i) $\mathrm{Tr}_{\mathbb{G}/\mathbb{F}} = \mathrm{Tr}_{\mathbb{E}/\mathbb{F}} \circ \mathrm{Tr}_{\mathbb{G}/\mathbb{E}}$
(ii) $\mathrm{N}_{\mathbb{G}/\mathbb{F}} = \mathrm{N}_{\mathbb{E}/\mathbb{F}} \circ \mathrm{N}_{\mathbb{G}/\mathbb{E}}$.

Proof. By virtue of Theorem 6.5.6 and Proposition 6.6.1, there exists $h, m \in \mathbb{N}$ such that $\mathbb{F} = \mathbb{F}_q$, $\mathbb{E} = \mathbb{F}_{q^h}$ and $\mathbb{G} = \mathbb{F}_{q^{hm}}$. For every $\alpha \in \mathbb{G}$ we have:

$$[\mathrm{Tr}_{\mathbb{E}/\mathbb{F}} \circ \mathrm{Tr}_{\mathbb{G}/\mathbb{E}}](\alpha) = \sum_{k=0}^{h-1} \left[\mathrm{Tr}_{\mathbb{G}/\mathbb{E}}(\alpha) \right]^{q^k}$$

$$= \sum_{k=0}^{h-1} \left[\sum_{j=0}^{m-1} \alpha^{q^{jh}} \right]^{q^k}$$

$$\text{(the map } \beta \mapsto \beta^{q^k} \text{ belongs to } \mathrm{Aut}(\mathbb{G})) = \sum_{k=0}^{h-1} \sum_{j=0}^{m-1} \alpha^{q^{jh+k}}$$

$$\text{(setting } r = hj + k) = \sum_{r=0}^{hm-1} \alpha^{q^r}$$

$$= \mathrm{Tr}_{\mathbb{G}/\mathbb{F}}(\alpha).$$

Analogously,

$$[\mathrm{N}_{\mathbb{E}/\mathbb{F}} \circ \mathrm{N}_{\mathbb{G}/\mathbb{E}}](\alpha) = \prod_{k=0}^{h-1} \left(\prod_{j=0}^{m-1} \alpha^{q^{jh}} \right)^{q^k}$$

$$= \prod_{k=0}^{h-1} \prod_{j=0}^{m-1} \alpha^{q^{jh+k}}$$

$$\text{(setting } r = hj + k) = \prod_{r=0}^{hm-1} \alpha^{q^r}$$

$$= \mathrm{N}_{\mathbb{G}/\mathbb{F}}(\alpha). \qquad \square$$

Theorem 6.7.2 (Hilbert Satz 90)

(i) $\mathrm{Tr}_{\mathbb{E}/\mathbb{F}}$ *is a surjective* $\mathbb{F}$-*linear map from* $\mathbb{E}$ *onto* $\mathbb{F}$ *and*

$$\mathrm{KerTr}_{\mathbb{E}/\mathbb{F}} = \{\alpha - \sigma(\alpha) : \alpha \in \mathbb{E}\}.$$

(ii) $N_{\mathbb{E}/\mathbb{F}}$ *yields (by restriction) a surjective homomorphism from the multiplicative group* $\mathbb{E}^*$ *of* $\mathbb{E}$ *into the multiplicative group* $\mathbb{F}^*$ *of* $\mathbb{F}$ *and*

$$\mathrm{Ker}N_{\mathbb{E}/\mathbb{F}} = \{\alpha\sigma(\alpha)^{-1} : \alpha \in \mathbb{E}\}.$$

Proof.

(i) The map $\mathrm{Tr}_{\mathbb{E}/\mathbb{F}}$ is $\mathbb{F}$-linear since

$$\mathrm{Tr}_{\mathbb{E}/\mathbb{F}}(\alpha_1\beta_1 + \alpha_2\beta_2) = \sum_{k=1}^{h} \sigma^k(\alpha_1\beta_1 + \alpha_2\beta_2)$$

$$(\text{since } \sigma^k \in \mathrm{Gal}(\mathbb{E}/\mathbb{F})) = \sum_{k=1}^{h} \alpha_1\sigma^k(\beta_1) + \alpha_2\sigma^k(\beta_2)$$

$$= \alpha_1\mathrm{Tr}_{\mathbb{E}/\mathbb{F}}(\beta_1) + \alpha_2\mathrm{Tr}_{\mathbb{E}/\mathbb{F}}(\beta_2)$$

for all $\alpha_i \in \mathbb{F}$ and $\beta_i \in \mathbb{E}$, $i = 1, 2$. As a consequence, $\mathrm{Im}\mathrm{Tr}_{\mathbb{E}/\mathbb{F}}$ is an $\mathbb{F}$-vector subspace of $\mathbb{F}$ and therefore (being $\mathbb{F}$ a field) it is either equal to $\{0\}$ or to the whole $\mathbb{F}$. But the first possibility implies that $\mathrm{Tr}_{\mathbb{E}/\mathbb{F}}$ is identically zero, which leads to a contradiction since it is the sum of $\mathbb{F}$-linearly independent $\mathbb{F}$-linear transformations of $\mathbb{E}$ (cf. Proposition 6.2.12). This shows that $\mathrm{Tr}_{\mathbb{E}/\mathbb{F}}$ is surjective. As a consequence,

$$|\mathrm{Ker}\mathrm{Tr}_{\mathbb{E}/\mathbb{F}}| = \frac{|\mathbb{E}|}{|\mathbb{F}|} = q^{h-1}.$$

Moreover, every element of the form $\alpha - \sigma(\alpha)$, with $\alpha \in \mathbb{E}$, clearly belongs to $\mathrm{Ker}\mathrm{Tr}_{\mathbb{E}/\mathbb{F}}$. Also, for α and β in $\mathbb{E}$ we have $\alpha - \sigma(\alpha) = \beta - \sigma(\beta)$ if and only if $\alpha - \beta = \sigma(\alpha - \beta)$, equivalently $\alpha - \beta \in \mathbb{F}$. We deduce that the set

$$\{\alpha - \sigma(\alpha) : \alpha \in \mathbb{E}\},$$

which consists of exactly q^{h-1} elements, coincides with $\mathrm{Ker}\mathrm{Tr}_{\mathbb{E}/\mathbb{F}}$.

(ii) As for (i), it is easy to check that $N_{\mathbb{E}/\mathbb{F}}$ is a group homomorphism between $\mathbb{E}^*$ and $\mathbb{F}^*$: we leave the details to the reader. Moreover, we have

$$N_{\mathbb{E}/\mathbb{F}}(\alpha) = \prod_{k=1}^{h} \sigma^k(\alpha) = \alpha^q\alpha^{q^2}\cdots\alpha^{q^{h-1}}\alpha = \alpha^{\sum_{k=0}^{h-1} q^k} = \alpha^{(q^h-1)/(q-1)}$$

for all $\alpha \in \mathbb{E}$. In particular, if α is a generator of $\mathbb{E}^*$, so that it has order $q^h - 1$, then $N_{\mathbb{E}/\mathbb{F}}(\alpha)$ has order $q - 1$ and therefore generates $\mathbb{F}^*$. It follows that $N_{\mathbb{E}/\mathbb{F}}$ is surjective. As a consequence,

$$|\mathrm{Ker}N_{\mathbb{E}/\mathbb{F}}| = \frac{|\mathbb{E}^*|}{|\mathbb{F}^*|} = \frac{q^h - 1}{q - 1}. \tag{6.25}$$

Moreover, every element of the form $\alpha\sigma(\alpha)^{-1}$, with $\alpha \in \mathbb{E}^*$, clearly belongs to $\mathrm{KerN}_{\mathbb{E}/\mathbb{F}}$. Also, for α and β in $\mathbb{E}^*$ we have $\alpha\sigma(\alpha)^{-1} = \beta\sigma(\beta)^{-1}$ if and only if $\alpha\beta^{-1} = \sigma(\alpha\beta^{-1})$, equivalently $\alpha\beta^{-1} \in \mathbb{F}^*$. We deduce that the set $\{\alpha\sigma(\alpha)^{-1} : \alpha \in \mathbb{E}^*\}$ has $(q^h - 1)/(q - 1)$ elements and therefore (cf.(6.25)) equals $\mathrm{KerN}_{\mathbb{E}/\mathbb{F}}$.

$\square$

Proposition 6.7.3 *Let* $\mathbb{F} \subseteq \mathbb{E}$ *be finite fields. Let* $\alpha \in \mathbb{E}$ *and suppose that* $\mathbb{E} = \mathbb{F}[\alpha]$. *Then, denoting by* $f(x) = x^h + a_{h-1}x^{h-1} + \cdots + a_1 x + a_0 \in \mathbb{F}[x]$ *the minimal polynomial of* α, *we have*

$$- a_{h-1} = \sum_{k=1}^{h} \sigma^k(\alpha) \equiv \mathrm{Tr}_{\mathbb{E}/\mathbb{F}}(\alpha) \tag{6.26}$$

and

$$(-1)^h a_0 = \prod_{k=1}^{h} \sigma^k(\alpha) \equiv \mathrm{N}_{\mathbb{E}/\mathbb{F}}(\alpha). \tag{6.27}$$

Proof. By virtue of Corollary 6.6.4 and Corollary 6.6.5 it follows that f is factorizable over $\mathbb{E}$ and its roots are precisely the elements $\sigma^k(\alpha), k = 1, 2, \ldots, h$. That is, $f(x) = (x - \alpha)(x - \sigma(\alpha)) \cdots (x - \sigma^{h-1}(\alpha))$, so that (6.26) and (6.27) follow. $\square$

Since, by definition, $f(\alpha) = 0$, we have $f(\sigma^k(\alpha)) = \sigma^k(f(\alpha)) = 0$ for all $k = 1, 2, \ldots, h$; moreover the elements $\sigma^k(\alpha) \in \mathbb{E}, k = 1, 2, \ldots, h$ are distinct.

Theorem 6.7.4 *Let* $\mathbb{F} \subseteq \mathbb{E}$ *be finite fields and let* $\alpha \in \mathbb{E}$. *Consider the* $\mathbb{F}$-*linear transformation* $\lambda(\alpha) \colon \mathbb{E} \to \mathbb{E}$ *defined by setting*

$$\lambda(\alpha)\beta = \alpha\beta$$

for all $\beta \in \mathbb{E}$. *Then we have*

$$\mathrm{Tr}\lambda(\alpha) = \mathrm{Tr}_{\mathbb{E}/\mathbb{F}}(\alpha)$$

and

$$\det \lambda(\alpha) = \mathrm{N}_{\mathbb{E}/\mathbb{F}}(\alpha).$$

Proof. Set $h = [\mathbb{E} : \mathbb{F}]$.

We first prove the statement under the hypothesis that $\mathbb{E} = \mathbb{F}[\alpha]$. In this case (see Proposition 6.2.4), we have that the elements

$$1, \alpha, \alpha^2, \ldots, \alpha^{h-1} \tag{6.28}$$

constitute a basis for the vector space $\mathbb{E}$ over $\mathbb{F}$ and the minimal polynomial $f \in \mathbb{F}[x]$ of α has degree h. We denote it by

$$f(x) = x^h + a_{h-1}x^{h-1} + \cdots + a_1 x + a_0. \tag{6.29}$$

Since $f(\lambda(\alpha)) = \lambda(f(\alpha))$, we have that f is the minimal polynomial of $\lambda(\alpha) \in \mathrm{End}_{\mathbb{F}}(\mathbb{E})$. Since the characteristic polynomial

$$p_{\lambda(\alpha)}(x) = \det(xI - \lambda(\alpha))$$

of $\lambda(\alpha)$ also has degree h, from the Cayley-Hamilton theorem, we deduce that, in fact, $f = p_{\lambda(\alpha)}$.

Keeping in mind (6.29), we have that the matrix $M_{\lambda(\alpha)}$ representing $\lambda(\alpha)$ in the basis (6.28) is the so-called *companion matrix* of f, namely

$$M_{\lambda(\alpha)} = \begin{pmatrix} 0 & 0 & 0 & \cdots & 0 & 0 & 0 & -a_0 \\ 1 & 0 & 0 & \cdots & 0 & 0 & 0 & -a_1 \\ 0 & 1 & 0 & \cdots & 0 & 0 & 0 & -a_2 \\ & & & \ddots & & & & \\ 0 & 0 & 0 & \cdots & 0 & 1 & 0 & -a_{h-2} \\ 0 & 0 & 0 & \ddots & 0 & 0 & 1 & -a_{h-1} \end{pmatrix}. \tag{6.30}$$

From this we deduce that

$$\mathrm{Tr}\lambda(\alpha) = \mathrm{Tr}M_{\lambda(\alpha)} = -a_{h-1} \quad \text{and} \quad \det\lambda(\alpha) = \det M_{\lambda(\alpha)} = (-1)^h a_0. \tag{6.31}$$

Comparing (6.26) and (6.27) with (6.31), the statement follows in the case $\mathbb{F}[\alpha] = \mathbb{E}$.

Suppose now that $\mathbb{F}[\alpha]$ is a proper subfield of $\mathbb{E}$. Then $m = [\mathbb{F}[\alpha] : \mathbb{F}]$ divides h (cf. Proposition 6.6.1). Let $\{u_j : j = 1, 2, \ldots, h/m\}$ be a vector space basis of $\mathbb{E}$ over $\mathbb{F}[\alpha]$. Moreover, as before, the elements α^k, $k = 1, 2, \ldots, m$, constitute a basis of $\mathbb{F}[\alpha]$ over $\mathbb{F}$. As a consequence of these facts,

$$\{\alpha^k u_j : k = 1, 2, \ldots, m; j = 1, 2, \ldots, h/m\}$$

is a vector space basis of $\mathbb{E}$ over $\mathbb{F}$. Thus, setting $U_j = \mathrm{span}_{\mathbb{F}}\{\alpha^k u_j : k = 1, 2, \ldots m\}$ for $j = 1, 2, \ldots, h/m$, we have the direct sum decomposition

$$\mathbb{F} = \bigoplus_{j=1}^{h/m} U_j$$

into $\lambda(\alpha)$-invariant subspaces. Moreover, $\lambda(\alpha)|_{U_j}$ is represented by an $m \times m$ matrix $M_{\lambda(\alpha)|_{U_j}}$ (in fact, independent of j) with coefficients in $\mathbb{F}$ as in (6.30)

$$
M_{\lambda(\alpha)|_{U_j}} = \begin{pmatrix}
0 & 0 & 0 & \cdots & 0 & 0 & 0 & -a_0 \\
1 & 0 & 0 & \cdots & 0 & 0 & 0 & -a_1 \\
0 & 1 & 0 & \cdots & 0 & 0 & 0 & -a_2 \\
& & & \ddots & & & & \\
0 & 0 & 0 & \cdots & 0 & 1 & 0 & -a_{m-2} \\
0 & 0 & 0 & \ddots & 0 & 0 & 1 & -a_{m-1}
\end{pmatrix},
$$

namely the companion matrix of the minimal polynomial $f(x) = x^m + a_{m-1}x^{m-1} + \cdots + a_1 x + a_0 \in \mathbb{F}[x]$ of α. Then, on the one hand, we have

$$
\mathrm{Tr}\lambda(\alpha) = \sum_{j=1}^{h/m} \mathrm{Tr}\left(\lambda(\alpha)|_{U_j}\right) = \sum_{j=1}^{h/m} \mathrm{Tr}\left(M_{\lambda(\alpha)|_{U_j}}\right) = \frac{h}{m}(-a_{m-1})
$$

and

$$
\det\lambda(\alpha) = \prod_{j=1}^{h/m} \det\left(\lambda(\alpha)|_{U_j}\right) = \prod_{j=1}^{h/m} \det\left(M_{\lambda(\alpha)|_{U_j}}\right) = ((-1)^m a_0)^{h/m}.
$$

On the other hand,

$$
\mathrm{Tr}_{\mathbb{E}/\mathbb{F}}(\alpha) = \sum_{k=1}^{h} \sigma^k(\alpha) =_* \frac{h}{m} \sum_{k=1}^{m} \sigma^k(\alpha) = \frac{h}{m}\mathrm{Tr}_{\mathbb{F}[\alpha]/\mathbb{F}}(\alpha) = \frac{h}{m}(-a_{m-1})
$$

where the last equality follows from (6.26), and

$$
\mathrm{N}_{\mathbb{E}/\mathbb{F}}(\alpha) = \prod_{k=1}^{h} \sigma^k(\alpha) =_* \left(\prod_{k=1}^{m} \sigma^k(\alpha)\right)^{h/m} = \left(\mathrm{N}_{\mathbb{F}[\alpha]/\mathbb{F}}(\alpha)\right)^{h/m} = ((-1)^m a_0)^{h/m}
$$

where the last equality follows from (6.27), and $=_*$ both follow from the equality $\mathrm{Gal}(\mathbb{E}, \mathbb{F}[\alpha]) = \langle \sigma^m \rangle$ (cf. Proposition 6.6.3). Thus, the general case follows as well. $\qquad\square$

6.8 Quadratic extensions

We now concentrate on the case of quadratic extensions. We split the analysis according to the parity of the characteristic p of the fields. Our purpose is to produce matrix representations of quadratic extensions similar to the well known matrix representation of the complex numbers $z = a + ib \mapsto \begin{pmatrix} a & -b \\ b & a \end{pmatrix}$, for all $a, b \in \mathbb{R}$. We begin with some general considerations.

Let p be a prime number, h a positive integer, and set $q = p^h$. Then $\mathrm{Gal}(\mathbb{F}_{q^2}/\mathbb{F}_q)$ is a cyclic group of order two. More precisely, it is generated by the automorphism σ defined by $\sigma(\alpha) = \alpha^q$ for all $\alpha \in \mathbb{F}_{q^2}$, which clearly fixes every element $\alpha \in \mathbb{F}_q$, and is involutory (cf. Corollary 6.4.3 and Proposition 6.6.3). By virtue of Proposition 6.5.4, the polynomial ring $\mathbb{F}_q[x]$ contains $(q^2 - q)/2$ irreducible monic polynomials of degree 2 and $\mathbb{F}_{q^2}$ may be obtained, abstractly, by adjoining one of the roots of any of these. Moreover, if $x^2 + ax + b \in \mathbb{F}_q[x]$ is irreducible over $\mathbb{F}_q$ and α, β are its roots, then $\sigma(\alpha) = \beta$ (and $\sigma(\beta) = \alpha$). Indeed, since σ fixes $\mathbb{F}_q$ pointwise, we have

$$\sigma(\alpha^2 + a\alpha + b) = \sigma(\alpha)^2 + a\sigma(\alpha) + b$$

so that $\sigma(\alpha)$ is still a root. But σ fixes exactly the elements in $\mathbb{F}_q$ so that, since $\alpha \notin \mathbb{F}_q$, we necessarily have $\sigma(\alpha) \neq \alpha$ and therefore $\sigma(\alpha) = \beta$.

We first examine the case when p is odd.

Theorem 6.8.1 *Suppose p is odd. Let η be a generator of the cyclic group $\mathbb{F}_q^*$ (cf. Theorem 6.3.3) and denote by $\pm i$ the square roots of η. Then $\pm i \notin \mathbb{F}_q$ and $\{1, i\}$ is a vector space basis for $\mathbb{F}_{q^2}$ over $\mathbb{F}_q$. Moreover, $\mathbb{F}_{q^2}$ is isomorphic (as an $\mathbb{F}_q$-algebra) to the algebra $\mathfrak{M}_2(\mathbb{F}_q, \eta) \subseteq \mathfrak{M}_2(\mathbb{F}_q)$ consisting of all matrices of the form*

$$\begin{pmatrix} \alpha & \eta\beta \\ \beta & \alpha \end{pmatrix}$$

with $\alpha, \beta \in \mathbb{F}_q$. The isomorphism is provided by the map $\mathfrak{M}_2(\mathbb{F}_q, \eta) \to \mathbb{F}_{q^2}$ given by

$$\begin{pmatrix} \alpha & \eta\beta \\ \beta & \alpha \end{pmatrix} \mapsto \alpha + i\beta \tag{6.32}$$

for all $\alpha, \beta \in \mathbb{F}_q$. Moreover

$$\sigma(\alpha + i\beta) = \alpha - i\beta$$

for all $\alpha, \beta \in \mathbb{F}_q$.

Proof. First observe that, under our assumptions on the parity of p, the order $q - 1$ of the cyclic group $\mathbb{F}_q^*$ is *even*. If we had $i \in \mathbb{F}_q$ then we would have

$$\eta^{\frac{q-1}{2}} = \left(i^2\right)^{\frac{q-1}{2}} = i^{q-1} = 1$$

which is impossible (since η has order $q - 1$).

Alternatively, η cannot be a square in $\mathbb{F}_q^*$ since, otherwise, every other element in $\mathbb{F}_q^*$ would also be a square, contradicting Proposition 6.4.4.

As a consequence, the polynomial $x^2 - \eta \in \mathbb{F}_q[x]$ is irreducible and therefore $\mathbb{F}_{q^2} = \mathbb{F}_q[i]$ so that, by Proposition 6.2.5, $\{1, i\}$ is a vector space basis of $\mathbb{F}_{q^2}$ over $\mathbb{F}_q$. We thus have

$$\mathbb{F}_{q^2} = \{\alpha + i\beta : \alpha, \beta \in \mathbb{F}_q\}$$

with addition given by

$$(\alpha_1 + i\beta_1) + (\alpha_2 + i\beta_2) = (\alpha_1 + \alpha_2) + i(\beta_1 + \beta_2)$$

and multiplication given by

$$(\alpha_1 + i\beta_1)(\alpha_2 + i\beta_2) = (\alpha_1\alpha_2 + \eta\beta_1\beta_2) + i(\alpha_1\beta_2 + \alpha_2\beta_1)$$

for all $\alpha_1, \alpha_2, \beta_1, \beta_2 \in \mathbb{F}_q$. Moreover, since $\sigma(i) = -i$ we also have

$$\sigma(\alpha + i\beta) = \alpha - i\beta$$

for all $\alpha, \beta \in \mathbb{F}_q$. Finally, as

$$\begin{pmatrix} \alpha_1 & \eta\beta_1 \\ \beta_1 & \alpha_1 \end{pmatrix} \begin{pmatrix} \alpha_2 & \eta\beta_2 \\ \beta_2 & \alpha_2 \end{pmatrix} = \begin{pmatrix} \alpha_1\alpha_2 + \eta\beta_1\beta_2 & \eta(\alpha_1\beta_2 + \alpha_2\beta_1) \\ \alpha_1\beta_2 + \alpha_2\beta_1 & \alpha_1\alpha_2 + \eta\beta_1\beta_2 \end{pmatrix}$$

we deduce that the map (6.32) is indeed an isomorphism. $\square$

Corollary 6.8.2 *Suppose p is odd. Then the $(q^2 - q)/2$ irreducible monic quadratic polynomials in $\mathbb{F}_q[x]$ (cf. Proposition 6.5.4) are exactly the polynomials*

$$p_{\alpha,\beta}(x) = x^2 - 2\alpha x + (\alpha^2 - \beta^2\eta)$$

where $\alpha \in \mathbb{F}_q$ and $\beta \in \mathbb{F}_q^*$.

Proof. Any irreducible monic quadratic polynomial over $\mathbb{F}_q$ is necessarily of the form $[x - (\alpha + i\beta)][x - \sigma(\alpha + i\beta)]$, with $\alpha, \beta \in \mathbb{F}_q$ and $\beta \neq 0$. Since $\sigma(\alpha + i\beta) = \alpha - i\beta$, the statement follows. (Note that $p_{\alpha,-\beta} = p_{\alpha,\beta}$.) $\square$

We now examine the case $p = 2$. Recall (cf. Proposition 6.4.4) that, in this case, all elements in $\mathbb{F}_{2^h}$ are squares.

Theorem 6.8.3 *There exists $j \in \mathbb{F}_{2^{2h}} \setminus \mathbb{F}_{2^h}$ and $\omega \in \mathbb{F}_{2^h}$ such that*

$$j^2 + j + \omega = 0 \quad (\text{equivalently,} \quad j^2 = j + \omega)$$

and

$$\mathbb{F}_{2^{2h}} = \mathbb{F}_{2^h}[j].$$

Moreover, the polynomial $x^2 + x + \omega \in \mathbb{F}_{2^h}[x]$ is irreducible *and the map*

$$\begin{pmatrix} \alpha & \omega\beta \\ \beta & \alpha + \beta \end{pmatrix} \mapsto \alpha + j\beta \tag{6.33}$$

yields an ($\mathbb{F}_{2^h}$-algebra) isomorphism of the algebra $\mathfrak{M}_2(\mathbb{F}_{2^h}, \omega) \subseteq \mathfrak{M}_2(\mathbb{F}_{2^h})$ consisting of all the matrices of the form

$$\begin{pmatrix} \alpha & \omega\beta \\ \beta & \alpha + \beta \end{pmatrix}$$

where $\alpha, \beta \in \mathbb{F}_{2^h}$, onto the field $\mathbb{F}_{2^{2h}}$. Finally,

$$\sigma(\alpha + j\beta) = (\alpha + \beta) + j\beta$$

for all $\alpha, \beta \in \mathbb{F}_{2^h}$.

Proof. Since $\mathbb{F}_{2^{2h}}$ is a quadratic extension of $\mathbb{F}_{2^h}$, there exists an irreducible polynomial $f(x) = x^2 + \alpha x + \beta \in \mathbb{F}_{2^h}[x]$ such that $\mathbb{F}_{2^{2h}} = \mathbb{F}_{2^h}[j]$, where $j \in \mathbb{F}_{2^{2h}} \setminus \mathbb{F}_{2^h}$ is a root of f. Note that $\alpha \neq 0$: otherwise, the polynomial $f(x) = x^2 + \beta$ would be reducible since every element in $\mathbb{F}_{2^h}$ is a square.

Thus, setting $y = x\alpha^{-1}$ and $\omega = \beta\alpha^{-2} \in \mathbb{F}_{2^h}$, the equation $x^2 + \alpha x + \beta = 0$ becomes $\alpha^2 y^2 + \alpha^2 y + \beta = 0$, equivalently, $y^2 + y + \omega = 0$.

Let then $j, j' \in \mathbb{F}_{2^{2h}}$ be the roots of $x^2 + x + \omega$, so that $(x - j)(x - j') = x^2 + x + \omega$, yielding $j + j' = 1$ and $jj' = \omega$. Thus $j' = 1 + j = \omega j^{-1}$ and $j^2 = \omega + j$. As a consequence, in the basis $\{1, j\}$ of $\mathbb{F}_{2^{2h}}$ over $\mathbb{F}_{2^h}$, addition and multiplication are given by

$$(\alpha_1 + j\beta_1) + (\alpha_2 + j\beta_2) = (\alpha_1 + \alpha_1) + j(\beta_1 + \beta_2)$$

and

$$(\alpha_1 + j\beta_1)(\alpha_2 + j\beta_2) = (\alpha_1\alpha_2 + \omega\beta_1\beta_2) + j(\alpha_1\beta_2 + \alpha_2\beta_1 + \beta_1\beta_2) \tag{6.34}$$

for all $\alpha_1, \alpha_2, \beta_1, \beta_2 \in \mathbb{F}_{2^h}$. Clearly, $\sigma(j) = j' = 1 + j = \omega j^{-1}$ and therefore

$$\sigma(\alpha + j\beta) = (\alpha + \beta) + j\beta$$

for all $\alpha, \beta \in \mathbb{F}_{2^h}$. Finally, we have

$$\begin{pmatrix} \alpha_1 & \omega\beta_1 \\ \beta_1 & \alpha_1 + \beta_1 \end{pmatrix} \begin{pmatrix} \alpha_2 & \omega\beta_2 \\ \beta_2 & \alpha_2 + \beta_2 \end{pmatrix}$$
$$= \begin{pmatrix} \alpha_1\alpha_2 + \omega\beta_1\beta_2 & \omega(\alpha_1\beta_2 + \alpha_2\beta_1 + \beta_1\beta_2) \\ \alpha_1\beta_2 + \alpha_2\beta_1 + \beta_1\beta_2 & \alpha_1\alpha_2 + \omega\beta_1\beta_2 + \alpha_1\beta_2 + \alpha_2\beta_1 + \beta_1\beta_2 \end{pmatrix}$$

for all $\alpha_1, \alpha_2, \beta_1, \beta_2 \in \mathbb{F}_{2^h}$. From (6.34) we deduce that the map (6.33) yields the desired isomorphism. $\qquad\square$

Corollary 6.8.4 *The $2^{2h-1} - 2^{h-1}$ irreducible monic quadratic polynomials in* $\mathbb{F}_{2^h}[x]$ *(cf. Proposition 6.5.4) are exactly the polynomials*

$$q_{\alpha,\beta}(x) = x^2 + \beta x + (\alpha^2 + \alpha\beta + \beta^2\omega)$$

where $\beta \in \mathbb{F}_{2^h}^*$ *and* $\alpha \in \mathbb{F}_{2^h}$.

Proof. Any irreducible monic quadratic polynomial over $\mathbb{F}_{2^h}$ is necessarily of the form $(x + (\alpha + j\beta))(x + \sigma(\alpha + j\beta))$ with $\alpha, \beta \in \mathbb{F}_q$ and $\beta \neq 0$. Since $\sigma(\alpha + j\beta) = (\alpha + \beta) + j\beta$, the statement follows. (Note that $q_{\alpha,\beta} = q_{\alpha',\beta'}$ if and only if $\beta' = \beta$ and $\alpha' \in \{\alpha, \alpha + \beta\}$). $\qquad\square$

In view of the next chapters, we set

$$\overline{\alpha} = \sigma(\alpha)$$

and call it the *conjugate* of $\alpha \in \mathbb{F}_{q^2}$. Explicit expressions are given in Theorem 6.8.1 and Theorem 6.8.3. Note also that

$$N_{\mathbb{F}_{q^2}/\mathbb{F}_q}(\alpha) = \alpha\overline{\alpha}$$

and

$$\text{Tr}_{\mathbb{F}_{q^2}/\mathbb{F}_q}(\alpha) = \alpha + \overline{\alpha}$$

for all $\alpha \in \mathbb{F}_{q^2}$. Moreover, $\alpha = \overline{\alpha}$ if and only if $\alpha \in \mathbb{F}_q$ (see also [86]).

7

Character theory of finite fields

In this chapter we give an introduction to the character theory of finite fields. Our exposition is mainly based on the books by Ireland and Rosen [79], Winnie Li [95], and by Lidl and Niederreiter [96]. Actually, one of the main goals is to present the generalized Kloosterman sums from Piatetski-Shapiro's monograph [123], which will play a fundamental role in Chapter 14 on the representation theory of GL$(2, \mathbb{F}_q)$. We also introduce the reader to the study of the number of solutions of equations over finite fields. This is quite a vast and difficult subject, which culminates with very deep results such as the Weil conjecture, proved by Deligne (see [95]). Finally, Section 7.8, devoted to the FFT over finite fields, is based on the book by Tolimieri, An, and Lu [160].

7.1 Generalities on additive and multiplicative characters

Let p be a prime number, n a positive integer, and consider $\mathbb{F}_q$, the finite field of order $q = p^n$. An *additive character* of $\mathbb{F}_q$ is a character of the finite Abelian group $(\mathbb{F}_q, +)$ (cf. Definition 2.3.1), that is, a map

$$\chi : \mathbb{F}_q \to \mathbb{T}$$

such that $\chi(x + y) = \chi(x)\chi(y)$ for all $x, y \in \mathbb{F}_q$ (here, as usual, $\mathbb{T} = \{z \in \mathbb{C} : |z| = 1\}$ is the (multiplicative) circle group). We observe (cf. Definition 2.3.1) that the additive characters constitute a (multiplicative) Abelian group, denoted by $\widehat{\mathbb{F}_q}$, called the *dual group* of $\mathbb{F}_q$. Clearly, if χ is an additive character, then

$$\overline{\chi(x)} = \chi(x)^{-1} = \chi(-x) = \chi^{-1}(x)$$

for all $x \in \mathbb{F}_q$. Moreover, for $\chi, \xi \in \widehat{\mathbb{F}_q}$, the orthogonality relations (cf. Proposition 2.3.5) are:

$$\langle \chi, \xi \rangle = \sum_{x \in \mathbb{F}_q} \chi(x)\overline{\xi(x)} = \begin{cases} q & \text{if } \chi = \xi \\ 0 & \text{if } \chi \neq \xi. \end{cases} \tag{7.1}$$

In particular, taking $\xi = \mathbf{1}$, we have

$$\sum_{x \in \mathbb{F}_q^*} \chi(x) = -1 \text{ for all } \chi \neq \mathbf{1}, \tag{7.2}$$

since $\sum_{x \in \mathbb{F}_q} \chi(x) = 0$ and $\chi(0) = 1$.

The *principal* (or *canonical*) *additive character* of $\mathbb{F}_q$ is defined by setting, for all $x \in \mathbb{F}_q$,

$$\chi_{princ}(x) = \exp[2\pi i \mathrm{Tr}(x)/p], \tag{7.3}$$

where $\mathrm{Tr} = \mathrm{Tr}_{\mathbb{F}_q/\mathbb{F}_p}$ denotes the trace (cf. (6.22)) and, as usual, we identify $\mathbb{F}_p$ with $\{0, 1, \ldots, p-1\}$ to compute the exponential. Since Tr is a surjective $\mathbb{F}_p$-linear map from $\mathbb{F}_q$ onto $\mathbb{F}_p$ (so that, in particular, $\mathrm{Tr}(x+y) = \mathrm{Tr}(x) + \mathrm{Tr}(y)$ for all $x, y \in \mathbb{F}_q$) by Hilbert Satz 90 (cf. Theorem 6.7.2), χ_{princ} is indeed a nontrivial additive character.

In the following we present another explicit isomorphism between $(\mathbb{F}_q, +)$ and its dual group $\widehat{\mathbb{F}_q}$ (cf. Corollary 2.3.4).

Proposition 7.1.1 *Let* χ *be a nontrivial additive character of* $\mathbb{F}_q$*. For each* $y \in \mathbb{F}_q$ *define* $\chi_y \colon \mathbb{F}_q \to \mathbb{T}$ *by setting*

$$\chi_y(x) = \chi(xy)$$

for all $x \in \mathbb{F}_q$*. Then* χ_y *is also an additive character of* $\mathbb{F}_q$*, and the map*

$$\Psi \colon \mathbb{F}_q \to \widehat{\mathbb{F}_q}$$
$$y \mapsto \chi_y$$

is a group isomorphism.

Proof. The fact that χ_y is an additive character and that Ψ is a group homomorphism follow immediately from the distributivity law in $\mathbb{F}_q$. Indeed,

$$\chi_y(x_1 + x_2) = \chi(y(x_1 + x_2)) = \chi(yx_1 + yx_2) = \chi(yx_1)\chi(yx_2) = \chi_y(x_1)\chi_y(x_2)$$

and

$$\chi_{y+z}(x) = \chi((y+z)x) = \chi(yx + zx) = \chi(yx)\chi(zx) = \chi_y(x)\chi_z(x)$$

for all x, x_1, x_2, y, and z in $\mathbb{F}_q$.

Suppose now that $y \neq 0$. Since χ is nontrivial, we can find $\bar{x} \in \mathbb{F}_q$ such that $\chi(\bar{x}) \neq 1$. Let $x = y^{-1}\bar{x}$, then $\chi_y(x) = \chi(yx) = \chi(\bar{x}) \neq 1$. Thus $y \notin \mathrm{Ker}(\Psi)$. This shows that Ψ is injective. Since $|\widehat{\mathbb{F}_q}| = |\mathbb{F}_q| = q$ (cf. Corollary 2.3.4), we deduce that Ψ is also surjective. $\qquad\square$

Exercise 7.1.2 Show that $\widehat{\mathbb{F}_q^2} = \{\chi_{s,t} : s, t \in \mathbb{F}_q\}$, where

$$\chi_{s,t}(x, y) = \chi_{princ}(sx + ty) \qquad (7.4)$$

for all $s, t, x, y, \in \mathbb{F}_q$.

Corollary 7.1.3 *Let $\chi \in \widehat{\mathbb{F}_q}$ be a nontrivial additive character. Then for all $z \in \mathbb{F}_q$ we have*

$$\sum_{x \in \mathbb{F}_q^*} \chi(xz) = \begin{cases} q-1 & \text{if } z = 0 \\ -1 & \text{if } z \neq 0. \end{cases}$$

Proof. It is an immediate consequence of Proposition 7.1.1 and (7.2). $\qquad\square$

If we choose $\chi = \chi_{princ}$, we get the *canonical isomorphism* between $\mathbb{F}_q$ and $\widehat{\mathbb{F}_q}$:

$$\chi_y(x) = \exp[2\pi i \mathrm{Tr}(xy)/p]; \qquad (7.5)$$

in particular, $\chi_1 = \chi = \chi_{princ}$, where 1 is the (multiplicative) identity element in the field $\mathbb{F}_q$, and $\chi_0 = \mathbf{1}$, the trivial character.

A *multiplicative character* of $\mathbb{F}_q$ is a character of the finite cyclic group $(\mathbb{F}_q^*, \cdot)$ (cf. Theorem 6.3.3 and Definition 2.3.1), that is, a map

$$\psi : \mathbb{F}_q^* \to \mathbb{T}$$

such that $\psi(xy) = \psi(x)\psi(y)$ for all $x, y \in \mathbb{F}_q^*$. We observe (cf. Definition 2.3.1) that the set $\widehat{\mathbb{F}_q^*}$ of all multiplicative characters is a (multiplicative) cyclic (cf. Remark 2.3.2) group, called the *dual group* of $\mathbb{F}_q^*$.

We can extend a multiplicative character $\psi \in \widehat{\mathbb{F}_q^*}$ to a map $\mathbb{F}_q \to \mathbb{T} \cup \{0\}$ (still denoted by ψ), by setting

$$\psi(0) = \begin{cases} 0 & \text{if } \psi \text{ is nontrivial} \\ 1 & \text{if } \psi = \mathbf{1}. \end{cases} \tag{7.6}$$

Clearly, if ψ is a multiplicative character, then

$$\overline{\psi(x)} = \psi(x)^{-1} = \psi(x^{-1}) = \psi^{-1}(x)$$

for all $x \in \mathbb{F}_q^*$.

Let $\psi \in \widehat{\mathbb{F}_q^*}$. In the following, we shall often encounter the quantity $\psi(-1)$: since $\psi(-1)^2 = \psi[(-1)^2] = \psi(1) = 1$, we necessarily have $\psi(-1) = \pm 1$. The *order* of ψ is the smallest positive integer m such that $\psi^m = \mathbf{1}$: clearly, m divides $q - 1$, since $\psi(x)^{q-1} = \psi(x^{q-1}) = \psi(1) = 1$ (alternatively, this is an immediate consequence of Lagrange's theorem; see Proposition 1.2.12). We recall (cf. Definition 6.3.4), that $x \in \mathbb{F}_q^*$ is called a primitive element of $\mathbb{F}_q$ if it generates $\mathbb{F}_q^*$.

Lemma 7.1.4 *Let ψ be a nontrivial multiplicative character of $\mathbb{F}_q$ and denote by m its order. Then $\psi(-1) = -1$ if and only if m is even and $\frac{q-1}{m}$ is odd.*

Proof. Since $\psi(x)^m = \psi^m(x) = 1$ for all $x \in \mathbb{F}_q^*$, all the values of ψ are m-th roots of unity. Let also x be a primitive element of $\mathbb{F}_q$. Then $\psi(x)$ is a primitive m-th root of 1, so that $\psi(x)^h \neq 1$ for $1 \leq h \leq m - 1$.

If m is odd, then -1 is not an m-th root of unity and therefore $\psi(-1)$ is necessary equal to 1.

Suppose now that m is even. Then $\psi(x)^h = -1$ if and only if $h \equiv \frac{m}{2} \mod m$. Moreover (note that $q - 1$ is even, because it is divisible by m), $x^{\frac{q-1}{2}} = -1$ (since $x^{q-1} = 1$ but $x^{\frac{q-1}{2}} \neq 1$). It follows that

$$\psi(-1) = \psi(x^{\frac{q-1}{2}}) = \psi(x)^{\frac{q-1}{2}}$$

so that

$$\psi(-1) = -1 \Leftrightarrow \frac{q-1}{2} \equiv \frac{m}{2} \mod m$$

$$\Leftrightarrow \frac{q-1}{m} \equiv 1 \mod 2$$

$$\Leftrightarrow \frac{q-1}{m} \text{ is odd.}$$

$\square$

Exercise 7.1.5 Fill in the details of the above equivalence $\frac{q-1}{2} \equiv \frac{m}{2} \mod m \Leftrightarrow \frac{q-1}{m} \equiv 1 \mod 2$.

Let $\psi, \phi \in \widehat{\mathbb{F}_q^*}$. The orthogonality relations are (cf. Proposition 2.3.5):

$$\langle \psi, \phi \rangle = \sum_{x \in \mathbb{F}_q^*} \psi(x)\overline{\phi(x)} = \begin{cases} q - 1 & \text{if } \psi = \phi \\ 0 & \text{if } \psi \neq \phi. \end{cases} \tag{7.7}$$

As a consequence, if ψ is nontrivial (taking ϕ the trivial character) we have

$$\sum_{x \in \mathbb{F}_q^* \setminus \{-1\}} \psi(x) = -\psi(-1) \tag{7.8}$$

so that, keeping in mind (7.6),

$$\sum_{x \in \mathbb{F}_q} \psi(x) = 0. \tag{7.9}$$

The dual orthogonal relations (cf. (2.13)) are

$$\sum_{\psi \in \widehat{\mathbb{F}_q^*}} \psi(x)\overline{\psi(y)} = \begin{cases} q - 1 & \text{if } x = y \\ 0 & \text{if } x \neq y. \end{cases} \tag{7.10}$$

Let x be a primitive element of $\mathbb{F}_q$. The *principal multiplicative character* of $\mathbb{F}_q^*$ associated with x is the multiplicative character ψ_{princ} defined by setting

$$\psi_{princ}(x^k) = \exp\left(\frac{2\pi i k}{q - 1}\right) \tag{7.11}$$

for all $k = 1, 2, \ldots, q - 1$.

Exercise 7.1.6 Show that ψ_{princ} is a generator of $\widehat{\mathbb{F}_q^*}$.

7.2 Decomposable characters

We fix $q = p^n$ and consider the field $\mathbb{F}_q$ together with its quadratic extension $\mathbb{F}_{q^2}$. We use the notation at the end of Section 6.8. In particular, if $\alpha \in \mathbb{F}_{q^2}^*$ then its conjugate is the element $\bar{\alpha} = \sigma(\alpha) \in \mathbb{F}_{q^2}^*$ and we have $\alpha\bar{\alpha} = N_{\mathbb{F}_{q^2}/\mathbb{F}_q}(\alpha)$ and $\alpha + \bar{\alpha} = \text{Tr}_{\mathbb{F}_{q^2}/\mathbb{F}_q}(\alpha) \in \mathbb{F}_q$.

Definition 7.2.1 Let ν be a character of $\mathbb{F}_{q^2}^*$.

One says that ν is *decomposable* if there exists a character ψ of $\mathbb{F}_q^*$ such that

$$\nu(\alpha) = \psi(\alpha\bar{\alpha}) \tag{7.12}$$

for all $\alpha \in \mathbb{F}_{q^2}^*$. If this is not the case, ν is called *indecomposable*.

Moreover, the *conjugate* of ν is the character $\overline{\nu}$ defined by

$$\overline{\nu}(\alpha) = \nu(\overline{\alpha}) \tag{7.13}$$

for all $\alpha \in \mathbb{F}_{q^2}^*$.

Proposition 7.2.2 *A character* $\nu \in \widehat{\mathbb{F}_{q^2}^*}$ *is decomposable if and only if* $\nu = \overline{\nu}$.

Proof. Suppose first that ν is decomposable. Then, by virtue of (7.12), we have, for all $\alpha \in \mathbb{F}_{q^2}^*$,

$$\overline{\nu}(\alpha) = \nu(\overline{\alpha}) = \psi(\overline{\alpha}\overline{\overline{\alpha}}) = \psi(\alpha\overline{\alpha}) = \nu(\alpha).$$

This shows that $\nu = \overline{\nu}$.

Conversely, if $\nu = \overline{\nu}$, we may set

$$\psi(\alpha\overline{\alpha}) = \nu(\alpha) \tag{7.14}$$

for all $\alpha \in \mathbb{F}_{q^2}^*$. Note that this is well defined since, by virtue of Hilbert satz 90 (Theorem 6.7.2), the map $N_{\mathbb{F}_{q^2}/\mathbb{F}_q} : \mathbb{F}_{q^2}^* \to \mathbb{F}_q^*$ is surjective with kernel $\mathrm{Ker}N_{\mathbb{F}_{q^2}/\mathbb{F}_q} = \{\alpha\overline{\alpha}^{-1} : \alpha \in \mathbb{F}_{q^2}^*\}$. Indeed, if $\alpha, \beta \in \mathbb{F}_{q^2}^*$ and $\alpha\overline{\alpha} = \beta\overline{\beta}$, then $N_{\mathbb{F}_{q^2}/\mathbb{F}_q}(\alpha) = N_{\mathbb{F}_{q^2}/\mathbb{F}_q}(\beta)$, that is, $\alpha\beta^{-1} \in \mathrm{Ker}N_{\mathbb{F}_{q^2}/\mathbb{F}_q}$ since the norm is a group homomorphism. Then there exists $\gamma \in \mathbb{F}_{q^2}^*$ such that $\alpha\beta^{-1} = \gamma\overline{\gamma}^{-1}$ so that $\nu(\alpha\beta^{-1}) = \nu(\gamma\overline{\gamma}^{-1}) = \nu(\gamma)\nu(\overline{\gamma})^{-1} = 1$ (recall that $\nu = \overline{\nu}$), showing that $\nu(\alpha) = \nu(\beta)$.

We leave it to the reader to check that ψ is indeed a character of $\mathbb{F}_q^*$. By construction, (7.12) follows from (7.14). $\qquad\square$

Proposition 7.2.3 *Let* $\nu \in \widehat{\mathbb{F}_{q^2}^*}$ *and suppose that it is not decomposable. Then*

$$\sum_{\substack{\beta \in \mathbb{F}_{q^2}^*: \\ \beta\overline{\beta}=\alpha}} \nu(\beta) = 0 \tag{7.15}$$

for all $\alpha \in \mathbb{F}_q^*$.

Proof. First of all, we show that there exists $\gamma \in \mathbb{F}_{q^2}$ such that $\gamma\overline{\gamma} = 1$ for which $\nu(\gamma) \neq 1$. Indeed, otherwise, if $\alpha, \beta \in \mathbb{F}_{q^2}^*$ satisfy $\alpha\overline{\alpha} = \beta\overline{\beta}$, then $\alpha\beta^{-1}\overline{\alpha\beta^{-1}} = 1$ and therefore $\nu(\alpha\beta^{-1}) = 1$ so that $\nu(\alpha) = \nu(\beta)$. We may then define a character ψ of $\mathbb{F}_q^*$ as in (7.14) and this would contradict our assumptions on the indecomposability of ν.

Thus, for all $\alpha \in \mathbb{F}_q^*$

$$\sum_{\substack{\beta \in \mathbb{F}_{q^2}^*: \\ \beta\bar{\beta}=\alpha}} \nu(\beta) = \sum_{\substack{\beta \in \mathbb{F}_{q^2}^*: \\ \beta\bar{\beta}=\alpha}} \nu(\gamma\beta) = \nu(\gamma) \sum_{\substack{\beta \in \mathbb{F}_{q^2}^*: \\ \beta\bar{\beta}=\alpha}} \nu(\beta),$$

where the first equality follows from the fact that $\gamma\bar{\gamma} = 1$. Since $\nu(\gamma) \neq 1$, (7.15) follows. $\qquad\square$

7.3 Generalized Kloosterman sums

In this section we introduce and study a family of *generalized Kloosterman sums*, that we shall use (cf. Section 14.6), following Piatetski-Shapiro [123], to describe the cuspidal representations of $GL(2, \mathbb{F}_q)$ and their associated Bessel functions, a finite analogue of the classical Bessel functions.

Let $q = p^n$ and consider the quadratic extension $\mathbb{F}_{q^2}$ of the field $\mathbb{F}_q$.

Let also χ be a nontrivial character of $\mathbb{F}_q$ and ν an indecomposable character of $\mathbb{F}_{q^2}^*$.

We use the notation in Section 6.8 and Section 7.2.

The *generalized Kloosterman sum* associated with the pair (χ, ν) is the map $j = j_{\chi,\nu} \colon \mathbb{F}_q^* \to \mathbb{C}$ defined by setting

$$j(x) = \frac{1}{q} \sum_{\substack{w \in \mathbb{F}_{q^2}^*: \\ w\bar{w}=x}} \chi(w + \bar{w})\nu(w) \qquad (7.16)$$

for all $x \in \mathbb{F}_q^*$.

We need a few technical formulas involving these sums: we begin with two results on additive characters.

Lemma 7.3.1 *Let $z \in \mathbb{F}_{q^2}^*$ and $\chi \in \widehat{\mathbb{F}_q}$. Then*

$$\sum_{t \in \mathbb{F}_{q^2}^*} \chi[tz + \bar{t}\bar{z}] = \begin{cases} q^2 - 1 & \text{if } \chi \text{ is trivial and/or } z = 0 \\ -1 & \text{otherwise.} \end{cases}$$

Proof. We first observe that the map $\tilde{\chi} \colon \mathbb{F}_{q^2} \to \mathbb{C}$ defined by

$$\tilde{\chi}(t) = \chi[tz + \bar{t}\bar{z}] \qquad (7.17)$$

for all $t \in \mathbb{F}_{q^2}$, is a character of $\mathbb{F}_{q^2}$.

Now, if χ is trivial and/or $z = 0$, then $\tilde{\chi}$ is the trivial character and therefore,

$$\sum_{t \in \mathbb{F}_q^*} \chi[tz + \bar{t}\bar{z}] = \sum_{t \in \mathbb{F}_{q^2}^*} \tilde{\chi}(t) = \sum_{t \in \mathbb{F}_q^*} 1 = |\mathbb{F}_q^*| = q^2 - 1.$$

Suppose now that χ is nontrivial and $z \neq 0$. We claim that the map $\mathbb{F}_{q^2} \ni t \mapsto tz + \bar{t}\bar{z} \in \mathbb{F}_q$ is surjective. Indeed, the map $t \mapsto tz$ is a bijection of $\mathbb{F}_{q^2}$ and the map $s \mapsto s + \bar{s} = \text{Tr}_{\mathbb{F}_{q^2}/\mathbb{F}_q}(s)$ is surjective by Hilbert Satz 90 (Theorem 6.7.2). It follows that the character (7.17) is nontrivial and, by the orthogonality relations of characters (cf. (7.1)),

$$\sum_{t \in \mathbb{F}_{q^2}} \chi[tz + \bar{t}\bar{z}] = \sum_{t \in \mathbb{F}_{q^2}} \tilde{\chi}(t) = \langle \tilde{\chi}, 1 \rangle = 0.$$

Since

$$\chi[tz + \bar{t}\bar{z}]_{t=0} = \chi(0) = 1,$$

the result follows. $\qquad\square$

Lemma 7.3.2 *Let $\chi \in \widehat{\mathbb{F}_q}$ be a nontrivial character. Let also $z \in \mathbb{F}_{q^2}^*$ and $y \in \mathbb{F}_q^*$. Then*

$$\sum_{t \in \mathbb{F}_{q^2}^*} \chi[y^{-1}(t + y + z)(\overline{t + y + z})] = -q - \chi[y^{-1}(y + z)(y + \bar{z})].$$

Proof. We have

$$\sum_{t \in \mathbb{F}_{q^2}^*} \chi[y^{-1}(t + y + z)(\overline{t + y + z})] = \sum_{\substack{s \in \mathbb{F}_{q^2}: \\ s \neq y + z}} \chi(y^{-1}s\bar{s})$$

$$= \sum_{s \in \mathbb{F}_{q^2}^*} \chi(y^{-1}s\bar{s}) + 1 - \chi[y^{-1}(y + z)(y + \bar{z})]$$

$$\text{(by (6.25) and setting } r = s\bar{s}) = (q + 1) \sum_{r \in \mathbb{F}_q^*} \chi(y^{-1}r)$$

$$+ 1 - \chi[y^{-1}(y + z)(y + \bar{z})]$$

$$\text{(by (7.1))} = -(q + 1) + 1 - \chi[y^{-1}(y + z)(y + \bar{z})]$$

$$= -q - \chi[y^{-1}(y + z)(y + \bar{z})]. \qquad\square$$

Proposition 7.3.3 *For every $x \in \mathbb{F}_q^*$ we have*

$$\overline{j(x)} = \overline{\nu(-x)} j(x).$$

Proof. Let $x \in \mathbb{F}_q^*$. Then, by definition of the Kloosterman sum j (cf. (7.16)),

$$\overline{j(x)} = \frac{1}{q} \sum_{\substack{y \in \mathbb{F}_{q^2}^*: \\ y\bar{y}=x}} \chi(-y - \bar{y})v(y^{-1})$$

$$=_* \frac{1}{q} \sum_{\substack{t \in \mathbb{F}_{q^2}^*: \\ t\bar{t}=x}} \chi[x(t^{-1} + \bar{t}^{-1})]v(-x^{-1}t)$$

$$=_{**} \overline{v(-x)} \frac{1}{q} \sum_{\substack{t \in \mathbb{F}_{q^2}^*: \\ t\bar{t}=x}} \chi(t + \bar{t})v(t)$$

$$= \overline{v(-x)} j(x),$$

where equality $=_*$ follows by setting $t = -xy^{-1}$ (so that $t\bar{t} = x$ and $y = -xt^{-1}$), and equality $=_{**}$ follows from $x(t^{-1} + \bar{t}^{-1}) = x\frac{t+\bar{t}}{t\bar{t}} = t + \bar{t}$. $\square$

Proposition 7.3.4 *For all* $x, y \in \mathbb{F}_q^*$ *we have*

$$\sum_{w \in \mathbb{F}_q^*} j(xw)j(yw)v(w^{-1})\chi(w) = -\chi(-x - y)v(-1)j(xy).$$

Proof. We have

$$\sum_{w \in \mathbb{F}_q^*} j(xw)j(yw)v(w^{-1})\chi(w)$$

$$= \frac{1}{q^2} \sum_{w \in \mathbb{F}_q^*} \sum_{\substack{t \in \mathbb{F}_{q^2}^*: \\ t\bar{t}=xw}} \sum_{\substack{s \in \mathbb{F}_{q^2}^*: \\ s\bar{s}=yw}} \chi(t + \bar{t} + s + \bar{s} + w)v(tsw^{-1}) \quad (7.18)$$

Let us set $z = yt(\bar{s})^{-1}$. First note that from $s\bar{s} = yw$ we get

$$tsw^{-1} = yt(\bar{s})^{-1} = z.$$

From $t\bar{t} = xw$ we then deduce

$$z\bar{z} = yt(\bar{s})^{-1}y\bar{t}s^{-1} = yt\bar{t}y(s\bar{s})^{-1} = yxwyw^{-1}y^{-1} = yx.$$

Moreover,

$$y^{-1}(s + y + z)(\overline{s + y + z}) = (y^{-1}s + 1 + y^{-1}z)(\bar{s} + y + \bar{z})$$
$$= y^{-1}s\bar{s} + s + y^{-1}s\bar{z} + \bar{s} + y$$
$$\quad + \bar{z} + y^{-1}z\bar{s} + z + y^{-1}z\bar{z} \quad (7.19)$$
$$= w + s + \bar{t} + \bar{s} + y + \bar{z} + t + z + x$$
$$= w + s + \bar{s} + t + \bar{t} + y + z + \bar{z} + x$$

and

$$y^{-1}(y+z)(y+\bar{z}) = (y+z)(1+y^{-1}\bar{z})$$
$$= y+z+\bar{z}+y^{-1}z\bar{z} \qquad (7.20)$$
$$= y+z+\bar{z}+x.$$

Then the calculation (7.18) continues as follows:

$$=_{(i)} \frac{1}{q^2} \sum_{w\in\mathbb{F}_q^*} \sum_{\substack{s\in\mathbb{F}_{q^2}^*:\\ s\bar{s}=yw}} \sum_{\substack{z\in\mathbb{F}_{q^2}^*:\\ z\bar{z}=xy}} \chi[y^{-1}(s+y+z)\overline{(s+y+z)} - x-y-z-\bar{z}]v(z)$$

$$=_{(ii)} \frac{1}{q^2} \sum_{s\in\mathbb{F}_{q^2}^*} \sum_{\substack{z\in\mathbb{F}_{q^2}^*:\\ z\bar{z}=xy}} \chi[y^{-1}(s+y+z)\overline{(s+y+z)} - x-y-z-\bar{z}]v(z)$$

$$= \frac{1}{q^2} \sum_{\substack{z\in\mathbb{F}_{q^2}^*:\\ z\bar{z}=xy}} \chi[-x-y-z-\bar{z}]v(z) \sum_{s\in\mathbb{F}_{q^2}^*} \chi[y^{-1}(s+y+z)\overline{(s+y+z)}]$$

$$=_{(iii)} \frac{1}{q^2} \sum_{\substack{z\in\mathbb{F}_{q^2}^*:\\ z\bar{z}=xy}} \chi[-x-y-z-\bar{z}]v(z) \left\{-q - \chi[y^{-1}(y+z)(y+\bar{z})]\right\}$$

$$= -\frac{1}{q} \sum_{\substack{z\in\mathbb{F}_{q^2}^*:\\ z\bar{z}=xy}} \chi[-x-y-z-\bar{z}]v(z)$$

$$- \frac{1}{q^2} \sum_{\substack{z\in\mathbb{F}_{q^2}^*:\\ z\bar{z}=xy}} \chi[-x-y-z-\bar{z}+y^{-1}(y+z)(y+\bar{z})]v(z)$$

$$=_{(iv)} -\frac{1}{q}\chi(-x-y)v(-1) \sum_{\substack{z\in\mathbb{F}_{q^2}^*:\\ z\bar{z}=xy}} \chi(z+\bar{z})v(z) - \frac{1}{q^2} \sum_{\substack{z\in\mathbb{F}_{q^2}^*:\\ z\bar{z}=xy}} v(z)$$

$$=_{(v)} -\chi(-x-y)v(-1)j(xy)$$

where $=_{(i)}$ follows from (7.19), $=_{(ii)}$ follows from Hilbert Satz 90, $=_{(iii)}$ follows from Lemma 7.3.2, $=_{(iv)}$ is obtained by changing z to $-z$ and because $-x-y-z-\bar{z}+y^{-1}(y+z)(y+\bar{z}) = 0$, by (7.20), and $=_{(v)}$ follows from Proposition 7.2.3 and the definition of j (cf. (7.16)). $\qquad \square$

Proposition 7.3.5 (Orthogonality relations) $\sum_{w\in\mathbb{F}_q^*} j(xw)\overline{j(yw)} = \delta_{x,y}$ *for all* $x,y \in \mathbb{F}_q^*$.

Proof. By definition of j, we have

$$\sum_{w \in \mathbb{F}_q^*} j(xw)\overline{j(yw)} = \frac{1}{q^2} \sum_{w \in \mathbb{F}_q^*} \sum_{\substack{t \in \mathbb{F}_{q^2}^*: \\ t\bar{t}=xw}} \sum_{\substack{s \in \mathbb{F}_{q^2}^*: \\ s\bar{s}=yw}} \chi(t + \bar{t} - s - \bar{s})\nu(ts^{-1})$$

$$(\text{setting } z = ts^{-1}) \quad = \frac{1}{q^2} \sum_{w \in \mathbb{F}_q^*} \sum_{\substack{s \in \mathbb{F}_{q^2}^*: \\ s\bar{s}=yw}} \sum_{\substack{z \in \mathbb{F}_{q^2}^*: \\ z\bar{z}=xy^{-1}}} \chi(zs + \bar{z}\bar{s} - s - \bar{s})\nu(z)$$

$$(\text{by (6.25)}) \quad = \frac{1}{q^2} \sum_{\substack{z \in \mathbb{F}_{q^2}^*: \\ z\bar{z}=xy^{-1}}} \left(\sum_{s \in \mathbb{F}_{q^2}^*} \chi((z-1)s + (\bar{z}-1)\bar{s}) \right) \nu(z).$$

If $x \neq y$, then $z \neq 1$ and, by virtue of Lemma 7.3.1, $\sum_{s \in \mathbb{F}_{q^2}^*} \chi((z-1)s + (\bar{z}-1)\bar{s}) = -1$, so that

$$\frac{1}{q^2} \sum_{\substack{z \in \mathbb{F}_{q^2}^*: \\ z\bar{z}=xy^{-1}}} \left(\sum_{s \in \mathbb{F}_{q^2}^*} \chi((z-1)s + (\bar{z}-1)\bar{s}) \right) \nu(z) = -\frac{1}{q^2} \sum_{\substack{z \in \mathbb{F}_{q^2}^*: \\ z\bar{z}=xy^{-1}}} \nu(z) = 0,$$

where the last equality follows from Proposition 7.2.3.

If $x = y$, then $z = 1$ is admissible and, again by virtue of Lemma 7.3.1,

$$\frac{1}{q^2} \sum_{\substack{z \in \mathbb{F}_{q^2}^*: \\ z\bar{z}=xy^{-1}}} \left(\sum_{s \in \mathbb{F}_{q^2}^*} \chi((z-1)s + (\bar{z}-1)\bar{s}) \right) \nu(z) = \frac{1}{q^2}[(q^2-1) - \sum_{\substack{z \in \mathbb{F}_{q^2}^* \setminus \{1\}: \\ z\bar{z}=1}} \nu(z)]$$

$$= \frac{1}{q^2}[(q^2-1) - (-1)]$$

$$= 1,$$

where the last but one equality follows from Proposition 7.2.3. $\qquad\square$

Corollary 7.3.6 *For every $x \in \mathbb{F}_q^*$ we have*

$$\sum_{y \in \mathbb{F}_q^*} j(xy)j(y)\nu(y^{-1}) = \begin{cases} \nu(-1) & \text{if } x = 1 \\ 0 & \text{if } x \neq 1. \end{cases}$$

Proof. Let $x \in \mathbb{F}_q^*$. Then

$$\sum_{y \in \mathbb{F}_q^*} j(xy)j(y)\nu(y^{-1}) = \sum_{y \in \mathbb{F}_q^*} j(xy)j(y)\nu(-y^{-1})\nu(-1)$$

$$\text{(by Proposition 7.3.3)} = \left(\sum_{y \in \mathbb{F}_q^*} j(xy)\overline{j(y)} \right) \nu(-1)$$

$$\text{(by Proposition 7.3.5)} = \delta_{x,1}\nu(-1). \qquad \square$$

In the following (see also Section 14.6), in order to emphasize the dependance of the map j from ν, we shall write j_ν (clearly j also depends on χ). Note that, from (7.16) it follows immediately that

$$j_{\bar{\nu}} = j_\nu, \tag{7.21}$$

where $\bar{\nu}$ is the conjugate character of ν (cf. (7.13)).

Theorem 7.3.7 *Let μ and ν be two indecomposable characters of $\mathbb{F}_{q^2}^*$. Suppose that $j_\mu = j_\nu$ and*

$$\mu|_{\mathbb{F}_q^*} = \nu|_{\mathbb{F}_q^*}. \tag{7.22}$$

Then $\mu = \nu$ or $\mu = \bar{\nu}$.

Proof. Our first assumption yields

$$\sum_{\substack{y \in \mathbb{F}_{q^2}^*: \\ y\bar{y}=x}} \chi(y + \bar{y})\mu(y) = qj_\mu(x) = qj_\nu(x) = \sum_{\substack{y \in \mathbb{F}_{q^2}^*: \\ y\bar{y}=x}} \chi(y + \bar{y})\nu(y)$$

for all $x \in \mathbb{F}_q^*$. Moreover, for $y \in \mathbb{F}_{q^2}^*$ and $\delta \in \mathbb{F}_q^*$, we set $z = \delta^{-1}y$ (i.e. $y = \delta z$) and $t = z\bar{z} = \delta^{-2}y\bar{y}$, so that, taking into account (7.22), from the above formula we deduce

$$\sum_{\substack{z \in \mathbb{F}_{q^2}^*: \\ z\bar{z}=t}} \chi[\delta(z + \bar{z})]\mu(z) = \sum_{\substack{z \in \mathbb{F}_{q^2}^*: \\ z\bar{z}=t}} \chi[\delta(z + \bar{z})]\nu(z) \tag{7.23}$$

for all $t \in \mathbb{F}_q^*$ and $\delta \in \mathbb{F}_q$ (the case $\delta = 0$ follows from Proposition 7.2.3).

Fix $t \in \mathbb{F}_q^*$. Then the solutions of the equation $z\bar{z} = t$ may be partitioned into sets of the form $\{z, \bar{z}\}$. Choose a complete system C_t of representatives for such sets, that is,

$$\{z \in \mathbb{F}_{q^2}^* : z\bar{z} = t\} = \coprod_{z \in C_t}\{z, \bar{z}\}.$$

Note (recall Proposition 6.4.4) that if t is a square in $\mathbb{F}_q$, say $t = u^2$, $u \in \mathbb{F}_q^*$, then also the singletons $\{u\}$ and $\{-u\}$ must be considered (and they coincide if q is even). We may then write (7.23) in the form

$$\sum_{z \in C_t \setminus \mathbb{F}_q} \chi[\delta(z + \bar{z})][\mu(z) + \mu(\bar{z}) - \nu(z) - \nu(\bar{z})]$$

$$+ \sum_{z \in C_t \cap \mathbb{F}_q} \chi[\delta(z + \bar{z})][\mu(z) - \nu(z)] = 0, \quad (7.24)$$

where $C_t \cap \mathbb{F}_q$ is empty if t is not a square. In any case, the second sum in the left hand side vanishes by virtue of (7.22).

We now set $\tilde{C}_t = \{z + \bar{z} : z \in C_t\}$. Since $z + \bar{z} = \mathrm{Tr}_{\mathbb{F}_{q^2}/\mathbb{F}_q}(z) \in \mathbb{F}_q$ for all $z \in \mathbb{F}_{q^2}$, we have $\tilde{C}_t \subseteq \mathbb{F}_q$. Moreover, every $x \in \tilde{C}_t$ corresponds to a unique set $\{z, \bar{z}\}$ (possibly $z = \bar{z}$) because the system

$$\begin{cases} z + \bar{z} = x \\ z\bar{z} = t \end{cases}$$

is equivalent to the equation $z^2 - xz + t = 0$. In other words, $x \in \tilde{C}_t$ determines $\{z, \bar{z}\}$ and the map

$$\begin{aligned} C_t &\to \tilde{C}_t \\ z &\mapsto z + \bar{z} \end{aligned}$$

is a bijection. Then we may define a function $f_t : \tilde{C}_t \to \mathbb{C}$ by setting

$$f_t(x) = \begin{cases} \mu(z) + \mu(\bar{z}) - \nu(z) - \nu(\bar{z}) & \text{if } z\bar{z} = t, z + \bar{z} = x, \text{ and } z \neq \bar{z} \\ \mu(z) - \nu(z) \equiv 0 & \text{if } z^2 = t, z \in \mathbb{F}_q, \text{ and } 2z = x. \end{cases}$$

Therefore (7.24) may be written in the form

$$\sum_{x \in \tilde{C}_t} \chi(\delta x) f_t(x) = 0 \quad (7.25)$$

for all $t \in \mathbb{F}_q^*$ and $\delta \in \mathbb{F}_q$. By Proposition 7.1.1, the functions $\psi_x \in L(\mathbb{F}_q)$, $x \in \tilde{C}_t$, defined by $\psi_x(\delta) = \chi(\delta x)$ for all $\delta \in \mathbb{F}_q$, are distinct characters of $\mathbb{F}_q$, and the left hand side of (7.25) may be considered as a linear combination of distinct characters. Since the characters are linearly independent, it follows that $f_t = 0$ for all $t \in \mathbb{F}_q^*$, that is,

$$\mu(z) + \mu(\bar{z}) = \nu(z) + \nu(\bar{z}) \quad (7.26)$$

for all $z \in \mathbb{F}_{q^2} \setminus \mathbb{F}_q$. Moreover, since μ and ν are multiplicative, and $z\bar{z} = \mathbb{N}_{\mathbb{F}_{q^2}/\mathbb{F}_q}(z) \in \mathbb{F}_q$ for all $z \in \mathbb{F}_{q^2}$, keeping in mind (7.22), we have

$$\mu(z)\mu(\bar{z}) = \nu(z)\nu(\bar{z}). \tag{7.27}$$

From (7.26) and (7.27) we deduce that the sets $\{\mu(z), \mu(\bar{z})\}$ and $\{\nu(z), \nu(\bar{z})\}$ solve the same quadratic equation, namely,

$$\lambda^2 - [\mu(z) + \mu(\bar{z})]\lambda + \mu(z)\mu(\bar{z}) = 0.$$

It follows that $\{\mu(z), \mu(\bar{z})\} = \{\nu(z), \nu(\bar{z})\}$, that is, $\mu(z) = \nu(z)$ or $\mu(z) = \nu(\bar{z})$, for each $z \in \mathbb{F}_{q^2} \setminus \mathbb{F}_q$.

Let z_0 be a generator of the cyclic group $\mathbb{F}_{q^2}^*$ (cf. Theorem 6.3.3). Then $\mu(z_0) = \nu(z_0)$ yields $\mu = \nu$, while $\mu(z_0) = \nu(\bar{z_0})$ yields $\mu = \bar{\nu}$. □

The (ordinary) *Kloosterman sums* are defined by

$$K(\chi; a, b) = \sum_{c \in \mathbb{F}_q^*} \chi(ac + bc^{-1}),$$

where χ is a nontrivial element of $\widehat{\mathbb{F}_q}$ and $a, b \in \mathbb{F}_q$. For more on these sums we refer to [96] and the references therein. We limit ourselves to a couple of elementary identities.

Exercise 7.3.8 Let $a, b \in \mathbb{F}_q$.

(a) Show that $K(\chi; a, b) = K(\chi; b, a)$;
(b) show that if $a \in \mathbb{F}_q^*$ then $K(\chi; a, b) = K(\chi; 1, ab)$.

7.4 Gauss sums

Definition 7.4.1 Let $\chi \in \widehat{\mathbb{F}_q}$ and $\psi \in \widehat{\mathbb{F}_q^*}$. We define the *Gauss sum* of the multiplicative character ψ and the additive character χ as the complex number

$$g(\psi, \chi) = \sum_{x \in \mathbb{F}_q^*} \psi(x)\chi(x). \tag{7.28}$$

Note that, by virtue of (4.18) and (4.22), the Gauss sum $G(n, p) = \tau(p, n)$ coincides with $g(\ell_p, \chi_n)$, where ℓ_p and χ_n are the multiplicative and additive characters defined in Section 4.4, respectively.

Proposition 7.4.2 *Denote by $\chi_0 = 1$ the trivial character of $\mathbb{F}_q$ (so that, by (7.6), it is also the trivial multiplicative character). Then for all $\chi \in \widehat{\mathbb{F}_q}$ and $\psi \in \widehat{\mathbb{F}_q^*}$ we have:*

(i) $g(\chi_0, \chi_0) = q - 1$;
(ii) $g(\chi_0, \chi) = -1$ *if $\chi \neq \chi_0$;*

(iii) $g(\psi, \chi_0) = 0$ *if* $\psi \neq \chi_0$;

(iv) $g(\psi, \chi) = \sum_{x \in \mathbb{F}_q} \psi(x) \chi(x) = \langle \psi, \overline{\chi} \rangle_{L(\mathbb{F}_p)}$ *if* $\psi \neq \chi_0$.

Proof. These are all elementary consequences of the orthogonality relations for the additive and multiplicative characters (in particular (7.2), (7.6), and (7.9)). We thus leave it to the reader to fill in the details of the proof. $\square$

Note that (iv) shows that for $\psi \neq \chi_0$, the Gauss sum $g(\psi, \chi)$ equals the Fourier coefficient (2.15), both of ψ with respect to $\overline{\chi}$ as well as of $\chi|_{\mathbb{F}_q^*}$ with respect to $\overline{\psi}$. We now present the basic properties of Gaussian sums.

Theorem 7.4.3 *Let* χ_y *be the additive character as in* (7.5), $\chi \in \widehat{\mathbb{F}_q}$ *and* $\psi \in \widehat{\mathbb{F}_q^*}$. *Then we have:*

(i) $g(\psi, \chi_y) = \overline{\psi(y)} g(\psi, \chi_1)$ *if* $y \neq 0$;

(ii) $g(\psi, \overline{\chi}) = \psi(-1) g(\psi, \chi)$;

(iii) $g(\overline{\psi}, \chi) = \psi(-1) \overline{g(\psi, \chi)}$;

(iv)

$$\psi = \frac{1}{q} \sum_{\substack{\chi \in \widehat{\mathbb{F}_q} \\ \chi \neq \chi_0}} g(\psi, \chi) \overline{\chi} = \frac{1}{q} g(\psi, \chi_1) \sum_{y \in \mathbb{F}_q^*} \overline{\psi(y)} \overline{\chi_y}$$

if $\psi \neq \chi_0$;

(v) $\chi|_{\mathbb{F}_q^*} = \frac{1}{q-1} \sum_{\psi \in \widehat{\mathbb{F}_q^*}} g(\psi, \chi) \overline{\psi}$;

(vi) $g(\psi, \chi) g(\overline{\psi}, \chi) = \psi(-1) q$ *if* $\psi, \chi \neq \chi_0$;

(vii) $|g(\psi, \chi)| = \sqrt{q}$ *if* $\psi, \chi \neq \chi_0$;

(viii) $g(\psi^p, \chi_y) = g(\psi, \chi_{\sigma(y)})$, *where* $\sigma(y) = y^p$ *is the Frobenius automorphism.*

Proof.

(i) Suppose $y \neq 0$. Then

$$g(\psi, \chi_y) = \sum_{x \in \mathbb{F}_q^*} \psi(x) \chi_1(xy)$$

$$(\text{setting } t = xy) = \sum_{t \in \mathbb{F}_q^*} \psi(ty^{-1}) \chi_1(t)$$

$$= \sum_{t \in \mathbb{F}_q^*} \psi(y^{-1}) \psi(t) \chi_1(t)$$

$$= \overline{\psi(y)} g(\psi, \chi_1).$$

(ii) We have:

$$g(\psi, \overline{\chi}) = \sum_{x \in \mathbb{F}_q^*} \psi(x)\overline{\chi(x)}$$

$$= \sum_{x \in \mathbb{F}_q^*} \psi(x)\chi(-x)$$

$$(\text{setting } y = -x) = \sum_{y \in \mathbb{F}_q^*} \psi(-y)\chi(y)$$

$$= \sum_{y \in \mathbb{F}_q^*} \psi(-1)\psi(y)\chi(y)$$

$$= \psi(-1)g(\psi, \chi).$$

(iii) By (ii) and recalling that $\psi(-1) = \pm 1$ (cf. Lemma 7.1.4), we have:

$$\overline{g(\psi, \chi)} = \psi(-1)\overline{g(\psi, \overline{\chi})}$$
$$= \psi(-1)g(\overline{\psi}, \chi).$$

(iv) and (v) are immediate consequences of Proposition 7.4.2 (iii) and (iv), the Fourier inversion formula (cf. (2.16)), and (i). We leave it to the reader to fill in the details.

(vi) We have:

$$g(\psi, \chi)g(\overline{\psi}, \chi) = \left[\sum_{x \in \mathbb{F}_q^*} \psi(x)\chi(x)\right] \cdot \left[\sum_{y \in \mathbb{F}_q^*} \overline{\psi(y)}\chi(y)\right]$$

$$= \sum_{x,y \in \mathbb{F}_q^*} \psi(xy^{-1})\chi(x+y)$$

$$(\text{setting } t = xy^{-1}) = \sum_{t \in \mathbb{F}_q^*} \psi(t) \sum_{y \in \mathbb{F}_q^*} \chi[y(t+1)]$$

$$(\text{by Corollary 7.1.3}) = (q-1)\psi(-1) - \sum_{t \in \mathbb{F}_q^* \setminus \{-1\}} \psi(t)$$

$$(\text{by (7.8)}) = (q-1)\psi(-1) - [-\psi(-1)]$$

$$= q\psi(-1).$$

(vii) Recalling, once more, that $\psi(-1) = \pm 1$, we have:

$$|g(\psi, \chi)|^2 = g(\psi, \chi)\overline{g(\psi, \chi)}$$
$$(\text{by (iii)}) = \psi(-1)g(\psi, \chi)g(\overline{\psi}, \chi)$$
$$(\text{by (vi)}) = q.$$

(viii) We have:

$$g(\psi^p, \chi_y) = \sum_{x \in \mathbb{F}_q^*} \psi^p(x) \chi_y(x)$$

$$= \sum_{x \in \mathbb{F}_q^*} \psi(x^p) \chi_y(x)$$

(setting $z = x^p$, and by bijectivity of σ) $= \sum_{z \in \mathbb{F}_q^*} \psi(z) \chi_y[\sigma^{-1}(z)]$

(by definition of χ_y) $= \sum_{z \in \mathbb{F}_q^*} \psi(z) \chi_1 \left(\sigma^{-1}[\sigma(y)z] \right)$

$$=_* \sum_{z \in \mathbb{F}_q^*} \psi(z) \chi_1[\sigma(y)z]$$

$$= g(\psi, \chi_{\sigma(y)}),$$

where $=_*$ follows from $\mathrm{Tr} \circ \sigma^{-1} = \mathrm{Tr}$ (cf. (7.3), (6.22), and (6.24)).

$\square$

Even if its module is given by Theorem 7.4.3.(vii), the exact evaluation of a Gauss sum $g(\psi, \chi)$ is a very difficult problem and only a few special values are known. See Gauss' original results in Theorem 4.4.15 for an important example. Other cases are in the books by Lidl and Niederreiter [96] and by Berndt, Evans, and Williams [20].

7.5 The Hasse-Davenport identity

In this section we reproduce Weil's proof [165] of the Hasse-Davenport identity [48], which relates the Gauss sums over a finite field and those over a finite extension. We split it into several preliminary results.

Let us fix $\psi \in \widehat{\mathbb{F}_q^*}$ and $\chi \in \widehat{\mathbb{F}_q}$, with ψ nontrivial. Moreover, for every monic polynomial $f(x) = x^n + a_{n-1}x^{n-1} + \cdots + a_0 \in \mathbb{F}_q[x]$, define the complex number $\lambda(f) = \lambda_{\psi, \chi}(f)$ by setting, keeping in mind (7.6),

$$\lambda(f) = \psi(a_0)\chi(a_{n-1}). \tag{7.29}$$

Notice that if $n = 1$ then $a_{n-1} = a_0$ and therefore $\lambda(f) = \psi(a_0)\chi(a_0)$. Since ψ is not trivial, we have $|\lambda(f)| = 1$ if $a_0 \neq 0$, while $\lambda(f) = 0$ if $a_0 = 0$. Moreover, if $g(x) = x^m + b_{m-1}x^{m-1} + \cdots + b_0 \in \mathbb{F}_q[x]$ then

$$f(x)g(x) = x^{n+m} + (a_{n-1} + b_{m-1})x^{n+m-1} + \cdots + a_0 b_0$$

so that

$$\lambda(f \cdot g) = \psi(a_0 b_0)\chi(a_{n-1} + b_{m-1}) = \lambda(f)\lambda(g),$$

that is, the map $\lambda\colon \mathbb{F}_q^{\mathrm{mon}}[x] \to \mathbb{C}$ is multiplicative (see Notation 6.6.6).

We define the formal power series $\ell(z) = \ell_{\psi,\chi}(z)$ by setting

$$\ell(z) = \sum_{f \in \mathbb{F}_q^{\mathrm{mon}}[x]} \lambda(f) z^{\deg f} \equiv \sum_{k=0}^{\infty} \left(\sum_{f \in \mathbb{F}_q^{\mathrm{mon},k}[x]} \lambda(f) \right) z^k. \qquad (7.30)$$

Proposition 7.5.1 *The series $\ell(z)$ converges for all $z \in \mathbb{C}$ and its sum is given by*

$$\ell(z) = 1 + g(\psi, \chi)z.$$

Proof. Clearly, $\mathbb{F}_q^{\mathrm{mon},0}[x] = \{1\}$. Moreover, $\mathbb{F}_q^{\mathrm{mon},1}[x] = \{x + a_0 : a_0 \in \mathbb{F}_q\}$ so that (recalling Proposition 7.4.2.(iv))

$$\sum_{f \in \mathbb{F}_q^{\mathrm{mon},1}[x]} \lambda(f) = \sum_{a_0 \in \mathbb{F}_q} \psi(a_0)\chi(a_0) = g(\psi, \chi).$$

Let $k \geq 2$. For every $a_0, a_{k-1} \in \mathbb{F}_q$ there are exactly q^{k-2} monic polynomials of the form $x^k + a_{k-1}x^{k-1} + \cdots + a_0$. Then we have

$$\sum_{f \in \mathbb{F}_q^{\mathrm{mon},k}[x]} \lambda(f) = q^{k-2} \sum_{a_{k-1},a_0} \psi(a_0)\chi(a_{k-1}) = 0,$$

since, being ψ nontrivial, $\sum_{a_0 \in \mathbb{F}_q} \psi(a_0) = 0$ (cf. (7.9)). $\qquad \square$

We have the following formal product development:

$$\ell(z) = \prod_{f \in \mathbb{F}_q^{\mathrm{mon,irr}}[x]} \frac{1}{1 - \lambda(f)z^{\deg f}}, \qquad (7.31)$$

where the right hand side must be seen as the product

$$\prod_{f \in \mathbb{F}_q^{\mathrm{mon,irr}}[x]} \left(\sum_{r=0}^{\infty} \lambda(f)^r z^{r \deg f} \right).$$

In other words, the coefficient of z^k in $\ell(z)$ is given by

$$\sum \lambda(f_1)^{r_1} \lambda(f_2)^{r_2} \cdots \lambda(f_s)^{r_s}, \qquad (7.32)$$

where the (finite) sum runs over all (distinct) $f_1, f_2, \ldots, f_s \in \mathbb{F}_q^{\mathrm{mon,irr}}[x]$ and $r_1, r_2, \ldots, r_s \in \mathbb{N}$ such that $r_1 \deg f_1 + r_2 \deg f_2 + \cdots + r_s \deg f_s = k$.

Indeed, (7.31) then amounts to saying that $\sum_{f \in \mathbb{F}_q^{\mathrm{mon},k}[x]} \lambda(f)$ equals the sum (7.32). But this simply follows from the fact that f may be written uniquely (up to reordering the factors) in the form $f = f_1^{r_1} f_2^{r_2} \cdots f_s^{r_s}$ with $f_1, f_2, \ldots, f_s \in \mathbb{F}_q^{\mathrm{mon,irr}}[x]$, $r_1, r_2, \ldots, r_s \in \mathbb{N}$, and, since λ is multiplicative, $\lambda(f_1^{r_1} f_2^{r_2} \cdots f_s^{r_s}) = \lambda(f_1)^{r_1} \lambda(f_2)^{r_2} \cdots \lambda(f_s)^{r_s}$.

Let now $h > 1$ and consider the field extension $\mathbb{F}_{q^h}$ of $\mathbb{F}_q$. We set

$$\Psi = \psi \circ N_{\mathbb{F}_{q^h}/\mathbb{F}_q} \quad \text{and} \quad X = \chi \circ \text{Tr}_{\mathbb{F}_{q^h}/\mathbb{F}_q} \tag{7.33}$$

and observe that $\Psi \in \widehat{\mathbb{F}_{q^h}^*}$ is nontrivial and $X \in \widehat{\mathbb{F}_{q^h}}$.

Also, in analogy with (7.29), we define $\Lambda = \Lambda_{\Psi,X} : \mathbb{F}_{q^h}^{\text{mon}}[x] \to \mathbb{C}$ by setting

$$\Lambda(F) = \Psi(A_0)X(A_{s-1})$$

for every monic polynomial $F(x) = x^s + A_{s-1}x^{s-1} + \cdots + A_1 x + A_0 \in \mathbb{F}_{q^h}[x]$.

Lemma 7.5.2 *Let* $f(x) = x^n + a_{n-1}x^{n-1} + \cdots + a_1 x + a_0$ *be an irreducible polynomial in* $\mathbb{F}_q[x]$. *Let also* $h > 1$ *and set* $d = \gcd(h, n)$. *Then, if* $F(x) \in \mathbb{F}_{q^h}[x]$ *is an irreducible and monic polynomial that divides* f, *we have*

$$\Lambda(F) = \lambda(f)^{\frac{h}{d}}.$$

Proof. We start by observing that, by (6.20), $s = \frac{n}{d}$ equals $\deg F$. Write $F(x) = x^s + A_{s-1}x^{s-1} + \cdots + A_1 x + A_0$. Let $\alpha \in \mathbb{F}_{q^n}$ be a root of f (see Corollary 6.6.4). Clearly, f is the minimal polynomial of α over $\mathbb{F}_q$ (see Corollary 6.6.5). Moreover, by virtue of (6.19), we may suppose that α is also a root of F (if necessary, we may replace α by $\sigma^{-\ell}(\alpha)$ for some $\ell \geq 1$). Since $hs = \frac{h}{d}n \geq n$, so that $\mathbb{F}_{q^{hs}} \supseteq \mathbb{F}_{q^n}$, we conclude that F is the minimal polynomial of $\alpha \in \mathbb{F}_{q^{hs}}$ over $\mathbb{F}_{q^h}$ (again by Corollary 6.6.5). By Proposition 6.7.3 (and the elementary fact that $\sigma(-1) = -1$), we have

$$A_0 = (-1)^s N_{\mathbb{F}_{q^{hs}}/\mathbb{F}_{q^h}}(\alpha) = N_{\mathbb{F}_{q^{hs}}/\mathbb{F}_{q^h}}(-\alpha) \tag{7.34}$$

$$A_{s-1} = -\text{Tr}_{\mathbb{F}_{q^{hs}}/\mathbb{F}_{q^h}}(\alpha) = \text{Tr}_{\mathbb{F}_{q^{hs}}/\mathbb{F}_{q^h}}(-\alpha) \tag{7.35}$$

$$a_0 = N_{\mathbb{F}_{q^n}/\mathbb{F}_q}(-\alpha) \tag{7.36}$$

$$a_{n-1} = \text{Tr}_{\mathbb{F}_{q^n}/\mathbb{F}_q}(-\alpha). \tag{7.37}$$

It follows that

$$\Lambda(F) = \Psi(A_0)X(A_{s-1})$$

$$\text{(by (7.34) and (7.35))} = \Psi[N_{\mathbb{F}_{q^{hs}}/\mathbb{F}_{q^h}}(-\alpha)] \cdot X[\text{Tr}_{\mathbb{F}_{q^{hs}}/\mathbb{F}_{q^h}}(-\alpha)]$$

$$\text{(by (7.33))} = \psi[N_{\mathbb{F}_{q^h}/\mathbb{F}_q} \circ N_{\mathbb{F}_{q^{hs}}/\mathbb{F}_{q^h}}(-\alpha)] \cdot$$
$$\cdot \chi[\text{Tr}_{\mathbb{F}_{q^h}/\mathbb{F}_q} \circ \text{Tr}_{\mathbb{F}_{q^{hs}}/\mathbb{F}_{q^h}}(-\alpha)]$$

$$\text{(by Proposition 6.7.1)} = \psi[N_{\mathbb{F}_{q^{hs}}/\mathbb{F}_q}(-\alpha)] \cdot \chi[\text{Tr}_{\mathbb{F}_{q^{hs}}/\mathbb{F}_q}(-\alpha)]$$

$$\text{(again by Proposition 6.7.1)} = \psi[N_{\mathbb{F}_{q^n}/\mathbb{F}_q} \circ N_{\mathbb{F}_{q^{hs}}/\mathbb{F}_{q^n}}(-\alpha)] \cdot$$
$$\cdot \chi[\text{Tr}_{\mathbb{F}_{q^n}/\mathbb{F}_q} \circ \text{Tr}_{\mathbb{F}_{q^{hs}}/\mathbb{F}_{q^n}}(-\alpha)]$$

$$\text{(since } \alpha \in \mathbb{F}_{q^n}) = \psi[N_{\mathbb{F}_{q^n}/\mathbb{F}_q}(-\alpha)^{h/d}] \cdot \chi\left[\frac{h}{d}\text{Tr}_{\mathbb{F}_{q^n}/\mathbb{F}_q}(-\alpha)\right]$$

$$= \{\psi[N_{\mathbb{F}_{q^n}/\mathbb{F}_q}(-\alpha)]\}^{\frac{h}{d}} \cdot \{\chi[\text{Tr}_{\mathbb{F}_{q^n}/\mathbb{F}_q}(-\alpha)]\}^{\frac{h}{d}}$$

$$\text{(by (7.36) and (7.37))} = [\psi(a_0)\chi(a_{n-1})]^{\frac{h}{d}}$$

$$= \lambda(f)^{\frac{h}{d}}. \qquad \qquad \square$$

Theorem 7.5.3 (Hasse-Davenport identity) *With the above notation (in particular, (7.33)) we have*

$$g(\Psi, X) = (-1)^{h-1}[g(\psi, \chi)]^h.$$

Proof. As in (7.30), with ψ and χ replaced by Ψ and X, respectively, we set

$$L(Z) = \sum_{F \in \mathbb{F}_{q^h}^{\text{mon}}[x]} \Lambda(F)Z^{\deg F}.$$

Then, Proposition 7.5.1 and (7.31) become

$$L(z) = 1 + g(\Psi, X)Z = \prod_{F \in \mathbb{F}_{q^h}^{\text{mon,irr}}[x]} \frac{1}{1 - \Lambda(F)Z^{\deg F}}$$

$$\text{(by Proposition 6.6.7)} = \prod_{f \in \mathbb{F}_q^{\text{mon,irr}}[x]} \prod_{\substack{F \in \mathbb{F}_{q^h}^{\text{mon,irr}}[x]: \\ F \mid f}} \frac{1}{1 - \Lambda(F)Z^{\deg F}}$$

$$=_* \prod_{f \in \mathbb{F}_q^{\text{mon,irr}}[x]} \frac{1}{[1 - \lambda(f)^{h/d}Z^{\deg f/d}]^d}$$

$$\text{(setting } Z = z^h) = \prod_{f \in \mathbb{F}_q^{\text{mon,irr}}[x]} [1 - \lambda(f)^{h/d}z^{\deg(f)\cdot h/d}]^{-d}$$

$$=_{**} \prod_{f \in \mathbb{F}_q^{\text{mon,irr}}[x]} \prod_{\ell=0}^{h/d-1} [1 - \lambda(f)\zeta^{d\ell}z^{\deg f}]^{-d}$$

$$=_{***} \prod_{f \in \mathbb{F}_q^{\text{mon,irr}}[x]} \prod_{j=0}^{h-1} [1 - \lambda(f)(\zeta^j z)^{\deg f}]^{-1}$$

$$\text{(by (7.31) and Proposition 7.5.1)} = \prod_{j=0}^{h-1} [1 + g(\psi, \chi)\zeta^j z]$$

$$=_{****} 1 - [-g(\psi, \chi)]^h z^h$$

$$= 1 - [-g(\psi, \chi)]^h Z,$$

where:

$=_*$ follows by Lemma 7.5.2 and recalling that $d = \gcd(\deg f, h)$;

$=_{**}$ follows by observing that, for $n \geq 1$, $z^n - 1 = \prod_{\ell=0}^{n-1}(z - \exp(2\ell\pi i/n))$ which yields (after dividing by z^n and setting $w = z^{-1}$) $1 - w^n = \prod_{\ell=0}^{n-1}(1 - \exp(2\ell\pi i/n)w)$ so that, setting $\zeta = \exp(2\pi i/h)$ and $n = h/d$, $1 - w^{h/d} = \prod_{\ell=0}^{h/d-1}(1 - \zeta^{d\ell}w)$;

$=_{***}$ the numbers

$$\zeta^{j\deg f}, \quad j = 0, 1, \ldots, h-1, \tag{7.38}$$

are the same as $\zeta^{d\ell}$, $\ell = 0, 1, \ldots, h/d$, with each number in (7.38) repeated d times. Indeed, $d = \gcd(\deg f, h)$ implies that the period of $\zeta^{\deg f}$ is h/d, and if $\deg f = md$ then $\zeta^{j\deg f} = \zeta^{mjd}$ (and $\gcd(m, h) = 1$);

$=_{****}$ finally follows from the equality $1 - w^h = \prod_{j=0}^{h-1}(1 - \zeta^j w)$ (cf. $=_{**}$).

Then the Hasse-Davenport identity follows from simplifying

$$1 + g(\Psi, X)Z = 1 - [-g(\psi, \chi)]^h Z. \qquad \square$$

7.6 Jacobi sums

Definition 7.6.1 For $a \in \mathbb{F}_q$ and $\psi_1, \psi_2, \ldots, \psi_n \in \widehat{\mathbb{F}_q^*}$, the associated *Jacobi sum* is the complex number

$$J_a(\psi_1, \psi_2, \ldots, \psi_n) = \sum_{\substack{b_1, b_2, \ldots, b_n \in \mathbb{F}_q: \\ b_1 + b_2 + \cdots + b_n = a}} \psi_1(b_1)\psi_2(b_2)\cdots\psi_n(b_n),$$

with the usual convention as in (7.6).

Note that this sum effectively depends only on $n - 1$ terms: we can choose $b_1, b_2, \ldots, b_{n-1}$ arbitrarily and then b_n is uniquely determined. Recall that **1** denotes the trivial character in $\widehat{\mathbb{F}_q^*}$.

Proposition 7.6.2 *Let* $a \in \mathbb{F}_q$ *and* $\psi_1, \psi_2, \ldots, \psi_n \in \widehat{\mathbb{F}_q^*}$. *Then the following holds.*

(i) $J_a(\underbrace{\mathbf{1}, \mathbf{1}, \ldots, \mathbf{1}}_{n \text{ times}}) = q^{n-1}$;

(ii) *if $a \neq 0$*

$$J_a(\psi_1, \psi_2, \ldots, \psi_n) = \psi_1(a)\psi_2(a) \cdots \psi_n(a)J_1(\psi_1, \psi_2, \ldots, \psi_n);$$

(iii) *if some but not all of the characters $\psi_1, \psi_2, \ldots, \psi_n$ are trivial, then*
$J_a(\psi_1, \psi_2, \ldots, \psi_n) = 0;$

(iv) *if ψ_n is nontrivial then*

$$J_0(\psi_1, \psi_2, \ldots, \psi_n)$$

$$= \begin{cases} 0 & \text{if } \psi_1\psi_2 \cdots \psi_n \neq \mathbf{1} \\ \psi_n(-1)(q-1)J_1(\psi_1, \psi_2, \ldots, \psi_{n-1}) & \text{if } \psi_1\psi_2 \cdots \psi_n = \mathbf{1}. \end{cases}$$

Proof.

(i) This is obvious: each term in the sum is equal to 1.

(ii) Setting $c_j = b_j a^{-1}$, for $j = 1, 2, \ldots, n$, from $b_1 + b_2 + \cdots + b_n = a$ we deduce that $c_1 + c_2 + \cdots + c_n = 1$ and therefore

$$J_a(\psi_1, \psi_2, \ldots, \psi_n) = \sum_{\substack{c_1, c_2, \ldots, c_n \in \mathbb{F}_q: \\ c_1 + c_2 + \cdots + c_n = 1}} \psi_1(ac_1)\psi_2(ac_2) \cdots \psi_n(ac_n)$$

$$= \psi_1(a)\psi_2(a) \cdots \psi_n(a) \sum_{\substack{c_1, c_2, \ldots, c_n \in \mathbb{F}_q: \\ c_1 + c_2 + \cdots + c_n = 1}} \psi_1(c_1)\psi_2(c_2) \cdots \psi_n(c_n)$$

$$= \psi_1(a)\psi_2(a) \cdots \psi_n(a)J_1(\psi_1, \psi_2, \ldots, \psi_n).$$

(iii) Up to reordering the characters, we may suppose that $\psi_1, \psi_2, \ldots, \psi_k$ are nontrivial and $\psi_{k+1}, \psi_{k+2}, \ldots, \psi_n$ are trivial for some $1 \leq k \leq n - 1$. Since for all $b_1, b_2, \ldots, b_k \in \mathbb{F}_q$ there exist q^{n-k-1} choices of $(b_{k+1}, b_{k+2}, \ldots, b_n)$ such that $b_{k+1} + b_{k+2} + \cdots + b_n = a - b_1 - b_2 - \cdots - b_k$, we have

$$J_a(\psi_1, \psi_2, \ldots, \psi_n) = \sum_{\substack{b_1, b_2, \ldots, b_n \in \mathbb{F}_q: \\ b_1 + b_2 + \cdots + b_n = a}} \psi_1(b_1)\psi_2(b_2) \cdots \psi_k(b_k)$$

$$= q^{n-k-1} \left(\sum_{b_1 \in \mathbb{F}_q} \psi_1(b_1) \right) \left(\sum_{b_2 \in \mathbb{F}_q} \psi_2(b_2) \right) \cdots \left(\sum_{b_k \in \mathbb{F}_q} \psi_k(b_k) \right)$$

(by (7.9)) $= 0.$

(iv) First note that we may assume $n \geq 2$ because, for $n = 1$ and $\psi_1 \neq \mathbf{1}$, the statement immediately follows from (7.6). Then

$$J_0(\psi_1, \psi_2, \ldots, \psi_n) = \sum_{a \in \mathbb{F}_q} \left(\sum_{\substack{b_1, b_2, \ldots, b_{n-1} \in \mathbb{F}_q: \\ b_1 + b_2 + \cdots + b_{n-1} = -a}} \psi_1(b_1)\psi_2(b_2) \cdots \psi_{n-1}(b_{n-1}) \right) \psi_n(a)$$

$$(\psi_n(0) = 0) \;=\; \sum_{a \in \mathbb{F}_q^*} \psi_n(a) J_{-a}(\psi_1, \psi_2, \ldots, \psi_{n-1})$$

$$\text{(by (ii))} \;=\; J_1(\psi_1, \psi_2, \ldots, \psi_{n-1})$$
$$\cdot \sum_{a \in \mathbb{F}_q^*} \psi_n(a)\psi_1(-a)\psi_2(-a) \cdots \psi_{n-1}(-a)$$

$$= J_1(\psi_1, \psi_2, \ldots, \psi_{n-1})\psi_1(-1)\psi_2(-1) \cdots \psi_{n-1}(-1)$$
$$\cdot \sum_{a \in \mathbb{F}_q^*} (\psi_1 \psi_2 \cdots \psi_n)(a).$$

Now, if $\psi_1 \psi_2 \cdots \psi_n$ is nontrivial, the statement follows from (7.9). If $\psi_1 \psi_2 \cdots \psi_n = \mathbf{1}$ then $\sum_{a \in \mathbb{F}_q^*} (\psi_1 \cdots \psi_n)(a) = q - 1$ and

$$\psi_1(-1)\psi_2(-1) \cdots \psi_{n-1}(-1) = \overline{\psi_n(-1)} = \psi_n(-1)$$

(recall that $\psi_n(-1) = \pm 1$; see Lemma 7.1.4). □

Corollary 7.6.3 *Suppose that* $\psi_1, \psi_2, \ldots, \psi_n \in \widehat{\mathbb{F}_q^*}$ *are nontrivial as well as their product. Then, setting* $\psi_0 = (\psi_1 \psi_2 \cdots \psi_n)^{-1}$, *one has*

$$J_1(\psi_1, \psi_2, \ldots, \psi_n) = \frac{\psi_0(-1)}{q - 1} J_0(\psi_0, \psi_1, \ldots, \psi_n)$$

and

$$J_{-1}(\psi_1, \psi_2, \ldots, \psi_n) = \frac{1}{q - 1} J_0(\psi_0, \psi_1, \ldots, \psi_n).$$

Proof. Applying Proposition 7.6.2.(iv) with ψ_n replaced by ψ_0, we get

$$J_0(\psi_0, \psi_1, \ldots, \psi_n) = (q - 1)\psi_0(-1)J_1(\psi_1, \psi_2, \ldots, \psi_n).$$

For the second identity, use 7.6.2.(ii). □

Actually, the term "Jacobi sum" is attributed to J_1 in [79] and [96], and to J_{-1} in [95].

Proposition 7.6.4 *Suppose that* $\psi_1, \psi_2, \ldots, \psi_n \in \widehat{\mathbb{F}_q^*}$ *are nontrivial as well as their product. Then, for every nontrivial* $\chi \in \widehat{\mathbb{F}_q}$, *we have:*

$$J_1(\psi_1, \psi_2, \ldots, \psi_n) = \frac{g(\psi_1, \chi)g(\psi_2, \chi)\cdots g(\psi_n, \chi)}{g(\psi_1\psi_2\cdots\psi_n, \chi)}.$$

Proof. Indeed, by Definition 7.4.1 and (7.6), we have

$$g(\psi_1, \chi)g(\psi_2, \chi)\cdots g(\psi_n, \chi)$$

$$= \left(\sum_{x_1\in\mathbb{F}_q}\psi_1(x_1)\chi(x_1)\right)\left(\sum_{x_2\in\mathbb{F}_q}\psi_2(x_2)\chi(x_2)\right)\cdots\left(\sum_{x_n\in\mathbb{F}_q}\psi_n(x_n)\chi(x_n)\right)$$

$$= \sum_{x_1,x_2,\ldots,x_n\in\mathbb{F}_q}\psi_1(x_1)\psi_2(x_2)\cdots\psi_n(x_n)\chi(x_1 + x_2 + \cdots + x_n)$$

$$= \sum_{a\in\mathbb{F}_q}\chi(a)\sum_{\substack{x_1,x_2,\ldots,x_n\in\mathbb{F}_q:\\x_1+x_2+\cdots+x_n=a}}\psi_1(x_1)\psi_2(x_2)\cdots\psi_n(x_n)$$

$$= \sum_{a\in\mathbb{F}_q}\chi(a)J_a(\psi_1, \psi_2, \ldots, \psi_n)$$

$$=_* J_1(\psi_1, \psi_2, \ldots, \psi_n)\sum_{a\in\mathbb{F}_q^*}(\psi_1\psi_2\cdots\psi_n)(a)\chi(a)$$

$$= J_1(\psi_1, \psi_2, \ldots, \psi_n)g(\psi_1\psi_2\cdots\psi_n, \chi),$$

where $=_*$ follows from Proposition 7.6.2.(ii) and (iv). By Theorem 7.4.3.(vii), $g(\psi_1\psi_2\cdots\psi_n, \chi) \neq 0$, and this observation ends the proof. □

Proposition 7.6.5 *Suppose that* $\psi_1, \psi_2, \ldots, \psi_n \in \widehat{\mathbb{F}_q^*}$ *are nontrivial while their product* $\psi_1\psi_2\cdots\psi_n$ *is trivial. Then*

$$J_1(\psi_1, \psi_2, \ldots, \psi_{n-1}) = \frac{\psi_n(-1)}{q}g(\psi_1, \chi)g(\psi_2, \chi)\cdots g(\psi_n, \chi),$$

for all nontrivial $\chi \in \widehat{\mathbb{F}_q}$. *Moreover,*

$$J_1(\psi_1, \psi_2, \ldots, \psi_n) = -\psi_n(-1)J_1(\psi_1, \psi_2, \ldots, \psi_{n-1}).$$

Proof. Since $\psi_n^{-1} = \psi_1\psi_2\cdots\psi_{n-1}$, by Theorem 7.4.3.(vi) we have

$$g(\psi_1\psi_2\cdots\psi_{n-1}, \chi)g(\psi_n, \chi) = \psi_n(-1)q$$

and therefore, by Proposition 7.6.4 (recall also that $\psi_n(-1) = \pm 1$; see Lemma 7.1.4),

$$
\begin{aligned}
J_1(\psi_1, \psi_2, \ldots, \psi_{n-1}) &= \frac{g(\psi_1, \chi)g(\psi_2, \chi) \cdots g(\psi_{n-1}, \chi)}{g(\psi_1 \psi_2 \cdots \psi_{n-1}, \chi)} \\
&= \frac{\psi_n(-1)}{q} g(\psi_1, \chi)g(\psi_2, \chi) \cdots g(\psi_n, \chi)
\end{aligned}
$$

and the first identity is proved.

Note now that the triviality of $\psi_1 \psi_2 \cdots \psi_n$ and Proposition 7.6.2.(ii) yield

$$
J_a(\psi_1, \psi_2, \ldots, \psi_n) = J_1(\psi_1, \psi_2, \ldots, \psi_n)
$$

for all $a \in \mathbb{F}_q^*$. Then

$$
J_0(\psi_1, \psi_2, \ldots, \psi_n) + (q-1)J_1(\psi_1, \psi_2, \ldots, \psi_n)
$$

$$
\begin{aligned}
&= \sum_{a \in \mathbb{F}_q} J_a(\psi_1, \psi_2, \ldots, \psi_n) \\
\text{(by Definition 7.6.1)} \quad &= \sum_{a \in \mathbb{F}_q} \sum_{\substack{b_1, b_2, \ldots, b_n \in \mathbb{F}_q: \\ b_1 + b_2 + \cdots + b_n = a}} \psi_1(b_1)\psi_2(b_2) \cdots \psi_n(b_n) \\
&= \sum_{c_1, c_2, \ldots, c_n \in \mathbb{F}_q} \psi_1(c_1)\psi_2(c_2) \cdots \psi_n(c_n) \\
&= \left(\sum_{c_1 \in \mathbb{F}_q} \psi_1(c_1) \right) \left(\sum_{c_2 \in \mathbb{F}_q} \psi_2(c_2) \right) \cdots \left(\sum_{c_n \in \mathbb{F}_q} \psi_n(c_n) \right) \\
\text{(by (7.9))} \quad &= 0.
\end{aligned}
$$

Therefore

$$
J_1(\psi_1, \psi_2, \ldots, \psi_n) = \frac{1}{1-q} J_0(\psi_1, \psi_2, \ldots, \psi_n)
$$

$$
\text{(by Proposition 7.6.2.(iv))} \quad = -\psi_n(-1)J_1(\psi_1, \psi_2, \ldots, \psi_{n-1}). \qquad \square
$$

Corollary 7.6.6 *Suppose that $\psi_1, \psi_2, \ldots, \psi_n \in \widehat{\mathbb{F}_q^*}$ are nontrivial. If their product $\psi_1 \psi_2 \cdots \psi_n$ is nontrivial then*

$$
|J_1(\psi_1, \psi_2, \ldots, \psi_n)| = q^{(n-1)/2}, \tag{7.39}
$$

while, if $\psi_1 \psi_2 \cdots \psi_n$ is trivial then

$$
|J_1(\psi_1, \psi_2, \ldots, \psi_n)| = q^{(n-2)/2}, \tag{7.40}
$$

and

$$|J_0(\psi_1, \psi_2, \ldots, \psi_n)| = (q-1)q^{(n-2)/2}. \qquad (7.41)$$

Proof. (7.39) follows from Theorem 7.4.3.(vii) and Proposition 7.6.4. Also, (7.40) follows from 7.4.3.(vii) and Proposition 7.6.5. Finally, (7.41) follows from Proposition 7.6.2.(iv) and (7.39). $\qquad \square$

Exercise 7.6.7 Let $\psi_1, \psi_2, \ldots, \psi_k \in \widehat{\mathbb{F}_q^*}$ and suppose that they are not all trivial. Denote by $\Psi_1, \Psi_2, \ldots, \Psi_k \in \widehat{\mathbb{F}_{q^h}^*}$ their corresponding extensions as in (7.33). Prove that

$$J_1(\Psi_1, \Psi_2, \ldots, \Psi_k) = (-1)^{(h-1)(k-1)} J_1(\psi_1, \psi_2, \ldots, \psi_k).$$

Hint. Use Proposition 7.6.2.(iii) if some character is trivial, then apply Proposition 7.6.4, Proposition 7.6.5, and Theorem 7.5.3.

For more on Jacobi sums we refer to the aforementioned book by Berndt, Evans, and Williams [20].

7.7 On the number of solutions of equations

This section is based on the original paper by Weil [165] and the monographs by Ireland and Rosen [79], Lidl and Niederreiter [96], and Winnie Li [95]. It contains very important results that led Weil (ibidem) to the statement of his celebrated conjecture, solved by Deligne [52] (see also [95]).

Let $r \in \mathbb{N}$ and $f(x_0, x_1, \ldots, x_r) \in \mathbb{F}_q[x_0, x_1, \ldots, x_r]$. We denote by N_f the number of solutions of the equation $f = 0$, that is,

$$N_f = |\{(x_0, x_1, \ldots, x_r) \in \mathbb{F}_q^{r+1} : f(x_0, x_1, \ldots, x_r) = 0\}|,$$

where $\mathbb{F}_q^{r+1}$ is the $(r+1)$-dimensional vector space over $\mathbb{F}_q$. Moreover, if $u \in \mathbb{F}_q$ and $n \in \mathbb{N}$, we denote by $N_n(u)$ the number of solutions of the equation $x^n = u$, that is,

$$N_n(u) = |\{x \in \mathbb{F}_q : x^n = u\}|.$$

Lemma 7.7.1

(i) *If $d = \gcd(n, q-1)$ then*

$$N_n(u) = \begin{cases} 1 & \text{if } u = 0 \\ d & \text{if } u \text{ is a } d\text{-th power in } \mathbb{F}_q^* \\ 0 & \text{otherwise.} \end{cases}$$

(ii) If $f(x_0, x_1, \ldots, x_r) = a_0 x_0^{n_0} + a_1 x_1^{n_1} + \cdots + a_r x_r^{n_r}$ with $a_i \in \mathbb{F}_q^*$ and integers $n_i > 0$, for $i = 1, 2, \ldots, r$, then

$$N_f = \sum_{\substack{u_0, u_1, \ldots, u_r \in \mathbb{F}_q: \\ \sum_{i=0}^r a_i u_i = 0}} N_{n_0}(u_0) N_{n_1}(u_1) \cdots N_{n_r}(u_r).$$

Proof.

(i) The case $u = 0$ is obvious; the remaining is just Remark 1.2.14.
(ii) Put $x_i^{n_i} = u_i$ for $i = 0, 1, \ldots, r$, and count the number of solutions of these equations. $\qquad \square$

Lemma 7.7.2 *With the same notation as in Lemma 7.7.1.(i) we have*

$$N_n(u) = \sum_{\substack{\psi \in \widehat{\mathbb{F}_q^*}: \\ \psi^d = 1}} \psi(u).$$

Proof. Suppose first that $u \in \mathbb{F}_q^*$ is a d-th power, say $u = v^d$, for some $v \in \mathbb{F}_q^*$. Then

$$\sum_{\substack{\psi \in \widehat{\mathbb{F}_q^*}: \\ \psi^d = 1}} \psi(u) = \sum_{\substack{\psi \in \widehat{\mathbb{F}_q^*}: \\ \psi^d = 1}} \psi(v^d)$$

$$= \sum_{\substack{\psi \in \widehat{\mathbb{F}_q^*}: \\ \psi^d = 1}} [\psi(v)]^d$$

$$= |\{\psi \in \widehat{\mathbb{F}_q^*} : \psi^d = 1\}|$$

$$= d,$$

where the last equality follows from Proposition 1.2.12 applied to the cyclic group $\widehat{\mathbb{F}_q^*}$ (recall also Corollary 2.3.4 and Exercise 7.1.6).

Suppose now that u is not a d-th power and let α be a generator of $\mathbb{F}_q^*$. Then we can find $k, r \in \mathbb{N}$ with $0 < r < d$ such that $u = \alpha^{dk+r}$. Thus, if $\psi^d = 1$, we have

$$\psi(u) = \psi(\alpha^r)$$

and we may think of ψ as a character of the quotient group $\mathbb{F}_q^*/H$, where

$$H = \{v^d : v \in \mathbb{F}_q^*\} = \{\alpha^{kd} : k = 0, 1, \ldots, \frac{q-1}{d}\}.$$

Since $\mathbb{F}_q^*/H$ is cyclic of order $q - 1/((q-1)/d) = d$, we conclude that $\{\psi \in \widehat{\mathbb{F}_q^*} : \psi^d = 1\}$ may be identified with $\widehat{\mathbb{F}_q^*/H}$, so that, using the dual orthogonal relations (7.10), we deduce that

$$\sum_{\substack{\psi \in \widehat{\mathbb{F}_q^*}: \\ \psi^d = 1}} \psi(u) = \sum_{\psi \in \widehat{\mathbb{F}_q^*}/H} \psi(1)\psi(\alpha^r) = 0.$$

To conclude, in both cases, we may invoke Lemma 7.7.1.(i). $\qquad\square$

In the following we use the notation

$$\Xi = \{(\psi_0, \psi_1, \ldots, \psi_r) \in (\widehat{\mathbb{F}_q^*})^{r+1} : \psi_i \neq \mathbf{1}, \psi_i^{d_i} = \mathbf{1}, i = 0, 1, \ldots, r\}$$

and

$$\Xi_1 = \{(\psi_0, \psi_1, \ldots, \psi_r) \in \Xi : \psi_0 \psi_1 \cdots \psi_r = \mathbf{1}\}.$$

Theorem 7.7.3 (Hua-Vandiver [77], Weil [165]: the homogeneous case)
*Let f be as in Lemma 7.7.1.(ii) and set $d_i = \gcd(n_i, q-1)$, for $i = 0, 1, \ldots, r$.
Then*

$$N_f = q^r + \sum_{(\psi_0, \psi_1, \ldots, \psi_r) \in \Xi_1} \psi_0(a_0^{-1})\psi_1(a_1^{-1}) \cdots \psi_r(a_r^{-1}) J_0(\psi_0, \psi_1, \ldots, \psi_r) \tag{7.42}$$

and

$$|N_f - q^r| \leq (q-1)q^{\frac{r-1}{2}} M, \tag{7.43}$$

where $M = |\Xi_1|$.

Proof. From Lemma 7.7.1 and Lemma 7.7.2 we deduce that

$$\begin{aligned}
N_f &= \sum_{\substack{u_0, u_1, \ldots, u_r \in \mathbb{F}_q: \\ \sum_{i=0}^r a_i u_i = 0}} \sum_{\substack{\psi_0, \psi_1, \ldots, \psi_r \in \widehat{\mathbb{F}_q^*}: \\ \psi_i^{d_i} = 1, \ i = 0, 1, \ldots, r}} \psi_0(u_0)\psi_1(u_1) \cdots \psi_r(u_r) \\
&= \sum_{\substack{\psi_0, \psi_1, \ldots, \psi_r \in \widehat{\mathbb{F}_q^*}: \\ \psi_i^{d_i} = 1, \ i = 0, 1, \ldots, r}} \psi_0(a_0^{-1})\psi_1(a_1^{-1}) \cdots \psi_r(a_r^{-1}) \\
&\qquad\qquad \cdot \sum_{\substack{u_0, u_1, \ldots, u_r \in \mathbb{F}_q: \\ \sum_{i=0}^r a_i u_i = 0}} \psi_0(a_0 u_0)\psi_1(a_1 u_1) \cdots \psi_r(a_r u_r) \\
&= \sum_{\substack{\psi_0, \psi_1, \ldots, \psi_r \in \widehat{\mathbb{F}_q^*}: \\ \psi_i^{d_i} = 1, \ i = 0, 1, \ldots, r}} \psi_0(a_0^{-1})\psi_1(a_1^{-1}) \cdots \psi_r(a_r^{-1}) J_0(\psi_0, \psi_1, \ldots, \psi_r).
\end{aligned}$$

Then (7.42) follows from Proposition 7.6.2.(i), (iii), (iv). Moreover, we deduce (7.43) from (7.42) and (7.41). $\qquad\square$

We now consider the equation

$$a_0 x_0^{n_0} + a_1 x_1^{n_1} + \cdots + a_r x_r^{n_r} = b, \tag{7.44}$$

where $n_0, n_1, \ldots, n_r$ are positive integers and $b \in \mathbb{F}_q^*$. We set

$$f(x_0, x_1, \ldots, x_r) = a_0 x_0^{n_0} + a_1 x_1^{n_1} + \cdots + a_r x_r^{n_r} - b$$

and

$$N_f = |\{(x_0, x_1, \ldots, x_r) \in \mathbb{F}_q^{r+1} : f(x_0, x_1, \ldots, x_r) = 0\}|.$$

Theorem 7.7.4 (Hua-Vandiver, Weil: the non-homogeneous case) *With the notation above, and setting again $d_i = \gcd(n_i, q - 1)$, $i = 0, 1, \ldots, r$, we have:*

$$N_f = q^r + \sum_{(\psi_0, \psi_1, \ldots, \psi_r) \in \Xi} (\psi_0 \psi_1 \cdots \psi_r)(b)$$
$$\cdot \psi_0(a_0^{-1}) \psi_1(a_1^{-1}) \cdots \psi_r(a_r^{-1}) J_1(\psi_0, \psi_1, \ldots, \psi_r) \tag{7.45}$$

and

$$|N_f - q^r| \le M q^{\frac{r-1}{2}} + M' q^{\frac{r}{2}} \tag{7.46}$$

where, as before, $M = |\Xi_1|$, and $M' = |\Xi \setminus \Xi_1|$.

Proof. Arguing as in the proof of Theorem 7.7.3 we get

$$N_f = \sum_{\substack{u_0, u_1, \ldots, u_r \in \mathbb{F}_q: \\ \sum_{i=0}^{r} a_i u_i = b}} \sum_{\substack{\psi_0, \psi_1, \ldots, \psi_r \in \widehat{\mathbb{F}_q^*}: \\ \psi_i^{d_i} = 1, \ i = 0, 1, \ldots, r}} \psi_0(u_0) \psi_1(u_1) \cdots \psi_r(u_r)$$

$$= \sum_{\substack{\psi_0, \psi_1, \ldots, \psi_r \in \widehat{\mathbb{F}_q^*}: \\ \psi_i^{d_i} = 1, \ i = 0, 1, \ldots, r}} \psi_0(a_0^{-1} b) \psi_1(a_1^{-1} b) \cdots \psi_r(a_r^{-1} b)$$

$$\cdot \sum_{\substack{u_0, u_1, \ldots, u_r \in \mathbb{F}_q: \\ \sum_{i=0}^{r} b^{-1} a_i u_i = 1}} \psi_0(b^{-1} a_0 u_0) \psi_1(b^{-1} a_1 u_1) \cdots \psi_r(b^{-1} a_r u_r)$$

$$= \sum_{\substack{\psi_0, \psi_1, \ldots, \psi_r \in \widehat{\mathbb{F}_q^*}: \\ \psi_i^{d_i} = 1, \ i = 0, 1, \ldots, r}} (\psi_0 \psi_1 \cdots \psi_r)(b) \psi_0(a_0^{-1}) \psi_1(a_1^{-1}) \cdots \psi_r(a_r^{-1}) J_1(\psi_0, \psi_1, \ldots, \psi_r)$$

and (7.45) follows again from Proposition 7.6.2.(i),(iii), while the estimate (7.46) follows easily from (7.39) and (7.40). $\qquad\square$

Corollary 7.7.5 *With the same notation as in Theorem 7.7.4 we have*

$$|N_f - q^r| \le (d_0 - 1)(d_1 - 1) \cdots (d_r - 1) q^{\frac{r}{2}}.$$

Proof. Just note that $M + M' = |\Xi| = (d_0 - 1)(d_1 - 1) \cdots (d_r - 1)$. $\square$

Remark 7.7.6 Note that, both in Theorem 7.7.3 and in Theorem 7.7.4, if $d_i = 1$ for some i, then $N_f = q^r$. This is obvious: for instance, suppose that $n_0 = 1$. Then, for any choice of $x_1, x_2, \ldots, x_r \in \mathbb{F}_q$, setting

$$x_0 = -\frac{1}{a_0}[a_1 x_1^{n_1} + a_2 x_2^{n_2} + \cdots + a_r x_r^{n_r} - b]$$

yields a solution of (7.44). Moreover, since the exact formulas and the estimates depend only on the numbers $d_0, d_1, \ldots, d_r$, one may assume that $n_0, n_1, \ldots, n_r$ are divisors of $q - 1$.

Corollary 7.7.7 *Let p be a prime number, $n_0, n_1, \ldots, n_r$ positive integers, and $a_0, a_1, \ldots, a_r, b \in \mathbb{Z}$. Then the number $N(p)$ of (non-congruent) solutions $(x_0, x_1, \ldots, x_r) \in \mathbb{Z}^{r+1}$ of the congruence*

$$a_0 x_0^{n_0} + a_1 x_1^{n_1} + \cdots + a_r x_r^{n_r} = b \mod p$$

satisfies the condition

$$|N(p) - p^r| \leq (n_0 - 1)(n_1 - 1) \cdots (n_r - 1)p^{r/2}.$$

In particular,

$$\lim_{\substack{p \to +\infty: \\ p \text{ prime}}} N(p) = +\infty.$$

Proof. This follows immediately from Corollary 7.7.5 after observing that $n_i \geq d_i$ for all $i = 0, 1, \ldots, r$. $\square$

We conclude this section with an exercise.

Exercise 7.7.8

(1) Prove that for every integer $k \geq 0$

$$\sum_{x \in \mathbb{F}_q} x^k = \begin{cases} 0 & \text{if } k = 0 \text{ or } (q-1) \nmid k \\ -1 & \text{if } k > 0 \text{ and } (q-1) | k \end{cases}$$

(here we assume $0^0 = 1$).
Hint: For $k > 0$ use a generator α of $\mathbb{F}_q^*$.

(2) Show that if $f \in \mathbb{F}_q[x_1, x_2, \ldots, x_n]$ and $\deg(f) < n(q-1)$ then

$$\sum_{\alpha_1, \alpha_2, \ldots, \alpha_n \in \mathbb{F}_q} f(\alpha_1, \alpha_2, \ldots, \alpha_n) = 0.$$

Hint: from (1) deduce the statement for a monomial.

(3) Show that if $f \in \mathbb{F}_q[x_1, x_2, \ldots, x_n]$ and $F = 1 - f^{q-1}$ then

$$N_f = \sum_{\alpha_1, \alpha_2, \ldots, \alpha_n \in \mathbb{F}_q} F(\alpha_1, \alpha_2, \ldots, \alpha_n),$$

where N_f is seen as an element of $\mathbb{F}_q$.

(4) **(Warning's Theorem** [164]) Prove that if $f \in \mathbb{F}_q[x_1, x_2, \ldots, x_n]$ and $\deg(f) < n$, then N_f is divisible by p.
Hint: from (2) and (3) it follows that $N_f = 0 \mod p$.

(5) **(Chevalley's Theorem** [39]) Show that if $f \in \mathbb{F}_q[x_1, x_2, \ldots, x_n]$ satisfies $f(0, 0, \ldots, 0) = 0$ and $\deg(f) < n$, then $N_f \geq 2$. In particular, $f = 0$ has a nontrivial solution.

Remark 7.7.9 Chevalley's theorem was conjectured by E. Artin in 1935 and immediately proved by Chevalley and generalized by Warning. The proof sketched in the above exercise is due to Ax [16]. Warning, actually, proved that $N_f \geq q^{n-\deg(f)}$; see the monograph by Lidl and Niederreiter [96], where these results are proved also for systems of polynomials.

7.8 The FFT over a finite field

In this section, following again [160], we describe the matrix form of several algorithms for the additive Fourier transform over $\mathbb{F}_q$, with $q = p^h$, $p \geq 3$ prime, and $h \geq 1$. We generalize Rader's algorithm discussed at the end of Section 5.4. The original sources are [2] and [14].

The Fourier Transform over $\mathbb{F}_q$ is defined as in (2.15) by setting

$$\widehat{f}(\chi) = \sum_{x \in \mathbb{F}_q} f(x)\overline{\chi(x)} \tag{7.47}$$

for all $f \in L(\mathbb{F}_q)$ and $\chi \in \widehat{\mathbb{F}_q}$. However, to keep notation similar to that in Section 5.4, we avoid conjugation for χ when describing the matrix representing (7.47). By means of Theorem 6.3.3, we fix a generator α of the cyclic group $\mathbb{F}_q^*$ and we introduce the following ordering for the elements of $\mathbb{F}_q$:

$$0, \alpha^0 = 1, \alpha, \alpha^2, \ldots, \alpha^{q-2}. \tag{7.48}$$

Then, using the representation (7.5), we define the *Fourier Matrix* $A_{\mathbb{F}_q}$ of $\mathbb{F}_q$ by setting

$$A_{\mathbb{F}_q} = \begin{pmatrix} 1 & 1 & \cdots & 1 \\ 1 & & & \\ \vdots & & C_{q-1} & \\ 1 & & & \end{pmatrix}, \tag{7.49}$$

where

$$C_{q-1} = \left(\exp[2\pi i\mathrm{Tr}(\alpha^{k+j})/p]\right)_{k,j=0}^{q-2} \qquad (7.50)$$

is the associated *core matrix*. C_{q-1} has the following property: its (k, j)-entry depends only upon $k + j \bmod q - 1$. A matrix with this property is called *skew circulant* mod $q - 1$. In particular, C_{q-1} is Hankel and therefore symmetric. Note also that in [51] it is given a different definition of skew-circulant matrices, but we follow the terminology in [160]. Clearly, (7.49) represents the matrix form of Rader's algorithm over $\mathbb{F}_q$. Now we describe three block decompositions of the core matrix C_{q-1}. First of all, we assume that $h \geq 2$ so that $q - 1 = p^h - 1$ is not a prime number (for instance, it is divisible by $p - 1$). We begin with a description of an analogue of the Cooley-Tukey algorithm due to Agarwal and Tukey. Suppose that $q - 1 = mn$ is a nontrivial (arbitrary) factorization of $q - 1$. Denote by

$$B = \langle \alpha^m \rangle \qquad (7.51)$$

the subgroup generated by α^m. Clearly, B is cyclic of order n and we have the coset decomposition

$$\mathbb{F}_q^* = \coprod_{k=0}^{m-1} \alpha^k B.$$

Now we choose a different ordering for $\mathbb{F}_q$ (in place of (7.48)): we first order B by setting

$$1, \alpha^m, \alpha^{2m}, \ldots, \alpha^{(n-1)m} \qquad (7.52)$$

and then we order $\mathbb{F}_q$:

$$0, B, \alpha B, \ldots, \alpha^{m-1}B. \qquad (7.53)$$

The core matrix corresponding to this ordering has the form

$$\begin{pmatrix} C(0,0) & C(0,1) & \cdots & C(0, m-1) \\ C(1,0) & C(1,1) & \cdots & C(1, m-1) \\ \vdots & \vdots & \ddots & \vdots \\ C(m-1,0) & C(m-1,1) & \cdots & C(m-1, m-1) \end{pmatrix} \qquad (7.54)$$

where $C(r, r')$, with $0 \leq r, r' \leq m - 1$, is the $n \times n$ matrix

$$C(r, r') = \left(\exp[2\pi i\mathrm{Tr}(\alpha^{r+r'+(s+s')m})/p]\right)_{s,s'=0}^{n-1}. \qquad (7.55)$$

Note that $C(r, r')$ is skew-circulant mod n. It follows that (7.54) is a Hankel (actually skew-circulant mod nm) matrix whose blocks are Hankel (actually skew-circulant mod n) matrices. A further property is presented in the following proposition.

Proposition 7.8.1 *Set*

$$S_n = \begin{pmatrix} 0 & 1 & 0 & \cdots & 0 \\ 0 & 0 & 1 & \cdots & 0 \\ \vdots & \vdots & & \ddots & \vdots \\ 0 & 0 & 0 & \cdots & 1 \\ 1 & 0 & 0 & \cdots & 0 \end{pmatrix}. \tag{7.56}$$

If $r + r' = r_1 + r'_1 \mod m$ and

$$\ell m = r + r' - r_1 - r'_1 \tag{7.57}$$

for some positive ℓ, then

$$C(r, r') = S_n^\ell C(r_1, r'_1).$$

Proof. From (7.57) we deduce that

$$r + r' + (s + s')m = r_1 + r'_1 + (\ell + s + s')m$$

so that

$$[C(r, r')]_{s,s'} = [C(r_1, r'_1)]_{s+\ell,s'} = [S_n^\ell C(r_1, r'_1)]_{s,s'},$$

where $s + \ell$ must be considered mod n. □

Remark 7.8.2 Clearly, the matrices (7.50) and (7.54) are similar and the similarity is realized by means of a permutation matrix (recall Corollary 5.3.2). More precisely, by means of the permutation of $\mathbb{F}_q^*$ that transforms the ordered sequence (7.48) into the ordered sequence (7.53). The easy details are left to the reader and the same remark holds true for the block decomposition (7.59).

Now we give an analogue of the Good formula (Corollary 5.4.13).

Suppose, as before, that $q - 1 = nm$. We now also require that $\gcd(n, m) = 1$. By Proposition 1.2.5 we have

$$\mathbb{Z}_{mn} \cong \mathbb{Z}_m \oplus \mathbb{Z}_n. \tag{7.58}$$

More precisely, the generator of $\mathbb{Z}_m$ is n and the generator of $\mathbb{Z}_n$ is m (for instance, take $a = 1$ in the proof of Proposition 1.2.5, or use Bezout's identity (1.2): $1 = um + vn \Rightarrow m = 1 \mod n$ and $n = 1 \mod m$). Setting $A = \langle \alpha^n \rangle$ and recalling that $B = \langle \alpha^m \rangle$ (cf. (7.51)), (7.58) yields the multiplicative decomposition

$$\mathbb{F}_q^* \cong A \times B$$

with A (respectively B) multiplicative cyclic of order m (respectively n). Then, we may replace the ordering (7.53) by

$$0, B, \alpha^n B, \alpha^{2n} B, \ldots, \alpha^{(m-1)n} B,$$

where B is ordered again as in (7.52). With this new ordering, the core matrix has the form

$$
\begin{pmatrix}
\widetilde{C}(0,0) & \widetilde{C}(0,1) & \cdots & \widetilde{C}(0, m-1) \\
\widetilde{C}(1,0) & \widetilde{C}(1,1) & \cdots & \widetilde{C}(1, m-1) \\
\vdots & \vdots & & \vdots \\
\widetilde{C}(m-1,0) & \widetilde{C}(m-1,1) & \cdots & \widetilde{C}(m-1, m-1)
\end{pmatrix}
\tag{7.59}
$$

where $\widetilde{C}(r, r')$, with $0 \le r, r' \le m-1$, is the $n \times n$ matrix

$$\widetilde{C}(r, r') = \left(\exp[2\pi i \mathrm{Tr}(\alpha^{(r+r')n+(s+s')m})/p] \right)_{s,s'=0}^{n-1}.$$

The $\widetilde{C}$s have the same properties of the Cs in (7.54). Moreover,

$$\widetilde{C}(r, r') = \widetilde{C}(r_1, r_1')$$

if $r + r' = r_1 + r_1' \bmod m$. Setting $T(r) = \widetilde{C}(r, 0)$, matrix (7.59) takes the form:

$$
\begin{pmatrix}
T(0) & T(1) & \cdots & T(m-1) \\
T(1) & T(2) & \cdots & T(0) \\
\vdots & \vdots & & \vdots \\
T(m-1) & T(0) & \cdots & T(m-2)
\end{pmatrix}.
$$

This matrix is block skew-circulant mod m and each block is skew-circulant mod n.

We consider now a particular case of (7.54). We take $m = \frac{p^h - 1}{p - 1}$ and $n = p - 1$. The matrix S_{p-1} is as in (7.56). Set also $\omega = e^{2\pi i/p}$ and $\varepsilon = \alpha^m$. Note that now $B \cong \mathbb{Z}_p^*$ and $\varepsilon \in \mathbb{F}_p^*$ is a generator of this group (recall Corollary 6.3.5).

Theorem 7.8.3 (Auslander, Feigh, and Winograd) *Define*

$$v \colon \{0, 1, \ldots, q-1\} \to \{0, 1, \ldots, p-2\} \cup \{-\infty\}$$

by means of the relation

$$
\begin{cases}
\mathrm{Tr}(\alpha^r) = \varepsilon^{v(r)} & \text{if } \mathrm{Tr}(\alpha^r) \neq 0 \\
v(r) = -\infty & \text{if } \mathrm{Tr}(\alpha^r) = 0
\end{cases}
$$

for $r = 0, 1, \ldots, q - 1$. Set also

$$S_{p-1}^{-\infty} = \begin{pmatrix} -1 & -1 & \cdots & -1 \\ -1 & -1 & \cdots & -1 \\ \vdots & \vdots & \ddots & \vdots \\ -1 & -1 & \cdots & -1 \end{pmatrix}.$$

Then the matrix (7.54) may be factorized as

$$[I_m \otimes C(p)]S,$$

where we use the notation in (5.27) and

$$S = \begin{pmatrix} S_{p-1}^{-\nu(0)} & S_{p-1}^{-\nu(1)} & \cdots & S_{p-1}^{-\nu(m-1)} \\ S_{p-1}^{-\nu(1)} & S_{p-1}^{-\nu(2)} & \cdots & S_{p-1}^{-\nu(m)} \\ \vdots & \vdots & & \vdots \\ S_{p-1}^{-\nu(m-1)} & S_{p-1}^{-\nu(m)} & \cdots & S_{p-1}^{-\nu(2m-2)} \end{pmatrix}$$

and

$$C(p) = \left(\omega^{\varepsilon^{k+j}} \right)_{k,j=0}^{p-2}.$$

Proof. First of all, recall that the trace is $\mathbb{F}_p$-linear by Hilbert Satz 90 (cf. Theorem 6.7.2). Therefore, since $\varepsilon = \alpha^m \in \mathbb{F}_p$ in (7.55) we have

$$\text{Tr}(\alpha^{r+r'+(s+s')m}) = \text{Tr}(\alpha^{r+r'}\varepsilon^{s+s'}) = \varepsilon^{s+s'}\text{Tr}(\alpha^{r+r'}). \qquad (7.60)$$

We consider two cases.

<u>First case:</u> $\text{Tr}(\alpha^{r+r'}) \neq 0$. Then $\text{Tr}(\alpha^{r+r'}) = \varepsilon^{\nu(r+r')}$ so that (7.55) becomes

$$[C(r, r')]_{s,s'} = \omega^{\varepsilon^{s+s'+\nu(r+r')}}.$$

On the other hand, since $S_{p-1}^{-\ell} = (\delta_{i-\ell,j})_{i,j=0}^{p-2}$, we have

$$[C(p)S_{p-1}^{-\nu(r+r')}]_{s,s'} = \sum_{t=0}^{p-1} \omega^{\varepsilon^{s+t}} \delta_{t-\nu(r+r'),s'} = \omega^{\varepsilon^{s+s'+\nu(r+r')}}.$$

<u>Second case:</u> $\text{Tr}(\alpha^{r+r'}) = 0$. Then, by means of (7.60), equation (7.55) becomes

$$C(r, r') = \begin{pmatrix} 1 & 1 & \cdots & 1 \\ 1 & 1 & \cdots & 1 \\ \vdots & \vdots & \ddots & \vdots \\ 1 & 1 & \cdots & 1 \end{pmatrix}.$$

Moreover, since $\sum_{x \in \mathbb{F}_p} \omega^x = \sum_{k=0}^{p-1} \omega^k = \frac{\omega^p - 1}{\omega - 1} = 0$, we have

$$[C(p)S_{p-1}^{-\infty}]_{s,s'} = -\sum_{t=0}^{p-2} \omega^{\varepsilon^{s+t}} = -\sum_{x \in \mathbb{F}_p^*} \omega^x = 1.$$

It follows that

$$C(p)S_{p-1}^{-\infty} = \begin{pmatrix} 1 & 1 & \cdots & 1 \\ 1 & 1 & \cdots & 1 \\ \vdots & \vdots & \ddots & \vdots \\ 1 & 1 & \cdots & 1 \end{pmatrix} = C(r, r').$$

$\square$

Part III
Graphs and expanders

8

Graphs and their products

This chapter is an introduction to (finite) graph theory with an emphasis on spectral analysis of k-regular graphs. In Section 8.2 we study strongly regular graphs with a description of the celebrated Petersen graph and the Clebsch graph: the latter, in particular, is also described in terms of number theory over the Galois field $\mathbb{F}_{16}$. In the subsequent sections, we describe bipartite graphs as well as three basic examples (the complete graph, the hypercube, and the discrete circle) based on the theories developed in Chapter 4. Other explicit examples can be found in Section 8.8, where we give a detailed exposition of various notions of graph products, culminating with the study of the lamplighter graph, of the replacement product, and of the zig-zag product, in Section 8.11, Section 8.12, and Section 8.13, respectively. See also our first monograph [29]. In Chapter 9 we shall focus on more advanced topics such as the Alon-Milman-Dodziuk theorem, the Alon-Boppana-Serre theorem, and explicit constructions of expanders.

8.1 Graphs and their adjacency matrix

An *undirected graph* is a triple $\mathcal{G} = (X, E, r)$, where X is a nonempty set of *vertices*, E is a set of *edges*, and $r \colon E \to \mathcal{P}(X)$ is a map from the edge set into the power set of X such that $0 < |r(e)| \leq 2$ (as usual, $|\cdot|$ denotes cardinality). If $e \in E$ satisfies $r(e) = \{x\}$, then we say that e is a *loop* based at x. We denote by $E_0 = \{e \in E : |r(e)| = 1\}$ the set of all loops of $\mathcal{G}$ and denote by $E_1 = E \setminus E_0 = \{e \in E : |r(e)| = 2\}$ the set of remaining edges. Moreover, if there exist distinct edges $e, e' \in E$ such that $r(e) = r(e')$, equivalently, if the map r is not injective, we say that the graph $\mathcal{G}$ has *multiple edges*. On the other hand, if the map r is injective, that is, $\mathcal{G}$ has no multiple edges, one says that the graph is *simple*.

Thus, a simple (undirected) graph without loops can be regarded just as a pair $\mathcal{G} = (X, E)$, where X is the set of vertices and any edge $e \in E$ is a two-subset $\{x, y\}$ of (distinct) elements of X (we identify e with $r(e)$).

A *directed graph* is a triple $\mathcal{G} = (X, E, \vec{r})$, where X is a set of *vertices*, E is a set of *(oriented) edges*, and $\vec{r}: E \to X \times X$, called an *orientation* of $\mathcal{G}$, is a map from the edge set into the Cartesian square of X. Writing $\vec{r}(e) = (e_-, e_+)$, we say that e_- (respectively e_+) is the *initial* (respectively *terminal*) vertex of the oriented edge $e \in E$. Note that a directed graph $\mathcal{G} = (X, E, \vec{r})$ can be viewed as an undirected graph $\mathcal{G} = (X, E, r)$ by setting

$$r(e) = \{e_-, e_+\} \tag{8.1}$$

for all $e \in E$. Clearly, $e \in E$ is a loop if and only if $e_- = e_+$. Conversely, given an undirected graph $\mathcal{G} = (X, E, r)$, for every $e \in E_1$ we may arbitrarily choose a labeling of the two elements in $r(e)$ and write $r(e) = \{e_-, e_+\}$. This defines an orientation $\vec{r}: E \to X \times X$ by setting

$$\vec{r}(e) = \begin{cases} (x, x) & \text{if } e \in E_0 \text{ and } r(e) = \{x\} \\ (e_-, e_+) & \text{if } e \in E_1 \text{ and } r(e) = \{e_-, e_+\}. \end{cases}$$

Note that there are exactly $2^{|E_1|}$ different orientations on $\mathcal{G}$. Moreover, the undirected graph associated (via (8.1)) with the newly defined directed graph $\mathcal{G} = (X, E, \vec{r})$ is the original undirected graph $\mathcal{G} = (X, E, r)$.

From now on, unless otherwise specified, all graphs will be undirected.

Let $\mathcal{G} = (X, E, r)$ be an (undirected) graph.

Two vertices x and y are called *neighbors* or *adjacent*, and we write $x \sim y$, provided there exists $e \in E$ such that $r(e) = \{x, y\}$. We then say that the edge e *joins* the vertices x and y. Given a vertex $x \in X$, we denote by

$$\mathcal{N}(x) = \{y \in X : y \sim x\} \subseteq X$$

the *neighborhood* of x, by $E_x = \{e \in E : r(e) \ni x\}$ the set of edges *incident* to x, and by $\deg x = |E_x|$, the number of edges incident to x, called the *degree* of x. Note that a vertex $x \in V$ belongs to $\mathcal{N}(x)$ if and only if there exists a loop $e \in E$ based at x (that is, $r(e) = \{x\}$). When $\deg(\cdot) = k$ is constant, we say that the graph is *regular* of *degree k*, or *k-regular*. Note that if $\mathcal{G}$ is simple then $|\mathcal{N}(x)| = |E_x| = \deg x$.

If X and E are both finite we say that $\mathcal{G}$ is *finite*. Note that a simple graph $\mathcal{G} = (X, E)$ without loops is finite if (and only if) X is finite.

Let $\mathcal{F} = (Y, F, s)$ be another (undirected) graph.

$\mathcal{F}$ is called a *subgraph* of $\mathcal{G}$ provided $Y \subseteq X$, $F \subseteq E$, and $r|_F = s$.

$\mathcal{F}$ is said to be *isomorphic* to $\mathcal{G}$ if there exists a pair $\Phi = (\phi, \varphi)$ of bijections $\phi \colon X \to Y$ and $\varphi \colon E \to F$ such that

$$s(\varphi(e)) = \phi(r(e))$$

for all $e \in E$. One then writes $\Phi \colon \mathcal{G} \to \mathcal{F}$ and calls it an *isomorphism* of the graphs $\mathcal{G}$ and $\mathcal{F}$. Moreover, if $\mathcal{G}$ and $\mathcal{F}$ are both directed, then $\Phi = (\phi, \varphi)$ is a *(directed graphs) isomorphism* of $\mathcal{G}$ and $\mathcal{F}$ if

$$(\phi(e_-), \phi(e_+)) = (\varphi(e)_-, \varphi(e)_+)$$

for all $e \in E$.

A (finite) *path* in $\mathcal{G}$ is a sequence $p = (x_0, e_1, x_1, e_2, x_2 \ldots, e_m, x_m)$, with $x_0, x_1, \ldots, x_m \in X$ and $e_1, e_2, \ldots, e_m \in E$ such that $r(e_i) = \{x_{i-1}, x_i\}$ for all $i = 1, \ldots, m$. The vertices x_0 and x_m are called the *initial* and *terminal* vertices of p, respectively, and one says that p connects them. The nonnegative number $|p| = m$ is called the *length* of the path p. When $m = 0$ one calls $p = (x_0)$ the *trivial path* based at x_0. If $x_0 = x_m$ one says that the path is *closed* and p is also called a *cycle*. The *inverse* of a path $p = (x_0, e_1, x_1, e_2, x_2 \ldots, e_m, x_m)$ is the path $p^{-1} = (x_m, e_m, x_{m-1}, \ldots, x_1, e_1, x_0)$; note that $|p^{-1}| = |p|$. Given two paths $p = (x_0, e_1, x_1, e_2, x_2 \ldots, e_m, x_m)$ and $p' = (x'_0, e'_1, x'_1, e'_2, x'_2 \ldots, e'_n, x'_n)$ with $x_m = x'_0$ we define their *composition* as the path $p \cdot p' = (x_0, e_1, x_1, e_2, x_2 \ldots, e_m, x_m = x'_0, e'_1, x'_1, e'_2, x'_2 \ldots, e'_n, x'_n)$; note that $|p \cdot p'| = |p| + |p'|$.

For $x, y \in X$ we write $x \approx y$ if there exists a path connecting them: clearly, $\approx$ is an equivalence relation on the set X of vertices of $\mathcal{G}$. The equivalence classes are called the *connected components* of $\mathcal{G}$. One says that $\mathcal{G}$ is *connected* if there exists a unique connected component, in other words, if any two vertices in X are connected by a path. If this is the case, the *geodesic distance* of two vertices $x, y \in X$, denoted $d(x, y)$, is the minimal length of a path connecting them.

The *diameter* of a finite connected graph $\mathcal{G}$, denoted $D(\mathcal{G})$, is the maximal distance of two vertices in $\mathcal{G}$, in formulæ,

$$D(\mathcal{G}) = \max\{d(x, y) : x, y \in X\}.$$

Proposition 8.1.1 *Let $\mathcal{G} = (X, E, r)$ be a finite connected k-regular graph. Then*

$$D(\mathcal{G}) \geq \log_k[(k-1)|X| + 1] - 1.$$

Proof. Fix a base vertex $x_0 \in X$ and set $X_j = \{x \in X : d(x, x_0) = j\}$ for $j = 0, 1, 2, \ldots, D = D(\mathcal{G})$ (note that we may have $X_{j_0} = \varnothing$ for some $0 < j_0 \leq D$; then $X_j = \varnothing$ for all $j_0 \leq j \leq D$). We have $|X_0| = |\{x_0\}| = 1$ and, since $\mathcal{G}$ is k-regular, $|X_1| \leq k$ and, recursively, $|X_j| \leq |X_{j-1}|(k-1) \leq k(k-1)^{j-1} < k^j$, for $j \geq 2$. Indeed, each $y \in X_j$ is joined with at least one vertex $x \in X_{j-1}$ and, in turn, each such $x \in X_{j-1}$ is joined with at most $k - 1$ vertices in X_j. It follows that

$$|X| = |X_0 \sqcup X_1 \sqcup X_2 \sqcup \cdots \sqcup X_D| \leq 1 + k + k^2 + \cdots + k^D = \frac{k^{D+1} - 1}{k - 1}.$$

We deduce that $k^{D+1} \geq (k - 1)|X| + 1$ and, finally, $D \geq \log_k[(k - 1)|X| + 1] - 1$. $\qquad\square$

Corollary 8.1.2 *Let* $(\mathcal{G}_n = (X_n, E_n, r_n))_{n \in \mathbb{N}}$ *be a family of finite connected k-regular graphs such that* $|X_n| \underset{n \to \infty}{\to} \infty$. *Then also* $D(X_n) \underset{n \to \infty}{\to} \infty$. $\qquad\square$

Let $\mathcal{G} = (X, E, r)$ be a finite graph. The *adjacency matrix* associated with $\mathcal{G}$ is the $X \times X$-matrix $A = (A(x, y))_{x, y \in X}$ defined by setting

$$A(x, y) = |r^{-1}(\{x, y\})|$$

for all $x, y \in X$. In other words, if $x \neq y$ we have $A(x, y) = |E_x \cap E_y|$ equals the number (possibly 0) of edges incident to both x and y, and $A(x, x)$ is the number (possibly 0) of loops based at x. Note that A is symmetric ($A(x, y) = A(y, x)$ for all $x, y \in X$), that $A(x, y) \neq 0$ if and only if $x \sim y$, and $\deg x = \sum_{y \in X} A(x, y)$. Thus, $\mathcal{G}$ is simple (respectively without loops) if and only if $A(x, y) \in \{0, 1\}$ for all $x, y \in X$ (respectively $A(x, x) = 0$ for all $x \in X$). Often, we shall identify the matrix A with the corresponding linear operator $A \colon L(X) \to L(X)$, called the *adjacency operator* associated with $\mathcal{G}$, defined by setting

$$[Af](x) = \sum_{y \in Y} A(x, y) f(y) = \sum_{y \in Y} A(y, x) f(y),$$

for all $f \in L(X)$ and $x \in X$. Note that $A \delta_x = \sum_{y \sim x} A(x, y) \delta_y$, for all $x \in X$.

Moreover, as A is symmetric, it is diagonalizable and its *spectrum* $\sigma(A) = \{\mu \in \mathbb{C} : A - \mu I$ is not invertible$\}$ (that is, the set of its eigenvalues) is real ($\sigma(A) \subseteq \mathbb{R}$), and there exists an orthogonal basis of $L(X)$ made up of real-valued eigenfunctions (see [91]). One refers to $\sigma(A)$ as to the *spectrum* of the graph $\mathcal{G}$.

Remark 8.1.3 Warning that if $\mathcal{G} = (X, E, \vec{r})$ is directed, in this book we define its adjacency matrix as the adjacency matrix A of the associated undirected graph $\mathcal{G} = (X, E, r)$ (cf. (8.1)). In other contexts, one sets $A(x, y) = |(\vec{r})^{-1}(x, y)|$ for all $x, y \in X$ and therefore, in general, A is not symmetric. On the contrary, in our setting, A is always symmetric!

We recall (cf. Proposition 2.1.1) that $W_0 \leq L(X)$ is the space of constant functions on X and $W_1 = \{f \in L(X) : \sum_{x \in X} f(x) = 0\}$, so that $L(X) = W_0 \bigoplus W_1$ (cf. (2.4)).

Proposition 8.1.4 *Let $\mathcal{G} = (X, E, r)$ be a finite graph, with adjacency matrix A. If $\mathcal{G}$ is k-regular, then the decomposition $L(X) = W_0 \oplus W_1$ is A-invariant and W_0 is the eigenspace corresponding to the eigenvalue k. Conversely, if W_0 is an eigenspace of A, then the graph is regular and the degree is given by the corresponding eigenvalue.*

Proof. Suppose first that $\mathcal{G}$ is k-regular and let us show that W_0 and W_1 are A-invariant. Let $f_0 \in W_0$ and $x \in X$. Then

$$[Af_0](x) = \sum_{y \in X} A(x, y)f_0(y) = \sum_{y \in X} A(x, y)f_0(x) = \deg x f_0(x), \qquad (8.2)$$

showing that $Af_0 = kf_0$. Similarly, if $f_1 \in W_1$ we have

$$\sum_{x \in X}[Af_1](x) = \sum_{x \in X}\sum_{y \in X} A(x, y)f_1(y)$$

$$= \sum_{y \in X}\sum_{x \in X} A(x, y)f_1(y)$$

$$(\text{since } \textstyle\sum_{x \in X} A(x, y) = \deg y = k) \quad = k\sum_{y \in X} f_1(y) = 0,$$

showing that $Af_1 \in W_1$.

Conversely, assume that a nontrivial constant function $f_0 \equiv c$ is an eigenvector of A, with eigenvalue α. Then, as in (8.2), $[Af_0](x) = (\deg x)c$, and as $Af_0 = \alpha f_0 \equiv \alpha c$ we deduce that $\deg x = \alpha$ for all $x \in X$. $\qquad\square$

Proposition 8.1.5 *Let $\mathcal{G} = (X, E, r)$ be a finite k-regular graph. Let $\mu_0 \geq \mu_1 \geq \cdots \geq \mu_{|X|-1}$ be the eigenvalues of the adjacency matrix A of $\mathcal{G}$. Then*

(i) *k is an eigenvalue and its multiplicity equals the number of connected components of $\mathcal{G}$; in particular, $\mathcal{G}$ is connected if and only if the multiplicity of k is equal to 1;*

(ii) *$|\mu_i| \leq k$ for $i = 0, 1, \ldots, |X| - 1$, so that $\mu_0 = k$.*

Proof.

(i) It follows from (8.2) that if $f \in L(X)$ is constant on each connected component of $\mathcal{G}$, then $Af = kf$. This shows that k is an eigenvalue of A and that its multiplicity is, at least, the number of connected components of $\mathcal{G}$ (the characteristic functions of these connected components are, clearly, linearly independent). Conversely, suppose that $Af = kf$ with $f \in L(X)$ non-identically zero and real-valued. Let $X_0 \subset X$ be a connected component of $\mathcal{G}$ and suppose that $|f|$, restricted to X_0, attains its maximum at the point $x_0 \in X_0$, i.e. $|f(x_0)| \geq |f(x)|$ for all $x \in X_0$. We may suppose, up to passing to $-f$, that $f(x_0) > 0$ so that $f(x_0) \geq f(x)$ for all $x \in X_0$. Then

$$\sum_{x \in X_0} A(x_0, x)[f(x_0) - f(x)] = \sum_{x \in X_0} A(x_0, x)f(x_0) - \sum_{x \in X_0} A(x_0, x)f(x)$$

$$= kf(x_0) - kf(x_0) = 0.$$

Since $A(x_0, x) \geq 0$ and $f(x_0) - f(x) \geq 0$ for all $x \in X_0$, we deduce that $f(x) = f(x_0)$ for all $x \sim x_0$. By induction on the geodesic distance from x_0, we deduce that $f(x) = f(x_0)$ for all $x \in X_0$, that is, f is constant on X_0. This shows that f is constant on the connected components of X. In particular, the multiplicity of k is at most, and therefore equal to, the number of connected components of $\mathcal{G}$.

(ii) Let μ be an eigenvalue and denote by $f \in L(X)$ a corresponding (non-trivial) real-valued eigenfunction. Suppose that $|f|$ attains its maximum at the point $x_0 \in X$, i.e. $|f(x_0)| \geq |f(x)|$ for all $x \in X$. As before, up to passing to $-f$, we may assume that $f(x_0) > 0$ so that $f(x_0) \geq |f(x)|$ for all $x \in X$. Then we have

$$|\mu|f(x_0) = |\mu f(x_0)| = |\sum_{x \in X} A(x_0, x)f(x)|$$

$$\leq \sum_{x \in X} A(x_0, x)|f(x)|$$

$$\leq \left(\sum_{x \in X} A(x_0, x)\right) f(x_0),$$

$$= kf(x_0),$$

so that $|\mu| \leq k$. $\qquad\qquad\square$

Proposition 8.1.6 *Let $\mathcal{G} = (X, E, r)$ be a finite graph and denote by $A = (A(x, y))_{x,y \in X}$ the associated adjacency matrix. Then, denoting by $A^\ell = \left(A^{(\ell)}(x, y)\right)_{x,y \in X}$, $\ell \in \mathbb{N}$, the ℓ-th power of A (with the convention that $A^0 = I$, the identity matrix; cf. Section 2.1), we have*

$$A^{(\ell)}(x, y) = \text{ the number of paths of length } \ell \text{ in } \mathcal{G} \text{ connecting } x \text{ and } y$$

for all $x, y \in X$.

Proof. Let $x, y \in X$. If $\ell = 0$ the statement follows from the fact that there is exactly one (respectively, no) path of length 0, the trivial path at x, connecting x and y for $x = y$ (respectively, $x \neq y$). Now, every path

$$p(x, y) = (x_0 = x, e_1, x_1, e_2, x_2, \ldots, e_\ell, x_\ell = z, e_{\ell+1}, x_{\ell+1} = y)$$

of length $\ell + 1$ in $\mathcal{G}$ connecting x to y is the composition of the path $p(x, z) = (x_0 = x, e_1, x_1, e_2, x_2, \ldots, e_\ell, x_\ell = z)$ of length ℓ connecting x to z, a neighbor of y, and the edge $e_{\ell+1} \equiv (z, e_{\ell+1}, y)$. By induction, the number of such paths $p(x, z)$ equals $A^{(\ell)}(x, z)$, and the number of edges $e \equiv (z, e, y)$ equals, by definition, $A(z, y)$. As a consequence, the number of paths of length $\ell + 1$ connecting x to y is given by

$$\sum_{\substack{e \in E: \\ r(e) = \{z, y\}}} A^{(\ell)}(x, z) = \sum_{z \in X} A^{(\ell)}(x, z) A(z, y) = A^{(\ell+1)}(x, y).$$

$\square$

8.2 Strongly regular graphs

This section contains a series of exercises on a remarkable family of regular graphs.

Definition 8.2.1 A finite simple graph $\mathcal{G} = (X, E)$ without loops is called *strongly regular* of parameters (v, k, λ, μ) if

   (i) it is regular of degree k and $|X| = v$;
   (ii) for all $\{x, y\} \in E$ there exist exactly λ vertices adjacent to both x and y;
   (iii) for all $x, y \in X$ with $x \neq y$ and $\{x, y\} \notin E$ there exist exactly μ vertices adjacent to both x and y.

Note that, in the above definition, $0 \leq \lambda \leq k - 1$ and $0 \leq \mu \leq k$. Moreover, if $\mu > 0$ then $\mathcal{G}$ is connected.

Exercise 8.2.2 Let $\mathcal{G} = (X, E)$ be a finite simple graph without loops and set $|X| = v$. Denote by A its adjacency matrix and set $J = \begin{pmatrix} 1 & \cdots & 1 \\ \vdots & & \vdots \\ 1 & \cdots & 1 \end{pmatrix}$ (the $v \times v$ matrix with all 1s). Show that $\mathcal{G}$ is strongly regular with parameters (v, k, λ, μ) if and only if A satisfies the equations:

$$AJ = kJ \tag{8.3}$$

and

$$A^2 + (\mu - \lambda)A + (\mu - k)I = \mu J. \tag{8.4}$$

Hint: (8.3) is equivalent to k-regularity; (8.4) may be written in the form

$$A^2 = kI + \lambda A + \mu(J - I - A)$$

and one may use Proposition 8.1.6.

Exercise 8.2.3 Let $\mathcal{G}$ be a connected, strongly regular graph with parameters (v, k, λ, μ).

(1) Show that the adjacency matrix A of $\mathcal{G}$ has exactly three eigenvalues, namely:
- k with multiplicity 1,
- $\theta = \frac{\lambda - \mu + \sqrt{\Delta}}{2}$,
- $\tau = \frac{\lambda - \mu - \sqrt{\Delta}}{2}$,

where $\Delta = (\lambda - \mu)^2 + 4(k - \mu)$.

Hint: use Proposition 8.1.6; apply (8.4) and use the fact that nonconstant eigenvectors f of A satisfy $Jf = 0$.

(2) Show that the multiplicities of θ and τ are

$$m_\theta = \frac{1}{2}\left[(v - 1) - \frac{2k + (v - 1)(\lambda - \mu)}{\sqrt{\Delta}}\right]$$

$$m_\tau = \frac{1}{2}\left[(v - 1) + \frac{2k + (v - 1)(\lambda - \mu)}{\sqrt{\Delta}}\right].$$

Hint: $m_\theta + m_\tau = v - 1$ and $0 = \text{Tr}(A) = \theta m_\theta + \tau m_\tau + k$.

Exercise 8.2.4 Let $m \geq 4$ and denote by X the set of all 2-subsets of $\{1, 2, \ldots, m\}$. The *triangular graph* $T(m)$ is the finite graph with vertex set X and such that two distinct vertices are adjacent if they are not disjoint.

Show that $T(m)$ is strongly regular with parameters $v = \binom{m}{2}$, $k = 2(m - 2)$, $\lambda = m - 2$, and $\mu = 4$.

Exercise 8.2.5 Let $\mathcal{G} = (X, E)$ be a finite simple graph without loops. The *complement* of $\mathcal{G}$ is the graph $\overline{\mathcal{G}}$ with vertex set X and edge set $\overline{E} = \{\{x, y\} : x, y \in X, x \neq y, \{x, y\} \notin E\}$.

(1) Show that if $\mathcal{G}$ is strongly regular with parameters (v, k, λ, μ), then $\overline{\mathcal{G}}$ (which is not necessarily connected!) is strongly regular with parameters $(v, v - k - 1, v - 2k + \mu - 2, v - 2k + \lambda)$.

(2) From (1) deduce that the parameters of a strongly regular graph satisfy the inequality $v - 2k + \mu - 2 \geq 0$.

(3) Suppose that $\mathcal{G}$ is strongly regular. Show that $\mathcal{G}$ and $\overline{\mathcal{G}}$ are both connected if and only if $0 < \mu < k < v - 1$. If this is the case, one says that $\mathcal{G}$ is *primitive*.

Hint: show that $\mu = 0$ implies $\lambda = k - 1$ and write $\mu < k$ in the form $v - 2k + \mu - 2 < (v - k - 1) - 1$.

The complement of the triangle graph $T(5)$ (see Exercise 8.2.4) is the celebrated *Petersen graph* (see Figure 8.1). The monograph [73] is entirely devoted to this graph, which turned out to serve as a counterexample to several important conjectures.

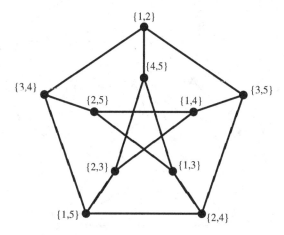

Figure 8.1. The Petersen graph.

Exercise 8.2.6 The *Clebsch graph* (see Figure 8.2) is defined as follows. The vertex set X consists of all subsets of even cardinality of $\{1, 2, 3, 4, 5\}$. Moreover, two vertices $A, B \in X$ are adjacent if $|A \bigtriangleup B| = 4$ (here $\bigtriangleup$ denotes the symmetric difference of two sets). Show that it is a $(16, 5, 0, 2)$ strongly regular graph.

In the following, we shall present another description of the Clebsch graph by using methods of number theory. An *edge coloring* of a graph $\mathcal{G} = (X, E)$ is a map $c \colon E \to C$, where C is a set of *colors*. A *monochromatic triangle* in $\mathcal{G}$ is a set of three vertices x, y, z such that $\{x, y\}, \{y, z\}, \{z, x\} \in E$ and have the same color. In the following exercise, we construct a very important coloring of the complete graph K_{16}, due to Greenwood and Gleason [68].

Exercise 8.2.7 Let $\mathbb{F}_{16}[x]$ denote the ring of polynomials in one indeterminate over the field $\mathbb{F}_{16}$.

(1) Show that

$$x^{15} + 1 = (x^4 + x + 1)(x^{11} + x^8 + x^7 + x^5 + x^3 + x^2 + x + 1).$$

(2) Show that the polynomial $p(x) = x^4 + x + 1$ is irreducible over $\mathbb{F}_2$. Let $\alpha \in \mathbb{F}_{16}^*$ be a root of p. Show that α is a generator of $\mathbb{F}_{16}^*$ and deduce from

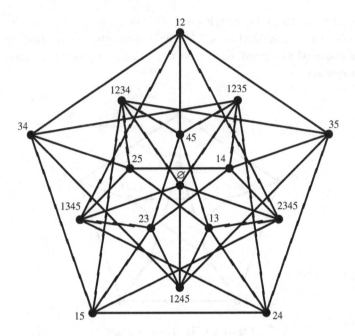

Figure 8.2. The Clebsch graph: for $1 \le i < j < h < k \le 5$ the string ij (respectively $ijhk$) indicates the subset $\{i, j\}$ (respectively $\{i, j, h, k, \}$). See also Figure 8.3.

Proposition 6.2.5 that every element of $\mathbb{F}_{16}$ may be uniquely represented in the form

$$\varepsilon_0 + \varepsilon_1 \alpha + \varepsilon_2 \alpha^2 + \varepsilon_3 \alpha^3, \tag{8.5}$$

where $\varepsilon_i \in \{0, 1\}$ for $0 \le i \le 3$.

(3) Let α be as in (2). Represent each element α^k, $k = 0, 1, \ldots, 14$, in the form (8.5) and show that the five cubes in $\mathbb{F}_{16}^*$ coincide with the elements

$$1, \ \alpha^3, \ \alpha^3 + \alpha^2, \ \alpha^3 + \alpha, \ \text{and } \alpha^3 + \alpha^2 + \alpha + 1.$$

Also show that the sum of two cubes in $\mathbb{F}_{16}^*$ is not a cube.
Hint: for instance, $1 + \alpha^3 = \alpha^{14}$ in $\mathbb{F}_{16}^*$.

(4) Consider the elements of $\mathbb{F}_{16}$ as the vertices of K_{16} (the complete graph on 16 vertices (see Section 8.4)). Color the edges of K_{16} in the following way: if $a, b \in \mathbb{F}_{16}$, $a \ne b$ and $a - b = \alpha^m$, then
- if $m \equiv 0 \mod 3$ (i.e. $a - b$ is a cube) the color of $\{a, b\}$ is *red*;
- if $m \equiv 1 \mod 3$ the color of $\{a, b\}$ is *blue*;
- if $m \equiv 2 \mod 3$ the color of $\{a, b\}$ is *green*.

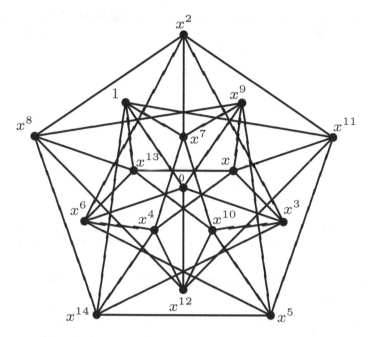

Figure 8.3. The Clebsch graph (cf. Figure 8.2): now the vertices are identified with the elements of $\mathbb{F}_{16}$. Moreover, $\mathbb{F}_{16} = \{0, 1, x, x^2, x^3, \ldots, x^{14}\}$, where x is a generator of the cyclic group $\mathbb{F}_{16}^*$.

Show that, with this coloring, K_{16} does not contain a monochromatic triangle.

Hint: show that if it contains a monochromatic triangle then it contains a *red* monochromatic triangle and then apply (3).

(5) Show that the graph $(\mathbb{F}_{16}, E)$, where E is the set of all *red* edges in (4), is isomorphic to the Clebsch graph (cf. Exercise 8.2.6).

Another important example of a strongly regular graph, namely the Paley graph, will be discussed in Exercise 9.4.5. For more on strongly regular graphs we refer to the monographs by van Lint and Wilson [97] and Godsil and Royle [65].

8.3 Bipartite graphs

Definition 8.3.1 A graph $\mathcal{G} = (X, E, r)$ is called *bipartite* if there exists a nontrivial partition $X = X_0 \coprod X_1$ of the set of vertices such that every edge $e \in E$ joins a (unique) vertex in X_0 with a (unique) vertex in X_1 (that is, $|r(e) \cap X_0| = 1 = |r(e) \cap X_1|$ for all $e \in E$). The sets X_0 and X_1 are called *partite sets* (cf. Figure 8.4).

Note that if a bipartite graph is connected, then the (nontrivial) partition of the set of vertices is unique. Moreover, any bipartite graph has necessarily no loops.

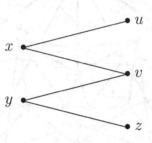

Figure 8.4. The bipartite graph $\mathcal{G} = (X, E)$ with vertex set $X = X_0 \coprod X_1$, where $X_0 = \{x, y\}$ and $X_1 = \{u, v, z\}$, and edge set $E = \{\{x, u\}, \{x, v\}, \{y, v\}, \{y, z\}\}$.

Exercise 8.3.2 Let $\mathcal{G} = (X, E, r)$ be a graph. Show that the following conditions are equivalent:

(a) $\mathcal{G}$ is bipartite;
(b) $\mathcal{G}$ is *bicolorable*, i.e. there exists a map $\phi \colon X \to \{0, 1\}$ such that $x \sim y$ infers $\phi(x) \neq \phi(y)$;
(c) $\mathcal{G}$ does not contain cycles of odd length.

Exercise 8.3.3 Let $\mathcal{G} = (X, E, r)$ be a finite bipartite graph with $X = X_0 \coprod X_1$ its partite sets partition. Consider the decomposition $L(X) = L(X_0) \oplus L(X_1)$. Show that if A denotes the adjacency matrix of $\mathcal{G}$ then we have:

(1) $A[L(X_0)] \subseteq L(X_1)$ (respectively $A[L(X_1)] \subseteq L(X_0)$);
(2) define $\varepsilon \colon L(X) \to L(X)$ (respectively $\tau \colon L(X) \to L(X)$) by setting

$$[\varepsilon f](x) = \begin{cases} f(x) & \text{if } x \in X_0 \\ -f(x) & \text{if } x \in X_1 \end{cases} \quad (\text{respectively, } \tau f = -\varepsilon f)$$

for all $f \in L(X)$ and $x \in X$. Show that (i) $A\varepsilon = \tau A$, (ii) $\varepsilon^2 = \tau^2 = I$, and (iii) $\tau\varepsilon = \varepsilon\tau = -I$.

The following provides another example of a structural (geometrical) property that reflects on the spectral theory of the graph.

Proposition 8.3.4 *Let $\mathcal{G} = (X, E, r)$ be a finite connected k-regular graph and denote by A the corresponding adjacency matrix. Then the following conditions are equivalent:*

(a) $\mathcal{G}$ is bipartite;

(b) the spectrum of A is symmetric with respect to 0;

(c) $-k$ is an eigenvalue of A.

Proof. Suppose that $\mathcal{G}$ is bipartite and let $X = X_0 \coprod X_1$ be the corresponding partite sets partition. Let $\lambda \in \sigma(A)$ and denote by $f \in L(X)$ a corresponding eigenfunction, so that, $Af = \lambda f$. Consider the function $g = \varepsilon f \in L(X)$ (cf. Exercise 8.3.3). Then we have (cf. Exercise 8.3.3):

$$Ag = A\varepsilon f = \tau A f = \lambda \tau f = -\lambda \varepsilon f = -\lambda g.$$

It follows that $-\lambda$ is an eigenvalue (with eigenfunction g). This shows that $\sigma(A)$ is symmetric with respect to 0, proving the implication (a) $\Rightarrow$ (b).

(b) $\Rightarrow$ (c) follows immediately from Proposition 8.1.5.(i).

(c) $\Rightarrow$ (a): suppose that $Af = -kf$ with $f \in L(X)$ nontrivial and real-valued. Denote by $x_0 \in X$ a maximum point for $|f|$; then, up to switching f to $-f$, we may suppose that $f(x_0) > 0$. Then the equality $-kf(x_0) = [Af](x_0) = \sum_{y \in X} A(x_0, y) f(y)$ may be rewritten $\sum_{y: y \sim x_0} A(x_0, y)[f(x_0) + f(y)] = 0$. Since $f(x_0) + f(y) \geq 0$, we deduce $f(y) = -f(x_0)$ for all $y \sim x_0$. Set $X_j = \{y \in X : f(y) = (-1)^j f(x_0)\}$ for $j = 0, 1$. Arguing as in the proof of Proposition 8.1.5.(i), and using induction on the geodesic distance from x_0, we deduce that indeed $X = X_0 \coprod X_1$ is a partite set decomposition, showing that X is bipartite. $\qquad\square$

Exercise 8.3.5 The *complete bipartite graph* $K_{n,m} = (X_{n,m}, E_{n,m})$ on $n + m$ vertices, $n, m \geq 1$, is the (finite, simple, and without loops) graph whose vertex set $X_{n,m} = X \sqcup Y$ is the disjoint union of a set X of cardinality n, and another set Y of cardinality m, and edge set $E_{n,m} = \{\{x, y\} : x \in X, y \in Y\}$. Show that the adjacency matrix of $K_{n,m}$ has the following eigenvalues:

- 0 with multiplicity $n + m - 2$
- $\sqrt{nm}$ with multiplicity 1
- $-\sqrt{nm}$ with multiplicity 1.

8.4 The complete graph

Definition 8.4.1 The *complete graph* on n vertices, $n \geq 1$, is the (finite, simple, and without loops) graph $K_n = (X_n, E_n)$ with vertex set $X_n = \{1, 2, \dots, n\}$ and edge set $E_n = \{\{x, y\} : x, y \in X_n, \ x \neq y\}$, that is, two vertices are connected if and only if they are distinct (cf. Figure 8.5).

Figure 8.5. The complete graphs K_2, K_3, K_4, and K_5.

Note that K_n is regular: indeed, each vertex has degree $n - 1$.
The adjacency matrix A of K_n is given by

$$A(x, y) = \begin{cases} 1 & \text{if } x \neq y \\ 0 & \text{if } x = y. \end{cases}$$

The graph K_n is always connected and it is bipartite if and only if $n = 2$.
Moreover (cf. Proposition 8.1.4), setting $W_0 = \{f \in L(X_n) : f \text{ is constant}\}$ and
$W_1 = \{f \in L(X_n) : \sum_{y \in X_n} f(y) = 0\}$, we have, for $f \in W_0$,

$$[Af](x) = \sum_{y \in X_n} A(x, y)f(y) = (n - 1)f(x)$$

and, for $f \in W_1$,

$$[Af](x) = \sum_{y \in X_n} A(x, y)f(y) = \sum_{\substack{y \in X_n \\ y \neq x}} f(y) = \left(\sum_{y \in X_n} f(y)\right) - f(x) = -f(x)$$

for all $x \in X_n$.
We deduce that (cf. Proposition 8.1.4):

- W_0 is an eigenspace for A corresponding to the eigenvalue $(n - 1)$, whose
  multiplicity is equal to $\dim W_0 = 1$;
- W_1 is an eigenspace for A corresponding to the eigenvalue -1, whose multiplicity is equal to $\dim W_1 = n - 1$.

8.5 The hypercube

Definition 8.5.1 The *n-dimensional hypercube*, $n \in \mathbb{N}$, is the (finite, simple,
and without loops) graph $Q_n = (X_n, E_n)$ with vertex set $X_n = \{0, 1\}^n$ and edge
set $E_n = \{\{x, y\} : d(x, y) = 1\}$, where

$$d(x, y) = |\{i : x_i \neq y_i, 1 \leq i \leq n\}|$$

is the *Hamming distance* of $x = (x_1, x_2, \ldots, x_n)$ and $y = (y_1, y_2, \ldots, y_n) \in X_n$
(cf. Figure 8.6).

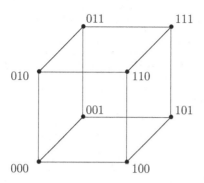

Figure 8.6. The 3-dimensional hypercube Q_3.

It is clear from the definition that the adjacency matrix $A = (A(x, y))_{x,y \in X_n}$ of Q_n is given by

$$A(x, y) = \begin{cases} 1 & \text{if } d(x, y) = 1 \\ 0 & \text{otherwise,} \end{cases}$$

for all $x, y \in X_n$.

We observe that X_n, equipped with the addition operation (that is, $(x + y)_i = x_i + y_i \bmod 2$, for all $x, y \in X_n$ and $1 \le i \le n$), is an Abelian group, with identity element $\mathbf{0} = (0, 0, \ldots, 0)$, isomorphic to $\mathbb{Z}_2^n$. The characters (cf. Definition 2.3.1) of such a group are given by (cf. Proposition 2.3.3) the functions $\chi_x \in L(X_n)$, $x \in X_n$, defined by setting

$$\chi_x(y) = (-1)^{x \cdot y} \tag{8.6}$$

for all $y \in X_n$, where $x \cdot y = \sum_{i=1}^n x_i y_i$.

Exercise 8.5.2 Show that $A \in \text{End}(L(X_n))$ satisfies the equivalent conditions in Theorem 2.4.10 (warning: the notation has changed), namely: A is $\mathbb{Z}_2^n$-invariant and it is the convolution operator with kernel $h \in L(X_n)$ defined by

$$h(x) = \begin{cases} 1 & \text{if } d(x, \mathbf{0}) = 1 \\ 0 & \text{otherwise,} \end{cases} \tag{8.7}$$

for all $x \in X_n$, so that its eigenfunctions are exactly the characters χ_x, $x \in \mathbb{Z}_2^n$.

For $x = (x_1, x_2, \ldots, x_n) \in X_n$ we define $w(x) = |\{j : x_j = 1\}|$ the *weight* of x. Note that $d(x, y) = w(x - y)$ for all $x, y \in X_n$.

Keeping in mind Corollary 2.4.11, the following provides a complete list of the eigenvalues of A.

Proposition 8.5.3 *The Fourier transform of the function $h \in L(X_n)$ in (8.7) is given by*

$$\widehat{h}(x) = n - 2w(x) \tag{8.8}$$

for all $x \in X_n$.

Proof. Let $x \in X_n$. Then we have

$$\widehat{h}(x) = \langle h, \chi_x \rangle$$

$$\text{(by (8.6))} \quad = \sum_{y \in X_n} h(y)(-1)^{x \cdot y}$$

$$\text{(by (8.7))} \quad = \sum_{j=1}^{n} (-1)^{x_j}$$

$$= \sum_{j:x_j=1} (-1)^{x_j} + \sum_{j:x_j=0} (-1)^{x_j}$$

$$= -w(x) + (n - w(x))$$

$$= n - 2w(x). \qquad \square$$

Note that, according to Proposition 8.3.4, the spectrum of A is symmetric with respect to 0, as Q_n is bipartite: its partite set partition is $X_n = \{x \in X_n : w(x) \text{ is odd}\} \coprod \{x \in X_n : w(x) \text{ is even}\}$.

We now determine the multiplicities of the eigenvalues (8.8) of A. It is clear that, for $0 \leq k \leq n$, the eigenspace associated with the eigenvalue $n - 2k$ is the subspace

$$V_k = \langle \chi_x : w(x) = k \rangle \leq L(X_n).$$

Moreover, its dimension is given by $\dim(V_k) = |\{x \in X_n : w(x) = k\}| = \binom{n}{k}$.

8.6 The discrete circle

Definition 8.6.1 The *discrete circle* (or *cycle graph*) on $n \geq 3$ *vertices*, is the (finite, simple, and without loops) graph $C_n = (X_n, E_n)$, where $X_n = \mathbb{Z}_n$ and $x, y \in X_n$ are adjacent if $x - y = \pm 1$ (cf. Figure 8.7).

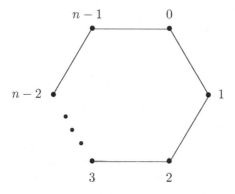

Figure 8.7. The discrete circle C_n.

Note that C_n is 2-regular and it is bipartite if and only if n is even. The associated adjacency matrix is circulant (see Exercise 2.4.16) and is given by

$$A = \begin{pmatrix} 0 & 1 & 0 & 0 & \cdots & 0 & 0 & 1 \\ 1 & 0 & 1 & 0 & \cdots & 0 & 0 & 0 \\ 0 & 1 & 0 & 1 & \cdots & 0 & 0 & 0 \\ \vdots & & & & \ddots & & & \vdots \\ 0 & 0 & 0 & 0 & \cdots & 1 & 0 & 1 \\ 1 & 0 & 0 & 0 & \cdots & 0 & 1 & 0 \end{pmatrix}.$$

Exercise 8.6.2 Show that $A \in \mathrm{End}(L(X_n))$ satisfies the equivalent conditions in Theorem 2.4.10 (warning: the notation has changed), namely: A is $\mathbb{Z}_n$-invariant and it is the convolution operator with kernel $h = \delta_1 + \delta_{-1} \in L(X_n)$, so that its eigenfunctions are exactly the characters χ_x, $x \in \mathbb{Z}_n$.

Recall (cf. Definition 2.2.1) that the characters of $\mathbb{Z}_n$ are the functions χ_x, $x \in \mathbb{Z}_n$, defined by

$$\chi_x(y) = \omega^{xy}$$

for all $y \in \mathbb{Z}_n$, where $\omega = \exp(\frac{2\pi i}{n})$. Moreover (cf. Exercise 2.4.4), the Fourier transform of a Dirac δ_x, $x \in \mathbb{Z}_n$, is given by $\widehat{\delta_x}(y) \equiv \widehat{\delta_x}(\chi_y) = \overline{\chi_y(x)}$ for all $y \in \mathbb{Z}_n$. By linearity we have, for all $y \in \mathbb{Z}_n$,

$$\widehat{h}(y) = \widehat{\delta_1}(y) + \widehat{\delta_{-1}}(y) = \overline{\chi_1}(y) + \overline{\chi_{-1}}(y)$$
$$= \exp(-\frac{2\pi yi}{n}) + \exp(\frac{2\pi yi}{n})$$
$$= 2\cos(\frac{2\pi y}{n}).$$

We remark that $\widehat{h}(y) = \widehat{h}(y')$, $y, y' \in \mathbb{Z}_n$, if and only if $y = \pm y'$. From Corollary 2.4.11 and the above remark, we deduce that the eigenvalues of A are exactly the numbers

$$2\cos(\frac{2\pi y}{n}), \quad y = 0, 1, \ldots, \left\lfloor \frac{n}{2} \right\rfloor. \tag{8.9}$$

We now determine their multiplicities, arguing separately on the parity of n.

If n is even, then the eigenvalues (8.9) corresponding to $y = 0$ and $y = \left\lceil \frac{n}{2} \right\rceil = \frac{n}{2}$ (these are 2 and -2, respectively) have multiplicity one, and all others have multiplicity two. Note that, according to Proposition 8.3.4, the spectrum of A is symmetric with respect to 0 as, in this case, $\mathcal{C}_n$ is bipartite.

If n is odd, then the eigenvalue (8.9) corresponding to $y = 0$ (namely, 2) has multiplicity one, and all others have multiplicity two. Moreover, in this case, $\mathcal{C}_n$ is not bipartite.

Exercise 8.6.3 (The 2-regular segment) For $n \geq 2$ let $\mathcal{G}_n = (X_n, E_n, r_n)$ denote the simple graph (with loops!) where: $X_n = \{0, 1, 2, \ldots, n-1\}$, $E_n = \sqcup_{i=0}^{n-2}\{i, i+1\} \sqcup \{0\} \sqcup \{n-1\}$, and $r_n \colon E_n \to \mathcal{P}(X_n)$ is the restriction to E_n of the identity map on $\mathcal{P}(X_n)$. This is called the *2-regular segment* on $n \geq 1$ vertices (cf. Figure 8.8).

$$0 \qquad 1 \qquad 2 \qquad\qquad\qquad\qquad n-2 \quad n-1$$

Figure 8.8. The 2-regular segment $\mathcal{G}_n$.

Show that the eigenvalues of $\mathcal{G}_n$ are

$$2\cos\frac{k\pi}{n}, \quad k = 0, 1, \ldots, n-1. \tag{8.10}$$

Hint: see [29, Exercise A1.0.15] as well as the books by Feller [61] and Karlin and Taylor [84].

8.7 Tensor products

In this section we introduce some notation and preliminary results that we shall use both in the present chapter as well as in other parts of the book. For a similar approach see also the beginning of [124, Chapter 5]. This section is in the same spirit of Section 2.1 and contains some complements to that section. It is also connected with Section 5.3, where the Kronecker products of matrices are introduced, and it will be generalized in Section 5.3, where we study tensor products of representations.

Let X and Y be two finite sets. The *tensor product* of two functions $f \in L(X)$ and $g \in L(Y)$ is the function $f \otimes g \in L(X \times Y)$ defined by setting

$$(f \otimes g)(x, y) = f(x)g(y) \tag{8.11}$$

for all $(x, y) \in X \times Y$. This way, we have the natural identification $\delta_{(x,y)} = \delta_x \otimes \delta_y$, so that the standard basis of $L(X \times Y)$ may be written in the form

$$\{\delta_x \otimes \delta_y : x \in X, y \in Y\}.$$

It is also easy to check that, for $f, f' \in L(X)$ and $g, g' \in L(Y)$, we have:

$$\langle f \otimes g, f' \otimes g' \rangle_{L(X \times Y)} = \langle f, f' \rangle_{L(X)} \cdot \langle g, g' \rangle_{L(Y)}. \tag{8.12}$$

If V is a subspace of $L(X)$ and W a subspace of $L(Y)$ then their tensor product $V \otimes W$ is the subspace of $L(X \times Y)$ generated by all products $f \otimes g$ with $f \in V$ and $g \in W$.

Now suppose that $A \in \text{End}(L(X))$ and $B \in \text{End}(L(Y))$ are linear operators. We define their *tensor product* $A \otimes B \in \text{End}(L(X \times Y))$ by setting

$$(A \otimes B)(f \otimes g) = Af \otimes Bg \tag{8.13}$$

for all $f \in L(X)$ and $g \in L(Y)$ (and then extending by linearity). It is easy to check that this definition is well posed. Indeed, we now derive the matrix representing $A \otimes B$. Suppose that $(a(x, x'))_{x,x' \in X}$ (respectively, $(b(y, y'))_{y,y' \in Y}$) is the matrix representing A (respectively, B) with respect to the standard basis of $L(X)$ (respectively, of $L(Y)$), see (2.2). Then, for all $x, x' \in X$ and $y, y' \in Y$,

$$\begin{aligned}
\left\{ [A \otimes B] \left(\delta_{x'} \otimes \delta_{y'} \right) \right\} (x, y) &= \left\{ (A\delta_{x'}) \otimes (B\delta_{y'}) \right\} (x, y) && \text{(by (8.13))} \\
&= (A\delta_{x'})(x) \cdot (B\delta_{y'})(y) && \text{(by (8.11))} \\
&= a(x, x')b(y, y') && \text{(by (2.1))}.
\end{aligned}$$

This shows that the matrix representing $A \otimes B$ with respect to the standard basis of $L(X \times Y)$ is

$$\left(a(x, x')b(y, y') \right)_{(x,y),(x',y') \in X \times Y}.$$

It is easy to see that this is a coordinate-free description of the Kronecker product introduced in Section 5.3: just take $X = \{1, 2, \ldots, n\}$ and $Y = \{1, 2, \ldots, m\}$. We leave it to the reader to check the details.

The *Kronecker sum* of A and B is the operator $A \otimes I_Y + I_X \otimes B \in \text{End}(L(X \times Y))$; see the monograph by Lancaster and Tismenetsky [91]. Clearly, this sum is represented by the matrix

$$\left(a(x, x')\delta_y(y') + \delta_x(x')b(y, y')\right)_{(x,y),(x',y')\in X\times Y}.$$

Now suppose that both A and B are symmetric, that is, $a(x, x') = a(x', x)$ and $b(y, y') = b(y', y)$ for all $x, x' \in X$ and $y, y' \in Y$. Then also $A \otimes B$ and $A \otimes I_Y + I_X \otimes B$ are symmetric. Recall that symmetric matrices are diagonalizable and have real eigenvalues. Let us denote by

• $\lambda_0, \lambda_1, \ldots, \lambda_{|X|-1}$ (respectively, $\mu_0, \mu_1, \ldots, \mu_{|Y|-1}$) the eigenvalues of A (respectively, of B);
• $\{f_0, f_1, \ldots, f_{|X|-1}\}$ (respectively, $\{g_0, g_1, \ldots, g_{|Y|-1}\}$) an orthonormal basis of (real-valued) eigenvectors for A (respectively, for B)

so that

$$Af_i = \lambda_i f_i \quad \text{and} \quad Bg_j = \mu_j g_j \tag{8.14}$$

for all $i = 0, 1, \ldots, |X| - 1$ and $j = 0, 1, \ldots, |Y| - 1$. The proof of the following proposition is immediate.

Proposition 8.7.1 *The set* $\{f_i \otimes g_j : i = 0, 1, \ldots, |X| - 1, j = 0, 1, \ldots, |Y| - 1\}$ *is an orthonormal basis of (real-valued) eigenvectors for both* $A \otimes B$ *and* $A \otimes I_Y + I_X \otimes B$. *Moreover, for all* $i = 0, 1, \ldots, |X| - 1$ *and* $j = 0, 1, \ldots, |Y| - 1$,

$$[A \otimes B](f_i \otimes g_j) = \lambda_i \mu_j (f_i \otimes g_j)$$

and

$$[A \otimes I_Y + I_X \otimes B](f_i \otimes g_j) = (\lambda_i + \mu_j)(f_i \otimes g_j);$$

in particular, the eigenvalues of $A \otimes B$ *are the* $\lambda_i \mu_j$s *while those of* $A \otimes I_Y + I_X \otimes B$ *are the* $(\lambda_i + \mu_j)$s.

More generally, if F is a two variable complex polynomial, then the eigenvalues of $F(A, B)$ (here the powers of matrices are the usual powers, while the other products (respectively, sums) involved are tensor products (respectively, Kronecker sums)) are $F(\lambda_i, \mu_j)$, $i = 0, 1, \ldots, |X| - 1$ and $j = 0, 1, \ldots, |Y| - 1$ (this is Stephanov's theorem [153]: see the monograph by Lancaster and Tismenetsky [91, Theorem 1, Section 12.2]).

Recall (cf. Proposition 2.1.1) that W_0 is the space of constant functions on X and $W_1 = \{f \in L(X) : \sum_{x \in X} f(x) = 0\}$. We also denote by J_Y the matrix

$(j(y, y'))_{y,y' \in Y}$ with $j(y, y') = 1$ for all $y, y' \in Y$. This way, for $f \in L(Y)$ we have

$$J_Y f = \left(\sum_{y \in Y} f(y) \right) \mathbf{1}_Y. \tag{8.15}$$

Proposition 8.7.2 *Let $A: L(X) \to L(X)$ and $B: L(Y) \to L(Y)$ be two linear operators and suppose that the decomposition $L(Y) = W_0(Y) \oplus W_1(Y)$ is B-invariant. Then the decomposition*

$$L(X \times Y) = [L(X) \otimes W_0(Y)] \oplus [L(X) \otimes W_1(Y)]$$

is invariant for $A \otimes J_Y + I_X \otimes B$. Moreover,

$$W_1(X \times Y) = [W_1(X) \otimes W_0(Y)] \oplus [L(X) \otimes W_1(Y)]. \tag{8.16}$$

Proof. Just note that $W_0(Y)$ and $W_1(Y)$ are J_Y-invariant ($J_Y - I_Y$ is the adjacency matrix of the complete graph with vertex set Y; see Section 8.4). Also, (8.16) follows immediately after observing that $W_0(X \times Y) = W_0(X) \otimes W_0(Y)$. $\quad\square$

Following [128] we introduce a notation for the decomposition (8.16) (see also the generalizations in [28] and [44]).

For $f \in W_1(X \times Y)$ we define $f^{\parallel} \in L(X \times Y)$ by setting

$$f^{\parallel}(x, y) = \frac{1}{|Y|} \sum_{z \in Y} f(x, z)$$

for all $(x, y) \in X \times Y$. Clearly, $f^{\parallel}$ does not depend on $y \in Y$, and $f^{\parallel} \in W_1(X) \otimes W_0(Y)$. Moreover, setting

$$f^{\perp} = f - f^{\parallel},$$

so that

$$f = f^{\parallel} + f^{\perp},$$

we have $f^{\perp} \in L(X) \otimes W_1(Y)$.

Another useful notation is the following. For $f \in L(X \times Y)$ and $x \in X$ we define $f_x \in L(Y)$ by setting

$$f_x(y) = f(x, y)$$

for all $y \in Y$. Then

$$f = \sum_{x \in X} \delta_x \otimes f_x. \tag{8.17}$$

Moreover, setting

$$f_x^{\|} = \frac{1}{|Y|} J_Y f_x \quad \text{and} \quad f_x^{\perp} = f_x - f_x^{\|} \tag{8.18}$$

we have

$$f^{\|} = \sum_{x \in X} \delta_x \otimes f_x^{\|} \tag{8.19}$$

and

$$f^{\perp} = \sum_{x \in X} \delta_x \otimes f_x^{\perp}. \tag{8.20}$$

Finally, following again [128], we define $C \colon L(X \times Y) \to L(X)$ by setting

$$[Cf](x) = \sum_{y \in Y} f(x, y) \tag{8.21}$$

for all $f \in L(X \times Y)$ and $x \in X$. Note the similarity between $f^{\|}$ and Cf: however, the former is a function of two variables (constant with respect to the second variable), while the latter is a function of a single variable. Moreover, $f^{\|}$ is normalized. Their relationship is expressed in (iv) of the following lemma.

Lemma 8.7.3

(i) $C(\delta_x \otimes \delta_y) = \delta_x$ *for all* $(x, y) \in X \times Y$;

(ii) $C|_{W_1(X \times Y)}$ *is a linear operator from* $W_1(X \times Y)$ *onto* $W_1(X)$;

(iii) $(Cf) \otimes \mathbf{1}_Y = (I_X \otimes J_Y)f$ *for all* $f \in L(X \times Y)$;

(iv) $Cf^{\|} = Cf$ *for all* $f \in L(X \times Y)$.

Proof.

(i) For $x, z \in X$ and $y \in Y$ we have

$$[C(\delta_x \otimes \delta_y)](z) = \sum_{t \in Y} (\delta_x \otimes \delta_y)(z, t) = \delta_x(z).$$

(ii) This is clear.

(iii) Using (8.17) we have, for all $f \in L(X \times Y)$,

$$(I_X \otimes J_Y)f = (I_X \otimes J_Y) \sum_{x \in X} (\delta_x \otimes f_x)$$

$$\text{(by (8.15))} \quad = \sum_{x \in X} \delta_x \otimes \left[\left(\sum_{y \in Y} f(x, y) \right) \mathbf{1}_Y \right]$$

$$= \sum_{x \in X} ([Cf](x)\delta_x) \otimes \mathbf{1}_Y$$

$$= (Cf) \otimes \mathbf{1}_Y.$$

(iv) It is a simple computation: for $f \in L(X \times Y)$ and $x \in X$ we have

$$[Cf^{\parallel}](x) = \sum_{y \in Y} f^{\parallel}(x, y) = \sum_{y \in Y} \frac{1}{|Y|} \sum_{z \in Y} f(x, z) = \sum_{z \in Y} f(x, z) = [Cf](x).$$

$\square$

Lemma 8.7.4 *Let* $f \in W_1(X \times Y)$. *Then*

$$f^{\parallel} = \frac{1}{|Y|}(I_X \otimes J_Y)f = \frac{1}{|Y|}(Cf) \otimes \mathbf{1}_Y.$$

Proof. Using again (8.17) we have

$$(I_X \otimes J_Y)f = (I_X \otimes J_Y) \sum_{x \in X} \delta_x \otimes f_x$$

$$= \sum_{x \in X} \delta_x \otimes (J_Y f_x)$$

$$(\text{by } (8.18)) = |Y| \sum_{x \in X} \delta_x \otimes f_x^{\parallel}$$

$$(\text{by } (8.19)) = |Y| f^{\parallel}.$$

The second equality follows from Lemma 8.7.3.(iii). $\square$

We use the notation Y^X to denote the space of all maps $f: X \to Y$ and refer to it as to an *exponential set*. Clearly,

$$Y^X = \underbrace{Y \times Y \times \cdots \times Y}_{|X| \text{ times}}.$$

We introduce a *coordinate-free* description of the tensor product

$$L(Y^X) = \underbrace{L(Y) \otimes L(Y) \otimes \cdots \otimes L(Y)}_{|X| \text{ times}}.$$

Given $\phi_x \in L(Y)$, $x \in X$, we define the tensor product of the family $(\phi_x)_{x \in X}$ as in (8.11) by setting:

$$\left(\bigotimes_{x \in X} \phi_x \right)(f) = \prod_{x \in X} \phi_x(f(x)),$$

for all $f \in Y^X$. Analogously, given $A_x \in \text{End}(L(Y))$, $x \in X$, the tensor product of the corresponding family of operators is defined as in (8.13) by setting:

$$\left(\bigotimes_{x \in X} A_x \right) \left(\bigotimes_{x \in X} \phi_x \right) = \bigotimes_{x \in X} A_x \phi_x, \tag{8.22}$$

for all tensors $\left(\bigotimes_{x \in X} \phi_x\right)$ (and then extended by linearity). Finally, note that for every $f \in Y^X$ we have

$$\delta_f = \bigotimes_{x \in X} \delta_{f(x)}. \tag{8.23}$$

8.8 Cartesian, tensor, and lexicographic products of graphs

In this section we give a detailed definition of three basic notions of graph products. See Remark 8.8.2 for a shorter description.

Recall that we use the symbol $\sim$ to denote the adjacency relation of vertices in a given graph.

Definition 8.8.1 Let $\mathcal{G} = (X, E, r)$ and $\mathcal{F} = (Y, F, s)$ be two finite graphs.

(i) The *Cartesian product* of $\mathcal{G}$ and $\mathcal{F}$ is the graph $\mathcal{G} \square \mathcal{F} = (X \times Y, E \square F, r \square s)$ where the edge set is

$$E \square F = (E \times Y) \sqcup (X \times F)$$

and $r \square s \colon E \square F \to \mathcal{P}(X \times Y)$ is defined by setting

$$[r \square s](e, y) = r(e) \times \{y\} \quad \text{and} \quad [r \square s](x, f) = \{x\} \times s(f)$$

for all $e \in E, y \in Y, x \in X$, and $f \in F$ (cf. Figure 8.9).

Note that if $\mathcal{G}$ and $\mathcal{F}$ are both directed, then $\mathcal{G} \square \mathcal{F}$ is also directed after defining the orientation $\vec{r} \square \vec{s} \colon E \square F \to X \times Y$ by setting

$$[\vec{r} \square \vec{s}](e, y) = ((e_-, y), (e_+, y)) \quad \text{and} \quad [\vec{r} \square \vec{s}](x, f) = ((x, f_-), (x, f_+))$$

for all $e \in E, y \in Y, x \in X$, and $f \in F$.

Finally note that if $\mathcal{G}$ and $\mathcal{F}$ are both simple (respectively, without loops), then $\mathcal{G} \square \mathcal{F}$ is also simple, with edge set

$$E \square F = \left\{ \{(x, y), (x', y')\} \subseteq X \times Y : \left[x \sim x' \text{ and } y = y'\right] \text{or} \left[x = x' \text{ and } y \sim y'\right] \right\}$$

(respectively, without loops).

(ii) Equip $\mathcal{G}$ and $\mathcal{F}$ with arbitrary orientations $\vec{r}$ and $\vec{s}$, respectively: different orientations will produce isomorphic graph products (exercise). Also, we denote, as usual, by $E_0 \subseteq E$ (respectively, $F_0 \subseteq F$) the set of all loops of $\mathcal{G}$ (respectively, $\mathcal{F}$) and $E_1 = E \setminus E_0$ (respectively, $F_1 = F \setminus F_0$). Let also $(E_1 \times F_1)_e$ and $(E_1 \times F_1)_o$ be two disjoint copies of the Cartesian product of the edge subsets E_1 and F_1 ("e" stands for *even* and "o" for *odd*).

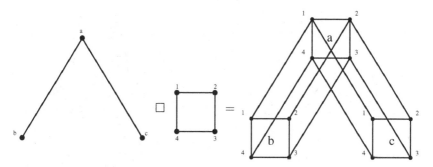

Figure 8.9. An example of Cartesian product of graphs.

The *tensor product* of $\mathcal{G}$ and $\mathcal{F}$ is the (undirected) graph $\mathcal{G} \otimes \mathcal{F} = (X \times Y, E \otimes F, \vec{r} \otimes \vec{s})$ where

$$E \otimes F = ((E \times F) \setminus (E_1 \times F_1)) \sqcup (E_1 \times F_1)_e \sqcup (E_1 \times F_1)_o$$
$$\equiv (E_0 \times F_0) \sqcup (E_0 \times F_1) \sqcup (E_1 \times F_0) \sqcup (E_1 \times F_1)_e \sqcup (E_1 \times F_1)_o$$

and

$$[\vec{r} \otimes \vec{s}](e, f) = \begin{cases} r(e) \times s(f) & \text{if } (e, f) \in (E \times F) \setminus (E_1 \times F_1) \\ \{(e_-, f_-), (e_+, f_+)\} & \text{if } (e, f) \in (E_1 \times F_1)_e \\ \{(e_-, f_+), (e_+, f_-)\} & \text{if } (e, f) \in (E_1 \times F_1)_o \end{cases}$$

for all $(e, f) \in E \otimes F$ (cf. Figure 8.10). Note that, if $\mathcal{G}$ and $\mathcal{F}$ have no loops, then one has $|E \otimes F| = 2|E| \cdot |F|$.

The tensor product $\mathcal{G} \otimes \mathcal{F}$ admits the natural orientation $\vec{t} \colon E \otimes F \to X \times Y$ defined by setting

$$\vec{t}(e, f) = \begin{cases} ((x, y), (x, y)) & \text{if } (e, f) \in E_0 \times F_0 \\ ((e_-, y), (e_+, y)) & \text{if } (e, f) \in E_1 \times F_0 \\ ((x, f_-), (x, f_+)) & \text{if } (e, f) \in E_0 \times F_1 \\ ((e_-, f_-), (e_+, f_+)) & \text{if } (e, f) \in (E_1 \times F_1)_e \\ ((e_-, f_+), (e_+, f_-)) & \text{if } (e, f) \in (E_1 \times F_1)_o \end{cases}$$

for all $(e, f) \in E \otimes F$.

Moreover, if $\mathcal{G}$ and $\mathcal{F}$ are both simple (respectively, without loops), then $\mathcal{G} \otimes \mathcal{F}$ is also simple, with edge set

$$E \otimes F = \left\{ \{(x, y), (x', y')\} \subseteq X \times Y : x \sim x' \text{ and } y \sim y' \right\}$$

(respectively, without loops).

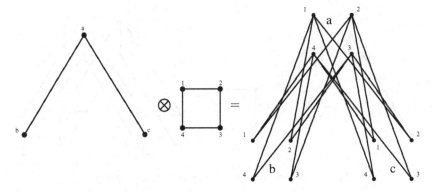

Figure 8.10. An example of tensor product of graphs.

(iii) Equip $\mathcal{G}$ with an arbitrary orientation $\vec{r}$: different orientations will produce isomorphic graph products (exercise). The *lexicographic product* (or *composition*) of $\mathcal{G}$ and $\mathcal{F}$ is the (undirected) graph $\mathcal{G} \circ \mathcal{F} = (X \times Y, E \circ F, \vec{r} \circ s)$ where

$$E \circ F = (E \times Y \times Y) \sqcup (X \times F)$$

and

$$[\vec{r} \circ s](e, y, y') = \{(e_-, y), (e_+, y')\} \quad \text{and} \quad [\vec{r} \circ s](x, f) = \{x\} \times s(f)$$

for all $e \in E$, $y, y' \in Y$, $x \in X$, and $f \in F$ (cf. Figure 8.11).

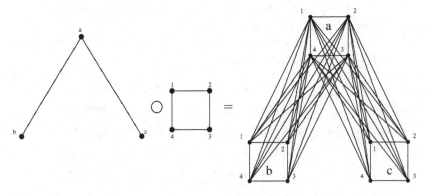

Figure 8.11. An example of lexicographic product of graphs.

Note that if also the second graph $\mathcal{F}$ is directed, say with an orientation $\vec{s}$, then $\mathcal{G} \circ \mathcal{F}$ admits the orientation $\vec{t}: E \circ F \to X \times Y$ defined by setting

$$\vec{t}(e, y, y') = \big((e_-, y), (e_+, y')\big) \quad \text{and} \quad \vec{t}(x, f) = ((x, f_-), (x, f_+))$$

for all $e \in E$, $y, y' \in Y$, $x \in X$, and $f \in F$.

Also, we may regard the Cartesian product $\mathcal{G}\square\mathcal{F}$ as a subgraph of $\mathcal{G}\circ\mathcal{F}$ (via the injection $E\times Y\ni(e,y)\mapsto(e,y,y)\in E\times Y\times Y$).

Finally note that if $\mathcal{G}$ and $\mathcal{F}$ are both simple (respectively, without loops), then $\mathcal{G}\circ\mathcal{F}$ is also simple, with edge set

$$E\circ F=\Big\{\{(x,y),(x',y')\}\subseteq X\times Y:\big[x\sim x'\big]\text{ or }\big[x=x'\text{and }y\sim y'\big]\Big\}$$

(respectively, without loops).

Remark 8.8.2 Summarizing, in all these products the vertex set is $X\times Y$. In the Cartesian product, two vertices (x,y) and (x',y') are adjacent if and only if one of the following two conditions is satisfied: $x\sim x'$ and $y=y'$, or $x=x'$ and $y\sim y'$. In the tensor product they are adjacent if and only if $x\sim x'$ and $y\sim y'$. Finally, in the lexicographic product they are adjacent if and only if one of the following two conditions is satisfied: $x\sim x'$ (*edge of the first type*), or $x=x'$ and $y\sim y'$ (*edge of the second type*). The more involved definitions given above are necessary in order to keep into account possible multiple edges and loops, as well as orientability.

Now denote by A (respectively, B) the adjacency matrix of $\mathcal{G}$ (respectively, $\mathcal{F}$) and suppose that $\lambda_0\geq\lambda_1\geq\cdots\geq\lambda_{|X|-1}$ (respectively, $\mu_0\geq\mu_1\geq\cdots\geq\mu_{|Y|-1}$) are the eigenvalues of A (respectively, of B). Let $\{f_0,f_1,\ldots,f_{|X|-1}\}\subset L(X)$ (respectively, $\{g_0,g_1,\ldots,g_{|Y|-1}\}\subset L(Y)$) be an orthonormal basis of eigenvectors, as in (8.14). Recall that J_Y denotes the matrix $(j(y,y'))_{y,y'\in Y}$ with $j(y,y')=1$ for all $y,y'\in Y$.

Proposition 8.8.3

(i) *The adjacency matrix of $\mathcal{G}\square\mathcal{F}$ is $A\otimes I_Y+I_X\otimes B$, and its eigenvalues are $\lambda_i+\mu_j$, $i=0,1,\ldots,|X|-1$; $j=0,1,\ldots,|Y|-1$.*

(ii) *The adjacency matrix of $\mathcal{G}\otimes\mathcal{F}$ is $A\otimes B$, and its eigenvalues are $\lambda_i\mu_j$, $i=0,1,\ldots,|X|-1$; $j=0,1,\ldots,|Y|-1$.*

(iii) *The adjacency matrix of $\mathcal{G}\circ\mathcal{F}$ is $A\otimes J_Y+I_X\otimes B$. Moreover, if $\mathcal{F}$ is k-regular, then its eigenvalues are:*
- $\lambda_i|Y|+k$, $i=0,1,\ldots,|X|-1$;
- μ_j, $j=1,\ldots,|Y|-1$, *each of them with multiplicity $|X|$.*

Proof.

(i) By definition, we have

$$A_{\mathcal{G}\square\mathcal{F}}\big((x,y),(x',y')\big)=A(x,x')\delta_{y,y'}+\delta_{x,x'}B(y,y')$$

for all $x,x'\in X$ and $y,y'\in Y$, proving the statement relative to the adjacency matrix. For the eigenvalues we apply Proposition 8.7.1.

(ii) We now have

$$A_{\mathcal{G} \otimes \mathcal{F}}\left((x, y), (x', y')\right) = A(x, x')B(y, y')$$

for all $x, x' \in X$ and $y, y' \in Y$, and Proposition 8.7.1 applies again.

(iii) In this final case we have

$$A_{\mathcal{G} \circ \mathcal{F}}\left((x, y), (x', y')\right) = A(x, x') + \delta_{x,x'}B(y, y')$$
$$= A(x, x')J_Y(y, y') + \delta_{x,x'}B(y, y')$$

for all $x, x' \in X$ and $y, y' \in Y$, proving the statement relative to the adjacency matrix. Suppose now that $\mathcal{F}$ is k-regular so that $\mu_0 = k$, $g_0 \in W_0(Y)$, and $g_j \in W_1(Y)$ for all $j = 1, 2, \ldots, |Y| - 1$. Then $J_Y g_0 = |Y|g_0$ while $J_Y g_j = 0$ for $j = 1, 2, \ldots, |Y| - 1$. Therefore,

$$[A \otimes J_Y + I_X \otimes B]\,(f_i \otimes g_0) = (\lambda_i|Y| + k)(f_i \otimes g_0)$$

for $i = 0, 1, 2 \ldots, |X| - 1$, while

$$[A \otimes J_Y + I_X \otimes B]\,(f_i \otimes g_j) = \mu_j(f_i \otimes g_j)$$

for $i = 0, 1, 2 \ldots, |X| - 1$ and $j = 1, 2, \ldots, |Y| - 1$.

$\square$

Remark 8.8.4 In [44], in the framework of the theory of Markov chains, the matrices $A \otimes I_Y + I_X \otimes B$ and $A \otimes J_Y + I_X \otimes B$ are called the *crossed* and *nested* products, respectively, and are combined to get a further generalization, called the *crested product* of the given Markov chains.

Corollary 8.8.5 *Suppose that $\mathcal{G}$ is h-regular and $\mathcal{F}$ is k-regular. Then*

(i) *$\mathcal{G} \square \mathcal{F}$ is $(h + k)$-regular, $\mathcal{G} \otimes \mathcal{F}$ is hk-regular, and $\mathcal{G} \circ \mathcal{F}$ is $(|Y|h + k)$-regular.*

(ii) *$\mathcal{G} \square \mathcal{F}$ is connected if and only if $\mathcal{G}$ and $\mathcal{F}$ are both connected; $\mathcal{G} \otimes \mathcal{F}$ is connected if and only if both factors are connected and at least one of them is nonbipartite; $\mathcal{G} \circ \mathcal{F}$ is connected if and only if $\mathcal{G}$ is connected.*

(iii) *Assuming that it is connected, the graph $\mathcal{G} \square \mathcal{F}$ is bipartite if and only if both $\mathcal{G}$ and $\mathcal{F}$ are bipartite. Similarly, assuming that it is connected, the graph $\mathcal{G} \otimes \mathcal{F}$ is bipartite if and only if at least one of the factors is bipartite. Finally, assuming that it is connected, the graph $\mathcal{G} \circ \mathcal{F}$ is not bipartite.*

Proof. We have $\lambda_0 = h$ (respectively, $\mu_0 = k$), $Af_0 = hf_0$, and f_0 is a nonzero constant function (respectively, $Bg_0 = kg_0$ and g_0 is a nonzero constant function).

(i) The function $f_0 \otimes g_0 \in L(X \times Y)$ is constant and it is a nontrivial eigenvector of
   - $A \otimes I_Y + I_X \otimes B$, with eigenvalue $h + k$;
   - $A \otimes B$, with eigenvalue hk;
   - $A \otimes J_Y + I_X \otimes B$, with eigenvalue $|Y|h + k$.

   In order to show regularity and determine the corresponding degree, we use the last statement in Proposition 8.1.4.

(ii) By virtue of Proposition 8.1.5, the graph $\mathcal{G} \square \mathcal{F}$ is connected if and only if $\lambda_0 + \mu_0 > \lambda_i + \mu_j$ for all $(i, j) \neq (0, 0)$, that is if and only if $\lambda_0 > \lambda_1$ and $\mu_0 > \mu_1$, and this is equivalent to saying that $\mathcal{G}$ and $\mathcal{F}$ are both connected.

   Similarly, $\mathcal{G} \otimes \mathcal{F}$ is connected if and only if

   $$\lambda_0 \mu_0 > \lambda_i \mu_j \text{ for all } (i, j) \neq (0, 0). \tag{8.24}$$

   If both factors are connected and at least one of them, say $\mathcal{G}$, is non-bipartite, by Proposition 8.3.4 we have $h = \lambda_0 > \lambda_1 \geq \cdots \lambda_{|X|-1} > -h$ and $k = \mu_0 > \mu_1 \geq \cdots \mu_{|Y|-1} \geq -k$; an elementary case-by-case analysis shows that (8.24) is satisfied. Conversely, if one of the graphs, say $\mathcal{G}$, is not connected then $\lambda_1 = \lambda_0 = h$ so that $\lambda_1 \mu_0 = \lambda_0 \mu_0$ and (8.24) is not verified; if both graphs are connected and bipartite then $\lambda_{|X|-1} = -h$ and $\mu_{|Y|-1} = -k$, so that $\lambda_{|X|-1} \mu_{|Y|-1} = (-h)(-k) = hk = \lambda_0 \mu_0$ and, again, (8.24) is not verified.

   Finally, observe that the eigenvalues of $\mathcal{G} \circ \mathcal{F}$ are

   $$h|Y| + k = \lambda_0 |Y| + \mu_0 \geq \lambda_1 |Y| + \mu_0 \geq \cdots \geq \lambda_{|X|-1}|Y|$$
   $$+ \mu_0 \geq \mu_1 \geq \mu_2 \geq \mu_{|Y|-1}$$

   and $\mathcal{G} \circ \mathcal{F}$ is connected if and only if the multiplicity of the eigenvalue $h|Y| + k$ is one, and this happens if and only if the multiplicity of $h = \lambda_0$ is one, that is, if and only if $\mathcal{G}$ is connected.

(iii) We again apply Proposition 8.3.4. The number $-(h + k)$ is an eigenvalue of the adjacency matrix of $\mathcal{G} \square \mathcal{F}$ if and only if $\lambda_{|X|-1} = -h$ *and* $\mu_{|Y|-1} = -k$. Similarly, $-hk$ is an eigenvalue of the adjacency matrix of $\mathcal{G} \otimes \mathcal{F}$ if and only if $\lambda_{|X|-1} = -h$ or $\mu_{|Y|-1} = -k$. Finally, since

   $$\mu_{|Y|-1} \geq -\mu_0 = -k > -(h|Y| + k),$$

   the number $-(h|Y| + k)$ cannot be an eigenvalue of the adjacency matrix of $\mathcal{G} \circ \mathcal{F}$. $\qquad \square$

Exercise 8.8.6 Give a direct combinatorial (i.e. not spectral) proof of Corollary 8.8.5.

Exercise 8.8.7 (The Hamming graph) Let n, m be two positive integers. The Hamming graph $\mathcal{H}_{n,m+1} = (X_{n,m+1}, E_{n,m+1})$ is the (finite simple without loops) graph with vertex set

$$X_{n,m+1} = \{0, 1, \ldots, m\}^n = \{(x_1, x_2, \ldots, x_n) : x_1, x_2, \ldots, x_n \in \{0, 1, \ldots, m\}\}$$

and two vertices $(x_1, x_2, \ldots, x_n)$ and $(y_1, y_2, \ldots, y_n) \in X_{n,m+1}$ are adjacent if there exists $1 \leq j \leq n$ such that $x_j \neq y_j$ and $x_i = y_i$ for all $i \neq j$. The *Hamming distance* between two vertices $(x_1, x_2, \ldots, x_n)$ and $(y_1, y_2, \ldots, y_n)$ is given by

$$d((x_1, x_2, \ldots, x_n), (y_1, y_2, \ldots, y_n)) = |\{j : x_j \neq y_j\}|.$$

Note that $\mathcal{H}_{n,2}$ (i.e. $m = 1$) coincides with the n-dimensional hypercube Q_n (cf. Section 8.5).

(1) Show that $\mathcal{H}_{n,m+1}$ is an nm-regular graph. Moreover, show that the Hamming distance coincides with the geodesic distance on the graph.

(2) Show that $\mathcal{H}_{n,m+1}$ is the Cartesian product of n copies of the complete graph K_{m+1} (with vertices $\{0, 1, \ldots, m\}$), that is, its adjacency matrix is

$$\sum_{j=1}^{n} I_{m+1} \otimes \cdots \otimes I_{m+1} \otimes A \otimes I_{m+1} \otimes \cdots \otimes I_{m+1},$$

where I_{m+1} is the $(m+1) \times (m+1)$ identity matrix and A (in the j-th position) is the adjacency matrix of K_{m+1}.

(3) For $\mathbf{i} = (i_1, i_2, \ldots, i_n) \in \{0, 1\}^n$ set $w(\mathbf{i}) = |\{k : i_k = 1\}|$ (the *weight* of $\mathbf{i}$). Recalling the spectral decomposition (see Proposition 8.1.4 and Section 8.4)

$$L(K_{m+1}) = W_0 \oplus W_1$$

for $0 \leq \ell \leq n$, we set

$$V_\ell = \bigoplus_{w(\mathbf{i})=\ell} W_{i_1} \otimes W_{i_2} \otimes \cdots \otimes W_{i_n}.$$

In other words, V_ℓ is the subspace spanned by all tensor products $f_1 \otimes f_2 \otimes \cdots \otimes f_n$ where ℓ (respectively, the remaining $n - \ell$) of the f_js belong to W_1 (respectively, W_0). Show that

$$L(X_{n,m+1}) = \oplus_{\ell=0}^{n} V_\ell$$

is the spectral decomposition relative to the adjacency matrix of $\mathcal{H}_{n,m+1}$, that the eigenvalue corresponding to V_ℓ is $nm - \ell(m+1)$, and that $\dim V_\ell = m^\ell \binom{n}{\ell}$.

8.9 Wreath product of finite graphs

This section is based on [45]: in particular, for simplicity, we only consider (finite) simple graphs without loops.

Let X be a finite set and $\mathcal{F} = (Y, F)$ a finite simple graph without loops. We endow the exponential set Y^X with a graph structure, denoted $\mathcal{F}^X$, by declaring that two vertices $f, f' \in Y^X$ are adjacent (and, as usual, we write $f \sim f'$) if there exists $x \in X$ such that $f(z) = f'(z)$ for all $z \in X \setminus \{x\}$ and $f(x) \sim f'(x)$ in $\mathcal{F}$. Note that $\mathcal{F}^X$ is simple and without loops; moreover, if $|X| = 2$ it coincides with the Cartesian square $\mathcal{F} \Box \mathcal{F}$. Denote by B the adjacency operator of $\mathcal{F}$ (that is, $B\delta_y = \sum_{y' \sim y} \delta_{y'} = \mathbf{1}_{N(y)}$ for all $y \in Y$) and by $\mathcal{B}$ the adjacency operator of $\mathcal{F}^X$ (that is, $\mathcal{B}\delta_f = \sum_{f' \sim f} \delta_{f'} = \mathbf{1}_{\mathcal{N}(f)}$ for all $f \in Y^X$). Also, for all $x, x' \in X$ we define the linear operator $B_{x,x'} : L(Y) \to L(Y)$ by setting

$$
B_{x,x'} = \begin{cases} B & \text{if } x = x' \\ I_Y & \text{if } x \neq x'. \end{cases}
$$

We now generalize Proposition 8.8.3.(i).

Proposition 8.9.1 *The adjacency operator $\mathcal{B}$ of $\mathcal{F}^X$ has the expression*

$$
\mathcal{B} = \sum_{x \in X} \bigotimes_{x' \in X} B_{x,x'}.
$$

Proof. Let $f \in Y^X$ and let us show that

$$
\mathcal{B}\delta_f = \left(\sum_{x \in X} \bigotimes_{x' \in X} B_{x,x'} \right) \delta_f. \tag{8.25}
$$

For $x, x' \in X$ define $\mathbf{1}_{x,x'} \in L(Y)$ by setting

$$
\mathbf{1}_{x,x'} = \begin{cases} \mathbf{1}_{\mathcal{N}(f(x))} & \text{if } x = x' \\ \delta_{f(x')} & \text{if } x \neq x'. \end{cases} \tag{8.26}
$$

Note that setting

$$
\mathcal{N}_x(f) = \left\{ f' \in Y^X : [f'(x') = f(x') \text{ for } x \neq x'] \text{ and } [f'(x) \sim f(x)] \right\} \tag{8.27}
$$

for all $x \in X$, in the graph $\mathcal{F}^X$ we have the partition

$$
\mathcal{N}(f) = \coprod_{x \in X} \mathcal{N}_x(f) \tag{8.28}
$$

and the map $\alpha \colon \mathcal{N}(f(x)) \to \mathcal{N}_x(f)$ defined by

$$\alpha(y)(x') = \begin{cases} y & \text{if } x' = x \\ f(x') & \text{if } x' \neq x \end{cases} \tag{8.29}$$

for all $y \in \mathcal{N}(f(x))$ and $x' \in X$, is a bijection. Then, on the one hand, we have

$$
\begin{aligned}
\mathcal{B}\delta_f &= \mathbf{1}_{\mathcal{N}(f)} \\
\text{(by (8.28))} &= \sum_{x \in X} \mathbf{1}_{\mathcal{N}_x(f)} \\
&= \sum_{x \in X} \sum_{f' \in \mathcal{N}_x(f)} \delta_{f'} \\
\text{(by (8.23))} &= \sum_{x \in X} \sum_{f' \in \mathcal{N}_x(f)} \bigotimes_{x' \in X} \delta_{f'(x')} \\
&= \sum_{x \in X} \sum_{f' \in \mathcal{N}_x(f)} \left[\left(\bigotimes_{x' \in X \setminus \{x\}} \delta_{f'(x')} \right) \otimes \delta_{f'(x)} \right] \\
\text{(by (8.27))} &= \sum_{x \in X} \sum_{f' \in \mathcal{N}_x(f)} \left[\left(\bigotimes_{x' \in X \setminus \{x\}} \delta_{f(x')} \right) \otimes \delta_{f'(x)} \right] \\
&= \sum_{x \in X} \left[\left(\bigotimes_{x' \in X \setminus \{x\}} \delta_{f(x')} \right) \otimes \left(\sum_{f' \in \mathcal{N}_x(f)} \delta_{f'(x)} \right) \right] \\
&= \sum_{x \in X} \left[\left(\bigotimes_{x' \in X \setminus \{x\}} \delta_{f(x')} \right) \otimes \left(\sum_{y \in \mathcal{N}(f(x))} \delta_{\alpha(y)(x)} \right) \right] \\
\text{(by (8.29))} &= \sum_{x \in X} \left[\left(\bigotimes_{x' \in X \setminus \{x\}} \delta_{f(x')} \right) \otimes \left(\sum_{y \in \mathcal{N}(f(x))} \delta_{y} \right) \right] \\
&= \sum_{x \in X} \left[\left(\bigotimes_{x' \in X \setminus \{x\}} \delta_{f(x')} \right) \otimes \mathbf{1}_{\mathcal{N}(f(x))} \right] \\
\text{(by (8.26))} &= \sum_{x \in X} \bigotimes_{x' \in X} \mathbf{1}_{x,x'}.
\end{aligned}
\tag{8.30}
$$

Moreover,

$$B_{x,x'}\delta_{f(x')} = \begin{cases} B\delta_{f(x)} & \text{if } x = x' \\ I_Y\delta_{f(x')} & \text{if } x \neq x' \end{cases} = \mathbf{1}_{x,x'}, \tag{8.31}$$

so that, on the other hand, keeping in mind (8.23), we have

$$\left(\bigotimes_{x' \in X} B_{x,x'}\right)\delta_f = \left[\left(\bigotimes_{x' \in X} B_{x,x'}\right)\left(\bigotimes_{x' \in X} \delta_{f(x')}\right)\right]$$

$$\text{(by (8.22))} \quad = \left[\left(\bigotimes_{x' \in X} B_{x,x'}\delta_{f(x')}\right)\right] \tag{8.32}$$

$$\text{(by (8.31))} \quad = \bigotimes_{x' \in X} \mathbf{1}_{x,x'}.$$

Summing up (over $x \in X$) in (8.32), and comparing it with (8.30), we finally deduce (8.25). □

Exercise 8.9.2 Show that the set of all eigenvalues of the adjacency operator $\mathcal{B}$ of $\mathcal{F}^X$ is given by

$$\left\{\sum_{x \in X} \mu_{\xi(x)} : \xi \in \{0, 1, \ldots, |Y| - 1\}^X\right\},$$

where $\mu_0, \mu_1, \ldots, \mu_{|Y|-1}$ are the eigenvalues of $\mathcal{F}$. Deduce, as a particular case, the set of all eigenvalues of the hypercube (cf. Section 8.5) and of the Hamming graph (cf. Exercise 8.8.7).

Let now $\mathcal{G} = (X, E)$ and $\mathcal{F} = (Y, F)$ be two finite simple graphs without loops.

Definition 8.9.3 The *wreath product* of $\mathcal{G}$ and $\mathcal{F}$ is the graph $\mathcal{G} \wr \mathcal{F} = (Y^X \times X, \mathcal{E})$ where the edge set is

$$\mathcal{E} = \left\{\{(f, x), (f', x')\} \subseteq Y^X \times X : [x = x' \text{ and } f' \in \mathcal{N}_x(f)]\right.$$
$$\left. \text{or } [x \sim x' \text{ and } f = f']\right\},$$

where $\mathcal{N}_x(f) \subseteq Y^X$ is as in (8.28). Moreover, $\{(f, x), (f', x')\} \in \mathcal{E}$ is called an *edge of the first type* (respectively, *edge of the second type*) provided $x = x'$ and $f' \in \mathcal{N}_x(f)$ (respectively, $x \sim x'$ and $f = f'$).

Remark 8.9.4 Note that, modulo the map $Y^X \times X \ni (f, x) \mapsto (x, f) \in X \times Y^X$, the wreath product $\mathcal{G} \wr \mathcal{F}$ can be viewed as a subgraph of the Cartesian product $\mathcal{G} \square \mathcal{F}^X$, and therefore of the lexicographic product $\mathcal{G} \circ \mathcal{F}^X$. Indeed, the set of all edges of the first type in $\mathcal{G} \wr \mathcal{F}$ forms a subset of those edges of the Cartesian product that are given by the less restrictive condition $x = x'$ and

$f \sim f'$; the set of all edges of the second type in $\mathcal{G} \wr \mathcal{F}$ are defined by the analogous condition in the Cartesian product (but they form a subset of the edges of the first type in the lexicographic product).

Theorem 8.9.5 *The adjacency operator of the wreath product* $\mathcal{G} \wr \mathcal{F}$ *has the expression*

$$\sum_{x \in X} \left[\left(\bigotimes_{x' \in X} B_{x,x'} \right) \bigotimes \Delta_x \right] + I_{Y^X} \otimes A, \tag{8.33}$$

where $\Delta_x \in \mathrm{End}(L(X))$ *is defined by setting* $\Delta_x(\delta_{x'}) = \delta_x(x')\delta_x$ *for all* $x, x' \in X$.

Proof. Let us show that the first summand in (8.33) takes into account all edges of the first type. Indeed, arguing as in the proof of Proposition 8.9.1, for $z \in X$ and $f \in Y^X$, we have:

$$\left\{ \sum_{x \in X} \left[\left(\bigotimes_{x' \in X} B_{x,x'} \right) \bigotimes \Delta_x \right] \right\} (\delta_f \otimes \delta_z) = \sum_{x \in X} \left[\left(\bigotimes_{x' \in X} B_{x,x'} \right) (\delta_f) \bigotimes \Delta_x(\delta_z) \right]$$

$$= \left(\bigotimes_{x' \in X} B_{z,x'} \right) (\delta_f) \bigotimes \delta_z$$

$$\text{(by (8.32))} = \left(\bigotimes_{x' \in X} 1_{z,x'} \right) \bigotimes \delta_z,$$

where the last expression is precisely the characteristic function of the set of all vertices adjacent to (f, z) by an edge of the first type.

Finally, the term $I_{Y^X} \otimes A$ takes into account all edges of the second type; compare it with the expression of the adjacency matrix of the Cartesian product in Proposition 8.8.3.(i). $\qquad \square$

In [45], D'Angeli and Donno introduced and used (8.33) as a definition of wreath product of matrices.

8.10 Lamplighter graphs and their spectral analysis

This section is based on our monograph [34] and the paper [136], but the version of the lamplighter that we analyze is the one described in [45, 58, 59].

Let $\mathcal{G} = (X, E)$ be a finite simple graph without loops.

Definition 8.10.1 The *lamplighter graph* associated with $\mathcal{G}$ is the finite graph $\mathcal{L} = (\mathcal{X}, \mathcal{E})$ with vertex set

$$\mathcal{X} = \{0, 1\}^X \times X = \left\{ (\omega, x) : \omega \in \{0, 1\}^X, x \in X \right\}$$

and edge set

$$\mathcal{E} = \Big\{ \{(\omega, x), (\theta, y)\} : \big[x = y, \omega(z) = \theta(z) \text{ for all } z \neq x \text{ and } \omega(x) \neq \theta(x) \big]$$
$$\text{or } \big[x \sim y \text{ and } \omega = \theta \big] \Big\}.$$

Clearly, $\mathcal{L}$ coincides with the wreath product $\mathcal{G} \wr K_2$, where K_2 is the complete graph on two vertices (cf. Figure 8.5).

Remark 8.10.2 Another description of the lamplighter graph is the following. We associate with each vertex $x \in X$ a lamp that may be either *on* or *off*. A *configuration* of the lamps is a map $\omega \colon X \to \{0, 1\}$: the value $\omega(x) = 1$ (respectively, $\omega(x) = 0$) indicates that the lamp at x is on (respectively, off). A vertex of the lamplighter is a pair (ω, x) consisting of a configuration of the lamps and a vertex of X. Two vertices (ω, x) and (θ, y) of the lamplighter graph are adjacent if and only if one of these two conditions are satisfied:

$$x \sim y \text{ and } \omega = \theta \qquad \text{(a } \textit{walk edge)};$$
$$x = y \text{ and } \omega \text{ and } \theta \text{ differ exactly in } x \qquad \text{(a } \textit{switch edge)}. \tag{8.34}$$

This is the so-called *walk or switch lamplighter*: the neighbors of the vertex (ω, x) may be obtained by either walking to a neighbor of x in $\mathcal{G}$ and leaving all the lamps at their current states, or remaining at x but changing the state of the lamp at x.

Finally note that two configurations ω and θ may be added: $(\omega + \theta)(x) = \omega(x) + \theta(x) \bmod 2$.

In the literature, several variations on this construction have been analyzed; see [136], and, for infinite lamplighters and their spectral computations [17, 69, 70, 94].

Let $\mathcal{A} \in \mathrm{End}(L(\mathcal{X}))$ denote the adjacency operator associated with the lamplighter graph $\mathcal{L}$, so that

$$[\mathcal{A}\Phi](\omega, x) = \sum_{(\theta, y) \sim (\omega, x)} \Phi(\theta, y),$$

for all $\Phi \in L(\mathcal{X})$ and $(\omega, x) \in \mathcal{X}$. Since $L(\mathcal{X}) \equiv L\big(\{0, 1\}^X\big) \otimes L(X)$, it is useful to determine the $\mathcal{A}$-image of a tensor product of functions: if $F \in L\big(\{0, 1\}^X\big)$ and $f \in L(X)$ we have

$$[\mathcal{A}(F \otimes f)](\omega, x) = F(\omega + \delta_x)f(x) + F(\omega) \sum_{y \sim x} f(y) \tag{8.35}$$

for all $(\omega, x) \in \{0, 1\}^X \times X$. Indeed, the first term corresponds to a switch at x (δ_x is regarded as the configuration with only the lamp at x on) and the second to a walk from x.

With each $\theta \in \{0, 1\}^X$ we associate the linear operator $A_\theta : L(X) \to L(X)$ defined by setting

$$[A_\theta f](x) = (-1)^{\theta(x)} f(x) + \sum_{y \sim x} f(y) \tag{8.36}$$

for all $f \in L(X)$ and $x \in X$, and the character $\chi_\theta \in \widehat{\mathbb{Z}_2^X} \equiv \widehat{\{0, 1\}^X} \subseteq L(\{0, 1\}^X)$ defined by setting

$$\chi_\theta(\omega) = (-1)^{\sum_{x \in X} \theta(x)\omega(x)}$$

for all $\omega \in \{0, 1\}^X$ (cf. Section 8.5).

Theorem 8.10.3 *For all $\theta \in \{0, 1\}^X$ and $f \in L(X)$ we have:*

$$\mathcal{A}(\chi_\theta \otimes f) = \chi_\theta \otimes A_\theta f. \tag{8.37}$$

Suppose also that $\lambda_{\theta,1}, \lambda_{\theta,2}, \ldots, \lambda_{\theta,h(\theta)}$ are the distinct eigenvalues of A_θ and $V_{\theta,j}$ is the eigenspace of A_θ corresponding to the eigenvalue $\lambda_{\theta,j}$, $j = 1, \ldots, h(\theta)$. Then

$$\left\{ \lambda_{\theta,j} : \theta \in \{0, 1\}^X, j = 1, 2, \ldots, h(\theta) \right\}$$

are the eigenvalues of $\mathcal{A}$ (not necessarily distinct) and $\mathcal{W}_{\theta,j} = \{\chi_\theta \otimes f : f \in V_{\theta,j}\}$ is the eigenspace of $\mathcal{A}$ corresponding to $\lambda_{\theta,j}$.

Proof. Applying (8.35) we get

$$[\mathcal{A}(\chi_\theta \otimes f)](\omega, x) = \chi_\theta(\omega + \delta_x) f(x) + \chi_\theta(\omega) \sum_{y \sim x} f(y)$$

$$= \chi_\theta(\omega) \left[(-1)^{\theta(x)} f(x) + \sum_{y \sim x} f(y) \right] \tag{8.38}$$

$$= [\chi_\theta \otimes A_\theta f](\omega, x).$$

The other statements follow easily from (8.37). $\qquad \square$

8.11 The lamplighter on the complete graph

This section is based on [45]. See also [34] and [136] for another version of the following construction.

Given a finite set X, we denote, as usual, by $W_0(X)$ the space of constant functions on X and $W_1(X) = \{f \in L(X) : \sum_{x \in X} f(x) = 0\}$. Then (cf. Proposition 2.1.1), we have the decomposition

$$L(X) = W_0(X) \oplus W_1(X). \tag{8.39}$$

Let now $K_n = (X, E)$ be the complete graph on n vertices so that $X = \{1, 2, \ldots, n\}$ and $E = \{\{x, y\} : x, y \in X, x \neq y\})$. The eigenspaces of the adjacency operator on the complete graph on n vertices are $W_0(X)$ and $W_1(X)$, with corresponding eigenvalues $n - 1$ and -1, respectively; see Section 8.4. Let $\mathcal{L} = (\mathcal{X}, \mathcal{E})$ be the associated lamplighter graph. Let $\theta \in \{0, 1\}^X$ and set

$$X_\theta = \{x \in X : \theta(x) = 0\}.$$

For $f \in L(X)$ and $x \in X$, equation (8.36) becomes:

$$[A_\theta f](x) = \begin{cases} f(x) + \displaystyle\sum_{\substack{y \in X_\theta: \\ y \neq x}} f(y) + \sum_{y \in X \setminus X_\theta} f(y) & \text{if } x \in X_\theta \\ -f(x) + \displaystyle\sum_{y \in X_\theta} f(y) + \sum_{\substack{y \in X \setminus X_\theta: \\ y \neq x}} f(y) & \text{if } x \in X \setminus X_\theta. \end{cases} \tag{8.40}$$

Let $f \in L(X)$. If

$$f|_{X_\theta} \in W_1(X_\theta) \text{ and } f|_{X \setminus X_\theta} \equiv 0 \tag{8.41}$$

then (8.40) becomes

$$[A_\theta f](x) = \begin{cases} f(x) + \displaystyle\sum_{\substack{y \in X_\theta: \\ y \neq x}} f(y) & \text{if } x \in X_\theta \\ \displaystyle\sum_{y \in X_\theta} f(y) & \text{if } x \in X \setminus X_\theta \end{cases}$$

$$= \sum_{y \in X_\theta} f(y) = 0 \quad \text{(in both cases)}.$$

Therefore, the space of all functions satisfying the conditions in (8.41) constitutes an A_θ-eigenspace with eigenvalue 0.

Similarly, if

$$f|_{X \setminus X_\theta} \in W_1(X \setminus X_\theta) \text{ and } f|_{X_\theta} \equiv 0 \tag{8.42}$$

then

$$
[A_\theta f](x) = \begin{cases} \displaystyle\sum_{y \in X \setminus X_\theta} f(y) & \text{if } x \in X_\theta \\ -f(x) + \displaystyle\sum_{\substack{y \in X \setminus X_\theta: \\ y \neq x}} f(y) & \text{if } x \in X \setminus X_\theta \end{cases}
$$

$$
= \begin{cases} 0 & \text{if } x \in X_\theta \\ -2f(x) & \text{if } x \in X \setminus X_\theta \end{cases}
$$

$$
= -2f(x) \quad \text{(in both cases)}.
$$

Therefore, the space of all functions satisfying the conditions in (8.42) constitutes an A_θ-eigenspace with eigenvalue -2.

Finally, suppose that $|X_\theta| = k$ with $0 \le k \le n$, and let $f = \alpha \mathbf{1}_{X_\theta} + \beta \mathbf{1}_{X \setminus X_\theta}$, for some $\alpha, \beta \in \mathbb{C}$. From (8.40) it follows that

$$
[A_\theta f](x) = \begin{cases} k\alpha + (n-k)\beta & \text{if } x \in X_\theta \\ k\alpha + (n-k-2)\beta & \text{if } x \in X \setminus X_\theta. \end{cases}
$$

Note that if $k = 0$ (respectively, $k = n$), that is, $X_\theta = \varnothing$ (respectively, $X_\theta = X$), then f is constant and is an A_θ-eigenvector with eigenvalue $n - 2$ (respectively, n). When $1 \le k \le n - 1$, elementary calculations show that the eigenvalues of the matrix $\left(\begin{smallmatrix} k & n-k \\ k & n-k-2 \end{smallmatrix} \right)$ are $\lambda_\pm^{(k)} = \frac{n-2 \pm \sqrt{(n-2)^2 + 8k}}{2}$ and the corresponding eigenvectors are $\left(1, \omega_\pm^{(k)}\right)^T$, where $\omega_\pm^{(k)} = \frac{\lambda_\pm^{(k)}}{2 + \lambda_\pm^{(k)}}$.

We then define the one-dimensional A_θ-eigenspaces (subspaces of $L(X)$)

$$
W_\theta^\pm = \{f = \alpha \mathbf{1}_{X_\theta} + \omega_\pm^{(k)} \alpha \mathbf{1}_{X \setminus X_\theta} : \alpha \in \mathbb{C}\},
$$

for $1 \le |X_\theta| \le n - 1$, and

$$
W_0 = \{f = \alpha \mathbf{1}_X : \alpha \in \mathbb{C}\},
$$

if $|X_\theta| = 0, n$.

We also define the following subspaces of $L(\mathcal{X})$:

$$
\mathcal{W}_{0;0} = \mathrm{span}(\mathbf{1} \otimes f : f \in W_0),
$$
$$
\mathcal{W}_{n;0} = \mathrm{span}((-\mathbf{1}) \otimes f : f \in W_0),
$$

where $\mathbf{1}(\omega) = 1$ and $[-\mathbf{1}](\omega) = (-1)^{\sum_{x \in X} \omega(x)}$, for all $\omega \in \{0, 1\}^X$, and, for $1 \le k \le n - 1$,

$$
\mathcal{W}_{k;0}^\pm = \mathrm{span}(\chi_\theta \otimes f : |X_\theta| = k, \ f \in W_\theta^\pm),
$$
$$
\mathcal{W}_{k;1} = \mathrm{span}(\chi_\theta \otimes f : |X_\theta| = k, \ f|_{X_\theta} \in W_1(X_\theta) \text{ and } f|_{X \setminus X_\theta} \equiv 0),
$$
$$
\mathcal{W}_{k;2} = \mathrm{span}(\chi_\theta \otimes f : |X_\theta| = k, f|_{X_\theta} \equiv 0 \text{ and } f|_{X \setminus X_\theta} \in W_1(X \setminus X_\theta)).
$$

Exercise 8.11.1 Show that

(1) $\mathcal{W}_{0;0}$ is the $\mathcal{A}$-eigenspace with eigenvalue $n - 2$;
(2) $\mathcal{W}_{n;0}$ is the $\mathcal{A}$-eigenspace with eigenvalue n;
(3) $\mathcal{W}_{k;0}^{\pm}$ is the $\mathcal{A}$-eigenspace with eigenvalue $\lambda_{\pm}^{(k)}$, for $k = 1, 2, \ldots, n - 1$;
(4) $\bigoplus_{k=1}^{n} \mathcal{W}_{k;1}$ is the $\mathcal{A}$-eigenspace with eigenvalue 0;
(5) $\bigoplus_{k=0}^{n-1} \mathcal{W}_{k;2}$ is the $\mathcal{A}$-eigenspace with eigenvalue -2.

8.12 The replacement product

In this section, based on [58], we introduce the replacement product. This is a natural construction but it is worthwhile to introduce specific notation in order to get a precise description of it. This notation will also be used for the zig-zag product (cf. Section 8.13).

Let $\mathcal{G} = (X, E, r)$ be a finite d-regular graph possibly with multiple edges and loops.

Let x and y be two distinct vertices in X. Recall that E_x denotes the set of edges incident to x. This way, $E_x \cap E_y$ is the set of edges joining x and y (note that $x \not\sim y$ if and only if $E_x \cap E_y = \varnothing$).

Set $[d] = \{1, 2, \ldots, d\}$. Then for each $x \in X$ we (arbitrarily) choose a bijective *labelling* of the edges incident to x using $[d]$ as the set of labels, that is, a bijection $h_x \colon E_x \to [d]$. We refer to $(h_x)_{x \in X}$ as to the (edge) *labelling* of $\mathcal{G}$ and we say that $\mathcal{G}$ is a *labelled* graph. Given a vertex $x \in X$ and an edge $e \in E$ such that $r(e) \ni x$, the label $h = h_x(e)$ is called the *color* of the edge e *near x* and we also say that e is the *h-edge near x*. Note that, unless otherwise specified, if $x, y \in X$ are distinct and adjacent, and $e \in E_x \cap E_y$, then there is no relation between the color $h_x(e)$ of e near x and the color $h_y(e)$ of e near y. Moreover, if $r(e) = \{x\}$, that is, e is a loop at x, then e has only the color $h_x(e)$ near x.

Definition 8.12.1 The *rotation map*

$$\mathrm{Rot}_{\mathcal{G}} \colon X \times [d] \longrightarrow X \times [d]$$

associated with the labelling $(h_x)_{x \in X}$ is the (bijective) map defined by setting

$$\mathrm{Rot}_{\mathcal{G}}(x, i) = (y, j) \text{ where } e = h_x^{-1}(i), \ r(e) = \{x, y\}, \text{ and } j = h_y(e), \quad (8.43)$$

for all $x \in X$ and $i \in [d]$.

In other words, if $e = h_x^{-1}(i) \in E$ is a loop at x, then $\mathrm{Rot}_{\mathcal{G}}(x, i) = (x, i)$, while if $r(e) = \{x, y\}$, with $y \neq x$, then $\mathrm{Rot}_{\mathcal{G}}(x, i) = (y, j)$, where j is the color of e near y. Note that

$$E = (X \times [d]) / \approx \quad (8.44)$$

where $\approx$ is the equivalence relation defined by setting $(x, i) \approx (x, i)$ and

$$(x, i) \approx (y, j) \text{ if } (y, j) = \text{Rot}_\mathcal{G}(x, i)$$

for all $x, y \in X$ and $i, j \in [d]$.

With the rotation map $\text{Rot}_\mathcal{G}$ we associate the permutation matrix $R_\mathcal{G}$ indexed by $X \times [d]$ defined by setting, for all $(x, i), (y, j) \in X \times [d]$,

$$R_\mathcal{G}\big((x, i), (y, j)\big) = \begin{cases} 1 & \text{if } \text{Rot}_\mathcal{G}(x, i) = (y, j) \\ 0 & \text{otherwise.} \end{cases} \tag{8.45}$$

In the following proposition, we show the connection between the permutation matrix $R_\mathcal{G}$ and the adjacency matrix $A = A_\mathcal{G}$ of $\mathcal{G}$. We use the operator C in (8.21) and we think of $R_\mathcal{G}$ (respectively, A) as a linear endomorphism of $L(X \times [d])$ (respectively, $L(X)$).

Proposition 8.12.2 *For all $f \in L(X)$ one has*

$$CR_\mathcal{G}(f \otimes \mathbf{1}_{[d]}) = Af.$$

Proof. Clearly, for $(x, i) \in X \times [d]$ we have

$$R_\mathcal{G}(\delta_y \otimes \delta_j) = \delta_x \otimes \delta_i$$

where $(y, j) = \text{Rot}_\mathcal{G}(x, i)$. Then

$$R_\mathcal{G}(\delta_x \otimes \mathbf{1}_{[d]}) = \sum_{i \in [d]} R_\mathcal{G}(\delta_x \otimes \delta_i) = \sum_{i \in [d]} \sum_{\substack{(y, j) \in X \times [d]: \\ \text{Rot}_\mathcal{G}(y, j) = (x, i)}} \delta_y \otimes \delta_j$$

so that

$$CR_\mathcal{G}(\delta_x \otimes \mathbf{1}_{[d]}) = \sum_{i \in [d]} \sum_{\substack{(y, j) \in X \times [d]: \\ \text{Rot}_\mathcal{G}(y, j) = (x, i)}} C(\delta_y \otimes \delta_j)$$

$$(\text{by Lemma 8.7.3.(ii)}) = \sum_{i \in [d]} \sum_{\substack{(y, j) \in X \times [d]: \\ \text{Rot}_\mathcal{G}(y, j) = (x, i)}} \delta_y$$

$$= \sum_{\substack{y \in X: \\ x \sim y \text{ in } \mathcal{G}}} \delta_y$$

$$= A\delta_x.$$

The general result follows by linearity. $\qquad\qquad\qquad\qquad\qquad\qquad\square$

Exercise 8.12.3 Show that if X is a finite nonempty set, then a map $\text{Rot}: X \times [d] \longrightarrow X \times [d]$ is the rotation map of a labelled d-regular graph with vertex set X if and only if Rot $\circ$ Rot is the identity map. Moreover, loops correspond to fixed-points of Rot.

Hint: Suppose Rot $\circ$ Rot is the identity map. For $x \in X$ set $\overline{E}_x = \{\text{Rot}(x, i) : i \in [d]\}$ and define $E = (\cup_{x \in X} \overline{E}_x) / \approx$, where $\approx$ is as in (8.44). Moreover, $r \colon E \to \mathcal{P}(X)$ is defined by setting $r[\text{Rot}(x, i)] = \{x, y\}$, where $\text{Rot}(x, i) = (y, j)$, for all $x \in X$ and $i \in [d]$.

Definition 8.12.4 Let $\mathcal{G} = (X, E, r_\mathcal{G})$ be a d-regular graph and $\mathcal{F} = (Y, F, r_\mathcal{F})$ a k-regular graph with $Y = [d]$. Assume that in both graphs we have defined a labelling and a rotation map as in Definition 8.12.1. Then their *replacement product* is the $(k + 1)$-regular graph $\mathcal{G}(\tau)\mathcal{F}$ with vertex set $X \times [d]$ and the rotation map defined by setting, for $x \in X$, $i \in [d]$, and $j \in [k + 1]$,

$$\text{Rot}_{\mathcal{G}(\tau)\mathcal{F}}((x, i), j) = \begin{cases} ((x, m), h) & \text{if } j \in [k] \text{ and } \text{Rot}_\mathcal{F}(i, j) = (m, h) \\ (\text{Rot}_\mathcal{G}(x, i), j) & \text{if } j = k + 1. \end{cases}$$

Exercise 8.12.5 Show that $\text{Rot}_{\mathcal{G}(\tau)\mathcal{F}} \circ \text{Rot}_{\mathcal{G}(\tau)\mathcal{F}}$ is the identity map so that, by Exercise 8.12.3, the definition of replacement product is well posed.

Remark 8.12.6 Actually, to define the replacement product it is not necessary to label $\mathcal{F}$. The definition may be modified by saying that $(x, i), (z, m) \in X \times [d]$ are adjacent in $\mathcal{G}(\tau)\mathcal{F}$ if

$$x \sim z \text{ and } \text{Rot}_\mathcal{G}(x, i) = (z, m) \quad \text{(edges of the first type)}$$

or $\hspace{10cm}$ (8.46)

$$x = z \text{ and } i \sim m \text{ in } \mathcal{F} \quad \text{(edges of the second type)}.$$

Clearly, each vertex is incident to exactly one edge of the first type and to k edges of the second type. Note also that the replacement product is a subgraph of the lexicographic product (cf. Definition 8.8.1). Indeed, the edges of the first type (respectively, second type) in (8.46) are a subset of the edges of the first type (respectively, precisely the set of all edges of the second type) in the lexicographic product.

Remark 8.12.7 A d-regular graph $\mathcal{G} = (X, E, r)$ is d-*edge-colorable* if there exists a map $\phi \colon E \to [d]$ such that the restriction of ϕ to E_x is a bijection for each $x \in X$. In other words, $\mathcal{G}$ is d-edge-colorable when we may assign a color to each edge in such a way that for each $x \in X$ and $j \in [d]$ there exists exactly one edge with color j incident to x. If such a map ϕ exists, we may use it to get a labelling of $\mathcal{G}$ such that if $x, y \in X$ and $e \in E_x \cap E_y$ then e has the same color $\phi(e)$ both near x and near y. This way, in (8.43) we always have $i = j$. If this condition is satisfied, we may write the first condition in (8.46) in the form:

$$x \sim z, \ i = m, \text{ and the label of the edge connecting } x \text{ and } z \text{ is } i. \quad (8.47)$$

Here is an informal description of the replacement product $\mathcal{G}(\Gamma)\mathcal{F}$; compare with the figures in Exercise 8.12.8. Replace each vertex of $\mathcal{G}$ by a copy of $\mathcal{F}$. The edges of each copy of $\mathcal{F}$ constitute the edges of the second type in (8.46). Then join the copies of $\mathcal{F}$ by means of the edges of $\mathcal{G}$, taking into account the labelling of $\mathcal{G}$, as in (8.46) (edges of the first type).

Exercise 8.12.8 Prove that the replacement products $K_5(\Gamma)C_4$ of the complete graph K_5 on five vertices (with the corresponding labellings) and the 4-circle C_4, are as in Figures 8.12 and 8.13. These examples, taken from [1], show that the replacement product does depend on the labelling of the first graph.

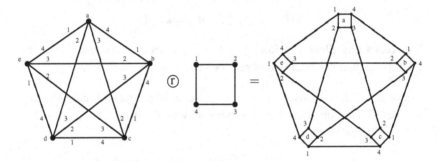

Figure 8.12. The replacement product $K_5(\Gamma)C_4$ (with a given labelling of K_5).

Proposition 8.12.9 *Let B be the adjacency matrix of $\mathcal{F}$ and $R_{\mathcal{G}}$ the permutation matrix in (8.45). Then the adjacency matrix of the replacement product $\mathcal{G}(\Gamma)\mathcal{F}$ is given by*

$$M_{\mathcal{G}(\Gamma)\mathcal{F}} = R_{\mathcal{G}} + I_X \otimes B.$$

Figure 8.13. The replacement product $K_5(\Gamma)C_4$ (with another labelling of K_5).

Proof. The matrix $R_{\mathcal{G}}$ (respectively, $I_X \otimes B$) takes into account all edges of the first type (respectively, second type) of $\mathcal{G}_{\textcircled{r}}\mathcal{F}$; compare with Proposition 8.8.3.(i). $\qquad\square$

We end this section by showing that the lamplighter construction in Section 8.10 may be obtained as a replacement product.

Let then $Q_n = (X, E)$ be the n-dimensional hypercube (see Section 8.5). Using the notation in both Section 8.10 and in the present section, we may identify X with $\{0, 1\}^{[n]}$. Moreover, two vertices $\omega, \theta \in \{0, 1\}^{[n]}$ are adjacent when there exists $j \in [n]$ such that: $\omega(j) \neq \theta(j)$ and $\omega(h) = \theta(h)$ for $h \neq j$. In this case, the edge $\{\omega, \theta\} \in E$ is labelled by the color j both near ω and near θ. This shows (cf. Remark 8.12.7) that the n-dimensional hypercube is n-edge-colorable.

Proposition 8.12.10 *Let $\mathcal{F} = ([n], E)$ be a simple graph without loops on n vertices. Then the product replacement $Q_n \textcircled{r} \mathcal{F}$ obtained by means of the labelling described above is isomorphic to the lamplighter $\mathcal{F} \wr K_2$.*

Proof. In the terminology of Remarks 8.10.2, 8.12.6, and 8.12.7, a switch edge in $\mathcal{F} \wr K_2$ corresponds to an edge of the first type in $Q_n \textcircled{r} \mathcal{F}$: both the switch condition in (8.34) and the conditions in (8.47) become: $i = m$, $\omega \sim \theta$, and the color of the edge connecting ω with θ is i.

Similarly, a walk edge in $\mathcal{F} \wr K_2$ corresponds to an edge of the second type in $Q_n \textcircled{r} \mathcal{F}$: for $(\omega, i), (\theta, m) \in Q_n \times [n]$ both the walk condition in (8.34) and the second condition in (8.46) become: $i \sim m$ and $\omega = \theta$. $\qquad\square$

8.13 The zig-zag product

This section is based on the exposition in [58]. The original sources are [74] and [128]. We assume all the notation in Section 8.12, in particular in Definition 8.12.4, so that $\mathcal{G} = (X, E, r_{\mathcal{G}})$ is a d-regular graph and $\mathcal{F} = (Y, F, r_{\mathcal{F}})$ a k-regular graph with $Y = [d]$.

Definition 8.13.1 The *zig-zag product* of $\mathcal{G}$ and $\mathcal{F}$ is the k^2-regular graph $\mathcal{G}\textcircled{z}\mathcal{F}$ with vertex set $X \times [d]$ and rotation map $\text{Rot}_{\mathcal{G}\textcircled{z}\mathcal{F}}$ described by the following conditions. We use the set $[k] \times [k]$ to label the edges of the graph and, for $x \in X$, $h \in [d]$, and $i, j \in [k]$,

$$\text{Rot}_{\mathcal{G}\textcircled{z}\mathcal{F}}((x, h), (i, j)) = ((y, l), (j', i')),$$

where $y \in X$, $l \in [d]$ and $i', j' \in [k]$ are determined by means of the following steps:

(i) $(h', i') = \mathrm{Rot}_{\mathcal{F}}(h, i)$;

(ii) $(y, l') = \mathrm{Rot}_{\mathcal{G}}(x, h')$;

(iii) $(l, j') = \mathrm{Rot}_{\mathcal{F}}(l', j)$.

Remark 8.13.2 Here is a more detailed description of these steps. We replace each vertex x of $\mathcal{G}$ with the vertices $(x, 1), (x, 2), \ldots, (x, d)$. Then the vertices $(x, h), (y, l) \in X \times [d]$ are adjacent in the zig-zag product $\mathcal{G}\textcircled{z}\mathcal{F}$ if it is possible to connect them in the replacement product $\mathcal{G}\textcircled{r}\mathcal{F}$ with a path of length three and of the following form.

(i) First of all, we choose an edge of the second type in $\mathcal{G}\textcircled{r}\mathcal{F}$ incident to (x, h), that is, we choose a label $i \in [k]$ so that the vertex (x, h') is determined by the rotation map: $\mathrm{Rot}_{\mathcal{F}}(h, i) = (h', i')$; this also yields the label $i' \in [k]$. We refer to this as to a *zig* move.

(ii) It is then determined the unique edge of the first type in $\mathcal{G}\textcircled{r}\mathcal{F}$ incident to (x, h'), that is, the vertex $(y, l') = \mathrm{Rot}_{\mathcal{G}}(x, h')$. We refer to this as to the *jump* move.

(iii) Finally, we choose an edge of the second type in $\mathcal{G}\textcircled{r}\mathcal{F}$ incident to (y, l'), that is, we choose a label $j \in [k]$ so that the vertex (y, l) is determined by the rotation map: $\mathrm{Rot}_{\mathcal{F}}(l', j) = (l, j')$, which also yields the label $j' \in [k]$. We refer to this as to a *zag* move.

Proposition 8.13.3 *Using the notation in Proposition 8.12.9, the adjacency matrix of the zig-zag product is:*

$$M_{\mathcal{G}\textcircled{z}\mathcal{F}} = (I_X \otimes B)R_{\mathcal{G}}(I_X \otimes B). \tag{8.48}$$

Moreover, there exists a $\left[(k + 1)^3 - k^2\right]$-regular graph $\mathcal{H}$ such that

$$M_{\mathcal{G}\textcircled{r}\mathcal{F}}^3 = M_{\mathcal{G}\textcircled{z}\mathcal{F}} + H,$$

where H is the adjacency matrix of $\mathcal{H}$.

Proof. Clearly, in (8.48) the two factors $(I_X \otimes B)$ take into account the zig and zag moves, while $R_{\mathcal{G}}$ is the jump move. Now consider the following graph $\mathcal{C}$. Its vertex set is again $X \times [d]$ and two vertices are adjacent in $\mathcal{C}$ if there is a path in $\mathcal{G}\textcircled{r}\mathcal{F}$ of length three connecting them. By Proposition 8.1.6, the adjacency matrix of $\mathcal{C}$ is $M_{\mathcal{G}\textcircled{r}\mathcal{F}}^3$. Moreover, $\mathcal{C}$ is regular of degree $(k + 1)^3$, possibly with multiple edges and loops. Finally, we conclude by noting that $\mathcal{G}\textcircled{z}\mathcal{F}$ is a subgraph of $\mathcal{C}$ so that, denoting by $\mathcal{H} = (X \times [d], E(\mathcal{H}))$ the subgraph of $\mathcal{C}$ with edge set $E(\mathcal{H}) = E(\mathcal{C}) \setminus E(\mathcal{G}\textcircled{z}\mathcal{F})$, we have, cf. Proposition 8.12.9,

$$H = \left[R_{\mathcal{G}} + (I_X \otimes B)\right]^3 - (I_X \otimes B)R_{\mathcal{G}}(I_X \otimes B). \qquad \square$$

Exercise 8.13.4 Using the first result in Exercise 8.12.8, prove that the zig-zag product of the complete graph K_5 on five vertices (with the given labeling) and the 4-circle C_4, is as in Figure 8.14.

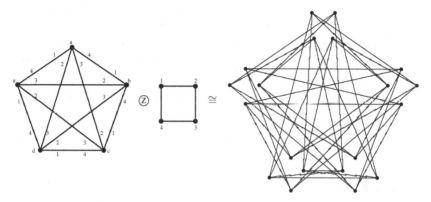

Figure 8.14. The zig-zag product $K_5 \textcircled{z} C_4$.

Remark 8.13.5 Proposition 8.13.3 and Exercise 8.13.4 show that it is not necessary to introduce a labelling in $\mathcal{F}$ in order to construct the zig-zag product. But the labelling of $\mathcal{F}$ is necessary to get a $[k] \times [k]$-labelling on the zig-zag graph.

Exercise 8.13.6 Assume the notation in Proposition 8.12.10. Define the *walk-switch-walk* lamplighter as follows: $(\omega, i), (\theta, m) \in Q_n \times [n]$ are adjacent if there exists $j \in [n]$ such that $i \sim j$, $j \sim m$, $\omega(h) = \theta(h)$ for $h \neq j$ and $\omega(j) \neq \theta(j)$. Show that this graph is isomorphic to the zig-zag product $Q_n \textcircled{z} \mathcal{F}$.

8.14 Cayley graphs, semidirect products, replacement products, and zig-zag products

In this section we introduce the concepts of a Cayley graph of a (finite) group (with respect to a given generating subset) and of a semidirect product of two (finite) groups. Then, by means of several exercises, we illustrate the connections between the Cayley graph of a semidirect product of two groups and a modified version of the replacement and of zig-zag products of the Cayely graphs of these groups (with respect to suitable generating subsets). They are based on the exposition in [58]. The original sources are [8] and [74].

Let G be a finite group. A subset $S \subseteq G$ is termed *generating* if every element $g \in G$ may be written as a product $g = s_1 s_2 \cdots s_m$ with $s_1, s_2, \ldots, s_m \in S \cup S^{-1}$ for some $m \geq 0$, where $S^{-1} = \{s^{-1} : s \in S\}$. A subset $S \subseteq G$ is said to be *symmetric* provided $S = S^{-1}$.

Let $S \subset G$ be a symmetric generating subset. Then the associated *Cayley graph* $\Gamma(G, S)$ is the graph with vertex set G and edge set $\{\{g, gs\} : s \in S, g \in G\}$. In other words, two vertices $g, g' \in G$ are adjacent if and only if $g^{-1}g' \in S$. Note that $\Gamma(G, S)$ is undirected since S is symmetric: $g^{-1}g' \in S$ if and only if $(g')^{-1}g = (g^{-1}g')^{-1} \in S$. Moreover, $\Gamma(G, S)$ has no multiple edges: if $gs = gs'$ for some $g \in G$ and $s, s' \in S$, then the cancellation property implies that $s = s'$. Moreover, $\Gamma(G, S)$ has loops if and only if S contains the identity element (and, if this is the case, then there is exactly one loop based at each vertex of $\Gamma(G, S)$). Finally, note that we may use the elements of S to get a labelling of $\Gamma(G, S)$: the rotation map (8.43) is then defined by setting

$$\mathrm{Rot}_{\Gamma(G,S)}(g, s) = (gs, s^{-1})$$

for all $g \in G$ and $s \in S$.

Exercise 8.14.1

(1) Show that the discrete circle C_n (cf. Definition 8.6.1) is the Cayley graph of the cyclic group $\mathbb{Z}_n$ with respect to the (symmetric) generating set $S = \{1, n - 1\}$.

(2) Show that the hypercube Q_n (cf. Definition 8.5.1) is the Cayley graph of the group $\mathbb{Z}_2^n$ with respect to the (symmetric) generating set $S = \{(1, 0, \ldots, 0), (0, 1, 0, \ldots, 0), \ldots, (0, 0, \ldots, 0, 1)\}$.

We now recall the well known construction of a semidirect product of two (finite) groups (see, for instance, [12, pp. 20–24], [148, pp. 6–8]).

Definition 8.14.2 (Semidirect product) Let G be a finite group and $N, H \le G$ two subgroups of G. Then G is the (internal) *semidirect product* of N by H and we write $G = N \rtimes H$, when the following conditions are satisfied:

(a) $N \trianglelefteq G$;
(b) $G = NH$;
(c) $N \cap H = \{1_G\}$.

Proposition 8.14.3 *Suppose that G is a semidirect product of N by H. Then*

(i) $G/N \cong H$;
(ii) *every $g \in G$ has a unique expression $g = nh$ with $n \in N$ and $h \in H$;*
(iii) *for any $h \in H$ and $n \in N$ set $\phi_h(n) = hnh^{-1}$. Then $\phi_h \in \mathrm{Aut}(N)$ for all $h \in H$ and the map*

$$H \longrightarrow \mathrm{Aut}(N)$$
$$h \longmapsto \phi_h$$

is a homomorphism (conjugation homomorphism);

(iv) *if nh, $n_1h_1 \in G$ are as in* (ii), *then their product is given by*

$$n_1 h_1 \cdot n_2 h_2 = [n_1 \cdot h_1 n_2 h_1^{-1}] h_1 h_2 = [n_1 \phi_{h_1}(n_2)] h_1 h_2. \quad (8.49)$$

Conversely, suppose that H and N are two (finite) groups and we are given a homomorphism

$$H \longrightarrow \mathrm{Aut}(N)$$
$$h \longmapsto \phi_h.$$

Set $G = \{(n, h) : n \in N, h \in H\}$ and define a product in G by setting

$$(n, h)(n_1, h_1) = (n\phi_h(n_1), hh_1)$$

for all $n, n_1 \in N$ and $h, h_1 \in H$ (compare with (8.49)). *Then G is a group and it is isomorphic to the (inner) semidirect product of $\widetilde{N} = \{(n, 1_H) : n \in N\} \cong N$ by $\widetilde{H} = \{(1_N, h) : H \in H\} \cong H$. The group G is called the external semidirect product of N by H with respect to ϕ and it is usually denoted by $N \rtimes_\phi H$. Moreover, with the above notation, the following conditions are equivalent:*

(a) *G is isomorphic to the direct product $\widetilde{N} \times \widetilde{H}$;*
(b) *$\widetilde{H}$ is normal in G;*
(c) *ϕ_h is the trivial automorphism of N for all $h \in H$.*

Proof. The proof is just an easy exercise and it is left to the reader. $\square$

Clearly, the internal and external semidirect products are equivalent constructions and we shall make no distinction between them.

Suppose now that $G = N \rtimes H$ is a semidirect product. For $n \in H$ we denote by n^H its orbit under the action of H, that is $n^H = \{hnh^{-1} : h \in H\}$. Let S_H (respectively, S_N) be a symmetric generating subset for H (respectively, N) and suppose that $n^H \in S_N$ for all $n \in S_N$ (in other words, S_N is H-invariant). Let then $x_1, x_2, \ldots, x_k \in S_N$ form a set of representative elements for the orbits of S_N under the action of H, that is,

$$S_N = x_1^H \coprod x_2^H \coprod \cdots \coprod x_k^H,$$

and set $S_N' = \left\{ x_1^{\pm 1}, x_2^{\pm 1}, \ldots, x_k^{\pm 1} \right\}$. In the following exercises we ask the reader to investigate the connections between the construction in Sections 8.12 and 8.13 and the semidirect product of groups.

Exercise 8.14.4

(1) Show that

$$S = S_H \cup S_N'$$

is a symmetric generating subset for G.

(2) Prove that the Cayley graph $\Gamma(G, S)$ is the *modified replacement product*

$$\Gamma(N, S'_N) \textcircled{r} \Gamma(H, S_H)$$

defined as follows. The vertex set is $G \equiv NH$. Each $g = nh \in G$ is incident to $|S_H|$ edges of the second type, which connect it with the vertices $\{nhs : s \in S_H\}$; this is as in Remark 8.12.6. Moreover, nh is also incident to $2k$ edges of the first type, which connect it with the vertices

$$\left\{ nhx_j^{\pm 1} \equiv (n \cdot hx_j^{\pm 1}h^{-1})h : j = 1, 2, \ldots, k \right\}.$$

(3) Show that the set

$$\widetilde{S} = \left\{ sx_j^{\pm 1}t : s, t \in S_H, j = 1, 2, \ldots, k \right\}$$

is another symmetric generating subset for G.

(4) Prove that the Cayley graph $\Gamma(G, \widetilde{S})$ is the *modified zig-zag product*

$$\Gamma(N, S'_N) \textcircled{z} \Gamma(H, S_H)$$

that may be defined as in Remark 8.13.2 but using the modified replacement product in (2).

Remark 8.14.5 If $k = 1$ and $x_1 = x_1^{-1}$, then the modified replacement product in Exercise 8.14.4.(2) coincides with an ordinary replacement product. The same holds for the modified zig-zag product in Exercise 8.14.4.(4). In general, a modified product may be seen as a "union" of ordinary products.

9

Expanders and Ramanujan graphs

This chapter is an introduction to the theory of expanders and Ramanujan graphs. It is based mainly on the exposition in the monograph by Davidoff-Sarnak-Valette [49] and the paper [74]. First of all, we present the basic theorems of Alon-Milman and Dodziuk, and of Alon-Boppana-Serre, on the isoperimetric constant and the spectral gap of a (finite, undirected, connected) regular graph, and their connections. We discuss a few examples with explicit computations showing optimality of the bounds given by the above theorems. Then we give the basic definitions of expanders and describe three fundamental constructions due to Margulis, Alon-Schwartz-Schapira (based on the replacement product, cf. Section 8.12), and Reingold-Vadhan-Wigderson [128] (based on the zig-zag product, cf. Section 8.13). In these constructions, the harmonic analysis on finite Abelian groups (cf. Chapter 2) and finite fields (cf. Chapter 6) we developed so far, plays a crucial role.

The original motivation for expander graphs was to build economical robust networks (e.g. for phones or computers): an expander with bounded valence is precisely an asymptotic robust graph with the number of edges growing linearly with size (number of vertices), for all subsets. Since their definition, expanders have found extensive applications in several branches of science and technology, for instance: in computer science, in designing algorithms, error correcting codes, extractors, pseudorandom generators, sorting networks (Ajtai, Komlós, and Szemerédi, [6]), robust computer networks (as in their initial motivation), and in cryptography (in order to construct hash functions: these are used in hash tables to quickly locate a data record given its search key). From a more theoretical viewpoint, they have also been used in proofs of many important results in computational complexity theory, such as SL = L (Reingold, [126]) and the PCP theorem (Dinur, [56]).

9.1 The Alon-Milman-Dodziuk theorem

In this section we present the discrete analogues, due to Dodziuk [57] and Alon-Milman [9], of the well-known Cheeger-Buser inequalities in Riemannian geometry (cf. [38] and [26, 27]).

Let $\mathcal{G} = (X, E, r)$ be a finite (undirected) k-regular graph (possibly with multiple edges and loops). Recall that $E_0 = \{e \in E : |r(e)| = 1\}$ denotes the set of all loops of $\mathcal{G}$ and $E_1 = \{e \in E : |r(e)| = 2\} = E \setminus E_0$.

Definition 9.1.1 Let $F \subseteq X$ be a set of vertices of $\mathcal{G}$. The *boundary* of F is the set

$$\partial F = \{e \in E : r(e) \cap F \neq \varnothing \text{ and } r(e) \cap (X \setminus F) \neq \varnothing\} \subseteq E_1$$

of all edges in $\mathcal{G}$ joining (vertices in) F with (vertices in) its complement $X \setminus F$.

The *isoperimetric constant* (also called the *Cheeger constant*) of $\mathcal{G}$ is the non-negative number

$$h(\mathcal{G}) = \min\left\{ \frac{|\partial F|}{|F|} : F \subseteq X, 0 < |F| \leq \frac{|X|}{2} \right\}.$$

Note that one has

$$|\partial F| = \sum_{\substack{x \in F \\ y \in X \setminus F}} A(x, y) = \sum_{\{x,y\} \in r(\partial F)} A(x, y). \tag{9.1}$$

Moreover, $h(\mathcal{G})$ is strictly positive if and only if $\mathcal{G}$ is connected, and

$$h(\mathcal{G}) \leq k. \tag{9.2}$$

Indeed, if $\mathcal{G}$ is connected, then ∂F is nonempty for all $\varnothing \neq F \subsetneq X$, thus showing that $h(\mathcal{G}) > 0$. If $\mathcal{G}$ is not connected, then there exists a connected component whose vertex set F satisfies $0 < |F| \leq \frac{|X|}{2}$ and, clearly, $\partial F = \varnothing$, showing, in this case, that $h(\mathcal{G}) = 0$. Moreover, if $\varnothing \neq F \subseteq X$, since $\mathcal{G}$ is k-regular, the total number of edges incident to some vertices in F is at most $|F|k$, so that $|\partial F| \leq |F|k$, and (9.2) follows.

Finally, note that some papers (for instance [10]) use the *normalized isoperimetric constant* (or *edge expansion constant*) that is defined as $h'(\mathcal{G}) = \frac{h(\mathcal{G})}{k}$, and, by virtue of (9.2), satisfies $h'(\mathcal{G}) \leq 1$.

Let A be the adjacency operator of $\mathcal{G}$ and set $\Delta = kI - A \in \text{End}(L(X))$, where, as usual, I denotes the identity map. Then, for $f \in L(X)$ and $x \in X$ we have that

$$[\Delta f](x) = kf(x) - \sum_{y \in X} A(x, y)f(y) = kf(x) - \sum_{y \sim x} A(x, y)f(y).$$

Moreover, keeping in mind Proposition 8.1.5 and the notation therein, we have that the eigenvalues of Δ are:

$$\lambda_0 = 0 \leq \lambda_1 = k - \mu_1 \leq \cdots \leq \lambda_{|X|-1} = k - \mu_{|X|-1}. \tag{9.3}$$

In the sequel we shall often use the following summation argument.

Remark 9.1.2 In our setting, for $a \in L(X \times X)$ symmetric (i.e. such that $a(x, y) = a(y, x)$ for all $x, y \in (X)$ and $b \in L(X)$, we have

$$\sum_{\{x,y\} \in r(E_1)} a(x, y) = \frac{1}{2} \sum_{x \in X} \sum_{\substack{y \in X: \\ y \sim x \\ y \neq x}} a(x, y) = \frac{1}{2} \sum_{y \in X} \sum_{\substack{x \in X: \\ x \sim y \\ x \neq y}} a(x, y) \tag{9.4}$$

and, by the regularity of $\mathcal{G}$ (namely, $\deg x = k$ for all $x \in X$),

$$\sum_{\{x,y\} \in r(E_1)} A(x, y) \, (b(x) + b(y)) = \sum_{x \in X} (k - A(x, x)) b(x). \tag{9.5}$$

In particular, taking $b = \mathbf{1}_X$ we get $2|E_1| = k|X| - |E_0|$, that is,

$$2|E_1| + |E_0| = k|X|. \tag{9.6}$$

Lemma 9.1.3 *Let $f \in L(X)$ be real valued. Then*

$$\langle \Delta f, f \rangle = \sum_{\{x,y\} \in r(E_1)} A(x, y) \, (f(x) - f(y))^2. \tag{9.7}$$

Proof. We have

$$\sum_{\{x,y\}\in r(E_1)} A(x,y)\,(f(x)-f(y))^2$$

$$= \sum_{\{x,y\}\in r(E_1)} A(x,y)\,\left(f(x)^2+f(y)^2\right)$$

$$-2\sum_{\{x,y\}\in r(E_1)} A(x,y)f(x)f(y)$$

$$=_* \sum_{x\in X}(k-A(x,x))f(x)^2 - \sum_{x\in X}\sum_{\substack{y\in X:\\y\sim x\\y\neq x}} A(x,y)f(x)f(y)$$

$$= k\sum_{x\in X} f(x)^2 - \sum_{x\in X}\sum_{\substack{y\in X:\\y\sim x}} A(x,y)f(x)f(y)$$

$$= k\sum_{x\in X} f(x)^2 - \sum_{x\in X}\sum_{y\in X} A(x,y)f(x)f(y)$$

$$= k\sum_{x\in X} f(x)^2 - \sum_{x\in X}[Af](x)f(x)$$

$$= k\langle f,f\rangle - \langle Af,f\rangle$$

$$= \langle \Delta f,f\rangle,$$

where $=_*$ follows from (9.5) with $b(x)=f(x)^2$, and from (9.4) with $a(x,y)=A(x,y)f(x)f(y)$. $\qquad\square$

Definition 9.1.4 The operator $\Delta \in \mathrm{End}(L(X))$ is called the *combinatorial Laplacian* and the right hand side of (9.7) the *Dirichlet form* on $\mathcal{G}$.

The terminology in the above definition is based on the classical mean-value property of harmonic functions on $\mathbb{R}^n$ (which constitute the kernel of the Euclidean Laplace operator $\Delta = \frac{\partial^2}{\partial x_1^2} + \frac{\partial^2}{\partial x_2^2} + \cdots + \frac{\partial^2}{\partial x_n^2}$).

Remark 9.1.5 Suppose that $\mathcal{G} = (X,E)$ is a finite simple graph without loops. Recall that we may identify the edge set E with the set of two-elements sets $\{x,y\} \subset X$ such that $x \sim y$. In this setting, the boundary of a subset $F \subset X$ is given by the set of edges

$$\partial F = \{\{x,y\} \in E : x \in F \text{ and } y \notin F\} \subseteq E.$$

Moreover, if $\mathcal{G}$ is k-regular, the combinatorial Laplacian and its associated Dirichlet form (9.7) can be expressed as

$$[\Delta f](x) = kf(x) - \sum_{y\sim x} f(y)$$

and

$$\langle \Delta f, f \rangle = \sum_{\{x,y\} \in E} (f(x) - f(y))^2 \, ,$$

respectively, for all $f \in L(X)$ and $x \in X$.

We recall (cf. Proposition 8.1.4) that if W_0 is the space of constant functions on X and $W_1 = \{f \in L(X) : \sum_{x \in X} f(x) = 0\}$, then $L(X) = W_0 \oplus W_1$.

Lemma 9.1.6 *Suppose that $\mathcal{G}$ is connected. Then we have*

$$\lambda_1 = k - \mu_1 = \min \left\{ \frac{\langle \Delta f, f \rangle}{\langle f, f \rangle} : f \in W_1, f \neq 0 \right\} \tag{9.8}$$

and

$$\mu_1 = k - \lambda_1 = \max \left\{ \frac{\langle A f, f \rangle}{\langle f, f \rangle} : f \in W_1, f \neq 0 \right\}. \tag{9.9}$$

Proof. Since $\mathcal{G}$ is connected, the multiplicity of the eigenvalue $\lambda_0 = 0$ of Δ is one: the corresponding eigenspace is W_0 (cf. Proposition 8.1.5). Therefore, the other eigenvalues of Δ, namely $\lambda_1 \leq \cdots \leq \lambda_{n-1}$ ($n = |X|$), are all positive with corresponding eigenfunctions $\phi_1, \ldots, \phi_{n-1}$ that can be chosen to be real valued and to constitute an orthonormal basis of W_1. Then, for every $f = \alpha_1 \phi_1 + \cdots + \alpha_{n-1} \phi_{n-1} \in W_1 \setminus \{0\}$ ($\alpha_1, \ldots, \alpha_{n-1} \in \mathbb{C}$) we have

$$
\begin{aligned}
\langle \Delta f, f \rangle &= \langle \Delta(\alpha_1 \phi_1 + \cdots + \alpha_{n-1} \phi_{n-1}), \alpha_1 \phi_1 + \cdots + \alpha_{n-1} \phi_{n-1} \rangle \\
&= \langle \lambda_1 \alpha_1 \phi_1 + \cdots + \lambda_{n-1} \alpha_{n-1} \phi_{n-1}, \alpha_1 \phi_1 + \cdots + \alpha_{n-1} \phi_{n-1} \rangle \\
&= \lambda_1 |\alpha_1|^2 + \cdots + \lambda_{n-1} |\alpha_{n-1}|^2 \\
\text{(by (9.3))} \quad &\geq \lambda_1 |\alpha_1|^2 + \cdots + \lambda_1 |\alpha_{n-1}|^2 \\
&= \lambda_1 (|\alpha_1|^2 + \cdots + |\alpha_{n-1}|^2) \\
&= \lambda_1 \langle f, f \rangle,
\end{aligned}
$$

showing that $\lambda_1 \leq \frac{\langle \Delta f, f \rangle}{\langle f, f \rangle}$. Since $\lambda_1 = \frac{\langle \Delta \phi_1, \phi_1 \rangle}{\langle \phi_1, \phi_1 \rangle}$, (9.8) follows. The proof of (9.9) is analogous and is left to the reader. $\qquad \square$

Theorem 9.1.7 (Alon-Milman) *Let $\mathcal{G} = (X, E, r)$ be a finite connected k-regular graph. Then*

$$\frac{k - \mu_1}{2} \leq h(\mathcal{G}).$$

Proof. We apply Lemma 9.1.6 to a suitable function in W_1. For $F \subseteq X$ such that $0 < |F| \leq \frac{|X|}{2}$, we define $f_F \in L(X)$ by setting

$$f_F(x) = \begin{cases} |X \setminus F| & \text{if } x \in F \\ -|F| & \text{if } x \in X \setminus F. \end{cases}$$

Then $\sum_{x \in X} f_F(x) = |X \setminus F| \cdot |F| - |F| \cdot |X \setminus F| = 0$, so that $f_F \in W_1$, and

$$\langle f_F, f_F \rangle = \sum_{x \in X} f_F(x)^2 = |X \setminus F|^2 \cdot |F| + |F|^2 \cdot |X \setminus F|$$

$$= |X \setminus F| \cdot |F| \cdot (|X \setminus F| + |F|) = |X \setminus F| \cdot |F| \cdot |X|.$$

Moreover,

$$f_F(x) - f_F(y) = \begin{cases} \pm |X| & \text{if } \{x, y\} \in r(\partial F) \\ 0 & \text{otherwise.} \end{cases}$$

Therefore, by virtue of Lemma 9.1.3 we have

$$\langle \Delta f_F, f_F \rangle = \sum_{\{x,y\} \in r(E_1)} A(x, y) \, (f_F(x) - f_F(y))^2$$

$$= |X|^2 \sum_{\{x,y\} \in r(\partial F)} A(x, y)$$

$$\text{(by (9.1))} \quad = |X|^2 \cdot |\partial F|.$$

Thus, from Lemma 9.1.6 we deduce that

$$\frac{|X|}{|X \setminus F|} \cdot \frac{|\partial F|}{|F|} = \frac{|X|^2 \cdot |\partial F|}{|X \setminus F| \cdot |F| \cdot |X|} = \frac{\langle \Delta f_F, f_F \rangle}{\langle f_F, f_F \rangle} \geq \lambda_1 = k - \mu_1.$$

Since $|F| \leq \frac{|X|}{2}$, we have $\frac{|X \setminus F|}{|X|} \geq \frac{1}{2}$ and therefore

$$\frac{|\partial F|}{|F|} \geq (k - \mu_1) \frac{|X \setminus F|}{|X|} \geq \frac{k - \mu_1}{2}. \tag{9.10}$$

As the isoperimetric constant $h(\mathcal{G})$ is, by definition, the minimum of the left hand side values (with $0 < |F| \leq \frac{|X|}{2}$) of (9.10), the statement follows. $\quad\square$

In the following theorem we give an upper bound for the isoperimetric constant.

Theorem 9.1.8 (Dodziuk) *Let $\mathcal{G} = (X, E, r)$ be a finite connected k-regular graph. Then*

$$h(\mathcal{G}) \leq \sqrt{2k(k - \mu_1)}.$$

Proof. Let $f \in L(X)$ be a non-negative function and denote by $\alpha_r > \alpha_{r-1} > \cdots > \alpha_1 > \alpha_0 \geq 0$ its values. Consider the map $j : X \to \{0, 1, \ldots, r\}$ defined by

$$f(x) = \alpha_{j(x)}$$

for all $x \in X$ (such a map j is clearly well defined). We also define the *level sets*

$$X_i = \{x \in X : f(x) \geq \alpha_i\} \equiv \{x \in X : j(x) \geq i\}$$

for $i = 0, 1, \ldots, r$. Clearly, $X_0 = X \supset X_1 \supset \cdots \supset X_r \neq \varnothing$. Finally, set

$$B_f = \sum_{\{x,y\} \in r(E)} A(x, y)|f(x)^2 - f(y)^2| = \sum_{\{x,y\} \in r(E_1)} A(x, y)|f(x)^2 - f(y)^2|.$$

Claim 1.

$$B_f = \sum_{h=1}^{r} |\partial X_h|(\alpha_h^2 - \alpha_{h-1}^2).$$

Proof of Claim 1. Given any $\{x, y\} \in r(E_1)$ we may suppose, up to exchanging x and y, that $f(x) \geq f(y)$, equivalently, $j(x) \geq j(y)$. This way, we have

$$r(\partial X_h) = \{\{x, y\} : j(y) < h \leq j(x)\} \tag{9.11}$$

for all $h = 1, 2, \ldots, r$. Moreover,

$$B_f = \sum_{\substack{\{x,y\} \in r(E_1): \\ j(x) > j(y)}} A(x, y) \left(\alpha_{j(x)}^2 - \alpha_{j(y)}^2\right) = \sum_{\substack{\{x,y\} \in r(E_1): \\ j(x) > j(y)}} A(x, y) \sum_{h=j(y)+1}^{j(x)} (\alpha_h^2 - \alpha_{h-1}^2).$$

In the last expression, each "telescopic" summand $(\alpha_h^2 - \alpha_{h-1}^2)$ appears exactly $A(x, y)$ times for every $\{x, y\} \in r(E_1)$ such that $j(x) \geq h > j(y)$, equivalently (cf. (9.11)), exactly $A(x, y)$ times for every $\{x, y\} \in r(\partial X_h)$. In other words, each "telescopic" summand appears exactly $|\partial X_h|$ times (cf. (9.1)). The claim follows. $\square$

Claim 2.

$$B_f \leq \sqrt{2k} \, \|f\| \, \langle \Delta f, f \rangle^{\frac{1}{2}}.$$

Proof of Claim 2. From the inequality $2ab \leq a^2 + b^2$, for all $a, b \in \mathbb{R}$, we deduce that

$$(f(x) + f(y))^2 = f(x)^2 + f(y)^2 + 2f(x)f(y) \leq 2[f(x)^2 + f(y)^2] \tag{9.12}$$

for all $x, y \in X$. Now,

$$B_f = \sum_{\{x,y\}\in r(E_1)} \sqrt{A(x,y)}|f(x) + f(y)| \cdot \sqrt{A(x,y)}|f(x) - f(y)|$$

$$\leq_{(*)} \left\{ \sum_{\{x,y\}\in r(E_1)} A(x,y)[f(x) + f(y)]^2 \right\}^{\frac{1}{2}} \left\{ \sum_{\{x,y\}\in r(E_1)} A(x,y)[f(x) - f(y)]^2 \right\}^{\frac{1}{2}}$$

$$\leq_{(**)} \sqrt{2} \left\{ \sum_{\{x,y\}\in r(E_1)} A(x,y)[f(x)^2 + f(y)^2] \right\}^{\frac{1}{2}} \langle \Delta f, f \rangle^{\frac{1}{2}}$$

$$=_{(***)} \sqrt{2} \left\{ \sum_{x\in X} (k - A(x,x))f(x)^2 \right\}^{\frac{1}{2}} \langle \Delta f, f \rangle^{\frac{1}{2}}$$

$$\leq \sqrt{2k} \left\{ \sum_{x\in X} f(x)^2 \right\}^{\frac{1}{2}} \langle \Delta f, f \rangle^{\frac{1}{2}},$$

where $\leq_{(*)}$ follows from the Cauchy-Schwarz inequality, $\leq_{(**)}$ follows from (9.12) and Lemma 9.1.3, and $=_{(***)}$ follows from (9.5). $\qquad\square$

We recall that the support of $f \in L(X)$ is the set

$$\mathrm{supp}(f) = \{x \in X : f(x) \neq 0\}.$$

Claim 3. *Suppose that*

$$|\mathrm{supp}(f)| \leq \frac{|X|}{2}.$$

Then

$$B_f \geq h(\mathcal{G})\|f\|^2.$$

Proof of Claim 3. By our hypothesis on f, we have $\alpha_0 = 0$, so that $X_1 = \mathrm{supp}(f)$, and $0 < |X_h| \leq \frac{|X|}{2}$ for every $h = 1, 2, \ldots, r$. Keeping in mind the definition of the isoperimetric constant, this implies that

$$|\partial X_h| \geq h(\mathcal{G})|X_h| \tag{9.13}$$

for every $h = 1, 2 \ldots, r$. From Claim 1 we deduce that

$$B_f = \sum_{h=1}^{r} |\partial X_h|(\alpha_h^2 - \alpha_{h-1}^2)$$

$$\text{(by (9.13))} \quad \geq h(\mathcal{G}) \sum_{h=1}^{r} |X_h|(\alpha_h^2 - \alpha_{h-1}^2)$$

$$= h(\mathcal{G}) \big[|X_r|(\alpha_r^2 - \alpha_{r-1}^2) + |X_{r-1}|(\alpha_{r-1}^2 - \alpha_{r-2}^2) +$$

$$+ \cdots + |X_2|(\alpha_2^2 - \alpha_1^2) + |X_1|\alpha_1^2 \big]$$

$$= h(\mathcal{G}) \big[|X_r|\alpha_r^2 + |X_{r-1} \setminus X_r|\alpha_{r-1}^2 +$$

$$+ |X_{r-2} \setminus X_{r-1}|\alpha_{r-2}^2 + \cdots + |X_1 \setminus X_2|\alpha_1^2 \big]$$

$$= h(\mathcal{G})\|f\|^2,$$

where the last equality follows from the fact that $X_{h-1} \setminus X_h$ is the set on which f takes the value α_{h-1}. □

Claim 4. *Let $1 \leq i \leq n - 1$. Denote by $\phi_i \in L(X)$ a real eigenfunction associated with the eigenvalue $\lambda_i = k - \mu_i$ and define $f_i \in L(X)$ by setting*

$$f_i(x) = \max\{\phi_i(x), 0\} = \frac{\phi_i(x) + |\phi_i(x)|}{2}$$

for all $x \in X$. Then

$$[\Delta f_i](x) \leq \lambda_i \phi_i(x)$$

for all $x \in X$ such that $\phi_i(x) > 0$. Moreover, we have

$$\langle \Delta f_i, f_i \rangle \leq \lambda_i \|f_i\|^2.$$

Proof of Claim 4. Let $x \in X$ such that $\phi_i(x) > 0$. Then we have $f_i(x) = \phi_i(x)$ and therefore

$$[\Delta f_i](x) = k f_i(x) - \sum_{y \in X} A(x, y) f_i(y)$$

$$= k\phi_i(x) - \sum_{\substack{y \in X: \\ \phi_i(y) > 0}} A(x, y)\phi_i(y)$$

$$\leq k\phi_i(x) - \sum_{y \in X} A(x, y)\phi_i(y)$$

$$= [\Delta \phi_i](x)$$

$$= \lambda_i \phi_i(x),$$

proving the first part of the claim. On the other hand,

$$\langle \Delta f_i, f_i \rangle = \sum_{x \in X} [\Delta f_i](x) f_i(x) = \sum_{\substack{x \in X: \\ \phi_i(x) > 0}} [\Delta f_i](x) \phi_i(x)$$

$$\leq \lambda_i \sum_{\substack{x \in X: \\ \phi_i(x) > 0}} \phi_i(x)^2 = \lambda_i \|f_i\|^2,$$

where the inequality follows from the first part of the claim. □

We are now in a position to complete the proof of Dodziuk's Theorem.

Let ϕ_1 be a real eigenfunction associated with the eigenvalue $\lambda_1 = k - \mu_1$. Switching ϕ_1 with $-\phi_1$, if necessary, we may suppose that the subset $X_+ = \{x \in X : \phi_1(x) > 0\}$ satisfies the condition $0 < |X_+| \leq \frac{|X|}{2}$ (observe that since $\phi_1 \in W_1$ and $\phi_1 \not\equiv 0$, the set $\{x \in X : \phi_1(x) > 0\}$ is nonempty). Taking into account, in order, Claim 3, Claim 2, and Claim 4 (and the notation therein), we deduce

$$h(\mathcal{G})\|f_1\|^2 \leq B_{f_1} \leq \sqrt{2k}\langle \Delta f_1, f_1 \rangle^{\frac{1}{2}} \|f_1\| \leq \sqrt{2k(k - \mu_1)}\|f_1\|^2,$$

and the statement follows after dividing by $\|f_1\|^2$. □

Definition 9.1.9 Let $\mathcal{G} = (X, E, r)$ be a finite connected k-regular graph. Denote by $k = \mu_0 > \mu_1 \geq \cdots \geq \mu_n$ the eigenvalues of the adjacency matrix of $\mathcal{G}$. The *spectral gap* of $\mathcal{G}$ is the positive number

$$\delta(\mathcal{G}) = \mu_0 - \mu_1 = k - \mu_1.$$

Remark 9.1.10 The theorem of Alon-Milman ensures that, in order to have a "large" isoperimetric constant $h(\mathcal{G})$, it suffices to have a "large" spectral gap $\delta(\mathcal{G})$. Conversely, the theorem of Dodziuk ensures that this is also a necessary condition. More specifically:

$$\delta(\mathcal{G}) \geq \delta \Rightarrow h(\mathcal{G}) \geq \frac{\delta}{2} \text{ (Alon-Milman)}$$

$$h(\mathcal{G}) \geq \varepsilon \Rightarrow \delta(\mathcal{G}) \geq \frac{\varepsilon^2}{2k} \text{ (Dodziuk)}.$$

In the remainder of this section we compare the exact values of the isoperimetric constant with the estimates provided by the theorems of Alon-Milman and Dodziuk for some graphs presented in Chapter 8.

Example 9.1.11 (The complete graph) Let K_n be the complete graph on $n \geq 1$ vertices (cf. Section 8.4). Recall that the graph K_n is regular of degree $k = n - 1$ and the eigenvalues of the associated adjacency matrix are $\mu_0 = n - 1$ (with multiplicity one) and $\mu_1 = -1$ (with multiplicity $n - 1$). As a consequence, by virtue of Theorem 9.1.7 and Theorem 9.1.8, the isoperimetric constant $h(K_n)$ satisfies

$$\frac{n}{2} = \frac{k - \mu_1}{2} \leq h(K_n) \leq \sqrt{2k(k - \mu_1)} = \sqrt{2(n - 1)n} \leq \sqrt{2}n.$$

Moreover, if $F_h = \{1, 2, \ldots, h\}$, $h = 1, 2, \ldots, n$, we have $|\partial F_h| = h(n - h)$ so that $\dfrac{|\partial F_h|}{|F_h|} = n - h$. It follows that

$$h(K_n) = \min_{1 \leq h \leq n/2} \frac{|\partial F_h|}{|F_h|} = \frac{|\partial F_{[n/2]}|}{|F_{[n/2]}|} = n - [n/2],$$

where, as usual, $[\cdot]$ denotes the integer part (floor function). It follows that $h(K_n) \approx n/2$ showing that the Alon-Milman inequality is asymptotically optimal; in fact, for n even we have $h(K_n) = n/2$ and, in this case, the Alon-Milman inequality is indeed an equality.

Example 9.1.12 (The hypercube) Let $Q_n = (X_n, E_n)$ be the n-dimensional hypercube, $n \geq 1$ (cf. Section 8.5). Recall that $X_n = \{0, 1\}^n$, the graph Q_n is regular of degree $k = n$, and that the second eigenvalue of the associated adjacency matrix is $\mu_1 = n - 2$. As a consequence, by virtue of Theorem 9.1.7 and Theorem 9.1.8, the isoperimetric constant $h(Q_n)$ satisfies

$$1 = \frac{k - \mu_1}{2} \leq h(Q_n) \leq \sqrt{2k(k - \mu_1)} = \sqrt{4n} = 2\sqrt{n}. \tag{9.14}$$

Moreover, if $F' = \{x \in X_n : x_1 = 0\}$ is the hyperplane $x_1 = 0$, we have $|F'| = |X_n|/2 = 2^{n-1}$ and, for every $x \in F'$, there exists exactly one edge in $\partial F'$ issuing from the vertex x, namely $\{x, x'\}$, where $x_1' = 1$ and $x_i' = x_i$ for $i = 2, 3, \ldots, n$. It follows that $|\partial F'| = |F'|$ and therefore from the Left Hand Side estimate in (9.14) we deduce

$$1 \leq h(Q_n) = \min_{0 < |F| \leq 2^{n-1}} \frac{|\partial F|}{|F|} \leq \frac{|\partial F'|}{|F'|} = 1,$$

showing that $h(Q_n) = 1$. We remark that, as for the complete graph, the Alon-Milman inequality is indeed an equality.

Example 9.1.13 (The discrete circle) Let $\mathcal{C}_n = (X_n, E_n)$ be the discrete circle on $n \geq 3$ vertices (cf. Section 8.6). Recall that $X_n = \mathbb{Z}_n$, the graph $\mathcal{C}_n$ is regular of degree $k = 2$, and that the second eigenvalue of the associated adjacency matrix is $\mu_1 = 2\cos(2\pi/n)$. As a consequence, by virtue of Theorem 9.1.7 and Theorem 9.1.8, the isoperimetric constant $h(\mathcal{C}_n)$ satisfies

$$1 - \cos(2\pi/n) = \frac{k - \mu_1}{2} \leq h(\mathcal{C}_n) \leq \sqrt{2k(k - \mu_1)} = 2\sqrt{2(1 - \cos(2\pi/n))}.$$
(9.15)

Let $F_h = \{0, 1, \ldots, h\}$, $h = 0, 1, \ldots, [n/2] - 1$. Then $0 < |F_h| = h + 1 \leq [n/2]$ and ∂F_h consists of the two edges $\{n - 1, 0\}$ and $\{h, h + 1\}$, so that $|\partial F_h| = 2$ and $\frac{|\partial F_h|}{|F_h|} = \frac{2}{h}$. It is also clear that if $F \subseteq X_n$, $0 < |F| \leq [n/2]$ is not connected (as a subgraph of $\mathcal{C}_n$), then $|\partial F| > 2$. It follows that

$$h(\mathcal{C}_n) = \min_{0 < |F| \leq [n/2]} \frac{|\partial F|}{|F|} = \min_{0 < h \leq [n/2] - 1} \frac{2}{h + 1} = \frac{2}{[n/2]} \approx \frac{4}{n}.$$

Comparing with (9.15), since

$$1 - \cos(2\pi/n) = 2\sin^2(\pi/n) \approx \frac{2\pi^2}{n^2}$$

and

$$2\sqrt{2(1 - \cos(2\pi/n))} = 4\sin(\pi/n) \approx \frac{4\pi}{n},$$

we deduce that in this case the upper bound provided by Dodziuk (Theorem 9.1.8) is asymptotically better than the lower bound provided by Alon-Milman (Theorem 9.1.7).

Example 9.1.14 (The 2-regular segment) Let $\mathcal{G}_n = (X_n, E_n, r_n)$ be the 2-regular segment on $n \geq 2$ vertices (cf. Exercise 8.6.3). Recall that $X_n = \{0, 1, 2, \ldots, n - 1\}$ and that the second eigenvalue of the associated adjacency matrix is $\mu_1 = 2\cos(\pi/n)$ (cf. (8.10)). The isoperimetric constant $h(\mathcal{G}_n)$ then satisfies the inequalities

$$1 - \cos(\pi/n) = \frac{k - \mu_1}{2} \leq h(\mathcal{G}_n) \leq \sqrt{2k(k - \mu_1)} = 2\sqrt{2(1 - \cos(\pi/n))}.$$
(9.16)

For $0 \leq h \leq k \leq [n/2] - 1$ we set $F_{h,k} = \{h, h + 1, \ldots, k\}$. Then $|F_{h,k}| = k - h + 1 \leq [n/2]$ and

$$\partial F_{h,k} = \begin{cases} \{\{h - 1, h\}, \{k, k + 1\}\} & \text{if } h > 0 \\ \{\{k, k + 1\}\} & \text{if } h = 0. \end{cases}$$

We then have

$$\frac{|\partial F_{h,k}|}{|F_{h,k}|} \geq \frac{|\partial F_{0,k}|}{|F_{0,k}|} = \frac{1}{k + 1} \geq \frac{1}{[n/2]}$$

so that

$$h(\mathcal{G}_n) = \min_{0 < |F| \le [n/2]} \frac{|\partial F|}{|F|} = \min_{0 < k \le [n/2]-1} \frac{1}{k+1} = \frac{1}{[n/2]} \approx \frac{2}{n}.$$

Comparing with (9.16), since

$$1 - \cos(\pi/n) = 2\sin^2(\pi/2n) \approx \frac{\pi^2}{2n^2}$$

and

$$2\sqrt{2(1 - \cos(\pi/n))} = 4\sin(\pi/2n) \approx \frac{2\pi}{n},$$

we deduce that, as for the discrete circle, the upper bound provided by Dodziuk is asymptotically better than the lower bound provided by Alon-Milman.

9.2 The Alon-Boppana-Serre theorem

In this section we present the Alon-Boppana-Serre Theorem. A weaker version (cf. Corollary 9.2.7) was originally proved by Alon and Boppana [7]. The present statement (cf. Theorem 9.2.6) is due to J.P. Serre [146] who studied eigenvalues of Hecke operators and their distribution. Our proof closely follows the presentation in the monograph by Davidoff, Sarnak, and Valette [49]. For another proof, due to Alon Nilli, we refer to the next section.

Let $\mathcal{G} = (X, E, r)$ be a finite connected k-regular graph.

Definition 9.2.1 (Hecke operators) A path $p = (x_0, e_1, x_1, e_2, \ldots, e_r, x_r)$ in $\mathcal{G}$ is said to be *non-backtracking* if $e_{i+1} \ne e_i$ for all $i = 1, 2, \ldots, r-1$.

(a) For $r \ge 1$ define the $X \times X$ matrix A_r by setting

$$A_r(x, y) = |\{\text{non-backtracking paths of length } r \text{ from } x \text{ to } y\}|$$

for all $x, y \in X$.

(b) For $m \ge 1$ set

$$T_m = \sum_{0 \le r \le [m/2]} A_{m-2r}.$$

We also set $T_0 = A_0 = I$ the identity matrix.

Clearly, $A_1 = T_1$ equals the adjacency matrix A of $\mathcal{G}$. Moreover, $T_2 = A_0 + A_2$ and, more generally, for $h \ge 1$,

$$T_{2h} = A_0 + A_2 + \cdots + A_{2h} \text{ and } T_{2h+1} = A_1 + A_3 + \cdots + A_{2h+1}. \quad (9.17)$$

Proposition 9.2.2 (Hecke relations I) *The matrices A_j's satisfy the following relations:*

(i) $A_1^2 = A_2 + kI$;
(ii) $A_1 A_r = A_r A_1 = A_{r+1} + (k-1)A_{r-1}$ *for all* $r \geq 2$.

Proof. Let $x, y \in X$ and $r \in \mathbb{N}$. We first recall (cf. Proposition 8.1.6) that $A_1^r(x, y)$ equals the number of all paths of length r connecting x and y, in particular, $A_1(x, y) \neq 0$ if and only if $x \sim y$.

(i) If x and y are distinct, then a path of length 2 connecting x and y is necessarily non-backtracking. Therefore, $A_1^2(x, y) = A_2(x, y)$.

Suppose now that $x = y$. For every neighbor $z \sim x$ (possibly, $z = x$) there are exactly $A(x, z)$ edges connecting x and z. Thus, among all the $A(x, z)^2$ paths $p = (x, e_1, z, e_2, x)$ of length 2 starting at x, passing by z, and returning at x (note that $A(x, z)^2 = A(x, z)A(z, x)$), there are exactly $A(x, z)$ which are backtracking ($e_1 = e_2$) and $A(x, z)(A(x, z) - 1)$ which are non-backtracking ($e_1 \neq e_2$). Altogether we have

$$A^2(x, x) = \sum_{z \sim x} A(x, z)^2 = \sum_{z \sim x} A(x, z) + \sum_{z \sim x} A(x, z)(A(x, z) - 1)$$
$$= k + A_2(x, x),$$

showing that $A_1^2 = A^2 = A_2 + kI$.

(ii) By definition we have

$$[A_1 A_r](x, y) = \sum_{z \in X} A_1(x, z) A_r(z, y). \tag{9.18}$$

Now, $A_r(z, y)$ counts the number of non-backtracking paths of length r connecting z and y. If $(z = x_0, e_1, x_1, e_2, \ldots, x_{r-1}, e_r, x_r = y)$ is one of these paths, we have two possibilities:

(a) $x \neq x_1$: then for every $e \in E$ such that $r(e) = \{x, z\}$, we have that $(x, e, z = x_0, e_1, x_1, e_2, \ldots, x_{r-1}, e_r, x_r = y)$ is a non-backtracking path of length $r + 1$ connecting x and y, and it contributes to the count of $A_{r+1}(x, y)$;

(b) $x = x_1$: then $(x = x_1, e_2, x_2, e_3, \ldots, x_{r-1}, e_r, x_r = y)$ is a non-backtracking path of length $r - 1$ connecting x and y: it contributes to the count of $A_{r-1}(x, y)$ and it appears exactly $(k - 1)$ times in (9.18) since e_1 can be any of the $(k - 1)$-edges such that $r(e_1) \ni x$ and $e_1 \neq e_2$.

This shows the equality $A_1 A_r = A_{r+1} + (k-1)A_{r-1}$. The proof that $A_r A_1 = A_{r+1} + (k-1)A_{r-1}$ (thus yielding also $A_1 A_r = A_r A_1$) is similar and it is left to the reader.   □

Corollary 9.2.3 (Hecke relations II) *For all $m \geq 1$ we have*

$$T_{m+1} = T_m T_1 - (k-1)T_{m-1}.$$

Proof. By Proposition 9.2.2.(i) we have

$$T_1^2 = A_1^2 = A_? + kI = T_2 + (k-1)T_0$$

and the case $m = 1$ immediately follows. In order to prove the general case observe that, for $h \geq 1$,

$$T_{2h} T_1 = T_{2h} A_1$$
$$\text{(by (9.17))} = A_0 A_1 + A_2 A_1 + \cdots + A_{2h} A_1$$
$$\text{(by Proposition 9.2.2.(ii))} = A_1 + A_3 + \cdots + A_{2h+1}$$
$$+ (k-1)(A_1 + A_3 + \cdots + A_{2h-1})$$
$$\text{(again by (9.17))} = kT_{2h-1} + A_{2h+1},$$

and, similarly,

$$T_{2h+1} T_1 = A_1^2 + A_3 A_1 + \cdots + A_{2h+1} A_1$$
$$= A_2 + kA_0 + A_4 + \cdots + A_{2h+2} + (k-1)(A_2 + A_4 + \cdots + A_{2h})$$
$$= kA_0 + A_{2h+2} + k(A_2 + \cdots + A_{2h})$$
$$= kT_{2h} + A_{2h+2}.$$

In other words,

$$T_m T_1 = kT_{m-1} + A_{m+1}$$

for all $m \geq 2$. From this we deduce

$$T_{m+1} - [T_m T_1 - (k-1)T_{m-1}] = T_{m+1} - kT_{m-1} - A_{m+1} + (k-1)T_{m-1}$$
$$= T_{m+1} - T_{m-1} - A_{m+1}$$
$$= 0,$$

and the statement follows.   □

Let P_m denote the modified Chebyshev polynomial as in (A.4).

Theorem 9.2.4 *For every $m \in \mathbb{N}$ we have*

$$T_m = P_m(A).$$

Proof. We proceed by induction on m. Clearly, $P_0 = 1$ so that $P_0(A) = I = T_0$, while $P_1(x) = x$ so that $P_1(A) = A = T_1$. Moreover,

$$
\begin{aligned}
P_{m+1}(A) &= P_m(A)A - (k-1)P_{m-1}(A) \\
&= T_m T_1 - (k-1)T_{m-1} \\
&= T_{m+1},
\end{aligned}
$$

where the first equality follows from Lemma A.9, the second one from the inductive hypothesis, and the last one from Corollary 9.2.3. □

Theorem 9.2.5 (Trace formula) *Denoting by $\mu_0 \geq \mu_1 \geq \cdots \geq \mu_{n-1}$ the eigenvalues of A, we have*

$$
\sum_{x \in X} \sum_{0 \leq r \leq [m/2]} A_{m-2r}(x, x) = \sum_{j=0}^{n-1} P_m(\mu_j)
$$

for all $m \geq 1$.

Proof. First note that

$$
\mathrm{Tr} A^\ell = \mu_0^\ell + \mu_1^\ell + \cdots + \mu_{n-1}^\ell \tag{9.19}
$$

for all $\ell \in \mathbb{N}$. Then we compute $\mathrm{Tr} T_m$ in two different ways. By definition of T_m (cf. Definition 9.2.1) we have

$$
\mathrm{Tr} T_m = \sum_{0 \leq r \leq [m/2]} \mathrm{Tr} A_{m-2r} = \sum_{0 \leq r \leq [m/2]} \sum_{x \in X} A_{m-2r}(x, x).
$$

On the other hand, from Theorem 9.2.4 we deduce that

$$
\mathrm{Tr} T_m = \mathrm{Tr} P_m(A) = \sum_{j=0}^{n-1} P_m(\mu_j),
$$

where the last equality follows from (9.19) and linearity of the trace. □

Theorem 9.2.6 (Alon-Boppana-Serre) *For every $\varepsilon > 0$ and $k \geq 3$ there exists a positive constant $C(\varepsilon, k)$ such that for every finite connected k-regular graph $\mathcal{G} = (X, E, r)$ the number of eigenvalues of the corresponding adjacency matrix belonging to the interval $[(2 - \varepsilon)\sqrt{k-1}, k]$ is at least $C(\varepsilon, k)|X|$. Note that $C(\varepsilon, k)$ does not depend on $|X|$ but only on ε and k.*

Proof. Let $\mathcal{G} = (X, E, r)$ be a finite connected k-regular graph with $|X| = n$ vertices and denote by $\mu_0 \geq \mu_1 \geq \cdots \geq \mu_{n-1}$ the eigenvalues of the associated adjacency matrix. From Theorem 9.2.5 and (A.6) we then deduce that

$$\sum_{j=0}^{n-1} X_m \left(\frac{\mu_j}{\sqrt{k-1}} \right) \geq 0 \qquad (9.20)$$

for all $m \in \mathbb{N}$. Let Z_ε be as in Corollary A.14. Then, by (9.20) and Corollary A.14.(i), we have

$$\sum_{j=0}^{n-1} Z_\varepsilon \left(\frac{\mu_j}{\sqrt{k-1}} \right) \geq 0.$$

Set $q = q(\varepsilon, k) = \max_{[2-\varepsilon, k/\sqrt{k-1}]} Z_\varepsilon$ and observe that, by virtue of Corollary A.14.(iii), we have $q > 0$ (since $k \geq 3$ implies $\frac{k}{\sqrt{k-1}} > 2$). If $\mu_j \geq (2 - \varepsilon)\sqrt{k-1}$ for all $j = 0, 1, \ldots, n-1$ there is nothing to prove. Otherwise, there exists $0 < j_0 \leq n-1$ such that

$$\mu_j \geq (2 - \varepsilon)\sqrt{k-1} \text{ for } 0 \leq j < j_0$$
$$\mu_j < (2 - \varepsilon)\sqrt{k-1} \text{ for } j_0 \leq j \leq n-1.$$

Then

$$\sum_{j=0}^{j_0-1} Z_\varepsilon \left(\frac{\mu_j}{\sqrt{k-1}} \right) \leq q j_0$$

while, by virtue of Corollary A.14.(ii),

$$\sum_{j=j_0}^{n-1} Z_\varepsilon \left(\frac{\mu_j}{\sqrt{k-1}} \right) \leq -(n - j_0).$$

Therefore

$$0 \leq \sum_{j=0}^{n-1} Z_\varepsilon \left(\frac{\mu_j}{\sqrt{k-1}} \right) \leq q j_0 - (n - j_0) = -n + j_0(q + 1)$$

so that the number j_0 of eigenvalues in $[(2 - \varepsilon)\sqrt{k-1}, k]$ satisfies

$$j_0 \geq \frac{n}{q+1} = \frac{1}{q+1}|X|,$$

and the proof is achieved by taking $C(\varepsilon, k) = \frac{1}{q+1}$. $\qquad \square$

Corollary 9.2.7 (Alon-Boppana) *Let $\mathcal{G}_n = (X_n, E_n, r_n)$, $n \in \mathbb{N}$, be a family of finite connected k-regular graphs, $k \geq 2$, such that $\lim_{n\to\infty} |X_n| = +\infty$. Then*

$$\liminf_{n\to\infty} \mu_1(\mathcal{G}_n) \geq 2\sqrt{k-1}.$$

Proof. For $k = 2$, each $\mathcal{G}_n$ is either a cycle or a 2-regular segment (cf. Exercise 8.6.3), and the result follows from (8.9) and Exercise 8.6.3, respectively. For $k \geq 3$, the statement follows from the previous theorem (since

$$\liminf_{n \to \infty} \mu_1(\mathcal{G}_n) \geq (2 - \varepsilon)\sqrt{k - 1}$$

for all $\varepsilon > 0$). □

9.3 Nilli's proof of the Alon-Boppana-Serre theorem

We now give an alternative proof of the Alon-Boppana-Serre theorem given by Alon Nilli [122] (a pseudonym of Noga Alon: Nilli Alon is his daughter; see [5] for a picture of Nilli Alon when she was a child). Our proof extends the original proof in [122] to graphs with multiple edges but with no loops. See also the discussion in [74].

We begin with an elementary lemma.

Lemma 9.3.1 *Let k and h be positive integers with $k \geq 3$. Set $\alpha = \frac{\pi}{2h}$ and*

$$\beta_i = \frac{\cos[(i - h)\alpha]}{(k - 1)^{i/2}}$$

for $i = 0, 1, \ldots, 2h$. Then the sequence $\beta_0, \beta_1, \ldots, \beta_{2h}$ is unimodal, *that is, there exists $0 \leq i_0 \leq 2h$ such that*

$$\beta_0 < \beta_1 \cdots < \beta_{i_0} < \beta_{i_0+1} \geq \beta_{i_0+2} \geq \cdots \geq \beta_{2h}.$$

More precisely:

- *for $k = 3$, $i_0 = 2$*
- *for $k = 4$, $i_0 = 1$*
- *for $k \geq 5$, $i_0 = 0$.*

Proof. First of all, note that (recall that $\alpha = \frac{\pi}{2h}$)

$$\cos[(i - h)\alpha] = \cos\left(\frac{i\pi}{2h} - \frac{\pi}{2}\right) = \sin\frac{i\pi}{2h} = \sin(i\alpha). \qquad (9.21)$$

Therefore, for $1 \leq i \leq 2h - 1$,

$$\frac{\beta_{i+1}}{\beta_i} = \frac{\sin[(i + 1)\alpha]}{\sqrt{k - 1}\sin(i\alpha)}. \qquad (9.22)$$

The function

$$g(\alpha) = i\sin[(i + 1)\alpha] - (i + 1)\sin(i\alpha)$$

satisfies $g(0) = 0$ and

$$g'(\alpha) = i(i+1)\left(\cos[(i+1)\alpha] - \cos(i\alpha)\right) \leq 0$$

for $0 \leq i\alpha \leq (i+1)\alpha \leq \pi$. This is the case since $(i+1)\alpha \leq 2h\frac{\pi}{2h} = \pi$. Then $0 = g(0) \geq g(\alpha)$ and therefore, from (9.22), it follows that

$$\frac{\beta_{i+1}}{\beta_i} \leq \frac{i+1}{i\sqrt{k-1}}, \tag{9.23}$$

for $1 \leq i \leq 2h - 1$. On the other hand, by the addition formulas for the sine function applied to the numerator of (9.22), we get

$$\frac{\beta_{i+1}}{\beta_i} = \frac{\cos\alpha + \cot(i\alpha)\sin\alpha}{\sqrt{k-1}}, \tag{9.24}$$

so that $\frac{\beta_{i+1}}{\beta_i}$ is decreasing for $1 \leq i \leq 2h - 1$. Moreover, from (9.21) $\beta_0 = 0 < \beta_1 = \frac{1}{\sqrt{k-1}}\sin\frac{\pi}{2h}$. Then we can take $i_0 + 1$ as the smallest $1 \leq i \leq 2h - 1$ such that the quantity in (9.24) is smaller than 1: this exists because for $i = h$ the quantity in (9.24) is equal to $\frac{\cos\alpha}{\sqrt{k-1}} < 1$ (recall that $k \geq 3$).

We now determine the values of i_0 for all $k \geq 3$.

<u>Case $k = 3$.</u> For $i = 3$, from (9.23) we get $\frac{\beta_4}{\beta_3} \leq \frac{4}{3\sqrt{2}} < 1$. Note that for $i = 2$, from (9.22) we get

$$\frac{\beta_3}{\beta_2} = \frac{\sin 3\alpha}{\sqrt{2}\sin 2\alpha} \underset{h\to+\infty}{\longrightarrow} \frac{3}{2\sqrt{2}} > 1,$$

so that $i_0 = 2$ is the correct index that works for *all* h.

<u>Case $k = 4$.</u> Again from (9.23) for $i = 2$ we get $\frac{\beta_3}{\beta_2} \leq \frac{3}{2\sqrt{3}} < 1$. For $i = 1$ we have $\frac{\beta_2}{\beta_1} = \frac{\sin 2\alpha}{\sqrt{3}\sin\alpha} \underset{h\to+\infty}{\longrightarrow} \frac{2}{\sqrt{3}}$. Therefore, $i_0 = 1$.

<u>Case $k \geq 5$.</u> From (9.22), for $i = 1$ we get

$$\frac{\beta_2}{\beta_1} = \frac{\sin 2\alpha}{\sqrt{k-1}\sin\alpha} = \frac{2\cos\alpha}{\sqrt{k-1}} \leq 1.$$

Then we have $i_0 = 0$.   □

Let now $\mathcal{G} = (X, E, r)$ be a finite graph. Given two subsets $Y, Z \subseteq X$ we set

$$A(Y, Z) = \sum_{(y,z)\in Y\times Z} A(y, z). \tag{9.25}$$

In other words, $A(Y, Z)$ equals the number of edges that join a vertex in Y with a vertex in Z. Note that $A(\{y\}, \{z\}) = A(y, z)$, so that we shall also write $A(y, Z)$ instead of $A(\{y\}, Z)$, for all $y, z \in X$ and $Z \subseteq X$. Moreover, if $x_1, x_2 \in Y \cap Z$ are distinct and adjacent, then in the sum (9.25) the equal summands $A(x_1, x_2)$ and

$A(x_2, x_1)$ both appear, giving altogether a contribution of $2A(x_1, x_2)$; in other words, the edges in $r^{-1}(\{x_1, x_2\})$ are counted twice.

For k and h positive integers, with $k \geq 3$, we set

$$\gamma_i = \beta_{i+i_0} \quad \text{for } 0 \leq i \leq 2h - i_0, \tag{9.26}$$

where the β_i's and i_0 are as in Lemma 9.3.1. Note that $\gamma_{2h-i_0} = \beta_{2h} = \cos \frac{\pi}{2} = 0$.

We now give a second lemma, of a pure combinatorial nature, which is the core of the proof of the main theorem of this section.

Lemma 9.3.2 *Let $\mathcal{G} = (X, E, r)$ be a finite connected k-regular graph, with $k \geq 3$, and denote by A its adjacency matrix. Suppose there exists a vertex $x_0 \in X$ with no loops based at it, and define $f \in L(X)$ by setting*

$$f(x) = \begin{cases} \gamma_i & \text{if } 0 \leq d(x, x_0) = i < 2h - i_0 \\ 0 & \text{if } d(x, x_0) \geq 2h - i_0, \end{cases}$$

where the γ_i's are as in (9.26). Then

$$\langle Af, f \rangle_{L(X)} \geq \langle f, f \rangle_{L(X)} 2\sqrt{k-1} \cos \alpha.$$

Proof. Set $X_i = \{x \in X : d(x, x_0) = i\}$ and $n_i = |X_i|$. By our assumption on x_0 we have $A(x_0, x_0) = 0$ and therefore

$$A(x_0, X_1) = k = |X_1| = n_1. \tag{9.27}$$

Moreover, for $i \geq 1$,

$$A(X_{i-1}, X_i) \geq |X_i| = n_i \tag{9.28}$$

and

$$A(X_{i-1}, X_i) + A(X_i, X_i) + A(X_{i+1}, X_i) = kn_i \tag{9.29}$$

because the left hand side counts all edges with a vertex in X_i (and the edges with both vertices in X_i, but which are not loops, are counted twice). Then

$$\langle f, f \rangle_{L(X)} = \sum_{x \in X} f(x)^2 = \sum_{i=0}^{2h-i_0-1} \sum_{x \in X_i} f(x)^2 = \sum_{i=0}^{2h-i_0-1} n_i \gamma_i^2 \tag{9.30}$$

and

$$\langle Af, f \rangle_{L(X)} = \sum_{x \in X} \sum_{\substack{y \in X: \\ y \sim x}} A(x, y) f(x) f(y)$$

$$= \sum_{i=0}^{2h-i_0-1} \gamma_i \sum_{x \in X_i} \sum_{\substack{y \in X: \\ y \sim x}} A(x, y) f(y)$$

$$= \gamma_0 \gamma_1 A(x_0, X_1) +$$

$$+ \sum_{i=1}^{2h-i_0-1} \gamma_i \left[\gamma_{i-1} A(X_{i-1}, X_i) + \gamma_i A(X_i, X_i) + \gamma_{i+1} A(X_{i+1}, X_i) \right].$$

$$(9.31)$$

In order to give a lower bound for (9.31), we first note that from $0 \leq \gamma_0 < \gamma_1$ (cf. Lemma 9.3.1) and (9.27) we deduce that

$$\gamma_0 \gamma_1 A(x_0, X_1) \geq \gamma_0^2 A(x_0, X_1) = \gamma_0^2 k \geq [2\sqrt{k-1} \cos \alpha] \gamma_0^2 \qquad (9.32)$$

(the last inequality follows immediately from $(k-2)^2 \geq 0$ and $\cos \alpha \leq 1$).

In the last line of (9.31), for the first term of the sum, corresponding to $i = 1$, keeping in mind (9.27) and $\gamma_1 \geq \gamma_2$, we have

$$\gamma_0 A(X_0, X_1) + \gamma_1 A(X_1, X_1) + \gamma_2 A(X_2, X_1)$$

$$\geq \gamma_0 A(X_0, X_1) + \gamma_2 [A(X_1, X_1) + A(X_2, X_1)]$$

$$\text{(by (9.29))} \quad = \gamma_0 A(X_0, X_1) + \gamma_2 [kn_1 - A(X_0, X_1)]$$

$$\text{(by (9.27))} \quad = \gamma_0 k + \gamma_2 [k^2 - k]$$

$$= k[\gamma_0 + (k-1)\gamma_2]$$

$$\text{(by (9.27))} \quad = n_1 [\gamma_0 + (k-1)\gamma_2].$$

As far as the terms corresponding to $i \geq 2$ are concerned, keeping in mind that $\gamma_{i-1} \geq \gamma_i \geq \gamma_{i+1}$, from (9.28) and (9.29) we deduce that

$$\gamma_{i-1} A(X_{i-1}, X_i) + \gamma_i A(X_i, X_i) + \gamma_{i+1} A(X_{i+1}, X_i)$$

$$\geq \gamma_{i-1} A(X_{i-1}, X_i) + \gamma_{i+1} [A(X_i, X_i) + A(X_{i+1}, X_i)]$$

$$= \gamma_{i-1} [n_i - n_i + A(X_{i-1}, X_i)] + \gamma_{i+1} [A(X_i, X_i) + A(X_{i+1}, X_i)]$$

$$= \gamma_{i-1} n_i + \gamma_{i-1} [A(X_{i-1}, X_i) - n_i] + \gamma_{i+1} [A(X_i, X_i) + A(X_{i+1}, X_i)]$$

$$\geq \gamma_{i-1} n_i + \gamma_{i+1} [A(X_{i-1}, X_i) - n_i] + \gamma_{i+1} [A(X_i, X_i) + A(X_{i+1}, X_i)]$$

$$= \gamma_{i-1} n_i + \gamma_{i+1} [-n_i + A(X_{i-1}, X_i) + A(X_i, X_i) + A(X_{i+1}, X_i)]$$

$$= \gamma_{i-1} n_i + \gamma_{i+1} (k-1) n_i$$

$$= n_i [\gamma_{i-1} + (k-1)\gamma_{i+1}].$$

Moreover, for all $i \geq 1$ we have

$$n_i[\gamma_{i-1} + (k-1)\gamma_{i+1}]$$

$$= \frac{\sqrt{k-1}}{(k-1)^{(i+i_0)/2}} n_i \{\cos[(i+i_0-h-1)\alpha] + \cos[(i+i_0-h+1)\alpha]\}$$

$$= \frac{2\sqrt{k-1}}{(k-1)^{(i+i_0)/2}} n_i \cos\alpha \cos[(i+i_0-h)\alpha]$$

$$= [2\sqrt{k-1}\cos\alpha]n_i\frac{\cos[(i+i_0-h)\alpha]}{(k-1)^{(i+i_0)/2}}$$

$$= [2\sqrt{k-1}\cos\alpha]n_i\gamma_i,$$

where the first equality follows from (9.26).

Using the above estimates, we get the desired lower bound for (9.31):

$$\langle Af, f\rangle_{L(X)} \geq [2\sqrt{k-1}\cos\alpha] \sum_{i=0}^{2h-i_0-1} n_i\gamma_i^2$$

$$(\text{by } (9.30)) \quad = [2\sqrt{k-1}\cos\alpha]\langle f, f\rangle_{L(X)}.$$

$\square$

To derive the main result of this section, we need to recall the *Courant-Fischer min-max formula* for the eigenvalues of a Hermitian operator.

Exercise 9.3.3 (Courant-Fischer min-max formula) Let W be an n-dimensional vector space and $T: W \to W$ a Hermitian operator. Denote by $\mu_0 \geq \mu_1 \geq \cdots \geq \mu_{n-1}$ the (real) eigenvalues of T and by $\{u_0, u_1, \ldots, u_{n-1}\}$ a corresponding orthonormal basis of eigenvectors. Let $0 \leq s \leq n-1$. Denote by $\mathbb{G}(W, s)$ the Grassmann variety of all s-dimensional subspaces of W and set $U_s = \langle u_s, u_{s+1}, \ldots, u_{n-1}\rangle$.

(1) Prove that for each $V \in \mathbb{G}(W, s+1)$ one has $\dim(V \cap U_s) \geq 1$.
   Hint: use the Grassmann identity.
(2) Show that

$$\max\{\langle Tw, w\rangle : w \in U_s, \|w\| = 1\} = \mu_s.$$

(3) From (1) and (2) deduce that for each $V \in \mathbb{G}(W, s+1)$

$$\min\{\langle Tv, v\rangle : v \in V, \|v\| = 1\} \leq \mu_s.$$

(4) Show that if $V = \langle u_0, u_1, \ldots, u_s\rangle$ then

$$\min\{\langle Tv, v\rangle : v \in V, \|v\| = 1\} = \mu_s.$$

(5) From (3) and (4) deduce the Courant-Fischer min-max formula

$$\max_{V \in \mathbb{G}(W, s+1)} \min\{\langle Tv, v\rangle : v \in V, \|v\| = 1\} = \mu_s.$$

We are now in position to present some fundamental estimates for the eigenvalues of a k-regular graph.

Theorem 9.3.4 *Let* $\mathcal{G} = (X, E, r)$ *be a finite connected k-regular graph, $k \geq 3$, with no loops. Suppose that there exist a positive integer h and $s + 1$ vertices* $x_1, x_2, \ldots, x_{s+1} \in X$ *such that $d(x_i, x_j) \geq 4h$, for $i \neq j$. Then*

$$\mu_s(\mathcal{G}) \geq 2\sqrt{k-1} \cos \frac{\pi}{2h}. \tag{9.33}$$

Proof. For $j = 1, 2, \ldots, s + 1$ define $f_j \in L(X)$ by setting

$$f_j(x) = \begin{cases} \gamma_i & \text{if } 0 \leq d(x, x_j) = i \leq 2h - i_0 \\ 0 & \text{if } d(x, x_j) > 2h - i_0, \end{cases}$$

where the γ_i's are as in (9.26). Then $\langle f_j, f_k\rangle_{L(X)} = 0$ (because f_j and f_k have disjoint supports) for $1 \leq j \neq k \leq s + 1$, so that $U = \langle f_1, f_2, \ldots, f_{s+1}\rangle$ is an $(s + 1)$-dimensional subspace of $L(X)$. Moreover, from Lemma 9.3.2 (where x_0 therein is replaced time after time by $x_1, x_2, \ldots, x_{s+1}$) we deduce that

$$\langle Af, f\rangle_{L(X)} \geq \langle f, f\rangle_{L(X)} 2\sqrt{k-1} \cos \frac{\pi}{2h} \tag{9.34}$$

for all $f \in U$.

From Exercise 9.3.3 (the Courant-Fischer min-max formula) and with the notation therein we deduce

$$\mu_s = \max_{V \in \mathbb{G}(L(X), s+1)} \min\{\langle Af, f\rangle_{L(X)} : f \in V, \|f\|_{L(X)} = 1\}. \tag{9.35}$$

Then (9.33) follows from (9.35) and (9.34). $\square$

Corollary 9.3.5 *Let* $\mathcal{G}$ *be a finite connected k-regular graph, $k \geq 3$, with no loops. Suppose that the diameter of $\mathcal{G}$ satisfies that $D(\mathcal{G}) \geq 4h$ for some positive integer h. Then*

$$\mu_1(\mathcal{G}) \geq 2\sqrt{k-1}\left(1 - \frac{\pi^2}{8h^2}\right).$$

Proof. Apply Theorem 9.3.4 with $s = 1$ and the estimate $\cos\theta \geq 1 - \frac{\theta^2}{2}$. $\square$

Corollary 9.3.6 (Alon-Boppana-Serre: II proof) *Let $\varepsilon > 0$ and $k \geq 3$. Then there exists a positive constant $C(\varepsilon, k)$ such that the following holds. For every finite connected k-regular graph $\mathcal{G} = (X, E, r)$ with no loops, the number of* *eigenvalues of the corresponding adjacency matrix belonging to the interval* $[(2 - \varepsilon)\sqrt{k-1}, k]$ *is at least* $C(\varepsilon, k)|X|$. *Explicitly, we may choose*

$$C(\varepsilon, k) = \begin{cases} 2^{-\frac{2\pi}{\sqrt{\varepsilon}} - 5} & \text{if } k = 3 \\ (k-1)^{-\frac{2\pi}{\sqrt{\varepsilon}} - 4} & \text{if } k \geq 4. \end{cases}$$

Proof. We start by denoting h as the (positive) integer satisfying

$$h \geq \frac{\pi}{2\sqrt{\varepsilon}} > h - 1, \tag{9.36}$$

so that $\varepsilon \geq \frac{\pi^2}{4h^2}$ and therefore (recall that $\cos\theta \geq 1 - \frac{\theta^2}{2}$)

$$2\sqrt{k-1}\cos\frac{\pi}{2h} \geq 2\sqrt{k-1}(1 - \frac{\pi^2}{8h^2}) \geq \sqrt{k-1}(2 - \varepsilon).$$

(We want to use the inequality in Theorem 9.3.4, that is,

$$\mu_0, \mu_1, \ldots, \mu_s \geq 2\sqrt{k-1}\cos\frac{\pi}{2h} \geq \sqrt{k-1}(2 - \varepsilon) \tag{9.37}$$

with the best possible, that is, the smallest, h.) According to Theorem 9.3.4, choose the largest s such that the hypotheses therein are satisfied, and let $x_1, x_2, \ldots, x_{s+1} \in X$ be the corresponding points. Then, for every $x \in X$ there exists $1 \leq j \leq s + 1$ such that $d(x, x_j) \leq 4h - 1$. Arguing as in the proof of Proposition 8.1.1, we conclude that

$$|X| \leq (s+1)[1 + k + k(k-1) + \cdots + k(k-1)^{4h-2}]$$
$$= (s+1)\left[1 + k\frac{(k-1)^{4h-1} - 1}{k-2}\right].$$

From (9.37) we deduce that such a constant $C(\varepsilon, k)$ exists and satisfies

$$C(\varepsilon, k) \geq \frac{s+1}{|X|} \geq \left[1 + k\frac{(k-1)^{4h-1} - 1}{k-2}\right]^{-1}. \tag{9.38}$$

Now, for $k \geq 4$ we have

$$1 + k\frac{(k-1)^{4h-1} - 1}{k-2} \leq (k-1)^{4h} \tag{9.39}$$

because this is equivalent to

$$-2 + k(k-1)^{4h-1} \leq (k-2)(k-1)^{4h}$$

which is certainly satisfied as $k \leq (k-2)(k-1)$ for $k \geq 4$. Therefore,

$$C(\varepsilon, k) \geq (k-1)^{-4h} \geq (k-1)^{-2\pi/\sqrt{\varepsilon} - 4}$$

where the first inequality follows from (9.38) and (9.39), and the second from (9.36). Finally, for $k = 3$ we may use

$$1 + k\frac{(k-1)^{4h-1}-1}{k-2}\Big|_{k=3} = 3 \cdot 2^{4h-1} - 2 \leq 2^{4h+1}$$

in place of (9.39). □

9.4 Ramanujan graphs

Definition 9.4.1 Let $\mathcal{G} = (X, E, r)$ be a finite connected k-regular graph. Denote by $k = \mu_0 > \mu_1 \geq \cdots \geq \mu_{n-1}$ the eigenvalues of the adjacency matrix of $\mathcal{G}$. Setting

$$\mu(\mathcal{G}) = \max\{|\mu_i| : |\mu_i| \neq k, i = 1, 2, \ldots, n-1\} \qquad (9.40)$$

one says that $\mathcal{G}$ is a *Ramanujan graph* provided

$$\mu(\mathcal{G}) \leq 2\sqrt{k-1}.$$

Note that if $\mathcal{G}$ is bipartite then (cf. Proposition 8.3.4) $\mathcal{G}$ is Ramanujan if and only if

$$\mu_1 \leq 2\sqrt{k-1}.$$

Exercise 9.4.2 (see [99]) Let $\mathcal{G}$ be a connected strongly regular graph with parameters (v, k, λ, μ) (cf. Definition 8.2.1). Show that $\mathcal{G}$ is Ramanujan if and only if

$$2|\lambda - \mu|\sqrt{k-1} \leq 3k + \mu - 4.$$

In the remainder of this section, we apply methods and results on finite fields established in Section 7.1 to introduce and describe the Paley graph, which constitutes an interesting example of a Ramanujan graph. We follow the approach in the monograph by van Lint and Wilson [97].

Let p be an odd prime and $q = p^n$. The *Legendre symbol* on $\mathbb{F}_q$ may be defined, as in Definition 4.4.7, by setting

$$\eta(y) = \begin{cases} 1 & \text{if } y \neq 0 \text{ is a square in } \mathbb{F}_q \\ -1 & \text{if } y \neq 0 \text{ is not a square in } \mathbb{F}_q \\ 0 & \text{if } y = 0 \end{cases}$$

(see also Proposition 6.4.4).

Exercise 9.4.3 Let $q = p^n$ with p an odd prime.

(1) Show that η is a multiplicative character of $\mathbb{F}_q$ and that, in the notation of (7.11), we have

$$\eta(x^k) = \exp(\pi i k) \quad \text{for } k = 0, 1, \ldots, q - 1.$$

(2) Prove that, for $z \neq 0$,

$$\sum_{y \in \mathbb{F}_q} \eta(y)\eta(y + z) = -1$$

Hint: for $y \neq 0$, $\eta(y)\eta(y + z) = \eta(y^2)\eta(1 + y^{-1}z)$.

(3) Prove that -1 is a square in $\mathbb{F}_q$ if and only if $q \equiv 1 \mod 4$.

(4) Define a matrix $R = (r(x, y))_{x,y \in \mathbb{F}_q}$ by setting

$$r(x, y) = \eta(x - y) \quad \text{for all } x, y \in \mathbb{F}_q.$$

Prove that

- R is symmetric (resp. antisymmetric) if $q \equiv 1 \mod 4$ (resp. $q \equiv 3 \mod 4$).
- $RJ = JR = 0$, where J is as in Exercise 8.2.2.(1).
- $RR^T = qI - J$
  Hint: Use (2).

Example 9.4.4 (The Paley graph) Let p be an odd prime and $q = p^n$. Suppose that $q \equiv 1 \mod 4$. The *Paley Graph* $P(q)$ has vertex set $\mathbb{F}_q$ and two distinct vertices $x, y \in \mathbb{F}_q$ are joined if $x - y$ is a square. Note that, by virtue of Exercise 9.4.3.(3), $x - y$ is a square if and only if $y - x$ is a square. We deduce that $P(q)$ is an undirected simple graph without loops.

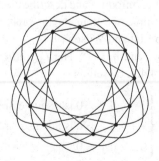

Figure 9.1. The Paley graph $P(13)$.

Exercise 9.4.5 We use the same notation as in Exercise 9.4.3 and Example 9.4.4.

(1) Show that the adjacency matrix of $P(q)$ is

$$A = \frac{1}{2}(R + J - I).$$

(2) Deduce that $P(q)$ is a strongly regular graph with parameters $(q, \frac{1}{2}(q - 1), \frac{1}{4}(q - 5), \frac{1}{4}(q - 1))$
   Hint: See Exercise 8.2.2 and Exercise 9.4.3.(4).
(3) [99, 161] Show that $P(q)$ is a Ramanujan graph
   Hint: Use Exercise 9.4.2.

9.5 Expander graphs

Definition 9.5.1 Let $\mathcal{G}_n = (X_n, E_n, r_n)$, $n \in \mathbb{N}$, be a sequence of finite (undirected) graphs. Suppose that there exist and integer $k \geq 2$ and $\varepsilon > 0$ such that

- $\mathcal{G}_n$ is k-regular for all $n \in \mathbb{N}$;
- $|X_n| \to +\infty$ as $n \to +\infty$;
- $h(\mathcal{G}_n) \geq \varepsilon$ for all $n \in \mathbb{N}$,

where $h(\cdot)$ denotes the isoperimetric constant (cf. Definition 9.1.1). Then we say that $(\mathcal{G}_n)_{n \in \mathbb{N}}$ is a family of *expander graphs* (briefly, *expanders*).

Remark 9.5.2 From (9.6)[†] we deduce that if $(\mathcal{G}_n)_{n \in \mathbb{N}}$ is a family of k-regular graphs, then

$$\frac{k}{2}|X_n| \leq |E_n| \leq k|X_n|$$

for all $n \in \mathbb{N}$, that is, the number of edges grows linearly with the size, i.e. with the number of vertices, of the graphs $\mathcal{G}_n$ (because k is fixed).

Also, the condition $h(\mathcal{G}_n) \geq \varepsilon$ ensures a *good connectivity* of the graph $\mathcal{G}_n$ in the following sense: if $A_n \subseteq X_n$ is a subset such that $|A_n| \leq \frac{|X_n|}{2}$, then, in order to "disconnect" A_n from its complement $X_n \setminus A_n$, that is, to remove ∂A_n, we need to "cut" at least $\varepsilon|A_n|$ edges of $\mathcal{G}_n$. Note that if $|A_n| \approx |X_n|$, then the quantity $\varepsilon|A_n|$ grows linearly with $|X_n|$. In other words, expanders provide a solution to the following min-max problem: to minimize the number of edges and to maximize the connectivity of the graphs.

[†] Note that in (9.6), E_0 (respectively, E_1) is not the edge set of $\mathcal{G}_0$ (respectively, $\mathcal{G}_1$), but denotes the loops (respectively, $E \setminus E_0$) of a generic graph $\mathcal{G} = (X, E, r)$.

Moreover, keeping k fixed and letting $|X_n| \to +\infty$ for $n \to +\infty$, the graphs $\mathcal{G}_n$ become more and more "sparse," that is, they have a large number $|X_n|$ of vertices, but each vertex has a "small" fixed number k of neighbors.

Recalling Remark 9.1.10, we immediately have the following equivalent definition of expanders.

Definition 9.5.3 (Spectral definition of expanders) Let $\mathcal{G}_n = (X_n, E_n, r_n)$, $n \in \mathbb{N}$, be a sequence of finite connected graphs. Suppose that there exist and integer $k \geq 2$ and $\delta > 0$ such that

- $\mathcal{G}_n$ is k-regular for all $n \in \mathbb{N}$;
- $|X_n| \to +\infty$ as $n \to +\infty$;
- $\delta(\mathcal{G}_n) \geq \delta$ for all $n \in \mathbb{N}$,

where $\delta(\cdot)$ denotes the spectral gap (cf. Definition 9.1.9). Then $(\mathcal{G}_n)_{n \in \mathbb{N}}$ is a family of *expanders*.

Remark 9.5.4 We may reformulate Corollary 9.2.7 as follows:

$$\limsup_{n \to \infty} \delta(\mathcal{G}_n) = k - \liminf_{n \to \infty} \mu_1(G) \leq k - 2\sqrt{k-1}. \qquad (9.41)$$

As a consequence, if $\delta(\mathcal{G}_n) \geq \delta$ for all $n \in \mathbb{N}$, then necessarily

$$\delta \leq k - 2\sqrt{k-1}. \qquad (9.42)$$

Example 9.5.5 Let $(\mathcal{G}_n)_{n \in \mathbb{N}}$ be a sequence of finite connected k-regular Ramanujan graphs. Suppose that $|X_n| \to +\infty$ as $n \to +\infty$. Then $(\mathcal{G}_n)_{n \in \mathbb{N}}$ is a family of expanders with $\delta = k - 2\sqrt{k-1}$ (cf. Definition 9.5.3). It follows from Remark 9.5.4 that a sequence of Ramanujan graphs is asymptotically *optimal* within the sequences of expanders.

The construction of a *single* Ramanujan graph is not difficult (see Exercise 9.4.2 and Exercise 9.4.5). On the contrary, the construction of a *sequence* of Ramanujan graphs of a fixed degree (and increasing size) requires very deep results from number theory. One of these results is the so-called *Ramanujan conjecture*, eventually proved by several mathematicians including Deligne and Drinfeld. For this reason, although Ramanujan never worked in graph theory, these expanders were named after him.

The first explicit construction of a sequence of Ramanujan graphs (of constant degree k and increasing size) were given for the following values of k:

- $k = p + 1$, with p an odd prime, by Lubotzky, Phillips, and Sarnak [101], and Margulis [112] in 1988;
- $k = 3$, by Chiu [40] in 1992;

- $k = q + 1$, with $q = p^r$, p prime and $r \geq 1$, by Morgenstern [116] in 1994.

An elementary account of the Lubotzky-Phillips-Sarnak graphs and Margulis graphs is in the monograph by Davidoff, Sarnak, and Valette [49] where, however, the authors do not provide a full proof of the Ramanujan property but only a weaker explicit estimate of the spectral gap (the construction of these graphs is relatively easy, but the proof of the Ramanujan property is indeed the difficult point). See also the monographs by Winnie Li [95], Lubotzky [99], and Sarnak [135, Chapter 3].

Very recently, in 2015, Marcus, Spielman, and Srivastava [109] proved that there exist infinite families of regular bipartite Ramanujan graphs of every degree $k \geq 3$. Later, in [110] they proved the existence of regular bipartite Ramanujan graphs of every degree and every number of vertices. With respect to the previous work, this is more elementary (although based on the *probabilistic method*, cf. [11]), but it does not provide an explicit construction. On the other hand, however, the construction of expanders is much more elementary: in the following sections we shall give several examples.

9.6 The Margulis example

In 1973 Margulis constructed the first example of a family of expanders [111]. His approach was quite abstract, based on the notion of Kazhdan property (T) (cf. the monograph by Bekka, de la Harpe, and Valette [19]). In 1981, Gabber and Galil [64], using classical Fourier analysis, were able to simplify Margulis example and to provide a lower bound of the spectral gap. Similar improvements were obtained in 1987 by Jimbo and Marouka [83] who used Fourier analysis on the finite group $\mathbb{Z}_n \oplus \mathbb{Z}_n$. Further simplifications were made by Hoory, Linial, and Wigderson [74], although they attributed the merit to Boppana. Our exposition is strictly based on this last reference.

We start by introducing some basic notation taken from Chapter 1 and Chapter 2. Let $n \geq 1$. Write the group $A = \mathbb{Z}_n \oplus \mathbb{Z}_n$ as a set of column vectors:

$$A = \left\{ \begin{pmatrix} x_1 \\ x_2 \end{pmatrix} : x_1, x_2 \in \mathbb{Z}_n \right\},$$

equipped with the usual componentwise addition, and denote by $0 = \begin{pmatrix} 0 \\ 0 \end{pmatrix}$ the zero of A. We also consider 2×2 matrices with entries in $\mathbb{Z}_n$. Clearly, a matrix $\begin{pmatrix} a & b \\ c & d \end{pmatrix}$, with $a, b, c, d \in \mathbb{Z}_n$, is invertible if and only if its determinant

$\det \begin{pmatrix} a & b \\ c & d \end{pmatrix} = ad - bc$ is invertible in $\mathbb{Z}_n$. Moreover, if this is the case, we have the usual formula

$$\begin{pmatrix} a & b \\ c & d \end{pmatrix}^{-1} = \begin{pmatrix} d(ad - bc)^{-1} & -b(ad - bc)^{-1} \\ -c(ad - bc)^{-1} & a(ad - bc)^{-1} \end{pmatrix}.$$

Let us set $\omega = e^{2\pi i/n}$ and, for $x = \begin{pmatrix} x_1 \\ x_2 \end{pmatrix}$ and $y = \begin{pmatrix} y_1 \\ y_2 \end{pmatrix} \in A$, write $\langle x, y \rangle = x_1 y_1 + x_2 y_2$. Arguing as in Section 2.4, we can write the Fourier transform of a function $f \in L(A)$ as

$$\widehat{f}(y) = \sum_{x \in A} f(x) \omega^{-\langle x, y \rangle} \quad \forall y \in A.$$

Then, the inversion formula (cf. Theorem 2.4.2) takes the form

$$f(x) = \frac{1}{n^2} \sum_{y \in A} \widehat{f}(y) \omega^{\langle x, y \rangle} \quad \forall x \in A,$$

while the Plancherel and Parseval formulas (cf. Theorem 2.4.3) become respectively:

$$\sqrt{\sum_{y \in A} |\widehat{f}(y)|^2} = n \cdot \sqrt{\sum_{x \in A} |f(x)|^2} \quad \forall f \in L(A)$$

and

$$\sum_{y \in A} \widehat{f_1}(y) \overline{\widehat{f_2}(y)} = n^2 \sum_{x \in A} f_1(x) \overline{f_2(x)} \quad \forall f_1, f_2 \in L(A).$$

Note also that $\widehat{f}(0) = \sum_{x \in A} f(x)$ so that

$$\widehat{f}(0) = 0 \Leftrightarrow \sum_{x \in A} f(x) = 0. \tag{9.43}$$

The following result is elementary but new.

Proposition 9.6.1 *Let $f \in L(A)$, B a 2×2 invertible matrix with entries in $\mathbb{Z}_n$, and $b \in A$. Define $g \in L(A)$ by setting $g(x) = f(Bx + b)$ for all $x \in A$. Then,*

$$\widehat{g}(y) = \omega^{\langle B^{-1} b, y \rangle} \widehat{f}((B^{-1})^T y),$$

for all $y \in A$.

Proof. Let $y \in A$. Then we have

$$\widehat{g}(y) = \sum_{x \in A} f(Bx + b)\omega^{-\langle x, y \rangle}$$

$$(z = Bx + b) \quad = \sum_{z \in A} f(z)\omega^{-\langle B^{-1}z - B^{-1}b, y \rangle}$$

$$= \omega^{\langle B^{-1}b, y \rangle} \sum_{z \in A} f(z)\omega^{-\langle z, (B^{-1})^T y \rangle}$$

$$= \omega^{\langle B^{-1}b, y \rangle} \widehat{f}((B^{-1})^T y). \qquad \square$$

In what follows, a special role will be played by the following 2×2 matrices with entries in $\mathbb{Z}_n$:

$$T_1 = \begin{pmatrix} 1 & 2 \\ 0 & 1 \end{pmatrix} \quad \text{and} \quad T_2 = \begin{pmatrix} 1 & 0 \\ 2 & 1 \end{pmatrix}$$

whose inverses are

$$T_1^{-1} = \begin{pmatrix} 1 & -2 \\ 0 & 1 \end{pmatrix} \quad \text{and} \quad T_2^{-1} = \begin{pmatrix} 1 & 0 \\ -2 & 1 \end{pmatrix}.$$

Clearly,

$$T_1 \begin{pmatrix} x_1 \\ x_2 \end{pmatrix} = \begin{pmatrix} x_1 + 2x_2 \\ x_2 \end{pmatrix}, \quad T_1^{-1} \begin{pmatrix} x_1 \\ x_2 \end{pmatrix} = \begin{pmatrix} x_1 - 2x_2 \\ x_2 \end{pmatrix}$$

$$T_2 \begin{pmatrix} x_1 \\ x_2 \end{pmatrix} = \begin{pmatrix} x_1 \\ 2x_1 + x_2 \end{pmatrix}, \quad T_2^{-1} \begin{pmatrix} x_1 \\ x_2 \end{pmatrix} = \begin{pmatrix} x_1 \\ -2x_1 + x_2 \end{pmatrix}$$

(9.44)

(everything mod n). Moreover, we identify $\mathbb{Z}_n$ with the integral interval $[-\frac{n}{2}, \frac{n}{2}) = \{k \in \mathbb{Z} : -\frac{n}{2} \leq k < \frac{n}{2}\}$. Clearly,

$$\left[-\frac{n}{2}, \frac{n}{2}\right) = \begin{cases} [-m, m) & \text{if } n = 2m \text{ is even} \\ [-m, m] & \text{if } n = 2m + 1 \text{ is odd.} \end{cases}$$

Then we can identify A with the set

$$\left\{ \begin{pmatrix} x_1 \\ x_2 \end{pmatrix} : x_1, x_2 \in \left[-\frac{n}{2}, \frac{n}{2}\right) \right\}.$$

The *diamond* in A is the set (see Figure 9.2)

$$D = \left\{ \begin{pmatrix} x_1 \\ x_2 \end{pmatrix} \in A : |x_1| + |x_2| < \frac{n}{2} \right\}.$$

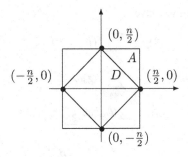

Figure 9.2. The diamond D in A.

We define a partial order in A by setting

$$\begin{pmatrix} x_1 \\ x_2 \end{pmatrix} > \begin{pmatrix} y_1 \\ y_2 \end{pmatrix} \text{ if } \begin{cases} |x_1| > |y_1| \text{ and } |x_2| \geq |y_2| \\ \text{or} \\ |x_1| \geq |y_1| \text{ and } |x_2| > |y_2|. \end{cases}$$

We now present a series of technical lemmas, which are essential for our subsequent calculations.

Lemma 9.6.2 *Let* $x = \begin{pmatrix} x_1 \\ x_2 \end{pmatrix} \in D \setminus \{0\}$.

(i) *If* $|x_1| = |x_2|$ *then two of the four points*

$$T_1 x, \quad T_1^{-1} x, \quad T_2 x, \quad T_2^{-1} x \tag{9.45}$$

are strictly greater than x *and the other two are incomparable with* x;

(ii) *if* $|x_1| \neq |x_2|$ *and* $x_1 \neq 0 \neq x_2$, *then three of the points in* (9.45) *are strictly greater than* x *and the other one is strictly smaller;*

(iii) *if* $|x_1| \neq |x_2|$ *but either* $x_1 = 0$ *or* $x_2 = 0$, *then two of the points in* (9.45) *are strictly greater than* x *and the other two are equal to* x.

Proof. (i) Suppose first that $x_1 = x_2$. Then

$$T_1^{-1} x = T_1^{-1} \begin{pmatrix} x_1 \\ x_1 \end{pmatrix} = \begin{pmatrix} -x_1 \\ x_1 \end{pmatrix} \text{ and } T_2^{-1} x = T_2^{-1} \begin{pmatrix} x_1 \\ x_1 \end{pmatrix} = \begin{pmatrix} x_1 \\ -x_1 \end{pmatrix}$$

are incomparable with x. Moreover,

$$|x_1| + |x_1| < \frac{n}{2} \Rightarrow |x_1| < \frac{n}{4},$$

and therefore

$$T_1 x = T_1 \begin{pmatrix} x_1 \\ x_1 \end{pmatrix} = \begin{pmatrix} 3x_1 \\ x_1 \end{pmatrix} \quad \text{and} \quad T_2 x = T_2 \begin{pmatrix} x_1 \\ x_1 \end{pmatrix} = \begin{pmatrix} x_1 \\ 3x_1 \end{pmatrix}.$$

First case: suppose $-\frac{n}{6} \le x_1 < \frac{n}{6}$. Then $-\frac{n}{2} \le 3x_1 < \frac{n}{2}$ and therefore $|3x_1| > |x_1|$ ensures that $T_1 x > x$ and $T_2 x > x$.

Second case: suppose $-\frac{n}{4} < x_1 < -\frac{n}{6}$. Then $-\frac{3}{4}n < 3x_1 < -\frac{n}{2}$ so that

$$\frac{n}{4} < 3x_1 + n < \frac{n}{2},$$

and we must take $3x_1 + n$ to represent $3x_1$ in the range $[-\frac{n}{2}, \frac{n}{2})$. This gives $|3x_1 + n| > \frac{n}{4} > |x_1|$ so that $T_1 x > x$ and $T_2 x > x$.

Third case: suppose $\frac{n}{6} \le x_1 < \frac{n}{4}$. Then $\frac{n}{2} \le 3x_1 < \frac{3}{4}n$ so that

$$-\frac{n}{2} \le 3x_1 - n < -\frac{n}{4},$$

and we must take $3x_1 - n$ to represent $3x_1$ in the range $[-\frac{n}{2}, \frac{n}{2})$. This gives $|3x_1 - n| > \frac{n}{4} > |x_1|$ and, again, $T_1 x > x$ and $T_2 x > x$.

When $x_1 = -x_2$ we may argue similarly: now $T_1^{-1} x > x$ and $T_2^{-1} x > x$, while $T_1 x$ and $T_2 x$ are incomparable with x. We leave the easy details to the reader.

(ii) By (9.44) it suffices to compare $|x_1 + 2x_2|$ and $|x_1 - 2x_2|$ with $|x_1|$, and $|x_2 + 2x_1|$ and $|x_2 - 2x_1|$ with $|x_2|$. It is easy to check (exercise) that, by means of the symmetries

$$x_1 \leftrightarrow -x_1, \quad x_2 \leftrightarrow -x_2, \quad \text{and } x_1 \leftrightarrow x_2,$$

we may reduce to the case

$$0 < x_2 < x_1.$$

Clearly, we also have $x_1 < \frac{n}{2}$, $x_1 + x_2 < \frac{n}{2}$, and $x_2 < \frac{n}{4}$.

First comparison: We have

$$|x_1 - 2x_2| = \begin{cases} x_1 - 2x_2 < x_1 & \text{if } x_2 < \frac{x_1}{2} \\ 2x_2 - x_1 = x_2 - (x_1 - x_2) < x_1 & \text{if } \frac{x_1}{2} \le x_2 < x_1, \end{cases}$$

and therefore $T_1^{-1} x < x$.

Second comparison:

If $x_1 + 2x_2 < \frac{n}{2}$ then $|x_1 + 2x_2| = x_1 + 2x_2 > x_1$.

If $x_1 + 2x_2 \ge \frac{n}{2}$ then $x_2 < \frac{n}{4}$ yields $\frac{n}{2} \le x_1 + 2x_2 \le \frac{3}{4}n$, which in turn implies that $-\frac{n}{2} \le -n + x_1 + 2x_2 < -\frac{n}{4}$, so that $-n + x_1 + 2x_2$ represents $x_1 + 2x_2$ in the range $[-\frac{n}{2}, \frac{n}{2})$ and $|-n + x_1 + 2x_2| = n - x_1 - 2x_2 > x_1$, since $x_1 + x_2 < \frac{n}{2}$.

In both cases, $T_1 x > x$.

Third comparison:

If $2x_1 + x_2 < \frac{n}{2}$ then $|2x_1 + x_2| = 2x_1 + x_2 > x_2$.

If $2x_1 + x_2 \geq \frac{n}{2}$ then, from $2x_1 + x_2 = (x_1 + x_2) + x_1 < \frac{n}{2} + \frac{n}{2} = n$ we deduce that $\frac{n}{2} \leq 2x_1 + x_2 < n$, which in turn implies that $-\frac{n}{2} \leq 2x_1 + x_2 - n < 0$, so that $2x_1 + x_2 - n$ represents $2x_1 + x_2$ in $[-\frac{n}{2}, \frac{n}{2})$ and $|2x_1 + x_2 - n| = n - 2x_1 - x_2 > x_2$, because $x_1 + x_2 < \frac{n}{2}$.

In both cases, $T_2 x > x$.

Fourth comparison:

If $-2x_1 + x_2 \geq -\frac{n}{2}$ then $|-2x_1 + x_2| = 2x_1 - x_2 > x_2$.

If $-2x_1 + x_2 < -\frac{n}{2}$, from $x_1 < \frac{n}{2}$ we deduce that $-2x_1 + x_2 > -n$, which in turn implies that $0 < -2x_1 + x_2 + n < \frac{n}{2}$, so that $-2x_1 + x_2 + n$ represents $-2x_1 + x_2$ in $[-\frac{n}{2}, \frac{n}{2})$ and $|-2x_1 + x_2 + n| = n - 2x_1 + x_2 > x_2$ (because $x_1 < \frac{n}{2}$).

In both cases, $T_2^{-1} x > x$.

(iii) Arguing as in (ii), we may reduce to the case $0 = x_2 < x_1$. Then $T_1^{\pm 1} x = x$, $T_2 x = (x_1, 2x_1)^T > x$, and $T_2^{-1} x = (x_1, -2x_1)^T > x$. $\qquad\square$

Lemma 9.6.3 *Let $\gamma : A \times A \to \mathbb{R}$ denote the function defined by setting*

$$\gamma(x, y) = \begin{cases} \frac{5}{4} & \text{if } x > y \\ \frac{4}{5} & \text{if } y > x \\ 1 & \text{otherwise,} \end{cases}$$

for all $x, y \in A$. Then

$$\gamma(x, y)\gamma(y, x) = 1 \tag{9.46}$$

and

$$\gamma(x, y) \leq \frac{5}{4} \tag{9.47}$$

for all $x, y \in A$. Moreover, if $x = \begin{pmatrix} x_1 \\ x_2 \end{pmatrix} \in A \setminus \{0\}$, we have

$$\left|\cos \frac{\pi x_1}{n}\right| \cdot [\gamma(x, T_2 x) + \gamma(x, T_2^{-1} x)]$$
$$+ \left|\cos \frac{\pi x_2}{n}\right| \cdot [\gamma(x, T_1 x) + \gamma(x, T_1^{-1} x)] \leq 3.65. \tag{9.48}$$

Proof. (9.46) and (9.47) are obvious. We divide the proof of (9.48) into two cases.

First case: x is outside the diamond D. By virtue of (9.47), the left hand side of (9.48) is bounded above by

$$\frac{5}{2}(|\cos\frac{\pi x_1}{n}| + |\cos\frac{\pi x_2}{n}|).$$ (9.49)

Since the cosine function is even, we may assume that $0 \le x_1, x_2 \le \frac{n}{2}$ so that $0 \le \frac{\pi x_2}{n} \le \frac{\pi}{2}$ and $x_2 \mapsto \cos\frac{\pi x_2}{n}$ is positive and decreasing. It follows that the maximum of (9.49) is achieved on the boundary of the diamond, and therefore (9.49) is bounded above by (here the max is over all $0 \le x_1 \le \frac{n}{2}$):

$$\max\left(\frac{5}{2}(|\cos\frac{\pi x_1}{n}| + |\cos\frac{\pi(n/2 - x_1)}{n}|)\right) = \max\left(\frac{5}{2}(\cos\frac{\pi x_1}{n} + \sin\frac{\pi x_1}{n})\right)$$

$$\le \frac{5\sqrt{2}}{2} < 3.65.$$

Second case: x is inside the diamond D. Now, using the trivial estimate $|\cos\theta| \le 1$ we get that the left hand side of (9.48) is bounded by

$$\gamma(x, T_1 x) + \gamma(x, T_1^{-1} x) + \gamma(x, T_2 x) + \gamma(x, T_2^{-1} x).$$ (9.50)

If $|x_1| = |x_2|$ by Lemma 9.6.2.(i) and the definition of γ we have that (9.50) is bounded above by $1 + 1 + \frac{4}{5} + \frac{4}{5} = 3.6 < 3.65$. Suppose now that $|x_1| \ne |x_2|$. If $x_1 \ne 0 \ne x_2$, then by Lemma 9.6.2.(ii) we have that (9.50) is bounded above by $3 \cdot \frac{4}{5} + \frac{5}{4} = 3.65$. If either x_1 or x_2 is equal to zero, then by Lemma 9.6.2.(iii), we again have that (9.50) is bounded above by $1 + 1 + \frac{4}{5} + \frac{4}{5} = 3.6 < 3.65$. $\square$

Lemma 9.6.4 *Let $G: A \to \mathbb{R}$ be a non-negative function such that $G(0) = 0$. Then*

$$\sum_{x \in A} 2G(x)\left[G(T_2^{-1} x)|\cos\frac{\pi x_1}{n}| + G(T_1^{-1} x)|\cos\frac{\pi x_2}{n}|\right] \le 3.65 \sum_{x \in A} G(x)^2.$$ (9.51)

Proof. Let $x, y \in A$. From (9.46) we deduce that

$$2G(x)G(y) \le \gamma(x, y)G(x)^2 + \gamma(y, x)G(y)^2.$$

Then, the left hand side of (9.51) is bounded above by

$$\sum_{x \in A}\left\{|\cos\frac{\pi x_1}{n}| \cdot \left[\gamma(x, T_2^{-1} x)G(x)^2 + \gamma(T_2^{-1} x, x)G(T_2^{-1} x)^2\right] + \right.$$

$$\left. +|\cos\frac{\pi x_2}{n}| \cdot \left[\gamma(x, T_1^{-1} x)G(x)^2 + \gamma(T_1^{-1} x, x)G(T_1^{-1} x)^2\right]\right\}.$$ (9.52)

Setting $x' = T_2^{-1}x$ and observing that $x'_1 = x_1$ (that is, T_2 and T_2^{-1} do not change x_1, see (9.44)), we get

$$\sum_{x \in A} |\cos \frac{\pi x_1}{n}| \gamma(T_2^{-1}x, x)G(T_2^{-1}x)^2 = \sum_{x' \in A} |\cos \frac{\pi x'_1}{n}| \gamma(x', T_2x')G(x')^2.$$

Similarly, with the change of variable $x'' = T_1^{-1}x$, we have $x''_2 = x_2$ and

$$\sum_{x \in A} |\cos \frac{\pi x_2}{n}| \gamma(T_1^{-1}x, x)G(T_1^{-1}x)^2 = \sum_{x'' \in A} |\cos \frac{\pi x''_2}{n}| \gamma(x'', T_1x'')G(x'')^2.$$

Therefore, recalling that $\cos \frac{\pi x'_1}{n} = \cos \frac{\pi x_1}{n}$ and $\cos \frac{\pi x''_2}{n} = \cos \frac{\pi x_2}{n}$, the upper bound (9.52) equals

$$\sum_{x \in A} G(x)^2 \left\{ |\cos \frac{\pi x_1}{n}| \cdot [\gamma(x, T_2x) + \gamma(x, T_2^{-1}x)] \right.$$
$$\left. + |\cos \frac{\pi x_2}{n}| \cdot [\gamma(x, T_1x) + \gamma(x, T_1^{-1}x)] \right\},$$

which, by virtue of Lemma 9.6.3 and the hypothesis $G(0) = 0$, is bounded above by $3.65 \sum_{x \in A} G(x)^2$. $\qquad\square$

Finally, we state a result, which is a consequence of the previous lemmas and that will quickly lead to the proof that the Margulis graphs are expanders. Recall that $W_1(A) = \{f \in L(A) : \sum_{x \in A} f(x) = 0\}$, and set $e_1 = (1, 0)^T$ and $e_2 = (0, 1)^T$.

Theorem 9.6.5 *For all real valued $f \in W_1(A)$ we have*

$$\sum_{x \in A} f(x)[f(T_1x) + f(T_1x + e_1) + f(T_2x) + f(T_2x + e_2)] \le 3.65 \|f\|^2_{L(A)}.$$

$$(9.53)$$

Proof. First of all, note that, by virtue of Proposition 9.6.1, if F denotes the Fourier transform of f, then the Fourier transform of the function

$$x \mapsto f(T_1x) + f(T_1x + e_1) + f(T_2x) + f(T_2x + e_2)$$

is the function

$$x = \begin{pmatrix} x_1 \\ x_2 \end{pmatrix} \mapsto F(T_2^{-1}x) + F(T_2^{-1}x)\omega^{x_1} + F(T_1^{-1}x) + F(T_1^{-1}x)\omega^{x_2},$$

because $(T_1^{-1})^T = T_2^{-1}$, $(T_2^{-1})^T = T_1^{-1}$, $T_1^{-1}e_1 = e_1$, and $T_2^{-1}e_2 = e_2$. Therefore, by the identities of Plancherel and Parseval, (9.53) is equivalent to

$$\sum_{x \in A} \overline{F(x)}[F(T_2^{-1}x)(1 + \omega^{x_1}) + F(T_1^{-1}x)(1 + \omega^{x_2})] \le 3.65 \|F\|^2_{L(A)}, \quad (9.54)$$

while condition $\sum_{x \in A} f(x) = 0$ is equivalent to $F(0) = 0$ (see (9.43)). Since

$$|1 + \omega^t|^2 = |1 + \cos \frac{2\pi t}{n} + i \sin \frac{2\pi t}{n}|^2$$
$$= 2(1 + \cos \frac{2\pi t}{n})$$
$$= 4 \cos^2 \frac{\pi t}{n},$$

then (9.54) follows from Lemma 9.6.4 and the triangular inequality (by setting $G = |F|$). $\qquad\square$

We now present the Gabber-Galil version of the Margulis construction.

Definition 9.6.6 (Margulis expanders) For every integer $n \geq 1$, we define the 8-regular graph $\mathcal{M}_n = (X_n, E, r_{\mathcal{M}_n})$, where $X_n = \mathbb{Z}_n^2$, equipped with the rotation map (cf. Exercise 8.12.3) $\mathrm{Rot}_{\mathcal{M}_n} : X_n \times [8] \to X_n \times [8]$ defined by setting

$$\mathrm{Rot}_{\mathcal{M}_n}(x, i) = (y_i, i + 4 \bmod 8)$$

for all $x \in X_n$ and $i \in [8]$, where

$$\begin{aligned} y_1 &= T_1 x, \quad y_2 = T_2 x, \quad y_3 = T_1 x + e_1, \quad y_4 = T_2 x + e_2, \\ y_5 &= T_1^{-1} x, \quad y_6 = T_2^{-1} x, \quad y_7 = T_1^{-1} x - e_1, \quad y_8 = T_2^{-1} x - e_2, \end{aligned} \quad (9.55)$$

for all $x \in X$.

Observe that the second line of (9.55) can be rewritten as

$$x = T_1 y_5 = T_2 y_6 = T_1 y_7 + e_1 = T_2 y_8 + e_2, \quad (9.56)$$

showing, in particular, that $\mathrm{Rot}_{\mathcal{M}_n}$ is indeed a rotation map (cf. Exercise 8.12.3).

Note also that $\mathcal{M}_n$ may have multiple edges and loops. For instance, $T_1 0 = T_2 0 = T_1^{-1} 0 = T_2^{-1} 0 = 0$ so that there are (exactly) four loops at 0.

Exercise 9.6.7

(1) Show that, if n is divisible by 4, then there are two (distinct) edges connecting $\begin{pmatrix} x_1 \\ n/4 \end{pmatrix}$ and $\begin{pmatrix} x_1 + n/2 \\ n/4 \end{pmatrix}$.

(2) Show that, if $n - 1$ is divisible by 4, then there are two (distinct) edges connecting $\begin{pmatrix} x_1 \\ (n-1)/4 \end{pmatrix}$ and $\begin{pmatrix} x_1 + (n-1)/2 \\ (n-1)/4 \end{pmatrix}$.

Theorem 9.6.8 *The 8-regular graphs $\mathcal{M}_n = (X_n, E_n, r_{\mathcal{M}_n})$ satisfy:*

$$\mu_1(\mathcal{M}_n) \leq 7.3$$

for all $n \in \mathbb{N}$. In particular, $(\mathcal{M}_n)_{n \geq 1}$ is a family of expanders.

Proof. Let $n \geq 5$. For $f \in W_1$ real valued we have

$$\langle A_{\mathcal{M}_n} f, f \rangle = \sum_{x \in X_n} [A_{\mathcal{M}_n} f](x) f(x)$$

$$= \sum_{x \in X_n} f(x)[f(T_1 x) + f(T_2 x) + f(T_1 x + e_1) + f(T_2 x + e_2)$$

$$+ f(T_1^{-1} x) + f(T_2^{-1} x) + f(T_1^{-1} x - e_1) + f(T_2^{-1} x - e_2)]$$

$$\text{(by (9.56))} = 2 \sum_{x \in X_n} f(x)[f(T_1 x) + f(T_2 x) + f(T_1 x + e_1) + f(T_2 x + e_2)]$$

$$\leq 7.3 \, \|f\|^2_{L(X_n)},$$

where the inequality follows from Theorem 9.6.5. From (9.9) we deduce that $\mu_1(\mathcal{M}_n) \leq 7.3$ and therefore

$$0.7 = 8 - 7.3 \leq k - \mu_1(\mathcal{M}_n) = \delta(\mathcal{M}_n).$$

Thus, in accordance with Definition 9.5.3, $(\mathcal{M}_n)_{n \in \mathbb{N}}$ is a family of expanders with spectral gap $\delta \geq 0.7$. $\qquad \square$

9.7 The Alon-Schwartz-Shapira estimate

This section is an exposition of the main result in [10], where the authors – using, however, a slightly different definition of a replacement product – give a lower bound for the isoperimetric constant of a replacement product. This result is interesting because it does not rely on spectral techniques but on a direct combinatorial argument.

We use the notation of Definition 8.12.4.

Theorem 9.7.1 *Let $\mathcal{G} = (X, E, r_{\mathcal{G}})$ be a d-regular graph and $\mathcal{F} = (Y, F, r_{\mathcal{F}})$ a k-regular graph with $Y = [d]$. Assume that in both graphs we have defined a labelling and a rotation map as in Definition 8.12.1. Then:*

$$h(\mathcal{G}_{\textcircled{r}} \mathcal{F}) \geq \min \left\{ \frac{1}{40} \left[\frac{h(\mathcal{G})}{d} \right]^2 h(\mathcal{F}), \frac{1}{8} \frac{h(\mathcal{G})}{d} \right\}. \tag{9.57}$$

Proof. First of all, for $x \in X$ we set $\Xi_x = \{x\} \times [d]$, so that we can regard the vertex set $X \times [d]$ of $\mathcal{G}_{\textcircled{r}} \mathcal{F}$ as the disjoint union $\coprod_{x \in X} \Xi_x$ (observe that each Ξ_x is a copy of $\mathcal{F}$ and the Ξ_xs are joined according to the structure of $\mathcal{G}$, as explained in Definition 8.12.4 and Remark 8.12.6).

Let now $\Gamma \subseteq X \times [d]$ such that

$$|\Gamma| \leq \frac{1}{2} |X \times [d]| = \frac{|X| d}{2}. \tag{9.58}$$

Set

- $\Gamma_x = \Gamma \cap \Xi_x$;
- $X' = \{x \in X : |\Gamma_x| \leq d - \frac{h(\mathcal{G})}{4}\}$ and $X'' = X \setminus X'$;
- $\Gamma' = \coprod_{x \in X'} \Gamma_x$ and $\Gamma'' = \coprod_{x \in X''} \Gamma_x$ (clearly, $\Gamma' \coprod \Gamma'' = \Gamma$).

We distinguish two cases.
First case:

$$|\Gamma'| \geq \frac{1}{10}\frac{h(\mathcal{G})}{d}|\Gamma|. \tag{9.59}$$

Note that, by definition, for $x \in X'$ we have

$$|\Xi_x \setminus \Gamma_x| = d - |\Gamma_x| \geq \frac{h(\mathcal{G})}{4}$$

so that (observing that $|\Gamma_x| \leq d$)

$$|\Xi_x \setminus \Gamma_x| \geq \frac{h(\mathcal{G})}{4d}d \geq \frac{h(\mathcal{G})}{4d}|\Gamma_x|.$$

Similarly, from (9.2) we deduce that $\frac{h(\mathcal{G})}{4d} \leq \frac{1}{4} < 1$, so that

$$|\Gamma_x| \geq \frac{h(\mathcal{G})}{4d}|\Gamma_x|.$$

Then, both in the case $|\Gamma_x| \leq \frac{d}{2}$ and in the case $|\Xi \setminus \Gamma_x| \leq \frac{d}{2}$, by definition of $h(\mathcal{F})$, we deduce that there are at least $\frac{h(\mathcal{G})}{4d}|\Gamma_x|h(\mathcal{F})$ edges (of the second kind) connecting Γ_x and its complement $\Xi_x \setminus \Gamma_x$ (a copy of $\mathcal{F}$). Then, by our assumption (9.59), the edges connecting Γ and its complement are at least

$$\frac{1}{10}\frac{h(\mathcal{G})}{d}\frac{h(\mathcal{G})}{4d}|\Gamma|h(\mathcal{F}) = \frac{1}{40}\left(\frac{h(\mathcal{G})}{d}\right)^2|\Gamma|h(\mathcal{F}).$$

After dividing by $|\Gamma|$, this yields the first term in the minimum in (9.57).
Second case:

$$|\Gamma'| < \frac{1}{10}\frac{h(\mathcal{G})}{d}|\Gamma|. \tag{9.60}$$

Since $\Gamma = \Gamma' \coprod \Gamma''$, this gives

$$|\Gamma''| > \left(1 - \frac{h(\mathcal{G})}{10d}\right)|\Gamma|. \tag{9.61}$$

Moreover, since, by definition, $|\Gamma_x| > d - \frac{h(\mathcal{G})}{4}$ for each $x \in X''$, summing up over X'' we get $|\Gamma''| > |X''| \left(d - \frac{h(\mathcal{G})}{4}\right)$, and therefore

$$|X''| < \frac{|\Gamma''|}{d - \frac{h(\mathcal{G})}{4}} \leq \frac{|\Gamma|}{d - \frac{h(\mathcal{G})}{4}} \leq \frac{\frac{1}{2}d|X|}{d - \frac{h(\mathcal{G})}{4}}$$

(where the last inequality follows from (9.58)). From the inequality $h(\mathcal{G}) \leq d$ we deduce that

$$\frac{\frac{d}{2}}{d - \frac{h(\mathcal{G})}{4}} \leq \frac{2}{3},$$

so that

$$|X''| \leq \frac{2}{3}|X|. \tag{9.62}$$

Note also that

$$|X''| \geq \frac{1}{d}|\Gamma''| \tag{9.63}$$

simply because

$$|\Gamma''| = |\coprod_{x \in X''} \Gamma_x| \leq |\coprod_{x \in X''} \Xi_x| = d|X''|.$$

We claim that

$$\min\{|X'|, |X''|\} \geq \frac{|X''|}{2}.$$

Indeed, from (9.62) we deduce that

$$|X'| = |X| - |X''| \geq \frac{1}{3}|X| \geq \frac{1}{2}|X''|.$$

By definition of $h(\mathcal{G})$, it follows that there exists a set F of edges of $\mathcal{G}$ such that

$$|F| \geq \frac{1}{2}h(\mathcal{G})|X''|$$

and F connects X' with X''. Denote by Φ the corresponding set of edges (of the first type) in $\mathcal{G}_{\textcircled{T}}\mathcal{F}$ (so that they connect vertices in $\coprod_{x \in X'} \Xi_x$ with vertices in $\coprod_{x \in X''} \Xi_x$). Since for $x \in X''$ we have $|\Gamma_x| > d - \frac{h(\mathcal{G})}{4}$ then $|\Xi_x \setminus \Gamma_x| < \frac{h(\mathcal{G})}{4}$ so that at most $\frac{h(\mathcal{G})}{4}|X''|$ of the edges in Φ connect a vertex in $\coprod_{x \in X''}(\Xi_x \setminus \Gamma_x)$ with a vertex in $\coprod_{x \in X'} \Xi_x$ (recall that each vertex is incident to exactly one edge of the first type, cf. Remark 8.12.6). Therefore, if we denote by Φ_2 the subset of Φ of all edges that connect vertices of Γ'' with vertices in $\coprod_{x \in X'} \Xi_x$, then

$$|\Phi_2| \geq |\Phi| - \frac{h(\mathcal{G})}{4}|X''| \geq \frac{1}{4}h(\mathcal{G})|X''|. \tag{9.64}$$

Consider the decomposition $\Phi_2 = \Phi_3 \coprod \Phi_4$, where Φ_3 are the edges that connect vertices of Γ'' with vertices in Γ' and Φ_4 its complement (so that an edge of Φ_4 connects a vertex of $\Gamma'' \subseteq \Gamma$ with a vertex in the complement of Γ). Then

$$|\Phi_3| \leq |\Gamma'|$$

$$\text{(by (9.60))} \quad \leq \frac{1}{10}\frac{h(\mathcal{G})}{d}|\Gamma|$$

$$\text{(by (9.61))} \quad \leq \frac{h(\mathcal{G})/10d}{1 - h(\mathcal{G})/10d}|\Gamma''|$$

$$\text{(because } h(\mathcal{G}) \leq d) \quad \leq \frac{h(\mathcal{G})}{9d}|\Gamma''|$$

$$\text{(by (9.63))} \quad \leq \frac{h(\mathcal{G})}{9}|X''|.$$

It follows that

$$|\Phi_4| = |\Phi_2| - |\Phi_3|$$

$$\text{(by (9.64))} \quad \geq \left(\frac{1}{4} - \frac{1}{9}\right)h(\mathcal{G})|X''|$$

$$\text{(by (9.63))} \quad \geq \frac{5}{36}\frac{h(\mathcal{G})}{d}|\Gamma''|$$

$$\text{(by (9.61))} \quad \geq \frac{5}{36}\frac{h(\mathcal{G})}{d}\left(1 - \frac{h(\mathcal{G})}{10d}\right)|\Gamma|$$

$$\text{(because } 1 - \frac{h(\mathcal{G})}{10d} \geq \frac{9}{10}) \quad \geq \frac{1}{8}\frac{h(\mathcal{G})}{d}|\Gamma|.$$

This computation yields the second term in the min of (9.57), ending the proof of the theorem. $\qquad\square$

Theorem 9.7.1 applies to situations where we have a lower bound for the normalized isoperimetric constant of $\mathcal{G}$. Here, we give an example.

Corollary 9.7.2 *Let* $(\mathcal{G}_n)_{n\in\mathbb{N}}$ *be a family of regular graphs such that*

- *the degree of* $\mathcal{G}_n$ *is* d_n *and* $d_n \to +\infty$ *as* $n \to +\infty$;
- $\mathcal{G}_n$ *has* a_n *vertices and* $a_n \to +\infty$ *as* $n \to +\infty$;
- *there exists* $\delta > 0$ *such that* $\frac{h(\mathcal{G}_n)}{d_n} \geq \delta$ *for all* $n \in \mathbb{N}$.

Let also $(\mathcal{F}_n)_{n\in\mathbb{N}}$ *be a family of k-degree expander graphs. Suppose that* $\mathcal{F}_n$ *has* d_n *vertices and there exists* $\epsilon > 0$ *such that* $h(\mathcal{F}_n) \geq \epsilon$ *for all* $n \in \mathbb{N}$. *Then* $\mathcal{G}_n \textcircled{\scriptsize\tau} \mathcal{F}_n$ *is a family of* $(k+1)$-*degree expanders with* $a_n d_n$ *vertices and*

$$h(\mathcal{G}_n \textcircled{\scriptsize\tau} \mathcal{F}_n) \geq \min\left(\frac{\delta^2\epsilon}{40}, \frac{\delta}{8}\right)$$

for all $n \in \mathbb{N}$.

Proof. It is an immediate consequence of Theorem 9.7.1.   □

Following [10], we construct a family of graphs that may take the role of $\mathcal{G}_n$ in Corollary 9.7.2.

Let p be a prime number, $q = p^t$ for some positive integer t, and denote by $\mathbb{F}_q$ the field of order q. Given a positive integer r, we define the finite graph LD(q, r) as follows. The vertex set is $\mathbb{F}_q^{r+1} = \{(a_0, a_1, \ldots, a_r) : a_j \in \mathbb{F}, j = 0, 1, \ldots, r\}$. For each $a = (a_0, a_1, \ldots, a_r) \in \mathbb{F}_q^{r+1}$ and for each $(x, y) \in \mathbb{F}_q^2$ there is a edge connecting a with

$$a + y(1, x, x^2, \ldots, x^r) = (a_0 + y, a_1 + yx, \ldots, a_r + yx^r).$$

This way, there are q loops at each vertex (these correspond to the case $y = 0$), and all other edges are simple. It follows that LD(q, r) is regular of degree q^2 and has q^{r+1} vertices.

Theorem 9.7.3 *Suppose that $1 \le r \le q$. Then*

$$\mu_1(\mathrm{LD}(q, r)) \le qr.$$

Proof. We give a complete spectral analysis of the graph LD(q, r), by exhibiting an orthonormal set of eigenvectors and by computing the relative eigenvalues. Actually, the eigenvectors are the character of the additive Abelian group $\mathbb{F}_q^{r+1}$, but, in our exposition, we prefer to follow the original sources and derive their properties from scratch. Fix a nontrivial linear map $L : \mathbb{F}_q \to \mathbb{F}_p$. For instance, thinking of $\mathbb{F}_q$ as a t-dimensional vector space over $\mathbb{F}_p$ (i.e., $\mathbb{F}_q = \{(\alpha_1, \alpha_2, \ldots, \alpha_t) : \alpha_i \in \mathbb{F}_p, i = 1, 2, \ldots, t\}$), then we can take $L(\alpha_1, \alpha_2, \ldots, \alpha_t) = \alpha_1$. Another choice could be the trace map $\mathrm{Tr}_{\mathbb{F}_q/\mathbb{F}_p}$ (cf. Section 6.7). Also, for $a = (a_0, a_1, \ldots, a_r)$ and $b = (b_0, b_1, \ldots, b_r) \in \mathbb{F}_q^{r+1}$ we set

$$a \cdot b = \sum_{j=0}^{r} a_j b_j.$$

Let $\omega = e^{2\pi i/p}$ be a primitive p-th root of the unity and, for $a \in \mathbb{F}_q^{r+1}$, define $v_a : \mathbb{F}_q^{r+1} \to \mathbb{C}$ by setting

$$v_a(b) = \omega^{L(a \cdot b)}$$

for all $b \in \mathbb{F}_q^{r+1}$.

Note that

$$\overline{v_a} = v_{-a}, \quad v_a(b) = v_b(a), \quad \text{and} \quad v_a(b + c) = v_a(b) v_a(c) \qquad (9.65)$$

for all $a, b, c \in \mathbb{F}_q^{r+1}$.

We claim that for $a \neq (0, 0, \ldots, 0)$

$$\sum_{b \in \mathbb{F}_q^{r+1}} v_a(b) = 0. \tag{9.66}$$

Indeed,

$$\begin{aligned}
\sum_{b \in \mathbb{F}_q^{r+1}} v_a(b) &= \sum_{b \in \mathbb{F}_q^{r+1}} \omega^{L(a \cdot b)} \\
&= \sum_{h \in \mathbb{F}_p} \sum_{\substack{b \in \mathbb{F}_q^{r+1} \\ L(a \cdot b) = h}} \omega^h \\
&= K \sum_{h \in \mathbb{F}_p} \omega^h \\
&= 0,
\end{aligned}$$

where

$$K = |\{b \in \mathbb{F}_q^{r+1} : L(a \cdot b) = h\}| = \frac{|\mathbb{F}_q^{r+1}|}{|\mathbb{F}_q|} \cdot \frac{|\mathbb{F}_q|}{p}$$

is independent of h, and the last equality follows from the fact that ω is a primitive p-th root of the unity (recall Lemma 2.2.3). The claim is proved.

As a consequence, for $a, b \in \mathbb{F}_q^{r+1}$ from (9.65) and (9.66) we deduce

$$\langle v_a, v_b \rangle_{L(\mathbb{F}_q^{r+1})} = \sum_{c \in \mathbb{F}_q^{r+1}} v_a(c) \overline{v_b(c)} = \sum_{c \in \mathbb{F}_q^{r+1}} v_{a-b}(c) = \delta_{a,b} |\mathbb{F}_q^{r+1}|,$$

that is, the set $\{v_a : a \in \mathbb{F}_q^{r+1}\}$ is an orthogonal basis in $L(\mathbb{F}_q^{r+1})$. More precisely, $(v_a)_{a \in \mathbb{F}_q^{r+1}}$ constitutes a parameterization of the characters of $\mathbb{F}_q^{r+1}$.

We now show that the functions $v_a \in L(\mathbb{F}_q^{r+1})$ are eigenvectors of the adjacency matrix A of the graph $LD(q, r)$. Indeed, for $a, b \in \mathbb{F}_q^{r+1}$ we have

$$[Av_a](b) = \sum_{x,y \in \mathbb{F}_q} v_a(b + y(1, x, x^2, \ldots, x^r))$$

$$(\text{by (9.65)}) \quad = \left(\sum_{x,y \in \mathbb{F}_q} v_a(y(1, x, x^2, \ldots, x^r)) \right) v_a(b)$$

so that, setting $p_a(x) = \sum_{j=0}^{r} a_j x^j$, we have that v_a is an eigenvector whose corresponding eigenvalue μ_a is given by

$$\begin{aligned}
\mu_a &= \sum_{x,y \in \mathbb{F}_q} v_a(y(1, x, x^2, \dots, x^r)) \\
&= \sum_{x,y \in \mathbb{F}_q} \omega^{L(y p_a(x))} \\
&= \sum_{\substack{x \in \mathbb{F}_q \\ p_a(x)=0}} \sum_{y \in \mathbb{F}_q} \omega^{L(y p_a(x))} \\
&= |\mathbb{F}_q| \cdot |\{x \in \mathbb{F}_q : p_a(x) = 0\}|,
\end{aligned} \tag{9.67}$$

where the two last equalities follow from the identity $\sum_{y \in \mathbb{F}_q} \omega^{L(y p_a(x))} = |\mathbb{F}_q| \delta_{0, p_a(x)}$. Now, if $a = (0, 0, \dots, 0)$, then $\mu_a = |\mathbb{F}_q|^2 = q^2$: this is the largest eigenvalue (recall that $\mathrm{LD}(q, r)$ is q^2-regular). If $a \neq (0, 0, \dots, 0)$, then the polynomial $p_a(x)$ has at most r roots in $\mathbb{F}_q$ and therefore $\mu_a \leq |\mathbb{F}_q| r = qr$. $\square$

Corollary 9.7.4 *Suppose that* $1 \leq r \leq q/2$. *Then*

$$h(\mathrm{LD}(q, r)) \geq \frac{q^2}{4}.$$

Proof. This follows from Theorem 9.7.3 and the Alon-Milman theorem (Theorem 9.1.7):

$$h(\mathrm{LD}(q, r)) \geq \frac{q^2 - \mu_1(\mathrm{LD}(q, r))}{2} \geq \frac{q^2 - qr}{2} \geq \frac{q^2}{4}. \qquad \square$$

Example 9.7.5 For $n \in \mathbb{N}$ let

$$\mathcal{G}_n = \mathrm{LD}(2^n, 2^{n-1})$$

and

$$\mathcal{F}_n = \mathcal{M}_{2^n}$$

the Margulis graph (cf. Definition 9.6.6). Recall that $\mathcal{G}_n$ has $2^{n(2^{n-1}+1)}$ vertices and degree $d(\mathcal{G}_n) = 2^{2n}$. Moreover, by Corollary 9.7.4, $h(\mathcal{G}_n) \geq \frac{2^{2n}}{4}$ so that

$$\frac{h(\mathcal{G}_n)}{d(\mathcal{G}_n)} \geq \frac{1}{4}.$$

Also, $\mathcal{F}_n$ has 2^{2n} vertices and constant degree $d(\mathcal{F}_n) = 8$. Moreover, by virtue of Theorem 9.6.8 and the Alon-Milman theorem (Theorem 9.1.7), we have

$$h(\mathcal{F}_n) \geq \frac{8 - \mu_1(\mathcal{F}_n)}{2} \geq \frac{8 - 7.3}{2} = \frac{7}{20}.$$

Then by Corollary 9.7.2 (with $\varepsilon = \frac{7}{20}$ and $\delta = \frac{1}{4}$) we have that $\{\mathcal{G}_n \textcircled{z} \mathcal{F}_n\}_{n \in \mathbb{N}}$ is a family of 9-degree expanders. In fact, for every $n \in \mathbb{N}$, the graph $\mathcal{G}_n \textcircled{z} \mathcal{F}_n$ has $2^{n(2^{n-1}+1)} \cdot 2^{2n} = 2^{n(2^{n-1}+3)}$ vertices and its isoperimetric constant satisfies

$$h(\mathcal{G}_n \textcircled{z} \mathcal{F}_n) \geq \min\left(\frac{1}{40} \cdot \frac{1}{16} \cdot \frac{7}{20}, \frac{1}{8} \cdot \frac{1}{4}\right) = \frac{7}{12800}.$$

9.8 Estimates of the first nontrivial eigenvalue for the Zig-Zag product

In this section, following [128], we give an upper bound for the first nontrivial eigenvalue of a zig-zag product in terms of the first nontrivial eigenvalues of its factors.

We first need to introduce a slightly modified version of $\mu_1(\mathcal{G})$. Keeping the notation of Proposition 8.1.5, for a connected k-regular graph $\mathcal{G}$ we set

$$\tilde{\mu}_1(\mathcal{G}) = \max\{|\mu_1|, |\mu_{n-1}|\}.$$

In other words, $\tilde{\mu}_1(\mathcal{G})$ is the largest (in absolute value) eigenvalue of the adjacency matrix of $\mathcal{G}$ different from $\mu_0 = k$. Note that, if $\mathcal{G}$ is bipartite, then, by Proposition 8.3.4, $\tilde{\mu}_1(\mathcal{G}) = k$. Moreover, $\mu_1(\mathcal{G}) \leq \tilde{\mu}_1(\mathcal{G})$ and, by replacing μ_1 by $\tilde{\mu}_1$, we obtain a variant of the spectral definition of expanders (cf. Definition 9.1.9 and Definition 9.5.3).

In the notation of Proposition 8.1.4 and Lemma 9.1.6 we have

$$\tilde{\mu}_1(\mathcal{G}) = \max_{f \in W_1, f \neq 0} \frac{\|Af\|}{\|f\|} = \max_{f \in W_1, f \neq 0} \frac{|\langle Af, f \rangle|}{\|f\|^2}. \tag{9.68}$$

Indeed, if $v_0, v_1, \ldots, v_{n-1}$ is an orthonormal basis of $L(X)$ such that $Av_j = \mu_j v_j$ for $j = 0, 1, \ldots, n-1$, then $v_1, \ldots, v_{n-1}$ is an orthonormal basis of W_1. Thus, if $f = \sum_{j=1}^{n-1} \alpha_j v_j$ we have $Af = \sum_{j=1}^{n-1} \alpha_j \mu_j v_j$ and

$$\langle Af, Af \rangle = \sum_{j=1}^{n-1} |\alpha_j|^2 \mu_j^2 \leq \tilde{\mu}_1(\mathcal{G})^2 \|f\|^2$$

so that

$$\frac{\langle Af, Af \rangle}{\|f\|^2} \leq \tilde{\mu}_1(\mathcal{G})^2.$$

On the other hand, if $|\mu_j| = \tilde{\mu}_1(\mathcal{G})$ ($j = 1$ or $j = n - 1$) then

$$\frac{\langle Av_j, Av_j \rangle}{\|v_j\|^2} = \tilde{\mu}_1(\mathcal{G})^2.$$

The proof of the other equality is similar.

Remark 9.8.1 It is important to notice that since the adjacency matrix A of $\mathcal{G}$ is real and symmetric, we can select the orthonormal basis $\{v_0, v_1, \ldots, v_{n-1}\}$ of $L(X)$ made up of real-valued functions. Thus, denoting by $L_{\mathbb{R}}(X)$ the space of all real-valued functions on X, in (9.68) we can replace W_1 by $W_1 \cap L_{\mathbb{R}}(X)$.

In the following we shall use the notation in Sections 8.7, 8.12, and 8.13.

Lemma 9.8.2 *Let $f \in W_1(X \times [d])$. Then*

$$(I_X \otimes B)f^{\perp} \in L(X) \otimes W_1([d]) \tag{9.69}$$

and

$$\|(I_X \otimes B)f^{\perp}\| \leq \tilde{\mu}_1(\mathcal{F})\|f^{\perp}\|.$$

Proof. First of all, using (8.20) we have

$$(I_X \otimes B)f^{\perp} = (I_X \otimes B)\left(\sum_{x \in X} \delta_x \otimes f_x^{\perp}\right) = \sum_{x \in X} \delta_x \otimes Bf_x^{\perp}.$$

Then, using again (8.20) and the B-invariance of $W_1([d])$ (cf. Proposition 8.1.4), (9.69) follows. Moreover, by (9.68)

$$\|Bf_x^{\perp}\|_{L([d])} \leq \tilde{\mu}_1(\mathcal{F})\|f_x^{\perp}\|_{L([d])}$$

for all $x \in X$ so that

$$\|(I_X \otimes B)f^{\perp}\|_{L(X \times [d])}^2 \leq \sum_{x \in X} \|\delta_x \otimes Bf_x^{\perp}\|_{L(X \times [d])}^2$$

$$\text{(by (8.12))} = \sum_{x \in X} \|Bf_x^{\perp}\|_{L([d])}^2$$

$$\leq \tilde{\mu}_1(\mathcal{F})^2 \sum_{x \in X} \|f_x^{\perp}\|_{L([d])}^2$$

$$= \tilde{\mu}_1(\mathcal{F})^2 \|f^{\perp}\|_{L(X \times [d])}^2. \qquad \square$$

Lemma 9.8.3 *Let $f \in W_1(X \times [d])$. Then*

$$|\langle R_{\mathcal{G}}f^{\|}, f^{\|}\rangle| \leq \frac{\tilde{\mu}_1(\mathcal{G})}{d}\|f^{\|}\|^2.$$

Proof. First of all, note that, from Lemma 8.7.4 and Proposition 8.12.2, it follows that

$$CR_{\mathcal{G}}f^{\|} = \frac{1}{d}CR_{\mathcal{G}}\left[(Cf) \otimes \mathbf{1}_{[d]}\right] = \frac{1}{d}ACf. \tag{9.70}$$

Then, again by Lemma 8.7.4, we have

$$\langle R_{\mathcal{G}} f^{\parallel}, f^{\parallel} \rangle = \frac{1}{d} \langle R_{\mathcal{G}} f^{\parallel}, (Cf) \otimes \mathbf{1}_{[d]} \rangle_{L(X \times [d])}$$

$$= \frac{1}{d} \sum_{(x,i) \in X \times [d]} (R_{\mathcal{G}} f^{\parallel})(x, i) \overline{[Cf](x)}$$

(by (8.21))$\quad = \frac{1}{d} \langle C R_{\mathcal{G}} f^{\parallel}, Cf \rangle_{L(X)}$

(by (9.70))$\quad = \frac{1}{d^2} \langle ACf, Cf \rangle_{L(X)}.$

Now, by Lemma 8.7.3.(ii), $Cf \in W_1(X)$ and therefore

$$|\langle R_{\mathcal{G}} f^{\parallel}, f^{\parallel} \rangle| = \frac{1}{d^2} |\langle ACf, Cf \rangle|$$

(by (9.68))$\quad \leq \frac{\tilde{\mu}_1(\mathcal{G})}{d^2} \|Cf\|_{L(X)}^2$

(by (8.12))$\quad = \frac{\tilde{\mu}_1(\mathcal{G})}{d^3} \|(Cf) \otimes \mathbf{1}_{[d]}\|_{L(X \times [d])}^2$

(by Lemma 8.7.4)$\quad = \frac{\tilde{\mu}_1(\mathcal{G})}{d} \|f^{\parallel}\|^2.$ $\qquad \square$

Recall that $L_{\mathbb{R}}(X \times [d])$ denotes the space of all real valued functions defined on $X \times [d]$. Since $R_{\mathcal{G}}$ is a symmetric matrix, $L_{\mathbb{R}}(X \times [d])$ decomposes into eigenspaces of $R_{\mathcal{G}}$ and, since $R_{\mathcal{G}}^2 = I_{X \times [d]}$, we deduce that $R_{\mathcal{G}}$ has only 1 and -1 as eigenvalues. Set

- $V_1 = \{f \in L_{\mathbb{R}}(X \times [d]) : R_{\mathcal{G}} f = f\}$
- $V_2 = \{f \in L_{\mathbb{R}}(X \times [d]) : R_{\mathcal{G}} f = -f\}.$

Then

$$L_{\mathbb{R}}(X \times [d]) = V_1 \oplus V_2$$

is the orthogonal decomposition of $L_{\mathbb{R}}(X \times [d])$ into eigenspaces of $R_{\mathcal{G}}$.

Lemma 9.8.4 *Let $f \in L_{\mathbb{R}}(X \times [d])$. Then we have*

$$\langle R_{\mathcal{G}} f, f \rangle = \cos(2\theta) \|f\|^2,$$

where θ is the angle between f and V_1.

Proof. Write $f = f_1 + f_2$, with $f_1 \in V_1$ and $f_2 \in V_2$, so that

$$\|f_1\| = \cos\theta \|f\| \quad \text{and} \quad \|f_2\| = \sin\theta \|f\|,$$

as shown in Figure 9.3.

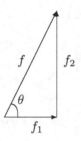

Figure 9.3. The decomposition $f = f_1 + f_2$, with $f_1 \in V_1$ and $f_2 \in V_2$.

Then

$$\langle R_g f, f \rangle = \langle f_1 - f_2, f_1 + f_2 \rangle$$
$$= \|f_1\|^2 - \|f_2\|^2$$
$$= (\cos^2 \theta - \sin^2 \theta)\|f\|^2$$
$$= \cos(2\theta)\|f\|^2. \qquad \square$$

We now introduce an auxiliary function: for $0 \leq \alpha, \beta \leq 1$ we set

$$\Phi(\alpha, \beta) = \frac{1}{2}(1 - \beta^2)\alpha + \frac{1}{2}\sqrt{(1 - \beta^2)^2\alpha^2 + 4\beta^2}.$$

The elementary properties of this function are described in the next lemma.

Lemma 9.8.5 *Let $0 \leq \alpha, \beta \leq 1$. Then the following holds.*

  (i) $\Phi(\alpha, 0) = \alpha$, $\Phi(0, \beta) = \beta$, *and* $\Phi(\alpha, 1) = \Phi(1, \beta) = 1$.
  (ii) *For $\beta < 1$ fixed, the function $\alpha \mapsto \Phi(\alpha, \beta)$ is strictly increasing.*
  (iii) *For $\alpha < 1$ fixed, the function $\beta \mapsto \Phi(\alpha, \beta)$ is strictly increasing.*
  (iv) *If $\alpha, \beta < 1$ then $\Phi(\alpha, \beta) < 1$.*
  (v) $\Phi(\alpha, \beta) \leq (1 - \beta^2)\alpha + \beta \leq \alpha + \beta$ (First upper bound).
  (vi) $\Phi(\alpha, \beta) \leq 1 - \frac{1}{2}(1 - \alpha)(1 - \beta^2)$ (Second upper bound).
  (vii) $\Phi(\alpha, \beta) \geq \frac{2\beta^2}{1 - \alpha + \beta^2(1 + \alpha)}$ (Lower bound).

Proof. (i) and (ii) are obvious. (iii) requires some elementary algebra. For the moment, suppose that $0 \leq \alpha < 1$ and $0 \leq \beta_1 < \beta_2 \leq 1$. Set $A_1 = (1 - \beta_1^2)\alpha$ and $A_2 = (1 - \beta_2^2)\alpha$. We have to prove that

$$A_1 + \sqrt{A_1^2 + 4\beta_1^2} < A_2 + \sqrt{A_2^2 + 4\beta_2^2}. \qquad (9.71)$$

First of all, note that $A_1 > A_2$ and

$$A_1^2 - A_2^2 = \alpha^2(\beta_2^2 - \beta_1^2)(2 - \beta_1^2 - \beta_2^2)$$
$$\leq 2(\beta_2^2 - \beta_1^2)$$
$$< 4(\beta_2^2 - \beta_1^2)$$

so that

$$A_1^2 + 4\beta_1^2 < A_2^2 + 4\beta_2^2. \tag{9.72}$$

We then write (9.71) in the form

$$A_1 - A_2 < \sqrt{A_2^2 + 4\beta_2^2} - \sqrt{A_1^2 + 4\beta_1^2}$$

which, by virtue of (9.72), is equivalent to (by squaring both sides)

$$\sqrt{A_2^2 + 4\beta_2^2}\sqrt{A_1^2 + 4\beta_1^2} < A_1 A_2 + 2\beta_1^2 + 2\beta_2^2.$$

Squaring again both sides, with some elementary calculations, (9.71) is in turn equivalent to

$$A_1^2\beta_2^2 + A_2^2\beta_1^2 < (\beta_1^2 - \beta_2^2)^2 + A_1 A_2(\beta_1^2 + \beta_2^2). \tag{9.73}$$

Now recalling that $A_j = (1 - \beta_j^2)\alpha$ for $j = 1, 2$ one easily checks that

$$A_1^2\beta_2^2 + A_2^2\beta_1^2 = \alpha^2(\beta_1^2 - \beta_2^2)^2 + A_1 A_2(\beta_1^2 + \beta_2^2)$$

and (9.73) follows. This shows (9.71).

(iv) follows from (i) and (ii) (or (iii)), but we give a straightforward direct proof. If $0 \leq \alpha, \beta < 1$ then $(1 - \beta^2)\alpha < 1 - \beta^2$ so that

$$\Phi(\alpha, \beta) = \frac{1}{2}(1 - \beta^2)\alpha + \frac{1}{2}\sqrt{(1 - \beta^2)^2\alpha^2 + 4\beta^2}$$
$$< \frac{1}{2}(1 - \beta^2) + \frac{1}{2}\sqrt{(1 - \beta^2)^2 + 4\beta^2}$$
$$= \frac{1}{2}(1 - \beta^2) + \frac{1}{2}(1 + \beta^2) = 1.$$

(v) Completing the square inside the square root we have

$$\Phi(\alpha, \beta) \leq \frac{1}{2}(1 - \beta^2)\alpha + \frac{1}{2}\sqrt{(1 - \beta^2)^2\alpha^2 + 4\beta^2 + 4\beta(1 - \beta^2)\alpha}$$
$$= \frac{1}{2}(1 - \beta^2)\alpha + \frac{1}{2}[(1 - \beta^2)\alpha + 2\beta]$$
$$= (1 - \beta^2)\alpha + \beta.$$

(vi) The inequality

$$\Phi(\alpha, \beta) \le 1 - \frac{1}{2}(1 - \alpha)(1 - \beta^2)$$

is equivalent to

$$\sqrt{(1 - \beta^2)^2 \alpha^2 + 4\beta^2} \le 1 + \beta^2.$$

Squaring both sides this becomes

$$\alpha^2(1 - \beta^2)^2 \le (1 - \beta^2)^2$$

which is satisfied since $\alpha^2 \le 1$.

(vii) If $\frac{2\beta^2}{1 - \alpha + \beta^2(1+\alpha)} \le \frac{1}{2}(1 - \beta^2)\alpha$ then there is nothing to prove. Otherwise, we can write the inequality in the form

$$\frac{2\beta^2}{1 - \alpha + \beta^2(1 + \alpha)} - \frac{1}{2}(1 - \beta^2)\alpha \le \frac{1}{2}\sqrt{(1 - \beta^2)^2 \alpha^2 + 4\beta^2}$$

and squaring both sides (the left hand side is positive) we get

$$4\beta^2 \le 2\alpha(1 - \beta^2)[(1 - \alpha) + \beta^2(1 + \alpha)] + [(1 - \alpha) + \beta^2(1 + \alpha)]^2,$$

that is,

$$\begin{aligned} 4\beta^2 + \alpha^2(1 - \beta^2)^2 &\le \{\alpha(1 - \beta^2) + [(1 - \alpha) + \beta^2(1 + \alpha)]\}^2 \\ &= (1 + \beta^2)^2 \end{aligned}$$

which becomes

$$\alpha^2(1 - \beta^2)^2 \le (1 - \beta^2)^2.$$

This is clearly satisfied since $\alpha^2 \le 1$. $\qquad\qquad\square$

Remark 9.8.6 The first upper bound is useful when α and β are small, while the second upper bound is useful when α and β are close to one. Moreover, it is an easy exercise to show that if $\beta < 1$ then

$$\alpha(1 - \beta^2) + \beta \le 1 - \frac{1}{2}(1 - \alpha)(1 - \beta^2)$$

if and only if

$$\alpha \le \frac{1 - \beta}{1 + \beta}.$$

In [127] the authors use the function $\Psi(\alpha, \beta) = 1 - (1 - \alpha)(1 - \beta)^2$ in place of Φ. This is also useful when α and β are close to one. We just note that

$$1 - \frac{1}{2}(1 - \alpha)(1 - \beta^2) \leq 1 - (1 - \alpha)(1 - \beta)^2$$

if and only if $\beta \geq \frac{1}{3}$. As a consequence, as soon as $\beta \geq \frac{1}{3}$, the second upper bound in Lemma 9.8.5 yields a better estimate than the one provided by Ψ in [127].

We are now in position to state and prove the main result of this section.

Theorem 9.8.7 (Reingold-Vadhan-Wigderson) *In the notation of Section 8.13 we have the following inequality for the first nontrivial eigenvalue of a zig-zag product:*

$$\tilde{\mu}_1(\mathcal{G}\textcircled{z}\mathcal{F}) \leq k^2 \Phi\left(\frac{\tilde{\mu}_1(\mathcal{G})}{d}, \frac{\tilde{\mu}_1(\mathcal{F})}{k}\right),$$

where Φ is the function in Lemma 9.8.5.

Proof. Let $0 \neq f \in W_1(X \times [d]) \cap L_{\mathbb{R}}(X \times [d])$ (cf. Remark 9.8.1). By virtue of Lemma 8.7.4 (recall that B is the adjacency matrix of $\mathcal{F}$) we have

$$(I_X \otimes B)f^{\parallel} = \frac{1}{d}(I_X \otimes B)[(Cf) \otimes \mathbf{1}_{[d]}]$$

$$(\text{as } B\mathbf{1}_{[d]} = k\mathbf{1}_{[d]}) = \frac{1}{d}[(Cf) \otimes k\mathbf{1}_{[d]}]$$

$$= kf^{\parallel}.$$

Therefore,

$$(I_X \otimes B)f = (I_X \otimes B)(f^{\parallel} + f^{\perp}) = kf^{\parallel} + (I_X \otimes B)f^{\perp}. \qquad (9.74)$$

Setting $\tilde{B} = \frac{1}{k}(I_X \otimes B)$ and recalling Proposition 8.13.3 we have

$$\langle M_{\mathcal{G}\textcircled{z}\mathcal{F}}f, f\rangle = \langle (I_X \otimes B)R_{\mathcal{G}}(I_X \otimes B)f, f\rangle$$

$$= \langle R_{\mathcal{G}}(I_X \otimes B)f, (I_X \otimes B)f\rangle$$

$$(\text{by } (9.74)) = k^2\langle R_{\mathcal{G}}(f^{\parallel} + \tilde{B}f^{\perp}), f^{\parallel} + \tilde{B}f^{\perp}\rangle$$

$$(\text{by Lemma } 9.8.4) = k^2 \cos 2\theta \|f^{\parallel} + \tilde{B}f^{\perp}\|^2$$

(where $\theta \in [0, \pi/2]$ is the angle between $f^{\parallel} + \tilde{B}f^{\perp}$ and V_1) so that

$$\frac{\langle M_{\mathcal{G}\textcircled{z}\mathcal{F}}f, f\rangle}{\|f\|^2} = k^2 \cos 2\theta \frac{\|f^{\parallel} + \tilde{B}f^{\perp}\|^2}{\|f^{\parallel} + f^{\perp}\|^2}. \qquad (9.75)$$

By virtue of (9.68), the remaining part of the proof is devoted to get an upper bound for the modulus of the right hand side of the above equality. We introduce three further angles:

- $\varphi \in [0, \pi/2]$ is the angle between $f^{\parallel}$ and $f = f^{\parallel} + f^{\perp}$ (see Figure 9.4);

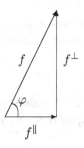

Figure 9.4. $\varphi \in [0, \pi/2]$ is the angle between $f^{\parallel}$ and $f = f^{\parallel} + f^{\perp}$.

- φ' is the angle between $f^{\parallel}$ and $f^{\parallel} + \tilde{B}f^{\perp}$ (see Figure 9.5);

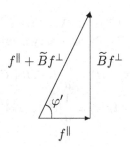

Figure 9.5. φ' is the angle between $f^{\parallel}$ and $f^{\parallel} + \tilde{B}f^{\perp}$.

- $\psi \in [0, \pi/2]$ is the angle between $f^{\parallel}$ and V_1.

By (9.69) we have that $f^{\parallel} \perp \tilde{B}f^{\perp}$ so that $\varphi' \in [0, \pi/2]$. We claim that

$$\theta \in [\psi - \varphi', \psi + \varphi'].$$

By symmetry, it suffices to prove that $\theta \leq \psi + \varphi'$, because by switching the role of ψ and θ (that is, switching $f^{\parallel}$ with $f^{\parallel} + \tilde{B}f^{\perp}$, see Figure 9.5) the inequality $\psi \leq \varphi' + \theta$ follows. Let h be the orthogonal projection of $f^{\parallel}$ into V_1 and denote by $\tilde{\theta}$ the angle between $f^{\parallel} + \tilde{B}f^{\perp}$ and h. Then, ψ is the angle between h and $f^{\parallel}$, $\tilde{\theta} \leq \psi + \varphi'$, by virtue of the triangular inequality for angles in a three dimensional real space and $\theta \leq \tilde{\theta}$ because θ is the minimal angle between $f^{\parallel} + \tilde{B}f^{\perp}$ and a vector $\tilde{h} \in V_1$.

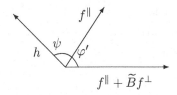

Figure 9.6. $\psi \in [0, \pi/2]$ is the angle between $f^{\parallel}$ and V_1.

Keeping in mind Figure 9.4 and Figure 9.5, and by virtue of Lemma 9.8.2, we have

$$\frac{\tan \varphi'}{\tan \varphi} = \frac{\|f^{\parallel}\| \tan \varphi'}{\|f^{\parallel}\| \tan \varphi} = \frac{\|\tilde{B} f^{\perp}\|}{\|f^{\perp}\|} \leq \frac{1}{k} \tilde{\mu}_1(\mathcal{F}). \qquad (9.76)$$

By Lemma 9.8.3, Lemma 9.8.4, and the definition of ψ

$$\cos 2\psi = \frac{\langle R_{\mathcal{G}} f^{\parallel}, f^{\parallel} \rangle}{\|f^{\parallel}\|^2} \leq \frac{\tilde{\mu}_1(\mathcal{G})}{d}. \qquad (9.77)$$

By Figure 9.4 and Figure 9.5,

$$\frac{\|f^{\parallel} + \tilde{B} f^{\perp}\|^2}{\|f^{\parallel} + f^{\perp}\|^2} = \frac{\frac{1}{\cos^2 \varphi'} \|f^{\parallel}\|^2}{\frac{1}{\cos^2 \varphi} \|f^{\parallel}\|^2} = \frac{\cos^2 \varphi}{\cos^2 \varphi'}.$$

In conclusion (see equation (9.75) and the observation following it), we have to maximize

$$k^2 |\cos 2\theta| \frac{\|f^{\parallel} + \tilde{B} f^{\perp}\|^2}{\|f^{\parallel} + f^{\perp}\|^2} = k^2 |\cos 2\theta| \frac{\cos^2 \varphi}{\cos^2 \varphi'}$$

subject to the constraints:

(1) $\varphi, \varphi', \psi \in [0, \frac{\pi}{2}]$;
(2) $\theta \in [\psi - \phi', \psi + \varphi']$;
(3) $\beta = \frac{\tan \varphi'}{\tan \varphi} \leq \frac{\tilde{\mu}_1(\mathcal{F})}{k}$ (cf. (9.76));
(4) $\alpha = |\cos 2\psi| \leq \frac{\tilde{\mu}_1(\mathcal{G})}{d}$ (cf. (9.77)).

We distinguish two cases, namely

$$0, \frac{\pi}{2} \notin [\psi - \varphi', \psi + \varphi'] \Leftrightarrow \varphi' < \min\{\psi, \frac{\pi}{2} - \psi\} \Leftrightarrow \varphi' < \psi < \frac{\pi}{2} - \varphi'$$

(this condition ensures that $\cos 2\psi < 1$) and

$$\varphi' \geq \min\{\psi, \frac{\pi}{2} - \psi\}$$

(now $\cos 2\psi = 1$ is possible).

Case I: $\varphi' < \min\{\psi, \frac{\pi}{2} - \psi\}$.

First of all, note that since

$$0 < \psi - \varphi' \le \theta \le \psi + \varphi' < \frac{\pi}{2}$$

we have

$$
\begin{aligned}
|\cos 2\theta| &\le \max\{|\cos 2(\psi + \varphi')|, |\cos 2(\psi - \varphi')|\} \\
&= \max\{|\cos 2\psi \cos 2\varphi' - \sin 2\psi \sin 2\varphi'|, \\
&\quad |\cos 2\psi \cos 2\varphi' + \sin 2\psi \sin 2\varphi'|\} \\
&=_* \begin{cases} \cos 2\psi \cos 2\varphi' + \sin 2\psi \sin 2\varphi' & \text{if } \cos 2\psi \cos 2\varphi' \ge 0 \\ -\cos 2\psi \cos 2\varphi' + \sin 2\psi \sin 2\varphi' & \text{if } \cos 2\psi \cos 2\varphi' < 0 \end{cases} \\
&= |\cos 2\psi \cos 2\varphi'| + \sin 2\psi \sin 2\varphi',
\end{aligned}
$$

where $=_*$ follows from $\sin 2\psi \sin 2\varphi' \ge 0$. Therefore

$$
\begin{aligned}
|\cos 2\theta| \frac{\cos^2 \varphi}{\cos^2 \varphi'} &\le \left| \frac{\cos^2 \varphi}{\cos^2 \varphi'} \cos 2\varphi' \cos 2\psi \right| + \frac{\cos^2 \varphi}{\cos^2 \varphi'} \sin 2\psi \sin 2\varphi' \\
&= \frac{1}{2} |(1 - \beta^2) \cos 2\psi + (1 + \beta^2) \cos 2\psi \cos 2\varphi| \\
&\quad + \beta \sin 2\psi \sin 2\varphi
\end{aligned}
\tag{9.78}
$$

where $\beta = \frac{\tan \varphi'}{\tan \varphi}$ as in (3), and the last equality follows from two elementary trigonometric identities, namely

$$\frac{\cos^2 \varphi}{\cos^2 \varphi'} \cos 2\varphi' = \frac{1}{2}[1 - \beta^2 + (1 + \beta^2) \cos 2\varphi],$$

which has a long but elementary proof, left to the reader, and

$$\frac{\cos^2 \varphi}{\cos^2 \varphi'} \sin 2\varphi' = \frac{\frac{\sin 2\varphi}{2\tan\varphi}}{\frac{\sin 2\varphi'}{2\tan\varphi'}} \sin 2\varphi' = \frac{\frac{1}{\tan\varphi}}{\frac{1}{\tan\varphi'}} \sin 2\varphi = \beta \sin 2\varphi.$$

Finally, the triangular inequality applied to (9.78) (recall that $|\beta| < 1$ by (9.76)) yields

$$
|\cos 2\theta| \frac{\cos^2 \varphi}{\cos^2 \varphi'} \leq \frac{1}{2}(1 - \beta^2)|\cos 2\psi|
$$

$$
+ \frac{1}{2}(1 + \beta^2)|\cos 2\psi| \cdot |\cos 2\varphi| + \frac{1}{2} \cdot 2\beta \sin 2\psi \cdot \sin 2\varphi
$$

$$
\leq_{**} \frac{1}{2}(1 - \beta^2)|\cos 2\psi|
$$

$$
+ \frac{1}{2}\sqrt{(1 + \beta^2)^2(\cos 2\psi)^2 + 4\beta^2(\sin 2\psi)^2}
$$

$$
\text{(by (4))} = \frac{1}{2}(1 - \beta^2)\alpha + \frac{1}{2}\sqrt{(1 - \beta^2)^2\alpha^2 + 4\beta^2}
$$

$$
= \Phi(\alpha, \beta),
$$

where $\leq_{**}$ follows by applying the Cauchy-Schwarz inequality. We then conclude by invoking Lemma 9.8.5.(ii) and (iii), and keeping in mind the inequalities in (3) and (4).

<u>Case II:</u> $\varphi' \geq \min\{\psi, \frac{\pi}{2} - \psi\}$.
We now have $\psi - \varphi' \leq 0$ or $\psi + \varphi' \geq \frac{\pi}{2}$ so that

$$
|\cos 2\theta| \frac{\cos^2 \varphi}{\cos^2 \varphi'} \leq \frac{\cos^2 \varphi}{\cos^2 \varphi'}
$$

$$
= \frac{\tan^2 \varphi'}{\tan^2 \varphi} + (1 - \frac{\tan^2 \varphi'}{\tan^2 \varphi})\cos^2 \varphi \qquad (9.79)
$$

$$
\text{(by (3))} = \beta^2 + (1 - \beta^2)\cos^2 \varphi,
$$

where the first equality is an elementary trigonometric identity, whose proof is left to the reader. Now, since $\varphi' \geq \min\{\psi, \frac{\pi}{2} - \psi\}$, we have

$$
\begin{cases} 2\varphi' \geq 2\psi \\ \text{or} \\ 2\varphi' \geq \pi - 2\psi \end{cases} \Rightarrow \begin{cases} \cos 2\varphi' \leq \cos 2\psi \\ \text{or} \\ \cos 2\varphi' \leq -\cos 2\psi \end{cases} \Rightarrow \cos 2\varphi' \leq |\cos 2\psi| = \alpha.
$$

Since

$$
\cos 2\varphi' = \frac{(1 + \beta^2)\cos^2 \varphi - \beta^2}{(1 - \beta^2)\cos^2 \varphi + \beta^2}
$$

(another trigonometric identity whose proof is left as an exercise) we get

$$
\frac{(1 + \beta^2)\cos^2 \varphi - \beta^2}{(1 - \beta^2)\cos^2 \varphi + \beta^2} \leq \alpha,
$$

which is equivalent to

$$\cos^2 \varphi \leq \frac{\beta^2(1+\alpha)}{\beta^2(1+\alpha)+1-\alpha}.$$

Applying this inequality to (9.79) we get

$$|\cos 2\theta| \frac{\cos^2 \varphi}{\cos^2 \varphi'} \leq \beta^2 + (1-\beta^2)\frac{\beta^2(1+\alpha)}{\beta^2(1+\alpha)+1-\alpha}$$
$$= \frac{2\beta^2}{1-\alpha+\beta^2(1+\alpha)}$$
$$\leq \Phi(\alpha, \beta),$$

where the last inequality follows from Lemma 9.8.5.(vii). The statement then follows, as in the previous case, from Lemma 9.8.5.(ii) and (iii), together with the inequalities in (3) and (4). $\qquad\square$

9.9 Explicit construction of expanders via the Zig-Zag product

In this section, we present the basic recursive construction that uses the estimates in Theorem 9.8.7 to construct a family of expander graphs. Let $\mathcal{G} = (X, E, r)$ be a finite connected graph. We define the *non-oriented square* of $\mathcal{G}$ as the graph $\mathcal{G}^2 = (X, F, s)$ with the same vertex set of $\mathcal{G}$, edge set defined as

$$F = \{\{x, e_1, y, e_2, z\} : x, y, z \in X, e_1, e_2 \in E, r(e_1) = \{x, y\}, r(e_2) = \{y, z\}\},$$

where $\{x, e_1, y, e_2, z\}$ should be thought of as the pair of paths (x, e_1, y, e_2, z) and (z, e_2, y, e_1, x), and $s(\{x, e_1, y, e_2, z\}) = \{x, z\}$ for all $\{x, e_1, y, e_2, z\} \in F$ (note that x, y, z are not necessarily distinct). In other words, F is the set of all (non-oriented) paths of length two in $\mathcal{G}$.

Clearly, if A is the adjacency matrix of $\mathcal{G}$, then A^2 is the adjacency matrix of $\mathcal{G}^2$ (see Proposition 8.1.6). Moreover, it is immediate to see that $\mathcal{G}^2$ is connected if and only if $\mathcal{G}$ is not bipartite: the reader is invited to find a direct proof of this fact and, in the case $\mathcal{G}$ is k-regular, to deduce it from Proposition 8.3.4 and Proposition 8.1.5.

Finally, if $\mathcal{G}$ is k-regular we clearly have

$$\tilde{\mu}_1(\mathcal{G}^2) = \tilde{\mu}_1(\mathcal{G})^2. \tag{9.80}$$

Theorem 9.9.1 *Let $\mathcal{G}$ be a d-regular graph with d^4 vertices, $d \geq 2$ and suppose that*

$$\tilde{\mu}_1(\mathcal{G}) \leq \frac{d}{4}.$$

Set

$$\mathcal{G}_1 = \mathcal{G}^2 \quad \text{and} \quad \mathcal{G}_{n+1} = \mathcal{G}_n^2 \textcircled{z} \mathcal{G} \text{ for } n \geq 1.$$

Then $\mathcal{G}_n$ is a d^2-regular graph with d^{4n} vertices and

$$\tilde{\mu}_1(\mathcal{G}_n) \leq \frac{d^2}{2}. \qquad (9.81)$$

In particular, the sequence $(\mathcal{G}_n)_{n\in\mathbb{N}}$ is a family of expanders.

Proof. By construction, $\mathcal{G}_1$ has d^4 vertices, is regular of degree d^2, and satisfies $\tilde{\mu}_1(\mathcal{G}_1) \leq \frac{d^2}{16}$ (by (9.80)). We proceed by induction. Suppose that $\mathcal{G}_n$ is a d^2-regular graph with d^{4n} vertices and (9.81) holds. Then $\mathcal{G}_n^2$ has d^{4n} vertices, is regular of degree d^4, and satisfies $\tilde{\mu}_1(\mathcal{G}_n^2) \leq \frac{d^4}{4}$. Therefore $\mathcal{G}_{n+1}$ has $d^{4n} \cdot d^4 = d^{4(n+1)}$ vertices, is regular of degree d^2 (by Definition 8.13.1), and, by Theorem 9.8.7 and Lemma 9.8.5.(v),

$$\tilde{\mu}_1(\mathcal{G}_{n+1}) \leq d^2 \left(\frac{1}{4} + \frac{1}{4}\right) = \frac{d^2}{2}.$$

The sequence $(\mathcal{G}_n)_{n\in\mathbb{N}}$ then forms a family of expanders (cf. Definition 9.5.3). □

Example 9.9.2 Consider the graph $LD(q, r)$ introduced in Section 9.7, where $q = p^t$ with p prime, and t, r positive integers. We use the notation in the proof of Theorem 9.7.3. Recall (cf. (9.67)) that the eigenvalues of $LD(q, r)$ are given by

$$\mu_a = \sum_{\substack{x\in\mathbb{F}_q \\ p_a(x)=0}} \sum_{y\in\mathbb{F}_q} \omega^{L(yp_a(x))},$$

$a \in \mathbb{F}_q^{r+1}$. Now, for $a \neq (0, 0, \ldots, 0)$, the polynomial $p_a(x)$ has at most r roots in $\mathbb{F}_q$ and therefore (cf. the end of the proof of Theorem 9.7.3)

$$|\mu_a| \leq qr. \qquad (9.82)$$

Then, for $r = 7$ and $q \geq 4r$ the graph $\mathcal{G} = LD(q, 7)$ satisfies the hypotheses of Theorem 9.9.1. Indeed, $\mathcal{G}$ is d-regular of degree $d = q^2$, the number of its vertices is $q^{r+1} = q^8 = d^4$, and (cf. (9.82))

$$\tilde{\mu}_1(\mathcal{G}) \leq r\sqrt{d} \leq \frac{d}{4}.$$

Part IV
Harmonic analysis on finite linear groups

10

Representation theory of finite groups

In this chapter we give a concise but quite detailed and complete exposition of the basic representation theory of finite groups. This may be considered as a noncommutative analogue of Chapter 2. Indeed, we emphasize the harmonic analytic point of view, focusing on unitary representations and Fourier transforms. Our exposition is based on our previous books [29], [33]. We also refer to the useful monographs by: Alperin and Bell [12], Diaconis [53], Fulton and Harris [63], Naimark and Stern [119], Serre [145], Simon [148], and Sternberg [154].

10.1 Representations, irreducibility, and equivalence

Let G be a finite group and V a finite dimensional vector space over $\mathbb{C}$. We denote by $\text{End}(V)$ the algebra (see Section 10.3) of all linear maps $T: V \to V$ and by $\text{GL}(V)$ the *general linear group* of V consisting of all invertible elements in $\text{End}(V)$.

Definition 10.1.1 A *representation* of G *over* V is a homomorphism $\rho: G \to \text{GL}(V)$. In other words, we have:

- $\rho(g): V \to V$ is linear and invertible for all $g \in G$;
- $\rho(g_1 g_2) = \rho(g_1)\rho(g_2)$ for all $g_1, g_2 \in G$;
- $\rho(g^{-1}) = \rho(g)^{-1}$ for all $g \in G$;
- $\rho(1_G) = I_V$ where 1_G is the identity element in G and $I_V: V \to V$ is the identity map (and thus the identity element in $\text{GL}(V)$).

We shall denote a representation by a pair (ρ, V). Note also that ρ may be seen as an action $(g, v) \mapsto \rho(g)v$ of G on V, where $\rho(g)$ is an invertible linear map for all $g \in G$. Denoting by n the dimension $\dim(V)$ of V, since $\text{GL}(V)$ is isomorphic to $GL(n, \mathbb{C})$, the group of invertible n-by-n complex matrices, we can regard a representation of G as a group homomorphism $\rho \colon G \to GL(n, \mathbb{C})$. Then n is the *dimension* or *degree* of ρ and it will be usually denoted by d_ρ.

The *kernel* of the representation (ρ, V) is $\mathrm{Ker}\rho = \{g \in G : \rho(g) = I_V\}$. The representation (ρ, V) is called *faithful* if $\mathrm{Ker}\rho = \{1_G\}$. In other words, ρ is faithful if and only if it is an isomorphism between G and a subgroup of $GL(V)$.

Let (ρ, V) be a representation of G and suppose that $W \leq V$ is a subspace. We say that W is *G-invariant* (or ρ-*invariant*) if $\rho(g)w \in W$ for all $g \in G$ and $w \in W$. Then, denoting by $\rho_W(g)$ the restriction of $\rho(g)$ to the subspace W, that is, $\rho_W(g)w = \rho(g)w$ for all $g \in G$ and $w \in W$, we say that (ρ_W, W) is the *restriction* of ρ to the (invariant) subspace W and call it a *sub-representation* of (ρ, V). We also say that ρ_W is *contained* in ρ and write $(\rho_W, W) \preceq (\rho, V)$, or simply $\rho_W \preceq \rho$. One also says that an element $v \in V$ is a *G-invariant vector*, provided $\rho(g)v = v$ for all $g \in G$. It is clear that the set of G-invariant vector is a G-invariant subspace $V^G \leq V$, which we call the *subspace of G-invariant vectors*. Clearly, every representation is a sub-representation of itself.

Let $K \leq G$ be a subgroup of G. Then setting $[\mathrm{Res}_K^G \rho](k) = \rho(k)$ for all $k \in K$, yields a K-representation $(\mathrm{Res}_K^G \rho, V)$. This is called the *restriction* of ρ to the subgroup K.

The representation (ρ, V) is called *irreducible* if the only G-invariant subspaces are trivial: $W \leq V$ and $\rho(g)W \leq W$ for all $g \in G$ implies that either $W = \{0\}$ or $W = V$.

The *direct sum* of given G-representations (ρ_j, W_j), $j = 1, 2, \ldots, k$, is the G-representation (ρ, V) where $V = W_1 \oplus W_2 \oplus \cdots \oplus W_k$ is the direct sum of the corresponding spaces, and $\rho = \rho_1 \oplus \rho_2 \oplus \cdots \oplus \rho_k$ is defined by setting

$$\rho(g)v = \rho_1(g)w_1 + \rho_2(g)w_2 + \cdots + \rho_k(g)w_k$$

for all $v = w_1 + w_2 + \cdots + w_k \in V$, $w_i \in W_i$, and $g \in G$. Conversely, if (ρ, V) is a G-representation and

$$V = W_1 \oplus W_2 \oplus \cdots \oplus W_k \tag{10.1}$$

is a direct sum decomposition into G-invariant subspaces, then $\rho = \rho_1 \oplus \rho_2 \oplus \cdots \oplus \rho_k$, where $\rho_j = \rho_{W_j}$, $j = 1, 2, \ldots, k$; we then say that (10.1) constitutes a (direct sum) *decomposition* of ρ.

Let (ρ, V) and (θ, W) be two representations of the same group G. Suppose that there exists a linear isomorphism $T \colon V \to W$ such that, for all $g \in G$,

$$\theta(g)T = T\rho(g). \tag{10.2}$$

Then one says that the two representations are *equivalent* and we write $\rho \sim \theta$. Note that $\sim$ is an equivalence relation and that two equivalent representations have the same degree. We write $\rho \nsim \theta$ to denote that ρ and θ are not equivalent.

We will also use the notation $V \cong W$ to indicate that the representations of G on V and W are equivalent. However, in expressions as (10.1) we prefer to use equality to emphasize that we have a concrete decomposition on V into direct sum of subspaces.

Suppose now that the complex vector space V is *unitary*, that is, it is endowed with an inner product that we shall denote by $\langle \cdot, \cdot \rangle_V$ (with associated norm $\| \cdot \|_V$); the subscript will usually be omitted when the space V is clear from the context. We recall (see [93, 91, 75]) that the *adjoint* of a linear operator $T \colon W \to V$ between two unitary spaces W, V is the unique linear operator $T^* \colon V \to W$ such that $\langle Tw, v \rangle_V = \langle w, T^*v \rangle_W$, for all $w \in W$, $v \in V$. Moreover, an endomorphism $U \colon V \to V$ is *unitary* if $U^*U = I = UU^*$ and this is equivalent to the condition $\langle Uv_1, Uv_2 \rangle = \langle v_1, v_2 \rangle$ for all $v_1, v_2 \in V$. Moreover, if U is a *unitary matrix* then $U^* = \overline{U}^T$, the conjugate transpose of U.

A representation (ρ, V) is called *unitary* if $\rho(g)$ is unitary for all $g \in G$, that is, $\langle \rho(g)v_1, \rho(g)v_2 \rangle = \langle v, w \rangle$ for all and $v_1, v_2 \in V$. We shall then say that the inner product $\langle \cdot, \cdot \rangle$ is *ρ-invariant* (or *G-invariant*).

Exercise 10.1.2 Show that every one-dimensional representation is unitary. *Hint:* Show that every inner product on $\mathbb{C}$ is of the form $\langle z_1, z_2 \rangle = \alpha z_1 \overline{z_2}$, where $\alpha > 0$, for all $z_1, z_2 \in \mathbb{C}$.

Given an arbitrary representation (ρ, V) of a finite group G it is always possible to endow V with an inner product making ρ unitary. If $\langle \cdot, \cdot \rangle$ is an arbitrary inner product on V, we define, for all v_1 and v_2 in V,

$$(v, w) = \sum_{g \in G} \langle \rho(g)v, \rho(g)w \rangle . \tag{10.3}$$

Proposition 10.1.3 *The representation (ρ, V) is unitary with respect to the scalar product $(\cdot, \cdot)$. In particular, every representation of G is equivalent to a unitary representation.*

Proof. First of all, it is easy to see that (10.3) defines an inner product on V. Moreover, for all $v_1, v_2 \in V$ and $h \in G$ we have

$$(\rho(h)v_1, \rho(h)v_2) = \sum_{g \in G} \langle \rho(g)\rho(h)v_1, \rho(g)\rho(h)v_2 \rangle$$

$$= \sum_{g \in G} \langle \rho(gh)v_1, \rho(gh)v_2 \rangle$$

$$(t = gh) \quad = \sum_{t \in G} \langle \rho(t)v_1, \rho(t)v_2 \rangle$$

$$= (v, w).$$

This shows that the inner product $(\cdot, \cdot)$ is G-invariant. $\qquad\square$

We are mostly interested in equivalence classes of representations, thus we might confine ourselves to unitary representations. Thus, from now on, given a G-representation (ρ, V), it is understood that V is a finite dimensional (complex) vector space endowed with a G-invariant inner product and that $\rho(g)$ is unitary for all $g \in G$: note that, under these assumptions, we thus have

$$\rho(g^{-1}) = \rho(g)^{-1} = \rho(g)^* \tag{10.4}$$

for all $g \in G$. Also, we shall use the *polar decomposition* of a linear operator (see any book of linear algebra, for instance [75]) in the following form: if $T: V \to W$ is a linear invertible map between two unitary spaces V and W then there exist a unique *positive, self-adjoint* operator $|T|: V \to V$ (that is, $\langle |T|v, v \rangle_V > 0$ and $\langle |T|v_1, v_2 \rangle_V = \langle v_1, |T|v_2 \rangle_V$ for all $v, v_1, v_2 \in V, v \neq 0$) and a unique unitary isomorphism $U: V \to W$ such that $T = U|T|$. We also recall that $|T|$ is the unique *positive square root* of the positive operator T^*T: this means that $|T|^2 = T^*T$ and $|T|$ is positive.

Lemma 10.1.4 *Let (ρ, V) and (θ, W) be two unitary representations of a finite group G and suppose that they are equivalent. Then they are also unitarily equivalent, that is, there exists a unitary isomorphism $U: V \to W$ such that $\rho(g) = U^{-1}\theta(g)U$ for all $g \in G$.*

Proof. Let $g \in G$. Since ρ and θ are equivalent, we write (10.2) in the form

$$\rho(g) = T^{-1}\theta(g)T. \tag{10.5}$$

Taking adjoints, using (10.4), and replacing g by g^{-1}, we have

$$\rho(g) = T^*\theta(g)(T^*)^{-1}.$$

From (10.5) we then deduce that $T^*T\rho(g)(T^*T)^{-1} = T^*\theta(g)(T^*)^{-1} = \rho(g)$, equivalently,

$$\rho(g)^{-1}(T^*T)\rho(g) = T^*T. \tag{10.6}$$

Now we use the polar decomposition of T: since $|T|^2 = T^*T$ we have, $\rho(g)^{-1}|T|^2\rho(g) = |T|^2$, that is, $\left[\rho(g)^{-1}|T|\rho(g)\right]^2 = |T|^2$, and $\rho(g)^{-1}|T|\rho(g)$ is positive:

$$\langle \rho(g)^{-1}|T|\rho(g)v, v \rangle = \langle |T|\rho(g)v, \rho(g)v \rangle > 0$$

for all $v \in V$, $v \neq 0$. Since $\rho(g)^{-1}|T|\rho(g)$ is the square root of the left hand side of (10.6), by the uniqueness of the positive square root we have $\rho(g)^{-1}|T|\rho(g) = |T|$, that we write in the form

$$|T|\rho(g)|T|^{-1} = \rho(g). \tag{10.7}$$

Then, if $T = U|T|$ is the polar decomposition of T, we have

$$U^{-1}\theta(g)U = |T|T^{-1}\theta(g)T|T|^{-1}$$

(by (10.5)) $= |T|\rho(g)|T|^{-1}$

(by (10.7)) $= \rho(g).$

This shows that ρ is unitarily equivalent to θ.   □

The assumption that the representation ρ is unitary has a simple but fundamental consequence: if W is a G-invariant subspace of V then $W^\perp = \{v \in V : \langle v, w \rangle = 0, \ \forall w \in W\}$, the *orthogonal complement* of W, is also G-invariant. Indeed, if $v \in W^\perp$ and $g \in G$ one has $\langle \rho(g)v, w \rangle = \langle v, \rho(g^{-1})w \rangle = 0$ for all $w \in W$. Moreover, V can be expressed as the *direct sum* of the orthogonal subspaces W and $W^\perp$, namely $V = W \oplus W^\perp$ and $\rho = \rho_W \oplus \rho_{W^\perp}$.

Lemma 10.1.5 *Every representation of G is the orthogonal direct sum of a finite number of irreducible representations.*

Proof. Let (ρ, V) be a representation of G. If ρ is irreducible there is nothing to prove. If not, as above we get a nontrivial orthogonal decomposition into G-invariant subspaces of the form $V = W \oplus W^\perp$. Then the proof follows by an easy inductive argument on the dimension of V, because $\dim W < \dim V$ and $\dim W^\perp < \dim V$.   □

Definition 10.1.6 (Dual) Let G be a finite group. We denote by $\widehat{G}$, called the *dual* of G, a complete set of irreducible pairwise non-equivalent (unitary) representations of G (in other words, $\widehat{G}$ contains exactly one element in each equivalence class of irreducible G-representations).

We will also use the following notation: if $\rho, \theta \in \widehat{G}$ then

$$\delta_{\rho,\theta} = \begin{cases} 1 & \text{if } \rho = \theta \\ 0 & \text{if } \rho \neq \theta, \end{cases}$$

(note that if $\rho, \theta \in \widehat{G}$ then $\rho \not\sim \theta$ is the same thing as $\rho \neq \theta$). We end this section by illustrating some fundamental examples.

Example 10.1.7 For any finite group G we define the *trivial representation* $(\iota, \mathbb{C})$ as the one-dimensional representation of G defined by setting $\iota(g) = \text{Id}_\mathbb{C}$ for all $g \in G$. As it is one-dimensional, it is also unitary (cf. Exercise 10.1.2) and irreducible.

Example 10.1.8 Let G be a finite group. Denote by $L(G) = \{f : G \to \mathbb{C}\}$ the space of all complex valued functions on G; it is a vector space with respect to the pointwise linear combinations: $(\alpha_1 f_1 + \alpha_2 f_2)(g) = \alpha_1 f_1(g) + \alpha_2 f_2(g)$ for all $f_1, f_2 \in L(G)$, $\alpha_1, \alpha_2 \in \mathbb{C}$, and $g \in G$. Introduce in $L(G)$ the inner product

$$\langle f_1, f_2 \rangle = \sum_{g \in G} f_1(g)\overline{f_2(g)} \tag{10.8}$$

for all $f_1, f_2 \in L(G)$. Then the representation $(\lambda_G, L(G))$ defined by

$$[\lambda_G(g)f](h) = f(g^{-1}h) \tag{10.9}$$

for all $g, h \in G$ and $f \in L(G)$, is called the *left regular representation* of G. It is easy to show that it is indeed a representation: if $g_1, g_2, g \in G$ and $f \in L(G)$ then we have

$$[\lambda_G(g_1)\lambda_G(g_2)f](g) = \{\lambda_G(g_1)[\lambda_G(g_2)f]\}(g)$$
$$= [\lambda_G(g_2)f](g_1^{-1}g)$$
$$= f(g_2^{-1}g_1^{-1}g)$$
$$= [\lambda_G(g_1g_2)f](g),$$

that is, $\lambda_G(g_1)\lambda_G(g_2) = \lambda_G(g_1g_2)$. Moreover, λ_G is unitary: if $g \in G$ and $f_1, f_2 \in L(G)$ then we have

$$\langle \lambda_G(g)f_1, \lambda_G(g)f_2 \rangle = \sum_{h \in G} f_1(g^{-1}h)\overline{f_2(g^{-1}h)}$$
$$(t = g^{-1}h) \qquad = \sum_{t \in G} f_1(t)\overline{f_2(t)}$$
$$= \langle f_1, f_2 \rangle.$$

Analogously, the representation $(\rho_G, L(G))$ defined by

$$[\rho_G(g)f](h) = f(hg) \tag{10.10}$$

for all $g, h \in G$ and $f \in L(G)$, is again a unitary representation, called the *right regular representation*. Note that these two representations commute: $\lambda_G(g)\rho_G(h) = \rho_G(h)\lambda_G(g)$ for all $g, h \in G$.

As in Section 2.1, we denote by $\delta_g \in L(G)$ the *Dirac function* at $g \in G$, defined by

$$\delta_g(h) = \begin{cases} 1 & \text{if } h = g \\ 0 & \text{otherwise.} \end{cases}$$

It is clear that $\{\delta_g : g \in G\}$ is an orthonormal basis in $L(G)$. Note also that $\lambda_G(h)\delta_g = \delta_{hg}$, for all $h, g \in G$ so that we may represent every $f \in L(G)$ in the form

$$f = \sum_{g \in G} f(g)\delta_g = \sum_{g \in G} f(g)\lambda_G(g)\delta_{1_G}. \tag{10.11}$$

Remark 10.1.9 In many books, the inner product (10.8) is normalized, that is $\langle f_1, f_2 \rangle_{L(G)} = \frac{1}{|G|} \sum_{g \in G} f_1(g)\overline{f_2(g)}$ and this changes many formulæ given in the following chapters by a factor of $1/|G|$. Our choice makes the Dirac functions an orthonormal basis. The normalized scalar product comes from the theory of compact groups, where the Haar measure is normalized in order to be a probability measure; see the monographs by Bump [23] and Simon [148].

Example 10.1.10 Let $G = S_n$ be the *symmetric group of degree n*, that is, the group of all *permutations* on n elements. The *sign representation* is the one-dimensional representation $(\varepsilon, \mathbb{C})$ defined by setting $\varepsilon(g) = (-1)^{\mathrm{sign}(g)}\mathrm{Id}_{\mathbb{C}}$, where $\mathrm{sign}(g)$, the *sign* of the permutation $g \in S_n$, is defined to be 1 if g is an *even permutation* (that is, g is the product of an even number of transpositions, equivalently $g \in A_n$, the *alternating group*), and -1 if g is an *odd permutation* (that is, $g \in S_n \setminus A_n$). As the map sign: $G \to S_n/A_n \equiv C_2$ is a group homomorphism, we have $\varepsilon(g_1 g_2) = \varepsilon(g_1)\varepsilon(g_2)$ for all $g_1, g_2 \in S_n$, so that ε is indeed a representation. As it is one-dimensional, it is also unitary (cf. Exercise 10.1.2) and irreducible.

Example 10.1.11 Let A be an Abelian group. Then its characters (see Section 2.3) are unitary representations of A and its dual $\widehat{A}$ is itself a group (cf. Definition 2.3.1; see also Corollary 10.2.7 and Example 10.2.27).

10.2 Schur's lemma and the orthogonality relations

Given two finite dimensional vector spaces V and W, recall that $\mathrm{Hom}(V, W)$ (respectively, $\mathrm{End}(V)$) denotes the vector space of all linear maps $T : V \to W$ (respectively, $T : V \to V$). Let G be a finite group and suppose that (ρ, V) and (θ, W) are two representations of G.

Definition 10.2.1 One says that $L \in \mathrm{Hom}(V, W)$ *intertwines* ρ and θ if

$$L\rho(g) = \theta(g)L,$$

for all $g \in G$. We will denote by $\mathrm{Hom}_G(V, W)$ (or $\mathrm{Hom}_G(\rho, \theta)$) the space of all such intertwiners; it is called the *commutant* of ρ and θ. When $W = V$ and $\theta = \rho$ it is denoted by $\mathrm{End}_G(V)$ (or $\mathrm{End}_G(\rho)$), and it is simply called the *commutant* of ρ.

We begin with an elementary but useful property.

Proposition 10.2.2 *A linear map* $L: V \to W$ *belongs to* $\mathrm{Hom}_G(V, W)$ *if and only if* L^* *belongs to* $\mathrm{Hom}_G(W, V)$.

Proof. For all $g \in G$ we have

$$L^*\theta(g) = L^*\theta(g^{-1})^* = (\theta(g^{-1})L)^* \text{ and } \rho(g)L^* = \rho(g^{-1})^*L^* = (L\rho(g^{-1}))^*,$$

so that $L^*\theta(g) = \rho(g)L^*$ if and only if $\theta(g^{-1})L = L\rho(g^{-1})$. $\square$

The map $L \to L^*$ is an antilinear isomorphism between $\mathrm{Hom}_G(V, W)$ and $\mathrm{Hom}_G(W, V)$: indeed, $(\alpha T_1 + \beta T_2)^* = \overline{\alpha} T_1^* + \overline{\beta} T_2^*$, for $\alpha, \beta \in \mathbb{C}$, $T_1, T_2 \in \mathrm{Hom}_G(V, W)$.

We now illustrate the fundamental results that relate the notion of reducibility of a representation with the existence of intertwiners.

Lemma 10.2.3 (Schur) *Let* (ρ, V) *and* (θ, W) *be two irreducible representations of* G. *If* $L \in \mathrm{Hom}_G(V, W)$ *then either* L *is the zero homomorphism, or it is an isomorphism.*

Proof. Consider the kernel $\mathrm{Ker} L = \{v \in V : Lv = 0\} \leq V$ and the range $\mathrm{Ran} L = \{Lv : v \in V\} \leq W$ of L. If L intertwines ρ and θ then $\mathrm{Ker} L$ and $\mathrm{Ran} L$ are ρ- and θ-invariant, respectively:

$$v \in \mathrm{Ker} L \;\Rightarrow\; Lv = 0 \;\Rightarrow\; L\rho(g)v = \theta(g)Lv = 0 \;\Rightarrow\; \rho(g)v \in \mathrm{Ker} L$$

and

$$w \in \mathrm{Ran} L \;\Rightarrow\; \exists v \in V : w = Lv \;\Rightarrow\; \theta(g)w = L\rho(g)v \in \mathrm{Ran} L.$$

By irreducibility, either $\mathrm{Ker} L = V$ (and necessarily $\mathrm{Ran} L = \{0\}$) or $\mathrm{Ker} L = \{0\}$ (and necessarily $\mathrm{Ran} L = W$). In the first case L vanishes, in the second case it is an isomorphism. $\square$

Corollary 10.2.4 *Let* (ρ, V) *be an irreducible representation of* G *and suppose that* $L \in \mathrm{End}_G(V)$ *(that is, L intertwines ρ with itself: $L\rho(g) = \rho(g)L$ for all $g \in G$). Then L is a multiple of the identity: there exists $\lambda \in \mathbb{C}$ such that $L = \lambda I_V$.*

Proof. Let λ be an eigenvalue of L (which exists because V is a complex vector space and $\mathbb{C}$ is algebraically closed). Then $(L - \lambda I_V) \in \mathrm{End}_G(V)$ and, by Schur's lemma, it is either an isomorphism or the zero homomorphism. But, by definition of an eigenvalue, it cannot be invertible, and therefore necessarily $L = \lambda I_V$. $\square$

The last corollary may be expressed in the form: if V is G-irreducible then

$$\mathrm{End}_G(V) = \{\lambda I_V : \lambda \in \mathbb{C}\} \equiv \mathbb{C} I_V.$$

Corollary 10.2.5 *Suppose that* (ρ, V) *and* (θ, W) *are irreducible equivalent G-representations. Then* $\dim \mathrm{Hom}_G(V, W) = 1$.

Proof. Let $T_1, T_2 \in \mathrm{Hom}_G(V, W) \setminus \{0\}$. Then, by Proposition 10.2.2 $T_2^* T_1 \in \mathrm{End}_G(V)$ so that, by Corollary 10.2.4, there exists $\lambda \in \mathbb{C}$ such that $T_2^* T_1 = \lambda I_V$, equivalently, $T_1 = \lambda T_2$. $\square$

Corollary 10.2.6 *Suppose that* (ρ, V) *and* (η, U) *are G-representations. Then* $\mathrm{Hom}_G(V, U)$ *is nontrivial if and only if* ρ *and* η *contain a common isomorphic irreducible G-representation.*

Proof. Suppose that $T \in \mathrm{Hom}_G(V, U)$ is nontrivial. Then $(\mathrm{Ker}\, T)^\perp \leq V$ is nontrivial, ρ-invariant, and therefore it contains an irreducible representation $W \leq V$ (recall Lemma 10.1.5). Clearly, $T|_W$ is an isomorphism intertwining W and $T(W) \leq U$. The proof of the converse is left as an exercise (see also Exercise 10.6.9). $\square$

Corollary 10.2.7 *Let G be a (finite) Abelian group. A representation* (ρ, V) *of G is irreducible if and only if it is one dimensional (so that it is a character).*

Proof. Let us use multiplicative notation for G. Then, for all $g, h \in G$ we have $\rho(g)\rho(h) = \rho(h)\rho(g)$, so that $\rho(g) \in \mathrm{End}_G(\rho)$. By Corollary 10.2.4, there exists a function $\chi : G \to \mathbb{C}$ such that $\rho(g) = \chi(g) I_V$, $\forall g \in G$. Then every subspace of V is ρ-invariant so that ρ is irreducible if and only if $\dim V = 1$. We leave it to the reader to check that χ is indeed a character. $\square$

Exercise 10.2.8 Show that if $\rho \in \widehat{G}$ and g is in the center $Z(G) = \{z \in G : zh = hz$ for all $h \in G\}$ of G, then there exists $\lambda \in \mathbb{C}$ such that $\rho(g) = \lambda I_V$.

Exercise 10.2.9 (Converse to Schur's lemma) Suppose that the commutant of a G-representation (ρ, V) is trivial, that is, $\mathrm{End}_G(V) = \mathbb{C} I_V$. Show that ρ is irreducible (see also Corollary 10.6.4).

Let (ρ, V) be a representation of G. Given $v, w \in V$ the element $u_{v,w}^\rho \in L(G)$ defined by $u_{v,w}^\rho(g) = \langle \rho(g)w, v \rangle$ for all $g \in G$, is called a *(matrix) coefficient* of the representation ρ; we will omit the superscript "ρ" when the representation ρ is clear from the context. If $\{v_1, v_2, \ldots, v_n\}$ is an orthonormal basis for V, then $\rho(g)$, viewed as an n-by-n matrix, coincides with the matrix $(u_{v_i, v_j}^\rho(g))_{i,j=1}^n$ (see Lemma 10.2.13.(ii)).

Note that if $f \in L(G)$ and $g \in G$, then (cf. (10.11)) one has

$$f(g) = \langle \lambda_G(g)\delta_{1_G}, \overline{f} \rangle = u_{\delta_{1_G}, \overline{f}}^{\lambda_G}(g),$$

where λ_G is the left regular representation of G and δ_{1_G} is the Dirac function at the identity element 1_G of G. This shows that any $f \in L(G)$ may be realized as a coefficient of a (unitary) representation.

Lemma 10.2.10 *Let (ρ, V) and (θ, W) be two irreducible, non-equivalent representations of G. Then all coefficients of ρ are orthogonal to all coefficients of θ.*

Proof. Let $v_1, v_2 \in V$ and $w_1, w_2 \in W$. Our goal is to show that the functions $u^\rho_{v_2,v_1}(g) = \langle \rho(g)v_1, v_2 \rangle_V$ and $u^\theta_{w_2,w_1}(g) = \langle \theta(g)w_1, w_2 \rangle_W$ are orthogonal in $L(G)$. Consider the linear transformation $L: V \to W$ defined by

$$Lv = \langle v, v_2 \rangle_V \, w_2, \qquad (10.12)$$

for all $v \in V$. Then, the linear transformation $\tilde{L}: V \to W$ defined by

$$\tilde{L} = \sum_{g \in G} \theta(g^{-1}) L \rho(g)$$

belongs to $\mathrm{Hom}_G(\rho, \theta)$. Indeed, for every $g \in G$,

$$\tilde{L}\rho(g) = \sum_{h \in G} \theta(h^{-1}) L \rho(hg)$$

$$(k = hg) \quad = \sum_{k \in G} \theta(gk^{-1}) L \rho(k)$$

$$= \theta(g)\tilde{L}.$$

Thus, by virtue of Schur's lemma, we have that either $\tilde{L}$ is invertible or $\tilde{L} = 0$. As $\rho \not\sim \theta$, necessarily the second possibility occurs and therefore

$$0 = \langle \tilde{L}v_1, w_1 \rangle_W = \sum_{g \in G} \langle L\rho(g)v_1, \theta(g)w_1 \rangle_W$$

$$(\text{by } (10.12)) \quad = \sum_{g \in G} \langle \rho(g)v_1, v_2 \rangle_V \cdot \langle w_2, \theta(g)w_1 \rangle_W$$

$$= \sum_{g \in G} \langle \rho(g)v_1, v_2 \rangle_V \cdot \overline{\langle \theta(g)w_1, w_2 \rangle_W}$$

$$= \sum_{g \in G} u^\rho_{v_2,v_1}(g) \overline{u^\theta_{w_2,w_1}(g)}$$

$$= \langle u^\rho_{v_2,v_1}, u^\theta_{w_2,w_1} \rangle_{L(G)}. \qquad \square$$

Theorem 10.2.11 *Let G be a finite group. Then there exist only finitely many pairwise inequivalent irreducible unitary representations. In other words,* $|\widehat{G}| < \infty$.

Proof. The space $L(G)$ is finite dimensional and contains only finitely many distinct pairwise orthogonal functions, and the statement follows from previous lemma. $\qquad\square$

Let now (ρ, V) be an irreducible G-representation, $d = \dim V$, and choose an orthonormal basis $\{v_1, v_2, \ldots, v_d\}$ of V. Recall that the *trace* of a linear operator $T \colon V \to V$ is given by $\mathrm{Tr}(T) = \sum_{j=1}^{n} \langle Tv_j, v_j \rangle$. It is easy to check that $\mathrm{Tr} \colon \mathrm{End}(V) \to \mathbb{C}$ is a linear map, that it does not depend on the choice of the basis, and that it satisfies the following central properties:

$$\mathrm{Tr}(TS) = \mathrm{Tr}(ST) \qquad \text{for all } S, T \in \mathrm{End}(V);$$

$$\mathrm{Tr}(T^{-1}ST) = \mathrm{Tr}(S) \qquad \text{for all } S \in \mathrm{End}(V) \text{ and } T \in \mathrm{GL}(V). \qquad (10.13)$$

Lemma 10.2.12 *The coefficients*

$$u_{i,j}^{\rho}(g) = \langle \rho(g)v_j, v_i \rangle_V, \quad i, j = 1, 2, \ldots, d, \qquad (10.14)$$

are pairwise orthogonal in $L(G)$. In formulæ,

$$\left\langle u_{i,j}^{\rho}, u_{k,h}^{\rho} \right\rangle_{L(G)} = \frac{|G|}{d} \delta_{ik} \delta_{jh}$$

for all $i, j, h, k = 1, 2, \ldots, d$.

Proof. Fix indices $1 \le i, k \le d$ and define $L_{ik} \in \mathrm{End}(V)$ by setting

$$L_{ik}(v) = \langle v, v_i \rangle \, v_k,$$

for all $v \in V$. It is easy to check that $\mathrm{Tr}(L_{ik}) = \delta_{ik}$. Now set

$$\widetilde{L}_{ik} = \frac{1}{|G|} \sum_{g \in G} \rho(g^{-1}) L_{ik} \rho(g)$$

and observe that $\widetilde{L}_{ik} \in \mathrm{End}_G(\rho)$ (see the proof of Lemma 10.2.10). As ρ is irreducible, from Corollary 10.2.4 we deduce that $\widetilde{L}_{ik} = \alpha I_V$, for a suitable $\alpha \in \mathbb{C}$. Indeed, $\alpha = \delta_{ik}/d$:

$$d\alpha = \mathrm{Tr}(\widetilde{L}_{ik})$$

$$= \frac{1}{|G|} \sum_{g \in G} \mathrm{Tr} \left[\rho(g^{-1}) L_{ik} \rho(g) \right]$$

$$\text{(by (10.13))} \quad = \mathrm{Tr}(L_{ik}).$$

It follows that $\widetilde{L}_{ik} = (1/d)\delta_{ik}I_V$ and therefore $\langle \widetilde{L}_{ik}v_j, v_h \rangle_V = (1/d)\delta_{jh}\delta_{ik}$. Since

$$\langle \widetilde{L}_{ik}v_j, v_h \rangle_V = \frac{1}{|G|} \sum_{g \in G} \langle L_{ik}\rho(g)v_j, \rho(g)v_h \rangle_V$$

$$= \frac{1}{|G|} \sum_{g \in G} \langle \rho(g)v_j, v_i \rangle_V \cdot \langle v_k, \rho(g)v_h \rangle_V$$

$$= \frac{1}{|G|} \left\langle u^\rho_{i,j}, u^\rho_{k,h} \right\rangle_{L(G)},$$

this ends the proof. □

The following lemma presents further properties of the matrix coefficients; these do not require the irreducibility of ρ.

Lemma 10.2.13 *Let* (ρ, V) *be a G-representation and let* $\{v_1, v_2, \ldots, v_d\}$ *be an orthonormal basis of V. With the notation in* (10.14) *one has:*

(i) $u^\rho_{i,j}(g^{-1}) = \overline{u^\rho_{j,i}(g)}$;

(ii) $\rho(g)v_j = \sum_{i=1}^d v_i u^\rho_{i,j}(g)$;

(iii) $u^\rho_{i,j}(g_1g_2) = \sum_{h=1}^d u^\rho_{i,h}(g_1)u^\rho_{h,j}(g_2)$;

(iv) $\sum_{j=1}^d \overline{u^\rho_{i,j}(g)}u^\rho_{k,j}(g) = \delta_{i,k}$ *and* $\sum_{i=1}^d \overline{u^\rho_{i,j}(g)}u^\rho_{i,k}(g) = \delta_{j,k}$ *(dual orthogonality relations)*

for all $g, g_1, g_2 \in G$ *and* $i, j, k = 1, 2, \ldots, d$.

Proof.

(i) This follows immediately from $\rho(g)^* = \rho(g^{-1})$ and $\langle v, w \rangle = \overline{\langle w, v \rangle}$ for all $g \in G$ and $v, w \in V$.

(ii) This is obvious, since for all $v \in V$ one has $v = \sum_{h=1}^n v_h\langle v, v_h \rangle$.

(iii) From (ii) we deduce that

$$\sum_{h=1}^d v_h u^\rho_{h,j}(g_1g_2) = \rho(g_1g_2)v_j = \rho(g_1)\rho(g_2)v_j = \sum_{h=1}^d \rho(g_1)v_h u^\rho_{h,j}(g_2)$$

and taking the scalar product with v_i we get the desired equality.

(iv) This is an immediate consequence of the unitarity of $\rho(g)$, that is, of the relation $\rho(g)\rho(g)^* = \rho(g)^*\rho(g) = I_V$, for all $g \in G$. □

In the following, we shall refer to $\left(u^\rho_{i,j}(g) \right)^n_{i,j=1}$ as a *matrix realization* of the representation ρ.

Definition 10.2.14 Let (ρ, V) be a G-representation. Then the map $\chi^\rho \in L(G)$ defined by setting

$$\chi^\rho(g) = \mathrm{Tr}[\rho(g)] \text{ for all } g \in G$$

is called the *character* of ρ.

Note that, for every $g \in G$, we have that $\rho(g)$, being unitary, is diagonalizable and therefore its trace $\mathrm{Tr}[\rho(g)] = \chi^\rho(g)$ coincides with the sum of its eigenvalues. From (10.13) it follows that two equivalent representations have the same character: indeed, $\mathrm{Tr}[T\rho(g)T^{-1}] = \mathrm{Tr}[\rho(g)]$ for every invertible operator T. Therefore, with each equivalence class of irreducible representations is associated a character. Clearly, using a matrix realization of ρ, one has $\mathrm{Tr}[\rho(g)[= \sum_{i=1}^n u_{i,i}^\rho(g)$ and this sum does not depend on the particular choice of the orthonormal system $\{v_1, v_2, \ldots, v_d\}$ in V. We observe that if ρ is one-dimensional, then $\rho(g) = \chi^\rho(g)I_V$ for all $g \in G$ and, by abuse of language, we say that the representation ρ coincides with its character and write $\rho = \chi^\rho$.

Proposition 10.2.15 *Let (ρ, V) be a G-representation. Then we have:*

(i) $\chi^\rho(1_G) = \dim V$;

(ii) $\chi^\rho(s^{-1}) = \overline{\chi^\rho(s)}$, *for all $s \in G$;*

(iii) $\chi^\rho(t^{-1}st) = \chi^\rho(s)$, *for all $s, t \in G$;*

(iv) *if $\rho = \rho_1 \oplus \rho_2$ then $\chi^\rho = \chi^{\rho_1} + \chi^{\rho_2}$;*

(v) *with the notation as in Lemma 10.2.13 we have:*

$$\chi^\rho = \sum_{i=1}^d u_{i,i}^\rho. \tag{10.15}$$

Proof.

(i) This is easy: $\rho(1_G) = I_V$ and $\mathrm{Tr}(I_V) = \dim V = d$.

(ii) We have

$$\chi^\rho(s^{-1}) = \mathrm{Tr}[\rho(s^{-1})] = \mathrm{Tr}[\rho(s)^*] = \overline{\chi^\rho(s)}$$

since $\rho(s)$ is unitary and $\mathrm{Tr}(A^*) = \overline{\mathrm{Tr}(A)}$ for all $A \in GL(V)$.

(iii) This follows again from the central property of the trace.

(iv) This is easy and left as an exercise.

(v) This is obvious. $\qquad\square$

Exercise 10.2.16 Let ρ be a G-representation and let $n = |G|$.

(1) Show that the eigenvalues of $\rho(g), g \in G$ are n-th roots of unity;

(2) deduce that $|\chi^\rho(g)| \le d_\rho$ for all $g \in G$.

Proposition 10.2.17 (Orthogonality relations for characters) *Let $\rho, \theta \in \widehat{G}$. Then*

$$\langle \chi^\rho, \chi^\theta \rangle_{L(G)} = |G| \delta_{\rho,\theta}. \tag{10.16}$$

In particular, two non-equivalent irreducible G-representations have different characters.

Proof. From (10.15), Lemma 10.2.10 and Lemma 10.2.12 we get

$$\langle \chi^\rho, \chi^\theta \rangle_{L(G)} = \sum_{i=1}^{d_\rho} \sum_{j=1}^{d_\theta} \langle u_{i,i}^\rho, u_{j,j}^\theta \rangle_{L(G)} = \sum_{i=1}^{d_\rho} \sum_{j=1}^{d_\theta} \delta_{\rho,\theta} \delta_{i,j} \frac{|G|}{d_\rho} = |G| \delta_{\rho,\theta}.$$

$\square$

We thus have that the characters of irreducible representations constitute an orthogonal system in $L(G)$ (in general not complete: see Theorem 10.3.13). Therefore they are finitely many and their cardinality equals the number of equivalence classes of irreducible representations (cf. Proposition 10.2.17 and the comments after Definition 10.2.14).

Proposition 10.2.18 *Let ρ and θ be two G-representations. Suppose that $\rho = \rho_1 \oplus \rho_2 \oplus \cdots \oplus \rho_k$ is a decomposition of ρ into irreducible subrepresentations and that θ is irreducible. Then, setting $m_\theta = |\{j : \rho_j \sim \theta\}|$, one has*

$$m_\theta = \frac{1}{|G|} \langle \chi^\rho, \chi^\theta \rangle_{L(G)}. \tag{10.17}$$

In particular, m_θ does not depend on the particular decomposition of ρ.

Proof. From Proposition 10.2.15.(iv) it follows that $\chi^\rho = \sum_{j=1}^k \chi^{\rho_j}$. Therefore, from Proposition 10.2.17 we deduce that

$$\langle \chi^\rho, \chi^\theta \rangle_{L(G)} = \sum_{j=1}^k \langle \chi^{\rho_j}, \chi^\theta \rangle_{L(G)} = \sum_{j=1}^k |G| \delta_{\rho_j,\theta} = |G| m_\theta. \qquad \square$$

Corollary 10.2.19 *Let (ρ, V) be a representation of G. Then, with the notation as in Proposition 10.2.18, one has*

$$\rho \sim \bigoplus_{\theta \in \widehat{G}} m_\theta \theta,$$

where $m_\theta \theta = \theta \oplus \theta \oplus \cdots \oplus \theta$ is the direct sum of m_θ copies of θ, and

$$V \cong \bigoplus_{\theta \in \widehat{G}} m_\theta W_\theta,$$

where $m_\theta W_\theta = W_\theta \oplus W_\theta \oplus \cdots \oplus W_\theta$ is the direct sum of m_θ copies of W_θ, the representation space of θ. Moreover,

$$\chi^\rho = \sum_{\theta \in \widehat{G}} m_\theta \chi^\theta.$$

Definition 10.2.20 The number m_θ in (10.17) is called the *multiplicity* of θ as a sub-representation of ρ. If θ is not contained in ρ then clearly $m_\theta = 0$. The subspace (of V which is isomorphic to) $m_\theta W_\theta$ is called the θ-*isotypic component* of V.

Example 10.2.21 Let (ρ, V) be a G-representation. Then the dimension $\dim(V^G)$ of the subspace of G-invariant vectors equals the multiplicity m_ι of the trivial representation ι of G as a sub-representation of ρ.

Corollary 10.2.22 *Let ρ, η be two representations of G. Suppose that $\rho = \oplus_{\theta \in \widehat{G}} m_\theta \theta$ and $\eta = \oplus_{\theta \in \widehat{G}} n_\theta \theta$ are their decompositions into irreducible subrepresentations, so that the numbers m_θ's and n_θ's are the corresponding multiplicities. Then, denoting by J the set of common irreducible representations, that is, $J = \{\theta \in \widehat{G} : m_\theta > 0$ and $n_\theta > 0\}$, we have*

$$\frac{1}{|G|} \langle \chi^\rho, \chi^\eta \rangle = \sum_{\theta \in J} m_\theta n_\theta.$$

Corollary 10.2.23 *A G-representation ρ is irreducible if and only if $\|\chi^\rho\|_{L(G)} = \sqrt{|G|}$.*

Corollary 10.2.24 *Two G-representations ρ and θ are equivalent if and only if $\chi^\rho = \chi^\theta$.*

Theorem 10.2.25 (Peter-Weyl) *Let G be a finite group and denote by $(\lambda_G, L(G))$ its left regular representation (see Example 10.1.8). Then the following hold:*

(i) *Every irreducible representation $\theta \in \widehat{G}$ appears in the decomposition of λ_G with multiplicity equal to its dimension d_θ, that is,*

$$L(G) \cong \bigoplus_{\theta \in \widehat{G}} d_\theta W_\theta, \tag{10.18}$$

where W_θ denotes the representation space of θ. Moreover,

$$\sum_{\theta \in \widehat{G}} d_\theta \chi^\theta = |G| \delta_{1_G}; \tag{10.19}$$

(ii) $|G| = \sum_{\theta \in \widehat{G}} d_\theta^2$;

(iii) *denoting by $u^{\theta}_{i,j}$ the matrix coefficient of $\theta \in \widehat{G}$ with respect to an orthonormal basis (see* (10.14)*), then the set*

$$\left\{ \sqrt{\frac{d_{\theta}}{|G|}} u^{\theta}_{i,j} : i, j = 1, \ldots, d_{\theta}, \theta \in \widehat{G} \right\}$$

is a complete orthonormal system in $L(G)$.

Proof.

(i) Denote by

$$\lambda_G \sim \bigoplus_{\theta \in \widehat{G}} m_{\theta} \theta \tag{10.20}$$

the decomposition of λ_G into irreducibles, as in Corollary 10.2.19, so that the integer m_{θ} denotes the multiplicity of the irreducible representation $\theta \in \widehat{G}$ in λ_G. Using the complete orthonormal system $\{\delta_g : g \in G\}$ of Dirac deltas in $L(G)$ and the identity $\lambda_G(h)\delta_g = \delta_{hg}$, we immediately obtain that

$$\chi^{\lambda_G}(g) = \sum_{h \in G} \langle \lambda_G(g)\delta_h, \delta_h \rangle = \sum_{h \in G} \langle \delta_{gh}, \delta_h \rangle = \begin{cases} |G| & \text{if } g = 1_G \\ 0 & \text{if } g \neq 1_G, \end{cases} \tag{10.21}$$

for all $g \in G$; in other words,

$$\chi^{\lambda_G} = |G|\delta_{1_G}. \tag{10.22}$$

From Proposition 10.2.18, (10.22), and Proposition 10.2.15, we deduce

$$m_{\theta} = \frac{1}{|G|} \langle \chi^{\lambda_G}, \chi^{\theta} \rangle = \chi^{\theta}(1_G) = d_{\theta}. \tag{10.23}$$

Then, (10.18) follows from (10.20) and (10.23), while (10.19) follows from, in order, (10.22), (10.20), and (10.23).

(ii) Taking dimensions in (10.18), we deduce that $|G| \equiv \dim L(G) = \sum_{\theta \in \widehat{G}} d_{\theta}^2$.

(iii) From Lemma 10.2.10 and Lemma 10.2.12 we have that the functions

$$\sqrt{\frac{d_{\theta}}{|G|}} u^{\theta}_{i,j} : i, j = 1, 2, \ldots, d_{\theta}, \theta \in \widehat{G}$$

constitute an orthonormal system in $L(G)$. This system is indeed complete since its cardinality $\sum_{\theta \in \widehat{G}} d_{\theta}^2 = |G|$ equals the dimension of the space $L(G)$. $\qquad \square$

The structure of the Peter-Weyl theorem will be examined further in Sections 10.3 and 10.5. For future reference, it is convenient to state explicitly the orthogonality relations for matrix coefficients in the following form, which immediately follows from Lemma 10.2.10 and Lemma 10.2.12. Let (θ, W) and (ρ, U) be two irreducible G-representations. Then

$$\langle u_{i,j}^\theta, u_{h,k}^\rho \rangle = \frac{|G|}{d_\theta} \delta_{\theta,\rho} \delta_{i,h} \delta_{j,k}. \tag{10.24}$$

We now present a useful formula for irreducible characters.

Proposition 10.2.26 *Let $(\theta, W) \in \widehat{G}$, $w \in W$ be a vector of norm 1, and $\phi(g) = \langle \theta(g)w, w \rangle$ the diagonal matrix coefficient associated with w. Then*

$$\chi^\theta(g) = \frac{d_\theta}{|G|} \sum_{h \in G} \phi(h^{-1}gh) \tag{10.25}$$

for all $g \in G$.

Proof. Let $\{v_1 = w, v_2, \ldots, v_{d_\theta}\}$ be an orthonormal basis of W and let $u_{i,j}^\theta$ be as in (10.14) (note that $\phi = u_{1,1}^\theta$). Then

$$\sum_{h \in G} \phi(h^{-1}gh) = \sum_{h \in G} \langle \theta(g)\theta(h)v_1, \theta(h)v_1 \rangle$$

$$(\text{by Lemma 10.2.13.(ii)}) \quad = \sum_{i,j=1}^{d_\theta} \sum_{h \in G} u_{i,1}^\theta(h)\overline{u_{j,1}^\theta(h)}\langle \theta(g)v_i, v_j \rangle$$

$$(\text{by (10.24) and (10.15)}) \quad = \frac{|G|}{d_\theta} \chi^\theta(g). \qquad \square$$

Example 10.2.27 Let A be a finite Abelian group. In Corollary 10.2.7 we have shown that its irreducible representations coincide with its characters. Now we can also deduce that A has exactly $|A|$ distinct characters: this agrees with Proposition 2.3.3.

Example 10.2.28 Let $D_n = \langle a, b : a^n = b^2 = 1, bab = a^{-1} \rangle$ denote the *dihedral* group of degree n, i.e. the group of isometries of a regular polygon with n vertices. Recall that $|D_n| = 2n$ and that any element of D_n may be written uniquely in the form $a^k b^\epsilon$, where $0 \leq k \leq n - 1$ and $\epsilon \in \{0, 1\}$. Moreover, the product of two elements is given by the following rule:

$$a^h b^\delta \cdot a^k b^\epsilon = a^h (b^\delta a^k b^\delta) b^{\delta+\epsilon} = \begin{cases} a^{h-k} b^{1+\epsilon} & \text{if } \delta = 1 \\ a^{h+k} b^\epsilon & \text{if } \delta = 0 \end{cases}$$

for all $h, k = 0, 1, \ldots, n - 1$ and $\delta, \epsilon \in \{0, 1\}$. Alternatively, D_n may be seen as the group of matrices generated by

$$a = \begin{pmatrix} \omega & 0 \\ 0 & \omega^{-1} \end{pmatrix} \text{ and } b = \begin{pmatrix} 0 & 1 \\ 1 & 0 \end{pmatrix},$$

where $\omega = e^{2\pi i/n} \equiv \cos\frac{2\pi}{n} + i \sin\frac{2\pi}{n}$ (compare with the representation ρ_1 below).

In the following, we determine $\widehat{D_n}$. We consider first the case when $\underline{n \text{ is even}}$. We have four one-dimensional representations (we identify these with the corresponding characters), $\chi^i, i = 1, 2, 3, 4$, defined by

$$\begin{aligned}
\chi^1(a^k b^\epsilon) &= 1 \\
\chi^2(a^k b^\epsilon) &= (-1)^\epsilon \\
\chi^3(a^k b^\epsilon) &= (-1)^k \\
\chi^4(a^k b^\epsilon) &= (-1)^{k+\epsilon}
\end{aligned} \tag{10.26}$$

for all $\epsilon = 0, 1$ and $k = 0, 1, \ldots, n - 1$. Setting $\omega = e^{2\pi i/n}$ as above, we also define the two-dimensional representations $\rho_t, t = 0, 1, \ldots, n$, by setting

$$\rho_t(a^k) = \begin{pmatrix} \omega^{tk} & 0 \\ 0 & \omega^{-tk} \end{pmatrix} \text{ and } \rho_t(a^k b) = \begin{pmatrix} 0 & \omega^{tk} \\ \omega^{-tk} & 0 \end{pmatrix}$$

for all $k = 0, 1, \ldots, n - 1$.

Exercise 10.2.29

(1) Show that each ρ_t is indeed a representation.
(2) Show that $\rho_t \sim \rho_{n-t}$.
(3) Show that $\chi^{\rho_0} = \chi^1 + \chi^2$ and $\chi^{\rho_{n/2}} = \chi^3 + \chi^4$.
(4) Show that ρ_t, with $1 \leq t \leq \frac{n}{2} - 1$, are pairwise non-equivalent irreducible representations in two different ways, namely:
   (i) by inspecting the invariant subspaces and intertwining operators;
   (ii) by computing the characters and their inner products.
(5) Conclude that $\chi^1, \chi^2, \chi^3, \chi^4, \rho_t$, with $1 \leq t < n/2$, constitute a complete list of irreducible representations of D_n.

Solution of (2): $\rho_{n-t}(g) = \rho_t(b)\rho_t(g)\rho_t(b)$ for all $g \in D_n$.

Exercise 10.2.30 Determine a complete list of irreducible representations of D_n in the case $\underline{n \text{ is odd}}$.
Solution: $\widehat{D_n}$ consists of χ^1, χ^2, and ρ_t with $t = 1, 2, \ldots, \frac{n-1}{2}$.

Exercise 10.2.31 The *generalized quaternion group* is $Q_n = \langle a, b : b^2 = a^n, b^{-1}ab = a^{-1} \rangle$. Note that Q_2 is the classical quaternionic group.

(1) Show that $b^2 = a^{-n}, b^4 = 1, a^{2n} = 1$ and that every element $g \in Q_n$ may be written in the form $g = a^k b^h$ with $0 \leq k \leq 2n - 1$ and $h \in \{0, 1\}$.

(2) Show that Q_n may be seen as the group of matrices generated by $a = \begin{pmatrix} \omega & 0 \\ 0 & \omega^{-1} \end{pmatrix}$ and $b = \begin{pmatrix} 0 & -1 \\ 1 & 0 \end{pmatrix}$, where $\omega = e^{\pi i/n}$. Deduce that the expression $g = a^k b^h$ is unique and that Q_n has $4n$ elements.

(3) Show that if n is even then $Q_n/\langle a^2 \rangle \cong C_2 \times C_2$ while if n is odd then $Q_n/\langle a^2 \rangle \cong C_4$.

(4) Denote by $\pi: Q_n \to Q_n/\langle a^2 \rangle$ the canonical quotient map. For every $\psi \in \widehat{Q_n/\langle a^2 \rangle}$ set $\overline{\psi} = \psi \circ \pi$: this is called the *inflation* of ψ (cf. Section 11.6). Show that the inflations $\overline{\psi}$, with $\psi \in \widehat{Q_n/\langle a^2 \rangle}$, are four one-dimensional, non-equivalent representations of Q_n.

(5) For $t = 0, 1, \ldots, n - 1$ set

$$\rho_t(a) = \begin{pmatrix} \omega^t & 0 \\ 0 & \omega^{-t} \end{pmatrix} \quad \text{and} \quad \rho_t(b) = \begin{pmatrix} 0 & (-1)^t \\ 1 & 0 \end{pmatrix}. \quad (10.27)$$

Show that (10.27) define $n - 1$ irreducible, non-equivalent representations of Q_n which, added to the four one-dimensional representations determined in (4), form a complete list for $\widehat{Q_n}$.

10.3 The group algebra and the Fourier transform

An *(associative) algebra* over $\mathbb{C}$ (or *complex algebra*) is a vector space $\mathcal{A}$ over $\mathbb{C}$ endowed with a multiplication operation, the *product*, such that $\mathcal{A}$ is a ring with respect to the sum and the product, and the following associative law holds for the product and multiplication by a scalar:

$$\alpha(AB) = (\alpha A)B = A(\alpha B)$$

for all $\alpha \in \mathbb{C}$ and $A, B \in \mathcal{A}$. The basic example is $\text{End}(V)$, where V is a finite-dimensional vector space over $\mathbb{C}$, with the usual operations of sum and product of operators, and of multiplication by scalars.

Let $\mathcal{A}$ be a complex algebra. A *subalgebra* of $\mathcal{A}$ is a subspace $\mathcal{B} \leq \mathcal{A}$, which is closed under multiplication. For instance, if V is a finite-dimensional vector space over $\mathbb{C}$, fix a basis $\mathcal{B} = \{v_1, v_2, \ldots, v_d\}$ of V. An operator $T \in \text{End}(V)$ is called $\mathcal{B}$-*diagonal* provided there exist scalars $\alpha_1, \alpha_2, \ldots, \alpha_d \in \mathbb{C}$ such that $Tv_i = \alpha_i v_i$ for all $i = 1, 2, \ldots, d$. Then the $\mathcal{B}$-diagonal operators constitute a subalgebra of $\text{End}(V)$.

An *involution* in $\mathcal{A}$ is a bijective map $A \mapsto A^*$ such that

- $(A^*)^* = A$
- $(\alpha A + \beta B)^* = \overline{\alpha} A^* + \overline{\beta} B^*$
- $(AB)^* = B^* A^*$ (*anti-multiplicative* property)

for all $\alpha, \beta \in \mathbb{C}$ and $A, B \in \mathcal{A}$. For instance, if $\mathcal{A} = \text{End}(V)$, then the map $T \mapsto T^*$ (where T^* is the adjoint of T) is an involution on $\mathcal{A}$; similarly for $\text{End}_G(V)$ (see Proposition 10.2.2). An algebra with involution is called an *involutive algebra* or $*$-*algebra*. An element A in a $*$-algebra $\mathcal{A}$ such that $A = A^*$ is called *self-adjoint*.

$\mathcal{A}$ is *unital* if it has a *unit*, that is, there exists an element $I \in \mathcal{A}$ such that $AI = IA = A$ for all $A \in \mathcal{A}$. Note that a unit is necessarily unique and self-adjoint. Indeed, if I and I' are units in $\mathcal{A}$, then $I = II' = I'$. Moreover, if $A \in \mathcal{A}$

$$I^*A = ((I^*A)^*)^* = (A^*(I^*)^*)^* = (A^*I)^* = (A^*)^* = A$$

and, similarly, $AI^* = A$. Thus $I = I^*$, by uniqueness of the unit.

The *dimension* of $\mathcal{A}$ is simply its dimension as a complex vector space.

In the following, we shall consider only finite-dimensional, unital, involutive, complex algebras.

The algebra $\mathcal{A}$ is *commutative* (or *Abelian*) if it is commutative as a ring, namely if $AB = BA$ for all $A, B \in \mathcal{A}$. A basic example is the following: let J be a finite set and denote by $\mathbb{C}^J$ the space of all functions $f: J \to \mathbb{C}$ with multiplication and involution given respectively by:

$$(f_1 f_2)(j) = f_1(j)f_2(j) \text{ and } f^*(j) = \overline{f(j)}, \tag{10.28}$$

for all $f, f_1, f_2 \in \mathbb{C}^J$ and $j \in J$. Clearly, $\mathbb{C}^J$ is isomorphic to the subalgebra of $\mathcal{B}$-diagonal operators in $\text{End}(V)$ (for any basis $\mathcal{B}$ of V and) for any vector space V with $\dim V = |J|$, as well as to the direct sum $\underbrace{\mathbb{C} \oplus \mathbb{C} \oplus \cdots \oplus \mathbb{C}}_{|J|-\text{times}}$.

The *center* $\mathcal{Z}(\mathcal{A})$ of $\mathcal{A}$ is the commutative subalgebra

$$\mathcal{Z}(\mathcal{A}) = \{B \in \mathcal{A} : AB = BA \text{ for all } A \in \mathcal{A}\}.$$

The direct sum $\mathcal{A} \oplus \mathcal{B}$ of two algebras $\mathcal{A}, \mathcal{B}$ is the vector space direct sum with the product defined componentwise: $(a_1, b_1)(a_2, b_2) = (a_1 a_2, b_1 b_2)$, for all $a_1, a_2 \in \mathcal{A}, b_1, b_2 \in \mathcal{B}$.

Let $\mathcal{A}_1$ and $\mathcal{A}_2$ be two involutive algebras and let $\phi: \mathcal{A}_1 \to \mathcal{A}_2$ be a map. One says that ϕ is a $*$-*homomorphism* provided that

- $\phi(\alpha A + \beta B) = \alpha \phi(A) + \beta \phi(B)$ (linearity)
- $\phi(AB) = \phi(A)\phi(B)$ (multiplicative property)
- $\phi(A^*) = [\phi(A)]^*$ (preservation of involution)

for all $\alpha, \beta \in \mathbb{C}$ and $A, B \in \mathcal{A}_1$. If in addition ϕ is a bijection, then it is called a *-isomorphism* between $\mathcal{A}_1$ and $\mathcal{A}_2$ and one says that $\mathcal{A}_1$ and $\mathcal{A}_2$ are *-isomorphic*. On the other hand, ϕ is a *-anti-homomorphism* if the multiplicative property is replaced by

$$\phi(AB) = \phi(B)\phi(A) \quad \text{(anti-multiplicative property)}$$

for all $A, B \in \mathcal{A}_1$. Finally, ϕ is a *-anti-isomorphism* if it is a bijective *-anti-homomorphism. If such a *-anti-isomorphism exists, then one says that $\mathcal{A}_1$ and $\mathcal{A}_2$ are *-anti-isomorphic*.

Let G be a finite group. Recall that $L(G)$ denotes the vector space of all functions $f: G \to \mathbb{C}$.

Definition 10.3.1 Let $f, f_1, f_2 \in L(G)$. We define the *convolution* of f_1 and f_2 and the adjoint of f as the functions $f_1 * f_2 \in L(G)$ and $f^* \in L(G)$ given by setting

$$[f_1 * f_2](g) = \sum_{h \in G} f_1(gh^{-1})f_1(h) \tag{10.29}$$

and

$$f^*(g) = \overline{f(g^{-1})} \tag{10.30}$$

for all $g \in G$, respectively.

Note that the convolution (10.29) may be also written in the following equivalent ways:

$$[f_1 * f_2](g) = \sum_{s,t \in G: st=g} f_1(s)f_2(t)$$

$$= \sum_{h \in G} f_1(h)f_2(h^{-1}g) = \sum_{h \in G} f_1(h)[\lambda_G(h)f_2](g). \tag{10.31}$$

Proposition 10.3.2 *The vector space $L(G)$ endowed with the convolution product* (10.29) *and the involution* (10.30) *is a unital, involutive algebra, with unit* δ_{1_G}. *It is called the* group algebra *of G.*

Proof. We leave it as an exercise to prove that the convolution is distributive with respect to the sum, and that δ_{1_G} is the unit.

Let $f_1, f_2, f_3 \in L(G)$ and $g \in G$. Then we have:

$$[f_1 * (f_2 * f_3)](g) = \sum_{h \in G} f_1(gh^{-1})(f_2 * f_3)(h)$$

$$= \sum_{h \in G} \sum_{t \in G} f_1(gh^{-1})f_2(ht^{-1})f_3(t)$$

(setting $h = st$) $\quad = \sum_{t \in G} \sum_{s \in G} f_1(gt^{-1}s^{-1})f_2(s)f_3(t)$

$$= \sum_{t \in G} (f_1 * f_2)(gt^{-1})f_3(t) = [(f_1 * f_2) * f_3](g).$$

This shows associativity of the convolution product. Finally,

$$[f_1^* * f_2^*](g) = \sum_{s \in G} f_1^*(gs)f_2^*(s^{-1})$$

$$= \sum_{s \in G} \overline{f_1(s^{-1}g^{-1})}\overline{f_2(s)}$$

$$= \overline{[f_2 * f_1](g^{-1})}$$

$$= [f_2 * f_1]^*(g),$$

which shows the anti-multiplicative property of the involution. $\qquad\square$

Proposition 10.3.3

(i) *For $s, t \in G$ we have $\delta_s * \delta_t = \delta_{st}$.*

(ii) *For $s \in G$, $f \in L(G)$ we have: $\delta_s * f = \lambda_G(s)f$ and $f * \delta_s = \rho_G(s^{-1})f$.*

(iii) *The center $\mathcal{Z}[L(G)]$ of the group algebra coincides with the set of all functions $f \in L(G)$ that are constant on each conjugacy class of G, that is, $f(s^{-1}ts) = f(t)$ for all $s, t \in G$. Such functions are termed* central *or* class functions.

(iv) *$L(G)$ is commutative if and only if G is Abelian.*

Proof. Let $g, s, t \in G$ and $f \in L(G)$.

(i) $(\delta_s * \delta_t)(g) = \sum_{h \in G} \delta_s(gh^{-1})\delta_t(h) = \delta_s(gt^{-1}) = \delta_{st}(g).$

(ii)

$$(\delta_s * f)(g) = \sum_{h \in G} \delta_s(h)f(h^{-1}g) = f(s^{-1}g) = [\lambda_G(s)f](g)$$

and similarly $(f * \delta_s)(g) = f(gs^{-1}) = [\rho_G(s^{-1})f](g).$

(iii) f belongs to the center if and only if $f * \delta_s = \delta_s * f$ for all $s \in L(G)$, that is if and only if $\delta_s * f * \delta_{s^{-1}} = f$ and this is equivalent to saying that f is central since, by (ii), $\delta_s * f * \delta_{s^{-1}}(t) = f(s^{-1}ts).$

(iv) $L(G)$ is commutative if and only if $\delta_{st} = \delta_s * \delta_t = \delta_t * \delta_s = \delta_{ts}$ for all $s, t \in G$, that is, if and only if G is Abelian. Alternatively, $L(G)$ is commutative if and only if it coincides with its center, that is, by (iii), if and only if each conjugacy class consists of one single element, and this is again equivalent to saying that G is Abelian. $\qquad\square$

Exercise 10.3.4 Show that $f \in L(G)$ is a class function if and only if $f(g_1g_2) = f(g_2g_1)$ for all $g_1, g_2 \in G$.

Given $f \in L(G)$ the *convolution operator* with *kernel f* is the linear operator $T_f \in \operatorname{End}(L(G))$ defined by setting:

$$T_f f' = f' * f, \qquad (10.32)$$

for all $f' \in L(G)$.

Proposition 10.3.5 $T_f \in \operatorname{End}_G(L(G))$ *for every* $f \in L(G)$; *here,* $\operatorname{End}_G(L(G))$ *is the commutant (cf. Definition* 10.2.1*) of the left regular representation of G. Moreover, the map*

$$\begin{array}{ccc} L(G) & \longrightarrow & \operatorname{End}_G(L(G)) \\ f & \longmapsto & T_f \end{array} \qquad (10.33)$$

is a ∗-anti-isomorphism of algebras, that is

$$T_{f_1 * f_2} = T_{f_2} T_{f_1} \qquad and \qquad T_{f^*} = (T_f)^* \qquad (10.34)$$

for all $f_1, f_2, f \in L(G)$.

Proof. First of all, for $f, f' \in L(G)$ and $g, g_0 \in G$ we have:

$$\begin{aligned}
[T_f \lambda_G(g) f'](g_0) &= \big([\lambda_G(g) f'] * f\big)(g_0) \\
&= \sum_{h \in G} [\lambda_G(g) f'](g_0 h) f(h^{-1}) \\
&= \sum_{h \in G} f'(g^{-1} g_0 h) f(h^{-1}) \\
&= [T_f f'](g^{-1} g_0) \\
&= \big(\lambda_G(g)[T_f f']\big)(g_0)
\end{aligned}$$

so that $T_f \lambda_G(g) = \lambda_G(g) T_f$. This shows that $T_f \in \operatorname{End}_G(L(G))$. Moreover, if $f, f_1, f_2 \in L(G)$ then, by associativity of the convolution product,

$$T_{f_1}(T_{f_2} f) = (f * f_2) * f_1 = f * (f_2 * f_1) = T_{f_2 * f_1} f,$$

so that $T_{f_1}T_{f_2} = T_{f_2 * f_1}$. Moreover,

$$\langle T_f f_1, f_2 \rangle_{L(G)} = \sum_{g \in G} \sum_{s \in G} f_1(gs)f(s^{-1})\overline{f_2(g)}$$

$$\text{(setting } g = ts^{-1}) \quad = \sum_{t \in G} \sum_{s \in G} f_1(t)f(s^{-1})\overline{f_2(ts^{-1})}$$

$$= \sum_{t \in G} \sum_{s \in G} f_1(t)\overline{f^*(s)f_2(ts^{-1})}$$

$$= \langle f_1, T_{f^*} f_2 \rangle_{L(G)},$$

that is, $(T_f)^* = T_{f^*}$. We now prove that the map $f \mapsto T_f$ is a bijection by showing that if $T \in \text{End}_G(L(G))$, then there exists a unique element $f \in L(G)$ such that $T = T_f$ and that, indeed, $f = T\delta_{1_G}$. Uniqueness is clear: let $f_1, f_2 \in L(G)$ and suppose that $T_{f_1} = T_{f_2}$. Then, recalling that δ_{1_G} is the unit in $L(G)$, we deduce that $f_1 = \delta_{1_G} * f_1 = T_{f_1}\delta_{1_G} = T_{f_2}\delta_{1_G} = \delta_{1_G} * f_2 = f_2$. Finally, if $f' \in L(G)$, then, using (10.11), we have

$$Tf' = T\left[\sum_{g \in G} f'(g)\lambda_G(g)\delta_{1_G}\right]$$

$$\text{(since } T \in \text{End}_G(L(G)) \quad = \sum_{g \in G} f'(g)\lambda_G(g)T\delta_{1_G}$$

$$\text{(by (10.31))} \quad = f' * (T\delta_{1_G}). \qquad \square$$

We now compute the convolution of matrix coefficients and characters. From now on, for each $\theta \in \widehat{G}$ we fix an orthonormal basis $\{v_j^\theta : j = 1, 2, \ldots, d_\theta\}$ in the representation space V_θ and denote by $u_{i,j}^\theta$, $i, j = 1, 2, \ldots, d_\theta$, the corresponding matrix coefficients (as in (10.14)).

Proposition 10.3.6 *For all $\theta, \sigma \in \widehat{G}$ we have:*

$$u_{i,j}^\theta * u_{h,k}^\sigma = \frac{|G|}{d_\theta}\delta_{\theta,\sigma}\delta_{j,h}u_{i,k}^\theta \tag{10.35}$$

for all $1 \leq i, j \leq d_\theta$ and $1 \leq h, k \leq d_\sigma$. Moreover,

$$\chi^\theta * \chi^\sigma = |G|\delta_{\theta,\sigma}\chi^\theta. \tag{10.36}$$

Proof. For all $g \in G$ we have

$$\left[u_{i,j}^{\theta} * u_{h,k}^{\sigma}\right](g) = \sum_{s \in G} u_{i,j}^{\theta}(gs)u_{h,k}^{\sigma}(s^{-1})$$

$$\text{(by (i) and (iii) in Proposition 10.2.13)} = \sum_{\ell=1}^{d_\theta} u_{i,\ell}^{\theta}(g) \sum_{s \in G} u_{\ell,j}^{\theta}(s)\overline{u_{k,h}^{\sigma}(s)}$$

$$\text{(by (10.24))} = \sum_{\ell=1}^{d_\theta} u_{i,\ell}^{\theta}(g)\delta_{\theta,\sigma}\delta_{\ell,k}\delta_{j,h}\frac{|G|}{d_\theta}$$

$$= \frac{|G|}{d_\theta}\delta_{\theta,\sigma}\delta_{j,h}u_{i,k}^{\theta}(g).$$

The convolutional property of the characters (10.36) then follows from (10.15) and (10.35). □

Definition 10.3.7 Let $f \in L(G)$ and $(\theta, W_\theta) \in \widehat{G}$. The *Fourier transform* of f with respect to θ is the linear operator $\widehat{f}(\theta) \in \text{End}(W_\theta)$ defined by setting

$$\widehat{f}(\theta) = \sum_{g \in G} f(g)\theta(g).$$

Proposition 10.3.8 *Let $f_1, f_2, f \in L(G)$ and $\theta \in \widehat{G}$. Then we have*

$$\widehat{f_1 * f_2}(\theta) = \widehat{f_1}(\theta)\widehat{f_2}(\theta) \tag{10.37}$$

and

$$\widehat{f^*}(\theta) = \widehat{f}(\theta)^*. \tag{10.38}$$

Proof. We have

$$\widehat{f_1 * f_2}(\theta) = \sum_{g \in G}\left[\sum_{h \in G} f_1(gh^{-1})f_2(h)\right]\theta(g)$$

$$= \sum_{g \in G}\sum_{h \in G} f_1(gh^{-1})f_2(h)\theta(gh^{-1})\theta(h)$$

$$= \sum_{h \in G}\left[\sum_{g \in G} f_1(gh^{-1})\theta(gh^{-1})\right] f_2(h)\theta(h)$$

$$= \widehat{f_1}(\theta)\widehat{f_2}(\theta).$$

This shows (10.37). For $v, w \in W_\theta$ we have:

$$\langle \widehat{f}^*(\theta)v, w \rangle = \sum_{g \in G} \overline{f(g^{-1})} \langle \theta(g)v, w \rangle$$

$$= \left\langle v, \sum_{g \in G} f(g^{-1})\theta(g)^* w \right\rangle$$

$$= \left\langle v, \sum_{g \in G} f(g^{-1})\theta(g^{-1})w \right\rangle$$

$$= \langle v, \widehat{f}(\theta)w \rangle$$

and (10.38) follows as well. $\qquad \square$

Proposition 10.3.9 *Let* $f \in \mathcal{Z}(L(G))$ *and* $(\theta, W_\theta) \in \widehat{G}$. *Then the Fourier transform of* f *with respect to* θ *is a scalar multiple of the identity, more precisely,*

$$\widehat{f}(\theta) = \lambda I_W \quad \text{with} \quad \lambda = \frac{1}{d_\theta} \sum_{g \in G} f(g)\chi^\theta(g) = \frac{1}{d_\theta}\langle f, \overline{\chi^\theta} \rangle.$$

Proof. Observe that

$$\theta(g)\widehat{f}(\theta)\theta(g^{-1}) = \sum_{h \in G} f(h)\theta(g)\theta(h)\theta(g^{-1}) = \sum_{h \in G} f(h)\theta(ghg^{-1})$$

(by Proposition 10.3.3.(iii)) $\quad = \sum_{h \in G} f(ghg^{-1})\theta(ghg^{-1}) = \widehat{f}(\theta),$

so that $\widehat{f}(\theta) \in \text{End}_G(W_\theta)$. By Corollary 10.2.4 we deduce that $\widehat{f}(\theta) = \lambda I_W$. Computing the trace, we obtain

$$\lambda d_\theta = \text{Tr}(\lambda I_W) = \text{Tr}\left[\widehat{f}(\theta)\right] = \sum_{h \in G} f(h)\chi^\theta(h) = \langle f, \overline{\chi^\theta} \rangle,$$

which yields the desired value of λ. $\qquad \square$

Theorem 10.3.10 (Fourier's inversion formula) *For* $f \in L(G)$ *one has*

$$f(g) = \frac{1}{|G|} \sum_{\theta \in \widehat{G}} d_\theta \text{Tr}\left[\theta(g^{-1})\widehat{f}(\theta)\right] \qquad (10.39)$$

for all $g \in G$. *In particular, if* $f_1, f_2 \in L(G)$ *satisfy the condition* $\widehat{f}_1(\theta) = \widehat{f}_2(\theta)$ *for every* $\theta \in \widehat{G}$, *then one has* $f_1 = f_2$.

Proof. Let $\{v_1^\theta, v_2^\theta, \ldots, v_{d_\theta}^\theta\}$ be an orthonormal basis for W_θ for all $\theta \in \widehat{G}$. By virtue of Theorem 10.2.25, the corresponding (normalized) coefficients $\frac{\sqrt{d_\theta}}{|G|} u_{i,j}^\theta$,

$i, j = 1, 2, \ldots, d_\theta, \theta \in \widehat{G}$, constitute an orthonormal basis in $L(G)$. As a consequence, also their conjugates $\frac{\sqrt{d_\theta}}{|G|} \overline{u^\theta_{i,j}}$ constitute an orthonormal basis and thus for every function $f \in L(G)$ we have

$$f(g) = \frac{1}{|G|} \sum_{\theta \in \widehat{G}} d_\theta \sum_{i,j=1}^{d_\theta} \left\langle f, \overline{u^\theta_{i,j}} \right\rangle \overline{u^\theta_{i,j}(g)}, \qquad (10.40)$$

for all $g \in G$. Now, recalling that $\widehat{f}(\theta) = \sum_{g \in G} f(g)\theta(g)$ we have

$$\langle f, \overline{u^\theta_{i,j}} \rangle - \sum_{g \in G} f(g) u^\theta_{i,j}(g) = \sum_{g \in G} f(g) \langle \theta(g) n^\theta_j, v^\theta_i \rangle = \langle \widehat{f}(\theta) v^\theta_j, v^\theta_i \rangle \quad (10.41)$$

and

$$\begin{aligned}
\sum_{i,j=1}^{d_\theta} \left\langle f, \overline{u^\theta_{i,j}} \right\rangle \overline{u^\theta_{i,j}(g)} &= \sum_{i,j=1}^{d_\theta} \langle \widehat{f}(\theta) v^\theta_j, v^\theta_i \rangle \langle v^\theta_i, \theta(g) v^\theta_j \rangle \\
&= \sum_{j=1}^{d_\theta} \langle \widehat{f}(\theta) v^\theta_j, \theta(g) v^\theta_j \rangle \\
&= \sum_{j=1}^{d_\theta} \langle \theta(g^{-1}) \widehat{f}(\theta) v^\theta_j, v^\theta_j \rangle \\
&= \mathrm{Tr} \left[\theta(g^{-1}) \widehat{f}(\theta) \right].
\end{aligned}$$

Thus, replacing this expression in (10.40), we deduce (10.39). $\qquad\square$

Exercise 10.3.11 Deduce the Fourier inversion formula (10.39) from (10.19), first in the case $f = \delta_g$, $g \in G$, and then, using linearity, in the general case (cf. (10.11)).

The Fourier inversion theorem shows that every function in $L(G)$ is uniquely determined by its Fourier transforms $\widehat{f}(\theta)$, $\theta \in \widehat{G}$. Note that although the expression of f, with respect to an orthonormal system made up of matrix coefficients is not unique but depends on the choice of an orthonormal basis in each representation space W_θ, $\theta \in \widehat{G}$, the Fourier inversion formula, however, does not depend on the choice of such bases.

Finally, from this analysis we deduce that the algebra $L(G)$ is isomorphic to a direct sum of matrix algebras, namely, $L(G) \cong \oplus_{\theta \in \widehat{G}} \mathfrak{M}_{d_\theta}(\mathbb{C})$, where $\mathfrak{M}_{d_\theta}(\mathbb{C}) \cong \mathrm{End}(W_\theta)$ is the algebra of d_θ-by-d_θ matrices over $\mathbb{C}$. In order to formulate more explicitly the properties of the Fourier transform as a linear map, we define the complex algebra

$$C(\widehat{G}) = \bigoplus_{\theta \in \widehat{G}} \mathrm{End}(W_\theta).$$

Clearly, $C(\widehat{G})$ is a direct sum of algebras and every element $T \in C(\widehat{G})$ will be written in the form $T = \oplus_{\theta \in \widehat{G}} T(\theta)$, where $T(\theta) \in \mathrm{End}(W_\theta)$ for each $\theta \in \widehat{G}$. It is also involutive with respect to the map $T \mapsto T^* = \oplus_{\theta \in \widehat{G}} T(\theta)^*$.

Corollary 10.3.12 *The Fourier transform*

$$
\begin{aligned}
L(G) &\longrightarrow C(\widehat{G}) \\
f &\longmapsto \widehat{f}
\end{aligned}
$$

is a $*$-isomorphism of $*$-algebras and its inverse is given by the map (inverse Fourier transform)

$$
\begin{aligned}
C(\widehat{G}) &\longrightarrow L(G) \\
T &\longmapsto T^\vee,
\end{aligned}
$$

where $T^\vee(g) = \frac{1}{|G|} \sum_{\theta \in \widehat{G}} d_\theta \mathrm{Tr}\left[\theta(g^{-1}) T(\theta)\right]$.

Theorem 10.3.13 *The Fourier inversion formula for a central function f has the form*

$$
f = \frac{1}{|G|} \sum_{\theta \in \widehat{G}} \langle f, \overline{\chi^\theta} \rangle_{L(G)} \overline{\chi^\theta}.
$$

In particular:

 (i) *the characters χ^θ, $\theta \in \widehat{G}$, constitute an orthogonal basis for the subspace of central functions;*
 (ii) *$|\widehat{G}|$ equals the number of conjugacy classes in G.*

Proof. The inversion formula follows from Proposition 10.3.9, taking into account that $\mathrm{Tr}\,\theta(g^{-1}) = \overline{\chi^\theta(g)}$ for all $g \in G$. Note also that from Proposition 10.2.15 and Proposition 10.2.17 it follows that the characters of irreducible representations form an orthogonal system in the space of central functions; the inversion formula ensures that it is also *complete*. Since the dimension of the space of central functions is equal to the number of conjugacy classes (recall Proposition 10.3.3.(iii)), this dimension must also equal the number of irreducible representations of G. $\qquad\square$

Corollary 10.3.14 (Dual orthogonality relations for characters) *Let $\mathcal{L} \subseteq G$ be a set of representatives for the conjugacy classes of G and denote by $\mathcal{C}(t) = \{g^{-1}tg : g \in G\}$ the conjugacy class of $t \in \mathcal{L}$. Then*

$$
\sum_{\theta \in \widehat{G}} \chi^\theta(t)\overline{\chi^\theta(t')} = \frac{|G|}{|\mathcal{C}(t)|}\delta_{t,t'} \tag{10.42}
$$

for all $t, t' \in \mathcal{L}$.

Proof. We begin by observing that (10.16) may be rewritten in the form

$$\sum_{t \in \mathcal{L}} \frac{|\mathcal{C}(t)|}{|G|} \chi^{\theta_1}(t) \overline{\chi^{\theta_2}(t)} = \delta_{\theta_1, \theta_2},$$

thus showing that the square (recall that $|\mathcal{L}| = |\widehat{G}|$) matrix $U = \left(U_{\theta, t}\right)_{\theta \in \widehat{G}, t \in \mathcal{L}}$, with $U_{\theta, t} = \sqrt{\frac{|\mathcal{C}(t)|}{|G|}} \chi^{\theta}(t)$, is unitary. Therefore

$$\sum_{\theta \in \widehat{G}} \sqrt{\frac{|\mathcal{C}(t_1)|}{|G|}} \chi^{\theta}(t_1) \cdot \sqrt{\frac{|\mathcal{C}(t_2)|}{|G|}} \overline{\chi^{\theta}(t_2)} = \delta_{t_1, t_2}$$

and the statement follows. □

Exercise 10.3.15 Deduce (10.42) from the dual orthogonality relations for matrix coefficients (cf. Lemma 10.2.13).

Exercise 10.3.16 Let G be a finite group.

(1) Use Theorem 10.3.13 to prove that G is Abelian if and only if its irreducible representations are all one-dimensional.
(2) More generally, prove that if G contains an Abelian subgroup A, then $d_\theta \leq |G/A|$ for all $\theta \in \widehat{G}$.

Solution of (2): Let $(\theta, V) \in \widehat{G}$. Consider the restriction $(\text{Res}_A^G \theta, V)$ and let $W \leq V$ be a nontrivial $\text{Res}_A^G \theta$-irreducible subspace. By (1) we have that W is one-dimensional. Set $H = \{g \in G : \theta(g)W \subseteq W\}$ and denote by $\mathcal{T} \subset G$ a complete set of representatives for the left cosets of H in G, so that $G = \bigsqcup_{t \in \mathcal{T}} tH$. Clearly $A \leq H$, $\theta(g)W \in \{\theta(t)W : t \in \mathcal{T}\}$ for all $g \in G$, and $\dim \theta(t)W = 1$ for all $t \in \mathcal{T}$. Since, by irreducibility, $V = \oplus_{t \in \mathcal{T}} \theta(t)W$, we deduce that $d_\theta = |\mathcal{T}| = |G/H| \leq |G/A|$.

Theorem 10.3.17 (Plancherel formula) *For all $f_1, f_2 \in L(G)$ we have:*

$$\langle f_1, f_2 \rangle_{L(G)} = \frac{1}{|G|} \sum_{\theta \in \widehat{G}} d_\theta \text{Tr} \left[\widehat{f_1}(\theta) \widehat{f_2}(\theta)^* \right]. \tag{10.43}$$

Proof. From Theorem 10.2.25.(iii) we deduce that

$$\langle f_1, f_2 \rangle = \sum_{\theta \in \widehat{G}} \frac{d_\theta}{|G|} \sum_{i, j=1}^{d_\theta} \left\langle f_1, \overline{u_{i,j}^\theta} \right\rangle \left\langle \overline{u_{i,j}^\theta}, f_2 \right\rangle,$$

and then, applying (10.41), we get

$$
\langle f_1, f_2 \rangle = \frac{1}{|G|} \sum_{\theta \in \widehat{G}} d_\theta \sum_{i,j=1}^{d_\theta} \langle \widehat{f_1}(\theta) v_j^\theta, v_i^\theta \rangle \cdot \langle v_i^\theta, \widehat{f_2}(\theta) v_j^\theta \rangle =
$$
$$
= \frac{1}{|G|} \sum_{\theta \in \widehat{G}} d_\theta \operatorname{Tr} \left[\widehat{f_1}(\theta) \widehat{f_2}(\theta)^* \right].
$$

□

10.4 Group actions and permutation characters

In the present section we suppose that the finite group G acts on a finite set X. We recall that this means that we have a map

$$
G \times X \longrightarrow X
$$
$$
(g, x) \longmapsto gx
$$

such that

- for each $g \in G$ the map $x \mapsto gx$ is a bijection (a permutation) of X, that we denote by $\pi(g)$;
- the map $g \mapsto \pi(g)$ is a homomorphism between G and $\operatorname{Sym}(X)$, the group of all permutations of X.

This is equivalent to saying that $(g_1 g_2)x = g_1(g_2 x)$ and $1_G x = x$ so that, in particular, $x \mapsto g^{-1}x$ is the inverse permutation $\pi(g)^{-1}$, for all $g_1, g_2 \in G$ and $x \in X$. We usually call gx the *g-image* of x.

For $x \in X$ denote by $\operatorname{Stab}_G(x) = \{g \in G : gx = x\}$ (or G_x) and $\operatorname{Orb}_G(x) = \{gx : g \in G\}$ (or Gx) the *stabilizer* and the *G-orbit* of x. It is easy to see that the orbits form a partition of X (see Exercise 10.4.1); the action is *transitive* if there is a single orbit, that is $\operatorname{Orb}_G(x) = X$ (and this clearly holds for all $x \in X$). Equivalently, it is transitive if and only if for all $x_1, x_2 \in X$ there exists $g \in G$ such that $gx_1 = x_2$. If G acts transitively on X we also say that X is a (*homogeneous*) *G-space*.

Exercise 10.4.1 Let X be a G-space.

(1) Show that $\operatorname{Stab}_G(gx) = g\operatorname{Stab}_G(x)g^{-1}$, for all $g \in G$ and $x \in X$.
(2) Show that for $x, x' \in X$, the relation $x \sim x'$ if x and x' belong to the same G-orbit is an equivalence relation on X, so that the G-orbits on X constitute the corresponding partition of X.

Lemma 10.4.2 *Let X be a G-space. Then*

$$
|G| = |\operatorname{Stab}_G(x)| \cdot |\operatorname{Orb}_G(x)| \tag{10.44}
$$

for all $x \in X$. Moreover,

$$\frac{1}{|G|} \sum_{x \in X} |\mathrm{Stab}_G(x)| = \textit{number of } G\textit{-orbits in } X.$$

Proof. Let $x \in X$ and consider the map $\phi \colon G \to \mathrm{Orb}_G(x)$ that maps g to gx. By definition it is surjective; moreover one has $\phi^{-1}(x) = \mathrm{Stab}_G(x)$ and, more generally, $\phi^{-1}(gx) = \{gk : k \in \mathrm{Stab}_G(x)\} = g\mathrm{Stab}_G(x)$ so that, in particular, $|\phi^{-1}(x')| = |\phi^{-1}(x)| = |\mathrm{Stab}_G(x)|$ for all $x' \in \mathrm{Orb}_G(x)$. Thus ϕ is a surjective $|\mathrm{Stab}_G(x)|$-to-one map and (10.44) follows. Moreover, if $X_1, X_2, \ldots, X_h$ are the orbits of G on X then

$$\frac{1}{|G|} \sum_{x \in X} |\mathrm{Stab}_G(x)| = \frac{1}{|G|} \sum_{i=1}^{h} \sum_{x \in X_i} |\mathrm{Stab}_G(x)|$$

$$\text{(by (10.44))} \quad = \frac{1}{|G|} \sum_{i=1}^{h} \sum_{x \in X_i} \frac{|G|}{|X_i|}$$

$$= \sum_{i=1}^{h} \frac{1}{|X_i|} \cdot |X_i|$$

$$= h. \qquad \square$$

Example 10.4.3 Let X be a G-space. As in Section 2.1, let $L(X)$ denote the vector space of all complex valued functions defined on X endowed with the inner product defined by $\langle f_1, f_2 \rangle_{L(X)} = \sum_{x \in X} f_1(x)\overline{f_2(x)}$, for all $f_1, f_2 \in L(X)$. The *permutation representation* of G on X is the G-representation $(\lambda, L(X))$ defined by

$$[\lambda(g)f](x) = f(g^{-1}x)$$

for all $f \in L(X)$, $g \in G$ and $x \in X$. As in Example 10.1.8 (which is actually a particular case of the present construction), it is easy to check that this is a unitary representation and that the Dirac functions δ_x, $x \in X$, form an orthonormal basis (now, $\delta_x(x) = 1$ and $\delta_x(y) = 0$ if $y \neq x$). Moreover, $\lambda(g)\delta_x = \delta_{gx}$ for all $g \in G$, $x \in X$, and $f = \sum_{x \in X} f(x)\delta_x$ for all $f \in L(X)$. Let now $X = \coprod_{j=1}^{h} X_j$ be the decomposition of X into G-orbits. Then

$$L(X) = \bigoplus_{j=1}^{h} L(X_j) \qquad (10.45)$$

is clearly a direct sum decomposition into G-invariant subspaces. Indeed, any $f \in L(X)$ may be written in the form $f = \sum_{j=1}^{h} f_j$, where $f_j \in L(X)$ is defined by setting

$$f_j(x) = \begin{cases} f(x) & \text{if } x \in X_j \\ 0 & \text{otherwise,} \end{cases} \qquad (10.46)$$

for all $j = 1, 2, \ldots, h$, so that f_j may be naturally identified with a function in $L(X_j)$. Moreover, (10.46) implies G-invariance of the decomposition (10.45). For this reason, it is customary, in representation theory, to consider only *transitive* actions (that is, the case $h = 1$). Note also that even in this case, a permutation representation on a set X with more than one element is not irreducible because the $(|X| - 1)$-dimensional space $W_1 = \{f \in L(X) : \sum_{x \in X} f(x) = 0\}$ is always G-invariant: if $f \in W_1$ and $g \in G$ then

$$\sum_{x \in X} [\lambda(g)f](x) = \sum_{x \in X} f(g^{-1}x) = \sum_{y \in X} f(y) = 0$$

so that $\lambda(g)f \in W_1$. Note also that, as in Section 2.1, we have the orthogonal decomposition $L(X) = W_0 \oplus W_1$, where $W_0 = \{f \in L(X) : f \text{ constant}\} = W_1^\perp$. More explicitly, for any $f \in L(X)$ we have

$$f = \frac{1}{|X|} \sum_{x \in X} f(x) + \left[f - \frac{1}{|X|} \sum_{x \in X} f(x) \right]$$

where the first summand (the *mean value*) belongs to W_0 and the second one to W_1. Another important consequence of transitivity is the following: the trivial representation of G is contained in $L(X)$ with multiplicity exactly one and coincides with W_0. Indeed, if $\lambda(g)f = f$ for all $g \in G$ then transitivity implies that f is constant (in general, the multiplicity of the trivial representation in $(\lambda, L(X))$ equals the number of G-orbits). In Exercise 10.4.16 we will give a necessary and sufficient condition for the irreducibility of W_1.

Example 10.4.4 Let $G = S_n$ be the symmetric group of degree n (cf. Example 10.1.10). The *natural permutation representation* of S_n is n-dimensional representation constructed as in Example 10.4.3, using the natural action of S_n on $X = 1, 2, \ldots, n$. See also Exercise 10.4.16.

Example 10.4.5 (The affine group over $\mathbb{F}_q$) Let $\mathbb{F}_q$ be the finite field with $q = p^m$ elements, where p is a prime number and $m \geq 1$ (see Chapter 6). The *(general) affine group (of degree one)* over $\mathbb{F}_q$ is the group of matrices

$$\mathrm{Aff}(\mathbb{F}_q) = \left\{ \begin{pmatrix} a & b \\ 0 & 1 \end{pmatrix} : a \in \mathbb{F}_q^*, b \in \mathbb{F}_q \right\}.$$

The terminology is due to the fact that $\mathrm{Aff}(\mathbb{F}_q)$ acts (transitively: this is an easy exercise) on $\mathbb{F}_q \equiv \left\{ \begin{pmatrix} x \\ 1 \end{pmatrix} : x \in \mathbb{F}_q \right\}$ by multiplication

$$\begin{pmatrix} a & b \\ 0 & 1 \end{pmatrix} \begin{pmatrix} x \\ 1 \end{pmatrix} = \begin{pmatrix} ax + b \\ 1 \end{pmatrix}$$

and the maps $x \mapsto ax + b$ (with $a \in \mathbb{F}_q^*$, $b \in \mathbb{F}_q$) are the *affine transformations* of $\mathbb{F}_q$. For this reason, one often also refers to $\mathrm{Aff}(\mathbb{F}_q)$ as to the *finite $ax + b$ group*.

This defines a permutation representation of $\mathrm{Aff}(\mathbb{F}_q)$, that will be examined in Exercise 10.4.7 and Exercise 10.4.16. In Section 12.1 we shall fully describe all irreducible representations of $\mathrm{Aff}(\mathbb{F}_q)$.

Consider the permutation representation of G on $L(X)$ defined in Example 10.4.3. The corresponding character χ^λ is called the *permutation character* of the action of G on X. In the following, we prove a basic formula for χ^λ.

Proposition 10.4.6 (Fixed point character formula) *Let* $g \in G$. *Then we have*

$$\chi^\lambda(g) = |\{x \in X : gx = x\}|, \tag{10.47}$$

that is, $\chi^\lambda(g)$ equals the number of points in X that are fixed by g.

Proof. Recall that the set $\{\delta_x : x \in X\}$ is an orthonormal basis in $L(X)$ and therefore

$$\chi^\lambda(g) = \sum_{x \in X} \langle \lambda(g)\delta_x, \delta_x \rangle_{L(X)} = \sum_{x \in X} \langle \delta_{gx}, \delta_x \rangle_{L(X)}.$$

This clearly counts the points in X that are fixed by g (compare with (10.21), which is just a special case). $\qquad\qquad\square$

Another formula for χ^λ, in the case of a transitive permutation representation, will be given in Corollary 11.1.14.

Example 10.4.7 Consider the permutation representation λ of the finite affine group $\mathrm{Aff}(\mathbb{F}_q)$ (cf. Example 10.4.5). The corresponding permutation character χ^λ is given by

$$\chi^\lambda \begin{pmatrix} a & b \\ 0 & 1 \end{pmatrix} = \begin{cases} 1 & \text{if } a \neq 1 \\ q & \text{if } a = 1 \text{ and } b = 0 \\ 0 & \text{otherwise} \end{cases}$$

for all $a \in \mathbb{F}_q^*$ and $b \in \mathbb{F}_q$. Indeed, solving the equation $ax + b = x$, that is, $(a - 1)x + b = 0$, we find:

- if $a \neq 1$ there is a unique solution given by $x = -\frac{b}{a-1}$;
- if $a = 1$ and $b = 0$ then each $x \in \mathbb{F}_q$ is a solution (the identity fixes every point);
- if $a = 1$ and $b \neq 0$ there are no solutions.

The following lemma is usually called "the Burnside lemma," but it was known already to Cauchy (see [21, 121, 169]).

Lemma 10.4.8 (Burnside's lemma) *Let G be a finite group acting on a finite set X and denote by $(\lambda, L(X))$ the corresponding permutation representation. Then we have:*

$$\frac{1}{|G|} \sum_{g \in G} \chi^\lambda(g) = \text{ number of } G\text{-orbits on } X.$$

Proof. We clearly have

$$\frac{1}{|G|} \sum_{g \in G} \chi^\lambda(g) = \frac{1}{|G|} \langle \chi^\lambda, 1_G \rangle_{L(G)}, \tag{10.48}$$

where $1_G = \chi^\iota$, the character of the trivial representation of G. By Proposition 10.2.18, the right hand side of (10.48) equals the multiplicity of the trivial representation as a sub-representation of the permutation representation λ. Since (cf. Example 10.4.3) $L(X)^G = \oplus_{i=1}^h \mathbb{C}1_{X_i}$, where 1_{X_i} denotes the characteristic function of the orbit X_i, $i = 1, 2, \ldots, h$, and (cf. Example 10.2.21) the multiplicity of the trivial representation in any G-representation V equals the dimension of the subspace V^G of G-invariant vectors, the right hand side of (10.48) is therefore equal to $\dim(L(X)^G) = h$, the number of G-orbits on X. $\quad\square$

Exercise 10.4.9 Deduce Burnside's Lemma from Lemma 10.4.2 and Proposition 10.4.6.

From now on, we assume that G acts transitively on X, that $K \leq G$ is the stabilizer of a fixed element $x_0 \in X$, and that $\mathcal{T}$ is a complete set of representatives for the left cosets of K in G, that is,

$$G = \coprod_{t \in \mathcal{T}} tK. \tag{10.49}$$

Then the map

$$\begin{aligned} \Psi \colon G/K &\to X \\ gK &\mapsto gx_0, \end{aligned} \tag{10.50}$$

where G/K is the set of all left cosets of K in G, is a bijection. Indeed, for $g_1, g_2 \in G$ we have $g_1 x_0 = g_2 x_0$ if and only if $g_1^{-1} g_2 \in K$, that is, $g_1 K = g_2 K$. Define an action of G on G/K by setting $g(g_0 K) = (g g_0) K$. It is easy to see that the map (10.50) is *G-equivariant* (or, that the *G*-spaces X and G/K are *isomorphic*), that is,

$$g\Psi(g_0 K) = \Psi(g(g_0 K)) \quad \forall g, g_0 \in G.$$

In other words, every transitive *G*-space is isomorphic to a *G*-space G/K (where K, as above, is the stabilizer of a point in X).

Exercise 10.4.10

   (1) Let $H, K \leq G$ be two subgroups. Show that G/H and G/K are isomorphic as *G*-spaces if and only if H and K are conjugate in G (there exists $g \in G$ such that $H = g^{-1} K g$).
   (2) Let X be a transitive *G*-space. Let $x_0, x_0' \in X$ and denote by $K, K' \leq G$ the corresponding stabilizers. Using Exercise 10.4.1 and (1) show that the *G*-spaces G/K and G/K' are isomorphic.

Given an action of a group G on a set X, the corresponding *diagonal action* of G on $X \times X$ is defined by setting

$$g(x_1, x_2) = (g x_1, g x_2), \quad g \in G, \ x_1, x_2 \in X.$$

We denote by $(\lambda^2, L(X \times X))$ the corresponding permutation representation.

Proposition 10.4.11 *Let X be a G-space and denote by $(\lambda, L(X))$ and $(\lambda^2, L(X \times X))$ the corresponding permutation representations. Then*

$$\chi^{\lambda^2} = (\chi^\lambda)^2.$$

Proof. Let $g \in G$. From the fixed point character formula (10.47) we deduce that

$$\left(\chi^\lambda(g)\right)^2 = |\{x \in X : gx = x\}|^2$$
$$= |\{x_1 \in X : g x_1 = x_1\}| \cdot |\{x_2 \in X : g x_2 = x_2\}|$$
$$= |\{(x_1, x_2) \in X \times X : g(x_1, x_2) = (x_1, x_2)\}|$$

(again by (10.47)) $= \chi^{\lambda^2}(g).$ $\qquad\square$

Proposition 10.4.12 *Let X be a G-space and denote, as usual, by $K \leq G$ the stabilizer of a fixed point $x_0 \in X$. Let $X = \Omega_0 \coprod \Omega_1 \coprod \cdots \coprod \Omega_n$ denote the decomposition of X into K-orbits (with $\Omega_0 = \{x_0\}$) and choose $x_i \in \Omega_i$, $i = 1, 2, \ldots, n$. Then the sets*

$$G(x_i, x_0) = \{(g x_i, g x_0) : g \in G\} \subseteq X \times X,$$

$i = 0, 1, 2, \ldots, n$, are the orbits of the diagonal action of G on $X \times X$.

Proof. First of all, note that if $(x, y) \in X \times X$ then there exist $g \in G, k \in K$, and $i \in \{0, 1, \ldots, n\}$ such that $gx_0 = y$ (G is transitive on X) and $gkx_i = x$ (let $Kx_i = \Omega_i$ be the K-orbit containing $g^{-1}x$). Therefore,

$$(x, y) = (gkx_i, gx_0) = (gkx_i, gkx_0) \in G(x_i, x_0).$$

This shows that

$$X \times X = \bigcup_{i=0}^{n} G(x_i, x_0). \tag{10.51}$$

It is also easy to show that $G(x_i, x_0) \cap G(x_j, x_0) = \varnothing$ if $i \neq j$: indeed if $g_1, g_2 \in G$ satisfy $g_1 x_i = g_2 x_j$ and $g_1 x_0 = g_2 x_0$ then, necessarily, $g_2^{-1}g_1 \in K$, and this forces $i = j$. Therefore (10.51) is in fact a disjoint union. □

Conversely, we may rephrase the above result as follows.

Corollary 10.4.13 *Let Θ be a G-orbit on $X \times X$. Then the set $\Omega = \{x \in X : (x, x_0) \in \Theta\}$ is an orbit of K on X and the map $\Theta \mapsto \Omega$ is a bijection between the set of orbits of G on $X \times X$ (with the diagonal action) and those of K on X.*

The following result was surely known to Schur and possibly even to Frobenius. Since a standard reference for it is the book by Wielandt [167], for convenience we refer to it as to "Wielandt's lemma." Another proof will be indicated in Exercise 11.4.9.

Lemma 10.4.14 (Wielandt) *Let X be a G-space. Suppose that $L(X) = \bigoplus_{i=0}^{N} m_i V_i$ is the decomposition of $L(X)$ into irreducible G-representations, where m_i denotes the multiplicity of V_i. Then*

$$\sum_{i=0}^{N} m_i^2 = number \ of \ G\text{-}orbits \ on \ X \times X = number \ of \ K\text{-}orbits \ on \ X. \tag{10.52}$$

Proof. Denote again by χ^λ the permutation character associated with the G-action on X. From Corollary 10.2.22 we deduce that:

$$\sum_{i=1}^{h} m_i^2 = \frac{1}{|G|} \langle \chi^\lambda, \chi^\lambda \rangle_{L(G)}$$

$$(\chi^\lambda = \overline{\chi^\lambda} \text{ by Proposition 10.4.6}) \quad = \frac{1}{|G|} \sum_{g \in G} \chi^\lambda(g)^2$$

$$(\text{by Proposition 10.4.11}) \quad = \frac{1}{|G|} \sum_{g \in G} \chi^{\lambda^2}(g)$$

$$(\text{by Lemma 10.4.8}) \quad = \text{number of } G\text{-orbits on } X \times X$$

$$(\text{by Corollary 10.4.13}) \quad = \text{number of } K\text{-orbits on } X.$$

In other words, by Proposition 10.2.18,

$$\frac{1}{|G|}\langle \chi^\lambda, \chi^\lambda \rangle = \frac{1}{|G|}\left\langle \left(\chi^\lambda\right)^2, \mathbf{1}_G \right\rangle = \frac{1}{|G|}\left\langle \chi^{\lambda^2}, \mathbf{1}_G \right\rangle$$

is equal to the multiplicity of the trivial representation in the permutation representation of G on $X \times X$. $\qquad\square$

The following is a slight but useful generalization of the previous result.

Exercise 10.4.15 Let G act transitively on two finite sets $X = G/K$ and $Y = G/H$. Define the diagonal action of G on $X \times Y$ by setting, for all $x \in X$, $y \in Y$, and $g \in G$

$$g(x, y) = (gx, gy).$$

(1) Show that the number of G-orbits on $X \times Y$ equals the number of H-orbits on X, which in turn equals the number of K-orbits on Y.
(2) Let $L(X) = \oplus_{i \in I} m_i V_i$ and $L(Y) = \oplus_{j \in J} n_j V_j$ denote the decomposition of the permutation representations $L(X)$ and $L(Y)$ into irreducible representations. Denoting by $I \cap J$ the set of indices corresponding to common (equivalent) sub-representations, show that the number of G-orbits on $X \times Y$ equals the sum $\sum_{i \in I \cap J} m_i n_i$.

An action of G on X is called *doubly transitive* if for all (x_1, x_2), $(y_1, y_2) \in (X \times X) \setminus \{(x, x) : x \in X\}$ there exists $g \in G$ such that $gx_i = y_i$ for $i = 1, 2$.

Exercise 10.4.16 Suppose that G acts transitively on X.

(1) Prove that G is doubly transitive on X if and only if K is transitive on $X \setminus \{x_0\}$.
(2) Let W_0 and W_1 be as in Example 10.4.3. Prove that $L(X) = W_0 \oplus W_1$ is the decomposition of the permutation representation into irreducibles if and only if G acts doubly transitively on X.
(3) Prove that if the action of G on $X = G/K$ is doubly transitive, then K is a maximal subgroup ($K < H \leq G$ infers $H = G$).
    Solution. Suppose that $K < H \leq G$ and let $h \in H \setminus K$ and $g \in G \setminus K$. By double transitivity applied to (K, hK), $(K, gK) \in (X \times X) \setminus \{(x, x) : x \in X\}$, there exists $g' \in G$ such that $g'K = K$ and $g'hK = gK$. But then $g' \in K$, $g'h \in H$ and therefore $g \in H$. This shows that $H = G$.
(4) Show that the action of S_n on $\{1, 2, \ldots, n\}$ is doubly transitive.
(5) Show that the action of $\mathrm{Aff}(\mathbb{F}_q)$ on $\mathbb{F}_q$ defined in Example 10.4.5 is doubly transitive. Deduce that the corresponding permutation representation decomposes into the sum of the trivial representation and of a $(q - 1)$-dimensional, irreducible representation. See also Section 12.1.

Exercise 10.4.17 Consider the dihedral group D_n in Example 10.2.28 and define an action of D_n on the additive cyclic group $\mathbb{Z}_n$ by setting $ah = h + 1$ and $bh = -h$ for all $h \in \mathbb{Z}_n$. Show that this coincides with the natural action of D_n on the regular polygon with n sides. Also show that the corresponding permutation representation λ decomposes as follows:

$$\lambda = \begin{cases} \chi_0 \oplus \chi_3 \oplus \left(\bigoplus_{j=1}^{\frac{n}{2}-1} \rho_j \right) & \text{if } n \text{ is even} \\[3mm] \chi_0 \oplus \left(\bigoplus_{j=1}^{\frac{n-1}{2}} \rho_j \right) & \text{if } n \text{ is odd.} \end{cases}$$

10.5 Conjugate representations and tensor products

The present section is devoted to two basic constructions in linear and multi-linear algebra, namely dual spaces and tensor products, in the framework of the representation theory of finite groups. We recall all basic notions but only for finite dimensional, complex unitary spaces.

Let V be a finite dimensional complex vector spaces. The *dual* V' of V is the space of all linear functionals $f : V \to \mathbb{C}$. If V is unitary, then the Riesz representation theorem ensures that for each $f \in V'$ there exists a unique vector $\xi(f) \in V$ such that:

$$f(v) = \langle v, \xi(f) \rangle, \quad \text{for all } v \in V. \tag{10.53}$$

The Riesz map $\xi = \xi_V : V' \to V$ is anti-linear, i.e. $\xi(\alpha f_1 + \beta f_2) = \overline{\alpha} \xi(f_1) + \overline{\beta} \xi(f_2)$, for all $\alpha, \beta \in \mathbb{C}$ and $f_1, f_2 \in V'$, and bijective. In V' we introduce an inner product by setting, for all f_1 and $f_2 \in V'$,

$$\langle f_1, f_2 \rangle_{V'} = \langle \xi(f_2), \xi(f_1) \rangle_V. \tag{10.54}$$

Thus, for $f \in V'$ and $v \in V$ one has

$$f(v) = \langle v, \xi(f) \rangle_V = \langle f, \xi^{-1}(v) \rangle_{V'}$$

which shows that $V'' = (V')'$, the bi-dual of V, is isometrically identified with V by means of ξ^{-1}.

Definition 10.5.1 Let G be a finite group and (ρ, V) a unitary representation of G. We define the *adjoint* or *conjugate representation* (ρ', V') of (ρ, V) by setting, for all $f \in V'$, $v \in V$ and $g \in G$

$$[\rho'(g)f](v) = f[\rho(g^{-1})v]. \tag{10.55}$$

It is easy to check that ρ' is a linear representation of G and ρ' is irreducible if and only if ρ is irreducible. This is an immediate consequence of the next proposition.

Proposition 10.5.2 *For all $g \in G$ we have:*

$$\rho'(g) = \xi^{-1}\rho(g)\xi. \tag{10.56}$$

Proof. For all $g \in G$, $v \in V$, and $f \in V'$ we have:

$$\begin{aligned}
\langle v, \xi[\rho'(g)f]\rangle &= [\rho'(g)f](v) && \text{by (10.53)} \\
&= f[\rho(g^{-1})v] && \text{by (10.55)} \\
&= \langle \rho(g^{-1})v, \xi(f)\rangle && \text{by (10.53)} \\
&= \langle v, \rho(g)[\xi(f)]\rangle
\end{aligned}$$

so that $\xi\rho'(g) = \rho(g)\xi$. $\qquad\square$

Remark 10.5.3 Note that, despite (10.56), in general $\rho' \not\sim \rho$: recall that the map ξ is *anti*-linear! However, the following result holds true (modulo the identification of V'' and V).

Corollary 10.5.4 *The double adjoint $(\rho')'$ coincides with ρ.*

Proof. We first observe that

$$\xi_{V'} = (\xi_V)^{-1}. \tag{10.57}$$

Thus, by applying Proposition 10.5.2 twice and (10.57), we obtain

$$(\rho')'(g) = \xi_{V'}^{-1}\rho'(g)\xi_{V'} = \xi_V\xi_V^{-1}\rho(g)\xi_V\xi_V^{-1} = \rho(g)$$

for all $g \in G$. $\qquad\square$

We now fix an orthonormal basis $\{v_1, v_2, \ldots, v_d\}$ of V and denote by $\{f_1, f_2, \ldots, f_d\}$ the orthonormal basis in V' which is dual to $\{v_1, v_2, \ldots, v_d\}$, that is, such that $f_i(v_j) = \delta_{i,j}$ (or, equivalently, $f_i = \xi^{-1}(v_i)$), for all $i, j = 1, 2, \ldots, d$.

Proposition 10.5.5 *The matrix coefficients $u'_{i,j}(g)$ of ρ' with respect to the dual basis $\{f_1, f_2, \ldots, f_d\}$ are the conjugates of those of ρ, in fomulæ:*

$$u'_{i,j}(g) = \overline{u_{i,j}(g)} \tag{10.58}$$

for all $g \in G$ and $i, j = 1, 2, \ldots, d$.

Proof. Keeping in mind (10.14), we have

$$u'_{i,j}(g) = \langle \rho'(g)f_j, f_i \rangle_{V'}$$

$$\text{(by (10.54))} \quad = \langle \xi(f_i), \xi[\rho'(g)f_j] \rangle_V$$

$$\text{(since } \xi(f_i) = v_i \text{ and by (10.56))} \quad = \langle v_i, \rho(g)v_j \rangle_V$$

$$= \overline{\langle \rho(g)v_j, v_i \rangle_V}$$

$$= \overline{u_{i,j}(g)}$$

for all $g \in G$ and $i, j = 1, 2, \ldots, d$. $\qquad\qquad\square$

Corollary 10.5.6 *The character of ρ' is the conjugate of the character of ρ:*

$$\chi^\rho(g) = \overline{\chi^{\rho'}(g)} \qquad\qquad (10.59)$$

for all $g \in G$.

For instance, if χ^k ($0 \leq k \leq n-1$) is a character of the cyclic group $\mathbb{Z}_n$ as in Section 2.2, then the character of the corresponding adjoint representation is χ^{-k}.

Exercise 10.5.7 (Fourier transform of a character) Prove that for θ and σ in $\widehat{G}$ we have $\widehat{\chi^\sigma}(\theta) = \delta_{\theta,\sigma} \frac{|G|}{d_\theta} I_{V_\theta}$.

Remark 10.5.8 A representation $\rho \in \widehat{G}$ is *self-conjugate* when ρ and ρ' are equivalent; it is *complex* when it is not self-conjugate. By virtue of (10.59), we may say that ρ is self-conjugate if and only if $\chi^\rho(g) \in \mathbb{R}$ for all $g \in G$, that is, its character is a real valued function. Similarly, ρ is complex if and only if $\chi^\rho(g) \in \mathbb{C} \setminus \mathbb{R}$ for some $g \in G$. The class of self-conjugate representations can be further split into two subclasses (*real* and *quaternionic*); we refer to [29, Section 9.7] for more details.

Now we apply the notion of a conjugate representation to the decomposition of the group algebra. Suppose that our choice of the elements of the dual $\widehat{G}$ of G makes it invariant under conjugation: for all $\theta \in \widehat{G}$, also $\theta' \in \widehat{G}$. Using the notation in Theorem 10.2.25, for each $\theta \in \widehat{G}$ we set:

$$M^\theta_{i,*} = \langle u^\theta_{i,j} : j = 1, 2, \ldots, d_\theta \rangle, \quad i = 1, 2, \ldots, d_\theta;$$

$$M^\theta_{*,j} = \langle u^\theta_{i,j} : i = 1, 2, \ldots, d_\theta \rangle, \quad j = 1, 2, \ldots, d_\theta;$$

$$M^\theta = \langle u^\theta_{i,j} : i, j = 1, 2, \ldots, d_\theta \rangle.$$

where $\langle \cdots \rangle$ indicates $\mathbb{C}$-linear span. Recall also the definition of the left (respectively right) regular representation in Example 10.1.8.

Theorem 10.5.9 *The following orthogonal decompositions hold:*

(i) $L(G) = \bigoplus_{\theta \in \widehat{G}} M^\theta$ *and each M^θ is both λ_G- and ρ_G-invariant;*

(ii) $M^\theta = \bigoplus_{i=1}^{d_\theta} M_{i,*}^\theta$; *each $M_{i,*}^\theta$ is ρ_G-invariant and the restriction of ρ_G to $M_{i,*}^\theta$ is equivalent to θ;*

(iii) $M^\theta = \bigoplus_{j=1}^{d_\theta} M_{*,j}^\theta$; *each $M_{*,j}^\theta$ is λ_G-invariant and the restriction of λ to $M_{*,j}^\theta$ is equivalent to θ'.*

Proof.

(i) The decomposition $L(G) = \bigoplus_{\theta \in \widehat{G}} M^\theta$ is just the Peter–Weyl theorem (Theorem 10.2.25); the λ_G- and ρ_G-invariance are proved below.

(ii) Let $g, g_1 \in G$ and $i, j \in \{1, 2, \ldots, d_\theta\}$. Then, by Lemma 10.2.13.(iii),

$$[\rho_G(g)u_{i,j}^\theta](g_1) = u_{i,j}^\theta(g_1 g) = \sum_{k=1}^{d_\theta} u_{i,k}^\theta(g_1)u_{k,j}^\theta(g),$$

i.e.

$$\rho_G(g)u_{i,j}^\theta = \sum_{k=1}^{d_\theta} u_{i,k}^\theta u_{k,j}^\theta(g).$$

Since, by Lemma 10.2.13.(ii), $\theta(g)v_j^\theta = \sum_{k=1}^{d_\theta} v_k^\theta u_{k,j}^\theta(g)$, we conclude that the map $v_j^\theta \mapsto u_{i,j}^\theta$, $j = 1, 2, \ldots, d_\theta$, extends to an invertible operator that intertwines θ with $\rho_G|_{M_{i,*}^\theta}$.

(iii) Let $g, g_1 \in G$ and $i, j \in \{1, 2, \ldots, d_\theta\}$. Then, by Lemma 10.2.13.(iii), Lemma 10.2.13.(i), and (10.58), we have

$$[\lambda_G(g)u_{i,j}^\theta](g_1) = u_{i,j}^\theta(g^{-1}g_1)$$

$$= \sum_{k=1}^{d_\theta} u_{i,k}^\theta(g^{-1})u_{k,j}^\theta(g_1)$$

$$= \sum_{k=1}^{d_\theta} \overline{u_{k,i}^\theta(g)}u_{k,j}^\theta(g_1)$$

$$= \sum_{k=1}^{d_\theta} u_{k,i}^{\theta'}(g)u_{k,j}^\theta(g_1),$$

i.e. $\lambda_G(g)u_{i,j}^\theta = \sum_{k=1}^{d_\theta} u_{k,j}^\theta u_{k,i}^{\theta'}(g)$. Again by Lemma 10.2.13.(ii) we have $\theta'(g)v_i^{\theta'} = \sum_{k=1}^{d_\theta} v_k^{\theta'} u_{k,i}^\theta(g)$, and this shows that the map $v_i^{\theta'} \mapsto u_{i,j}^\theta$, $i = 1, 2, \ldots, d_\theta$, extends to an invertible operator that intertwines θ' with $\lambda_G|_{M_{*,j}^\theta}$. $\square$

The representation M^θ is the θ-*isotypic component* of $L(G)$ (see Definition 10.2.20).

Exercise 10.5.10 Show that the orthogonal projection $E_\theta \colon L(G) \to M^\theta$ is given by $E_\theta f = \frac{1}{|G|} f * \chi^\theta$, for all $f \in L(G)$.

We now turn to the second fundamental construction in linear and multilinear algebra in the framework of representation theory of finite groups we alluded to above, namely tensor products. In Section 8.7 we have already given an elementary introduction to tensor products.

Let then V and W be two finite dimensional, complex, unitary spaces. A map $B \colon V \times W \to \mathbb{C}$ is said to be *bi-antilinear* provided

$$B(v_1 + v_2, w) = B(v_1, w) + B(v_2, w)$$
$$B(v, w_1 + w_2) = B(v, w_1) + B(v, w_2)$$
$$B(\alpha v, \beta w) = \overline{\alpha}\overline{\beta}B(v, w)$$

for all $v_1, v_2 \in V$, $w_1, w_2 \in W$, and $\alpha, \beta \in \mathbb{C}$. Clearly, the set of all such bi-antilinear maps is a complex vector space in a natural way; we denote it by $V \bigotimes W$ and call it the *tensor product* of V and W.

For $v \in V$ and $w \in W$ we denote by $v \otimes w$ the element in $V \bigotimes W$ defined by

$$[v \otimes w](v', w') = \langle v, v' \rangle_V \langle w, w' \rangle_W$$

for all $v' \in V$ and $w' \in W$. Elements of this kind are called *simple tensors*. Note that the map

$$V \times W \longrightarrow V \bigotimes W$$
$$(v, w) \longmapsto v \otimes w$$

is bilinear, that is,

$$(\alpha_1 v_1 + \alpha_2 v_2) \otimes (\beta_1 w_1 + \beta_2 w_2)$$
$$= \alpha_1 \beta_1 v_1 \otimes w_1 + \alpha_1 \beta_2 v_1 \otimes w_2 + \alpha_2 \beta_1 v_2 \otimes w_1 + \alpha_2 \beta_2 v_2 \otimes w_2,$$

for all $\alpha_i, \beta_i \in \mathbb{C}, v_i \in V$, and $w_i \in W, i = 1, 2$. We claim that the corresponding image spans the whole $V \bigotimes W$. Indeed, if $\{v_i\}_{i=1}^{d_V}$ and $\{w_j\}_{j=1}^{d_W}$ denote two bases for V and W, respectively, then for all $B \in V \bigotimes W$ we clearly have

$$B = \sum_{i=1}^{d_V} \sum_{j=1}^{d_W} B(v_i, w_j) v_i \otimes v_j.$$

This incidentally shows that the simple tensors $v_i \otimes w_j$, $i = 1, \ldots, d_V$ and $j = 1, \ldots, d_W$, generate $V \otimes W$. Since these are also linearly independent (exercise), they constitute a basis for $V \otimes W$, so that, in particular, $\dim(V \otimes W) = \dim(V) \cdot \dim(W)$.

We now endow $V \otimes W$ with a scalar product $\langle \cdot, \cdot \rangle_{V \otimes W}$ by setting

$$\langle v_1 \otimes w_1, v_2 \otimes w_2 \rangle_{V \otimes W} = \langle v_1, v_2 \rangle_V \langle w_1, w_2 \rangle_W \qquad (10.60)$$

and then extending by linearity. This way, if the bases $\{v_i\}_{i=1}^{d_V}$ and $\{w_j\}_{j=1}^{d_W}$ are orthonormal in V and W, respectively, then so is $\{v_i \otimes w_j\}_{\substack{i=1,\ldots,d_V \\ j=1,\ldots,d_W}}$ in $V \otimes W$.

Let now $A \in \operatorname{End}(V)$ and $B \in \operatorname{End}(W)$. Define $A \otimes B \in \operatorname{End}(V \otimes W)$ by setting, for all $C \in V \otimes W$,

$$\{[A \otimes B](C)\}\,(v', w') = C(A^* v', B^* w')$$

for all $v' \in V$ and $w' \in W$, where $A^* \in \operatorname{End}(V)$ and $B^* \in \operatorname{End}(W)$ are the adjoint operators. For $v, v' \in V$ and $w, w' \in W$ we then have

$$
\begin{aligned}
\{[A \otimes B](v \otimes w)\}\,(v', w') &= [v \otimes w](A^* v', B^* w') \\
&= \langle v, A^* v' \rangle_V \langle w, B^* w' \rangle_W \\
&= \langle Av, v' \rangle_V \langle Bw, w' \rangle_W \\
&= [(Av) \otimes (Bw)](v', w').
\end{aligned}
$$

This shows that

$$[A \otimes B](v \otimes w) = (Av) \otimes (Bw). \qquad (10.61)$$

Lemma 10.5.11 *Let* $A \in \operatorname{End}(V)$ *and* $B \in \operatorname{End}(W)$. *Then* $\operatorname{Tr}(A \otimes B) = \operatorname{Tr}(A)\operatorname{Tr}(B)$.

Proof. Let $\{v_i\}_{i=1}^{d_V}$ and $\{w_j\}_{j=1}^{d_W}$ be two orthonormal bases in V and W, respectively. Then

$$
\begin{aligned}
\operatorname{Tr}(A \otimes B) &= \sum_{\substack{i=1,\ldots,d_V \\ j=1,\ldots,d_W}} \langle [A \otimes B](v_i \otimes w_j), v_i \otimes w_j \rangle_{V \otimes W} \\
\text{(by (10.61))} \quad &= \sum_{\substack{i=1,\ldots,d_V \\ j=1,\ldots,d_W}} \langle (Av_i) \otimes (Bw_j), v_i \otimes w_j \rangle_{V \otimes W} \\
\text{(by (10.60))} \quad &= \sum_{\substack{i=1,\ldots,d_V \\ j=1,\ldots,d_W}} \langle Av_i, v_i \rangle_V \langle Bw_j, w_j \rangle_W \\
&= \operatorname{Tr}(A)\operatorname{Tr}(B). \qquad \square
\end{aligned}
$$

Exercise 10.5.12

(1) Show that the bilinear map

$$\phi : V \times W \to V \otimes W$$
$$(v, w) \mapsto v \otimes w$$

is universal in the sense that if Z is another complex vector space and $\psi : V \times W \to Z$ is bilinear, then there exists a unique linear map $\theta : V \otimes W \to Z$ such that $\theta(v \otimes w) = \phi(v, w)$, that is, such that the diagram

$$V \times W \xrightarrow{\phi} V \otimes W$$
$$\searrow \psi \qquad \swarrow \theta$$
$$Z$$

is commutative (i.e. $\psi = \theta \circ \phi$).

(2) Show that the above universal property characterizes the tensor product: let U be a complex vector space and let $\psi : V \times W \to U$ be a bilinear map such that

(a) $\psi(V \times W) = \{\psi(v, w) : v \in V, w \in W\}$ generates U;

(b) for any complex vector space Z and any bilinear map $\tau : V \times W \to Z$ there exists a unique linear map $\theta : U \to Z$ such that $\tau = \theta \circ \psi$.

Then there exists a linear isomorphism $\alpha : V \otimes W \to U$ such that $\psi = \alpha \circ \phi$.

Exercise 10.5.13 Let $V, W,$ and Z be finite dimensional, complex unitary spaces. Prove that the following natural isomorphisms hold:

(1) $V \otimes W \cong W \otimes V$;

(2) $\mathbb{C} \otimes V \cong V$;

(3) $(V \otimes W) \otimes Z \cong V \otimes (W \otimes Z)$;

(4) $(V \oplus W) \otimes Z \cong (V \otimes Z) \oplus (W \otimes Z)$.

Note that the third isomorphism, namely the associativity of the tensor product, may be recursively extended to the tensor product of k vector spaces: we then denote by $V_1 \otimes V_2 \otimes \cdots \otimes V_k$ the set of all k-antilinear maps $B : V_1 \times V_2 \times \cdots \times V_k \to \mathbb{C}$.

We now introduce and study two kinds of tensor product of representations.

Definition 10.5.14 Let G_1 and G_2 be two finite groups and let (ρ_1, V_1) and (ρ_2, V_2) be representations of G_1 and G_2, respectively. We define the *outer tensor product* of ρ_1 and ρ_2 as the representation $(\rho_1 \boxtimes \rho_2, V_1 \otimes V_2)$ of $G_1 \times G_2$

defined by setting

$$[\rho_1 \boxtimes \rho_2](g_1, g_2) = \rho_1(g_1) \otimes \rho_2(g_2) \in \text{End}\left(V_1 \bigotimes V_2\right)$$

for all $g_i \in G_i$, $i = 1, 2$.

When $G_1 = G_2 = G$ the *internal tensor product* of ρ_1 and ρ_2 is the G-representation $(\rho_1 \otimes \rho_2, V_1 \bigotimes V_2)$ defined by setting

$$[\rho_1 \otimes \rho_2](g) = \rho_1(g) \otimes \rho_2(g) \in \text{End}\left(V_1 \bigotimes V_2\right)$$

for all $g \in G$.

In the above definition, we have used the symbols "$\boxtimes$" and "$\otimes$" to make a distinction between these two notions of tensor product (compare with [63]). Note that, however, in both cases the space will be simply denoted by $V_1 \bigotimes V_2$. Moreover, it is obvious that, modulo the isomorphism between G and $\widetilde{G} = \{(g, g) : g \in G\} \leq G \times G$, the internal tensor product $\rho_1 \otimes \rho_2$ is unitarily equivalent to the restriction $\text{Res}_{\widetilde{G}}^{G \times G}(\rho_1 \boxtimes \rho_2)$.

Lemma 10.5.15 *Let ρ_1 and ρ_2 be two representations of two finite groups G_1 and G_2, respectively, and denote by χ^{ρ_1} and χ^{ρ_2} their characters. Then, the character of $\rho_1 \boxtimes \rho_2$ is given by*

$$\chi^{\rho_1 \boxtimes \rho_2}(g_1, g_2) = \chi^{\rho_1}(g_1)\chi^{\rho_2}(g_2) \tag{10.62}$$

for all $g_1 \in G_1$ and $g_2 \in G_2$. In particular, if both ρ_1 and ρ_2 are one-dimensional, so that they coincide with their characters, then one has that $\rho_1 \boxtimes \rho_2 = \chi^{\rho_1} \boxtimes \chi^{\rho_2} = \chi^{\rho_1}\chi^{\rho_2}$, the pointwise product of the characters. When $G_1 = G_2 = G$, as the internal tensor product is concerned, (10.62) becomes

$$\chi^{\rho_1 \otimes \rho_2}(g) = \chi^{\rho_1}(g)\chi^{\rho_2}(g) \tag{10.63}$$

for all $g \in G$.

Proof. This follows immediately from Definition 10.2.14 and Lemma 10.5.11. $\qquad\square$

Theorem 10.5.16 *Let G_1 and G_2 be two finite groups and let $\theta_1 \in \widehat{G_1}$ and $\theta_2 \in \widehat{G_2}$. Then $\theta_1 \boxtimes \theta_2$ is an irreducible representation of $G_1 \times G_2$. Moreover, if also $\sigma_1 \in \widehat{G_1}$ and $\sigma_2 \in \widehat{G_2}$ then $\theta_1 \boxtimes \theta_2 \sim \sigma_1 \boxtimes \sigma_2$ if and only if $\theta_1 = \sigma_1$ and $\theta_2 = \sigma_2$.*

Proof. By Proposition 10.2.17 and Corollary 10.2.23 it suffices to check that $\langle \chi^{\theta_1 \boxtimes \theta_2}, \chi^{\sigma_1 \boxtimes \sigma_2} \rangle$ is either $|G_1 \times G_2| \equiv |G_1| \cdot |G_2|$ if $\sigma_1 = \theta_1$ and $\sigma_2 = \theta_2$, or 0 otherwise. Now we have

$$\langle \chi^{\theta_1 \boxtimes \theta_2}, \chi^{\sigma_1 \boxtimes \sigma_2} \rangle = \sum_{(g_1, g_2) \in G_1 \times G_2} \chi^{\theta_1 \boxtimes \theta_2}(g_1, g_2)\overline{\chi^{\sigma_1 \boxtimes \sigma_2}(g_1, g_2)}$$

$$\text{(by Lemma 10.5.15)} \quad = \sum_{\substack{g_1 \in G_1 \\ g_2 \in G_2}} \chi^{\theta_1}(g_1)\chi^{\theta_2}(g_2)\overline{\chi^{\sigma_1}(g_1)}\,\overline{\chi^{\sigma_2}(g_2)}$$

$$= \sum_{g_1 \in G_1} \chi^{\theta_1}(g_1)\overline{\chi^{\sigma_1}(g_1)} \sum_{g_2 \in G_2} \chi^{\theta_2}(g_2)\overline{\chi^{\sigma_2}(g_2)}$$

$$= \langle \chi^{\theta_1}, \chi^{\sigma_1} \rangle \cdot \langle \chi^{\theta_2}, \chi^{\sigma_2} \rangle$$

$$\text{(by Proposition 10.2.17)} \quad = \begin{cases} |G_1| \cdot |G_2| & \text{if } \theta_1 = \sigma_1 \text{ and } \theta_2 = \sigma_2 \\ 0 & \text{otherwise.} \end{cases}$$

$\square$

Corollary 10.5.17 *Let G_1 and G_2 be two finite groups. Then the map*

$$\widehat{G_1} \times \widehat{G_2} \longrightarrow \widehat{G_1 \times G_2} \tag{10.64}$$
$$(\theta_1, \theta_2) \longmapsto \theta_1 \boxtimes \theta_2$$

is a bijection.

Proof. We first observe that every conjugacy class in $G_1 \times G_2$ is the Cartesian product of a conjugacy class in G_1 by one in G_2, and vice versa. Thus, keeping in mind Theorem 10.3.13, we have that $|\widehat{G_1 \times G_2}|$ equals the number of conjugacy classes in $G_1 \times G_2$, which in turn equals the product of the numbers of conjugacy classes in G_1 and G_2, and therefore, again by Theorem 10.3.13, equals $|\widehat{G_1}| \cdot |\widehat{G_2}|$. Therefore, by the previous theorem, the map (10.64) is indeed a bijection. Alternatively, it is immediate to check (exercise) that

$$\sum_{\theta_1 \in \widehat{G_1}} \sum_{\theta_2 \in \widehat{G_2}} (d_{\theta_1 \boxtimes \theta_2})^2 = |G_1 \times G_2|$$

and then we may invoke Theorem 10.2.25.(iii). $\square$

Exercise 10.5.18 Let G (respectively H) be a finite group and let X (respectively Y) be a finite homogenous G-space (respectively H-space). Let λ and μ denote the corresponding permutation representations. In Section 8.7 we showed that the map $\delta_x \otimes \delta_y \mapsto \delta_{(x,y)}$, $x \in X$, $y \in Y$, yields a natural isomorphism $L(X) \bigotimes L(Y) \cong L(X \times Y)$.

(1) Show that $\lambda \boxtimes \mu$ is equivalent to the permutation representation of $G \times H$ on $X \times Y$.

(2) Show that if $G = H$ and $X = Y$, then the internal tensor product $\lambda \otimes \mu$ is equivalent to the permutation representation associated with the diagonal action of G on $X \times X$.

By means of the two basic constructions (adjoints and tensor products), we now reinterpret the decomposition of the group algebra (cf. Theorem 10.5.9).

First of all, we recall that if V is a finite dimensional vector space and V' denotes its dual, then $\mathrm{End}(V) \cong V' \otimes V$. An explicit isomorphism is given by linearly extending to the whole of $V' \otimes V$ the map

$$
\begin{aligned}
V' \otimes V &\longrightarrow \mathrm{End}(V) \\
f \otimes v &\longmapsto T_{f,v}
\end{aligned}
\tag{10.65}
$$

where $T_{f,v}(w) = f(w)v$ for all $w \in V$.

Exercise 10.5.19 Fill up all the details relative to (10.65).

Now consider the action of $G \times G$ on G given by

$$
(g_1, g_2) \cdot g = g_1 g g_2^{-1}
$$

for all $g, g_1, g_2 \in G$, and the associated $(G \times G)$-permutation representation $(\eta, L(G))$ given by

$$
[\eta(g_1, g_2)f](g) = f(g_1^{-1} g g_2),
$$

for all $f \in L(G)$ and $g, g_1, g_2 \in G$. Note that, in terms of the left and right regular representations, we have $\eta(g_1, g_2) = \lambda_G(g_1)\rho_G(g_2) = \rho_G(g_2)\lambda_G(g_1)$, for all $g_1, g_2 \in G$. The stabilizer of the point 1_G is the diagonal subgroup $\widetilde{G} = \{(g, g) : g \in G\}$, clearly isomorphic to G, and in the present setting (10.50) yields:

$$
G = (G \times G)/\widetilde{G}.
$$

Theorem 10.5.20 *With the notation as in Theorem 10.5.9, the restriction of η to M^θ is equivalent to $\theta' \boxtimes \theta$. In particular, it is irreducible.*

Proof. For $f \in W_{\theta'}$ and $v \in W_\theta$ define $F_{f,v}^\theta \in L(G)$ by setting

$$
F_{f,v}^\theta(g) = f(\theta(g)v),
\tag{10.66}
$$

for all $g \in G$. Noticing that, for all $i, j = 1, 2, \ldots, d_\theta$ and $g \in G$, one has

$$
\begin{aligned}
u_{i,j}^\theta(g) &= \langle \theta(g)v_j^\theta, v_i^\theta \rangle \\
\text{(by (10.53))} \quad &= [\xi^{-1}(v_i^\theta)] \left(\theta(g)v_j^\theta \right) \\
&= F_{\xi^{-1}(v_i^\theta), v_j^\theta}^\theta(g),
\end{aligned}
$$

we deduce that the $F^\theta_{f,v}$s span the whole of M^θ. Moreover, if $(g_1, g_2) \in G \times G$ and $g \in G$, we have

$$[\eta(g_1, g_2)F^\theta_{f,v}](g) = F^\theta_{f,v}(g_1^{-1}gg_2)$$
$$\text{(by (10.66))} = f\left(\theta(g_1^{-1}gg_2)v\right)$$
$$\text{(by (10.53))} = \langle\theta(g_1^{-1}gg_2)v, \xi(f)\rangle$$
$$\text{(by (10.56))} = \langle\theta(g)\theta(g_2)v, \xi[\theta'(g_1)f]\rangle$$
$$= [\theta'(g_1)f](\theta(g)\theta(g_2)v)$$
$$= F^\theta_{\theta'(g_1)f,\theta(g_2)v}(g)$$

so that the surjective map

$$W_{\theta'} \otimes W_\theta \longrightarrow M^\theta$$
$$f \otimes v \longmapsto F^\theta_{f,v}$$

intertwines $\theta' \boxtimes \theta$ with $\eta|_{M^\theta}$. The irreducibility of $\theta' \boxtimes \theta$ follows from Theorem 10.5.16. $\qquad\square$

Recalling Corollary 10.3.12, the Fourier transform may be seen as an isomorphism between $L(G)$ and $\bigoplus_{\theta \in \widehat{G}} \left(W'_\theta \otimes W_\theta\right)$, if we identify $\text{End}(W_\theta)$ with $W'_\theta \otimes W_\theta$ as in (10.65).

Exercise 10.5.21 Using the notation in (10.65), (10.66), and in Corollary 10.3.12, show that the inverse Fourier transform of a tensor product $f \otimes v \in W'_\theta \otimes W_\theta$ is given by:

$$(f \otimes v)^\vee(g) = \frac{d_\theta}{|G|}F^\theta_{f,v}(g^{-1})$$

for all $g \in G$.

10.6 The commutant of a representation

In this section we study the commutant $\text{End}_G(V)$ of a G-representation (ρ, V). First of all, we recall some basic facts on projections (see any book on linear algebra, for instance [91]). Let V be finite dimensional unitary space. A linear transformation $E \in \text{End}(V)$ is called a *projection* if it is *idempotent*, that is, $E^2 = E$. If the range $W = \text{Ran}E$ is orthogonal to the null space $\text{Ker}E$, we say that E is an *orthogonal projection* of V onto W. It is easy to see that a projection E is orthogonal if and only if it is self-adjoint, that is, $E = E^*$.

Let now (V, ρ) be a representation of a finite group G and suppose that

$$V \cong \bigoplus_{\theta \in J} m_\theta W_\theta \qquad (10.67)$$

is the decomposition into irreducibles as in Corollary 10.2.19 (with $J = \{\theta \in \widehat{G} : m_\theta > 0\}$). We can decompose the isotypic component $m_\theta W_\theta$ by choosing suitable operators $I_{\theta,1}, I_{\theta,2}, \ldots, I_{\theta,m_\theta} \in \mathrm{Hom}_G(W_\theta, V)$, in such a way that

$$V = \bigoplus_{\theta \in J} \bigoplus_{j=1}^{m_\theta} I_{\theta,j} W_\theta \qquad (10.68)$$

is an orthogonal decomposition, and

$$\langle I_{\theta,i} w_1, I_{\sigma,j} w_2 \rangle_V = \delta_{\theta,\sigma} \delta_{i,j} \langle w_1, w_2 \rangle_{W_\theta} \qquad (10.69)$$

for all $\theta, \sigma \in J$, $i = 1, 2, \ldots, m_\theta$, $j = 1, 2, \ldots, m_\sigma$, $w_1 \in W_\theta$ and $w_2 \in W_\sigma$. In particular, each $I_{\theta,j}$ is an isometry and the $I_{\theta,j}$s are linearly independent in $\mathrm{Hom}_G(W, V)$. Then any vector $v \in V$ may be uniquely written in the form $v = \sum_{\theta \in J} \sum_{j=1}^{m_\theta} v_{\theta,j}$, with $v_{\theta,j} \in I_{\theta,j} W_\theta$. The operator $E_{\theta,j} \in \mathrm{End}(V)$, defined by setting $E_{\theta,j}(v) = v_{\theta,j}$ for all $v \in V$, is the orthogonal projection from V onto $I_{\theta,j} W_\theta$. In particular, $I_V = \sum_{\theta \in J} \sum_{j=1}^{m_\theta} E_{\theta,j}$.

Observe that if $v = \sum_{\theta \in J} \sum_{j=1}^{m_\theta} v_{\theta,j}$ then $\rho(g)v = \sum_{\theta \in J} \sum_{j=1}^{m_\theta} \rho(g)v_{\theta,j}$. As $\rho(g)v_{\theta,j} \in I_{\theta,j} W_\theta$, by the uniqueness of such a decomposition, we have that $E_{\theta,j}\rho(g)v = \rho(g)v_{\theta,j} = \rho(g)E_{\theta,j}v$. Therefore, $E_{\theta,j} \in \mathrm{End}_G(V)$.

Lemma 10.6.1 *With the above notation the following hold.*

(i) *The space* $\mathrm{Hom}_G(W_\theta, V)$ *is spanned by* $I_{\theta,1}, I_{\theta,2}, \ldots, I_{\theta,m_\theta}$. *In particular,* $m_\theta = \dim \mathrm{Hom}_G(W_\theta, V)$.

(ii) *We have*

$$I_{\theta,k}^* I_{\sigma,j} = \delta_{\sigma,\theta} \delta_{j,k} I_{W_\theta} \qquad (10.70)$$

for all $\theta, \sigma \in J$, $k = 1, 2, \ldots, m_\theta$, $j = 1, 2, \ldots, m_\sigma$; *in particular,* $I_{\theta,j}^*|_{I_{\theta,j} W_\theta}$ *is the inverse of* $I_{\theta,j} \colon W_\theta \to I_{\theta,j} W_\theta (\leq V)$.

Proof.

(i) If $T \in \mathrm{Hom}_G(W_\theta, V)$, then

$$T = I_V T = \sum_{\sigma \in J} \sum_{k=1}^{m_\sigma} E_{\sigma,k} T.$$

Since $\mathrm{Ran} E_{\sigma,k} = I_{\sigma,k} W_\sigma$, if follows from Lemma 10.2.3 that, if $\sigma \neq \theta$, then $E_{\sigma,k} T = 0$. Moreover, from Corollary 10.2.5, one deduces that $E_{\theta,k} T = \alpha_k I_{\theta,k}$ for some $\alpha_k \in \mathbb{C}$. Thus,

$$T = \sum_{k=1}^{m_\theta} E_{\theta,k} T = \sum_{k=1}^{m_\theta} \alpha_k I_{\theta,k}.$$

(ii) By Proposition 10.2.2, $I_{\theta,j}^* I_{\theta,j} \in \mathrm{End}_G(W_\theta)$ so that, by Schur's Lemma, $I_{\theta,j}^* I_{\theta,j} = \alpha I_{W_\theta}$ for some $\alpha \in \mathbb{C}$. Moreover, from (10.69) it follows that

$$\langle I_{\theta,j}^* I_{\theta,j} w, w \rangle_{W_\theta} = \langle I_{\theta,j} w, I_{\theta,j} w \rangle_V = \|w\|_{W_\theta}^2,$$

for all $w \in W_\theta$, so that necessarily $\alpha = 1$. On the other hand, if $(\sigma, j) \neq (\theta, k)$ then, again by means of (10.69), we deduce that

$$\langle I_{\theta,k}^* I_{\sigma,j} w, u \rangle_{W_\theta} = \langle I_{\sigma,j} w, I_{\theta,k} u \rangle_V = 0. \qquad \square$$

Clearly, the decomposition of the θ-isotypic component of V into irreducible sub-representations is not unique: it corresponds to the choice of a basis in $\mathrm{Hom}_G(W_\theta, V)$.

Now, for all $\theta \in J$ and $1 \leq j, k \leq m_\theta$, define $T_{k,j}^\theta \in \mathrm{End}_G(V)$ by setting

$$T_{k,j}^\theta v = \begin{cases} I_{\theta,k} I_{\theta,j}^* v & \text{if } v \in I_{\theta,j} W_\theta \\ 0 & \text{if } v \in V \ominus I_{\theta,j} W_\theta. \end{cases} \qquad (10.71)$$

where $V \ominus I_{\theta,j} W_\theta$ is the orthogonal complement of $I_{\theta,j} W_\theta$ in V.

Lemma 10.6.2 *With the above notation, we have:*

$$\mathrm{Ran} T_{k,j}^\theta = I_{\theta,k} W_\theta, \qquad \mathrm{Ker} T_{k,j}^\theta = V \ominus I_{\theta,j} W_\theta,$$

$$T_{k,j}^\sigma T_{s,t}^\theta = \delta_{\sigma,\theta} \delta_{j,s} T_{k,t}^\theta \qquad (10.72)$$

and

$$\left(T_{k,j}^\theta \right)^* = T_{j,k}^\theta. \qquad (10.73)$$

In particular,

$$T_{j,j}^\theta \equiv E_{\theta,j}.$$

and

$$\mathrm{Hom}_G(I_{\theta,j} W_\theta, I_{\theta,k} W_\theta) = \mathbb{C} T_{k,j}^\theta.$$

Proof. From (10.70) and (10.71) we deduce that, for all $w \in W_\theta$,

$$T_{k,j}^\theta I_{\theta,j} w = I_{\theta,k} I_{\theta,j}^* I_{\theta,j} w = I_{\theta,k} w \qquad (10.74)$$

so that $\operatorname{Ran} T_{k,j}^{\theta} = I_{\theta,k} W_{\theta}$. The same arguments yield $\operatorname{Ker} T_{k,j}^{\theta} = V \ominus I_{\theta,j} W_{\theta}$,

$$T_{k,j}^{\sigma} T_{s,t}^{\theta} v = \begin{cases} T_{k,j}^{\sigma} I_{\theta,s} I_{\theta,t}^{*} v & \text{if } v \in I_{\theta,t} W_{\theta} \\ 0 & \text{if } v \in V \ominus I_{\theta,t} W_{\theta}. \end{cases}$$

$$= \delta_{\sigma,\theta} \delta_{j,s} T_{k,t}^{\theta} v,$$

and

$$\langle T_{k,j}^{\theta} v_1, v_2 \rangle_V = \begin{cases} \langle I_{\theta,k} I_{\theta,j}^{*} v_1, v_2 \rangle_V & \text{if } v_1 \in I_{\theta,j} W_{\theta} \text{ and } v_2 \in I_{\theta,k} W_{\theta} \\ 0 & \text{otherwise} \end{cases}$$

$$= \langle v_1, T_{j,k}^{\theta} v_2 \rangle.$$

Finally, from (10.71) and (10.74) we deduce that $T_{j,j}^{\theta} I_{\sigma,k} w = \delta_{\sigma,\theta} \delta_{j,k} I_{\theta,j} w$, which yields $T_{j,j}^{\theta} \equiv E_{\theta,j}$, while Corollary 10.2.5 ensures that every operator $T \in \operatorname{Hom}_G(I_{\theta,k} W_{\theta}, I_{\theta,k} W_{\theta})$ is a scalar multiple of $T_{k,j}^{\theta}$. $\qquad \square$

Theorem 10.6.3 *With the above notation, the set*

$$\{T_{k,j}^{\theta} : \theta \in J, k, j = 1, 2, \ldots, m_{\theta}\} \qquad (10.75)$$

is a vector space basis for $\operatorname{End}_G(V)$. *Moreover, the map*

$$\operatorname{End}_G(V) \longrightarrow \bigoplus_{\theta \in J} \mathfrak{M}_{m_{\theta}}(\mathbb{C})$$

$$T \longmapsto \bigoplus_{\theta \in J} \left(\alpha_{k,j}^{\theta} \right)_{k,j=1}^{m_{\theta}}$$

where the $\alpha_{k,j}^{\theta}$s are the coefficients of T with respect to the basis (10.75), that is,

$$T = \sum_{\theta \in J} \sum_{k,j=1}^{m_{\theta}} \alpha_{k,j}^{\theta} T_{k,j}^{\theta},$$

is a $*$-isomorphism of algebras.

Proof. Let $T \in \operatorname{End}_G(V)$. We have

$$T = I_V T I_V = \left(\sum_{\sigma \in J} \sum_{k=1}^{m_{\sigma}} E_{\sigma,k} \right) T \left(\sum_{\theta \in J} \sum_{j=1}^{m_{\theta}} E_{\theta,j} \right)$$

$$= \sum_{\sigma,\theta \in J} \sum_{k=1}^{m_{\sigma}} \sum_{j=1}^{m_{\theta}} E_{\sigma,k} T E_{\theta,j}.$$

Observe that

- $\operatorname{Ran} E_{\sigma,k} T E_{\theta,j} \leq \operatorname{Ran} E_{\sigma,k} = I_{\sigma,k} W_{\sigma}$;

- $\mathrm{Ker}E_{\sigma,k}TE_{\theta,j} \geq \mathrm{Ker}E_{\theta,j} = V \ominus I_{\theta,j}W_\theta$;
- the restriction to $I_{\theta,j}W_\theta$ of $E_{\sigma,k}TE_{\theta,j}$ is in $\mathrm{Hom}_G(I_{\theta,j}W_\theta, I_{\sigma,k}W_\sigma)$.

From Lemma 10.2.3, it follows that $E_{\sigma,k}TE_{\theta,j} = 0$ if $\sigma \neq \theta$, while, if $\sigma = \theta$, by Corollary 10.2.5 one has that $E_{\theta,k}TE_{\theta,j}$ is a multiple of $T_{k,j}^\theta$, that is, there exist $\alpha_{k,j}^\theta \in \mathbb{C}$ such that

$$E_{\theta,k}TE_{\theta,j} = \alpha_{k,j}^\theta T_{k,j}^\theta.$$

This proves that the $T_{k,j}^\theta$s generate $\mathrm{End}_G(V)$. To prove independence, suppose that we can express the 0-operator as

$$0 = \sum_{\theta \in J} \sum_{k,j=1}^{m_\theta} \alpha_{k,j}^\theta T_{k,j}^\theta.$$

For $v \in I_{\theta,j}W_\theta$, $v \neq 0$, we obtain that $0 = \sum_{k=1}^{m_\theta} \alpha_{k,j}^\theta T_{k,j}^\theta v$ and this in turn implies that $\alpha_{k,j}^\theta = 0$ for all $k = 1, 2, \ldots, m_\theta$, as $T_{k',j}^\theta v$ and $T_{k,j}^\theta v$ belong to independent subspaces in V if $k \neq k'$.

The isomorphism of the algebras follows from (10.72):

$$\left(\sum_{\theta \in J} \sum_{k,j=1}^{m_\theta} \alpha_{k,j}^\theta T_{k,j}^\theta \right) \left(\sum_{\sigma \in J} \sum_{h,i=1}^{m_\sigma} \beta_{h,i}^\sigma T_{h,i}^\sigma \right) = \sum_{\theta,\sigma \in J} \sum_{k,j=1}^{m_\theta} \sum_{h,i=1}^{m_\sigma} \alpha_{k,j}^\theta \beta_{h,i}^\sigma \delta_{\sigma,\theta} \delta_{j,h} T_{k,i}^\theta$$

$$= \sum_{\theta \in J} \sum_{k,i=1}^{m_\theta} \left(\sum_{j=1}^{m_\theta} \alpha_{k,j}^\theta \beta_{j,i}^\theta \right) T_{k,i}^\theta.$$

The fact that it is also a $*$-isomorphism easily follows from (10.73). $\qquad\square$

Corollary 10.6.4 *With the above notation we have that*

$$\mathrm{dimEnd}_G(V) = \sum_{\theta \in J} m_\theta^2.$$

In particular, V is irreducible if and only if $\mathrm{dimEnd}_G(V) = 1$.

Definition 10.6.5 A representation (ρ, V) is *multiplicity-free* if $m_\theta = 1$ for all $\theta \in J$.

Corollary 10.6.6 *A representation (ρ, V) is multiplicity-free if and only if $\mathrm{End}_G(V)$ is commutative.*

Observe that

$$E_\theta = \sum_{j=1}^{m_\theta} E_{\theta,j} \equiv \sum_{j=1}^{m_\theta} T_{j,j}^\theta$$

is the projection from V onto the θ-isotypic component $m_\theta W_\theta$. It is called the *minimal central projection* associated with θ.

Recall the definition of the product in $\mathbb{C}^J$ in (10.28).

Corollary 10.6.7 *The center* $\mathcal{Z} = \mathcal{Z}(\mathrm{End}_G(V))$ *is isomorphic to* $\mathbb{C}^J$. *Moreover, the minimal central projections* E_θ, $\theta \in J$, *constitute a basis for* $\mathcal{Z}$.

Proof. The space $\mathrm{End}_G(V)$ is isomorphic to the direct sum $\bigoplus_{\theta \in J} \mathfrak{M}_{m_\theta}(\mathbb{C})$. But $A \in \mathfrak{M}_{m_\theta}(\mathbb{C})$ commutes with any other $B \in \mathfrak{M}_{m_\theta}(\mathbb{C})$ if and only if it is a scalar multiple of the identity: $A \in \mathbb{C}I_{m_\theta}$. $\square$

Exercise 10.6.8 Show that $E_\theta = \frac{d_\theta}{|G|} \sum_{g \in G} \rho(g) \chi^{\theta'}(g)$. Compare with Exercise 10.5.7 and Exercise 10.5.10.

Exercise 10.6.9 Let (ρ, V) and (η, U) be two G-representations. Suppose that $V \cong \bigoplus_{\theta \in J} m_\theta W_\theta$ and $U \cong \bigoplus_{\theta \in K} n_\theta W_\theta$, $J, K \subseteq \widehat{G}$, are the decompositions of V and U into irreducible representations. Show that we have an isomorphism

$$\mathrm{Hom}_G(U, V) \cong \bigoplus_{\theta \in K \cap J} \mathfrak{M}_{n_\theta, m_\theta}(\mathbb{C})$$

as vector spaces.

Exercise 10.6.10 Let V and W be two inner product vector spaces.

(1) Show that

$$\langle T_1, T_2 \rangle_{\mathrm{Hom}(W,V)} = \frac{1}{\dim W} \mathrm{Tr}(T_2^* T_1),$$

with $T_1, T_2 \in \mathrm{Hom}(W, V)$, defines an inner product in $\mathrm{Hom}(W, V)$ (called the *normalized Hilbert-Schmidt inner product*).

(2) Show that if $\dim W \leq \dim V$ and $T \in \mathrm{Hom}(W, V)$ is an isometry then $\|T\|_{\mathrm{Hom}(W,V)} = 1$.

Exercise 10.6.11 Let (ρ, V) and (θ, W) be two G-representations. Suppose that (θ, W) is irreducible and denote by $m = \dim \mathrm{Hom}_G(W, V)$ the multiplicity of θ in (ρ, V). Let also $T_1, T_2, \ldots, T_m \in \mathrm{Hom}_G(W, V)$. Show that the following facts are equivalent:

(a) $\langle T_i w_1, T_j w_2 \rangle_V = \langle w_1, w_2 \rangle_W \delta_{i,j}$, for all $w_1, w_2 \in W$ and $i, j = 1, 2, \ldots, m$;

(b) the W-isotypic component of V is equal to the orthogonal direct sum

$$T_1 W \oplus T_2 W \oplus \cdots \oplus T_m W,$$

and each operator T_j is an isometry from W onto $T_j W$;

(c) the operators $T_1, T_2, \ldots, T_m$ form an orthonormal basis for $\mathrm{Hom}_G(W, V)$ (with respect to the normalized Hilbert-Schmidt inner product);

(d) $T_j^* T_i = \delta_{i,j} I_W$, for all $i, j = 1, 2, \ldots, m$.

Exercise 10.6.12 In the notation of Corollary 10.3.12, see also Exercise 10.5.21,

(1) show that the Fourier transform is an isometric $*$-isomorphism between the group algebra $L(G)$ and $C(\widehat{G})$, where the scalar product is defined by setting

$$\langle T, S \rangle_{C(\widehat{G})} = \frac{1}{|G|} \sum_{\theta \in \widehat{G}} d_\theta \mathrm{Tr}[S(\theta)^* T(\theta)],$$

for all $S, T \in C(\widehat{G})$.

(2) Show that the Fourier transform and the inverse Fourier transform are one the adjoint of the other, that is, if we identify M^θ with $W'_\theta \otimes W_\theta$ by means of Theorem 10.5.20, then

$$\langle F, (f \otimes v)^\vee \rangle_{L(G)} = \langle \widehat{F}, f \otimes v \rangle_{C(\widehat{G})}$$

for all $F \in L(G)$, $v \in W_\theta$, $f \in W'_\theta$, and $\theta \in \widehat{G}$.

Solution. Fix $\theta \in \widehat{G}$ and let $\{v_1, v_2, \ldots, v_{d_\theta}\}$ be an orthonormal basis in W_θ. Then, for $v \in W_\theta$ and $f \in W'_\theta$ one has

$$\langle F, (f \otimes v)^\vee \rangle_{L(G)} = \frac{d_\theta}{|G|} \sum_{g \in G} F(g) \overline{f[\theta(g^{-1})v]}$$

$$= \frac{d_\theta}{|G|} \sum_{g \in G} F(g) f \overline{\left(\sum_{i=1}^{d_\theta} \langle \theta(g^{-1})v, v_i \rangle_{W_\theta} v_i \right)}$$

$$= \frac{d_\theta}{|G|} \sum_{i=1}^{d_\theta} \overline{f(v_i)} \sum_{g \in G} \langle F(g)\theta(g)v_i, v \rangle_{W_\theta}$$

$$= \frac{d_\theta}{|G|} \sum_{i=1}^{d_\theta} \overline{f(v_i)} \langle \widehat{F}(\theta)v_i, v \rangle_{W_\theta}$$

$$= \frac{d_\theta}{|G|} \sum_{i=1}^{d_\theta} \langle \widehat{F}(\theta)v_i, [f \otimes v](v_i) \rangle_{W_\theta}$$

$$= \langle \widehat{F}, f \otimes v \rangle_{C(\widehat{G})}.$$

10.7 A noncommutative FFT

The aim of this section is to present a noncommutative version of the FFT developed by Diaconis and Rockmore in [54]. Let G be a finite group, $K \leq G$ a subgroup, and $\mathcal{T} \subset G$ a complete set of representatives for the left cosets of K (cf. (10.49)). Given an irreducible G-representation (θ, W), we consider an orthogonal decomposition

$$\mathrm{Res}_K^G W = \bigoplus_{j=1}^{m} V_{\sigma_j} \tag{10.76}$$

of its restriction to K, into irreducible K-representations. Note that in (10.76) the K-representations (σ_j, V_{σ_j}), $j = 1, 2, \ldots, m$, are not necessarily pairwise inequivalent. Then, by choosing an orthonormal basis in each V_{σ_j} in (10.76), we get an orthonormal basis for W such that, identifying a linear operator with the associated matrix,

$$\theta(k) = \begin{pmatrix} \sigma_1(k) & & & \\ & \sigma_2(k) & & \\ & & \ddots & \\ & & & \sigma_m(k) \end{pmatrix}, \tag{10.77}$$

for all $k \in K$.

Exercise 10.7.1 Check the details of (10.77).

The orthogonal basis for W that leads to (10.77) is called an *adapted basis* to the decomposition in (10.76). Then, for $f \in L(G)$, its Fourier transform evaluated at θ is given by

$$\begin{aligned} \widehat{f}(\theta) &= \sum_{g \in G} f(g)\theta(g) \\ &= \sum_{t \in \mathcal{T}} \theta(t) \sum_{k \in K} f_t(k)\theta(k), \end{aligned} \tag{10.78}$$

where $f_t \in L(K)$, $t \in \mathcal{T}$, is defined by $f_t(k) = f(tk)$ for all $k \in K$. By virtue of (10.77), we have, for all $t \in \mathcal{T}$,

$$\sum_{k \in K} f_t(k)\theta(k) = \begin{pmatrix} \widehat{f_t}(\sigma_1) & & & \\ & \widehat{f_t}(\sigma_2) & & \\ & & \ddots & \\ & & & \widehat{f_t}(\sigma_m) \end{pmatrix}. \tag{10.79}$$

By combining (10.78) and (10.79), we get an algorithm that reduces the computation of $\widehat{f}(\theta)$ to the computation of smaller (dimension) Fourier transforms (the $\widehat{f_t}(\sigma_j)$s) and then to multiplications of these by the matrices $\theta(t)$s.

Exercise 10.7.2 Denote by $T(G)$ (respectively $T(K)$) the number of operations required to compute the Fourier transform of a given $f \in L(G)$ at each irreducible representation of G (respectively of K), and by $M(d)$ the number of operations needed to compute the product of two $(d \times d)$-matrices. Show that

$$T(G) = |\mathcal{T}| \cdot T(K) + (|\mathcal{T}| - 1) \sum_{\sigma \in \widehat{K}} M(d_\sigma).$$

Exercise 10.7.3 Show that the Cooley-Tukey algorithm in (5.62) is a particular case of the algorithm considered in this section.

Hint. Just observe that $G = \mathbb{Z}_{nm}$ and $K = \mathbb{Z}_m$.

Diaconis and Rockmore also considered recursive applications of this basic algorithm when a chain

$$G = G_0 \geq G_1 \geq G_2 \geq \cdots \geq G_m \geq G_{m+1} = \{1_G\}$$

of subgroups is available, providing several specific examples.

11

Induced representations and Mackey theory

In this chapter we introduce the theory of induced representations. This is a central topic in the representation theory of finite groups. We emphasize the analytic approach and include a detailed treatment of Mackey's theory, which will play a fundamental role in the following chapters, and of the little group method, due to Mackey and Wigner, that will be used extensively in Chapter 12. Other treatments of these topics are in the books by Naimark and Stern [119], Sternberg [154], Simon [148], Serre [145], Curtis and Reiner [42, 43], Huppert [78], Shaw [147], and Bump [23]. See also our previous monographs [33, 34] and the expository paper [30].

11.1 Induced representations

Throughout this section, G is a finite group, K a subgroup of G and (σ, V) a finite dimensional unitary representation of K. We suppose that $\mathcal{T}$ is a system of representatives for the set G/K of left cosets of K in G as in (10.49). We also assume that $1_G \in \mathcal{T}$ is the representative of K. We denote by $V[G]$ the vector space of all functions $f: G \to V$.

Definition 11.1.1 (Induced representation) The *induced representation* of a K-representation (σ, V) is the G-representation $(\lambda, \mathrm{Ind}_K^G V)$ whose representation space is

$$\mathrm{Ind}_K^G V = \{f \in V[G] : f(gk) = \sigma(k^{-1})f(g), \text{ for all } g \in G, k \in K\}, \quad (11.1)$$

with the action λ given by

$$[\lambda(g_1)f](g_2) = f(g_1^{-1}g_2), \qquad \text{for all } g_1, g_2 \in G \text{ and } f \in \mathrm{Ind}_K^G V. \quad (11.2)$$

Note that $\lambda(g)f \in \text{Ind}_K^G V$ for all $g \in G$ and $f \in \text{Ind}_K^G V$, and that λ is indeed a representation (compare with the definition of the left regular representation in (10.9)). Sometimes we shall denote λ by $\text{Ind}_K^G \sigma$.

In $\text{Ind}_K^G V$ we can define an invariant scalar product by setting

$$\langle f_1, f_2 \rangle_{\text{Ind}_K^G V} = \frac{1}{|K|} \sum_{g \in G} \langle f_1(g), f_2(g) \rangle_V \tag{11.3}$$

for $f_1, f_2 \in \text{Ind}_K^G V$; it is easy to check that $(\lambda, \text{Ind}_K^G V)$ is unitary with respect to this scalar product. We also use the following reduced form of (11.3):

$$\langle f_1, f_2 \rangle_{\text{Ind}_K^G V} = \sum_{t \in \mathcal{T}} \langle f_1(t), f_2(t) \rangle_V. \tag{11.4}$$

Indeed, if $g \in G$ and $g = tk$, $k \in K, t \in \mathcal{T}$, then from (11.1) and the unitarity of σ we deduce that $\langle f_1(g), f_2(g) \rangle_V = \langle \sigma(k^{-1})f_1(t), \sigma(k^{-1})f_2(t) \rangle_V = \langle f_1(t), f_2(t) \rangle_V$.

Now we explore the structure of an induced representation. For every $v \in V$ define the function $f_v \in V[G]$ by setting

$$f_v(g) = \begin{cases} \sigma(g^{-1})v & \text{if } g \in K \\ 0 & \text{otherwise.} \end{cases} \tag{11.5}$$

It is easy to check that $f_v \in \text{Ind}_K^G V$ and that the subspace $\widetilde{V} = \{f_v : v \in V\}$ of $\text{Ind}_K^G V$ is K-invariant and K-isomorphic to V; indeed,

$$\lambda(k)f_v = f_{\sigma(k)v} \tag{11.6}$$

for all $k \in K$.

Proposition 11.1.2 *With the same notation as in* (10.49), *we have the direct sum decomposition*

$$\text{Ind}_K^G V = \bigoplus_{t \in \mathcal{T}} \lambda(t)\widetilde{V}. \tag{11.7}$$

Proof. Take $f \in \text{Ind}_K^G V$ and set $v_t = f(t) \in V$ for every $t \in \mathcal{T}$. Then, for $t_0 \in \mathcal{T}$ and $k \in K$, we have $t^{-1}t_0 k \in K$ if and only if $t = t_0$, and therefore

$$\sum_{t \in \mathcal{T}} \lambda(t)f_{v_t}(t_0 k) = \sum_{t \in \mathcal{T}} f_{v_t}(t^{-1}t_0 k) = f_{v_{t_0}}(k)$$

$$= \sigma(k^{-1})v_{t_0} = \sigma(k^{-1})f(t_0) = f(t_0 k)$$

that is, since $t_0 k \in G$ is arbitrary,

$$f = \sum_{t \in \mathcal{T}} \lambda(t)f_{v_t}. \tag{11.8}$$

Note also that such an expression is unique: indeed, from (11.1) it follows that every $f \in \text{Ind}_K^G V$ is uniquely determined by its values on $\mathcal{T}$. $\qquad\square$

Conversely, we have:

Lemma 11.1.3 *Let (τ, W) be a representation of G and V a K-invariant subspace such that the direct decomposition*

$$W = \bigoplus_{t \in \mathcal{T}} \tau(t)V \tag{11.9}$$

holds. Then the G-representations W and $\text{Ind}_K^G V$ are isomorphic.

Proof. If we define $\tilde{V}$ as in (11.7) it follows that $\text{Ind}_K^G V$ and W are G-isomorphic. The easy details are left as an exercise. $\qquad\square$

Remark 11.1.4 In some books, as for instance Serre's monograph [145], induced representations are defined by means of the property in Lemma 11.1.3.

We observe that the dimension of the induced representation is given by

$$\dim(\text{Ind}_K^G V) = [G : K] \cdot \dim(V) \tag{11.10}$$

as it immediately follows from (11.7) and observing that $|\mathcal{T}| = [G : K]$. We now prove that induction is transitive.

Proposition 11.1.5 (Induction in stages) *Let $K \leq H \leq G$ be finite groups and (σ, V) a K-representation.*

(i) *The map $f \mapsto F$ given by $F(g, h) = [f(g)](h)$, for all $f \in (V[H])[G]$, $F \in V[G \times H]$, $g \in G$, and $h \in H$, yields a vector space isomorphism between $(V[H])[G]$ and $V[G \times H]$. By restriction, it yields an isomorphism between the G-representations $\text{Ind}_H^G(\text{Ind}_K^H V)$ and*

$$\{F \in V[G \times H] : F(gh, h'k) = \sigma(k^{-1})F(g, hh'),$$
$$\forall g \in G, h, h' \in H, k \in K\}. \tag{11.11}$$

(ii) *The map $F \mapsto \tilde{F}$, where F is in the space (11.11) and $\tilde{F} \in V[G]$ is defined by setting $\tilde{F}(g) = F(g, 1_G)$, for all $g \in G$, yields an isomorphism between the G-representations (11.11) and $\text{Ind}_K^G V$. The corresponding inverse map is given by $\tilde{F} \mapsto F$, where $F(g, h) = \tilde{F}(gh)$, for all $h \in H$, $g \in G$.*

(iii) *The following isometric isomorphism of G-representations holds:*

$$\text{Ind}_H^G(\text{Ind}_K^H V) \cong \text{Ind}_K^G V. \tag{11.12}$$

Proof.

(i) The isomorphism $(V[H])[G] \cong V[G \times H]$ induced by the map $f \mapsto F$ is obvious. Moreover, from the definition of an induced representation, we get

$$\text{Ind}_K^H V = \{f' \in V[H] : f'(hk) = \sigma(k^{-1})f'(h), \ \forall h \in H, k \in K\}$$

and, setting $\theta = \text{Ind}_K^H \sigma$,

$$\text{Ind}_H^G(\text{Ind}_K^H V) = \{f \in (\text{Ind}_K^H V)[G] :$$
$$f(gh) = \theta(h^{-1})f(g), \ \forall g \in G, h \in H\}.$$

We deduce that if $f \in \text{Ind}_H^G(\text{Ind}_K^H V)$ then we have

$$\begin{aligned}
F(gh, h'k) &= [f(gh)](h'k) \\
&= \sigma(k^{-1})\left([f(gh)](h')\right) \\
&= \sigma(k^{-1})[\theta(h^{-1})f(g)](h') \\
&= \sigma(k^{-1})[f(g)](hh') \\
&= \sigma(k^{-1})F(g, hh'),
\end{aligned}$$

for all $g \in G$, $h, h' \in H$, and $k \in K$. This shows that F belongs to (11.11). By means of the same arguments, it is easy to check that each F in (11.11) is the image of some $f \in \text{Ind}_H^G(\text{Ind}_K^H V)$.

(ii) Let F be in the space (11.11). It is immediate to check that $F(g, h) = F(gh, 1_G)$, for all $g \in G$ and $h \in H$, so that F is uniquely determined by its values on $G \times \{1_G\}$. As a consequence, we have

$$\widetilde{F}(gk) = F(gk, 1_G) = F(g, k) = \sigma(k^{-1})F(g, 1_G) = \sigma(k^{-1})\widetilde{F}(g),$$

for all $g \in G$ and $k \in K$, so that $\widetilde{F} \in \text{Ind}_K^G V$.

(iii) The isomorphism follows immediately from (i) and (ii). Finally, it is immediate to check that, modulo the identifications in (i) and (ii), one has $\|\widetilde{F}\|_{\text{Ind}_K^G V} = \|F\|_{\text{Ind}_H^G \text{Ind}_K^H V}$. $\square$

Example 11.1.6 (Permutation representation) Let G be a finite group acting transitively on a finite set X. Choose $x_0 \in X$ and let $K = \{g \in G : gx_0 = x_0\}$ be its stabilizer. As in Example 10.4.3, we denote by $(\lambda, L(X))$ the corresponding permutation representation of G. Let now $(\iota_K, \mathbb{C})$ denote the *trivial* (one dimensional) representation of K. Then

$$\text{Ind}_K^G \mathbb{C} = \{f \in L(G) : f(gk) = f(g), \forall g \in G, k \in K\} = L(G)^K$$

(the space of all right-K-invariant functions on G). The latter is isomorphic to $L(X)$: the map $f \mapsto \tilde{f}$, where $f \in L(X)$ and $\tilde{f} \in L(G)^K$ is given by

$$\tilde{f}(g) = f(gx_0) \tag{11.13}$$

for all $g \in G$, yields the desired G-isomorphism. We can rephrase the above discussion by saying that the permutation representation λ and the induced representation $\mathrm{Ind}_K^G \iota_K$ are equivalent. Recalling the identification $X = G/K$ as G-spaces, we can thus write:

$$(\lambda, L(G/K)) \sim (\mathrm{Ind}_K^G \iota_K, L(G)^K). \tag{11.14}$$

Exercise 11.1.7 Suppose that $K \leq H \leq G$, set $X = G/K$, $Y = G/H$, $Z = H/K$, and suppose that $x_0 \in X$ (respectively, $y_0 \in Y$) is the point stabilized by K (respectively H).

(1) Show that there exists a unique surjective map $\pi : X \to Y$ such that $\pi(x_0) = y_0$ and $\pi(gx) = g\pi(x)$ for all $x \in X$ and $g \in G$ (that is, π is G-equivariant).
(2) Show that, in the present setting, transitivity of induction has the following more explicit form: $L(X) \cong \mathrm{Ind}_H^G L(Z) \cong \oplus_{y \in Y} L(\pi^{-1}(y))$.

See [138] for some examples and applications of these simple facts.

Example 11.1.8 Let G be a finite group and $N \leq G$ a normal subgroup. Denote by $\lambda_{G/N}$ the left regular representation of G/N and by $\bar{\lambda}$ the permutation representation of G on G/N (note that the corresponding representation spaces are the same, namely $L(G/N)$). Then

$$\bar{\lambda}(g) = \lambda_{G/N}(gN) \tag{11.15}$$

for all $g \in G$. Indeed, if $f \in L(G/N)$ and $g, g_0 \in G$, one has

$$[\lambda_{G/N}(gN)f](g_0N) = f[(gN)^{-1}(g_0N)] = f(g^{-1}g_0N) = [\bar{\lambda}(g)f](g_0N).$$

Example 11.1.9 Let G be a finite group and $K \leq G$ a subgroup. Let also χ be a one-dimensional representation of K. Recall that $\chi : K \to \mathbb{C}$ satisfies: $|\chi(k_1)| = 1$, $\chi(k_1 k_2) = \chi(k_1)\chi(k_2)$, so that $\chi(k^{-1}) = \chi(k)^{-1} = \overline{\chi(k)}$, for all $k_1, k_2, k \in G$, and $\chi(1_K) = 1$. Then the representation space of $\mathrm{Ind}_K^G \chi$, that we denote by $\mathrm{Ind}_K^G \mathbb{C}$, is made up of all $f \in L(G)$ such that

$$f(gk) = \overline{\chi(k)}f(g) \tag{11.16}$$

for all $k \in K$ and $g \in G$. The corresponding G-action is again given by left translation:

$$[\mathrm{Ind}_K^G \chi(g)f](g') = f(g^{-1}g')$$

for all $f \in \mathrm{Ind}_K^G \mathbb{C}$ and $g, g' \in G$.

Now (11.7) becomes

$$\text{Ind}_K^G \mathbb{C} = \bigoplus_{t \in \mathcal{T}} \lambda(t)(\mathbb{C}\overline{\chi}),$$ (11.17)

where χ is extended to the whole G by setting $\chi(g) = 0$ for all $g \in G \setminus K$ (note that, this way, $f = \overline{\chi} \in L(G)$ satisfies (11.16)).

Exercise 11.1.10 Suppose that A, B are finite Abelian groups, $B \leq A$ and let χ be a character of B. Show that a character ψ of A is contained in $\text{Ind}_B^A \chi$ if and only if $\psi(b) = \chi(b)$ for all $b \in B$ and, if this is the case, its multiplicity is equal to 1.

Now we give a formula for the matrix coefficients and the character of an induced representation.

Theorem 11.1.11 *Let G be a finite group, $K \leq G$ a subgroup, and $\mathcal{T} \subseteq G$ a complete set of representatives for the left cosets of K in G. Let also (σ, V) be a K-representation, $\{e_1, e_2, \ldots, e_d\}$ an orthonormal basis for V and denote by $\lambda = \text{Ind}_K^G \sigma$ the corresponding induced representation. Define $f_{e_j} \in \text{Ind}_K^G V$ as in (11.5) and $f_{t,j} = \lambda(t) f_{e_j} \in \text{Ind}_K^G V$ for all $t \in \mathcal{T}$ and $j = 1, 2, \ldots, d$. Then $\{f_{t,j} : t \in \mathcal{T}, j = 1, 2, \ldots, d\}$ is an orthonormal basis for $\text{Ind}_K^G V$ with respect to the scalar product (11.3) and the corresponding matrix coefficients of λ are given by the formula*

$$\langle \lambda(g) f_{t,j}, f_{s,i} \rangle_{\text{Ind}_K^G V} = \begin{cases} \langle \sigma(s^{-1}gt)e_j, e_i \rangle_V & \text{if } s^{-1}gt \in K \\ 0 & \text{otherwise} \end{cases}$$

for all $s, t \in \mathcal{T}$ and $i, j = 1, 2, \ldots, d$.

Proof. The fact that $\{f_{t,j} : t \in \mathcal{T}, j = 1, 2, \ldots, n\}$ is an orthonormal basis easily follows from (11.4) and the formula $f_{t,j}(s) = \delta_{st} e_j$, for $s, t \in \mathcal{T}$. Now suppose that $g \in G$ and $r \in \mathcal{T}$. Then there exist $t_1 \in \mathcal{T}$ and $k \in K$ such that $g^{-1}r = t_1 k$ and therefore

$$[\lambda(g) f_{t,j}](r) = f_{t,j}(g^{-1}r)$$
$$= f_{t,j}(t_1 k)$$
$$= \delta_{t,t_1} \sigma(k^{-1}) e_j.$$

Since $k = t_1^{-1} g^{-1} r$ and

$$t = t_1 \iff g^{-1}r \in t K \iff r^{-1}gt \in K,$$

we deduce that

$$[\lambda(g)f_{t,j}](r) = \begin{cases} \sigma(r^{-1}gt)e_j & \text{if } r^{-1}gt \in K \\ 0 & \text{otherwise.} \end{cases}$$

We can use this formula and (11.4) to compute the matrix coefficients of the induced representation λ: for $s, t \in \mathcal{T}$ and $i, j = 1, 2, \ldots, d$, we have

$$\langle \lambda(g)f_{t,j}, f_{s,i} \rangle_{\mathrm{Ind}_K^G V} = \sum_{r \in \mathcal{T}} \langle [\lambda(g)f_{t,j}](r), f_{s,i}(r) \rangle_V$$

$$= \begin{cases} \langle \sigma(s^{-1}gt)e_j, e_i \rangle_V & \text{if } s^{-1}gt \in K \\ 0 & \text{otherwise.} \end{cases}$$

$\square$

Corollary 11.1.12 (Frobenius character formula) *Let G be a finite group, $K \leq G$ a subgroup, and (σ, V) a K-representation. Then the character of the induced representation $\mathrm{Ind}_K^G \sigma$ is given by*

$$\chi^{\mathrm{Ind}_K^G \sigma}(g) = \sum_{\substack{t \in \mathcal{T}: \\ t^{-1}gt \in K}} \chi^{\sigma}(t^{-1}gt). \tag{11.18}$$

Proof. Let $u_{i,j}^{\sigma}$ denote the matrix coefficients of σ and $u_{s,i;t,j}^{\lambda}$ those of λ. Then Theorem 11.1.11 yields:

$$u_{s,i;t,j}^{\lambda}(g) = \begin{cases} u_{i,j}^{\sigma}(s^{-1}gt) & \text{if } s^{-1}gt \in K \\ 0 & \text{otherwise,} \end{cases} \tag{11.19}$$

that is, if $U(k) = \left(u_{i,j}^{\sigma}(k) \right)_{i,j=1}^{d}$, then the matrix $\left(u_{t,i;s,j}^{\lambda}(g) \right)_{\substack{i,j=1,2,\ldots,d \\ t,s \in \mathcal{T}}}$ is given in block form by $\left(U(t^{-1}gs) \right)_{t,s \in \mathcal{T}}$, where $U(t^{-1}gs) = 0$ whenever $t^{-1}gs \notin K$. By taking the trace of this block matrix, we immediately get the expression for the character of λ in terms of the character of σ. $\square$

There is another useful way to write Frobenius character formula. If $\mathcal{C}$ is a conjugacy class in G, then $\mathcal{C} \cap K$ is invariant under conjugation by elements of K so that it is partitioned as

$$\mathcal{C} \cap K = \coprod_{i=1}^{m} \mathcal{D}_i, \tag{11.20}$$

where the $\mathcal{D}_i$'s are conjugacy classes in K.

Proposition 11.1.13 *Let G be a finite group, $K \leq G$ a subgroup, and (σ, V) a K-representation. Then we have:*

$$\chi^{\operatorname{Ind}_K^G \sigma}(\mathcal{C}) = \frac{|G|}{|K| \cdot |\mathcal{C}|} \sum_{i=1}^{m} |\mathcal{D}_i| \chi^{\sigma}(\mathcal{D}_i), \tag{11.21}$$

where $\chi(\mathcal{C})$ denotes the value $\chi(c)$ of the character χ at each $c \in \mathcal{C}$.

Proof. If $c, c' \in \mathcal{C}$, then

$$|\{g \in G : g^{-1}cg = c'\}| = \frac{|G|}{|\mathcal{C}|}. \tag{11.22}$$

Indeed, G acts transitively on $\mathcal{C}$ by conjugation ($c \mapsto g^{-1}cg$, for all $c \in \mathcal{C}$ and $g \in G$), and the stabilizer of c coincides with its centralizer, whose order is $|G|/|\mathcal{C}|$; see Lemma 10.4.2. Therefore, by Frobenius character formula, for $c \in \mathcal{C}$ we have

$$\chi^{\operatorname{Ind}_K^G \sigma}(\mathcal{C}) = \sum_{\substack{t \in \mathcal{T}: \\ t^{-1}ct \in K}} \chi^{\sigma}(t^{-1}ct)$$

$$= \frac{1}{|K|} \sum_{k \in K} \sum_{\substack{t \in \mathcal{T}: \\ t^{-1}ct \in K}} \chi^{\sigma}(k^{-1}t^{-1}ctk)$$

$$(g = tk) = \frac{1}{|K|} \sum_{\substack{g \in G: \\ g^{-1}cg \in K}} \chi^{\sigma}(g^{-1}cg)$$

$$(\text{by } (11.22)) = \frac{1}{|K|} \sum_{i=1}^{m} \frac{|G|}{|\mathcal{C}|} \sum_{k \in \mathcal{D}_i} \chi^{\sigma}(k)$$

$$= \frac{|G|}{|K| \cdot |\mathcal{C}|} \sum_{i=1}^{m} |\mathcal{D}_i| \chi^{\sigma}(\mathcal{D}_i). \qquad \square$$

Corollary 11.1.14 *For a permutation representation $(\lambda, L(X))$ (cf. Example 11.1.6), formula (11.21) becomes:*

$$\chi^{\lambda}(\mathcal{C}) = \frac{|X|}{|\mathcal{C}|} |\mathcal{C} \cap K|.$$

Exercise 11.1.15 Deduce the fixed point character formula (Proposition 10.4.6) from Frobenius character formula.

In the last part of this section, we illustrate two fundamental results that connect tensor products (cf. Section 10.5) and induced representations.

Theorem 11.1.16 *Let G be a finite group and $K \leq G$ a subgroup. Let (θ, W) be a G-representation and (σ, V) a K-representation. Then the map*

$$\phi \colon W \bigotimes \operatorname{Ind}_K^G V \to \operatorname{Ind}_K^G[(\operatorname{Res}_K^G W) \bigotimes V] \qquad (11.23)$$

defined by setting

$$[\phi(w \otimes f)](g) = \theta(g^{-1})w \otimes f(g),$$

for all $w \in W, f \in \operatorname{Ind}_K^G V$, and $g \in G$, is an isometric isomorphism of G-representations, so that, in particular,

$$\phi \in \operatorname{Hom}_G\left(\theta \otimes \operatorname{Ind}_K^G \sigma, \operatorname{Ind}_K^G[\operatorname{Res}_K^G \theta \otimes \sigma]\right).$$

Proof. The space $W \bigotimes \operatorname{Ind}_K^G V$ is spanned by all products $w \otimes f$ where $w \in W$ and $f \in V[G]$ satisfies $f(gk) = \sigma(k^{-1})f(g)$, for all $k \in K$ and $g \in G$. Let us set, as usual, $\lambda = \operatorname{Ind}_K^G \sigma$. The space $\operatorname{Ind}_K^G[(\operatorname{Res}_K^G W) \bigotimes V]$ is made up of all functions $F \in (W \bigotimes V)[G]$ such that

$$F(gk) = [\theta(k^{-1}) \otimes \sigma(k^{-1})]F(g), \qquad (11.24)$$

for all $k \in K$ and $g \in G$, and it is spanned by all functions of the form $\lambda_1(g)F_{w \otimes v}$, for $g \in G$, $w \in W$, $v \in V$, where $\lambda_1 = \operatorname{Ind}_K^G[(\operatorname{Res}_K^G \theta) \otimes \sigma]$ is as in (11.2) and $F_{w \otimes v}$ is given by (11.5). First of all, observe that $\phi(w \otimes f) \in \operatorname{Ind}_K^G[(\operatorname{Res}_K^G W) \bigotimes V]$. Indeed, $\phi(w \otimes f) \in (W \bigotimes V)[G]$ and satisfies (11.24):

$$\begin{aligned}
[\phi(w \otimes f)](gk) &= \theta(k^{-1}g^{-1})w \otimes f(gk) \\
&= [\theta(k^{-1}) \otimes \sigma(k^{-1})]\left(\theta(g^{-1})w \otimes f(g)\right) \\
&= [\theta(k^{-1}) \otimes \sigma(k^{-1})]\left(\phi(w \otimes f)\right)(g).
\end{aligned}$$

Let us show that the map (11.23) is G-equivariant: for all $g, g_0 \in G$ we have

$$\begin{aligned}
(\phi\{[\theta(g)w] \otimes [\lambda(g)f]\})(g_0) &= \theta(g_0^{-1}g)w \otimes f(g^{-1}g_0) \\
&= [\phi(w \otimes f)](g^{-1}g_0) \\
&= [\lambda_1(g)\phi(w \otimes f)](g_0),
\end{aligned}$$

that is, ϕ intertwines $\theta \otimes \lambda$ and λ_1. Now we prove that the map ϕ is surjective. For $w \in W, v \in V$, and $k \in K$ we have

$$[\phi(w \otimes f_v)](k) = \theta(k^{-1})w \otimes f_v(k) = \theta(k^{-1})w \otimes \sigma(k^{-1})v = F_{w \otimes v}(k)$$

and $[\phi(w \otimes f_v)](g) = 0 = F_{w \otimes v}(g)$ if $g \in G \setminus K$, so that $\phi(w \otimes f_v) = F_{w \otimes v}$. Since the functions of the form $\lambda_1(g)F_{w \otimes v}$ span $\operatorname{Ind}_K^G[(\operatorname{Res}_K^G W) \bigotimes V]$, we conclude that ϕ is surjective. Since

$$\dim\left[W \bigotimes \operatorname{Ind}_K^G V\right] = \dim W \dim V |G/K| = \dim\left\{\operatorname{Ind}_K^G[(\operatorname{Res}_K^G W) \bigotimes V]\right\}$$

it is also injective, so that it is an isomorphism. We leave it to the reader to check that ϕ is indeed an isometry. □

Corollary 11.1.17 *Let G be a finite group, $K \leq G$ a subgroup, and $x_0 \in X = G/K$ be the point stabilized by K. Let (θ, W) (respectively, $(\lambda, L(X))$) be a representation (respectively, the corresponding permutation representation) of G. Then the map*

$$\phi: W \bigotimes L(X) \to \mathrm{Ind}_K^G \mathrm{Res}_K^G W$$

defined by setting

$$[\phi(w \otimes f)](g) = f(gx_0)\theta(g^{-1})w,$$

for all $f \in L(X)$, $w \in W$, and $g \in G$, is an isometric isomorphism.

Proof. Apply Theorem 11.1.16 with $\sigma = \iota_K$ the trivial representation of K. In this case $\mathrm{Ind}_K^G V \cong L(X)$ (see Example 11.1.6, in particular (11.14)) and $(\mathrm{Res}_K^G W) \bigotimes V = (\mathrm{Res}_K^G W) \bigotimes \mathbb{C} \cong \mathrm{Res}_K^G W$. □

In the last corollary, we have shown that $\mathrm{Ind}_K^G \mathrm{Res}_K^G W$ is isomorphic to $W \bigotimes L(X)$. This is the first elementary result that connects induction and restriction. Sections 11.2, 11.4, and 11.5 are devoted to deeper results of this kind. In particular, Mackey's lemma in Section 11.5 examines the structure of $\mathrm{Res}_H^G \mathrm{Ind}_K^G V$, where V is a K-representation and $H \leq G$ is another subgroup.

Another property of the induction operation is additivity.

Proposition 11.1.18 *Let G be a finite group and $K \leq G$ a subgroup. Let (σ_1, V_1) and (σ_2, V_2) be two representations of K. Then*

$$\mathrm{Ind}_K^G \left(\rho_1 \bigoplus \rho_2 \right) \sim \mathrm{Ind}_K^G(\rho_1) \bigoplus \mathrm{Ind}_K^G(\rho_2).$$

Proof. We leave it to the reader to check that the map

$$\Phi: \left(\mathrm{Ind}_K^G V_1 \oplus \mathrm{Ind}_K^G V_2 \right) \to \mathrm{Ind}_K^G (V_1 \oplus V_2),$$

defined by $[\Phi(f_1 + f_2)](g) = f_1(g) + f_2(g)$, for all $f_i \in \mathrm{Ind}_K^G V_i$, $i = 1, 2$ and $g \in G$ is a bijective map in $\mathrm{Hom}_G(\mathrm{Ind}_K^G(\rho_1) \bigoplus \mathrm{Ind}_K^G(\rho_2), \mathrm{Ind}_K^G \left(\rho_1 \bigoplus \rho_2 \right))$. □

Exercise 11.1.19 Let G be a finite group and $K \leq G$ a subgroup. Let (σ, V) be a K-representation. Consider the tensor product $L(G) \bigotimes V$, its subspace $\mathcal{V}$ spanned by $\{\delta_{gk} \otimes v - \delta_g \otimes \sigma(k)v : g \in G, k \in K, v \in V\}$, and the G-representation $(\gamma, L(G) \bigotimes V)$ given by

$$\gamma(g)(\delta_{g'} \otimes v) = \delta_{gg'} \otimes v$$

for all $g, g' \in G$ and $v \in V$. Show that $\mathcal{V}$ is γ-invariant and that $\mathrm{Ind}_K^G V \cong [L(G) \bigotimes V]/\mathcal{V}$ as G-representations.

The above yields a classical, more algebraic, definition of an induced representation; see the monograph by Alperin and Bell [12].

11.2 Frobenius reciprocity

This section is devoted to the first fundamental result, due to Frobenius, that relates the operations of induction and restriction for group representations. We assume all the notation in Section 11.1; in particular, we suppose that (θ, W) is a G-representation (with $d_\theta = \dim W$) and (σ, V) is a K-representation. For a more detailed analysis of Frobenius reciprocity, we refer to [137, 140, 37].

Theorem 11.2.1 (Frobenius reciprocity) *For each $T \in \mathrm{Hom}_G(W, \mathrm{Ind}_K^G V)$ define* $\hat{T} : W \to V$ *by setting, for every $w \in W$,*

$$\hat{T} w = [Tw](1_G). \tag{11.25}$$

Then $\hat{T} \in \mathrm{Hom}_K(\mathrm{Res}_K^G W, V)$ *and the map*

$$\mathrm{Hom}_G(W, \mathrm{Ind}_K^G V) \longrightarrow \mathrm{Hom}_K(\mathrm{Res}_K^G W, V)$$
$$T \longmapsto \hat{T}$$

is an isomorphism of vector spaces. Its inverse is the map $L \mapsto \overset{\vee}{L}$ where, for $L \in \mathrm{Hom}_K(\mathrm{Res}_K^G W, V)$,

$$\left[\overset{\vee}{L} w \right](g) = L\theta(g^{-1})w, \tag{11.26}$$

for all $w \in W$ and $g \in G$.

Proof. First of all, we show that $\hat{T} \in \mathrm{Hom}_K(\mathrm{Res}_K^G W, V)$:

$$\begin{aligned}
\hat{T}\theta(k)w &= \{T[\theta(k)w]\}(1_G) \\
\left(T \in \mathrm{Hom}_G(W, \mathrm{Ind}_K^G V) \right) \quad &= [\lambda(k)(Tw)](1_G) \\
\text{(by (11.2))} \quad &= [Tw](k^{-1}) \\
\text{(by (11.1))} \quad &= \sigma(k)[Tw](1_G) \\
&= \sigma(k)\hat{T}w
\end{aligned}$$

for all $k \in K$ and $w \in W$.

Conversely, if $L \in \mathrm{Hom}_K(\mathrm{Res}_K^G W, V)$ then from (11.26) we deduce that

$$\left[\overset{\vee}{L w}\right](gk) = L\theta(k^{-1})\theta(g^{-1})w = \sigma(k^{-1})L\theta(g^{-1})w = \sigma(k^{-1})\left[\overset{\vee}{L w}\right](g),$$

for all $w \in W$, $k \in K$ and $g \in G$, so that $\overset{\vee}{L w} \in \mathrm{Ind}_K^G V$. Moreover, if $g_0 \in G$ we have

$$\left[\overset{\vee}{L\theta(g)w}\right](g_0) = L\theta(g_0^{-1})\theta(g)w = L\theta[(g^{-1}g_0)^{-1})]w$$

$$= \left[\overset{\vee}{L w}\right](g^{-1}g_0) = \left[\lambda(g)\overset{\vee}{L w}\right](g_0),$$

and this shows that $\overset{\vee}{L} \in \mathrm{Hom}_G(W, \mathrm{Ind}_K^G V)$. Finally,

$$\left[\left(\hat{T}\right)^{\vee} w\right](g) = \hat{T}\theta(g^{-1})w = [T\theta(g^{-1})w](1_G)$$

$$= \left[\lambda(g^{-1})(Tw)\right](1_G) = [Tw](g)$$

and

$$\left(\overset{\vee}{L}\right)^{\wedge} w = \left[\overset{\vee}{L w}\right](1_G) = Lw,$$

for all $w \in W$ and $g \in G$, that is, $(\hat{T})^{\vee} = T$ and $(\overset{\vee}{L})^{\wedge} = L$. It follows that the linear maps $T \mapsto \hat{T}$ and $L \mapsto \overset{\vee}{L}$ are one inverse to the other, and therefore are isomorphisms. $\qquad \square$

From Theorem 11.2.1, Lemma 10.6.1.(i), and Lemma 10.6.2 we deduce the following:

Corollary 11.2.2 *Suppose that W and V are irreducible. Then the multiplicity of W in $\mathrm{Ind}_K^G V$ equals the multiplicity of V in $\mathrm{Res}_K W$.*

Corollary 11.2.3 *Suppose that W and V are irreducible, and that W is contained in $\mathrm{Ind}_K^G V$ with multiplicity m. Then*

$$\dim W \geq m \dim V.$$

In particular, if $\dim W = 1$ one has $\dim V = 1$ and $m = 1$.

Proof. $\mathrm{Res}_K^G W$ contains m copies of V and $\dim \mathrm{Res}_K^G W = \dim W$. $\qquad \square$

From the point of view of character theory, Frobenius reciprocity may be formulated in the following form:

Proposition 11.2.4

$$\frac{1}{|G|}\langle \chi^\theta, \chi^{\mathrm{Ind}_K^G \sigma}\rangle_{L(G)} = \frac{1}{|K|}\langle \chi^{\mathrm{Res}_K^G \theta}, \chi^\sigma\rangle_{L(K)}.$$

Proof. Although this may be deduced from Corollary 11.2.2 (see Exercise 11.2.5), we reproduce the easy proof based on Frobenius character formula. Let $\mathcal{C}_j, j = 1, 2, \ldots, n$ be the conjugacy classes of G and suppose that $\mathcal{C}_j \cap K = \bigsqcup_{i=1}^{m_j} \mathcal{D}_{i,j}$ (with $\mathcal{D}_{i,j} \subset \mathcal{C}_j$ a K-equivalence class) as in (11.20). Then we have:

$$\frac{1}{|G|}\langle \chi^\theta, \chi^{\mathrm{Ind}_K^G \sigma}\rangle_{L(G)} = \frac{1}{|G|}\sum_{j=1}^{n}|\mathcal{C}_j|\chi^\theta(\mathcal{C}_j)\overline{\chi^{\mathrm{Ind}_K^G \sigma}(\mathcal{C}_j)}$$

$$\text{(by (11.21))} \quad = \frac{1}{|K|}\sum_{j=1}^{n}\sum_{i=1}^{m_j}|\mathcal{D}_{i,j}|\chi^\theta(\mathcal{D}_{i,j})\overline{\chi^\sigma(\mathcal{D}_{i,j})}$$

$$= \frac{1}{|K|}\langle \chi^{\mathrm{Res}_K^G \theta}, \chi^\sigma\rangle_{L(K)}. \qquad \square$$

Exercise 11.2.5 Deduce Proposition 11.2.4 from Proposition 10.2.18 and Corollary 11.2.2.

Exercise 11.2.6 With the notation as in Theorem 11.2.1, show that the map $T \mapsto \sqrt{|G/K|}\hat{T}$ is an isometry with respect to the scalar product in Exercise 10.6.10.

Exercise 11.2.7 (The other side of Frobenius reciprocity) For each $T \in \mathrm{Hom}_G(\mathrm{Ind}_K^G V, W)$ define $\overset{\circ}{T} \in \mathrm{Hom}(V, W)$ by setting $\overset{\circ}{T}v = Tf_v$, for all $v \in V$ (f_v is as in (11.5)).

(1) Show that $\overset{\circ}{T} \in \mathrm{Hom}_K(V, \mathrm{Res}_K^G W)$.
(2) Show that $(T^*)^\circ = \left(\hat{T}\right)^*$.
(3) Show that the map

$$\mathrm{Hom}_G(\mathrm{Ind}_K^G V, W) \longrightarrow \mathrm{Hom}_K(V, \mathrm{Res}_K^G W)$$
$$T \longmapsto \overset{\circ}{T}$$

is an isometric isomorphism of vector spaces and that its inverse is the map $L \mapsto \overset{\circ}{L}$ defined by setting $\overset{\circ}{L}f = \sum_{t \in \mathcal{T}} \theta(t)Lf(t)$ for all $L \in \mathrm{Hom}_K(V, \mathrm{Res}_K^G W)$ and $f \in \mathrm{Ind}_K^G V$.

We now examine Frobenius reciprocity in a particular case: from now on, the K-representation (σ, V) is one-dimensional and we shall identify it with its character $\chi = \chi^\sigma$. We then denote by $\text{Ind}_K^G \mathbb{C}$ the representation space of $\lambda = \text{Ind}_K^G \chi$ (see also Example 11.1.9).

We denote by $W^{K,\chi}$ the χ-isotypic component in $\text{Res}_K^G W$, that is,

$$W^{K,\chi} = \{w \in W : \theta(k)w = \chi(k)w \text{ for all } k \in K\}. \qquad (11.27)$$

Note that when $\chi = \iota_K$ is the trivial K-representation, then

$$W^{K,\iota_K} = W^K = \{w \in W : \theta(k)w = w \text{ for all } k \in K\}$$

is the subspace of K-invariant vectors in W.

Proposition 11.2.8 *Suppose that $W^{K,\chi}$ is nontrivial. With each $u \in W^{K,\chi}$ we associate a linear map $T_u \colon W \to L(G)$ defined by setting*

$$[T_u w](g) = \sqrt{\frac{d_\theta}{|G/K|}} \langle w, \theta(g)u \rangle_W, \qquad (11.28)$$

for all $w \in W$ and $g \in G$. Then:

(i) *for all $u \in W^{K,\chi}$ we have $T_u \in \text{Hom}_G(\theta, \text{Ind}_K^G \chi)$;*
(ii) *if (θ, W) is irreducible and $\|u\|_W = 1$ then $T_u \colon W \to \text{Ind}_K^G \mathbb{C}$ is isometric.*

Proof.

(i) Let $u \in W^{K,\chi}$ and define a linear functional $L \colon W \to \mathbb{C}$ by setting $Lw = \langle w, u \rangle_W$, for all $w \in W$ (that is, in the notation of (10.53), $L = \xi^{-1}(u)$). Then $L \in \text{Hom}_K(\text{Res}_K^G W, \chi)$:

$$L\theta(k)w = \langle \theta(k)w, u \rangle_W = \langle w, \theta(k^{-1})u \rangle_W = \chi(k)\langle w, u \rangle_W = \chi(k)Lw,$$

for all $w \in W$, $k \in K$. Since $T_u = \sqrt{\frac{d_\theta}{|G/K|}} \overset{\vee}{L}$, from Theorem 11.2.1 we deduce that $T_u \in \text{Hom}_G(\theta, \text{Ind}_K^G \chi)$.

(ii) Suppose that $\{u_i : i = 1, 2, \ldots, d_\theta\}$ is an orthonormal basis in W with $u_1 = u$. Then, for every $w = \sum_{i=1}^{d_\theta} \alpha_i u_i \in W$, $\alpha_i \in \mathbb{C}$, we have (cf. (11.3)):

$$\|T_u w\|^2_{\mathrm{Ind}_K^G \chi} = \frac{1}{|K|} \cdot \frac{d_\theta}{|G/K|} \sum_{g \in G} \langle w, \sigma(g)u \rangle_W \overline{\langle w, \sigma(g)u \rangle_W}$$

$$= \frac{d_\theta}{|G|} \sum_{i,j=1}^{d_\theta} \alpha_i \overline{\alpha_j} \sum_{g \in G} \langle u_i, \sigma(g)u_1 \rangle_W \overline{\langle u_j, \sigma(g)u_1 \rangle_W}$$

(by(10.24)) $$= \sum_{i=1}^{d_\theta} |\alpha_i|^2$$

$$= \|w\|^2_W.$$

This shows that T_u is an isometry. $\qquad\square$

11.3 Preliminaries on Mackey's theory

In the present and next two sections, we use all the notation of Section 11.1. We also suppose that H is another subgroup of G and that (v, U) is an H-representation. We set $\lambda_1 = \mathrm{Ind}_H^G v$. Moreover, we assume that S is a set of representatives for the set $H \backslash G / K$ of all H-K double cosets in G, so that

$$G = \coprod_{s \in S} HsK, \qquad (11.29)$$

with $1_G \in S$ (this is the representative of HK). For each $s \in S$, we set

$$G_s = H \cap sKs^{-1}. \qquad (11.30)$$

Clearly, G_s is a subgroup of H while $s^{-1}G_s s$ is a subgroup of K. We start with a simple but useful Lemma.

Lemma 11.3.1 *Let $h, h_1 \in H$, $k, k_1 \in K$, and $s \in S$. Then we have*

$$hsk = h_1 s k_1 \Leftrightarrow \exists x \in G_s \text{ such that } h_1 = hx \text{ and } k_1 = s^{-1}x^{-1}sk.$$

Proof. We have $hsk = h_1 s k_1$ if and only if $skk_1^{-1}s^{-1} = h^{-1}h_1$. By (11.30), this holds if and only if $h_1 = hx$ and $k_1 = s^{-1}x^{-1}sk$ with $x = h^{-1}h_1 (= skk_1^{-1}s^{-1}) \in G_s$. $\qquad\square$

Remark 11.3.2 From the lemma above it follows that

$$|HsK| = \frac{|H||K|}{|G_s|}.$$

Indeed, for each $g \in HsK$ there exist exactly $|G_s|$ pairs $(h, k) \in H \times K$ such that $g = hsk$. Observe also that $H \backslash G / K$ can be interpreted as the set of H-orbits on

$X = G/K$: if $x_0 \in X$ is the point stabilized by K, then these orbits are

$$\{Hsx_0 : s \in \mathcal{S}\}.$$

Moreover, the subgroup G_s can be identified with the stabilizer in H of the point sx_0.

We leave it as an exercise to check the above statements.

For all $s \in \mathcal{S}$, we denote by (σ_s, V_s) the representation of G_s on $V_s = V$ defined by setting

$$\sigma_s(x) = \sigma(s^{-1}xs) \tag{11.31}$$

for all $x \in G_s$. We also define

$$\mathcal{S}_0 = \{s \in \mathcal{S} : \mathrm{Hom}_{G_s}(\mathrm{Res}_{G_s}^{H} v, \sigma_s) \text{ is nontrivial}\}. \tag{11.32}$$

11.4 Mackey's formula for invariants

In this section, we expose a series of results of Mackey on the space of intertwining operators between two induced representations. The particular case of the commutant of the representation obtained by inducing a one dimensional representation will be analyzed more closely in Chapter 13. See also [140] and [37].

We assume the notation from the previous section.

Definition 11.4.1 We denote by $\mathcal{V} = \mathcal{V}(G, H, K, v, \sigma)$ the set of all maps $F: G \to \mathrm{Hom}(U, V)$ such that

$$F(hgk) = \sigma(k^{-1})F(g)v(h^{-1})$$

for all $g \in G$, $h \in H$, and $k \in K$.

Lemma 11.4.2

(i) *For $s \in \mathcal{S}_0$ and $T \in \mathrm{Hom}_{G_s}(\mathrm{Res}_{G_s}^{H} v, \sigma_s)$ define $\mathcal{L}_T: G \to \mathrm{Hom}(U, V)$ by setting*

$$\mathcal{L}_T(g) = \begin{cases} \sigma(k^{-1})Tv(h^{-1}) & \text{if } g = hsk \in HsK \\ 0 & \text{otherwise.} \end{cases} \tag{11.33}$$

Then $\mathcal{L}_T$ is well defined and belongs to $\mathcal{V}$.

(ii) *Let $F \in \mathcal{V}$. Then $F(s) \in \mathrm{Hom}_{G_s}(\mathrm{Res}_{G_s}^{H} v, \sigma_s)$ for all $s \in \mathcal{S}$.*

(iii) *Let $F \in \mathcal{V}$. Then $F = \sum_{s \in \mathcal{S}_0} \mathcal{L}_{F(s)}$ and the nontrivial elements in this sum are linearly independent.*

(iv) *The map*

$$\begin{aligned} \mathcal{V} &\longrightarrow \bigoplus_{s \in \mathcal{S}_0} \mathrm{Hom}_{G_s}(\mathrm{Res}^H_{G_s} v, \sigma_s) \\ F &\longmapsto \bigoplus_{s \in \mathcal{S}_0} F(s) \end{aligned} \qquad (11.34)$$

is an isomorphism of vector spaces.

Proof.

(i) It suffices to show that $\mathcal{L}_T$ is well defined. Indeed, if $hsk = h_1 s k_1$, then, by Lemma 11.3.1, $h_1 = hx$ and $k_1 = s^{-1}x^{-1}sk$ with $x \in G_s$, so that

$$\begin{aligned} \sigma(k_1^{-1})Tv(h_1^{-1}) &= \sigma(k^{-1}s^{-1}xs)Tv(x^{-1}h^{-1}) \\ \text{(by (11.31))} \qquad &= \sigma(k^{-1})\sigma_s(x)Tv(x^{-1}h^{-1}) \\ (T \in \mathrm{Hom}_{G_s}(\mathrm{Res}^H_{G_s} v, \sigma_s)) \qquad &= \sigma(k^{-1})Tv(x)v(x^{-1}h^{-1}) \\ &= \sigma(k^{-1})Tv(h^{-1}). \end{aligned}$$

(ii) For all $x \in G_s$, by definition of $\mathcal{V}$, we have

$$\begin{aligned} F(s)v(x) &= F(x^{-1}s) \\ &= F(s \cdot s^{-1}x^{-1}s) \\ &= \sigma(s^{-1}xs)F(s) \\ &= \sigma_s(x)F(s) \end{aligned}$$

that is, $F(s) \in \mathrm{Hom}_{G_s}(\mathrm{Res}^H_{G_s} v, \sigma_s)$.

(iii) Clearly, F is determined by its values on $\mathcal{S}$: indeed if $g = hsk$, with $h \in H$, $k \in K$, and $s \in \mathcal{S}$, we have

$$F(g) = F(hsk) = \sigma(k^{-1})F(s)v(h^{-1}) = \mathcal{L}_{F(s)}(g).$$

Moreover, this vanishes on the cosets HsK with $s \notin \mathcal{S}_0$. As a consequence, $F = \sum_{s \in \mathcal{S}_0} \mathcal{L}_{F(s)}$ and the nontrivial elements in this sum are linearly independent because they are supported on different double cosets.

(iv) Surjectivity of the map follows from (11.33). Indeed, T is the image of $\mathcal{L}_T$. Injectivity is a consequence of (iii). $\qquad \square$

For $F \in \mathcal{V}$ define the operator $\xi(F) \in \mathrm{Hom}(\mathrm{Ind}^G_H U, \mathrm{Ind}^G_K V)$ by setting

$$[\xi(F)f](g) = \sum_{r \in G} F(r^{-1}g)f(r), \qquad (11.35)$$

for all $f \in \mathrm{Ind}^G_H U$ and $g \in G$. It is then immediate to check that $\xi(F)f \in \mathrm{Ind}^G_K V$.

Also, for $T \in \text{Hom}_G(\text{Ind}_H^G U, \text{Ind}_K^G V)$ define the map $F_T : G \to \text{Hom}(U, V)$ by setting

$$F_T(g)u = \frac{1}{|H|}[T f_u](g) \tag{11.36}$$

for all $u \in U$ and $g \in G$, where f_u is as in (11.5) (but with K, V now replaced by H, U, respectively).

Theorem 11.4.3 *We have* $\xi(F) \in \text{Hom}_G(\text{Ind}_H^G U, \text{Ind}_K^G V)$ *for all* $F \in \mathcal{V}$ *and the map*

$$\xi : \mathcal{V} \longrightarrow \text{Hom}_G(\text{Ind}_H^G U, \text{Ind}_K^G V)$$

is an isomorphism of vector spaces. The corresponding inverse map is given by $T \mapsto F_T$.

Proof. Let $F \in \mathcal{V}$, $f \in \text{Ind}_H^G U$ and $g_0, g \in G$. Then we have

$$
\begin{aligned}
[\lambda(g)\xi(F)f](g_0) &= [\xi(F)f](g^{-1}g_0) \\
&= \sum_{r \in G} F(r^{-1}g^{-1}g_0)f(r) \\
(\text{setting } q = gr) \quad &= \sum_{q \in G} F(q^{-1}g_0)f(g^{-1}q) \\
&= \sum_{q \in G} F(q^{-1}g_0)[\lambda_1(g)f](q) \\
&= [\xi(F)\lambda_1(g)f](g_0),
\end{aligned}
$$

that is, $\lambda(g)\xi(F) = \xi(F)\lambda_1(g)$. This shows that $\xi(F) \in \text{Hom}_G(\text{Ind}_H^G U, \text{Ind}_K^G V)$.

Let now $h \in H, k \in K, g \in G, u \in U$ and $T \in \text{Hom}_G(\text{Ind}_H^G U, \text{Ind}_K^G V)$. Then we have

$$
\begin{aligned}
F_T(hgk)u &= \frac{1}{|H|}[T f_u](hgk) \\
(T \in \text{Hom}_G(\text{Ind}_H^G U, \text{Ind}_K^G V)) \quad &= \sigma(k^{-1})\left\{ \frac{1}{|H|}[T \lambda_1(h^{-1})f_u](g) \right\} \\
(\text{by (11.6)}) \quad &= \sigma(k^{-1})\left\{ \frac{1}{|H|}[T f_{\nu(h^{-1})u}](g) \right\} \\
&= \sigma(k^{-1})F_T(g)\nu(h^{-1})u.
\end{aligned}
$$

This shows that $F_T \in \mathcal{V}$.

We now prove that ξ is a bijection. Let $T \in \text{Hom}_G(\text{Ind}_H^G U, \text{Ind}_K^G V)$ and $F \in \mathcal{V}$. Since the functions $\lambda_1(g)f_u$, $g \in G$ and $u \in U$, span $\text{Ind}_H^G U$ (cf. Proposition

11.1.2), we have that $\xi(F) = T$ if and only if

$$\xi(F)\lambda_1(g)f_u = T\lambda_1(g)f_u \qquad (11.37)$$

for all $g \in G$ and $u \in U$.

We have

$$
\begin{aligned}
[T\lambda_1(g)f_u](g_0) &= [\lambda(g)Tf_u](g_0) \qquad (T \in \mathrm{Hom}_G(\mathrm{Ind}_H^G U, \mathrm{Ind}_K^G V)) \\
&= [Tf_u](g^{-1}g_0) \\
&= |H|F_T(g^{-1}g_0)u \qquad\qquad \text{(by (11.36))}
\end{aligned}
$$

and

$$
\begin{aligned}
[\xi(F)\lambda_1(g)f_u](g_0) &= \sum_{r \in G} F(r^{-1}g_0)f_u(g^{-1}r) \qquad\qquad \text{(by (11.35))} \\
&= \sum_{h \in H} F(h^{-1}g^{-1}g_0)v(h^{-1})u \quad \text{(by (11.5) with } g^{-1}r = h) \\
&= |H| \cdot F(g^{-1}g_0)u. \qquad\qquad \text{(by Definition 11.4.1)}
\end{aligned}
$$

for all $u \in U$, $g, g_0 \in G$. From (11.37) we then deduce that $\xi(F) = T$ if and only if $F = F_T$. $\qquad \square$

From Lemma 11.4.2.(iv) and Theorem 11.4.3 we deduce the following:

Corollary 11.4.4 (Mackey's formula for invariants) *The map*

$$
\begin{array}{ccc}
\mathrm{Hom}_G(\mathrm{Ind}_H^G v, \mathrm{Ind}_K^G \sigma) & \longrightarrow & \bigoplus_{s \in \mathcal{S}_0} \mathrm{Hom}_{G_s}(\mathrm{Res}_{G_s}^H v, \sigma_s) \\
T & \longmapsto & \bigoplus_{s \in \mathcal{S}_0} F_T(s),
\end{array} \qquad (11.38)
$$

is an isomorphism of vector spaces.

Proof. This map is nothing but the composition of the isomorphisms ξ^{-1} and (11.34). $\qquad \square$

By taking dimensions we deduce:

Corollary 11.4.5 (Mackey's intertwining number theorem)

$$\mathrm{dimHom}_G(\mathrm{Ind}_H^G v, \mathrm{Ind}_K^G \sigma) = \sum_{s \in \mathcal{S}} \mathrm{dimHom}_{G_s}(\mathrm{Res}_{G_s}^H v, \sigma_s)$$

Note that in the above sum the only contribution to the right hand side comes from the elements $s \in \mathcal{S}_0$. The following is one of the most useful results in Mackey's theory.

Corollary 11.4.6 (Mackey's irreducibility criterion) *Suppose $H = K$ and $v = \sigma$. Then $\mathrm{Ind}_K^G \sigma$ is irreducible if and only if the following conditions are both met:*

(a) (σ, V) is irreducible;

(b) for every $s \in \mathcal{S} \setminus \{1_G\}$, the G_s-representations $\mathrm{Res}^K_{G_s} \sigma$ and σ_s contain no common irreducible subrepresentations.

Proof. First of all, note that $G_{1_G} = K$ and $\sigma_{1_G} = \sigma$, so that Mackey's intertwining number theorem (Corollary 11.4.5) yields

$$\dim\mathrm{Hom}_G(\mathrm{Ind}^G_K\sigma, \mathrm{Ind}^G_K\sigma) = \dim\mathrm{Hom}_K(\sigma, \sigma) + \sum_{s\in\mathcal{S}\setminus\{1_G\}}\dim\mathrm{Hom}_{G_s}(\mathrm{Res}^K_{G_s}\sigma, \sigma_s).$$

We conclude by recalling that from Corollary 10.6.4 it follows that $\mathrm{Ind}^G_K\sigma$ is irreducible if and only if $\dim\mathrm{Hom}_G(\mathrm{Ind}^G_K\sigma, \mathrm{Ind}^G_K\sigma) = 1$ and then invoking Corollary 10.2.6 (see also Problem 10.6.9). □

Remark 11.4.7 Now we explain the terminology for "invariant" in Corollary 11.4.4. If (θ, W) is a G-representation, its invariant subspace is $\{w \in W : \theta(g)w = w, \ \forall g \in G\}$, that is, the isotypic component of the trivial representation ι_G in θ. If (ξ, Z) is another representation of G, then, defining a G-representation $(\eta, \mathrm{Hom}(W, Z))$ by setting

$$\eta(g)T = \xi(g)T\theta(g^{-1}),$$

for all $g \in G$ and $T \in \mathrm{Hom}(W, Z)$, we have that $\mathrm{Hom}_G(W, Z)$ is exactly the invariant subspace of η.

Exercise 11.4.8 Show that, for $H = G$ and $(\nu, U) = (\theta, W)$, Mackey's formula for invariants (11.38) reduces to Frobenius reciprocity (Theorem 11.2.1). More precisely, show that the maps (11.25) and (11.26) and their properties may be deduced from (11.33), (11.35), and (11.36). Examine the connections between the case $K = G$ and the other side of Frobenius reciprocity in Exercise 11.2.7.

Exercise 11.4.9 Deduce Lemma 10.4.14 from Corollary 11.4.5, taking into account Remark 11.3.2.

Remark 11.4.10 We now examine the case in which $\sigma = \chi$ and $\nu = \psi$ are one-dimensional (see Example 11.1.9). We have $U = V = \mathbb{C}$ and $\mathcal{S}_0 = \{s \in \mathcal{S} : \mathrm{Res}^H_{G_s}\psi = \chi_s\}$. Moreover, in the map (11.38), we have $F_T(s) = \frac{1}{|H|}[T\overline{\psi}](s) \in \mathbb{C}$, and the intertwining number theorem (Corollary 11.4.5) is just the formula

$$\dim\mathrm{Hom}_G(\mathrm{Ind}^G_H\psi, \mathrm{Ind}^G_K\chi) = |\mathcal{S}_0|.$$

Finally, in the case $H = K$ and $\psi = \chi$, the representation $\mathrm{Ind}^G_K\chi$ is irreducible if and only if $\mathrm{Res}^K_{G_s}\chi \neq \chi_s$ for all $s \in \mathcal{S} \setminus \{1_G\}$ (equivalently, $\mathcal{S}_0 = \{1_G\}$).

Exercise 11.4.11 Suppose that $H = K$ and $v = \sigma$. Define a multiplication operation in $\mathcal{V} = \mathcal{V}(G, K, K, \sigma, \sigma)$ (cf. Definition 11.4.1) by setting $[F_1 * F_2](g) = \sum_{g_1 \in G} F_1(g_1^{-1}g)F_2(g_1)$ for all $F_1, F_2 \in \mathcal{V}$ and $g \in G$. Also define the map $F \mapsto F^*$ by setting $F^*(f) = [F(g^{-1})]^*$, for all $F \in \mathcal{V}, g \in G$.

(1) Show that $\mathcal{V}$ is an involutive algebra.
(2) Show that if $\xi : \mathcal{V} \to \operatorname{Hom}_G(\operatorname{Ind}_K^G \sigma, \operatorname{Ind}_K^G \sigma)$ is as in (11.35), then we gave $\xi(F_1 * F_2) = \xi(F_1)\xi(F_2)$ and $\xi(F^*) = \xi(F)^*$. Taking into account Theorem 11.4.3, deduce that ξ is a $*$-isomorphism.
(3) With the notation in (10.49) and (11.5), show that $[\xi(F)\lambda(t)f_v](g) = |K| \cdot F(t^{-1}g)v$, for all $F \in \mathcal{V}, v \in V, t \in \mathcal{T}$ and $g \in G$.
(4) Deduce that $\operatorname{Tr}[\xi(F)] = |G| \cdot \operatorname{Tr}[F(1_G)]$.

Exercise 11.4.12 Let $\xi : \mathcal{V}(G, H, K, v, \sigma) \to \operatorname{Hom}_G(\operatorname{Ind}_H^G v, \operatorname{Ind}_K^G \sigma)$ and $\widetilde{\xi} : \mathcal{V}(G, H, H, v, v) \to \operatorname{Hom}_G(\operatorname{Ind}_H^G v, \operatorname{Ind}_H^G v)$ be as in (11.35).

(1) Let $F_1, F_2 \in \mathcal{V}(G, H, K, v, \sigma)$ and define $F : G \to \operatorname{Hom}(\operatorname{Ind}_H^G v, \operatorname{Ind}_H^G v)$ by setting

$$F(g) = \frac{|H|}{|K|} \sum_{g_1 \in G} [F_2(g^{-1}g_1)]^* F_1(g_1),$$

for all $g \in G$. Show that $F \in \mathcal{V}(G, H, H, v, v)$ and $\xi(F_2)^*\xi(F_1) = \widetilde{\xi}(F)$.
(2) Given two finite-dimensional vector spaces $\widetilde{U}$ and $\widetilde{V}$ and $T_1, T_2 \in \operatorname{Hom}(\widetilde{U}, \widetilde{V})$, set

$$\langle T_1, T_2 \rangle_{\operatorname{Hom}(\widetilde{U}, \widetilde{V})} = \operatorname{Tr}(T_2^* T_1).$$

Taking into account Exercise 11.4.11, deduce that

$$\langle \xi(F_1), \xi(F_2) \rangle_{\operatorname{Hom}(\operatorname{Ind}_H^G U, \operatorname{Ind}_K^G V)} = \frac{|H|^2}{|K|} \sum_{g \in G} \langle F_1(g), F_2(g) \rangle_{\operatorname{Hom}(U,V)}$$

$$\equiv |H|^3 \sum_{s \in \mathcal{S}} \frac{1}{|G_s|} \langle F_1(s), F_2(s) \rangle_{\operatorname{Hom}(U,V)}.$$

11.5 Mackey's lemma

In Corollary 11.1.17 we have examined the composition $\operatorname{Ind} \circ \operatorname{Res}$. The following famous lemma, due to Mackey, considers the inverse composition, namely $\operatorname{Res} \circ \operatorname{Ind}$. It essentially constitutes a representation theoretic analogue of the decomposition (11.29).

We assume the notation from Section 11.3.

Theorem 11.5.1 (Mackey's lemma) *The map*

$$
\begin{aligned}
\operatorname{Res}_H^G \operatorname{Ind}_K^G V &\longrightarrow \bigoplus_{s \in \mathcal{S}} \operatorname{Ind}_{G_s}^H V_s \\
F &\longmapsto \bigoplus_{s \in \mathcal{S}} f_s,
\end{aligned}
\tag{11.39}
$$

where $f_s \in \operatorname{Ind}_{G_s}^H V_s$ is defined by setting $f_s(h) = F(hs)$ for all $h \in H$, is an isomorphism of vector spaces. Moreover, the subspace Z_s of $\operatorname{Res}_H^G \operatorname{Ind}_K^G V$ isomorphic to $\operatorname{Ind}_{G_s}^H V_s$ is given by

$$
Z_s = \{F \in V[G] : F(hs'k) = \delta_{s,s'} \sigma(k^{-1}) F(hs), \ \forall h \in H, k \in K \text{ and } s' \in \mathcal{S}\},
$$

that is, it is made up of all functions in $\operatorname{Ind}_K^G V$ that vanish outside HsK.

Proof. By definition of $\operatorname{Ind}_K^G V$ and Z_s, it is clear that

$$
\operatorname{Ind}_K^G V = \bigoplus_{s \in \mathcal{S}} Z_s.
\tag{11.40}
$$

Suppose that $F \in Z_s$ and $f_s : H \to V$ is as in the statement. Then, if $x \in G_s$ we have

$$
f_s(hx) = F(hxs) = F(hss^{-1}xs) = \sigma(s^{-1}x^{-1}s)F(hs) = \sigma_s(x^{-1})f_s(h)
$$

so that $f_s \in \operatorname{Ind}_{G_s}^H V_s$. Vice versa, given $f \in \operatorname{Ind}_{G_s}^H V_s$ consider the map $F_s : G \to V_s$ defined by $F_s(hs'k) = \delta_{s,s'}\sigma(k^{-1})f(h)$ for $k \in K$, $h \in H$ and $s' \in \mathcal{S}$. We claim that F_s is well defined: indeed if $hsk = h_1 s k_1$, by Lemma 11.3.1 we have $h_1 = hx$ and $k_1 = s^{-1}x^{-1}sk$ with $x \in G_s$, so that

$$
\begin{aligned}
\sigma(k_1^{-1})f(h_1) &= \sigma(k^{-1})[\sigma(s^{-1}xs)f(h_1)] \\
&= \sigma(k^{-1})[\sigma_s(x)f(h_1)] \\
&= \sigma(k^{-1})f(h_1 x^{-1}) \\
&= \sigma(k^{-1})f(h).
\end{aligned}
$$

Moreover,

$$
F_s(hs'k) = \delta_{s,s'}\sigma(k^{-1})f(h) = \sigma(k^{-1})F_s(hs),
$$

so that $F_s \in Z_s$. This shows that the map $F \mapsto f_s$ is an isomorphism between Z_s and $\operatorname{Ind}_{G_s}^H V_s$; since H acts on both spaces by left translation, we deduce that this map is also an intertwiner. Recalling (11.40), this ends the proof. $\square$

Exercise 11.5.2 Show that the isomorphism in Corollary 11.4.4 may be deduced from the isomorphism in Exercise 11.2.7.(3), from Mackey's lemma (Theorem 11.5.1), and Frobenius reciprocity (Theorem 11.2.1). Deduce also the explicit form of the isomorphism (11.34).

Theorem 11.5.3 (Mackey's tensor product theorem)

$$\mathrm{Ind}_H^G \nu \otimes \mathrm{Ind}_K^G \sigma \sim \bigoplus_{s \in S} \mathrm{Ind}_{G_s}^G \left[\mathrm{Res}_{G_s}^H \nu \otimes \sigma_s \right].$$

Proof. We have:

$$\mathrm{Ind}_H^G \nu \otimes \mathrm{Ind}_K^G \sigma \sim \mathrm{Ind}_H^G \left[\nu \otimes \mathrm{Res}_H^G (\mathrm{Ind}_K^G \sigma) \right] \text{ (by Theorem 11.1.16)}$$

$$\sim \mathrm{Ind}_H^G \left[\nu \otimes \left(\bigoplus_{s \in S} \mathrm{Ind}_{G_s}^H \sigma_s \right) \right] \text{ (by Mackey's lemma)}$$

$$\sim \mathrm{Ind}_H^G \left\{ \bigoplus_{s \in S} \mathrm{Ind}_{G_s}^H \left[\mathrm{Res}_{G_s}^H \nu \otimes \sigma_s \right] \right\} \text{ (by Theorem 11.1.16)}$$

$$\sim \bigoplus_{s \in S} \mathrm{Ind}_{G_s}^G \left[\mathrm{Res}_{G_s}^H \nu \otimes \sigma_s \right],$$

where the last equivalence follows from Proposition 11.1.5 and Proposition 11.1.18. $\qquad\square$

11.6 The Mackey-Wigner little group method

In this section we present a powerful method to construct irreducible representations (sometimes exhausting the whole dual) for a class of finite groups. We actually examine a particular case that will suffice for our subsequent purposes. For a more general treatment, we refer to our monograph [34] (see also [31]).

Let G be a finite group and suppose that $A \leq G$ is an *Abelian normal* subgroup. We assume the notation in Section 2.3.

There is a natural action of G on the dual of A: if $\chi \in \widehat{A}$ and $g \in G$ we define the *g-conjugate* ${}^g\chi \in \widehat{A}$ of χ by setting

$$ {}^g\chi(a) = \chi(g^{-1}ag) \tag{11.41}$$

for all $a \in A$. It is easy to check that ${}^{g_1}({}^{g_2}\chi) = {}^{g_1 g_2}\chi$ for all $g_1, g_2 \in G$ and that ${}^{1_G}\chi = \chi$, so that G-conjugation is indeed an action on $\widehat{A}$. The stabilizer of an element $\chi \in \widehat{A}$ is the subgroup

$$K_\chi = \mathrm{Stab}_G(\chi) = \{ g \in G : {}^g\chi = \chi \},$$

which is called the *inertia group* of χ. Note that $A \leq K_\chi$ since A is Abelian. We say that $\chi \in \widehat{A}$ has an *extension* to K_χ if there exists a one-dimensional representation $\widetilde{\chi}$ of K_χ such that $\widetilde{\chi}(a) = \chi(a)$ for all $a \in A$, that is, $\mathrm{Res}_A^{K_\chi} \widetilde{\chi} = \chi$. Now consider the quotient group K_χ / A. Given $\psi \in \widehat{K_\chi / A}$ we define its *inflation* to K_χ as the irreducible representation $\overline{\psi}$ of K_χ given by setting $\overline{\psi}(h) = \psi(hA)$

for all $h \in K_\chi$ (compare with (11.15)). Clearly, this is just the composition of the canonical homomorphism $K_\chi \to K_\chi/A$ with $\psi \colon K_\chi/A \to \mathrm{GL}(V_\psi)$, where V_ψ is the representation space of ψ.

Theorem 11.6.1 *Let* $\chi \in \widehat{A}$ *and suppose that* χ *has an extension* $\widetilde{\chi}$ *to* K_χ. *Then*

$$\mathrm{Ind}_A^{K_\chi} \chi = \bigoplus_{\psi \in \widehat{K_\chi/A}} d_\psi (\widetilde{\chi} \otimes \overline{\psi}), \tag{11.42}$$

where, as usual, d_ψ *denotes the dimension of* $\psi \in \widehat{K_\chi/A}$. *Moreover, the G-representations*

$$\mathrm{Ind}_{K_\chi}^G (\widetilde{\chi} \otimes \overline{\psi}), \qquad \psi \in \widehat{K_\chi/A}, \tag{11.43}$$

are irreducible and pairwise inequivalent.

Proof. From (11.23) we deduce that

$$\mathrm{Ind}_A^{K_\chi} \chi = \mathrm{Ind}_A^{K_\chi} (\chi \otimes \iota_A) = \mathrm{Ind}_A^{K_\chi} \left[\left(\mathrm{Res}_A^{K_\chi} \widetilde{\chi} \right) \otimes \iota_A \right] = \widetilde{\chi} \otimes \mathrm{Ind}_A^{K_\chi} \iota_A = \widetilde{\chi} \otimes \overline{\lambda},$$

where ι_A denotes the trivial representation of A and $\overline{\lambda}$ is the inflation of the regular representation λ of K_χ/A (cf. Example 11.1.8). Since $\lambda = \oplus_{\psi \in \widehat{K_\chi/A}} d_\psi \psi$, we have $\overline{\lambda} = \oplus_{\psi \in \widehat{K_\chi/A}} d_\psi \overline{\psi}$, from which (11.42) immediately follows.

Now suppose that $\mathcal{S}$ is a complete set of representatives for the double K_χ cosets in G (with $1_G \in \mathcal{S}$) and, as in (11.30) and (11.31) (with $H = K = K_\chi$), set $G_s = K_\chi \cap s K_\chi s^{-1}$ and

$$(\widetilde{\chi} \otimes \overline{\psi})_s(x) = (\widetilde{\chi} \otimes \overline{\psi})(s^{-1} x s),$$

for all $x \in G_s$ and $s \in \mathcal{S}$. Since $s^{-1} a s \in A$ for all $a \in A$, we have $\psi(s^{-1} a s A) = \psi(A)$, and therefore

$$(\widetilde{\chi} \otimes \overline{\psi})_s(a) = {}^s\chi(a)\psi(A)$$

for all $a \in A$, so that (recalling Proposition 10.2.15.(i))

$$\mathrm{Res}_A^{G_s} (\widetilde{\chi} \otimes \overline{\psi})_s \sim d_\psi {}^s\chi.$$

In particular, for $s \neq 1_G$ the G_s-representations $\mathrm{Res}_{G_s}^{K_\chi} (\widetilde{\chi} \otimes \overline{\psi})$ and $(\widetilde{\chi} \otimes \overline{\psi})_s$ cannot have common irreducible subrepresentations because these would lead to common subrepresentations between their restrictions to A, but ${}^s\chi \neq \chi$ because $s \notin K_\chi$. From Corollary 11.4.6 we deduce that $\mathrm{Ind}_{K_\chi}^G (\widetilde{\chi} \otimes \overline{\psi})$ is irreducible.

Finally, denote now by μ the representation $\mathrm{Ind}_{K_\chi}^G (\widetilde{\chi} \otimes \overline{\psi})$ and by Z its representation space. If $f \in Z$ and $a \in A$ then, for all $g \in G$, we have:

$$[\mu(a)f](g) = f(a^{-1}g) = f(g \cdot g^{-1}a^{-1}g) = (\widetilde{\chi} \otimes \overline{\psi})(g^{-1}ag)f(g) = {}^g\chi(a)f(g).$$

It follows that, in the notation as in Theorem 11.5.1 (with $\nu = \sigma = \widetilde{\chi} \otimes \overline{\psi}$) we have: $Z_{1_G} = \{f \in Z : \mu(a)f = \chi(a)f, \forall a \in A\}$. Indeed, Z_{1_G} is the space of all $f \in Z$ supported on K_χ. Moreover, in the decomposition

$$\mathrm{Res}_{K_\chi}^G \mathrm{Ind}_{K_\chi}^G (\widetilde{\chi} \otimes \overline{\psi}) \cong \bigoplus_{s \in \mathcal{S}} \mathrm{Ind}_{G_s}^{K_\chi} (\widetilde{\chi} \otimes \overline{\psi})_s,$$

Z_{1_G} is the representation space of $\widetilde{\chi} \otimes \overline{\psi}$ (because $G_{1_G} = K_\chi$). This means that the action of G on the χ-isotypic component of $\mathrm{Res}_A^G \mathrm{Ind}_{K_\chi}^G (\widetilde{\chi} \otimes \overline{\psi})$ corresponds exactly to $\widetilde{\chi} \otimes \overline{\psi}$, and this implies that the representations in (11.43) are pairwise inequivalent, because different representations come from different ψs. In other words, $\mathrm{Ind}_{K_\chi}^G (\widetilde{\chi} \otimes \overline{\psi})$ uniquely determines ψ. $\qquad\square$

Theorem 11.6.2 (The little group method) *Suppose that every $\chi \in \widehat{A}$ has an extension $\widetilde{\chi}$ to its inertia group K_χ. Define on $\widehat{A}$ an equivalence relation $\approx$ by setting $\chi_1 \approx \chi_2$ if there exists $g \in G$ such that ${}^g\chi_1 = \chi_2$. Let X be a complete set of representatives of the corresponding quotient space $\widehat{A}/\approx$. Then*

$$\widehat{G} = \left\{ \mathrm{Ind}_{K_\chi}^G (\widetilde{\chi} \otimes \overline{\psi}) : \chi \in X, \psi \in \widehat{K_\chi/A} \right\}. \tag{11.44}$$

More precisely, the right hand side in (11.44) is a complete list of all irreducible G-representations and, for different values of χ and ψ, the corresponding representations are inequivalent.

Proof. From Theorem 11.6.1 it follows that the representations in the list are irreducible. Moreover, from (11.42) and transitivity of induction (cf. Proposition 11.1.5), for any $\chi \in X$ we deduce that

$$\mathrm{Ind}_A^G \chi \cong \bigoplus_{\psi \in \widehat{K_\chi/A}} d_\psi \mathrm{Ind}_{K_\chi}^G (\widetilde{\chi} \otimes \overline{\psi}). \tag{11.45}$$

Suppose that $\mathcal{T}$ is a complete set of left (in this case, also right and double) cosets of A if G. Set $\lambda = \mathrm{Ind}_A^G \chi$ and denote by $\mathrm{Ind}_A^G \mathbb{C}$ the corresponding representation space (cf. Example 11.1.9). For $t \in \mathcal{T}$ and $g \in G$, we have $[\lambda(t)\overline{\chi}](g) \neq 0$ only if $g = a_t t \in At = tA$ and

$$[\lambda(a)\lambda(t)\overline{\chi}](g) = \overline{\chi}(t^{-1}a^{-1}g) = \overline{\chi}(t^{-1}g \cdot g^{-1}a^{-1}g)$$
$$= {}^g\chi(a)\overline{\chi}(t^{-1}g) = {}^t\chi(a)[\lambda(t)\overline{\chi}](g).$$

Thus,

$$\lambda(a)\,[\lambda(t)\overline{\chi}] = {}^t\chi(a)\,[\lambda(t)\overline{\chi}]\,,$$

and (11.17) now implies that

$$\mathrm{Res}_A^G \mathrm{Ind}_A^G \chi \sim \bigoplus_{t \in T} {}^t\chi, \tag{11.46}$$

which is clearly a particular case of (11.39). It follows that if $\chi_1, \chi_2 \in X$ are distinct, then two irreducible representations of the form $\mathrm{Ind}_{K_{\chi_1}}^G (\widetilde{\chi_1} \otimes \overline{\psi_1})$ and $\mathrm{Ind}_{K_{\chi_2}}^G (\widetilde{\chi_2} \otimes \overline{\psi_2})$ as in (11.44) cannot be equivalent because, by virtue of (11.45) and (11.46), their restrictions to A contain inequivalent representations (the G-conjugates of χ_1 and χ_2, respectively). The inequivalence of two representations of the form $\mathrm{Ind}_{K_\chi}^G (\widetilde{\chi} \otimes \overline{\psi_1})$ and $\mathrm{Ind}_{K_\chi}^G (\widetilde{\chi} \otimes \overline{\psi_2})$, with the same χ but $\psi_1 \neq \psi_2$, has been already proved in Theorem 11.6.1.

Now suppose that (θ, W) is a G-irreducible representation. Then $\mathrm{Res}_A^G \theta$ decomposes into the direct sum of characters of A. If $\xi \in \widehat{A}$ is contained in $\mathrm{Res}_A^G \theta$ then there exists $w \in W$, $w \neq 0$, such that $\theta(a)w = \xi(a)w$. For any $g \in G$ we have:

$$\theta(a)[\theta(g)w] = \theta(g \cdot g^{-1}ag)w = \theta(g)\theta(g^{-1}ag)w$$
$$= \xi(g^{-1}ag)\theta(g)w = {}^g\xi(a)[\theta(g)w],$$

that is, $\mathrm{Res}_A^G \theta$ contains all the g-conjugates of ξ and, in particular, an element $\chi \in X$. By Frobenius reciprocity, θ is contained in $\mathrm{Ind}_A^G \chi$. Keeping in mind (11.45), this implies that θ equals one of the representations in (11.44). $\quad\square$

11.7 Semidirect products with an Abelian group

In this section we apply the little group method to an important class of semidirect products (cf. Section 8.14), namely we suppose that the normal subgroup is Abelian.

Theorem 11.7.1 *Let G be a finite group and suppose that $G = A \rtimes H$ with A an Abelian (normal) subgroup. Given $\chi \in \widehat{A}$, its inertia group K_χ coincides with $A \rtimes H_\chi$, where $H_\chi = \mathrm{Stab}_H(\chi) = \{h \in H : {}^h\chi = \chi\}$. Moreover, any $\chi \in \widehat{A}$ may be extended to a one-dimensional representation $\widetilde{\chi} \in \widehat{A \rtimes H_\chi}$ by setting*

$$\widetilde{\chi}(ah) = \chi(a) \qquad \forall a \in A, \; h \in H_\chi. \tag{11.47}$$

Finally, with the notation used in Theorem 11.6.2, we have:

$$\widehat{G} = \{\mathrm{Ind}_{A \rtimes H_\chi}^G (\widetilde{\chi} \otimes \overline{\psi}) : \chi \in X, \psi \in \widehat{H_\chi}\}.$$

Proof. For $a, a_1 \in A$ and $h \in H$ we have

$$^{ah}\chi(a_1) = \chi(h^{-1}a^{-1}a_1 ah) = \chi(h^{-1}a^{-1}h)\chi(h^{-1}a_1 h)\chi(h^{-1}ah)$$
$$= \chi(h^{-1}a_1 h) = {}^h\chi(a_1)$$

thus showing that the inertia subgroup of χ coincides with $A \rtimes H_\chi$. Let $\chi \in \widehat{A}$ and let us show that the extension of χ defined by (11.47) is a representation. By definition of H_χ, we have that χ is invariant by conjugation with elements in H_χ so that, if $a_1, a_2 \in A$ and $h_1, h_2 \in H_\chi$, we have

$$\widetilde{\chi}(a_1 h_1 \cdot a_2 h_2) = \widetilde{\chi}(a_1 h_1 a_2 h_1^{-1} \cdot h_1 h_2) = \chi(a_1 h_1 a_2 h_1^{-1})$$
$$= \chi(a_1)\chi(a_2) = \widetilde{\chi}(a_1 h_1)\widetilde{\chi}(a_2 h_2).$$

Finally, the last statement is just an application of Theorem 11.6.2.   □

12

Fourier analysis on finite affine groups
and finite Heisenberg groups

In this chapter we study the representation theory of two finite matrix groups, the affine group (or $ax + b$ group) and the Heisenberg group, with entries in a finite field or in the finite ring $\mathbb{Z}/n\mathbb{Z}$.

We consider specific problems of Harmonic Analysis: our main results (taken from [15]), consist in a revisitation of the Discrete Fourier Transform and of the Fast Fourier Transform from the point of view of the representation theory of the Heisenberg group. Other sources are the monograph by Terras [159], our book on the representation theory of wreath products of finite groups [34], and [142]. The results of Section 12.1 will play a fundamental role in Chapter 14.

We closely follow Notation 1.1.17, that is, we use $\mathbb{Z}_n$ when we want to emphasize that our arguments are based *only* on the structure of the additive Abelian group of the integers mod n, while we use $\mathbb{Z}/n\mathbb{Z}$ when the whole structure of a finite ring is used, that is, multiplication enters the picture. We think that this distinction is important in view of possible generalizations of some of our arguments, for instance to more general Abelian (or even noncommutative) groups, and to other rings.

12.1 Representation theory of the affine group $\mathrm{Aff}(\mathbb{F}_q)$

Let q be a power of a prime number and denote by $\mathbb{F}_q$ the field with q elements (as in Chapter 6). Recall, cf. Example 10.4.5, that the (*general*) *affine group* (*of degree one*) over $\mathbb{F}_q$ is the subgroup $\mathrm{Aff}(\mathbb{F}_q)$ of $\mathrm{GL}(2, \mathbb{F}_q)$ defined by

$$\mathrm{Aff}(\mathbb{F}_q) = \left\{ \begin{pmatrix} a & b \\ 0 & 1 \end{pmatrix} : a \in \mathbb{F}_q^*,\ b \in \mathbb{F}_q \right\}.$$

Note that Aff($\mathbb{F}_q$) acts doubly transitively (cf. Exercise 10.4.16.(5)) on $\mathbb{F}_q \equiv \left\{ \begin{pmatrix} x \\ 1 \end{pmatrix} : x \in \mathbb{F}_q \right\}$ by multiplication:

$$\begin{pmatrix} a & b \\ 0 & 1 \end{pmatrix} \begin{pmatrix} x \\ 1 \end{pmatrix} = \begin{pmatrix} ax + b \\ 1 \end{pmatrix}. \tag{12.1}$$

We begin with some elementary algebraic properties and use the notion of a semidirect product of groups (cf. Definition 8.14.2). Consider the following Abelian subgroups of Aff($\mathbb{F}_q$):

$$A = \left\{ \begin{pmatrix} a & 0 \\ 0 & 1 \end{pmatrix} : a \in \mathbb{F}_q^* \right\} \cong \mathbb{F}_q^* \text{ and } U = \left\{ \begin{pmatrix} 1 & b \\ 0 & 1 \end{pmatrix} : b \in \mathbb{F}_q \right\} \cong \mathbb{F}_q. \tag{12.2}$$

Lemma 12.1.1

(i) *The inverse of* $\begin{pmatrix} a & b \\ 0 & 1 \end{pmatrix} \in$ Aff($\mathbb{F}_q$) *is* $\begin{pmatrix} a & b \\ 0 & 1 \end{pmatrix}^{-1} = \begin{pmatrix} a^{-1} & -a^{-1}b \\ 0 & 1 \end{pmatrix}$;

(ii) *the subgroup U is normal and one has*

$$\text{Aff}(\mathbb{F}_q) \cong U \rtimes A \equiv \mathbb{F}_q \rtimes \mathbb{F}_q^*; \tag{12.3}$$

(iii) *the conjugacy classes of the group* Aff($\mathbb{F}_q$) *are the following:*

- $C_0 = \left\{ \begin{pmatrix} 1 & 0 \\ 0 & 1 \end{pmatrix} \right\}$;

- $C_1 = \left\{ \begin{pmatrix} 1 & b \\ 0 & 1 \end{pmatrix} : b \in \mathbb{F}_q^* \right\}$;

- $C_a = \left\{ \begin{pmatrix} a & b \\ 0 & 1 \end{pmatrix} : b \in \mathbb{F}_q \right\}$, *where* $a \in \mathbb{F}_q^*$, $a \neq 1$.

Proof.

(i) This is a trivial calculation. From this, one easily deduces the identity

$$\begin{pmatrix} u & v \\ 0 & 1 \end{pmatrix} \begin{pmatrix} a & b \\ 0 & 1 \end{pmatrix} \begin{pmatrix} u & v \\ 0 & 1 \end{pmatrix}^{-1} = \begin{pmatrix} a & (1-a)v + bu \\ 0 & 1 \end{pmatrix} \tag{12.4}$$

for all $u, a \in \mathbb{F}_q^*$ and $v, b \in \mathbb{F}_q$.

(ii) The normality of U follows from (12.4), after taking $a = 1$. Since

$$\begin{pmatrix} a & b \\ 0 & 1 \end{pmatrix} = \begin{pmatrix} a & 0 \\ 0 & 1 \end{pmatrix} \begin{pmatrix} 1 & a^{-1}b \\ 0 & 1 \end{pmatrix}$$

for all $a \in \mathbb{F}_q^*$ and $b \in \mathbb{F}_q$, we deduce that Aff($\mathbb{F}_q$) $= AU$. Then (12.3) follows from the fact that $A \cap U = \left\{ \begin{pmatrix} 1 & 0 \\ 0 & 1 \end{pmatrix} \right\} = \{1_{\text{Aff}(\mathbb{F}_q)}\}$.

(iii) This is a case-by-case analysis by means of (12.4). $\qquad\square$

Since $\mathrm{Aff}(\mathbb{F}_q)$ is a semidirect product with an Abelian normal subgroup (cf. (12.3)), we can apply the little group method (Theorem 11.7.1) in order to get a complete list of all irreducible representations of $\mathrm{Aff}(\mathbb{F}_q)$. As usual, $\widehat{\mathbb{F}_q}$ (respectively $\widehat{\mathbb{F}_q^*}$) will denote the dual of the additive group $\mathbb{F}_q$ (respectively of the multiplicative group $\mathbb{F}_q^*$).

From Lemma 12.1.1.(ii) and (12.4), after identifying A with the multiplicative group $\mathbb{F}_q^*$ (via the map $\begin{pmatrix} a & 0 \\ 0 & 1 \end{pmatrix} \mapsto a$) and U with the additive group $\mathbb{F}_q$ (via the map $\begin{pmatrix} 1 & b \\ 0 & 1 \end{pmatrix} \mapsto b$), it follows that the conjugacy action (cf. (11.41)) of $A \equiv \mathbb{F}_q^*$ on $\widehat{U} \equiv \widehat{\mathbb{F}_q}$ is given by

$$^a\chi(b) = \chi(a^{-1}b) \tag{12.5}$$

for all $\chi \in \widehat{U}, b \in \mathbb{F}_q$, and $a \in \mathbb{F}_q^*$.

Denote by $\chi_0 \equiv 1$ the trivial character of U.

Lemma 12.1.2 *The action of A on $\widehat{U}$ has exactly two orbits, namely $\{\chi_0\}$ and $\widehat{\mathbb{F}_q} \setminus \{\chi_0\}$. Moreover, the stabilizer of $\chi \in \widehat{U}$ is given by*

$$\mathrm{Stab}_A(\chi) = \begin{cases} \{1_A\} & \text{if } \chi \neq \chi_0 \\ A & \text{if } \chi = \chi_0. \end{cases}$$

Proof. It is clear that χ_0 is a fixed point. From now on, let $\chi \in \widehat{U}$ be a nontrivial character. For $a \in \mathbb{F}_q$ let us set

$$^a\chi^* = \begin{cases} ^{a^{-1}}\chi & \text{if } a \in \mathbb{F}_q^* \\ \chi_0 & \text{if } a = 0, \end{cases}$$

that is, $^a\chi^*(x) = \chi(ax)$ for all $x \in \mathbb{F}_q$. We claim that the map $a \mapsto \,^a\chi^*$ yields an isomorphism from $\mathbb{F}_q$ onto $\widehat{\mathbb{F}_q}$. Indeed, it is straightforward to check that $^{(a+b)}\chi^*(x) = \,^a\chi^*(x)^b\chi^*(x)$ for all $a, b, x \in \mathbb{F}_q$. Moreover, if $a \neq 0$ we have $^a\chi^* \neq \chi_0$ since the map $x \mapsto ax$ is a bijection of $\mathbb{F}_q$. This shows that the homomorphism $a \mapsto \,^a\chi^*$ is injective. Since $|\mathbb{F}_q| = |\widehat{\mathbb{F}_q}|$, it is in fact bijective. As a consequence, we have that $\{^a\chi : a \neq 0\} = \{^a\chi^* : a \neq 0\}$ coincides with the set of all nontrivial characters. $\quad\square$

Theorem 12.1.3 *The group $\mathrm{Aff}(\mathbb{F}_q)$ has exactly $q - 1$ one-dimensional representations and one $(q - 1)$-dimensional irreducible representation. The first ones are obtained by associating with each $\psi \in \widehat{A}$ the group homomorphism*

Ψ: Aff($\mathbb{F}_q$) $\rightarrow$ $\mathbb{T}$ *defined by*

$$\Psi \begin{pmatrix} a & b \\ 0 & 1 \end{pmatrix} = \psi(a) \tag{12.6}$$

for all $\begin{pmatrix} a & b \\ 0 & 1 \end{pmatrix}$ $\in$ Aff($\mathbb{F}_q$). *The* $(q-1)$-*dimensional irreducible representation is given by*

$$\pi = \mathrm{Ind}_U^{\mathrm{Aff}(\mathbb{F}_q)} \chi, \tag{12.7}$$

where χ *is any nontrivial character of* U. *Moreover, the character* χ^π *of* π *is given by:*

$$\chi^\pi \begin{pmatrix} a & b \\ 0 & 1 \end{pmatrix} = \begin{cases} q-1 & \textit{if } a = 1 \textit{ and } b = 0 \\ -1 & \textit{if } a = 1 \textit{ and } b \neq 0 \\ 0 & \textit{otherwise.} \end{cases} \tag{12.8}$$

Proof. This is just an application of the little group method (Theorem 11.7.1). Indeed, by Lemma 12.1.2, the inertia group of the trivial character $\chi_0 \in \widehat{U}$ is Aff($\mathbb{F}_q$). This provides the $q-1$ one-dimensional representations simply by taking any character $\psi \in \widehat{A}$. Moreover, the inertia group of any nontrivial character $\chi \in \widehat{U}$ is U since, by Lemma 12.1.2, $\mathrm{Stab}_A(\chi) = \{1_A\}$.

Finally, from (11.18) with $\mathcal{T} = A$, and using again (12.5), we immediately get

$$\chi^\pi \begin{pmatrix} a & b \\ 0 & 1 \end{pmatrix} = \begin{cases} \sum_{\alpha \in \mathbb{F}_q^*} \chi(\alpha^{-1}b) & \text{if } a = 1 \\ 0 & \text{otherwise.} \end{cases}$$

Then (12.8) follows from Corollary 7.1.3. $\square$

We now give a concrete realization of π.

Proposition 12.1.4 *Fix* $\chi \in \widehat{\mathbb{F}_q} \setminus \{\chi_0\}$ *and set*

$$\left[\pi^\sharp \begin{pmatrix} a & b \\ 0 & 1 \end{pmatrix} f \right](x) = \chi(bx^{-1}) f(a^{-1}x), \tag{12.9}$$

for all $f \in L(\mathbb{F}_q^*)$, $\begin{pmatrix} a & b \\ 0 & 1 \end{pmatrix}$ $\in$ Aff($\mathbb{F}_q$) *and* $x \in \mathbb{F}_q^*$. *Then* $(\pi^\sharp, L(\mathbb{F}_q^*))$ *is a representation of* Aff($\mathbb{F}_q$) *and*

$$\pi^\sharp \sim \pi = \mathrm{Ind}_U^{\mathrm{Aff}(\mathbb{F}_q)} \chi.$$

Proof. From Definition 11.1.1 it follows that the representation space of π is

$$W = \left\{ \widetilde{f} \colon \mathrm{Aff}(\mathbb{F}_q) \to \mathbb{C} : \widetilde{f}(gu) = \overline{\chi(u)} \widetilde{f}(g), \forall g \in \mathrm{Aff}(\mathbb{F}_q), u \in U \right\}.$$

Then for $\tilde{f} \in W$ and $\begin{pmatrix} a & b \\ 0 & 1 \end{pmatrix} \in \mathrm{Aff}(\mathbb{F}_q)$ we have

$$\tilde{f}\begin{pmatrix} a & b \\ 0 & 1 \end{pmatrix} = \tilde{f}\left[\begin{pmatrix} a & 0 \\ 0 & 1 \end{pmatrix}\begin{pmatrix} 1 & ba^{-1} \\ 0 & 1 \end{pmatrix}\right] = \overline{\chi(ba^{-1})}\tilde{f}\begin{pmatrix} a & 0 \\ 0 & 1 \end{pmatrix}$$

so that the map $W \ni \tilde{f} \mapsto f \in L(\mathbb{F}_q^*)$, where $f(x) = \tilde{f}\begin{pmatrix} x & 0 \\ 0 & 1 \end{pmatrix}$ for all $x \in \mathbb{F}_q^*$, is a well defined isomorphism of vector spaces. Moreover,

$$\left[\pi\begin{pmatrix} a & b \\ 0 & 1 \end{pmatrix}\tilde{f}\right]\begin{pmatrix} x & 0 \\ 0 & 1 \end{pmatrix} = \tilde{f}\begin{pmatrix} a^{-1}x & -a^{-1}b \\ 0 & 1 \end{pmatrix}$$
$$= \overline{\chi(-bx^{-1})}f(a^{-1}x)$$
$$= \chi(bx^{-1})f(a^{-1}x). \qquad \square$$

Corollary 12.1.5

$$\mathrm{Res}_A^{\mathrm{Aff}(\mathbb{F}_q)} \pi \sim \bigoplus_{\psi \in \widehat{A}} \psi.$$

Proof. If $\psi \in \widehat{A} (\cong \widehat{\mathbb{F}_q^*})$, then $\psi \in L(\mathbb{F}_q^*)$ satisfies

$$\left[\pi\begin{pmatrix} a & 0 \\ 0 & 1 \end{pmatrix}\psi\right](x) = \psi(a^{-1}x) = \overline{\psi(a)}\psi(x)$$

for all $a, x \in \mathbb{F}_q^*$. $\qquad \square$

Exercise 12.1.6 Check that $\pi^\sharp$, defined by (12.9), is an irreducible representation of $\mathrm{Aff}(\mathbb{F}_q)$ without using the theory of induced representations.

Corollary 12.1.7

$$\mathrm{Res}_U^{\mathrm{Aff}(\mathbb{F}_q)} \pi = \bigoplus_{\chi \in \widehat{U} \setminus \{\chi_0\}} \chi.$$

Proof. Since $\pi = \mathrm{Ind}_U^{\mathrm{Aff}(\mathbb{F}_q)}\chi$ for any nontrivial character $\chi \in \widehat{U}$ and $\dim \pi = q - 1$ equals the cardinality of the set of all nontrivial characters of U, the statement follows from Frobenius reciprocity. $\qquad \square$

Exercise 12.1.8 Recalling the notation in (12.6) and (12.8), directly prove the following:

(1) $\mathrm{Res}_U^{\mathrm{Aff}(\mathbb{F}_q)}\Psi = \chi_0$ and $\mathrm{Res}_A^{\mathrm{Aff}(\mathbb{F}_q)}\Psi = \psi$.
(2) Deduce (by using Frobenius reciprocity and Corollary 12.1.5) that

$$\mathrm{Ind}_U^{\mathrm{Aff}(\mathbb{F}_q)}\chi_0 = \oplus_{\psi \in \widehat{A}}\Psi \qquad \text{and} \qquad \mathrm{Ind}_A^{\mathrm{Aff}(\mathbb{F}_q)}\psi = \pi \oplus \Psi.$$

(3) Show a connection between (12.8) and the character formula in Example 10.4.7, taking into account Exercise 10.4.16.

Exercise 12.1.9 Consider $\mathbb{F}_q$ as a subfield of $\mathbb{F}_{q^m}$, $m \geq 2$; see Section 6.6.

(1) Denote by π_q (resp. π_{q^m}) the $(q - 1)$-dimensional irreducible representation of $\mathbb{F}_q$ (resp. the $(q^m - 1)$-dimensional of $\mathbb{F}_{q^m}$). Prove that $\mathrm{Ind}_{\mathrm{Aff}(\mathbb{F}_q)}^{\mathrm{Aff}(\mathbb{F}_{q^m})} \pi_q = q^{m-1} \pi_{q^m}$.
Hint: the restrictions of the one-dimensional representations of Aff($\mathbb{F}_{q^m}$) cannot contain π_q.

(2) For $\xi \in \widehat{\mathbb{F}_{q^m}}$, set $\xi^\natural = \mathrm{Res}_{\mathbb{F}_q}^{\mathbb{F}_{q^m}} \xi$ and denote by Ξ the corresponding one-dimensional representation of Aff($\mathbb{F}_{q^m}$). With the notation in Theorem 12.1.3, prove that

$$\mathrm{Ind}_{\mathrm{Aff}(\mathbb{F}_q)}^{\mathrm{Aff}(\mathbb{F}_{q^m})} \Psi = \frac{q^{m-1} - 1}{q - 1} \pi_{q^m} \oplus \left(\bigoplus_{\substack{\xi \in \widehat{\mathbb{F}_{q^m}}: \\ \xi^\natural = \psi}} \Xi \right).$$

Hint: Examine $\mathrm{Res}_{\mathrm{Aff}(\mathbb{F}_q)}^{\mathrm{Aff}(\mathbb{F}_{q^m})} \Xi$.

See [140] for a detailed analysis of the commutant of $\mathrm{Ind}_{\mathrm{Aff}(\mathbb{F}_q)}^{\mathrm{Aff}(\mathbb{F}_{q^m})} \pi_q$.

We end this section with a brief treatment of the automorphism group of Aff($\mathbb{F}_q$). First, we recall some elementary facts of group theory; see the monographs by Robinson [129], Rotman [132], and Machì [103], for more details.

Let G be a finite group and denote by Aut(G) its automorphism group. With each $g \in G$ we associate the *inner automorphism* given by: $\xi_g(h) = ghg^{-1}$, for all $h \in G$. The inner automorphisms form a subgroup of Aut(G), denoted Inn(G). If $g \in G$ and $\alpha \in$ Aut(G) then $\alpha \circ \xi_g \circ \alpha^{-1} = \xi_{\alpha(g)}$; in particular, Inn($G$) is normal in Aut($G$).

A subgroup N is *characteristic* if it is invariant with respect to every automorphism of G: $\alpha(N) = N$ for all $\alpha \in$ Aut(G). Clearly, a subgroup is normal if and only if it is invariant with respect to every inner automorphism and therefore a characteristic subgroup is also normal. Two particular characteristic groups are: the *center* $Z(G) = \{g \in G : gh = hg$ for all $h \in G\}$ and the *derived subgroup* G', which is the subgroup generated by all *commutators*, namely, the elements of the form $ghg^{-1}h^{-1}$, $g, h \in G$. Recall that if N is normal in G then the quotient group G/N is Abelian if and only if $G' \leq N$, and that, if $G' \leq H \leq G$, then H is normal in G. Finally, given $g \in G$, the inner automorphism ξ_g is trivial if and only if $g \in Z(G)$. As a consequence, Inn(G) $\cong G/Z(G)$.

Exercise 12.1.10 Verify all the statements in the last two paragraphs.

Exercise 12.1.11

(1) Prove that the center of $\mathrm{Aff}(\mathbb{F}_q)$ is trivial while its derived subgroup is U.

(2) For $u \in \mathbb{F}_q^*$ and $v \in \mathbb{F}_q$ denote by $\xi_{u,v}$ the inner automorphism of $\mathrm{Aff}(\mathbb{F}_q)$ associated with the element $\begin{pmatrix} u & v \\ 0 & 1 \end{pmatrix}$, that is, $\xi_{u,v} \begin{pmatrix} a & b \\ 0 & 1 \end{pmatrix} = \begin{pmatrix} a & (1-a)v + bu \\ 0 & 1 \end{pmatrix}$ for all $a \in \mathbb{F}_q^*$ and $b \in \mathbb{F}_q$. Prove that for all choices of $\begin{pmatrix} a & b \\ 0 & 1 \end{pmatrix} \in \mathrm{Aff}(\mathbb{F}_q)$, with $a \neq 1$ and $\begin{pmatrix} 1 & c \\ 0 & 1 \end{pmatrix} \in U$ with $c \neq 0$, there exists $\xi_{u,v}$ such that

$$\xi_{u,v} \begin{pmatrix} a & b \\ 0 & 1 \end{pmatrix} \in A \quad \text{and} \quad \xi_{u,v} \begin{pmatrix} 1 & c \\ 0 & 1 \end{pmatrix} = \begin{pmatrix} 1 & 1 \\ 0 & 1 \end{pmatrix}.$$

(3) Deduce the following fact: for each nontrivial $\alpha \in \mathrm{Aut}(\mathrm{Aff}(\mathbb{F}_q))$ there exists $\xi_{u,v} \in \mathrm{Inn}(\mathrm{Aff}(\mathbb{F}_q))$ such that:

$$\xi_{u,v} \circ \alpha(A) = A \quad \text{and} \quad \xi_{u,v} \circ \alpha \begin{pmatrix} 1 & 1 \\ 0 & 1 \end{pmatrix} = \begin{pmatrix} 1 & 1 \\ 0 & 1 \end{pmatrix}.$$

(4) Suppose that $q = p^n$, p prime number, and denote by σ the Frobenius automorphism of $\mathbb{F}_q$ (cf. Section 6.4). With the notation in (3), let us set $\beta = \xi_{u,v} \circ \alpha$. Prove that there exists $0 \leq k \leq n - 1$ such that
$$\beta \begin{pmatrix} a & b \\ 0 & 1 \end{pmatrix} = \begin{pmatrix} \sigma^k(a) & \sigma^k(b) \\ 0 & 1 \end{pmatrix} \text{ for all } \begin{pmatrix} a & b \\ 0 & 1 \end{pmatrix} \in \mathrm{Aff}(\mathbb{F}_q).$$
Hint: First of all, consider the restrictions $\beta|_A$ and $\beta|_U$. Then apply β to (12.4) with $a = b = 1$ and $v = 0$.

(5) Deduce that $\mathrm{Aut}\left(\mathrm{Aff}(\mathbb{F}_q)\right) \cong \mathrm{Aff}(\mathbb{F}_q) \rtimes \mathrm{Aut}(\mathbb{F}_q)$.

12.2 Representation theory of the affine group $\mathrm{Aff}(\mathbb{Z}/n\mathbb{Z})$

In this section we examine the representation theory of the group

$$\mathrm{Aff}(\mathbb{Z}/n\mathbb{Z}) = \left\{ \begin{pmatrix} a & b \\ 0 & 1 \end{pmatrix} : a \in \mathcal{U}(\mathbb{Z}/n\mathbb{Z}), b \in \mathbb{Z}/n\mathbb{Z} \right\},$$

that is, the affine group over the ring $\mathbb{Z}/n\mathbb{Z}$. As far as we know, most of the results presented here are new. We use the notation in Chapter 1. Clearly, for $n = p$ prime we have $\mathrm{Aff}(\mathbb{Z}/n\mathbb{Z}) = \mathrm{Aff}(\mathbb{F}_p)$.

First of all, in order to generalize the arguments in the proof of Lemma 12.1.2, we study the action γ of $\mathcal{U}(\mathbb{Z}/n\mathbb{Z})$ on $\mathbb{Z}/n\mathbb{Z}$ given by multiplication:

$$\gamma(a)b = ab,$$

for all $a \in \mathcal{U}(\mathbb{Z}/n\mathbb{Z})$ and $b \in \mathbb{Z}/n\mathbb{Z}$. From the results in Section 1.5 it follows that it coincides with the action of Aut($\mathbb{Z}_n$) on $\mathbb{Z}_n$. This action has been extensively studied in [4]. We limit ourselves to report some basic results, which form an interesting complement to Gauss' results in Proposition 1.1.20 and Proposition 1.2.13. We first introduce the following notation: for $n \in \mathbb{N}$, we denote by $D(n)$ the set of all *positive divisors* of n. Moreover for $r \in D(n)$ we set $A(r) = \{0 \le k \le n - 1 : \gcd(k, n) = n/r\}$ (cf. (1.6)), and regard $A(r)$ as a subset of $\mathbb{Z}/n\mathbb{Z}$. In particular, $A(n) \equiv \mathcal{U}(\mathbb{Z}/n\mathbb{Z})$ and $A(1) = \{0\}$.

Theorem 12.2.1 *The decomposition of* $\mathbb{Z}/n\mathbb{Z}$ *into the orbits of* γ *is*

$$\mathbb{Z}/n\mathbb{Z} = \coprod_{r \in D(n)} A(r). \tag{12.10}$$

Moreover, the stabilizer of $\frac{n}{r} \in A(r)$ *is*

$$\mathcal{U}_r(\mathbb{Z}/n\mathbb{Z}) = \{a \in \mathcal{U}(\mathbb{Z}/n\mathbb{Z}) : a \equiv 1 \bmod r\} \tag{12.11}$$

and

$$\frac{\mathcal{U}(\mathbb{Z}/n\mathbb{Z})}{\mathcal{U}_r(\mathbb{Z}/n\mathbb{Z})} \cong \mathcal{U}(\mathbb{Z}/r\mathbb{Z}). \tag{12.12}$$

Proof. For each $r \in D(n)$ let

$$\mathrm{Orb}(n/r) = \left\{a\frac{n}{r} \bmod n : a \in \mathcal{U}(\mathbb{Z}/n\mathbb{Z})\right\}$$

be the orbit containing n/r, Clearly, if $\gcd(a, n) = 1$ then also $\gcd(a, r) = 1$, so that $\gcd(an/r, n) = \gcd(a, r)n/r = n/r$, and this yields

$$\mathrm{Orb}(n/r) \subseteq A(r). \tag{12.13}$$

The solutions $a \in \mathbb{Z}$ of the congruence equation $a\frac{n}{r} \equiv \frac{n}{r} \bmod n$ are given by Proposition 1.2.13 (and its proof): selecting 1 as a fixed solution, they are:

$$1 + jr, \qquad j = 0, 1, \ldots, \frac{n}{r} - 1.$$

Among these numbers, we must select those belonging to $\mathcal{U}(\mathbb{Z}/n\mathbb{Z})$, and this proves (12.11).

Now consider the map

$$\Theta : \mathcal{U}(\mathbb{Z}/n\mathbb{Z}) \equiv A(n) \to \mathcal{U}(\mathbb{Z}/r\mathbb{Z})$$

given by $\Theta(a) = b$, if $a = b + jr$ with $0 \leq b \leq r - 1$ and $j \geq 0$, that is, b is the remainder of the division of a by r. Clearly, it is well defined: if $\gcd(a, n) = 1$ then $\gcd(b, r) = 1$. Indeed, $\gcd(b, r)|a$ and $r|n$ force $\gcd(b, r)|\gcd(a, n)$. Moreover, it is straightforward to check that it is a homomorphism, namely $\Theta(a_1 a_2) \equiv \Theta(a_1)\Theta(a_2) \bmod r$. Let us prove that it is surjective. Let $b \in \mathcal{U}(\mathbb{Z}/r\mathbb{Z})$, that is $0 \leq b \leq r - 1$ and $\gcd(b, r) = 1$. Consider the integer

$$a = b + p_1 p_2 \cdots p_m r,$$

where $p_1, p_2, \ldots, p_m$ are the (distinct) primes that divide n but not b. Now, if p is a prime and $p|n$ then we have two possibilities:

- if $p\,|b$ then $p \nmid p_1 p_2 \cdots p_m r$ and therefore p cannot divide a;
- if $p \nmid b$ then $p|p_1 p_2 \cdots p_m$ and therefore again p cannot divide a.

In conclusion, p does not divide a and we have proved that $\gcd(a, n) = 1$. As clearly, $b = \Theta(a)$, this ensures that Θ is surjective. Finally, from (12.11) we deduce that $\mathcal{U}_r(\mathbb{Z}/n\mathbb{Z}) = \mathrm{Ker}\Theta$ and this implies (12.12). In particular,

$$|\mathcal{U}_r(\mathbb{Z}/n\mathbb{Z})| = \frac{\varphi(n)}{\varphi(r)},$$

where φ is the Euler totient function (see Definition 1.1.18). Then we have:

$$
\begin{aligned}
\varphi(r) &= |A(r)| & \text{(by (1.8))} \\
&\geq |\mathrm{Orb}(n/r)| & \text{(by (12.13))} \\
&= \frac{|\mathcal{U}(\mathbb{Z}/n\mathbb{Z})|}{|\mathcal{U}_r(\mathbb{Z}/n\mathbb{Z})|} & \text{(by (10.44))} \\
&= \varphi(r),
\end{aligned}
$$

which forces the equality in (12.13), and (12.10) follows. $\qquad\square$

We recall (cf. Definition 1.1.6 and Exercise 1.1.5) that the *greatest common divisor* $\gcd(m, n, k)$ of three integers m, n, k is the largest positive integer that divides each of m, n, k and it equals the smallest positive integer that may be written in the form $um + vn + wk$, with $u, v, w \in \mathbb{Z}$; in fact $\{um + vn + wk : u, v, w \in \mathbb{Z}\}$ is the principal ideal in $\mathbb{Z}$ generated by $\gcd(m, n, k)$. Compare with Section 1.1. See also the monographs by Apostol [13] and Nathanson [118]. In the following, we consider the action of $\mathrm{Aff}(\mathbb{Z}/n\mathbb{Z})$ on $\mathbb{Z}/n\mathbb{Z}$, in analogy with (12.1), as well as the subgroups (cf. (12.2))

$$A = \left\{ \begin{pmatrix} a & 0 \\ 0 & 1 \end{pmatrix} : a \in \mathcal{U}(\mathbb{Z}/n\mathbb{Z}) \right\} \text{ and } U = \left\{ \begin{pmatrix} 1 & b \\ 0 & 1 \end{pmatrix} : b \in \mathbb{Z}/n\mathbb{Z} \right\}.$$

Lemma 12.2.2

(i) *The subgroup U is normal and one has*

$$\text{Aff}(\mathbb{Z}/n\mathbb{Z}) \cong U \rtimes A \equiv \mathbb{Z}_n \rtimes \mathcal{U}(\mathbb{Z}/n\mathbb{Z}); \quad (12.14)$$

(ii) *the conjugacy classes of the group* Aff($\mathbb{Z}/n\mathbb{Z}$) *are listed as follows:*

- $C_0 = \left\{ \begin{pmatrix} 1 & 0 \\ 0 & 1 \end{pmatrix} \right\};$

- $C_r = \left\{ \begin{pmatrix} 1 & b \\ 0 & 1 \end{pmatrix} : b \subset \Lambda(r) \right\}$, *where* $r \in D(n);$

- $C_{a,d} = \left\{ \begin{pmatrix} a & b \\ 0 & 1 \end{pmatrix} : b \in \mathbb{Z}/n\mathbb{Z} \text{ and } \gcd(a-1,n,b) = d \right\},$
  where $a \in \mathcal{U}(\mathbb{Z}/n\mathbb{Z})$, $a \neq 1$, *and* $d \in D(\gcd(a-1,n))$).

Proof.

(i) See the proof of the corresponding statement in Lemma 12.1.1.
(ii) By (12.4), for $a = 1$ the computation of the conjugacy orbits reduces to the computation of the γ-orbits in Theorem 12.2.1 and, this way, we determine the orbits C_r, $r \in D(n)$.

Now suppose that $a \in \mathcal{U}(\mathbb{Z}/n\mathbb{Z})$, $a \neq 1$, and $b \in \mathbb{Z}/n\mathbb{Z}$. Again by (12.4), we have to determine those $c \in \mathbb{Z}/n\mathbb{Z}$ such that the equation

$$v(1-a) + ub = c \quad (12.15)$$

has solutions $u \in \mathcal{U}(\mathbb{Z}/n\mathbb{Z})$ and $v \in \mathbb{Z}/n\mathbb{Z}$. First of all, note that if we think of a, b, c, u, v as integers, then this equation may be rewritten in the form

$$v(1-a) + ub + kn = c \text{ and } \gcd(u,n) = 1, \quad (12.16)$$

with $v, u, k \in \mathbb{Z}$ (k serves as another unknown). By the properties of the gcd, equation (12.16) has a solution only if $\gcd(1-a,b,n)|c$. Since we can switch the role of b and c in (12.15) (because u is invertible mod n), we conclude that this equation has a solution only if $\gcd(1-a,b,n) = \gcd(1-a,c,n)$.

Now suppose that $\gcd(1-a,b,n) = \gcd(1-a,c,n)$; we want to show that (12.16) has a solution. Set $r = \gcd(1-a,n)$, so that $\gcd(b,r) = \gcd(1-a,b,n) = \gcd(1-a,c,n) = \gcd(c,r)$. Then there exist $v, k \in \mathbb{Z}$ such that $r = v(1-a) + kn$. With this position, (12.16) becomes:

$$ub + r = c \text{ and } \gcd(u,n) = 1.$$

Moreover, in the last equation r may be replaced by any of its multiples hr, $h \in \mathbb{Z}$, because this corresponds to the replacement of v, k by vh, kh, respectively. Therefore, to solve (12.16) it suffices to solve $ub \equiv c \bmod r$, which, multiplied by $\frac{n}{r}$, yields the equivalent equation

$$u\frac{nb}{r} \equiv \frac{nc}{r} \bmod n \text{ and } \gcd(u, n) = 1.$$

By Theorem 12.2.1 the last equation has a solution because

$$\gcd\left(\frac{nb}{r}, n\right) = \gcd\left(\frac{n}{r}b, \frac{n}{r}r\right) = \frac{n}{r}\gcd(b, r) = \frac{n}{r}\gcd(c, r) = \gcd\left(\frac{nc}{r}, n\right).$$

□

Since $\mathrm{Aff}(\mathbb{Z}/n\mathbb{Z})$ is a semidirect product with an Abelian normal subgroup (cf. (12.14)), we can again apply Theorem 11.7.1 (the little group method) to get a complete list of all irreducible representations of $\mathrm{Aff}(\mathbb{Z}/n\mathbb{Z})$. As usual, $\widehat{\mathbb{Z}/n\mathbb{Z}}$ (respectively $\widehat{\mathcal{U}(\mathbb{Z}/n\mathbb{Z})}$) will denote the dual of the additive group $\mathbb{Z}/n\mathbb{Z}$ (respectively the multiplicative group $\mathcal{U}(\mathbb{Z}/n\mathbb{Z})$). After identifying A with the multiplicative group $\mathcal{U}(\mathbb{Z}/n\mathbb{Z})$ (via the map $\begin{pmatrix} a & 0 \\ 0 & 1 \end{pmatrix} \mapsto a$) and U with the additive group $\mathbb{Z}/n\mathbb{Z}$ (via the map $\begin{pmatrix} 1 & b \\ 0 & 1 \end{pmatrix} \mapsto b$), it follows from (12.4) that the conjugacy action (cf. (11.41)) of A on $\widehat{U} \equiv \widehat{\mathbb{Z}/n\mathbb{Z}}$ is given by

$$^a\chi(b) = \chi(a^{-1}b) \tag{12.17}$$

for all $\chi \in \widehat{U}$, $b \in \mathbb{Z}/n\mathbb{Z}$, and $a \in \mathcal{U}(\mathbb{Z}/n\mathbb{Z})$. For $0 \le k \le n - 1$, denote by χ_k the character of U given by: $\chi_k(b) = \exp\frac{2\pi kbi}{n}$, for all $0 \le b \le n - 1$, so that (12.17) becomes: $^a\chi_k = \chi_{a^{-1}k}$.

Lemma 12.2.3 *The orbits of the action of A on $\widehat{U}$ are:*

$$\Omega_r = \{\chi_k : k \in A(r)\}, \quad r \in D(n).$$

Moreover, the stabilizer of $\chi_{n/r} \in \Omega_r$ is the group $\mathcal{U}_r(\mathbb{Z}/n\mathbb{Z})$.

Proof. This is an immediate consequence of Theorem 12.2.1. □

Now we may apply the little group method.

Theorem 12.2.4

$$\widehat{\mathrm{Aff}(\mathbb{Z}/n\mathbb{Z})} = \left\{\pi_{r,\psi} = \mathrm{Ind}_{U \rtimes \mathcal{U}_r(\mathbb{Z}/n\mathbb{Z})}^{\mathrm{Aff}(\mathbb{Z}/n\mathbb{Z})}\left(\widetilde{\chi_{n/r}} \otimes \overline{\psi}\right) : r \in D(r), \psi \in \widehat{\mathcal{U}_r(\mathbb{Z}/n\mathbb{Z})}\right\}.$$

More precisely, the right hand side is a complete list of irreducible, pairwise inequivalent representations of $\mathrm{Aff}(\mathbb{Z}/n\mathbb{Z})$. *Moreover,*

$$\dim \pi_{r,\psi} = \varphi(r),$$

and $\mathrm{Aff}(\mathbb{Z}/n\mathbb{Z})$ *has* $\frac{\varphi(n)}{\varphi(r)}$ *irreducible, pairwise inequivalent representations of dimension* $\varphi(r)$.

Note that

$$\sum_{r \in D(n)} \sum_{\psi \in \widehat{\mathcal{U}_r(\mathbb{Z}/n\mathbb{Z})}} \left(\dim \pi_{r,\psi}\right)^2 = \sum_{r \in D(n)} \frac{\varphi(n)}{\varphi(r)} \cdot \varphi(r)^2 - \varphi(n) \sum_{r \in D(n)} \varphi(r)$$

$$\begin{aligned}
\text{(by Proposition 1.1.20)} \qquad &= \varphi(n)n \\
\text{(by (12.14))} \qquad &= |\mathrm{Aff}(\mathbb{Z}/n\mathbb{Z})|,
\end{aligned}$$

in agreement with Theorem 10.2.25.(iii).

12.3 Representation theory of the Heisenberg group $H_3(\mathbb{Z}/n\mathbb{Z})$

This section is based on [142]. A recent application of the material in this section is in [24].

The *Heisenberg group* over $\mathbb{Z}/n\mathbb{Z}$ is the matrix group

$$H_3(\mathbb{Z}/n\mathbb{Z}) = \left\{ \begin{pmatrix} 1 & x & z \\ 0 & 1 & y \\ 0 & 0 & 1 \end{pmatrix} : x, y, z \in \mathbb{Z}/n\mathbb{Z} \right\}.$$

Exercise 12.3.1 Show that $H_3(\mathbb{Z}/n\mathbb{Z})$ is isomorphic to the direct product $\mathbb{Z}/n\mathbb{Z} \times \mathbb{Z}/n\mathbb{Z} \times \mathbb{Z}/n\mathbb{Z}$ endowed with the multiplication

$$(x, y, z) \cdot (u, v, w) = (x + u, y + v, xv + w + z), \qquad (12.18)$$

for all $x, y, z, u, v, w \in \mathbb{Z}/n\mathbb{Z}$. In particular, check that

$$(x, y, z)^{-1} = (-x, -y, -z + xy), \qquad (12.19)$$

$$(x, y, z)^{-1}(u, v, w) = (u - x, v - y, w - z + xy - xv), \qquad (12.20)$$

$$(x, y, z)(u, v, w)(x, y, z)^{-1} = (u, v, w + xv - yu), \qquad (12.21)$$

and

$$(x, y, z) = (0, y, z)(x, 0, 0) = (0, 0, z) \cdot (0, y, 0) \cdot (x, 0, 0). \qquad (12.22)$$

In what follows, we use the notation in Exercise 12.3.1 rather than the matrix notation.

Proposition 12.3.2 *The conjugacy classes of $H_3(\mathbb{Z}/n\mathbb{Z})$ are:*

$$\mathcal{C}_{a,b,c} = \left\{ (a, b, c + k \gcd(a, b, n)) : k = 0, 1, \dots, \frac{n}{\gcd(a, b, n)} - 1 \right\},$$

$a, b \in \mathbb{Z}/n\mathbb{Z}$ and $c = 0, 1, \dots, \gcd(a, b, n) - 1$.

Proof. By (12.21), the conjugacy class containing a fixed element $(a, b, c) \in H_3(\mathbb{Z}/n\mathbb{Z})$ is

$$\{(a, b, c + xb - ya) : x, y \in \mathbb{Z}/n\mathbb{Z}\}.$$

We argue as in the proof of Lemma 12.2.2(ii). We fix an element $m \in \mathbb{Z}/n\mathbb{Z}$ and study the equation $xb - ya = m$ in the unknowns $x, y \in \mathbb{Z}/n\mathbb{Z}$. This is equivalent to

$$xb - ya + kn = m \tag{12.23}$$

in the unknowns $x, y, k \in \mathbb{Z}$ (we think of a, b, m as integers). Clearly, (12.23) has a solution if and only if $\gcd(a, b, n) | m$. Therefore, two elements $(a, b, c), (u, v, w) \in H_3(\mathbb{Z}/n\mathbb{Z})$ are conjugate if and only if $a = u, b = v$, and $c \equiv w \mod \gcd(a, b, n)$. $\square$

Proposition 12.3.3 *The Heisenberg group is the semidirect product*

$$H_3(\mathbb{Z}/n\mathbb{Z}) \cong \mathbb{Z}_n^2 \rtimes_\phi \mathbb{Z}_n, \tag{12.24}$$

where $\mathbb{Z}_n^2 = \{(0, v, w) : v, w \in \mathbb{Z}_n\}$ and $\mathbb{Z}_n = \{(x, 0, 0) : x \in \mathbb{Z}_n\}$ are viewed as additive groups, and ϕ is the $\mathbb{Z}_n$-action on $\mathbb{Z}_n^2$ given by

$$\phi_x(v, w) = (v, w + xv),$$

for all $x \in \mathbb{Z}_n$ and $(v, w) \in \mathbb{Z}_n^2$ (here x, v, w are viewed as elements in $\mathbb{Z}/n\mathbb{Z}$).

Proof. This follows from (12.21) and (12.22). Just note that, in particular, $(x, 0, 0)(0, v, w)(x, 0, 0)^{-1} = (0, v, w + xv)$. $\square$

We next apply Theorem 11.7.1, with

$$G = H_3(\mathbb{Z}/n\mathbb{Z}), \quad A = \mathbb{Z}_n^2, \quad \text{and} \quad H = \mathbb{Z}_n.$$

To this end, we need some preliminary results. Recall that the elements of $\widehat{A}$ are the characters $\chi_{s,t}, s, t = 0, 1, \dots, n - 1$, given by

$$\chi_{s,t}(v, w) = \exp\left(\frac{2\pi i}{n}(sv + tw) \right), \tag{12.25}$$

for all $u, v \in \mathbb{Z}_n$; see Section 2.3.

Proposition 12.3.4 *The orbits of H on $\widehat{A}$ are:*

$$\mathcal{R}_{k,t} = \left\{ \chi_{s,t} : s \equiv k \bmod \gcd(t,n) \right\},$$

for $0 \le t \le n-1$ and $0 \le k \le \gcd(t,n) - 1$. Moreover, the stabilizer of $\chi_{s,t} \in \mathcal{R}_{k,t}$ does not depend on the choice of s and it is given by

$$H_{\chi_{s,t}} = \left\{ (x,0,0) \in H : x \equiv 0 \bmod \frac{n}{\gcd(t,n)} \right\} \cong \mathbb{Z}_{\gcd(t,n)}.$$

Proof. The action of H on $\widehat{A}$ is given explicitly by:

$$\begin{aligned}
{}^{(x,0,0)}\chi_{s,t}(v,w) &= \chi_{s,t}(v, w - xv) \\
&= \exp\left[\frac{2\pi i}{n}[sv + t(w - xv)] \right] \\
&= \exp\left\{ \frac{2\pi i}{n}[(s - tx)v + tw] \right\} \\
&= \chi_{s-tx,t}(v,w).
\end{aligned}$$

Then χ_{s_1,t_1} and χ_{s_2,t_2} belong to the same H-orbit if and only if $t_1 = t_2 = t$ and there exists $x \in \mathbb{Z}$ such that $s_1 - tx \equiv s_2 \bmod n$. By Proposition 1.2.13 this equation has a solution if and only if $s_1 \equiv s_2 \bmod \gcd(t,n)$. Finally, we observe that the stabilizer of $\chi_{s,t}$ is made up of those $x \in H$ such that $xt = 0 \bmod n$. $\qquad\square$

In more explicit form,

$$\mathcal{R}_{k,t} = \left\{ \chi_{s,t} : s = k + j\gcd(t,n), 0 \le j \le \frac{n}{\gcd(n,t)} - 1 \right\}$$

and

$$H_{\chi_{s,t}} = \left\{ \left(j\frac{n}{\gcd(t,n)}, 0, 0 \right) : 0 \le j \le \gcd(n,t) - 1 \right\}.$$

Moreover, for a given t with $0 \le t \le n-1$ we have the following particular cases:

- If $t = 0$ then $\gcd(0,n) = n$ and $\mathcal{R}_{k,0} = \{\chi_{k,0}\}, k = 0, 1, \dots, n-1$: now each orbit consists of a single element and its stabilizer is $H_{\chi_{k,0}} = H$.
- If $\gcd(t,n) = 1$ then we have exactly one orbit of n elements, namely $\mathcal{R}_{0,t} = \{\chi_{s,t} : s = 0, 1, \dots, n-1\}$, and the stabilizer is trivial: $H_{\chi_{s,t}} = \{(0,0,0)\}$.

According to the preceding analysis, we can choose

$$X = \{\chi_{k,t} : 0 \le t \le n-1, 0 \le k \le \gcd(t,n) - 1\}$$

as a set of representatives of the quotient space $\widehat{A}/\approx$ (cf. Theorem 11.6.2). By (11.47) and (12.22) we deduce that the extension of these characters to $A \rtimes H_{\chi_{k,t}}$ is given by

$$\widetilde{\chi_{k,t}}(x, y, z) = \chi_{k,t}(y, z), \qquad (12.26)$$

for all $(x, y, z) \in A \rtimes H_{\chi_{k,t}}$. We also need a parameterization of the characters of the groups $H_{\chi_{k,t}} \cong \mathbb{Z}_{\gcd(t,n)}$: they are given by

$$\psi_{\gcd(t,n),h}(j) = \exp\left(\frac{2\pi i}{\gcd(t, n)} hj\right),$$

$h, j = 0, 1, \ldots, \gcd(t, n) - 1$. Their inflation to $A \rtimes H_{\chi_{k,t}}$ is given by

$$\overline{\psi_{\gcd(t,n),h}}(x, y, z) = \psi_{\gcd(t,n),h}\left(\frac{x \gcd(t, n)}{n}\right) \equiv \exp\left(\frac{2\pi i}{n} hx\right),$$

for all $(x, y, z) \in A \rtimes H_{\chi_{k,t}}$ (so that $\frac{n}{\gcd(t,n)}|x$). We now have all necessary tools needed to apply Theorem 11.7.1.

Theorem 12.3.5

$$\widehat{H_3(\mathbb{Z}/n\mathbb{Z})} = \left\{\pi_{k,t,h} = \mathrm{Ind}_{A \rtimes H_{\chi_{k,t}}}^{H_3(\mathbb{Z}/n\mathbb{Z})}\left(\widetilde{\chi_{k,t}} \otimes \overline{\psi_{\gcd(t,n),h}}\right) : \right.$$
$$\left. 0 \leq t \leq n - 1, 0 \leq h, k \leq \gcd(t, n) - 1\right\}. \quad (12.27)$$

More precisely, the right hand side is a complete list of irreducible, pairwise inequivalent representations of $H_3(\mathbb{Z}/n\mathbb{Z})$. Moreover,

$$\dim \pi_{k,t,h} = \frac{n}{\gcd(t, n)}$$

and, for each $d \in D(n)$, the group $H_3(\mathbb{Z}/n\mathbb{Z})$ has exactly $d^2\varphi(n/d)$ irreducible, pairwise inequivalent representations of dimension $\frac{n}{d}$. In particular, it has n^2 one-dimensional representations (case $d = n$) and $\varphi(n)$ irreducible representations of maximal dimension n (case $d = 1$).

As for $\mathrm{Aff}(\mathbb{Z}/n\mathbb{Z})$, note that

$$\sum_{\substack{d \in D(n) \\ \gcd(t,n)=d}} \sum_{0 \leq t \leq n-1;} \sum_{k,h=1}^{d-1} \left(\dim \pi_{k,t,h}\right)^2 = \sum_{d \in D(n)} \left(\frac{n}{d}\right)^2 \cdot d^2\varphi(n/d)$$

$$\text{(by Proposition 1.1.20)} \quad = n^3$$

$$= |H_3(\mathbb{Z}/n\mathbb{Z})|,$$

in agreement with Theorem 10.2.25.(iii).

Proposition 12.3.6 *Fix* $0 \leq t \leq n-1$ *and* $0 \leq h, k \leq d-1$, *where* $d = \gcd(t, n)$. *Then a matrix form of* $\pi_{k,t,h}$ *is given by the map*

$$H_3(\mathbb{Z}/n\mathbb{Z}) \ni (x, y, z) \rightarrow \Pi_{k,t,h}(x, y, z) = \left(\Pi_{k,t,h;r,s}(x, y, z) \right)_{r,s=0}^{\frac{n}{d}-1},$$

where $\Pi_{k,t,h;r,s}(x, y, z) = 0$ *if* $\frac{n}{d} \nmid (x+s-r)$ *and*

$$\Pi_{k,t,h;r,s}(x, y, z) = \exp\left(\frac{2\pi i}{n} [ky + t(z - ry) + h(x + s - r)] \right), \quad (12.28)$$

otherwise.

Proof. If $(x, y, z) \in H_3(\mathbb{Z}/n\mathbb{Z})$ we may compute the remainder of x modulo $\frac{n}{d}$, namely the integer $0 \leq r \leq \frac{n}{d} - 1$ given by the Euclidean division: $x = q\frac{n}{d} + r$. Therefore $(x, y, z) = (r, 0, 0)(q\frac{n}{d}, y, z - ry)$, where $(q\frac{n}{d}, y, z - ry) \in A \rtimes H_{\chi_{k,t}}$ and

$$H_3(\mathbb{Z}/n\mathbb{Z}) = \coprod_{r=0}^{\frac{n}{d}-1} (r, 0, 0) \left(A \rtimes H_{\chi_{k,t}} \right) \quad (12.29)$$

is the decomposition of $H_3(\mathbb{Z}/n\mathbb{Z})$ into left cosets of $A \rtimes H_{\chi_{k,t}}$; see (10.49). Moreover, if $0 \leq r, s \leq \frac{n}{d} - 1$ then

$$(r, 0, 0)^{-1}(x, y, z)(s, 0, 0) = (x + s - r, y, z - ry)$$

belongs to $A \rtimes H_{\chi_{k,t}}$ if and only if $\frac{n}{d} | (x + s - r)$. If this is the case, we have

$$\left(\widetilde{\chi_{k,t}} \otimes \overline{\psi_{d,h}} \right) (x + s - r, y, z - ry) = \chi_{k,t}(y, z - ry) \psi_{d,h}\left(\frac{(x + s - r)d}{n} \right).$$

Then (12.28) follows from (11.19), taking into account the explicit formulas for $\widetilde{\chi_{k,t}}$ and $\overline{\psi_{d,h}}$. $\qquad\qquad\square$

We now study some particular cases of (12.28).

• For $t = 0$ we get the n^2 one-dimensional representations, given by:

$$\Pi_{k,0,h}(x, y, z) = \exp\left[\frac{2\pi i}{n}(ky + hx) \right],$$

for $(x, y, z) \in H_3(\mathbb{Z}/n\mathbb{Z})$, $0 \leq k, h \leq n - 1$.

• Suppose that $x = 1$ and $y = z = 0$. Then the number $1 + s - r$ is divisible by $\frac{n}{d}$ in the following two cases: if $1 + s - r = 0$, and therefore the corresponding entry is equal to 1, and if $s = \frac{n}{d} - 1$, $r = 0$, so that the entry is equal to

$\exp(\frac{2\pi i}{d}h)$. Therefore,

$$\Pi_{k,t,h}(1,0,0) = \begin{pmatrix} 0 & 1 & 0 & \cdots & 0 \\ 0 & 0 & 1 & \cdots & 0 \\ \vdots & \vdots & \ddots & \ddots & \vdots \\ \vdots & \vdots & & \ddots & 1 \\ \exp(\frac{2\pi i}{d}h) & 0 & 0 & \cdots & 0 \end{pmatrix}.$$

- For $y = z = 0$ we have $(x,0,0) = (1,0,0)^x$ and therefore:

$$\Pi_{k,t,h}(x,0,0) = \begin{pmatrix} 0 & 1 & 0 & \cdots & 0 \\ 0 & 0 & 1 & \cdots & 0 \\ \vdots & \vdots & \ddots & \ddots & \vdots \\ \vdots & \vdots & & \ddots & 1 \\ \exp(\frac{2\pi i}{d}h) & 0 & 0 & \cdots & 0 \end{pmatrix}^x. \qquad (12.30)$$

- Suppose that $x = 0$. Then $\frac{n}{d}|(s-r)$ if and only if $s = r$, so that the matrix is *diagonal* and the r-th coefficient is

$$\exp\left(\frac{2\pi i}{n}[ky + t(z - ry)]\right) = \exp\left[\frac{2\pi i}{n}(ky + tz)\right]\exp\left(-rty\frac{2\pi i}{n}\right).$$

Therefore

$$\Pi_{k,t,h}(0,y,z) = \exp\left[\frac{2\pi i}{n}(ky + tz)\right]$$

$$\cdot \begin{pmatrix} 1 & 0 & 0 & \cdots & 0 \\ 0 & \exp\left(-ty\frac{2\pi i}{n}\right) & 0 & \cdots & 0 \\ 0 & 0 & \exp\left(-2ty\frac{2\pi i}{n}\right) & & \vdots \\ \vdots & \vdots & & \ddots & 0 \\ 0 & 0 & 0 & \cdots & \exp\left[-\left(\frac{n}{d}-1\right)ty\frac{2\pi i}{n}\right] \end{pmatrix}.$$

$$(12.31)$$

In particular, if also $y = 0$, then the matrix is scalar: $\Pi_{k,t,h}(0,0,z) = \exp\left(\frac{2\pi i}{n}tz\right)I_{n/d}$.

- Finally, we observe that we can use (12.22) to reduce the computation of $\Pi_{k,t,h}(x,y,z)$ to the cases (12.30) and (12.31), because $\Pi_{k,t,h}(x,y,z) = \Pi_{k,t,h}(0,y,z)\Pi_{k,t,h}(x,0,0)$.

Exercise 12.3.7 Prove the following explicit expression for the character $\chi_{k,t,h}$ of the representation $\pi_{k,t,h}$:

$$\chi_{k,t,h}(x, y, z) = \mathbf{1}_{n/d}(x)\mathbf{1}_{n/d}(y)\frac{n}{d}\exp\left[\frac{2\pi i}{n}(hx + ky + tz)\right], \quad (12.32)$$

where

$$\mathbf{1}_{n/d}(x) = \begin{cases} 1 & \text{if } \frac{n}{d}|x \\ 0 & \text{otherwise.} \end{cases}$$

Exercise 12.3.8

(1) By means of Proposition 10.2.18 and (12.32) prove that

$$\mathrm{Res}_H^{H_3(\mathbb{Z}/n\mathbb{Z})}\pi_{k,t,h} = \bigoplus_{\substack{0 \leq \ell \leq n-1: \\ \ell \equiv h \bmod d}} \chi_\ell$$

and

$$\mathrm{Res}_A^{H_3(\mathbb{Z}/n\mathbb{Z})}\pi_{k,t,h} = \bigoplus_{\substack{0 \leq s \leq n-1: \\ s \equiv k \bmod d}} \chi_{s,t},$$

where $\chi_\ell(x) = \exp\left(\frac{2\pi i}{n}\ell x\right)$ for all $0 \leq x \leq n-1$ (characters of $H \equiv \mathbb{Z}_n$) and $\chi_{s,t}$ is as in (12.25).

(2) By means of Frobenius reciprocity, deduce that

$$\mathrm{Ind}_H^{H_3(\mathbb{Z}/n\mathbb{Z})}\chi_\ell \sim \bigoplus_{\substack{0 \leq t \leq n-1 \\ 0 \leq k \leq \gcd(t,n)-1}} \pi_{k,t,h(t,\ell)},$$

where $h(t, \ell)$ is the remainder of the division of ℓ by $\gcd(t, n)$, and

$$\mathrm{Ind}_A^{H_3(\mathbb{Z}/n\mathbb{Z})}\chi_{s,t} \sim \bigoplus_{0 \leq h \leq d-1} \pi_{k,t,h},$$

where k is the remainder of the division of s by d.

12.4 The DFT revisited

The connection between classical Fourier analysis and the continuous Heisenberg group has been well studied and we refer to the expository paper [76], and Folland's monograph [62]. In one of our main sources, namely [15], this connection is extended to the finite case and our purpose is to give a clear exposition of these facts; see also [142]. We focus on the key point: by means of suitable realizations of the irreducible representation $\pi_{0,1,0}$, the Heisenberg group may be seen as a group of unitary transformations of $L(\mathbb{Z}/n\mathbb{Z})$, and the Fourier transform intertwines two different such realizations.

For the moment, we fix a positive integer n and we set $\chi(k) = \exp\left(\frac{2\pi i}{n}k\right)$, for $k \in \mathbb{Z}$. Also, to simplify notation, we set $G = H_3(\mathbb{Z}/n\mathbb{Z})$. Moreover, in the notation of (12.27), we set $\pi = \pi_{0,1,0}$ and we denote by V_π its representation space. From (11.16), and (12.18) with $u = 0$, it follows that V_π is made up of all functions $f: G \to \mathbb{C}$ such that

$$f(x, y + v, xv + z + w) = \chi(-w)f(x, y, z), \qquad (12.33)$$

for all $(x, y, z) \in G$ and $v, w \in \mathbb{Z}/n\mathbb{Z}$. Indeed, in (12.25) we have $\chi_{0,1}(v, w) = \chi(w)$, in (12.26) and (12.27) the subgroup $H_{\chi_{0,1}}$ is trivial, and, finally, $\pi = \mathrm{Ind}_A^G \chi_{0,1}$. From (12.33) and the identity $(x, y, z) = (x, 0, 0)(0, y, z - xy)$, it follows that $f \in L(G)$ belongs to V_π if and only if it satisfies the condition:

$$f(x, y, z) = \chi(-z + xy)f(x, 0, 0), \qquad (12.34)$$

for all $(x, y, z) \in G$, so that it is determined by its values on the subgroup H. In other words, in (11.17) $\mathcal{T} \equiv H$ (actually, this is a particular case of (12.29)). Finally, we observe that from (12.20) with $v = w = 0$ it follows that

$$[\pi(x, y, z)f](u, 0, 0) = f(u - x, -y, -z + xy). \qquad (12.35)$$

We now translate π into an equivalent representation on $L(\mathbb{Z}/n\mathbb{Z})$ showing its relevance to the DFT on a cyclic group. We need a series of notation and identities. First of all, invoking (12.34) we can define the linear operator $U: V_\pi \to L(\mathbb{Z}/n\mathbb{Z})$ by setting

$$[Uf](x) = f(x, 0, 0), \qquad (12.36)$$

for all $f \in V_\pi$ and $x \in \mathbb{Z}/n\mathbb{Z}$. Its inverse is given by

$$\left[U^{-1}f\right](x, y, z) = \chi(-z + xy)f(x), \qquad (12.37)$$

for all $f \in L(\mathbb{Z}/n\mathbb{Z})$ and $(x, y, z) \in G$. It is immediate to show that U (and therefore U^{-1}) is an isometric isomorphism; just recall the definition of scalar product in an induced representation (11.3). Then we set

$$\pi^\sharp(x, y, z) = U\pi(x, y, z)U^{-1} \qquad (12.38)$$

for all $(x, y, z) \in G$. Clearly, $\pi^\sharp$ is a unitary representation of G on $L(\mathbb{Z}/n\mathbb{Z})$, equivalent to π. But another description of $\pi^\sharp$ will reveal its importance. We introduce three unitary operators T_x (*translation operator*), M_y (*multiplier operator*), and S_z on $L(\mathbb{Z}/n\mathbb{Z})$ by setting:

$$[T_xf](u) = f(u - x), \quad [M_yf](u) = \chi(-yu)f(u), \quad [S_zf](u) = \chi(z)f(u),$$

for all $f \in L(\mathbb{Z}/n\mathbb{Z})$ and $x, y, z, u \in \mathbb{Z}/n\mathbb{Z}$. Note that T_x has already been defined in Section 2.4.

Lemma 12.4.1 *We have the following* commutation relation:

$$T_x M_y = S_{xy} M_y T_x, \tag{12.39}$$

for all $x, y \in \mathbb{Z}/n\mathbb{Z}$.

Proof. Let $f \in L(\mathbb{Z}/n\mathbb{Z})$ and $x, y, u \in \mathbb{Z}/n\mathbb{Z}$. Then

$$\begin{aligned}
\left[T_x M_y f\right](u) &= \left[M_y f\right](u - x) = \chi(-yu + xy)f(u - x) \\
&= \chi(-yu + xy)\left[T_x f\right](u) = \chi(xy)\left[M_y T_x f\right](u) = \left[S_{xy} M_y T_x f\right](u). \quad \square
\end{aligned}$$

The Fourier transform intertwines T_x and M_y: from Exercise 2.4.7 (see also Lemma 4.1.1) it follows that

$$\mathcal{F} T_x = M_x \mathcal{F} \qquad \text{and} \qquad \mathcal{F} M_y = T_{-y} \mathcal{F}. \tag{12.40}$$

We use the normalized Fourier transform, see Section 4.1. Note also the analogous identities for the inverse Fourier transform: $\mathcal{F}^{-1} T_x = M_{-x} \mathcal{F}^{-1}$ and $\mathcal{F}^{-1} M_y = T_y \mathcal{F}^{-1}$.

Theorem 12.4.2

(i) *The irreducible representation $\pi^{\sharp}$ defined in (12.38) may be expressed in the form:*

$$\pi^{\sharp}(x, y, z) = S_z M_y T_x, \tag{12.41}$$

$(x, y, z) \in G$. *Moreover, it is a faithful representation of G as a group of unitary operators on $L(\mathbb{Z}/n\mathbb{Z})$.*

(ii) *The map $J \colon G \to G$ defined by setting $J(x, y, z) = (-y, x, z - xy)$, for all $(x, y, z) \in G$, is an order four automorphism of G.*

(iii) *The G-representation $\pi^{\flat} = \pi^{\sharp} \circ J$ is equivalent to $\pi^{\sharp}$ and the equivalence is realized by the Fourier transform:*

$$\mathcal{F} \pi^{\sharp}(x, y, z) = \pi^{\flat}(x, y, z)\mathcal{F}, \tag{12.42}$$

for all $(x, y, z) \in G$.

Proof.

(i) For all $f \in L(\mathbb{Z}/n\mathbb{Z})$, $(x, y, z) \in G$, and $u \in \mathbb{Z}/n\mathbb{Z}$, we have:

$$\begin{aligned}
\left[\pi^{\sharp}(x, y, z)f\right](u) &= \left[U\pi(x, y, z)U^{-1}f\right](u) \\
(\text{by } (12.36)) &= \left[\pi(x, y, z)U^{-1}f\right](u, 0, 0) \\
(\text{by } (12.35)) &= \left[U^{-1}f\right](u - x, -y, -z + xy) \\
(\text{by } (12.37)) &= \chi(z - uy)f(u - x) \\
&= \left[S_z M_y T_x f\right](u).
\end{aligned}$$

Moreover, if $(x, y, z) \in \mathrm{Ker}\pi^{\sharp}$ then $\pi^{\sharp}(x, y, z)\delta_0 = \delta_0$, that is, $\chi(z - uy)\delta_x(u) = \delta_0(u)$ for all $u \in \mathbb{Z}/n\mathbb{Z}$. It follows that $x = 0 = y = z$.

(ii) This follows from easy calculations. For instance, $J^2(x, y, z) = (-x, -y, z)$ yields $J^4 = \mathrm{Id}_G$.

(iii) First of all, note that from (12.41) and (12.39) we deduce that:

$$\pi^{\flat}(x, y, z) = \pi^{\sharp}(-y, x, z - xy) = S_{z-xy}M_xT_{-y} = S_zT_{-y}M_x.$$

Therefore, using the identities in (12.40) we get:

$$\mathcal{F}\pi^{\sharp}(x, y, z) = \mathcal{F}S_zM_yT_x = S_zT_{-y}\mathcal{F}T_x = S_zT_{-y}M_x\mathcal{F} = \pi^{\flat}(x, y, z)\mathcal{F}.$$

$\square$

Note that, in the proof above, we have also obtained the following explicit form of $\pi^{\sharp}$:

$$[\pi^{\sharp}(x, y, z)f](u) = \chi(z - uy)f(u - x). \tag{12.43}$$

In other words, G may be seen as the group generated by the translation operators T_x and the multiplier operators M_y; then the operators S_z enter the picture by virtue of the commutation relation (12.39). The automorphism J switches the role of x and y, giving a different realization of G as a group of unitary operators. The Fourier transform intertwines the translation and multiplier operators and therefore also the different realizations of G. That is, J corresponds to the conjugation by $\mathcal{F}$, in formulæ $\pi^{\flat} = \mathcal{F}\pi^{\sharp}\mathcal{F}^{-1}$. Note also that the order of J as an automorphism of G coincides with the order of $\mathcal{F}$ as a unitary operator; see Proposition 4.1.2. We may also express all of this by saying that the diagram in Figure 12.1 is commutative

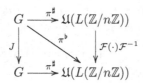

Figure 12.1. The commutative diagram showing that the Fourier transform $\mathcal{F}$ intertwines the representations $\pi^{\flat}$ and $\pi^{\sharp}$. Here, $\mathfrak{U}(L(\mathbb{Z}/n\mathbb{Z}))$ is the group of unitary operators on $L(\mathbb{Z}/n\mathbb{Z})$, and $\mathcal{F}(\cdot)\mathcal{F}^{-1}$ indicates conjugation by $\mathcal{F}$.

Finally, note that the J-image of the group $\mathbb{Z}_n^2$ in (12.24) is nothing but $\{(u, 0, w) : u, w \in \mathbb{Z}_n\}$.

Exercise 12.4.3 Define $\pi^{\sharp}$ by means of (12.41). Then, using the commutation relations (12.39), prove that $\pi^{\sharp}$ is a representation of G and, furthermore, using the converse to Schur's lemma (Exercise 10.2.9) and Theorem 2.4.10, prove that it is irreducible.

12.5 The FFT revisited

In this section, following again [15], we derive an operator form of the Fast Fourier Transform by means of intertwining operators between different realizations of the representation $\pi_{0,1,0}$. We begin by fixing two integers $m, n \geq 2$ and setting $G = H_3(\mathbb{Z}/nm\mathbb{Z})$. We introduce the subgroups

$$K_1 = \{(rn, sm, 0) : 0 \leq r \leq m - 1, 0 \leq s \leq n - 1\}$$

and

$$K_2 = \{(sm, rn, 0) : 0 \leq r \leq m - 1, 0 \leq s \leq n - 1\},$$

both isomorphic to $\mathbb{Z}_m \oplus \mathbb{Z}_n$. Clearly, an element $(x, y, z) \in G$ belongs to K_1 if and only if $z = 0$, $n|x$, and $m|y$, while it belongs to K_2 if and only if $z = 0$, $m|x$, and $n|y$. In what follows, we use some notation similar to that in Chapter 5. In particular, for $0 \leq u, v \leq nm - 1$ we set

$$u = \tilde{s} + rn, \quad v = \tilde{r} + sm, \quad \text{with } 0 \leq s, \tilde{s} \leq n - 1, \ 0 \leq r, \tilde{r} \leq m - 1. \tag{12.44}$$

We also use the notation $\chi(u) = \exp(\frac{2\pi i}{mn}u)$ and $\pi^\sharp, \pi^\flat$ as in Section 12.4, but now n is replaced with nm. Then we define Z_1 as the space of all $f \in L(G)$ such that:

$$f(u, v, w) = \chi(s\tilde{s}m - w)f(\tilde{s}, \tilde{r}, 0) \tag{12.45}$$

for all $(u, v, w) \in G$, where u, v are as in (12.44). Finally, we define the *Weil-Berezin map* $W_1 : L(\mathbb{Z}/nm\mathbb{Z}) \to L(G)$ by setting

$$[W_1 f](x, y, z) = \frac{1}{m\sqrt{n}}\chi(xy - z)\sum_{\ell=0}^{m-1} f(\ell n + x)\chi(\ell ny), \tag{12.46}$$

for all $f \in L(\mathbb{Z}/nm\mathbb{Z})$ and $(x, y, z) \in G$.

Proposition 12.5.1

(i) *In the notation of Example 11.1.6, $L(G/K_1)$ is the space of all $f \in L(G)$ such that:*

$$f(u, v, w) = f(\tilde{s}, \tilde{r}, w - s\tilde{s}m) \tag{12.47}$$

for all $(u, v, w) \in G$, where $s, \tilde{s}, r, \tilde{r}$ are as in (12.44).

(ii) Z_1 *is a subspace of* $L(G/K_1)$ *and it is invariant with respect to the left regular representation* λ *of* G.

(iii) *Denote by* λ_1 *the restriction of the left regular representation of* G *to* Z_1 *and endow this space with the norm of* $L(G/K_1)$ *(recall* (11.3)). *Then the* W_1-*image of* $L(\mathbb{Z}/nm\mathbb{Z})$ *is exactly* Z_1 *and* W_1 *is an isometry that intertwines* $\pi^\sharp$ *with* λ_1: *for all* $(x, y, z) \in G$

$$W_1 \pi^\sharp(x, y, z) = \lambda_1(x, y, z)W_1. \tag{12.48}$$

Proof.

(i) A function $f \in L(G)$ is right K_1-invariant if and only if

$$f(u + rn, v + sm, w + usm) = f(u, v, w), \tag{12.49}$$

for all $(u, v, w) \in G$ and $(rn, sm, 0) \in K_1$. Moreover, in the notation of (12.44), each element of G may be written uniquely in the form

$$(u, v, w) = (\widetilde{s}, \widetilde{r}, w - s\widetilde{s}m)(rn, sm, 0).$$

Therefore

$$\{(\widetilde{s}, \widetilde{r}, w) : 0 \leq \widetilde{s} \leq n - 1, 0 \leq \widetilde{r} \leq m - 1, 0 \leq w \leq mn - 1\}$$

is a set of representatives for the left cosets of K_1 in G and our assertion is a particular case of (11.7) and (11.17); see also Example 11.1.6.

(ii) If f satisfies (12.45), then it also satisfies (12.47). Indeed, (12.45), with $s = r = 0$ and w replaced with $w - s\widetilde{s}m$, yields

$$f(\widetilde{s}, \widetilde{r}, w - s\widetilde{s}m) = \chi(s\widetilde{s}m - w)f(\widetilde{s}, \widetilde{r}, 0), \tag{12.50}$$

and therefore, for arbitrary u, v, w,

$$
\begin{aligned}
f(u, v, w) &= \chi(s\widetilde{s}m - w)f(\widetilde{s}, \widetilde{r}, 0) && \text{(by (12.45))} \\
&= f(\widetilde{s}, \widetilde{r}, w - s\widetilde{s}m). && \text{(by (12.50))}
\end{aligned}
$$

It follows that $Z_1 \leq L(G/K_1)$. Note also that if $f \in Z_1$ then

$$f(u, v, w) = \chi(-w)f(u, v, 0), \tag{12.51}$$

because both sides are equal to $\chi(-w)\chi(s\widetilde{s}m)f(\widetilde{s}, \widetilde{r}, 0)$. Moreover, it is easy to check that Z_1 is exactly the set of all $f \in L(G)$ that verify both (12.47) and (12.51). Finally, by means of (12.20), we deduce that if f

satisfies (12.51) then

$$
\begin{aligned}
[\lambda(x, y, z)f](u, v, w) &= f(u - x, v - y, w - z + xy - xv) \\
&= \chi(-w)\chi(z - xy + xv)f(u - x, v - y, 0) \\
&= \chi(-w)f(u - x, v - y, -z + xy - xv) \\
&= \chi(-w)[\lambda(x, y, z)f](u, v, 0).
\end{aligned}
$$

That is, the space of all functions satisfying condition (12.51) is λ-invariant. Therefore, also Z_1 is λ-invariant, because it is the subspace of all functions in $L(G/K_1)$ satisfying (12.51).

(iii) For $f \in L(\mathbb{Z}/nm\mathbb{Z})$ and assuming (12.44), we have:

$$
\begin{aligned}
m\sqrt{n}[W_1 f](u, v, w) &= m\sqrt{n}[W_1 f](\widetilde{s} + rn, \widetilde{r} + sm, w) \\
&= \chi(-w + \widetilde{s}\widetilde{r} + \widetilde{s}sm + \widetilde{r}rn) \\
&\quad \cdot \sum_{\ell=0}^{m-1} f(\ell n + rn + \widetilde{s})\chi(\ell(\widetilde{r} + sm)n) \\
&= \chi(-w + \widetilde{s}\widetilde{r} + \widetilde{s}sm + \widetilde{r}rn) \sum_{\ell=0}^{m-1} f((\ell + r)n + \widetilde{s})\chi(\ell\widetilde{r}n) \\
(t = \ell + r) \quad &= \chi(-w + \widetilde{s}\widetilde{r} + \widetilde{s}sm) \sum_{t=0}^{m-1} f(tn + \widetilde{s})\chi(t\widetilde{r}n) \\
&= m\sqrt{n}\chi(-w + \widetilde{s}sm)[W_1 f](\widetilde{s}, \widetilde{r}, 0).
\end{aligned}
$$

Therefore, by (12.45), the image of W_1 is contained in Z_1. Moreover, for $f_1, f_2 \in L(\mathbb{Z}/nm\mathbb{Z})$ we have:

$$
\begin{aligned}
\langle W_1 f_1, W_1 f_2 \rangle_{Z_1} &= \frac{1}{nm} \sum_{(x,y,z) \in G} [W_1 f_1](x, y, z)\overline{[W_1 f_2](x, y, z)} \\
&= \frac{1}{n^2 m^3} \sum_{z \in \mathbb{Z}/nm\mathbb{Z}} \sum_{\ell_1, \ell_2 = 0}^{m-1} \sum_{x \in \mathbb{Z}/nm\mathbb{Z}} f_1(\ell_1 n + x)\overline{f_2(\ell_2 n + x)} \\
&\quad \cdot \sum_{y \in \mathbb{Z}/nm\mathbb{Z}} \chi(\ell_1 ny)\overline{\chi(\ell_2 ny)} \\
\text{(by (2.7))} \quad &= \frac{1}{m} \sum_{\ell_1 = 0}^{m-1} \sum_{x \in \mathbb{Z}/nm\mathbb{Z}} f_1(\ell_1 n + x)\overline{f_2(\ell_1 n + x)} \\
&= \langle f_1, f_2 \rangle_{L(\mathbb{Z}/nm\mathbb{Z})}.
\end{aligned}
$$

It follows that W_1 is an isometry. Finally, for $(x, y, z), (u, v, w) \in G$ and $f \in L(\mathbb{Z}/nm\mathbb{Z})$ we have:

$$[\lambda_1(x, y, z)W_1 f](u, v, w) = [W_1 f](u - x, v - y, -z + xy + w - xv)$$

$$= \frac{1}{m\sqrt{n}} \chi(z - w + uv - uy)$$

$$\cdot \sum_{\ell=0}^{m-1} f(\ell n + u - x)\chi(\ell n(v - y))$$

(by (12.43))
$$= \frac{1}{m\sqrt{n}} \chi(-w + uv)$$

$$\cdot \sum_{\ell=0}^{m-1} [\pi^\sharp(x, y, z)f](\ell n + u)\chi(\ell nv)$$

$$= \left[W_1 \pi^\sharp(x, y, z)f\right](u, v, w). \qquad \square$$

In Exercise 12.5.9 we outline a different proof of the fact that W_1 is an intertwining operator, also showing how to derive its expression.

Now we concentrate on K_2. First of all, we change the notation in (12.44): for $0 \le u, v \le nm - 1$ we set

$$u = \widetilde{r} + sm, \quad v = \widetilde{s} + rn, \quad \text{with } 0 \le s, \widetilde{s} \le n - 1, \ 0 \le r, \widetilde{r} \le m - 1. \tag{12.52}$$

Then we define Z_2 as the space of all $f \in L(G)$ such that

$$f(u, v, w) = \chi(\widetilde{r}\widetilde{r}n - w)f(\widetilde{r}, \widetilde{s}, 0) \tag{12.53}$$

for all $(u, v, w) \in G$, where u, v are as in (12.52). Moreover, we define $W_2: L(\mathbb{Z}/nm\mathbb{Z}) \to L(G)$ by setting

$$[W_2 f](x, y, z) = \frac{1}{n\sqrt{m}} \chi(xy - z) \sum_{t=0}^{n-1} f(tm - x)\chi(-tmy), \tag{12.54}$$

for all $f \in L(\mathbb{Z}/nm\mathbb{Z})$, $(x, y, z) \in G$. Finally, we define $M: L(G) \to L(G)$ by setting $Mf = f \circ J$, where J is as in Theorem 12.4.2(ii), that is,

$$[Mf](x, y, z) = f(-y, x, z - xy)$$

for all $f \in L(G)$ and $(x, y, z) \in G$.

Proposition 12.5.2

(i) Z_2 is a subspace of $L(G/K_2)$ and it is the M-image of Z_1.

(ii) *If we set $\lambda_2(x, y, z) = M\lambda_1(x, y, z)M^{-1}$, that is,*

$$M\lambda_1(x, y, z) = \lambda_2(x, y, z)M, \qquad (12.55)$$

then λ_2 is a representation of G on Z_2 equivalent to λ_1 (by means of (12.55)). Moreover,

$$[\lambda_2(x, y, z)f](u, v, w) = f(y - u, v + x, w - z - yv),$$

for all (x, y, z), $(u, v, w) \in G$ and $f \in Z_2$.

(iii) *Endow the space Z_2 with the norm of $L(G/K_2)$ (recall (11.3)). Then the W_2-image of $L(\mathbb{Z}/nm\mathbb{Z})$ is exactly Z_2 and W_2 is an isometry that intertwines $\pi^\flat$ with λ_2. Moreover, if $\mathcal{F}$ is the Fourier transform on $\mathbb{Z}_{nm}$ then*

$$W_2 = MW_1\mathcal{F}^{-1}. \qquad (12.56)$$

Proof.

(i) The proof that $Z_2 \leq L(G/K_2)$ is the same of that in Proposition 12.5.1(ii); see also Exercise 12.5.3. Moreover, using the notation in (12.52), for all $f \in Z_1$ we have:

$$[Mf](\widetilde{r} + sm, \widetilde{s} + rn, w) = f(-\widetilde{s} - rn, \widetilde{r} + sm, w - \widetilde{rs} - \widetilde{r}rn - \widetilde{s}sm)$$
$$\text{(by (12.51))} \qquad = \chi(-w + \widetilde{r}rn)f(-\widetilde{s} - rn, \widetilde{r} + sm, -\widetilde{rs} - \widetilde{s}sm)$$
$$\text{(by (12.49))} \qquad = \chi(-w + \widetilde{r}rn)f(-\widetilde{s}, \widetilde{r}, -\widetilde{rs})$$
$$= \chi(-w + \widetilde{r}rn)[Mf](\widetilde{r}, \widetilde{s}, 0),$$

so that $Mf \in Z_2$.

(ii) From its definition and the fact that M is an isometry between Z_1 and Z_2 it follows that λ_2 is a G-representation on Z_2. Moreover, for all (x, y, z), $(u, v, w) \in G$, we get

$$\left[M\lambda_1(x, y, z)M^{-1}f\right](u, v, w) = \left[\lambda_1(x, y, z)M^{-1}f\right](-v, u, w - uv)$$
$$\text{(by (12.20))} \qquad = \left[M^{-1}f\right](-v - x, u - y, w - uv - z + xy - xu)$$
$$= f(u - y, v + x, w - z - yv).$$

(iii) For all $(x, y, z) \in G$, we have:

$$MW_1\mathcal{F}^{-1}\pi^\flat(x, y, z) = MW_1\pi^\sharp(x, y, z)\mathcal{F}^{-1} \qquad \text{(by (12.42))}$$
$$= M\lambda_1(x, y, z)W_1\mathcal{F}^{-1} \qquad \text{(by (12.48))}$$
$$= \lambda_2(x, y, z)MW_1\mathcal{F}^{-1} \qquad \text{(by (12.55))}.$$

Therefore, it suffices to prove directly (12.56). Indeed, for every $f \in L(\mathbb{Z}/nm\mathbb{Z})$ we have:

$$\left[MW_1\mathcal{F}^{-1}f\right](x, y, z) = \left[W_1\mathcal{F}^{-1}f\right](-y, x, z - xy)$$

$$\text{(by (12.46))} \quad = \frac{\chi(-z)}{m\sqrt{n}} \sum_{\ell=0}^{m-1} \left[\mathcal{F}^{-1}f\right](\ell n - y)\chi(\ell nx)$$

$$= \frac{\chi(-z)}{nm\sqrt{m}} \sum_{\ell=0}^{m-1} \sum_{u=0}^{nm-1} f(u)\chi(u(\ell n - y))\chi(\ell nx)$$

$$= \frac{\chi(-z)}{nm\sqrt{m}} \sum_{u=0}^{nm-1} f(u)\chi(-uy) \sum_{\ell=0}^{m-1} \chi(\ell(x+u)n)$$

$$\text{(by (2.7))} \quad = \frac{\chi(-z)}{n\sqrt{m}} \sum_{\substack{u=0 \\ u \equiv -x \bmod m}}^{nm-1} f(u)\chi(-uy)$$

$$(u = -x + tm) \quad = \frac{\chi(xy - z)}{n\sqrt{m}} \sum_{t=0}^{n-1} f(tm - x)\chi(-tmy). \qquad \square$$

Exercise 12.5.3

(1) Let G be a finite group, J an automorphism of G, $K \subset G$ a subgroup, and set $[Mf](g) = f(J(g))$, for all $g \in G$ and $f \in L(G)$. Prove that the M-image of $L(G/K)$ is $L\left(G/J^{-1}(K)\right)$.

(2) Prove that $Z_2 \leq L(G/K_2)$ (cfr. Proposition 12.5.2.(i)) by showing that $J^{-1}(K_1) = K_2$.

As a direct consequence of (12.56), we get immediately the first formulation of the main result of this section.

Corollary 12.5.4 *The Discrete Fourier Transform on* $\mathbb{Z}_{nm}$ *has the following factorization:*

$$\mathcal{F} = W_2^{-1}MW_1. \tag{12.57}$$

In other words, the diagram in Figure 12.2 is commutative.

We now introduce some notation in order to give a second version of (12.57). We define the linear operators $C_1 \colon Z_1 \to L\left(\mathbb{Z}/n\mathbb{Z} \times \mathbb{Z}/m\mathbb{Z}\right)$ and $C_2 \colon Z_2 \to L\left(\mathbb{Z}/m\mathbb{Z} \times \mathbb{Z}/n\mathbb{Z}\right)$ by setting

$$[C_1f_1](\tilde{s}, \tilde{r}) = f_1(\tilde{s}, \tilde{r}, 0) \quad \text{and} \quad [C_2f_2](\tilde{r}, \tilde{s}) = f_1(\tilde{r}, \tilde{s}, 0),$$

for all $f_j \in Z_j$, $j = 1, 2$, $0 \leq \tilde{s} \leq n - 1$ and $0 \leq \tilde{r} \leq m - 1$. From (12.45) and (12.53) it follows that C_1 and C_2 are isomorphisms of vector spaces. Then we

$$L(\mathbb{Z}/nm\mathbb{Z}) \xrightarrow{W_1} Z_1$$

$$\mathcal{F} \downarrow \qquad \qquad \downarrow M$$

$$L(\mathbb{Z}/nm\mathbb{Z}) \xrightarrow{W_2} Z_2$$

Figure 12.2. The commutative diagram representing the factorization (12.57) of the Fourier transform $\mathcal{F}$. Compare it with the diagram in Figure 12.1: note that, in both cases, the DFT is connected with the action of the automorphism J.

set

$$\widetilde{W}_1 = C_1 W_1 \qquad \text{and} \qquad \widetilde{W}_2 = C_2 W_2.$$

That is, $[\widetilde{W}_1 f_1](\widetilde{s}, \widetilde{r}) = [W_1 f_1](\widetilde{s}, \widetilde{r}, 0)$, and similarly for $\widetilde{W}_2$. Finally, we define $\widetilde{M} \colon L(\mathbb{Z}/n\mathbb{Z} \times \mathbb{Z}/m\mathbb{Z}) \to L(\mathbb{Z}/m\mathbb{Z} \times \mathbb{Z}/n\mathbb{Z})$ by setting

$$[\widetilde{M} f](\widetilde{r}, \widetilde{s}) = \chi(\widetilde{rs}) f(-\widetilde{s}, \widetilde{r}).$$

Proposition 12.5.5

(i) *We have* $\widetilde{M} = C_2 M C_1^{-1}$, *that is, the diagram*

$$Z_1 \xrightarrow{\ C_1\ } L(\mathbb{Z}/n\mathbb{Z} \times \mathbb{Z}/m\mathbb{Z})$$

$$M \downarrow \qquad \qquad \downarrow \widetilde{M}$$

$$Z_2 \xrightarrow{\ C_2\ } L(\mathbb{Z}/m\mathbb{Z} \times \mathbb{Z}/n\mathbb{Z})$$

is commutative.

(ii) *The Discrete Fourier Transform on* $\mathbb{Z}_{mn}$ *may be factorized in the form:*

$$\mathcal{F} = \widetilde{W}_2^{-1} \widetilde{M} \widetilde{W}_1. \tag{12.58}$$

Proof.

(i) For $f \in L(\mathbb{Z}/n\mathbb{Z} \times \mathbb{Z}/m\mathbb{Z})$ and $(\widetilde{r}, \widetilde{s}) \in \mathbb{Z}/m\mathbb{Z} \times \mathbb{Z}/n\mathbb{Z}$ we have:

$$[C_2 M C_1^{-1} f](\widetilde{r}, \widetilde{s}) = [M C_1^{-1} f](\widetilde{r}, \widetilde{s}, 0) = [C_1^{-1} f](-\widetilde{s}, \widetilde{r}, -\widetilde{sr})$$

$$(\text{by } (12.51)) \qquad = \chi(\widetilde{sr}) f(-\widetilde{s}, \widetilde{r}).$$

(ii) From the definition of $\widetilde{W}_1, \widetilde{W}_2$, from (i) and from (12.57) it follows that

$$\widetilde{W}_2^{-1} \widetilde{M} \widetilde{W}_1 = W_2^{-1} C_2^{-1} \widetilde{M} C_1 W_1 = W_2^{-1} M W_1 = \mathcal{F}. \qquad \square$$

In other words, also the diagram in Figure 12.5 is commutative.

$$
\begin{array}{ccc}
L(\mathbb{Z}/nm\mathbb{Z}) & \xrightarrow{\;\widetilde{W}_1\;} & L\,(\mathbb{Z}/n\mathbb{Z} \times \mathbb{Z}/m\mathbb{Z}) \\
{\scriptstyle \mathcal{F}}\big\downarrow & & \big\downarrow{\scriptstyle \widetilde{M}} \\
L(\mathbb{Z}/nm\mathbb{Z}) & \xrightarrow{\;\widetilde{W}_2\;} & L\,(\mathbb{Z}/m\mathbb{Z} \times \mathbb{Z}/n\mathbb{Z})
\end{array}
$$

Figure 12.3. The commutative diagram representing the factorization (12.58) of the Fourier transform $\mathcal{F}$. Compare it with the diagram in Figure 12.2.

In order to give the third and final factorization of the DFT, we introduce the following five operators

$$D_1 : L(\mathbb{Z}/nm\mathbb{Z}) \longrightarrow L\,(\mathbb{Z}/n\mathbb{Z} \times \mathbb{Z}/m\mathbb{Z})$$
$$D_2 : L(\mathbb{Z}/nm\mathbb{Z}) \longrightarrow L\,(\mathbb{Z}/m\mathbb{Z} \times \mathbb{Z}/n\mathbb{Z})$$
$$R_1 : L\,(\mathbb{Z}/n\mathbb{Z} \times \mathbb{Z}/m\mathbb{Z}) \longrightarrow L\,(\mathbb{Z}/n\mathbb{Z} \times \mathbb{Z}/m\mathbb{Z})$$
$$R_2 : L\,(\mathbb{Z}/m\mathbb{Z} \times \mathbb{Z}/n\mathbb{Z}) \longrightarrow L\,(\mathbb{Z}/m\mathbb{Z} \times \mathbb{Z}/n\mathbb{Z})$$
$$T : L\,(\mathbb{Z}/n\mathbb{Z} \times \mathbb{Z}/m\mathbb{Z}) \longrightarrow L\,(\mathbb{Z}/m\mathbb{Z} \times \mathbb{Z}/n\mathbb{Z})$$

defined by setting

$$[D_1 f](\widetilde{s}, \widetilde{r}) = f(\widetilde{r}n + \widetilde{s})$$
$$[D_2 f](\widetilde{r}, \widetilde{s}) = f(\widetilde{s}m + \widetilde{r})$$
$$[R_1 f_1](\widetilde{s}, \widetilde{r}) = \chi(\widetilde{s}\widetilde{r})f_1(\widetilde{s}, -\widetilde{r})$$
$$[R_2 f_2](\widetilde{r}, \widetilde{s}) = \chi(-\widetilde{s}\widetilde{r})f_2(-\widetilde{r}, -\widetilde{s})$$
$$[T f_1](\widetilde{r}, \widetilde{s}) = \chi(-\widetilde{s}\widetilde{r})f_1(\widetilde{s}, \widetilde{r}),$$

for all $f \in L(\mathbb{Z}/nm\mathbb{Z})$, $f_1 \in L\,(\mathbb{Z}/n\mathbb{Z} \times \mathbb{Z}/m\mathbb{Z})$, $f_2 \in L\,(\mathbb{Z}/m\mathbb{Z} \times \mathbb{Z}/n\mathbb{Z})$, and $0 \le \widetilde{s} \le n - 1$, $0 \le \widetilde{r} \le m - 1$. Finally, we introduce the following notation: we denote by $\mathcal{F}_k$ (respectively $\mathcal{F}_k^{-1}$, I_k) the normalized Fourier transform, cf. Exercise 2.4.13, (respectively its inverse, the identity operator) on $\mathbb{Z}/k\mathbb{Z}$. Moreover, we identify $L\,(\mathbb{Z}/n\mathbb{Z} \times \mathbb{Z}/m\mathbb{Z})$ with $L\,(\mathbb{Z}/n\mathbb{Z}) \otimes L\,(\mathbb{Z}/m\mathbb{Z})$; see Section 8.7 and Section 10.5.

Proposition 12.5.6 *We have:*

$$(I_n \otimes \mathcal{F}_m)\,D_1 = \sqrt{nm}R_1\widetilde{W}_1,$$

$$\left(I_m \otimes \mathcal{F}_n^{-1}\right)D_2 = \sqrt{nm}R_2\widetilde{W}_2,$$

and

$$R_2\widetilde{M}R_1 = T.$$

Proof. Indeed, for $f \in L(\mathbb{Z}/nm\mathbb{Z})$, $(\widetilde{s}, \widetilde{r}) \in \mathbb{Z}/n\mathbb{Z} \times \mathbb{Z}/m\mathbb{Z}$, we have:

$$[(I_n \otimes \mathcal{F}_m) D_1 f] (\widetilde{s}, \widetilde{r}) = \frac{1}{\sqrt{m}} \sum_{\ell=0}^{m-1} [D_1 f](\widetilde{s}, \ell) \chi(-\ell n \widetilde{r})$$

$$= \frac{1}{\sqrt{m}} \sum_{\ell=0}^{m-1} f(\ell n + \widetilde{s}) \chi(-\ell n \widetilde{r})$$

$$\text{(by (12.46))} \quad = \sqrt{nm} \chi(\widetilde{s}\widetilde{r})[W_1 f](\widetilde{s}, -\widetilde{r}, 0)$$

$$= \sqrt{nm} \left[R_1 \widetilde{W}_1 f \right] (\widetilde{s}, \widetilde{r}).$$

Similarly,

$$\left[\left(I_m \otimes \mathcal{F}_n^{-1} \right) D_2 f \right] (\widetilde{r}, \widetilde{s}) = \frac{1}{\sqrt{n}} \sum_{t=0}^{n-1} [D_2 f](\widetilde{r}, t) \chi(tm\widetilde{s})$$

$$= \frac{1}{\sqrt{n}} \sum_{t=0}^{n-1} f(tm + \widetilde{r}) \chi(tm\widetilde{s})$$

$$= \sqrt{nm} \chi(-\widetilde{s}\widetilde{r})[W_2 f](-\widetilde{r}, -\widetilde{s}, 0)$$

$$= \sqrt{nm} \left[R_2 \widetilde{W}_2 f \right] (\widetilde{r}, \widetilde{s}).$$

Finally, for $f \in L(\mathbb{Z}/n\mathbb{Z} \times \mathbb{Z}/m\mathbb{Z})$,

$$\left[R_2 \widetilde{M} R_1 f \right] (\widetilde{r}, \widetilde{s}) = \chi(-\widetilde{s}\widetilde{r}) \left[\widetilde{M} R_1 f \right] (-\widetilde{r}, -\widetilde{s})$$

$$= [R_1 f] (\widetilde{s}, -\widetilde{r})$$

$$= \chi(-\widetilde{s}\widetilde{r}) f(\widetilde{s}, \widetilde{r}).$$

$\square$

Finally, we are in position to present the third version of (12.57), which is an operator version of the matrix factorizations in Section 5.5; see, in particular, the Vector Form in Exercise 5.5.1.

Theorem 12.5.7

$$\mathcal{F}_{nm} = D_2^{-1} (I_m \otimes \mathcal{F}_n) T (I_n \otimes \mathcal{F}_m) D_1. \tag{12.59}$$

Proof. From Proposition 12.5.5.(ii) and Proposition 12.5.6, noting also that $R_1^{-1} = R_1$, we get:

$$\mathcal{F} = \widetilde{W}_2^{-1} \widetilde{M} \widetilde{W}_1$$

$$= D_2^{-1} (I_m \otimes \mathcal{F}_n) R_2 \cdot \widetilde{M} \cdot R_1 (I_n \otimes \mathcal{F}_m) D_1$$

$$= D_2^{-1} (I_m \otimes \mathcal{F}_n) T (I_n \otimes \mathcal{F}_m) D_1.$$

$\square$

The factorization (12.59) is equivalent to the commutativity of the following diagram:

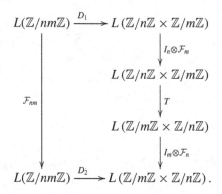

Clearly, the significance of the machinery developed in this section is *not* in the proof of (12.59) (see the following exercise), *but* in the group theoretic interpretation of each operator involved and of the various formulas obtained.

Exercise 12.5.8 Give a direct proof of (12.59), based only on the definition of the operators involved.

In the following exercise, we present an alternative approach to Proposition 12.5.1.(ii). In particular, we show how the machinery developed in Chapter 10 and Chapter 11 may be used to *derive* the exact form of the Weil-Berezin map (12.46).

Exercise 12.5.9

(1) Let d be a divisor of mn. Set $d_1 = \gcd(m, d)$, $m_1 = m/d_1$, $d_2 = d/d_1$, and $d_3 = \gcd(n, d)$, $n_1 = n/d_3$, $d_4 = d/d_3$. Prove that $d_2 | d_3$ and give an example in which $d_3 > d_2$.

(2) Arguing as in Exercise 12.3.8, and with the preceding notation, prove that the multiplicity of $\pi_{k,t,h}$ in the permutation representation $L(G/K_1)$ is equal to d_3/d_2 if $h \equiv 0 \bmod d_1$ and $k \equiv 0 \bmod d_3$, and, otherwise, it is equal to zero. In particular, $L(G/K_1)$ is not multiplicity-free, in general.

(3) Show that the multiplicity of $\pi_{0,1,0}$ in $L(G/K_1)$ is equal to 1 in two ways: (i) by using the results in (2); (ii) by showing that the space of K_1-invariant vectors in $L(\mathbb{Z}/mn\mathbb{Z})$ with respect to the representation $\pi^\sharp$ is one-dimensional and it is spanned by the function $\varphi = \frac{1}{\sqrt{m}} \sum_{r=0}^{m-1} \delta_{rn}$.

(4) Use Proposition 11.2.8 and (3) to prove Proposition 12.5.1.(iii), in particular to get the expression for W_1 in (12.46) (that is, $W_1 = T_\varphi$).

12.6 Representation theory of the Heisenberg group $H_3(\mathbb{F}_q)$

This section is based on Chapter 18 of Terras' monograph [159]; see also the exposition in [34]. Some details are similar to those in Section 12.3 so that they are omitted and/or left as exercises.

Let $\mathbb{F}_q$ be a finite field, $q = p^r$ with p a prime number. The *Heisenberg group* over $\mathbb{F}_q$ is the matrix group

$$H_3(\mathbb{F}_q) = \left\{ \begin{pmatrix} 1 & x & z \\ 0 & 1 & y \\ 0 & 0 & 1 \end{pmatrix} : x, y, z \in \mathbb{F}_q \right\}.$$

Clearly, all the identities in Exercise 12.3.1 still hold. In particular, we shall denote the elements of $H_3(\mathbb{F}_q)$ by $(x, y, z) \in \mathbb{F}_q \times \mathbb{F}_q \times \mathbb{F}_q \equiv \mathbb{F}_q^3$ with the multiplication as in (12.18).

Exercise 12.6.1

(1) From (12.21) deduce that the conjugacy classes of $H_3(\mathbb{F}_q)$ are:
   - $\mathcal{C}_w = \{(0, 0, w)\}$, $w \in \mathbb{F}_q$ (q one-element classes);
   - $\mathcal{C}_{u,v} = \{(u, v, w) : w \in \mathbb{F}_q\}$, $u, v \in \mathbb{F}_q$, $(u, v) \neq (0, 0)$
     ($q^2 - 1$ classes of q elements each).
(2) Prove also that

$$H_3(\mathbb{F}_q) \cong \mathbb{F}_q^2 \rtimes_\phi \mathbb{F}_q,$$

where $\mathbb{F}_q^2 = \{(0, v, w) : v, w \in \mathbb{F}_q\}$ and $\mathbb{F}_q = \{(x, 0, 0) : x \in \mathbb{F}_q\}$ are viewed as additive groups and ϕ is the $\mathbb{F}_q$-action on $\mathbb{F}_q^2$ given by

$$\phi_x(v, w) = (v, w + xv)$$

with $x, v, w \in \mathbb{F}_q$.

Using the notation from Theorem 11.7.1 (with $G = H_3(\mathbb{F}_q)$, $A = \mathbb{F}_q^2$ and $H = \mathbb{F}_q$), given $\chi_{s,t} \in \widehat{A}$ (cf. (7.4)), we have

$$H_{\chi_{s,t}} = \begin{cases} \{1_H\} & \text{if } t \neq 0 \\ H & \text{if } t = 0. \end{cases}$$

Indeed, from

$$^{(x,0,0)}\chi_{s,t}(v, w) = \chi_{s,t}(v, w - xv)$$
$$= \chi_{princ}(sv + t(w - xv))$$
$$= \chi_{princ}((s - tx)v + tw)$$
$$= \chi_{s-tx,t}(v, w)$$

we deduce that $^{(x,0,0)}\chi_{s,t} = \chi_{s,t}$ if and only if either $t = 0$ (in this case, the $\approx$ equivalence class of each $\chi_{s,0}$ reduces to the element $\chi_{s,0}$ itself, and therefore $H_{\chi_{s,0}} = H$), or $t \neq 0$ and $x = 0$ (so that $H_{\chi_{s,t}} = \{1_H\}$).

According to the preceding analysis, we can choose

$$X = \{\chi_{s,0} : s \in \mathbb{F}_q\} \cup \{\chi_{0,t} : t \in \mathbb{F}_q, t \neq 0\}$$

as a set of representatives of the quotient space $\widehat{A}/ \approx$ (cf. Theorem 11.6.2). Then, for every $s, u \in \mathbb{F}_q$ if we denote by $\psi_{s,u} \in \widehat{H_3(\mathbb{F}_q)}$ the character defined by

$$\psi_{s,u}(x, y, z) = \chi_{princ}(sy + ux)$$

recalling that $H_{\chi_{s,0}} = H$ (so that $A \rtimes H_{\chi_{s,0}} = H_3(\mathbb{F}_q)$) and that $\overline{\chi_u} \in \widehat{H_3(\mathbb{F}_q)}$ denotes the inflation of $\chi_u \in \widehat{H_3(\mathbb{F}_q)}/A = \widehat{H} = \mathbb{F}_q$, we have

$$\text{Ind}_{A \rtimes H_{\chi_{s,0}}}^{H_3(\mathbb{F}_q)} (\widetilde{\chi_{s,0}} \otimes \overline{\chi_u})(x, y, z) = (\widetilde{\chi_{s,0}} \otimes \overline{\chi_u})(x, y, z)$$
$$= \chi_{s,0}(y, z)\chi_u(x)$$
$$= \chi_{princ}(sy + ux)$$
$$= \psi_{s,u}(x, y, z)$$

so that

$$\text{Ind}_{A \rtimes H_{\chi_{s,0}}}^{H_3(\mathbb{F}_q)} (\widetilde{\chi_{s,0}} \otimes \overline{\chi_u}) = \psi_{s,u}.$$

On the other hand, if $t \neq 0$, then $H_{\chi_{0,t}} = \{1_H\}$ (so that $A \rtimes H_{\chi_{0,t}} = A$) and we may set

$$\pi_t := \text{Ind}_{A \rtimes H_{\chi_{0,t}}}^{H_3(\mathbb{F}_q)} (\widetilde{\chi_{0,t}}) = \text{Ind}_A^{H_3(\mathbb{F}_q)} \chi_{0,t} \in \widehat{H_3(\mathbb{F}_q)}. \tag{12.60}$$

From Theorem 11.7.1 we deduce that $\widehat{H_3(\mathbb{F}_q)}$ consists exactly of the q^2 one-dimensional representations $\psi_{s,u}$, $s, u \in \mathbb{F}_q$, and the $q - 1$ representations π_t, $t \in \mathbb{F}_q^*$, of dimension $[H_3(\mathbb{F}_q) : A] = |H| = |\mathbb{F}_q| = q$.

Exercise 12.6.2 Use (12.60) to show that a matrix realization of π_t, $t \in \mathbb{F}_q^*$, is given by

$$U(x, y, z) = \chi_{princ}(tz)D(ty)W(x),$$

for all $x, y, z \in \mathbb{F}_q$, where $D(ty)$ is the $q \times q$ diagonal matrix

$$D(ty) = \begin{pmatrix} 1 & 0 & 0 & 0 & \cdots & 0 \\ 0 & \chi(-ty) & 0 & 0 & \cdots & 0 \\ 0 & 0 & \chi(-\alpha ty) & 0 & \cdots & 0 \\ \vdots & \vdots & \vdots & & \ddots & \vdots \\ 0 & 0 & 0 & 0 & \cdots & \chi(-\alpha^{q-2}ty) \end{pmatrix}$$

α being a generator of the cyclic group $\mathbb{F}_q^*$, and $W(x)$ being the $q \times q$ permutation matrix defined by

$$W(x)_{i,j} = \delta_i(j + x),$$

for all $i, j \in \mathbb{F}_q$.

Hint: Use equation (12.22) and observe that $S = \{(i, 0, 0) : i \in \mathbb{F}_q\} = H = \mathbb{F}_q$ is a system of representatives for the left cosets of $A = \mathbb{F}_q^2$ in $G = H_3(\mathbb{F}_q)$. Use the identities

$$(-i, 0, 0)(0, 0, z)(j, 0, 0) = (j - i, 0, z)$$
$$(-i, 0, 0)(0, y, 0)(j, 0, 0) = (j - i, y, -iy)$$
$$(-i, 0, 0)(x, 0, 0)(j, 0, 0) = (j - i + x, 0, 0)$$

for all $i, j, x, y, z \in \mathbb{F}_q$. To get the matrix $D(ty)$ set $i, j = 0, 1, \alpha, \alpha^2, \ldots, \alpha^{q-2}$.

13

Hecke algebras and multiplicity-free triples

In this chapter we develop the basic theory of finite multiplicity-free triples. This is a subject that has not yet received the attention it deserves. As far as we know, the only book that treats this topic is Macdonald's [105]. The classical theory of finite Gelfand pairs, which constitutes a particular yet fundamental case, was essentially covered in our first monograph [29]. Other references on the material of this chapter include [139, 140], [37], [152], and [25].

13.1 Preliminaries and notation

Let G be a finite group and $K \leq G$ a subgroup. We assume all the basic notation in Section 11.1 and Section 11.3 (the latter with $H = K$). In addition, we suppose that χ is a one-dimensional representation of K. We consider the representation space $\text{Ind}_K^G \mathbb{C}$ of $\text{Ind}_K^G \chi$ as a subspace of the group algebra $L(G)$ (see Example 11.1.9) and we define $\psi \in L(K)$ by setting

$$\psi(k) = \frac{1}{|K|} \overline{\chi(k)} \equiv \frac{1}{|K|} \chi \left(k^{-1} \right) \tag{13.1}$$

for all $k \in K$. Then, regarding $L(K)$ as a subalgebra of $L(G)$, we define the convolution operator $P \colon L(G) \to L(G)$ by setting $Pf = f * \psi$, that is,

$$[Pf](g) = \frac{1}{|K|} \sum_{k \in K} f(gk) \chi(k)$$

for all $f \in L(G)$ and $g \in G$.

Proposition 13.1.1 The function ψ satisfies the identities

$$\psi * \psi = \psi \quad \text{and} \quad \psi^* = \psi. \tag{13.2}$$

460

Moreover, P is the orthogonal projection of $L(G)$ onto $\mathrm{Ind}_K^G \mathbb{C}$. In other words,

$$\mathrm{Ind}_K^G \mathbb{C} = \{f * \psi : f \in L(G)\} \equiv \{f \in L(G) : f * \psi = f\}. \qquad (13.3)$$

Proof. The first identity in (13.2) follows from (10.36) and, together with the first formula in (10.34), ensures that P is an idempotent. The second identity follows immediately from the analogous properties of characters (cf. Proposition 10.2.15.(ii)). This, together with the second formula in (10.34), implies that P is self-adjoint. This shows that P is an orthogonal projection. Moreover, from (11.16) we deduce that

$$[Pf](g) = [f * \psi](g) = \frac{1}{|K|} \sum_{k \in K} f(gk)\chi(k) = f(g)\frac{1}{|K|} \sum_{k \in K} 1 = f(g)$$

for all $f \in \mathrm{Ind}_K^G \mathbb{C}$ and $g \in G$, that is, $Pf = f$ (and, in particular, $\mathrm{Ran}P \supseteq \mathrm{Ind}_K^G \mathbb{C}$). Finally, let us show that the range of P is contained in (and therefore equals) $\mathrm{Ind}_K^G \mathbb{C}$. Indeed, for all $f \in L(G)$, $g \in G$ and $k_1 \in K$ we have

$$[Pf](gk_1) = \frac{1}{|K|} \sum_{k \in K} f(gk_1 k)\chi(k)$$

$$(k_2 = k_1 k) \qquad = \frac{1}{|K|} \sum_{k_2 \in K} f(gk_2)\chi(k_1^{-1} k_2)$$

$$= \overline{\chi(k_1)}[Pf](g),$$

that is, Pf satisfies (11.16) and therefore $Pf \in \mathrm{Ind}_K^G \mathbb{C}$. We conclude that $\mathrm{Ran}P = \mathrm{Ind}_K^G \mathbb{C}$. $\qquad \square$

Let now $J \subseteq \widehat{G}$ denote the set of all irreducible G-representations contained in $\mathrm{Ind}_K^G \chi$. For $(\theta, W_\theta) \in J$, denote by $m_\theta > 0$ its multiplicity in $\mathrm{Ind}_K^G \chi$, that is,

$$\mathrm{Ind}_K^G \chi \sim \bigoplus_{\theta \in J} m_\theta \theta. \qquad (13.4)$$

From Corollary 10.6.6 we deduce that $\mathrm{Ind}_K^G \chi$ is multiplicity free (that is, $m_\theta = 1$ for all $\theta \in J$) if and only if $\mathrm{End}_G(\mathrm{Ind}_K^G \chi)$ is commutative, and, if this is the case, Corollary 10.6.7 ensures that

$$\mathrm{End}_G(\mathrm{Ind}_K^G \mathbb{C}) \cong \mathbb{C}^J. \qquad (13.5)$$

Finally, note that now (11.30) becomes $G_s = K \cap sKs^{-1}$, and (11.32) becomes

$$\mathcal{S}_0 = \{s \in \mathcal{S} : \chi(x) = \chi(s^{-1}xs), \ \forall x \in G_s\}. \qquad (13.6)$$

13.2 Hecke algebras

Definition 13.2.1 The *Hecke algebra* $\mathcal{H}(G, K, \chi)$ associated with G, K, and χ, is

$$\mathcal{H}(G, K, \chi) = \left\{ f \in L(G) : f(k_1 g k_2) = \overline{\chi(k_1 k_2)} f(g), \text{ for all } g \in G, k_1, k_2 \in K \right\}.$$

Note that, in the notation of Definition 11.4.1, we have

$$\mathcal{H}(G, K, \chi) = \mathcal{V}(G, K, K, \chi, \chi).$$

Remark 13.2.2 When $\chi = \iota_K$ (see Example 11.1.6), the Hecke algebra $\mathcal{H}(G, K, \chi)$ equals the subalgebra of all *bi-K-invariant* functions

$$L(K \backslash G / K) = \{ f \in L(G) : f(k_1 g k_2) = f(g), \text{ for all } g \in G, k_1, k_2 \in K \}.$$

Note that, under the isomorphism (11.13), $L(K \backslash G / K)$ corresponds to the subspace $L(G/K)^K$ of all functions in $L(G/K)$ that are invariant under the action of K, that is, that are constant on the orbits of K on G/K.

Theorem 13.2.3 $\mathcal{H}(G, K, \chi)$ *is an involutive subalgebra of* $L(G)$. *Moreover,*

(i) $\mathcal{H}(G, K, \chi)$ *is contained in* $\mathrm{Ind}_K^G \mathbb{C}$ *and in fact*

$$\mathcal{H}(G, K, \chi) = \{ \psi * f * \psi : f \in L(G) \} \equiv \{ f \in L(G) : f = \psi * f * \psi \}.$$

(ii) *The map*

$$\begin{aligned}
\mathcal{H}(G, K, \chi) &\longrightarrow \mathrm{End}_G \left(\mathrm{Ind}_K^G \mathbb{C} \right) \\
f &\longmapsto T_f|_{\mathrm{Ind}_K^G \mathbb{C}}
\end{aligned} \tag{13.7}$$

is a $*$-anti-isomorphism of algebras and

$$\mathrm{Ker} T_f \supseteq \left[\mathrm{Ind}_K^G \mathbb{C} \right]^\perp = \mathrm{Ker} P$$

(see Proposition 13.1.1), for all $f \in \mathcal{H}(G, K, \chi)$.

Proof. We leave it to the reader the easy task to check that the vector space $\mathcal{H}(G, K, \psi)$ is closed under convolution and involution, thus showing that it is an involutive subalgebra of $L(G)$.

(i) Suppose that $f = \psi * f * \psi$, that is, $f = \frac{1}{|K|^2}\overline{\chi} * f * \overline{\chi}$. Then for all $k_1, k_2 \in K, g \in G$, we have

$$f(k_1 g k_2) = \frac{1}{|K|^2}[\overline{\chi} * f * \overline{\chi}](k_1 g k_2)$$

$$= \frac{1}{|K|^2} \sum_{\substack{r \in k_2^{-1}g^{-1}K \\ k \in K}} \overline{\chi(k_1 g k_2 r)} f(r^{-1}k)\overline{\chi(k^{-1})}$$

$$(u = k_2 r \text{ and } h = k_2 k) = \frac{1}{|K|^2} \sum_{\substack{u \in g^{-1}K \\ h \in K}} \overline{\chi(k_1 g u)} f(u^{-1}h)\overline{\chi(h^{-1}k_2)}$$

$$= \frac{1}{|K|^2} \sum_{\substack{u \in g^{-1}K \\ h \in K}} \overline{\chi(k_1)}\overline{\chi(gu)} f(u^{-1}h)\overline{\chi(h^{-1})}\overline{\chi(k_2)}$$

$$= \frac{1}{|K|^2}\overline{\chi(k_1)} \cdot [\overline{\chi} * f * \overline{\chi}](g) \cdot \overline{\chi(k_2)}$$

$$= \overline{\chi(k_1)}f(g)\overline{\chi(k_2)},$$

so that $f \in \mathcal{H}(G, K, \chi)$.

Vice versa, if $f \in \mathcal{H}(G, K, \chi)$ then, for all $g \in G$ and $k_1, k_2 \in K$, we have:

$$[\psi * f * \psi](g) = \frac{1}{|K|^2}[\overline{\chi} * f * \overline{\chi}](g)$$

$$= \frac{1}{|K|^2} \sum_{\substack{r \in g^{-1}K \\ k_2 \in K}} \overline{\chi(gr)} f(r^{-1}k_2)\overline{\chi(k_2^{-1})}$$

$$(\text{setting } k_1 = gr) \quad = \frac{1}{|K|^2} \sum_{k_1, k_2 \in K} \overline{\chi(k_1)} f(k_1^{-1}g k_2)\overline{\chi(k_2^{-1})}$$

$$(f \in \mathcal{H}(G, K, \chi)) \quad = f(g).$$

It is now easy to check that $\mathcal{H}(G, K, \psi)$ is contained in $\mathrm{Ind}_K^G\mathbb{C}$: indeed, if $f = \psi * f * \psi$ then

$$Pf = f * \psi = \psi * f * \psi * \psi = \psi * f * \psi = f, \qquad (13.8)$$

and we can invoke (13.3).

(ii) Let $f \in \mathcal{H}(G, K, \chi)$. Then if $f' \in \mathrm{Ker}P$ we have

$$T_f f' = f' * f = f' * \psi * f * \psi = [Pf'] * f * \psi = 0,$$

so that $f' \in \mathrm{Ker}T_f$. This shows the inclusion $\mathrm{Ker}P \subseteq \mathrm{Ker}T_f$.

Also, if $f'' \in \mathrm{Ind}_K^G\mathbb{C}$ we have

$$P(T_f f'') = P(f'' * f) = P(f'' * \psi * f) = f'' * \psi * f * \psi = f'' * f = T_f f'',$$

that is, $T_f(\mathrm{Ind}_K^G \mathbb{C}) \subseteq \mathrm{Ind}_K^G \mathbb{C}$. It follows that the restriction of the anti-isomorphism (10.33) to the subalgebra $\mathcal{H}(G, K, \chi)$ yields the desired anti-isomorphism (13.7). $\qquad\square$

The following is a useful computational rule.

Lemma 13.2.4 *For all $f_1 \in \mathcal{H}(G, K, \chi)$ and $f_2 \in L(G)$ we have*

$$[f_1 * \psi * f_2 * \psi](1_G) = [f_1 * f_2](1_G). \tag{13.9}$$

Proof. Indeed, from (13.8) we deduce $f_1 * \psi * f_2 * \psi = f_1 * f_2 * \psi$ so that

$$[f_1 * \psi * f_2 * \psi](1_G) = [f_1 * f_2 * \psi](1_G)$$
$$= \sum_{h \in G} \sum_{k \in K} f_1(kh) f_2(h^{-1}) \psi(k^{-1}) = [\psi * f_1 * f_2](1_G) = [f_1 * f_2](1_G).$$

$\qquad\square$

Definition 13.2.5 The *Curtis and Fossum basis* of $\mathcal{H}(G, K, \chi)$ is the set $\{a_s : s \in \mathcal{S}_0\}$ of functions in $L(G)$ defined by setting

$$a_s(g) = \begin{cases} \frac{1}{|K|} \overline{\chi(k_1)\chi(k_2)} & \text{if } g = k_1 s k_2 \text{ for some } k_1, k_2 \in K \\ 0 & \text{if } g \notin KsK \end{cases} \tag{13.10}$$

for all $g \in G$.

Note that (13.10) is well-defined: indeed, if $k_1 s k_2 = k_3 s k_4$ then by Lemma 11.3.1 there exists $x \in G_s$ such that $k_1 = k_3 x$ and $k_2 = s^{-1} x^{-1} s k_4$, and therefore

$$\chi(k_1)\chi(k_2) = \chi(k_3)\chi(k_4)\chi(x)\chi(s^{-1}x^{-1}s) = \chi(k_3)\chi(k_4),$$

because $s \in \mathcal{S}_0$ (see (13.6)). See also Lemma 13.2.6 below.

Clearly, for each $f \in \mathcal{H}(G, K, \chi)$ we have:

$$f = |K| \sum_{s \in \mathcal{S}_0} f(s) a_s. \tag{13.11}$$

Moreover, for $s, t \in \mathcal{S}_0$

$$\langle a_s, a_t \rangle_{L(G)} = \delta_{s,t} \frac{1}{|G_s|}. \tag{13.12}$$

Indeed, for $s \neq t$ the supports of a_s and a_t are disjoint, so that these functions are orthogonal. For $s = t$ we have: $\sum_{g \in KsK} |a_s(g)|^2 = \frac{|KsK|}{|K|^2} = \frac{1}{|G_s|}$ (see Remark 11.3.2). From (13.11) and (13.12) we deduce that

$$f(s) = \frac{|G_s|}{|K|} \langle f, a_s \rangle_{L(G)}. \tag{13.13}$$

Note also that changing the double cosets representatives will multiply each basis element by some root of 1 (if $\chi = \iota_K$, such a root is just 1). Finally, $a_{1_G} \equiv \psi$ and, more generally, $a_s(k_1 s k_2) = |K| \psi(k_1) \psi(k_2)$, for all $k_1, k_2 \in K$.

Lemma 13.2.6 *For all $s \in S_0$ we have*

$$a_s = \frac{|K|}{|G_s|} \psi * \delta_s * \psi.$$

Proof. Let $s \in S_0$. First of all, observe that

$$[\psi * \delta_s * \psi](g) = \frac{1}{|K|^2} \sum_{\substack{t \in g^{-1}K \\ k \in K}} \overline{\chi(gt)} \delta_s(t^{-1}k) \overline{\chi(k^{-1})} \tag{13.14}$$

for all $g \in G$. Moreover, $\delta_s(t^{-1}k) \neq 0$ only if $t^{-1}k = s$ and this forces

$$g = gt \cdot t^{-1} = gt \cdot s \cdot k^{-1} \in KsK$$

so that if $g \notin KsK$ then the above convolution is 0. Let $g = k_1 s k_2$ with $k_1, k_2 \in K$. Then (13.14) becomes (setting $t = ks^{-1}$)

$$
\begin{aligned}
[\psi * \delta_s * \psi](k_1 s k_2) &= \frac{1}{|K|^2} \sum_{k \in K} \overline{\chi(k_1 s k_2 k s^{-1})} \overline{\chi(k^{-1})} \\
(x = s k_2 k s^{-1}) \quad &= \frac{1}{|K|^2} \sum_{x \in G_s} \overline{\chi(k_1)} \overline{\chi(x)} \overline{\chi(s^{-1} x^{-1} s k_2)} \\
(\chi(x) = \chi_s(x)) \quad &= \frac{1}{|K|^2} \overline{\chi(k_1)} \overline{\chi(k_2)} \sum_{x \in G_s} \overline{\chi(x)} \chi(x) \\
&= \frac{|G_s|}{|K|^2} \overline{\chi(k_1)} \overline{\chi(k_2)} \\
&= a_s(k_1 s k_2).
\end{aligned}
$$
$\square$

For all $r, s \in S_0$ there exist complex numbers $\mu_{rst}, t \in S_0$, such that

$$a_r * a_s = \sum_{t \in S_0} \mu_{rst} a_t. \tag{13.15}$$

The numbers μ_{rst}, $r, s, t \in S_0$, are called the *structure constants* of the Hecke algebra $\mathcal{H}(G, K, \chi)$ relative to the basis $\{a_s : s \in S_0\}$.

Lemma 13.2.7 *The structure constants are given by the following formula:*

$$\mu_{rst} = |K| \sum_{g \in (KrK) \cap (tKs^{-1}K)} a_r(g) a_s(g^{-1}t),$$

for all $r, s, t \in S_0$.

Proof. On the one hand, from (13.10) and (13.15) we have

$$[a_r * a_s](t) = \frac{1}{|K|}\mu_{rst} \tag{13.16}$$

for all $r, s, t \in \mathcal{S}_0$. On the other hand, just computing the convolution, we get:

$$[a_r * a_s](t) = \sum_{g \in G} a_r(g)a_s(g^{-1}t)$$
$$= \sum_{g \in (KrK) \cap (tKs^{-1}K)} a_r(g)a_s(g^{-1}t). \tag{13.17}$$

Comparing (13.16) and (13.17), the lemma follows.					□

13.3 Commutative Hecke algebras

Definition 13.3.1 Let G be a finite group, $K \subset G$ a subgroup, and χ a one-dimensional K-representation. We say that (G, K, χ) is a *multiplicity-free triple* provided the Hecke algebra $\mathcal{H}(G, K, \chi)$ is commutative.

Moreover, we say that (G, K) is a *Gelfand pair* provided that (G, K, ι_K) is a multiplicity-free triple, that is, $\mathcal{H}(G, K, \iota_k)(\cong L(K \backslash G / K))$ is commutative.

Theorem 13.3.2 *The following conditions are equivalent.*

(a) (G, K, χ) *is a multiplicity-free triple;*
(b) *the induced representation* $\mathrm{Ind}_K^G \chi$ *decomposes without multiplicity;*
(c) $\dim W_\theta^{K,\chi} \leq 1$ *for each irreducible G-representation (θ, W_θ) (cf. Definition (11.27)).*

Moreover, if these equivalent conditions are satisfied, with the notation of Remark 11.4.10 (with $H = K$ and $v = \chi$) and (13.4), we have

$$\dim \mathcal{H}(G, K, \chi) = |J| = |\mathcal{S}_0|.$$

Proof. From Corollary 10.6.6 it follows that (G, K, χ) is a multiplicity-free triple if and only if $\mathrm{Ind}_K^G \chi$ decomposes without multiplicity; see also (13.5). Moreover, from Frobenius reciprocity (Theorem 11.2.1) this is equivalent to the fact that χ has multiplicity at most one in the restriction to K of each irreducible G-representation. Finally, if $\mathrm{Ind}_K^G \chi$ is multiplicity free, we may invoke Remark 11.4.10, (13.5), and (13.6) to conclude that $\dim \mathcal{H}(G, K, \chi) = \dim \mathbb{C}^J = |J| = |\mathcal{S}_0|$.					□

Now we examine a series of *sufficient conditions* for the commutativity of the Hecke algebra. An *anti-automorphism* of G is a bijective map $\tau : G \to G$

such that:

$$\tau(g_1 g_2) = \tau(g_2)\tau(g_1)$$

for all $g_1, g_2 \in G$. It is *involutive* if $\tau^2 = \mathrm{id}_G$, where id_G is the identity map on G. Clearly, $\tau(1_G) = 1_G$ and $\tau(g^{-1}) = \tau(g)^{-1}$ for all $g \in G$. Note that the map $\mathrm{inv} \colon G \to G$, defined by $\mathrm{inv}(g) = g^{-1}$ for all $g \in G$, is an involutory anti-automorphism, while if τ is as above, then $g \mapsto \tau(g^{-1})$ is an automorphism of G.

Let τ be an anti-automorphism of G. We define a linear map

$$L(G) \longrightarrow L(G)$$
$$f \longmapsto f^\tau$$

by setting

$$f^\tau(g) = f(\tau(g)) \qquad (13.18)$$

for all $f \in L(G)$, $g \in G$.

Given an algebra $\mathcal{A}$, a bijective linear map $\varphi \colon \mathcal{A} \to \mathcal{A}$ such that $\varphi(a_1 a_2) = \varphi(a_2)\varphi(a_1)$ for all $a_1, a_2 \in \mathcal{A}$, is called an *anti-automorphism* of $\mathcal{A}$. If in addition, $\varphi^2 = \mathrm{id}_{\mathcal{A}}$, where $\mathrm{id}_{\mathcal{A}}$ is the identity map on $\mathcal{A}$, then one says that φ is *involutive*.

Lemma 13.3.3 *Let τ be an (involutive) anti-automorphism of G. Then the map $f \mapsto f^\tau$ is an (involutive) anti-automorphism of $L(G)$.*

Proof. It is clear that the map $f \mapsto f^\tau$ is a linear isomorphism. Let $f_1, f_2, f \in L(G)$ and $g \in G$. We have

$$(f_1 * f_2)^\tau(g) = (f_1 * f_2)(\tau(g)) = \sum_{h \in G} f_1(\tau(g)h) f_2(h^{-1})$$

$$= \sum_{h \in G} f_1\left(\tau[\tau^{-1}(h)g]\right) f_2\left(\tau\left[\tau^{-1}(h)^{-1}\right]\right)$$

$$= \sum_{h \in G} f_2^\tau\left(\tau^{-1}(h)^{-1}\right) f_1^\tau\left(\tau^{-1}(h)g\right)$$

$$= \left(f_2^\tau * f_1^\tau\right)(g).$$

Moreover, if τ is involutive, so is the maps $f \mapsto f^\tau$. Indeed,

$$[(f^\tau)^\tau](g) = [f^\tau](\tau(g)) = f(\tau^2(g)) = f(g). \qquad \square$$

The next proposition is just a generalization of the following well known and easy fact: if $\mathcal{A}$ is a subalgebra of the full matrix algebra $\mathcal{M}_n(\mathbb{F})$, $n \in \mathbb{N}$ where $\mathbb{F}$ is any field, and each matrix $A \in \mathcal{A}$ is symmetric, then $\mathcal{A}$ is commutative.

Proposition 13.3.4 *Let τ be an anti-automorphism of G and $\mathcal{A}$ a subalgebra of $L(G)$ such that $f^\tau = f$ for all $f \in \mathcal{A}$. Then $\mathcal{A}$ is commutative.*

Proof. For all $f_1, f_2 \in \mathcal{A}$ we have:

$$f_1 * f_2 = (f_1 * f_2)^\tau = f_2^\tau * f_1^\tau = f_2 * f_1. \qquad \square$$

Remark 13.3.5 In Proposition 13.3.4, the anti-automorphism $f \mapsto f^\tau$ may be replaced by any anti-automorphism $\Phi \colon L(G) \to L(G)$.

Corollary 13.3.6 *Let τ be an anti-automorphism of G. Suppose that*

$$f^\tau = f \qquad \text{for all } f \in \mathcal{H}(G, K, \chi). \tag{13.19}$$

Then (G, K, χ) is a multiplicity-free triple.
   Moreover, condition (13.19) is satisfied if:

  (i) *(Bump and Ginzburg [25]) $\tau(K) = K$, $\chi^\tau = \chi$, and for every $s \in \mathcal{S}_0$ there exist $k_1, k_2 \in K$ such that $\tau(s) = k_1 s k_2$ and $\chi(k_1)\chi(k_2) = 1$;*
  (ii) *(symmetric Gelfand pairs) $\chi = \iota_K$, $\tau = \text{inv}$, and $g^{-1} \in KgK$ for all $g \in G$.*

Proof.

  (i) In this case, it is immediate to check that the elements in the Curtis-Fossum basis (Definition 13.2.5) satisfy $a_s^\tau = a_s$, for all $s \in \mathcal{S}_0$.
  (ii) This is just a particular case of (i). $\square$

Exercise 13.3.7 Assume the notation in Proposition 10.4.12 with $X = G/K$. Prove that (G, K) is a symmetric Gelfand pair (i.e. satisfies the conditions in (ii) of Corollary 13.3.6) if and only if the orbits of G on $X \times X$ are symmetric, that is, for all $x, y \in X$, the pairs (x, y) and (y, x) belong to the same G-orbit.

A group G is said to be *ambivalent* if g^{-1} is conjugate to g for all $g \in G$.

Exercise 13.3.8 Denote by $\widetilde{G}$ the diagonal subgroup of $G \times G$, that is, $\widetilde{G} = \{(g, g) : g \in G\} \cong G$.

  (1) Prove that $L(G) = \oplus_{\sigma \in \widehat{G}} M^\sigma$ (see Theorem 10.5.9) is the decomposition of $L(G)$ into irreducible $G \times G$-representations.
  (2) Deduce that $(G \times G, \widetilde{G})$ is a Gelfand pair.
  (3) Prove that the Gelfand pair $(G \times G, \widetilde{G})$ is symmetric if and only if G is ambivalent.

Exercise 13.3.9 (Weakly symmetric Gelfand pairs) Suppose that there exists $\xi \in \text{Aut}(G)$ such that $g^{-1} = K\xi(g)K$, for all $g \in G$. Show that (G, K) is a Gelfand pair; see [53].

Exercise 13.3.10 $\left(\text{Aff}(\mathbb{F}_q), U\right)$ is a Gelfand pair: this follows immediately from Exercise 12.1.8. Use the characterization of the automorphisms of $\text{Aff}(\mathbb{F}_q)$ in Exercise 12.1.11 to deduce that it is *not* weakly symmetric.

13.4 Spherical functions: intrinsic theory

In this section we introduce and develop the theory of spherical function (associated with a multiplicity-free triple) in an intrinsic way, that is, we consider and analyze all the properties of spherical functions without appealing to their explicit form as matrix coefficients (this will be treated in Section 13.5).

Let (G, K, χ) be a multiplicity-free triple.

Definition 13.4.1 An element $\phi \in \mathcal{H}(G, K, \chi)$ is called a *spherical function* if it satisfies the following conditions:

$$\phi(1_G) = 1 \tag{13.20}$$

and, for all $f \in \mathcal{H}(G, K, \chi)$ there exists $\lambda_{\phi,f} \in \mathbb{C}$ such that

$$\phi * f = \lambda_{\phi,f}\phi. \tag{13.21}$$

Condition (13.21) may be reformulated in the following way: ϕ is an eigenvector of the convolution operator T_f, for every $f \in \mathcal{H}(G, K, \chi)$. Moreover, by means of (13.20) and (13.21) we get $\lambda_{\phi,f} = [\phi * f](1_G)$. As a consequence, the following equivalent formulation of (13.21) holds (recall that, by definition of a multiplicity-free triple, the Hecke algebra $\mathcal{H}(G, K, \chi)$ is commutative):

$$\phi * f = [\phi * f](1_G)\phi = [f * \phi](1_G)\phi = f * \phi. \tag{13.22}$$

Now we give the basic functional identity satisfied by all spherical functions; it involves the function ψ defined in (13.1).

Theorem 13.4.2 *A function* $\phi \in L(G)$, $\phi \neq 0$, *is spherical if and only if it satisfies the functional identity*

$$\sum_{k \in K} \phi(gkh)\overline{\psi(k)} = \phi(g)\phi(h), \tag{13.23}$$

for all $g, h \in G$.

Proof. Suppose that $\phi \in L(G)$, $\phi \neq 0$, satisfies (13.23). Choose $h \in G$ such that $\phi(h) \neq 0$; writing (13.23) in the form $\phi(g) = \frac{1}{\phi(h)} \sum_{k \in K} \phi(gkh)\overline{\psi(k)}$

we get

$$[\phi * \psi](g) = \frac{1}{\phi(h)} \sum_{k,k_1 \in K} \phi(gk_1kh)\overline{\psi(k)}\psi(k_1^{-1})$$

$$(k_1k = k_2) \qquad = \frac{1}{\phi(h)} \sum_{k_2 \in K} \phi(gk_2h)\overline{[\psi * \psi](k_2)}$$

$$(\text{by } (13.2)) \qquad = \frac{1}{\phi(h)} \sum_{k_2 \in K} \phi(gk_2h)\overline{\psi(k_2)}$$

$$(\text{by } (13.23)) \qquad = \phi(g)$$

for all $g \in G$, showing that $\phi * \psi = \phi$. Similarly, one proves that $\psi * \phi = \phi$. As a consequence, $\psi * \phi * \psi = \psi * \phi = \phi$, that is, (cf. Theorem 13.2.3.(i)) $\phi \in \mathcal{H}(G, K, \chi)$. Then, taking $h = 1_G$ in (13.23) we get

$$\phi(g)\phi(1_G) = \sum_{k \in K} \phi(gk)\overline{\psi(k)} = [\phi * \psi](g) = \phi(g)$$

for all $g \in G$, and therefore (recall that $\phi \neq 0$) $\phi(1_G) = 1$. Finally, for all $f \in \mathcal{H}(G, K, \chi)$ and $g \in G$, we have

$$[\phi * f](g) = [\phi * f * \psi](g)$$

$$= \sum_{h \in G} \sum_{k \in K} \phi(gkh)f(h^{-1})\overline{\psi(k)}$$

$$(\text{by } (13.23)) \qquad = \phi(g) \sum_{h \in G} \phi(h)f(h^{-1})$$

$$= [\phi * f](1_G)\phi(g)$$

so that also (13.22) is satisfied. It follows that ϕ is spherical.

Conversely, suppose that ϕ is spherical. For all $g \in G$, define $F_g \in L(G)$ by setting

$$F_g(h) = \sum_{k \in K} \phi(gkh)\overline{\psi(k)},$$

for all $h \in G$. For $f \in \mathcal{H}(G, K, \chi)$ and $g, g_1 \in G$ we then have

$$[F_g * f](g_1) = \sum_{k \in K} \sum_{h \in G} \phi(gkg_1h)f(h^{-1})\overline{\psi(k)}$$

$$(\text{by } (13.22)) \qquad = [\phi * f](1_G) \sum_{k \in K} \phi(gkg_1)\overline{\psi(k)} \qquad (13.24)$$

$$= [\phi * f](1_G)F_g(g_1).$$

For all $g \in G$, we also define $J_g \in L(G)$ by setting

$$J_g(h) = \sum_{k \in K} f(hkg)\overline{\psi(k)}$$

for all $h \in G$. We claim that $J_g \in \mathcal{H}(G, K, \chi)$. Indeed,

$$[\psi * J_g * \psi](h) = \sum_{k,k_1,k_2 \in K} \psi(k_1) f(k_1^{-1}hk_2^{-1}kg)\psi(k_2)\overline{\psi(k)}$$

$$(k_3 = k_2^{-1}k) \quad = \sum_{k,k_3 \in K} [\psi * f](hk_3 g)\psi(kk_3^{-1})\psi(k^{-1})$$

$$= \sum_{k_3 \in K} f(hk_3 g)[\psi * \psi](k_3^{-1})$$

$$= \sum_{k_3 \in K} f(hk_3 g)\overline{\psi(k_3)}$$

$$= J_g(h).$$

This shows that $\psi * J_g * \psi = J_g$. Moreover, for $g_1 \in G$ we have

$$[\phi * J_{g_1}](1_G) = \sum_{h \in G} \phi(h^{-1}) \sum_{k \in K} f(hkg_1)\overline{\psi(k)}$$

$$(hk = t) \quad = \sum_{t \in G} \left[\sum_{k \in K} \psi(k^{-1})\phi(kt^{-1}) \right] f(tg_1)$$

$$= \sum_{t \in G} [\psi * \phi](t^{-1}) f(tg_1) \tag{13.25}$$

$$= \sum_{t \in G} \phi(t^{-1}) f(tg_1)$$

$$= [\phi * f](g_1)$$

$$(\text{by } (13.22)) \quad = [\phi * f](1_G)\phi(g_1).$$

It follows that, for $g, g_1 \in G$,

$$[F_g * f](g_1) = \sum_{h \in G} \sum_{k \in K} \phi(gkg_1 h)\overline{\psi(k)} f(h^{-1})$$

$$(kg_1 h = t) \quad = \sum_{t \in G} \phi(gt) \sum_{k \in K} \overline{\psi(k)} f(t^{-1}kg_1)$$

$$= [\phi * J_{g_1}](g) \tag{13.26}$$

$$(\text{by } (13.22)) \quad = [\phi * J_{g_1}](1_G)\phi(g)$$

$$(\text{by } (13.25)) \quad = [\phi * f](1_G)\phi(g_1)\phi(g).$$

From (13.24) and (13.26) we get

$$[\phi * f](1_G)F_g(g_1) = [\phi * f](1_G)\phi(g_1)\phi(g),$$

and taking $f \in \mathcal{H}(G, K, \chi)$ such that $[\phi * f](1_G) \neq 0$ this yields

$$\sum_{k \in K} \phi(gkg_1)\overline{\psi(k)} = F_g(g_1) = \phi(g_1)\phi(g),$$

which is exactly (13.23) with h replaced by g_1. In order to complete the proof, we are only left to show the existence of such an f. Since $\phi \neq 0$, we can find $f_1 \in L(G)$ such that $[\phi * f_1](1_G) \neq 0$. Then, keeping in mind (13.9), we have that $f = \psi * f_1 * \psi \in \mathcal{H}(G, K, \chi)$ satisfies $[\phi * f](1_G) \neq 0$. $\square$

Definition 13.4.3 A linear functional $\Phi \colon \mathcal{H}(G, K, \chi) \to \mathbb{C}$ is *multiplicative* if

$$\Phi(f_1 * f_2) = \Phi(f_1)\Phi(f_2)$$

for all $f_1, f_2 \in \mathcal{H}(G, K, \chi)$.

Theorem 13.4.4 *Let ϕ be a spherical function and set*

$$\Phi(f) = \sum_{g \in G} f(g)\phi(g^{-1}) \equiv [f * \phi](1_G) \tag{13.27}$$

for all $f \in \mathcal{H}(G, K, \chi)$. Then Φ is a linear multiplicative functional on $\mathcal{H}(G, K, \chi)$. Moreover, any nontrivial linear multiplicative functional on $\mathcal{H}(G, K, \chi)$ is of this form.

Proof. Let Φ as in (13.27). For $f_1, f_2 \in \mathcal{H}(G, K, \chi)$, by means of a repeated application of (13.22), we get:

$$\begin{aligned}
\Phi(f_1 * f_2) &= [(f_1 * f_2) * \phi](1_G) \\
&= [f_1 * (f_2 * \phi)](1_G) \\
&= [[f_2 * \phi](1_G)f_1 * \phi](1_G) \\
&= [f_1 * \phi](1_G)[f_2 * \phi](1_G) \\
&= \Phi(f_1)\Phi(f_2).
\end{aligned}$$

This shows that Φ is multiplicative. Conversely, suppose that Φ is a nontrivial multiplicative linear functional on $\mathcal{H}(G, K, \chi)$. We extend Φ to a linear functional on the whole $L(G)$ by considering the map $f_2 \mapsto \Phi(\psi * f_2 * \psi)$ for all $f_2 \in L(G)$. By Riesz theorem, we can find an element $\varphi \in L(G)$ such that

$$\Phi(\psi * f_2 * \psi) = \sum_{g \in G} f_2(g)\varphi(g^{-1}) \tag{13.28}$$

for all $f_2 \in L(G)$. From (13.9) we deduce that if $f_1 \in \mathcal{H}(G, K, \chi)$ then

$$\Phi(f_1) = [f_1 * \varphi](1_G) = [f_1 * \psi * \varphi * \psi](1_G).$$

Therefore, setting $\phi = \psi * \varphi * \psi \in \mathcal{H}(G, K, \chi)$, we then have

$$\Phi(f_1) = [\phi * f_1](1_G) \tag{13.29}$$

for all $f_1 \in \mathcal{H}(G, K, \chi)$. With this position, (13.9) also yields

$$\Phi(\psi * f_2 * \psi) = [\phi * \psi * f_2 * \psi](1_G) = [\phi * f_2](1_G) = \sum_{h \in G} \phi(h) f_2(h^{-1})$$

for all $f_2 \in L(G)$, and therefore in (13.28) the function φ may be replaced by the function ϕ. Since Φ is multiplicative, for $f_1 \in \mathcal{H}(G, K, \chi)$ and $f_2 \in L(G)$ the expression

$$
\begin{aligned}
\Phi(f_1 * \psi * f_2 * \psi) &= [\phi * f_1 * \psi * f_2 * \psi](1_G) \\
\text{(by (13.9))} \qquad &= [\phi * f_1 * f_2](1_G) \\
&= \sum_{h \in G} [\phi * f_1](h) f_2(h^{-1})
\end{aligned}
$$

must be equal to

$$\Phi(f_1)\Phi(\psi * f_2 * \psi) = \sum_{h \in G} \Phi(f_1)\phi(h) f_2(h^{-1}).$$

Since $f_2 \in L(G)$ was arbitrary, we get the equality $[\phi * f_1](h) = \Phi(f_1)\phi(h)$, so that, in particular, ϕ satisfies condition (13.21). Taking $h = 1_G$ and choosing $f_1 \in \mathcal{H}(G, K, \chi)$ such that $\Phi(f_1) \neq 0$ (recall that Φ is nontrivial), and keeping in mind (13.29), this gives $\Phi(f_1) = [\phi * f_1](1_G) = \Phi(f_1)\phi(1_G)$. It follows that $\phi(1_G) = 1$. In conclusion, ϕ is a spherical function. $\qquad \square$

Corollary 13.4.5 *The number of distinct spherical functions is equal to* $|J|$, *the number of irreducible G-representations contained in* $\mathrm{Ind}_K^G \chi$.

Proof. We have $\mathcal{H}(G, K, \chi) \cong \mathbb{C}^J$ (see (13.5)) and every linear multiplicative functional on $\mathbb{C}^J$ is of the form $\mathbb{C}^J \ni \lambda \mapsto \lambda(\theta)$, for a fixed $\theta \in J$. $\qquad \square$

In the following we use the notation in (10.9).

Proposition 13.4.6 *Let ϕ and μ be two distinct spherical functions. Then the following holds.*

  (i) $\phi(g^{-1}) = \overline{\phi(g)}$ *for all $g \in G$, that is, $\phi^* = \phi$;*
  (ii) $\phi * \mu = 0$;
  (iii) $\langle \lambda_G(g_1)\phi, \lambda_G(g_2)\mu \rangle_{L(G)} = 0$ *for all $g_1, g_2 \in G$, in particular ϕ and μ are orthogonal: $\langle \phi, \mu \rangle_{L(G)} = 0$.*

Proof.

(i) By definition of a spherical function, one has

$$\phi^* * \phi = [\phi^* * \phi](1_G)\phi = \|\phi\|^2\phi.$$

As a consequence, since $(\phi^* * \phi)^* = \phi^* * \phi$, we have

$$[\phi^* * \phi](g) = \overline{[\phi^* * \phi](g^{-1})} = \overline{[\phi^* * \phi](1_G)} \cdot \overline{\phi(g^{-1})} = \|\phi\|^2\overline{\phi(g^{-1})}$$

and therefore we must have $\phi = \phi^*$.

(ii) By commutativity,

$$[\phi * \mu](1_G)\phi(g) = [\phi * \mu](g) = [\mu * \phi](g) = [\mu * \phi](1_G)\mu(g).$$

Therefore, if $\phi \neq \mu$, necessarily $[\phi * \mu](1_G) = [\mu * \phi](1_G) = 0$ and this also yields $\phi * \mu = 0$.

(iii) Let $g_1, g_2 \in G$. Then

$$\langle \lambda_G(g_1)\phi, \lambda_G(g_2)\mu \rangle = \langle \phi, \lambda_G(g_1^{-1}g_2)\mu \rangle = \sum_{h \in G} \phi(h)\overline{\mu\left[(g_1^{-1}g_2)^{-1}h\right]}$$

$$= \sum_{h \in G} \phi(h)\mu^*(h^{-1}g_1^{-1}g_2) = [\phi * \mu^*](g_1^{-1}g_2) = [\phi * \mu](g_1^{-1}g_2) = 0,$$

where the last equality follows from (ii).　　　　□

Theorem 13.4.7 *For each spherical function ϕ define $U_\phi = \langle \lambda_G(g)\phi : g \in G \rangle$, the subspace of $L(G)$ spanned by all translates of ϕ. Then*

$$\mathrm{Ind}_K^G\mathbb{C} = \bigoplus_\phi U_\phi,$$

where the sum runs over all spherical functions, is the decomposition of $\mathrm{Ind}_K^G\mathbb{C}$ *into irreducible G-representations.*

Proof. Each subspace U_ϕ is clearly G-invariant and contained in $\mathrm{Ind}_K^G\mathbb{C}$ (recall Theorem 13.2.3). Moreover, by virtue of Lemma 13.4.6.(iii), if ϕ and μ are distinct then the spaces U_ϕ and U_μ are orthogonal. Finally, we can invoke Corollary 13.4.5 to conclude that each U_ϕ is irreducible and that the sum $\bigoplus_\phi U_\phi$ exhausts the whole $\mathrm{Ind}_K^G\mathbb{C}$.　　　　□

The space U_ϕ is called the *spherical representation* associated with the spherical function ϕ.

13.5 Harmonic analysis on the Hecke algebra $\mathcal{H}(G, K, \chi)$

The first purpose of this section is to present a different realization of spherical functions as matrix coefficients associated with spherical representations.

Suppose again that (G, K, χ) is a multiplicity-free triple. Let J be as in (13.5) (but now $m_\theta = 1$ for all $\theta \in J$). For each $\theta \in J$ choose a vector $w^\theta \in W^{K,\chi}$ of norm one (recall (11.27)). Such w^θ is unique up to a scalar multiple of modulus one (usually called a *phase factor*); see Theorem 13.3.2. Moreover, we are in the multiplicity free case of Theorem 10.6.3: for each $\theta \in J$ we may choose $T_\theta \in \mathrm{Hom}_G(W_\theta, \mathrm{Ind}_K^G \mathbb{C})$, which is also an isometry, so that $\mathrm{Hom}_G(W_\theta, \mathrm{Ind}_K^G \mathbb{C}) = \langle T_\theta \rangle$ and

$$\mathrm{Ind}_K^G \mathbb{C} = \bigoplus_{\theta \in J} T_\theta W_\theta \tag{13.30}$$

is an explicit orthogonal decomposition. Clearly, our choice of w^θ and (11.28) in Proposition 11.2.8 may be used to get an explicit form for $T_\theta = T_{w^\theta}$:

$$[T_\theta w](g) = \sqrt{\frac{d_\theta}{|G/K|}} \langle w, \theta(g) w^\theta \rangle_{W_\theta}, \tag{13.31}$$

for all $w \in W_\theta$ and $g \in G$. Again, T_θ is defined up to a phase factor. Note that now the map (13.7) is a $*$-isomorphism because the algebras involved are commutative.

Proposition 13.5.1 *Let* (13.30) *be an explicit decomposition of* $\mathrm{Ind}_K^G \mathbb{C}$ *into irreducible, inequivalent G-representation. Then for* $f \in \mathcal{H}(G, K, \chi)$ *the following hold:*

(i) *the decomposition of* $\mathrm{Ind}_K^G \mathbb{C}$ *into eigenspaces of the convolution operator* T_f *is given by* (13.30)*;*

(ii) *if* $\lambda_f(\theta)$ *denotes the eigenvalue of* T_f *associated with the subspace* $T_\theta W_\theta$ *then the map*

$$\mathcal{H}(G, K, \chi) \longrightarrow \mathbb{C}^J$$
$$f \longmapsto \lambda_f,$$

is an algebra isomorphism.

Proof.

(i) By Theorem 13.2.3.(ii) and multiplicity freeness of $\mathrm{Ind}_K^G \chi$, the convolution operator T_f intertwines each irreducible representation $T_\theta W_\theta$ with itself so that, by Schur's lemma, it is a multiple of the identity on each irreducible space.

(ii) If $f_1 \in \mathcal{H}(G, K, \chi)$, $f \in \mathrm{Ind}_K^G \mathbb{C}$, and $f = \sum_{\theta \in J} f_\theta$ with $f_\theta \in T_\theta W_\theta$, then $T_{f_1}(f) = \sum_{\theta \in J} \lambda_{f_1}(\theta) f_\theta$. Therefore $T_{f_1 * f_2} = T_{f_1} T_{f_2}$ yields

$$\lambda_{f_1 * f_2} = \lambda_{f_1} \lambda_{f_2}$$

for all $f_1, f_2 \in \mathcal{H}(G, K, \chi)$. $\qquad\square$

An explicit expression of λ_f will be given in Proposition 13.5.4.

For each θ define $\phi^\theta \in L(G)$ by setting

$$\phi^\theta(g) = \langle w^\theta, \theta(g)w^\theta \rangle_{W_\theta} \tag{13.32}$$

for all $g \in G$.

Theorem 13.5.2 *The function ϕ^θ is spherical and it is associated with W_θ, that is, in the notation of Theorem 13.4.7, we have $U_{\phi^\theta} = T_\theta W_\theta$. Moreover, the spherical functions satisfy the following orthogonality relations:*

$$\langle \phi^\theta, \phi^\rho \rangle_{L(G)} = \frac{|G|}{d_\theta} \delta_{\theta,\rho}, \tag{13.33}$$

for $\theta, \rho \in J$.

Proof. By (13.31) we have $\phi^\theta = \sqrt{\frac{|G/K|}{d_\theta}} T_\theta w^\theta$ and therefore, by Proposition 11.2.8, ϕ^θ belongs to the subspace of $\mathrm{Ind}_K^G \mathbb{C}$ isomorphic to W_θ, namely to $T_\theta W_\theta$ in (13.30). Now we use the functional identity (13.23) to show that ϕ^θ is a spherical function. We need to prove a preliminary identity. First of all, we choose an orthonormal basis $\{u_i : i = 1, 2, \ldots, d_\theta\}$ for W_θ in the following way. Let $\mathrm{Res}_K^G \theta = \chi \oplus \left(\oplus_\eta m_\eta \eta \right)$ be the decomposition of $\mathrm{Res}_K^G \theta$ into irreducible K-representations (the η's are pairwise distinct and each of them is distinct from χ; m_η is the multiplicity of η). We suppose that $u_1 = w^\theta$ and that each u_i, $2 \leq i \leq d_\theta$, belongs to some irreducible W_η. Then by (10.24) we have

$$\sum_{k \in K} \langle u_1, \theta(k)u_1 \rangle_{W_\theta} \langle \theta(k)u_i, u_j \rangle_{W_\theta} = |K| \delta_{1i} \delta_{1j}. \tag{13.34}$$

Since $\theta(k)u_1 = \chi(k)u_1$ we have $\psi(k) = \frac{1}{|K|} \langle u_1, \theta(k)u_1 \rangle$ and therefore (13.34) may be written in the form

$$\left\langle \sum_{k \in K} \psi(k)\theta(k)u_i, u_j \right\rangle_{W_\theta} = \delta_{1i} \delta_{1j}$$

and this yields

$$\sum_{k \in K} \psi(k)\theta(k)u_i = \delta_{i1} u_1 \tag{13.35}$$

for all $i = 1, 2, \ldots, d_\theta$. We are now in position to check (13.23):

$$\sum_{k \in K} \phi^\theta(gkh)\overline{\psi(k)} = \sum_{k \in K} \langle w^\theta, \theta(gkh)w^\theta \rangle_{W_\theta} \overline{\psi(k)}$$

$$=_* \sum_{i=1}^{d_\theta} \langle \theta(g^{-1})w^\theta, u_i \rangle_{W_\theta} \sum_{k \in K} \overline{\langle \theta(kh)u_1, u_i \rangle_{W_\theta} \psi(k)}$$

$$= \sum_{i=1}^{d_\theta} \langle \theta(g^{-1})w^\theta, u_i \rangle_{W_\theta} \overline{\langle \theta(h)u_1, \sum_{k \in K} \psi(k^{-1})\theta(k^{-1})u_i \rangle_{W_\theta}}$$

(by (13.35)) $= \phi^\theta(g)\phi^\theta(h),$

where equality $=_*$ follows from $\theta(kh)u_1 = \sum_{i=1}^{d_\theta} \langle \theta(kh)u_1, u_i \rangle_{W_\theta} u_i$: recall that $\{u_i : i = 1, 2, \ldots, d_\theta\}$ is an orthonormal basis. Finally, (13.33) is a particular case of (10.24). $\square$

Remark 13.5.3 Suppose that (G, K, χ) is a multiplicity-free triple. Then $(G, K, \overline{\chi})$ is also multiplicity-free. Indeed, $\mathcal{H}(G, K, \overline{\chi}) = \overline{\mathcal{H}(G, K, \chi)}$, that is, the functions in $\mathcal{H}(G, K, \overline{\chi})$ are the conjugates of the functions in $\mathcal{H}(G, K, \chi)$. Moreover, if $\{\phi^\theta : \theta \in J\}$ are the spherical functions with respect to χ then their conjugates $\{\overline{\phi^\theta} : \theta \in J\}$ are the spherical functions with respect to $\overline{\chi}$ (this may be deduced, for instance, directly from Definition 13.4.1). Finally, from (11.18) it follows that $\chi^{\mathrm{Ind}_K^G \overline{\chi}} = \overline{\chi^{\mathrm{Ind}_K^G \chi}}$ and therefore $\theta \in \widehat{G}$ is contained in $\mathrm{Ind}_K^G \chi$ if and only if its conjugate θ' (cf. Section 10.5) is contained in $\mathrm{Ind}_K^G \overline{\chi}$. Indeed, $\overline{\phi^\theta}$ equals the spherical function with respect to $\overline{\chi}$ associated with θ'.

Moreover, from (13.32) it follows that ϕ^θ is *not* a matrix coefficient of θ but of θ'. This happens because ϕ^θ belongs to the sub-representation of $\mathrm{Ind}_K^G \mathbb{C} \leq L(G)$ isomorphic to θ but, by Theorem 10.5.9, the restriction of the left regular representation λ to $M_{*,1}^\theta$ is isomorphic to θ', that is, $W_\theta \sim M_{*,1}^{\theta'}$.

The *spherical Fourier transform* is the linear map

$$\mathcal{F}: \mathcal{H}(G, K, \chi) \longrightarrow L(J)$$

defined by setting, for $f \in \mathcal{H}(G, K, \chi)$ and $\theta \in J$,

$$[\mathcal{F}f](\theta) = \sum_{g \in G} f(g)\overline{\phi^\theta(g)}.$$

From the orthogonality relations (13.33) we immediately deduce the *inversion formula*:

$$f = \frac{1}{|G|} \sum_{\theta \in J} d_\theta \mathcal{F}f(\theta)\phi^\theta$$

and the *Plancherel formula*:

$$\langle f_1, f_2 \rangle_{L(G)} = \frac{1}{|G|} \sum_{\theta \in J} d_\theta \mathcal{F}f_1(\theta)\overline{\mathcal{F}f_2(\theta)},$$

for all $f, f_1, f_2 \in \mathcal{H}(G, K, \chi)$. In particular, $\|f\|_{L(G)}^2 = \frac{1}{|G|}\sum_{\theta \in J} d_\theta |\mathcal{F}f(\theta)|^2$. Finally, the *convolution formula*

$$\mathcal{F}(f_1 * f_2) = (\mathcal{F}f_1)(\mathcal{F}f_2)$$

follows from the inversion formula and (10.35).

Now we are in position to give an explicit formula for the eigenvalues $\lambda_f(\theta)$, $\theta \in J$, in Proposition 13.5.1.(ii).

Proposition 13.5.4 *For all $f \in \mathcal{H}(G, K, \chi)$ we have*

$$\lambda_f = \mathcal{F}f.$$

Proof. Let $f \in \mathcal{H}(G, K, \chi)$ and $\theta \in J$. It suffices to compute $\lambda_f(\theta)$ for the eigenvector ϕ^θ:

$$[T_f\phi^\theta](g) = [f * \phi^\theta](g)$$
$$(\text{by (13.22)}) \qquad = [f * \phi^\theta](1_G)\phi^\theta(g)$$
$$= \sum_{h \in G} f(h)\phi^\theta(h^{-1})\phi^\theta(g)$$
$$(\text{by Proposition 13.4.6.(i)}) \qquad = [\mathcal{F}f](\theta)\phi^\theta(g). \qquad \square$$

Proposition 13.5.5 *The operator $E_\theta \colon \text{Ind}_K^G\mathbb{C} \longrightarrow L(G)$ defined by setting*

$$E_\theta f = \frac{d_\theta}{|G|} f * \phi^\theta,$$

for all $f \in \text{Ind}_K^G\mathbb{C}$, is the orthogonal projection from $\text{Ind}_K^G\mathbb{C}$ onto $T_\theta W_\theta$.

Proof. First of all, note that, for $g \in G$ and $f \in \text{Ind}_K^G\mathbb{C}$, we have:

$$[E_\theta f](g) = \frac{d_\theta}{|G|} \sum_{h \in G} f(h)\phi^\theta(h^{-1}g)$$
$$= \frac{d_\theta}{|G|} \sum_{h \in G} f(h)\overline{\phi^\theta(g^{-1}h)} = \frac{d_\theta}{|G|} \langle f, \lambda_G(g)\phi^\theta \rangle_{L(G)},$$

where λ_G is as in (10.9). Therefore, for $\eta \in J \setminus \{\theta\}$ and $h \in G$,

$$\left[E_\theta\lambda_G(h)\phi^\eta\right](g) = \frac{d_\theta}{|G|} \langle \lambda_G(h)\phi^\eta, \lambda_G(g)\phi^\theta \rangle_{L(G)} = 0$$

by Proposition 13.4.6.(iii), that is, $\displaystyle\bigoplus_{\eta \in J, \eta \neq \theta} T_\eta W_\eta \subseteq \mathrm{Ker} E_\theta$. Similarly,

$$
\begin{aligned}
\left[E_\theta \lambda_G(h)\phi^\theta\right](g) &= \frac{d_\theta}{|G|} \langle \lambda_G(h)\phi^\theta, \lambda_G(g)\phi^\theta \rangle_{L(G)} \\
&= \frac{d_\theta}{|G|} \langle \phi^\theta, \lambda_G(h^{-1}g)\phi^\theta \rangle_{L(G)} \\
&= \frac{d_\theta}{|G|} [\phi^\theta * \phi^\theta](h^{-1}g) \\
\text{(by (13.22))} \quad &= \frac{d_\theta}{|G|} [\phi^\theta * \phi^\theta](1_G)\phi^\theta(h^{-1}g) \\
(\phi^\theta * \phi^\theta(1_g) = \|\phi^\theta\|^2_{L(G)} = |G|/d_\theta) \quad &= \lambda_G(h)\phi^\theta(g).
\end{aligned}
$$

We then conclude by using Theorem 13.4.7. $\qquad\qquad\qquad\qquad\qquad\qquad$ □

We now show that the spherical function ϕ^θ and the character χ^θ may be expressed one in terms of the other.

Proposition 13.5.6 *For all $g \in G$ we have:*

$$
\chi^\theta(g) = \frac{d_\theta}{|G|} \sum_{h \in G} \overline{\phi^\theta(h^{-1}gh)} \tag{13.36}
$$

and

$$
\phi^\theta(g) = [\overline{\chi^\theta} * \psi](g). \tag{13.37}
$$

Proof. Clearly, (13.36) is just a particular case of (10.25), keeping into account (13.32). On the other hand, using the bases in (13.35) we have

$$
\begin{aligned}
[\overline{\chi^\theta} * \psi](g) &= \sum_{k \in K} \sum_{i=1}^{d_\theta} \overline{\langle \theta(gk^{-1})u_i, u_i \rangle} \psi(k) \\
&= \sum_{k \in K} \sum_{i=1}^{d_\theta} \overline{\langle \theta(g) \sum_{k \in K} \psi(k^{-1})\theta(k^{-1})u_i, u_i \rangle} \\
\text{(by (13.35))} \quad &= \phi^\theta(g).
\end{aligned}
$$

$\qquad\qquad\qquad\qquad\qquad\qquad\qquad\qquad\qquad\qquad\qquad\qquad\qquad\qquad$ □

In what follows, for $f \in L(G)$ and $\theta \in J$ we set

$$
\chi^\theta(f) = \sum_{g \in G} \chi^\theta(g)f(g) \equiv \langle \chi^\theta, \overline{f} \rangle
$$

and, similarly,

$$\phi^\theta(f) = \sum_{g \in G} \phi^\theta(g) f(g) \equiv \langle \phi^\theta, \overline{f} \rangle.$$

We use the Curtis-Fossum basis in Definition 13.2.5.

Proposition 13.5.7 (Curtis-Fossum) *Let $\theta \in J$. Then the following hold:*

(i) *The spherical function ϕ^θ can be expressed as*

$$\phi^\theta = \sum_{s \in \mathcal{S}_0} |G_s| \phi^\theta(\overline{a_s}) a_s.$$

(ii) *The orthogonality relations for the spherical functions may be written in the form:*

$$\sum_{s \in \mathcal{S}_0} |G_s| \phi^\theta(\overline{a_s}) \overline{\phi^\rho(\overline{a_s})} = \delta_{\theta,\rho} \frac{|G|}{d_\theta}, \qquad \rho \in J.$$

(iii) *The dimension d_θ is given by*

$$d_\theta = \frac{|G|}{\sum_{s \in \mathcal{S}_0} |G_s| \cdot |\phi^\theta(\overline{a_s})|^2}.$$

Proof.

(i) This is an immediate consequence of (13.11) and (13.13).

(ii) From (i) and (13.12) we have:

$$\langle \phi^\theta, \phi^\rho \rangle_{L(G)} = \sum_{s \in \mathcal{S}_0} |G_s|^2 \phi^\theta(\overline{a_s}) \overline{\phi^\rho(\overline{a_s})} \|a_s\|_{L(G)}^2 = \sum_{s \in \mathcal{S}_0} |G_s| \phi^\theta(\overline{a_s}) \overline{\phi^\rho(\overline{a_s})}.$$

Then we may invoke (13.33).

(iii) It follows immediately from (ii). □

Remark 13.5.8 When $\chi = \iota_K$ and (G, K) is a Gelfand pair, it is customary to use the isomorphism (11.13) to define the spherical functions as K-invariant functions on X (see Remark 13.2.2). That is, for $\theta \in J$ we define $\varphi^\theta \in L(X)$ by setting $\varphi^\theta(x) = \phi^\theta(g)$ if $gx_0 = x$. Then the orthogonality relations become: $\sum_{x \in X} \varphi^\theta(x) \overline{\varphi^\rho(x)} = \delta_{\theta,\rho} \frac{|X|}{d_\theta}$. We refer to [29] for an extensive treatment of this case.

Exercise 13.5.9 Prove that, in the setting of Exercise 13.3.8, the spherical function in M^θ is equal to $\frac{1}{d_\theta} \chi^\theta$.

Exercise 13.5.10 Let G be a finite group and suppose it acts doubly transitively on a set X. Denote by K the stabilizer of a fixed element $x_0 \in X$. Show that (G, K) is a symmetric Gelfand pair, that $L(X) = W_0 \oplus W_1$ (cf. Proposition

2.1.1) is the decomposition into spherical representations, and that the corresponding spherical functions are given by $\phi_0 \equiv 1$ and

$$\phi_1(x) = \begin{cases} 1 & \text{if } x = x_0 \\ -\frac{1}{1-|X|} & \text{otherwise} \end{cases}$$

for all $x \in X$.

Exercise 13.5.11 From Exercise 12.1.8 we deduce that $\left(\text{Aff}(\mathbb{F}_q), A, \psi\right)$ is a multiplicity-free triple for any character $\psi \in \widehat{A}$. By means of (13.31) and/or (13.37) applied to (12.8), show that the spherical functions are given by:

$$\phi^\pi \begin{pmatrix} a & b \\ 0 & 1 \end{pmatrix} = \begin{cases} \overline{\psi(a)} & \text{if } b = 0 \\ -\frac{1}{q-1}\overline{\psi(a)} & \text{otherwise,} \end{cases}$$

and $\phi^\Psi \begin{pmatrix} a & b \\ 0 & 1 \end{pmatrix} = \overline{\psi(a)}$, for all $\begin{pmatrix} a & b \\ 0 & 1 \end{pmatrix} \in \text{Aff}(\mathbb{F}_q)$.

14

Representation theory of $\mathrm{GL}(2, \mathbb{F}_q)$

This chapter is devoted to the representation theory of the general linear group $\mathrm{GL}(2, \mathbb{F}_q)$. It contains an exposition of all the results in Piatetski-Shapiro's monograph [123]. We have added some more details and reinterpreted the whole theory in terms of our "multiplicity-free triples" developed in the preceding chapter. Section 7.3, on generalized Kloosterman sums, also plays here a fundamental role. In the final sections, we present a complete set of formulas for the decomposition of induced representations $\mathrm{Ind}_{\mathbb{F}_q}^{\mathbb{F}_{q^m}}$ and of inner tensor products.

14.1 Matrices associated with linear operators

First of all, we need to study the conjugacy classes in $\mathrm{GL}(2, \mathbb{F})$. For this purpose, we recall some basic facts of linear algebra over an arbitrary field $\mathbb{F}$ and, subsequently, we concentrate on the finite case. If the field $\mathbb{F}$ is algebraically closed, we shall make use of the *Jordan canonical form*, while, in the general case, our standard tool will be the *rational canonical form*.

Let $\mathbb{F}$ be a field and denote by $\mathfrak{M}_n(\mathbb{F})$ the algebra of all $n \times n$ matrices with entries in $\mathbb{F}$. Then the multiplicative group $\mathrm{GL}(n, \mathbb{F}) = \mathcal{U}(\mathfrak{M}_n(\mathbb{F}))$, consisting of all *invertible* matrices, acts on $\mathfrak{M}_n(\mathbb{F})$ by conjugation. The action of an element $A \in \mathrm{GL}(n, \mathbb{F})$ on $\mathfrak{M}_n(\mathbb{F})$ is then given by:

$$B \mapsto ABA^{-1}$$

for all $B \in \mathfrak{M}_n(\mathbb{F})$. The orbits under this action are the *conjugacy classes* of $\mathfrak{M}_n(\mathbb{F})$ and the choice of a suitable canonical element in the conjugacy class of a matrix $B \in \mathfrak{M}_n(\mathbb{F})$ is called a *canonical form* for B.

We identify the n-dimensional vector space $\mathbb{F}^n$ with the vector space $\mathfrak{M}_{n,1}$ of n-dimensional column vectors. Also we fix an *(ordered) basis* $\mathbf{Y} = (Y_1, Y_2, \ldots, Y_n)$ of $\mathbb{F}^n$.

Let $L\colon \mathbb{F}^n \to \mathbb{F}^n$ be a linear operator. Then the *matrix* $C = C(L; \mathbf{Y}) = (c_{i,j})_{i,j=1}^n$ *representing* the operator L with respect to the basis $\mathbf{Y}$ is defined by

$$L(Y_j) = \sum_{i=1}^n c_{i,j} Y_i$$

for all $j = 1, 2, \ldots, n$.

Vice versa, with each $B \in \mathfrak{M}_n(\mathbb{F})$ we associate the linear operator $L_B \colon \mathbb{F}^n \to \mathbb{F}^n$ defined by setting $L_B(X) = BX$ for all $X \in \mathbb{F}^n$.

Let also $\mathbf{X} = (X_1, X_2, \ldots, X_n)$ denote the *canonical (ordered) basis* of $\mathbb{F}^n$, that is,

$$X_1 = \begin{pmatrix} 1 \\ 0 \\ 0 \\ \vdots \\ 0 \end{pmatrix}, \ X_2 = \begin{pmatrix} 0 \\ 1 \\ 0 \\ \vdots \\ 0 \end{pmatrix}, \ \ldots, \ X_n = \begin{pmatrix} 0 \\ 0 \\ 0 \\ \vdots \\ 1 \end{pmatrix}.$$

Then, for $j = 1, 2, \ldots, n$, the vector $L_B(X_j)$ equals the j-th column of the matrix B. In other words, the matrix $C(L_B; \mathbf{X})$ representing L_B with respect to the canonical basis is the matrix B itself.

Let $A = A(\mathbf{Y}) \in GL(n, \mathbb{F})$ denote the *change of basis matrix*, that is, the unique invertible matrix A such that $Y_j = A^{-1} X_j$, equivalently, $X_j = AY_j$, for all $j = 1, 2, \ldots, n$. Then the matrix $C = C(L_B; \mathbf{Y})$ representing the linear operator L_B in the basis $\mathbf{Y}$ is given by $C = ABA^{-1}$. Indeed, if

$$BY_j = L_B(Y_j) = \sum_{i=1}^n c_{i,j} Y_i,$$

then

$$ABA^{-1}X_j = ABY_j = A \sum_{i=1}^n c_{i,j} Y_i = \sum_{i=1}^n c_{i,j} AY_i = \sum_{i=1}^n c_{i,j} X_i = CX_j$$

for all $j = 1, 2, \ldots, n$.

This shows that finding a canonical form C for B corresponds to choosing a suitable basis $\mathbf{Y}$ in $\mathbb{F}^n$ such that $C = C(L_B; \mathbf{Y})$.

14.2 Canonical forms for $\mathfrak{M}_2(\mathbb{F})$

We now describe a canonical form for matrices in $\mathfrak{M}_2(\mathbb{F})$.

We denote by $\mathbb{F}[\lambda]$ the $\mathbb{F}$-vector space of all polynomials with coefficients in $\mathbb{F}$ and indeterminate λ.

Let $B = \begin{pmatrix} \alpha & \beta \\ \gamma & \delta \end{pmatrix} \in \mathfrak{M}_2(\mathbb{F})$.

Given $t(\lambda) = a_n\lambda^n + a_{n-1}\lambda^{n-1} + \cdots + a_1\lambda + a_0 \in \mathbb{F}[\lambda]$ we set $t(B) = a_nB^n + a_{n-1}B^{n-1} + \cdots + a_1B + a_0I \in \mathfrak{M}_2(\mathbb{F})$, where $I \in \mathfrak{M}_2(\mathbb{F})$ denotes the identity matrix.

The *characteristic polynomial* $q = q_B \in \mathbb{F}[\lambda]$ of the matrix B is defined as

$$q(\lambda) = \det(\lambda I - B) = \det \begin{pmatrix} \lambda - \alpha & \beta \\ \gamma & \lambda - \delta \end{pmatrix} = \lambda^2 - \lambda(\alpha + \delta) + (\alpha\delta - \beta\gamma).$$

Exercise 14.2.1 Show, by a direct calculation, that $q(B) = 0 \in \mathfrak{M}_2(\mathbb{F})$ (*Cayley-Hamilton theorem*). Moreover, given $\lambda_1 \in \mathbb{F}$ show that $q(\lambda_1) = 0$ if and only if λ_1 is an *eigenvalue* of B (i.e. there exists an *eigenvector* $Y \in \mathbb{F}^2 \setminus \{0\}$ such that $BY = \lambda_1Y$).

The *minimal polynomial* $p = p_B \in \mathbb{F}[\lambda]$ of B is the monic polynomial of least degree such that $p(B) = 0$. We clearly have two cases:

(a) $\deg(p) = 1$. Then $p(\lambda) = \lambda - \lambda_1$ for some $\lambda_1 \in \mathbb{F}$ and $p(B) = 0$ implies that $B = \lambda_1I$ is a scalar matrix.

(b) $\deg(p) = 2$. Then $p(\lambda) = q(\lambda)$ and B is not a scalar matrix. We further distinguish three subcases:

(b$_1$) $p(\lambda)$ has two distinct roots in $\mathbb{F}$: there exist $\lambda_1, \lambda_2 \in \mathbb{F}$, $\lambda_1 \neq \lambda_2$, such that $p(\lambda) = (\lambda - \lambda_1)(\lambda - \lambda_2)$, equivalently, B has two distinct eigenvalues. Let $Y_1, Y_2 \in \mathbb{F}^2$ be two corresponding eigenvectors: $BY_1 = \lambda_1Y_1$ and $BY_2 = \lambda_2Y_2$. Then Y_1 and Y_2 are linearly independent: if $\alpha_1Y_1 + \alpha_2Y_2 = 0$, with $\alpha_1, \alpha_2 \in \mathbb{F}$, by applying B to both sides we deduce that $\alpha_1\lambda_1Y_1 + \alpha_2\lambda_2Y_2 = 0$ so that

$$\alpha_2(\lambda_1 - \lambda_2)Y_2 = \lambda_1(\alpha_1Y_1 + \alpha_2Y_2) - (\alpha_1\lambda_1Y_1 + \alpha_2\lambda_2Y_2) = 0.$$

Since $\lambda_1 \neq \lambda_2$, we deduce that $\alpha_2 = 0$ and, in turn, $\alpha_1 = 0$.

The matrix $C = C(L_B; \mathbf{Y})$ representing L_B in the basis $\mathbf{Y} = (Y_1, Y_2)$ is then given by

$$C = \begin{pmatrix} \lambda_1 & 0 \\ 0 & \lambda_2 \end{pmatrix}$$

that is, C is a diagonal matrix with distinct diagonal terms.

Note also that the matrices $\begin{pmatrix} \lambda_1 & 0 \\ 0 & \lambda_2 \end{pmatrix}$ and $C(L_B; (Y_2, Y_1)) = \begin{pmatrix} \lambda_2 & 0 \\ 0 & \lambda_1 \end{pmatrix}$ are conjugate. Indeed:

$$\begin{pmatrix} 0 & 1 \\ 1 & 0 \end{pmatrix} \begin{pmatrix} \lambda_1 & 0 \\ 0 & \lambda_2 \end{pmatrix} \begin{pmatrix} 0 & 1 \\ 1 & 0 \end{pmatrix} = \begin{pmatrix} \lambda_2 & 0 \\ 0 & \lambda_1 \end{pmatrix}. \tag{14.1}$$

(b$_2$) $p(\lambda) = (\lambda - \lambda_1)^2$, where $\lambda_1 \in \mathbb{F}$. Then there exists an eigenvector Y_1 associated with λ_1, so that $BY_1 = \lambda_1 Y_1$. Moreover, there exists a vector $\widetilde{Y} \in \mathbb{F}^2$ (any vector that is not a scalar multiple of Y_1) such that $(B - \lambda_1 I)\widetilde{Y} \neq 0$, because $B - \lambda_1 I \neq 0$. Then $(B - \lambda_1 I)^2 \widetilde{Y} = p(B)\widetilde{Y} = 0$ implies (exercise) that there exists $\alpha' \in \mathbb{F} \setminus \{0\}$ such that

$$(B - \lambda_1 I)\widetilde{Y} = \alpha' Y_1. \tag{14.2}$$

Setting $Y_2 = \frac{1}{\alpha'}\widetilde{Y}$ equation (14.2) becomes

$$BY_2 = \lambda_1 Y_2 + Y_1$$

and, in the basis $\mathbf{Y} = (Y_1, Y_2)$, the operator L_B is represented by the matrix $C = C(L_B; \mathbf{Y})$ given by

$$C = \begin{pmatrix} \lambda_1 & 1 \\ 0 & \lambda_1 \end{pmatrix},$$

which constitutes the simplest (non-trivial) example of a *Jordan canonical form*.

(b$_3$) $p(\lambda) = \lambda^2 + \alpha'\lambda + \beta'$, where $\alpha', \beta' \in \mathbb{F}$, is irreducible over $\mathbb{F}$. Consider a vector $Y_1 \neq 0$. Then $Y_2 = BY_1$ is not a multiple of Y_1 (otherwise Y_1 would be an eigenvector) and therefore $\mathbf{Y} = (Y_1, Y_2)$ is a basis for $\mathbb{F}^2$. Since $B^2 + \alpha'B + \beta'I = 0$ (cf. Exercise 14.2.1), we have that $BY_2 = B^2 Y_1 = -\alpha' BY_1 - \beta' Y_1 = -\beta' Y_1 - \alpha' Y_2$, so that, in the basis $\mathbf{Y}$, the operator L_B is represented by the matrix $C = C(L_B; \mathbf{Y})$ given by

$$C = \begin{pmatrix} 0 & -\beta' \\ 1 & -\alpha' \end{pmatrix}. \tag{14.3}$$

This is the simplest (non-trivial) example of a *rational canonical form*.

From the previous case-by-case analysis we immediately deduce the following:

Theorem 14.2.2 *Two matrices in* $\mathfrak{M}_2(\mathbb{F})$ *are conjugate if and only if they have the same minimal and characteristic polynomials. For non-scalar matrices it suffices that they have the same characteristic polynomial.*

Remark 14.2.3 In $\mathfrak{M}_n(\mathbb{F})$ with $n > 2$, Theorem 14.2.2 is no longer true and the full machinery for the *rational canonical form* and the theory of *invariant factors* (or *invariant polynomials*, or *elementary divisors*) must be used to get a parameterization of the conjugacy classes, i.e. in the terminology of linear algebra, to establish if two matrices are similar.

If the field $\mathbb{F}$ is algebraically closed, the *Jordan canonical form* may be used in place of the rational canonical form. See, for instance, Herstein's book [71].

We now introduce four important subgroups of GL(2, $\mathbb{F}$), namely,

$$B = \left\{ \begin{pmatrix} \alpha & \beta \\ 0 & \delta \end{pmatrix} : \alpha, \delta \in \mathbb{F}^*, \beta \in \mathbb{F} \right\} \quad \text{(the *Borel* subgroup)}$$

$$D = \left\{ \begin{pmatrix} \alpha & 0 \\ 0 & \delta \end{pmatrix} : \alpha, \delta \in \mathbb{F}^* \right\} \quad \text{(the subgroup of *diagonal* matrices)}$$

$$U = \left\{ \begin{pmatrix} 1 & \beta \\ 0 & 1 \end{pmatrix} : \beta \in \mathbb{F} \right\} \quad \text{(the subgroup of *unipotent* matrices)}$$

$$Z = \left\{ \begin{pmatrix} \alpha & 0 \\ 0 & \alpha \end{pmatrix} : \alpha \in \mathbb{F}^* \right\} \quad \text{(the *center*)},$$

where, as usual, $\mathbb{F}^*$ denotes the multiplicative subgroup of $\mathbb{F}$ consisting of all nonzero elements.

Clearly, U is Abelian and isomorphic to the additive group of $\mathbb{F}$:

$$\begin{pmatrix} 1 & \beta_1 \\ 0 & 1 \end{pmatrix} \begin{pmatrix} 1 & \beta_2 \\ 0 & 1 \end{pmatrix} = \begin{pmatrix} 1 & \beta_1 + \beta_2 \\ 0 & 1 \end{pmatrix}$$

for all $\beta_1, \beta_2 \in \mathbb{F}$; see Section 12.1.

Moreover, U is a normal subgroup of B:

$$\begin{pmatrix} \alpha & \beta \\ 0 & \delta \end{pmatrix} \begin{pmatrix} 1 & \beta' \\ 0 & 1 \end{pmatrix} \begin{pmatrix} \alpha & \beta \\ 0 & \delta \end{pmatrix}^{-1} = \begin{pmatrix} \alpha & \alpha\beta' + \beta \\ 0 & \delta \end{pmatrix} \begin{pmatrix} \alpha^{-1} & -\beta\delta^{-1}\alpha^{-1} \\ 0 & \delta^{-1} \end{pmatrix}$$

$$= \begin{pmatrix} 1 & \alpha\delta^{-1}\beta' \\ 0 & 1 \end{pmatrix}$$

for all $\beta, \beta' \in \mathbb{F}$ and $\alpha, \delta \in \mathbb{F}^*$.

Recall that given a group G, the *derived subgroup* (or *commutator subgroup*) of G is the subgroup $G' = [G, G]$ generated by the commutators $[g, h] = g^{-1}h^{-1}gh$, with $g, h \in G$. Moreover, setting $G^{(0)} = G$ and $G^{(k)} = [G^{(k-1)}, G^{(k-1)}]$ for $k = 1, 2, \ldots$, one says that G is *solvable* provided there exists $k_0 \in \mathbb{N}$ such that $G^{(k_0)} = \{1_G\}$. Finally, given $g \in G$ and a subgroup $H \leq G$, the *centralizer* of g in H is the subgroup $\{h \in H : hg = gh\} \leq H$. See also Section 12.1.

Lemma 14.2.4

(i) *The centralizer in* $\mathrm{GL}(2, \mathbb{F})$ *of the matrix* $\begin{pmatrix} \lambda_1 & 0 \\ 0 & \lambda_2 \end{pmatrix}$, *with* $\lambda_1 \neq \lambda_2 \in \mathbb{F}$, *is the subgroup D.*

(ii) *The centralizer in* $\mathrm{GL}(2, \mathbb{F})$ *of the matrix* $\begin{pmatrix} \lambda_1 & 1 \\ 0 & \lambda_1 \end{pmatrix}$, *with* $\lambda_1 \in \mathbb{F}$, *is the subgroup ZU, which equals* $\left\{ \begin{pmatrix} \alpha & \beta \\ 0 & \alpha \end{pmatrix} : \alpha \in \mathbb{F}^*, \beta \in \mathbb{F} \right\}$.

(iii) $B = U \rtimes D$, *i.e. B is the semidirect product of U by D. Moreover, U is the derived subgroup of B, and B is solvable.*

(iv) *Setting* $w = \begin{pmatrix} 0 & 1 \\ 1 & 0 \end{pmatrix}$, *we have the* **Bruhat decomposition***:*

$$\mathrm{GL}(2, \mathbb{F}) = B \bigsqcup BwU \equiv B \bigsqcup UwB,$$

where $\bigsqcup$ *denotes a disjoint union. Moreover, every element* $g \in \mathrm{GL}(2, \mathbb{F}) \setminus B$ *may be uniquely written in the form* $g = uwb$ *with* $u \in U$ *and* $b \in B$.

Proof. The proof is nothing but easy calculations, which we leave to the reader as an exercise.

For instance, (iv) follows from the fact that if $\begin{pmatrix} \alpha & \beta \\ \gamma & \delta \end{pmatrix} \in \mathrm{GL}(2, \mathbb{F}) \setminus B$ (so that $\gamma \in \mathbb{F}^*$) then, as one easily checks,

$$\begin{pmatrix} \alpha & \beta \\ \gamma & \delta \end{pmatrix} = \begin{pmatrix} \beta - \alpha\gamma^{-1}\delta & \alpha \\ 0 & \gamma \end{pmatrix} \begin{pmatrix} 0 & 1 \\ 1 & 0 \end{pmatrix} \begin{pmatrix} 1 & \gamma^{-1}\delta \\ 0 & 1 \end{pmatrix}$$

$$= \begin{pmatrix} 1 & \alpha\gamma^{-1} \\ 0 & 1 \end{pmatrix} \begin{pmatrix} 0 & 1 \\ 1 & 0 \end{pmatrix} \begin{pmatrix} \gamma & \delta \\ 0 & \beta - \alpha\gamma^{-1}\delta \end{pmatrix},$$

and these factorizations are unique. $\qquad\qquad\qquad\qquad\qquad\qquad\square$

Another important subgroup is

$$\mathrm{Aff}(\mathbb{F}) = \left\{ \begin{pmatrix} \alpha & \beta \\ 0 & 1 \end{pmatrix} : \alpha \in \mathbb{F}^*, \beta \in \mathbb{F} \right\},$$

the *affine group* over $\mathbb{F}$ (cf. Example 10.4.5 and Section 12.1).

Exercise 14.2.5 Show the following:

(1) $Z \cap \mathrm{Aff}(\mathbb{F}) = \{I\}$ and $Z \cdot \mathrm{Aff}(\mathbb{F}) = \mathrm{Aff}(\mathbb{F}) \cdot Z = B$;
(2) $\mathrm{Aff}(\mathbb{F})$ is a normal subgroup of B and deduce that $B \cong \mathrm{Aff}(\mathbb{F}) \times Z$ (direct product);
(3) $\mathrm{Aff}(\mathbb{F}) = U \rtimes A$ (semi-direct product), where A is the subgroup $\left\{ \begin{pmatrix} \alpha & 0 \\ 0 & 1 \end{pmatrix} : \alpha \in \mathbb{F}^* \right\} \cong \mathbb{F}^*$; see Section 12.1.

14.3 The finite case

From now on, we concentrate on the finite case, that is, we consider the group GL(2, $\mathbb{F}_q$), where $\mathbb{F}_q$ is a finite field of order $q = p^h$, where p is a prime number and $h \geq 1$.

Proposition 14.3.1 GL(2, $\mathbb{F}_q$) *is a finite group of order*

$$|\mathrm{GL}(2, \mathbb{F}_q)| = (q^2 - 1)(q^2 - q) = q(q + 1)(q - 1)^2.$$

Proof. The first row of a matrix $\begin{pmatrix} \alpha & \beta \\ \gamma & \delta \end{pmatrix} \in \mathrm{GL}(2, \mathbb{F}_q)$ may be chosen in $q^2 - 1$ ways: it is an arbitrary ordered pair $(\alpha, \beta) \in (\mathbb{F}_q \times \mathbb{F}_q) \setminus \{(0, 0)\}$. Then the second row (γ, δ) is an arbitrary ordered pair in $(\mathbb{F}_q \times \mathbb{F}_q) \setminus \{(\lambda a, \lambda b) : \lambda \in \mathbb{F}_q\}$, and there are $q^2 - q$ such pairs.

Another proof is the following. Consider the projective line $\mathbb{P}(\mathbb{F}_q) = \left((\mathbb{F}_q \times \mathbb{F}_q) \setminus \{(0, 0)\} \right) / \sim$, where $\sim$ is the equivalence relation on $(\mathbb{F}_q \times \mathbb{F}_q) \setminus \{(0, 0)\}$ defined by $(x, y) \sim (u, v)$ if there exists $\lambda \in \mathbb{F}_q^*$ such that $(x, y) = (\lambda u, \lambda v)$. The action of GL(2, $\mathbb{F}_q$) on $\mathbb{F}_q \times \mathbb{F}_q$ fixes $(0, 0)$ and preserves $\sim$, and therefore induces an action of GL(2, $\mathbb{F}_q$) on $\mathbb{P}(\mathbb{F}_q)$. Moreover, it is easy to check that this induced action is transitive. The stabilizer of the $\sim$-class of $(1, 0)$ is the Borel subgroup B. Since $|\mathbb{P}(\mathbb{F}_q)| = \frac{q^2 - 1}{q - 1} = q + 1$ and $|B| = q(q - 1)^2$, we obtain again $|\mathrm{GL}(2, \mathbb{F}_q)| = |\mathbb{P}(\mathbb{F}_q)| \cdot |B| = (q + 1)q(q - 1)^2$; recall (10.44). $\square$

Using the notation (and results) of Section 6.8, we introduce another fundamental subgroup of GL(2, $\mathbb{F}_q$). The *Cartan* (or *non-split Cartan*) *subgroup* of GL(2, $\mathbb{F}_q$) is the subgroup C defined by

$$C = \left\{ \begin{pmatrix} \alpha & \omega\beta \\ \beta & \alpha + \beta \end{pmatrix} : \alpha, \beta \in \mathbb{F}_q, (\alpha, \beta) \neq (0, 0) \right\}$$

if $p = 2$, where $\omega \in \mathbb{F}_q$ is as in Theorem 6.8.3, and

$$C = \left\{ \begin{pmatrix} \alpha & \eta\beta \\ \beta & \alpha \end{pmatrix} : \alpha, \beta \in \mathbb{F}_q, (\alpha, \beta) \neq (0, 0) \right\}$$

if $p > 2$, where $\eta \in \mathbb{F}_q$ is as in Theorem 6.8.1.

In both cases, we have (cf. the just mentioned theorems) a group isomorphism

$$C \cong \mathbb{F}_{q^2}^*.$$

In the following theorem, we use the elements of $C \setminus Z$ to parameterize the conjugacy classes of type (b$_3$) in Section 14.2. Note that

$$C \setminus Z = \left\{ \begin{pmatrix} \alpha & \omega\beta \\ \beta & \alpha + \beta \end{pmatrix} : \alpha \in \mathbb{F}_q, \beta \in \mathbb{F}_q^* \right\}$$

if $p = 2$, and

$$C \setminus Z = \left\{ \begin{pmatrix} \alpha & \eta\beta \\ \beta & \alpha \end{pmatrix} : \alpha \in \mathbb{F}_q, \beta \in \mathbb{F}_q^* \right\}$$

if $p > 2$.

Theorem 14.3.2 *The following table describes the conjugacy classes of* $GL(2, \mathbb{F}_q)$

Table 14.1. *The conjugacy classes of* $GL(2, \mathbb{F}_q)$.

TYPE	RE	NC	NE	NAME	C(RE)
(a)	$\begin{pmatrix} \lambda & 0 \\ 0 & \lambda \end{pmatrix}, \lambda \neq 0$	$q - 1$	1	central	$GL(2, \mathbb{F}_q)$
(b$_1$)	$\begin{pmatrix} \lambda_1 & 0 \\ 0 & \lambda_2 \end{pmatrix}, \lambda_1 \neq \lambda_2$	$(q-1)(q-2)/2$	$q^2 + q$	hyperbolic	D
(b$_2$)	$\begin{pmatrix} \lambda & 1 \\ 0 & \lambda \end{pmatrix}, \lambda \neq 0$	$q - 1$	$q^2 - 1$	parabolic	ZU
(b$_3$)	$C \setminus Z$	$q(q-1)/2$	$q^2 - q$	elliptic	C

where

- *TYPE stands for* type *of the conjugacy class according to the classification in Section 14.2;*
- *RE stands for* representative element: *for each (conjugacy) class we indicate a representative element;*
- *NC stands for* number of conjugacy classes: *this equals the number of representative elements;*
- *NE stands for the* number of elements *in each class;*
- *NAME stands for the* denomination *of this type of class;*
- *C(RE) stands for the* centralizer *in* $GL(2, \mathbb{F}_q)$ *of the representative element.*

Moreover, the two matrices of type (b_1)

$$\begin{pmatrix} \lambda_1 & 0 \\ 0 & \lambda_2 \end{pmatrix} \quad and \quad \begin{pmatrix} \lambda_2 & 0 \\ 0 & \lambda_1 \end{pmatrix} \tag{14.4}$$

represent the same class. Similarly, the two matrices of type (b_3)

$$\begin{pmatrix} \alpha & \omega\beta \\ \beta & \alpha+\beta \end{pmatrix} \quad and \quad \begin{pmatrix} \alpha+\beta & \omega\beta \\ \beta & \alpha \end{pmatrix} \in C \setminus Z \tag{14.5}$$

when $p = 2$, *and*

$$\begin{pmatrix} \alpha & \eta\beta \\ \beta & \alpha \end{pmatrix} \quad and \quad \begin{pmatrix} \alpha & -\eta\beta \\ -\beta & \alpha \end{pmatrix} \in C \setminus Z \tag{14.6}$$

when $p > 2$, *represent the same class.*

Proof. The first row in the above table follows from Section 14.2.(a) and the trivial fact that any central element is fixed under conjugation.

The second row follows from Section 14.2.(b_1), Lemma 14.2.4.(i), and the fact that the number of elements in each conjugacy class is given by

$$\frac{|\mathrm{GL}(2, \mathbb{F}_q)|}{|D|} = \frac{q(q+1)(q-1)^2}{(q-1)^2} = q^2 + q.$$

Moreover, we have already observed (cf. (14.1)) that the matrices in (14.4) are conjugate. Similarly, the third row follows from Section 14.2.(b_2) and Lemma 14.2.4.(ii), noticing also that the number of elements in each conjugacy class now equals

$$\frac{|\mathrm{GL}(2, \mathbb{F}_q)|}{|ZU|} = \frac{q(q+1)(q-1)^2}{(q-1)q} = q^2 - 1,$$

where the first equality follows from Proposition 14.3.1.

Finally, to get the fourth row, we distinguish two cases according to the parity of p.

For $p = 2$ the characteristic polynomial of the representative $\begin{pmatrix} \alpha & \omega\beta \\ \beta & \alpha+\beta \end{pmatrix}$ is given by

$$\det \begin{pmatrix} \lambda+\alpha & \omega\beta \\ \beta & \lambda+(\alpha+\beta) \end{pmatrix} = \lambda^2 + \beta\lambda + (\alpha^2 + \alpha\beta + \beta^2\omega) \tag{14.7}$$

so that, by Corollary 6.8.4, it is irreducible.

Moreover, since the matrices $\begin{pmatrix} 0 & \alpha^2+\alpha\beta+\beta^2\omega \\ 1 & \beta \end{pmatrix}$ and $\begin{pmatrix} \alpha+\beta & \omega\beta \\ \beta & \alpha \end{pmatrix}$ have the same characteristic polynomial as in (14.7), we deduce that the matrix

$\begin{pmatrix} \alpha & \omega\beta \\ \beta & \alpha+\beta \end{pmatrix}$ belongs to the same conjugacy class of $\begin{pmatrix} 0 & \alpha^2+\alpha\beta+\beta^2\omega \\ 1 & \beta \end{pmatrix}$

and $\begin{pmatrix} \alpha+\beta & \omega\beta \\ \beta & \alpha \end{pmatrix}$. Since, by Corollary 6.8.4, all irreducible quadratic polynomials over $\mathbb{F}_q$ are as in (14.7), we deduce that the elements in $C \setminus Z$ parameterize all conjugacy classes of type (b_3). Finally, (recall that $\beta \neq 0$) we have

$$\begin{pmatrix} x & y \\ z & u \end{pmatrix}\begin{pmatrix} \alpha & \omega\beta \\ \beta & \alpha+\beta \end{pmatrix} = \begin{pmatrix} x\alpha+y\beta & x\omega\beta+y(\alpha+\beta) \\ z\alpha+u\beta & z\omega\beta+u(\alpha+\beta) \end{pmatrix}$$

equals

$$\begin{pmatrix} \alpha & \omega\beta \\ \beta & \alpha+\beta \end{pmatrix}\begin{pmatrix} x & y \\ z & u \end{pmatrix} = \begin{pmatrix} \alpha x+\omega\beta z & \alpha y+\omega\beta u \\ \beta x+z(\alpha+\beta) & \beta y+u(\alpha+\beta) \end{pmatrix}$$

if and only if $\omega z = y$ and $x+z = u$. As a consequence, the centralizer of any element in $C \setminus Z$ is the subgroup C. We deduce that the number of elements in each conjugacy class is given by

$$\frac{|GL(2,\mathbb{F}_q)|}{|C|} = \frac{q(q+1)(q-1)^2}{q^2-1} = q^2-q. \tag{14.8}$$

Suppose now that $p > 2$. The characteristic polynomial of the representative $\begin{pmatrix} \alpha & \eta\beta \\ \beta & \alpha \end{pmatrix}$ is given by

$$\det\begin{pmatrix} \lambda-\alpha & -\eta\beta \\ -\beta & \lambda-\alpha \end{pmatrix} = \lambda^2 - 2\alpha\lambda + \alpha^2 - \eta\beta^2 \tag{14.9}$$

which is again irreducible by virtue of Corollary 6.8.2.

As in the case $p = 2$, we deduce that the element $\begin{pmatrix} \alpha & \eta\beta \\ \beta & \alpha \end{pmatrix}$ belongs to the same conjugacy class of $\begin{pmatrix} 0 & \eta\beta^2-\alpha^2 \\ 1 & 2\alpha \end{pmatrix}$ (see Section 14.2.(b_3) or (14.3)). Again, since all irreducible quadratic polynomials are as in (14.9), the elements in $C \setminus Z$ parameterize the conjugacy classes of type (b_3). Moreover, $\begin{pmatrix} \alpha & \eta\beta \\ \beta & \alpha \end{pmatrix}$ and $\begin{pmatrix} \alpha & -\eta\beta \\ -\beta & \alpha \end{pmatrix}$ have the same characteristic polynomial, so that they are conjugate (by $\begin{pmatrix} 0 & -\eta \\ 1 & 0 \end{pmatrix}$, for instance).

Finally, (recall, once more, that $\beta \neq 0$) another simple computation shows that

$$\begin{pmatrix} x & y \\ z & u \end{pmatrix} \begin{pmatrix} \alpha & \eta\beta \\ \beta & \alpha \end{pmatrix} = \begin{pmatrix} \alpha & \eta\beta \\ \beta & \alpha \end{pmatrix} \begin{pmatrix} x & y \\ z & u \end{pmatrix}$$

if and only if $\eta z = y$ and $x = u$. As a consequence, the centralizer of an element in $C \setminus Z$ is again C and the number of elements in each conjugacy class is again expressed by (14.8). $\qquad\square$

Remark 14.3.3 From the discussion in Section 14.2 and from the proof of Theorem 14.3.2, it follows that the representatives of type (b$_3$) may also be taken of the form $\begin{pmatrix} 0 & -z\bar{z} \\ 1 & z+\bar{z} \end{pmatrix}$, with $z \in \mathbb{F}_{q^2} \setminus \mathbb{F}_q$.

14.4 Representation theory of the Borel subgroup

As in (12.6), we associate with each $\psi \in \widehat{\mathbb{F}_q^*}$ the function $\Psi \colon Z \to \mathbb{C}$ defined by

$$\Psi \begin{pmatrix} \alpha & 0 \\ 0 & \alpha \end{pmatrix} = \psi(\alpha) \tag{14.10}$$

for all $\alpha \in \mathbb{F}_q^*$. It is immediate to check that Ψ is a character of Z.

The representation theory of B may then easily be deduced from Theorem 12.1.3 and the isomorphism

$$B \cong \mathrm{Aff}(\mathbb{F}_q) \times Z \cong \mathrm{Aff}(\mathbb{F}_q) \times \mathbb{F}_q^*$$

that gives (see Corollary 10.5.17)

$$\widehat{B} \cong \widehat{\mathrm{Aff}(\mathbb{F}_q)} \times \widehat{Z} \cong \widehat{\mathrm{Aff}(\mathbb{F}_q)} \times \widehat{\mathbb{F}_q^*}.$$

Theorem 14.4.1 *The Borel subgroup B has exactly $(q-1)^2$ one-dimensional representations, namely $\Psi_1 \boxtimes \Psi_2$, where $\Psi_1 \in \widehat{\mathrm{Aff}(\mathbb{F}_q)}$ is one-dimensional and $\Psi_2 \in \widehat{Z}$, and $q-1$ irreducible $(q-1)$-dimensional representations, namely $\pi \boxtimes \Psi$, where $\pi \in \widehat{\mathrm{Aff}(\mathbb{F}_q)}$ is as in (12.7) and $\Psi \in \widehat{Z}$.*

Explicitly, these are given by

$$(\Psi_1 \boxtimes \Psi_2) \begin{pmatrix} \alpha & \beta \\ 0 & \delta \end{pmatrix} = \psi_1(\alpha\delta^{-1})\psi_2(\delta) \quad \text{for all } \begin{pmatrix} \alpha & \beta \\ 0 & \delta \end{pmatrix} \in B, \tag{14.11}$$

where $\Psi_1 \in \widehat{\text{Aff}(\mathbb{F}_q)}$ *(resp.* $\Psi_2 \in \widehat{Z}$*) is the character associated with* $\psi_1 \in \widehat{\mathbb{F}_q^*}$ *(resp.* $\psi_2 \in \widehat{\mathbb{F}_q^*}$*), and*

$$(\pi \boxtimes \Psi) \begin{pmatrix} \alpha & \beta \\ 0 & \delta \end{pmatrix} = \pi \begin{pmatrix} \alpha\delta^{-1} & \beta\delta^{-1} \\ 0 & 1 \end{pmatrix} \psi(\delta) \quad \text{for all } \begin{pmatrix} \alpha & \beta \\ 0 & \delta \end{pmatrix} \in B,$$

where $\Psi \in \widehat{Z}$ *is the character associated with* $\psi \in \widehat{\mathbb{F}_q^*}$.

Proof. Each irreducible representation of B is the tensor product of an irreducible representation of $\text{Aff}(\mathbb{F}_q)$ and an irreducible representation of Z (see Corollary 10.5.17). Moreover, for any $\begin{pmatrix} \alpha & \beta \\ 0 & \delta \end{pmatrix} \in B$ we have the unique decomposition

$$\begin{pmatrix} \alpha & \beta \\ 0 & \delta \end{pmatrix} = \begin{pmatrix} \alpha\delta^{-1} & \beta\delta^{-1} \\ 0 & 1 \end{pmatrix} \begin{pmatrix} \delta & 0 \\ 0 & \delta \end{pmatrix} \in \text{Aff}(\mathbb{F}_q)Z. \qquad \square$$

Remark 14.4.2 Given $\psi_1, \psi_2 \in \widehat{\mathbb{F}_q^*}$ let us set $\psi_2' := \psi_1^{-1}\psi_2 \in \widehat{\mathbb{F}_q^*}$. Then the irreducible one dimensional representation (14.11) can be expressed by

$$(\Psi_1 \boxtimes \Psi_2) \begin{pmatrix} \alpha & \beta \\ 0 & \delta \end{pmatrix} = \psi_1(\alpha)\psi_2'(\delta) \quad \text{for all } \begin{pmatrix} \alpha & \beta \\ 0 & \delta \end{pmatrix} \in B.$$

As a consequence, we shall rearrange the parameterization of the pairs (ψ_1, ψ_2) (equivalently, (ψ_1, ψ_2')) in $\widehat{\mathbb{F}_q^*} \times \widehat{\mathbb{F}_q^*}$ and denote by $\chi_{\psi_1,\psi_2} \in \widehat{B}$ the one-dimensional representation given by

$$\chi_{\psi_1,\psi_2} \begin{pmatrix} \alpha & \beta \\ 0 & \delta \end{pmatrix} = \psi_1(\alpha)\psi_2(\delta) \qquad (14.12)$$

for all $\begin{pmatrix} \alpha & \beta \\ 0 & \delta \end{pmatrix} \in B$. We deduce from (14.12) that restricting to D all one-dimensional representations of B provides us with all irreducible representations of its (Abelian) subgroup D. Also, for simplicity of notation, we shall identify $\text{Res}_D^B \chi_{\psi_1,\psi_2}$ and χ_{ψ_1,ψ_2}.

In the following, for every character χ of D we denote by ${}^w\chi$ (cf. (11.41)) the character of D defined by ${}^w\chi(d) = \chi(wdw)$ for all $d \in D$, where the element w is as in Lemma 14.2.4.(iv). We shall then say that χ is w-*invariant*, provided ${}^w\chi = \chi$.

We thus have

$$
\begin{aligned}
{}^w\chi_{\psi_1,\psi_2}\begin{pmatrix} \alpha & 0 \\ 0 & \delta \end{pmatrix} &= \chi_{\psi_1,\psi_2}\left(w\begin{pmatrix} \alpha & 0 \\ 0 & \delta \end{pmatrix}w\right) \\
&= \chi_{\psi_1,\psi_2}\begin{pmatrix} \delta & 0 \\ 0 & \alpha \end{pmatrix} \\
&= \psi_1(\delta)\psi_2(\alpha) \\
&= \chi_{\psi_2,\psi_1}\begin{pmatrix} \alpha & 0 \\ 0 & \delta \end{pmatrix}
\end{aligned}
\tag{14.13}
$$

for all $\begin{pmatrix} \alpha & 0 \\ 0 & \delta \end{pmatrix} \in D$.

It follows that χ_{ψ_1,ψ_2} is w-invariant if and only if $\psi_1 = \psi_2$.

Proposition 14.4.3 *Let* $\psi \in \widehat{\mathbb{F}_q^*}$. *Then*

$$
\chi_{\psi,\psi}(b) = \psi(\det(b))
$$

for all $b \in B$.

Proof. This is a simple calculation: indeed we have

$$
\chi_{\psi,\psi}\begin{pmatrix} \alpha & \beta \\ 0 & \delta \end{pmatrix} = \psi(\alpha)\psi(\delta) = \psi(\alpha\delta) = \psi(\det\begin{pmatrix} \alpha & \beta \\ 0 & \delta \end{pmatrix})
$$

for all $\alpha, \delta \in \mathbb{F}_q^*$ and $\beta \in \mathbb{F}_q$. $\qquad\square$

14.5 Parabolic induction

In this section we determine the irreducible representation of GL(2, $\mathbb{F}_q$) that may be obtained by inducing up the characters of the Borel subgroup B. First, we give a general principle.

Proposition 14.5.1 *Let* G *be a finite group and* $N \trianglelefteq G$ *a normal subgroup. Then the map* $(\rho, U) \mapsto (\widetilde{\rho}, U)$ *defined by*

$$
\widetilde{\rho}(gN)u = \rho(g)u
\tag{14.14}
$$

for all $g \in G$ *and* $u \in U$, *yields a bijection between the set of all* G-*representations* (ρ, U) *such that* $\mathrm{Res}_N^G\rho$ *is trivial and the set of all* G/N-*representations. Moreover, this bijection preserves irreducibility and direct-sums.*

Proof. Let (ρ, U) be a G-representation and suppose that $\mathrm{Res}_N^G\rho$ is trivial. We note that (14.14) is well defined. Indeed, if $g_1, g_2 \in G$ satisfy $g_1N = g_2N$, then $g_1^{-1}g_2 \in N$ so that $\rho(g_1^{-1}g_2)u = u$, equivalently, $\rho(g_1)u = \rho(g_2)u$, for all $u \in U$, showing that $\widetilde{\rho}(g_1 N) = \widetilde{\rho}(g_2 N)$. Vice versa, given a G/N-representation (σ, U), let $(\check{\sigma}, U)$ be the G-representation defined by

$$\check{\sigma}(g)u = \sigma(gN)u \tag{14.15}$$

for all $u \in U$. In other words, $\check{\sigma}$ is the composition of σ with the quotient map $G \to G/N$. Clearly, $\mathrm{Res}_N^G \check{\sigma}$ is trivial. Moreover, the map $\sigma \mapsto \check{\sigma}$ is the inverse of the map $\rho \mapsto \widetilde{\rho}$ given by (14.14). It is straightforward to check that if ρ is irreducible (resp. $\rho = \rho_1 \oplus \rho_2$) then $\widetilde{\rho}$ is irreducible (resp. $\widetilde{\rho} = \widetilde{\rho_1} \oplus \widetilde{\rho_2}$). $\quad\square$

The G-representation $(\check{\sigma}, U)$ defined in (14.15) is called the *inflation* of the G/N-representation (σ, U). See also Section 11.6.

Corollary 14.5.2 *Let H be a finite group and denote by H' its derived subgroup. Then there exists a bijective correspondence between the set of all (irreducible) one-dimensional H-representations and the characters of H/H'.*

Proof. We first observe that if $(\rho, U) \in \widehat{H}$ is one-dimensional, then $\mathrm{Ker}(\rho) = \{h \in H : \rho(h) = \mathrm{id}_U\}$ necessarily contains H': indeed $H/\mathrm{Ker}(\rho) \cong \rho(H) \leq \mathbb{T} = \{z \in \mathbb{C} : |z| = 1\}$ is Abelian. Then the corollary follows from Proposition 14.5.1 after noticing that H' is normal in H and that H/H' is Abelian so that its irreducible representations are all one-dimensional, i.e. characters. $\quad\square$

Proposition 14.5.3 *Let G be a finite group and $H \leq G$ a subgroup. Denote by H' the derived group of H. Let (ρ, V) be an irreducible G-representation. Then the following conditions are equivalent:*

(a) *the subspace $V^{H'}$ of H'-invariant vectors is nontrivial;*
(b) *there exists a one-dimensional representation χ of H such that ρ is contained in $\mathrm{Ind}_H^G \chi$.*

Proof. First of all, note that the subspace $V^{H'}$ is H-invariant. Indeed, H' is normal in H and therefore for $h \in H$ and $v \in V^{H'}$ we have

$$\rho(h')\rho(h)v = \rho(h \cdot h^{-1}h'h)v = \rho(h)\rho(h^{-1}h'h)v = \rho(h)v$$

for all $h' \in H'$, thus showing that $\rho(h)v \in V^{H'}$ (observe that, in fact, the H-invariance of $V^{H'}$ only depends on the normality of H' in H).

Consider the H-representation $(\mathrm{Res}_H^G \rho, V^{H'})$ and observe that its restriction to H' is trivial. By virtue of Proposition 14.5.1 we can identify it with a representation of the Abelian group H/H' and therefore, again by Proposition 14.5.1, it decomposes as a direct sum of one-dimensional H-representations.

Thus, if $V^{H'}$ is not trivial, we can find a character $\chi \in \widehat{H}$ such that $\chi \preceq (\text{Res}_H^G \rho, V^{H'}) \preceq (\text{Res}_H^G \rho, V)$. By Frobenius reciprocity we have that $\rho \preceq \text{Ind}_H^G \chi$.

Conversely, if ρ is contained in $\text{Ind}_H^G \chi$, for some character $\chi \in \widehat{H}$, then, again by Frobenius reciprocity, $\text{Res}_H^G \rho$ contains χ, which, by Corollary 14.5.2, is trivial on H'. It follows that V contains H'-invariant vectors. $\qquad\square$

The space $J(V) = V^{H'}$ is called the *Jacquet module* of the G-representation (ρ, V) relative to the subgroup $H \leq G$.

We now apply the above results in the case where $G = \text{GL}(2, \mathbb{F}_q)$ and $H = B$, so that $H' = B' = U$ (see Lemma 14.2.4).

Notation 14.5.4 From now on, unless otherwise specified, we simply denote $GL(2, \mathbb{F}_q)$ by G. Moreover if χ is a one-dimensional representation of B, we use the notation $(\widehat{\chi}, V)$ to denote $(\text{Ind}_B^G \chi, \text{Ind}_B^G \mathbb{C})$. Also, given the correspondence between the one-dimensional representations of B and the characters of its subgroup D, by abuse of notation (observe that B is not invariant by conjugation by w) we also denote by $^w\chi$ the one-dimensional representation of B corresponding to the character $^w\chi \in \widehat{D}$ (cf. (11.41)).

Proposition 14.5.5 *Let* χ *be a one-dimensional representation of* B. *Then*

$$(\text{Res}_B^G \widehat{\chi}, V^U) \sim (\chi \oplus {}^w\chi, \mathbb{C}^2).$$

Proof. First of all note that the space $V^U \leq \text{Ind}_B^G \mathbb{C}$ is made up of all functions $f : G \to \mathbb{C}$ such that

$$f(gb) = \overline{\chi(b)} f(g) \quad \text{for all } b \in B \text{ and } g \in G \qquad (14.16)$$

(by the definition of an induced representation) and

$$f(u^{-1}g) = f(g) \quad \text{for all } u \in U \text{ and } g \in G$$

(by U-invariance). Then, by the Bruhat decomposition (see Lemma 14.2.4), any function f satisfying these conditions is uniquely determined by its values at 1_G and w:

$$\begin{aligned} f(b) &= \overline{\chi(b)} f(1_G) & \text{for all } b \in B \\ f(uwb) &= \overline{\chi(b)} f(w) & \text{for all } b \in B \text{ and } u \in U. \end{aligned} \qquad (14.17)$$

As a consequence, $\dim V^U = 2$ and the functions f_0 and f_1 in V^U satisfying

$$f_0(1_G) = 1, \ f_0(w) = 0 \text{ and } f_1(1_G) = 0, \ f_1(w) = 1$$

constitute a basis for V^U.

Let us determine the corresponding matrix coefficients for the representation $(\mathrm{Res}_B^G \widehat{\chi}, V^U)$. We have

$$[\widehat{\chi}(b)f_0](1_G) = f_0(b^{-1}) = \chi(b)f_0(1_G) \text{ for all } b \in B.$$

Moreover, for every $b \in B$ there exist $b' \in B$ and $u \in U$ such that $b^{-1}w = uwb'$ so that

$$[\widehat{\chi}(b)f_0](w) = f_0(b^{-1}w) = f_0(uwb') = \overline{\chi(b')}f_0(w) = 0.$$

This shows that

$$\widehat{\chi}(b)f_0 = \chi(b)f_0.$$

We now consider the action of B on f_1. Let $b \in B$. Then we can find $\alpha_0, \alpha_1 \in \mathbb{C}$ such that

$$\widehat{\chi}(b)f_1 = \alpha_0 f_0 + \alpha_1 f_1.$$

Evaluating this expression at 1_G we get

$$\alpha_0 = [\widehat{\chi}(b)f_1](1_G) = f_1(b^{-1}) = \overline{\chi(b)}f_1(1_G) = 0$$

so that

$$\widehat{\chi}(b)f_1 = \alpha_1 f_1.$$

Since f_1 is U-invariant, arguing as in the proof of Proposition 14.5.3, the action of B on f_1 is given by the action of $D \cong B/U \equiv B/B'$. As a consequence, setting $d = bU \in B/U$ we have

$$[\widehat{\chi}(d)f_1](1_G) = f_1(d^{-1}) = 0 \quad \text{for all } d \in D$$

and

$$\begin{aligned}
[\widehat{\chi}(d)f_1](w) &= f_1(d^{-1}w) \\
&= f(w \cdot wd^{-1}w) \\
&= \chi(wdw)f_1(w) \\
&= {}^w\chi(d)f_1(w)
\end{aligned}$$

that is, $\widehat{\chi}(d)f_1 = {}^w\chi(d)f_1$, for all $d \in D$. This, in turn, implies $\widehat{\chi}(b)f_1 = {}^w\chi(b)f_1$, for all $b \in B$. $\qquad\square$

For the convenience of the reader, we now recall from Section 11.4 two basic facts on the theory of induced representations in the particular case when the representations that we are inducing are one-dimensional. See also Remark 11.4.10. Let G be a finite group, $K \leq G$ a subgroup, and $\mathcal{S} \ni 1_G$ a system of representatives for the double K-cosets, so that we have the decomposition $G = \bigsqcup_{s \in \mathcal{S}} KsK$. Let χ, ξ be one-dimensional representations of K. For $s \in \mathcal{S}$

let $K_s = sKs^{-1} \cap K$ and define a one-dimensional representation of K_s by setting

$$\xi_s(x) = \xi(s^{-1}xs) \quad \text{for all } x \in K_s.$$

Then we have Mackey's formula for invariants (cf. Corollary 11.4.4)

$$\mathrm{Hom}_G(\mathrm{Ind}_K^G \chi, \mathrm{Ind}_K^G \xi) \cong \bigoplus_{s \in \mathcal{S}} \mathrm{Hom}_{K_s}(\mathrm{Res}_{K_s}^K \chi, \xi_s)$$

and

$$\mathrm{Hom}_{K_s}(\mathrm{Res}_{K_s}^K \chi, \xi_s) \cong \begin{cases} \mathbb{C} & \text{if } \mathrm{Res}_{K_s}^K \chi = \xi_s \\ \{0\} & \text{otherwise.} \end{cases}$$

In particular, for $\xi = \chi$ we get Mackey's criterion for irreducibility (cf. Corollary 11.4.6): $\mathrm{Ind}_K^G \chi$ is irreducible if and only if

$$\mathrm{Res}_{K_s}^K \chi \neq \chi_s \quad \text{for all } s \in \mathcal{S} \setminus \{1_G\}.$$

Let again $G = \mathrm{GL}(2, \mathbb{F}_q)$ and, for each $\psi \in \widehat{\mathbb{F}_q^*}$, define a one-dimensional representation $\widehat{\chi}_\psi^0$ of G by setting

$$\widehat{\chi}_\psi^0(g) = \psi(\det g) \quad \text{for all } g \in G. \tag{14.18}$$

Theorem 14.5.6 *Keeping in mind* (14.12) *and Notation 14.5.4, we have:*

(i) *Let* $\psi_1, \psi_2, \xi_1, \xi_2 \in \widehat{\mathbb{F}_q^*}$. *If* $\psi_1 \neq \psi_2$ *then* $\widehat{\chi}_{\psi_1, \psi_2}$ *is an irreducible representation of G of dimension* $q + 1$. *Moreover,* $\widehat{\chi}_{\psi_1, \psi_2} \sim \widehat{\chi}_{\xi_1, \xi_2}$ *if and only if* $\{\psi_1, \psi_2\} = \{\xi_1, \xi_2\}$. *In particular,*

$$\left\{ \widehat{\chi}_{\psi_1, \psi_2}(= \widehat{\chi}_{\psi_2, \psi_1}) : \psi_1 \neq \psi_2 \in \widehat{\mathbb{F}_q^*} \right\}$$

consists of $\frac{(q-1)(q-2)}{2}$ *pairwise nonequivalent irreducible representations of G.*

(ii) *For each* $\psi \in \widehat{\mathbb{F}_q^*}$ *there exists an irreducible G-representation* $\widehat{\chi}_\psi^1$ *of dimension q such that*

$$\widehat{\chi}_{\psi, \psi} = \widehat{\chi}_\psi^0 \oplus \widehat{\chi}_\psi^1.$$

Moreover,

$$\left\{ \widehat{\chi}_\psi^1 : \psi \in \widehat{\mathbb{F}_q^*} \right\}$$

is a set of $(q - 1)$ *pairwise nonequivalent q-dimensional G-representations, while*

$$\left\{ \widehat{\chi}_\psi^0 : \psi \in \widehat{\mathbb{F}_q^*} \right\}$$

is the set of all *one-dimensional G-representations.*

Proof. First of all, note that the Bruhat decomposition in Lemma 14.2.4 may be also written in the form

$$G = B \coprod BwB$$

yielding a decomposition of G into double B-cosets. Moreover, $wBw \cap B = D$, so that, if χ, ξ are one-dimensional representations of B, Mackey's formula for invariants becomes

$$\mathrm{Hom}_G(\widehat{\chi}, \widehat{\xi}) \cong \mathrm{Hom}_B(\chi, \xi) \oplus \mathrm{Hom}_D(\mathrm{Res}_D^B \chi, {}^w\xi)$$
$$\doteq \mathrm{Hom}_B(\chi, \xi) \oplus \mathrm{Hom}_D(\chi, {}^w\xi). \qquad (14.19)$$

In particular, for $\xi = \chi$ and $\xi \neq {}^w\chi$ (more precisely, $\chi \neq {}^w\chi$) we get the irreducibility of $\widehat{\chi}$; for $\chi \neq {}^w\chi$, $\xi \neq {}^w\xi$, and $\{\chi, {}^w\chi\} \neq \{\xi, {}^w\xi\}$ we get the nonequivalence of the irreducible representations $\widehat{\chi}$ and $\widehat{\xi}$. Their dimension is just $[G : B] = q + 1$. Note that their nonequivalence also follows from Proposition 14.5.5. Finally, we can invoke Theorem 14.4.1 and (14.13).

Now suppose that $\chi = {}^w\chi$. From (14.19) we deduce that $\dim\mathrm{Hom}_G(\widehat{\chi}, \widehat{\chi}) = 2$, so that $\widehat{\chi}$ decomposes into the sum of two irreducible B-representations. Moreover, $\widehat{\chi}_\psi^0$ is contained in $\widehat{\chi}_{\psi,\psi}$. Indeed, setting $f(g) = \overline{\psi(\det g)}$, we have

$$f(gb) = \overline{\psi(\det(gb))} = \overline{\psi(\det g)} \cdot \overline{\psi(\det b)} = \overline{\chi_{\psi,\psi}(b)}f(g) \qquad (14.20)$$

for all $g \in G$ and $b \in B$, so that (14.16) is satisfied, and

$$[\widehat{\chi}_{\psi,\psi}(g)f](g_0) = f(g^{-1}g_0) = \widehat{\chi}_\psi^0(g)f(g_0) \qquad (14.21)$$

for all $g, g_0 \in G$. Therefore, there exists a second irreducible representation $\widehat{\chi}_\psi^1$ in $\widehat{\chi}$ with $\dim\widehat{\chi}_\psi^1 = (q+1) - 1 = q$. Again by (14.19), for different ψs we get nonequivalent representations (this also follows from Proposition 14.5.5). Finally, if ξ is a one-dimensional G-representation, then it is contained in $\mathrm{Ind}_B^G\chi$, where $\chi = \mathrm{Res}_B^G\xi$. This follows from computations as in (14.20) and (14.21). Alternatively, $\mathrm{Res}_U^G\xi \equiv 1$, because U is the commutator subgroup of B so that, by Proposition 14.5.3, ξ is contained in some $\mathrm{Ind}_B^G\chi$. In any case, we have proved that $\{\widehat{\chi}_\psi^0, \psi \in \widehat{\mathbb{F}_q}\}$ is the list of all one-dimensional G-representations. $\qquad \square$

As a byproduct, we deduce the following result of a purely algebraic flavor:

Corollary 14.5.7 *The commutator subgroup of* $\mathrm{GL}(2, \mathbb{F}_q)$ *is* $\mathrm{SL}(2, \mathbb{F}_q)$.

Proof. SL(2, $\mathbb{F}_q$) is normal and GL(2, $\mathbb{F}_q$)/SL(2, $\mathbb{F}_q$) is Abelian, because we have the homomorphism

$$\text{GL}(2, \mathbb{F}_q) \to \mathbb{F}_q^*$$
$$g \mapsto \det g$$

whose kernel is SL(2, $\mathbb{F}_q$). In particular, GL(2, $\mathbb{F}_q$)/SL(2, $\mathbb{F}_q$) $\cong \mathbb{F}_q^*$, so that SL(2, $\mathbb{F}_q$) $\supseteq$ GL(2, $\mathbb{F}_q$)$'$, and $|$GL(2, $\mathbb{F}_q$)/SL(2, $\mathbb{F}_q$)$| = q - 1$. But, for any finite group G, the quantity $|G/G'|$ equals the number of one-dimensional irreducible G-representations (see Corollary 14.5.2) and, by Theorem 14.5.6, this number is exactly $|\mathbb{F}_q^*| = q - 1$. This forces SL(2, $\mathbb{F}_q$) = GL(2, $\mathbb{F}_q$)$'$. $\quad\square$

Remark 14.5.8 From Proposition 14.5.3 and Proposition 14.5.5 it follows that for any one-dimensional representation χ of B, the induced representation $\widehat{\chi}$ decomposes as the sum of at most two irreducible G-representations. Indeed, if $\widehat{\chi} = \sigma_1 \oplus \sigma_2 \oplus \cdots \oplus \sigma_m$, by Proposition 14.5.3 each σ_i contains a nontrivial U-invariant vector, while, by Proposition 14.5.5, $\widehat{\chi}$ contains exactly a two-dimensional space of U-invariant vectors. This fact might be used to get an alternative proof of the fact that $\widehat{\chi}_{\psi,\psi}$ contains exactly two irreducible representations.

Proposition 14.5.9 *Let* $\psi, \psi_1, \psi_2 \in \widehat{\mathbb{F}_q^*}$ *and denote by* Ψ, Ψ_1, Ψ_2 *the corresponding representations of* Aff($\mathbb{F}_q$) *(cf. Theorem 12.1.3). Then*

$$\text{Res}^G_{\text{Aff}(\mathbb{F}_q)} \widehat{\chi}^1_\psi = \Psi \oplus \pi$$

and, if $\psi_1 \neq \psi_2$,

$$\text{Res}^G_{\text{Aff}(\mathbb{F}_q)} \widehat{\chi}_{\psi_1, \psi_2} = \Psi_1 \oplus \Psi_2 \oplus \pi,$$

where π *is the unique* $(q - 1)$-*dimensional irreducible representation of* Aff($\mathbb{F}_q$) *(cf. Theorem 12.1.3).*

Proof. We first note that the space V^U (with V as in Proposition 14.5.5) being B-invariant, it is also Aff($\mathbb{F}_q$)-invariant, and, moreover, dim$V^U = 2$. It is also clear that $\text{Res}^G_{\text{Aff}(\mathbb{F}_q)} \widehat{\chi}_{\psi_1, \psi_2} \succeq \Psi_1 \oplus \Psi_2$. Indeed, by (14.13) and Proposition 14.5.5, the B-representation on V^U is isomorphic to $\chi_{\psi_1, \psi_2} \oplus \chi_{\psi_2, \psi_1}$ and $\text{Res}^B_{\text{Aff}(\mathbb{F}_q)} \widehat{\chi}_{\psi_1, \psi_2} = \Psi_1$. Then, there exists an Aff($\mathbb{F}_q$)-invariant subspace W such that $V = V^U \oplus W$. The space W cannot contain a one-dimensional representation of Aff($\mathbb{F}_q$), otherwise it would contain U-invariant vectors (note that U is the commutator subgroup also of Aff($\mathbb{F}_q$)). Therefore, W necessarily coincides with the representation space of π.

The case $\psi_1 = \psi_2 = \psi$ is analogous. $\quad\square$

Exercise 14.5.10

(1) From Proposition 14.5.9 and Frobenius reciprocity, deduce that, for all $\psi \in \widehat{\mathbb{F}_q^*}$,

$$\mathrm{Ind}_{\mathrm{Aff}(\mathbb{F}_q)}^G \Psi = \widehat{\chi}_\psi^0 \oplus \widehat{\chi}_\psi^1 \oplus \left(\bigoplus_{\substack{\psi_1 \in \widehat{\mathbb{F}_q^*}: \\ \psi_1 \neq \psi}} \widehat{\chi}_{\psi_1, \psi} \right).$$

(2) From Exercise 12.1.8.(2) and transitivity of induction, deduce that

$$\mathrm{Ind}_U^G \chi_0 = \left(\bigoplus_{\psi \in \widehat{\mathbb{F}_q^*}} \widehat{\chi}_\psi^0 \right) \oplus \left(\bigoplus_{\psi \in \widehat{\mathbb{F}_q^*}} \widehat{\chi}_\psi^1 \right) \oplus 2 \left(\bigoplus_{\{\psi_1, \psi_2\}} \widehat{\chi}_{\psi_1, \psi_2} \right),$$

where $\{\psi_1, \psi_2\}$ runs over all two-subsets of $\widehat{\mathbb{F}_q^*}$ (in other words, in the last summand, the representation $\widehat{\chi}_{\psi_1, \psi_2} = \widehat{\chi}_{\psi_2, \psi_1}$ is counted once, but it appears with multiplicity 2 in the decomposition).

14.6 Cuspidal representations

This section is devoted to a close analysis of the cuspidal representations of G. The last part heavily relies on the material from Section 7.3. Let G be a finite group and $K \leq G$ a subgroup. Consider a one-dimensional K-representation $(\chi, \mathbb{C})$ that we identify with its character. As usual, we fix a complete set $S \subseteq G$ of representatives for the double K-cosets in G, so that $G = \bigsqcup_{s \in S} KsK$, and set $K_s = K \cap sKs^{-1}$. Also, cf. (11.32), we denote by S_0 the set of $s \in S$ such that $\mathrm{Hom}_{K_s}(\mathrm{Res}_{K_s}^K \chi, \chi_s)$ is not trivial.

For the convenience of the reader, in the following theorem we collect some results about the Hecke algebra $\mathcal{H}(G, K, \chi)$ from Chapter 13.

Theorem 14.6.1 *Let*

$$\mathcal{H}(G, K, \chi) = \{f \in L(G) : f(k_1 g k_2) = \overline{\chi(k_1)} f(g) \overline{\chi(k_2)}, \ \forall k_1, k_2 \in K, g \in G\}.$$

Then the following hold:

(i) $\mathrm{End}_G(\mathrm{Ind}_K^G \chi) \cong \mathcal{H}(G, K, \chi)$;
(ii) $S_0 = \{s \in S : \chi(s^{-1} x s) = \chi(x), \text{ for all } x \in K_s\}$;

(iii) *every function* $f \in \mathcal{H}(G, K, \chi)$ *only depends on its values on* $\mathcal{S}_0$, *namely,*

$$f(g) = \begin{cases} \overline{\chi(k_1)}f(s)\overline{\chi(k_2)} & \text{if } g = k_1 s k_2 \text{ with } s \in \mathcal{S}_0 \\ 0 & \text{otherwise.} \end{cases}$$

Definition 14.6.2 A $\mathrm{GL}(2, \mathbb{F}_q)$-representation (ρ, V) whose subspace V^U of U-invariant vectors is trivial is called a *cuspidal representation*. We denote by $\mathrm{Cusp} = \mathrm{Cusp}(\mathrm{GL}(2, \mathbb{F}_q)) \subset \widehat{\mathrm{GL}(2, \mathbb{F}_q)}$ a complete set of pairwise nonequivalent *irreducible* cuspidal representations.

Theorem 14.6.3 *Let* χ *be a non-trivial character of the (Abelian) group* U. *Then* $\mathrm{Ind}_U^G \chi$ *is multiplicity-free and does not depend on the particular choice of* χ. *Moreover*

$$\mathrm{Ind}_U^G \chi = \left[\bigoplus_{\psi \in \widehat{\mathbb{F}_q^*}} \widehat{\chi}_\psi^1 \right] \oplus \left[\bigoplus_{\psi_1 \neq \psi_2 \in \widehat{\mathbb{F}_q^*}} \widehat{\chi}_{\psi_1, \psi_2} \right] \oplus \left[\bigoplus_{\rho \in \mathrm{Cusp}} \rho \right]. \quad (14.22)$$

In other words, (G, U, χ) *is a multiplicity-free triple for every non-trivial character* $\chi \in \widehat{U}$ *(cf. Chapter 13) and* $\mathrm{Ind}_U^G \chi$ *contains all the irreducible G-representations of dimension greater than one.*

Proof. We present two proofs of (14.22): the first one is of a more theoretical flavor, the second one relies on the computation of the number of conjugacy classes of G.

First proof. We first observe that U is a normal subgroup of B and that one has $B = \coprod_{d \in D} dU = \coprod_{d \in D} UdU$. From the Bruhat decomposition (cf. Lemma 14.2.4) we then get

$$G = B \coprod UwB = \left(\coprod_{d \in D} UdU \right) \coprod \left(\coprod_{d \in D} UwdU \right).$$

As a consequence we can take $\mathcal{S} := D \coprod wD$ as a complete set of representatives for the double U-cosets in G. Moreover, it is easy to check that $dUd^{-1} \cap U = U$ and $wdUd^{-1}w \cap U = \{1_G\}$ for all $d \in D$. Thus (cf. Theorem 14.6.1.(ii)), we have that

$$\mathcal{S}_0 = Z \coprod wD = \mathcal{S} \setminus (D \setminus Z). \quad (14.23)$$

From Theorem 14.6.1.(iii) we deduce that every function $f \in \mathcal{H}(G, K, \psi)$ vanishes on $\coprod_{d \in D \setminus Z} dU$.

Consider now the map $\tau : G \to G$ defined by setting

$$\tau \begin{pmatrix} \alpha & \beta \\ \gamma & \delta \end{pmatrix} = \begin{pmatrix} \delta & \beta \\ \gamma & \alpha \end{pmatrix}$$

for all $\begin{pmatrix} \alpha & \beta \\ \gamma & \delta \end{pmatrix} \in G$. It is easy to check that τ is an involutive anti-automorphism of G, that is, $\tau(g_1 g_2) = \tau(g_2)\tau(g_1)$ and $\tau^2(g) = g$ for all $g_1, g_2, g \in G$. We claim that

$$f^\tau = f \quad \text{for all } f \in \mathcal{H}(G, U, \chi), \tag{14.24}$$

where $f^\tau \in L(G)$ is defined by setting $f^\tau(g) = f(\tau(g))$ for all $g \in G$ (cf. (13.18)). In order to show (14.24), we recall that every $f \in \mathcal{H}(G, U, \chi)$ is supported in $\coprod_{s \in Z} \coprod_{wD} UsU$ and observe that τ fixes pointwise the subgroup U. As a consequence, it suffices to show that τ also fixes all elements in $Z \coprod wD$. First of all, it is obvious that $\tau(z) = z$ for all $z \in Z$. The remaining part is a simple calculation:

$$\tau(wd) = \tau \left(\begin{pmatrix} 0 & 1 \\ 1 & 0 \end{pmatrix} \begin{pmatrix} \alpha & 0 \\ 0 & \beta \end{pmatrix} \right) = \tau \left(\begin{pmatrix} 0 & \beta \\ \alpha & 0 \end{pmatrix} \right) = \begin{pmatrix} 0 & \beta \\ \alpha & 0 \end{pmatrix} = wd$$

for all $d = \begin{pmatrix} \alpha & 0 \\ 0 & \beta \end{pmatrix} \in D$. The claim follows.

By Proposition 13.3.4, the algebra $\mathcal{H}(G, U, \chi)$ is commutative and therefore $\mathrm{Ind}_U^G \chi$ is multiplicity-free. By transitivity of induction and (12.7) we have

$$\mathrm{Ind}_U^G \chi = \mathrm{Ind}_{\mathrm{Aff}(\mathbb{F}_q)}^G \mathrm{Ind}_U^{\mathrm{Aff}(\mathbb{F}_q)} \chi = \mathrm{Ind}_{\mathrm{Aff}(\mathbb{F}_q)}^G \pi \tag{14.25}$$

so that also $\mathrm{Ind}_{\mathrm{Aff}(\mathbb{F}_q)}^G \pi$ is multiplicity-free.

The multiplicity of $\widehat{\chi}_\psi^1$ and $\widehat{\chi}_{\psi_1, \psi_2}$ in $\mathrm{Ind}_U^G \chi$ is equal to one by (14.25), Proposition 14.5.9, and Frobenius reciprocity. If ρ is cuspidal, then $\mathrm{Res}_{\mathrm{Aff}(\mathbb{F}_q)}^G \rho$ cannot contain a one-dimensional representation Ψ of $\mathrm{Aff}(\mathbb{F}_q)$, because otherwise it would contain also nontrivial U-invariant vectors (recall the proof of Proposition 14.5.9 and the fact that U is the commutator subgroup of $\mathrm{Aff}(\mathbb{F}_q)$). Then $\mathrm{Res}_{\mathrm{Aff}(\mathbb{F}_q)}^G \rho$ must be a multiple of π. Therefore,

$$1 \geq \text{multiplicity of a cuspidal representation } \rho \text{ in } \mathrm{Ind}_{\mathrm{Aff}(\mathbb{F}_q)}^G \pi$$

$$= \text{multiplicity of } \pi \text{ in } \mathrm{Res}_{\mathrm{Aff}(\mathbb{F}_q)}^G \rho \quad \text{(by Frobenius reciprocity)}$$

$$\geq 1$$

implies that all these multiplicities are equal to 1. Finally, from Corollary 11.2.3 and (14.25) it follows that $\mathrm{Ind}_U^G \chi$ cannot contain one-dimensional G-representations.

Second proof. In Theorem 14.5.6 we have determined:

- $q-1$ one-dimensional representations of G (the $\widehat{\chi}^0_\psi$s);
- $q-1$ irreducible q-dimensional representations of G (the $\widehat{\chi}^1_\psi$s);
- $\frac{(q-1)(q-2)}{2}$ irreducible $(q+1)$-dimensional representations of G (the $\widehat{\chi}_{\psi_1,\psi_2}$s).

Since G has $2(q-1) + \frac{(q-1)(q-2)}{2} + \frac{q(q-1)}{2}$ conjugacy classes (see Theorem 14.3.2), from Theorem 10.3.13.(ii) it follows that there exist exactly $\frac{q(q-1)}{2}$ irreducible representations missing in the above list: these are the cuspidal representations. Moreover (cf. (11.10) and Proposition 14.3.1)

$$\dim\mathrm{Ind}^G_U\chi = [G:U] = (q+1)(q-1)^2. \tag{14.26}$$

Invoking again Theorem 14.5.6 and using the last part of the first proof, we deduce that the $\widehat{\chi}^1_\psi$s and $\widehat{\chi}_{\psi_1,\psi_2}$s sum up in $\mathrm{Ind}^G_U\chi$ forming a subspace of dimension

$$\sum_{\psi\in\widehat{\mathbb{F}}^*_q}\dim\widehat{\chi}^1_\psi + \sum_{\psi_1\neq\psi_2\in\widehat{\mathbb{F}}^*_q}\dim\widehat{\chi}_{\psi_1,\psi_2} = q(q-1) + \frac{(q^2-1)(q-2)}{2}$$

$$= (q-1)\frac{q^2+q-2}{2}. \tag{14.27}$$

Denoting by $r_\rho \geq 1$ the multiplicity of π in $\mathrm{Res}^G_{\mathrm{Aff}(\mathbb{F}_q)}\rho \in \mathrm{Cusp}$, so that $\dim\rho = r_\rho\dim\pi = r_\rho(q-1)$ (cf. the first proof), by subtracting (14.27) from (14.26), we deduce

$$\sum_{\rho\in\mathrm{Cusp}} r_\rho(q-1) = (q-1)\frac{q(q-1)}{2},$$

that is, $\sum_{\rho\in\mathrm{Cusp}} r_\rho = \frac{q(q-1)}{2}$. Since this is a sum of $\frac{q(q-1)}{2}$ integers $r_\rho \geq 1$, we deduce that $r_\rho = 1$ for every cuspidal representation ρ. $\qquad\square$

Remark 14.6.4 Alternatively, from (14.23) and multiplicity freeness of $\mathrm{Ind}^G_U\chi$ one deduces that

$$\dim\mathrm{End}_G(\mathrm{Ind}^G_U\chi) = |Z| + |wD| = (q-1) + (q-1)^2 = q(q-1).$$

Since parabolic induction yields

$$q-1+\frac{(q-1)(q-2)}{2} = \frac{q(q-1)}{2}$$

irreducible representations in $\mathrm{Ind}^G_U\chi$, there are other $\frac{q(q-1)}{2}$ irreducible representations in $\mathrm{Ind}^G_U\chi$, and these must be exactly the $\frac{q(q-1)}{2}$ cuspidal representations.

Corollary 14.6.5 *A G-representation* (ρ, V) *(not necessarily irreducible) is a cuspidal representation if and only if* $\mathrm{Res}^G_{\mathrm{Aff}(\mathbb{F}_q)}\rho = \pi$. *In particular,* $\dim\rho = q - 1$ *for every cuspidal representation.*

Proof. The "only if" part can be immediately deduced from the proof of the previous theorem where we have shown that, if ρ is cuspidal, then $\mathrm{Res}^G_{\mathrm{Aff}(\mathbb{F}_q)}\rho = \pi$ and, in particular, $\dim\rho = q - 1$. The "if" part is trivial: if (ρ, V) is a G-representation and $\mathrm{Res}^G_{\mathrm{Aff}(\mathbb{F}_q)}\rho = \pi$ then ρ is G-irreducible, since π is $\mathrm{Aff}(\mathbb{F}_q)$-irreducible. Moreover, V cannot contain nontrivial U-invariant vectors because,

$$\mathrm{Res}^G_U\rho = \mathrm{Res}^{\mathrm{Aff}(\mathbb{F}_q)}_U\mathrm{Res}^G_{\mathrm{Aff}(\mathbb{F}_q)}\rho = \mathrm{Res}^{\mathrm{Aff}(\mathbb{F}_q)}_U\pi = \bigoplus_{\substack{\chi\in\widehat{U}\\\chi \text{ nontrivial}}} \chi,$$

where the last equality follows from Corollary 12.1.7. $\square$

We now introduce a special element in B:

$$b_0 = \begin{pmatrix} -1 & -1 \\ 0 & 1 \end{pmatrix}. \tag{14.28}$$

The following property is elementary, but useful: for all $b \in B \setminus D$ there exist $d_1, d_2 \in D$ such that

$$b = d_1 b_0 d_2. \tag{14.29}$$

Indeed, if $\begin{pmatrix} \alpha & \beta \\ 0 & \delta \end{pmatrix} \in B \setminus D$, that is $\beta \neq 0$, then

$$\begin{pmatrix} \alpha & \beta \\ 0 & \delta \end{pmatrix} = \begin{pmatrix} 1 & 0 \\ 0 & -\delta\beta^{-1} \end{pmatrix}\begin{pmatrix} -1 & -1 \\ 0 & 1 \end{pmatrix}\begin{pmatrix} -\alpha & 0 \\ 0 & -\beta \end{pmatrix}.$$

Also note that if $d = \begin{pmatrix} \alpha & 0 \\ 0 & \delta \end{pmatrix} \in D$ then

$$\tilde{d} = wdw = \begin{pmatrix} \delta & 0 \\ 0 & \alpha \end{pmatrix} \in D. \tag{14.30}$$

Exercise 14.6.6 From Exercise 14.5.10 and Exercise 12.1.8, deduce that, for $\psi \in \widehat{A}$,

$$\mathrm{Ind}^G_A\psi = \left(\mathrm{Ind}^G_U\chi\right) \oplus \left[\widehat{\chi}^0_\psi \oplus \widehat{\chi}^1_\psi \oplus \left(\bigoplus_{\substack{\psi_1\in\widehat{\mathbb{F}}^*_q:\\\psi_1\neq\psi}} \widehat{\chi}_{\psi_1,\psi}\right)\right],$$

where χ is any nontrivial character of U.

Proposition 14.6.7 *Let V be a finite dimensional vector space and $\rho\colon G \to$ End(V) a map such that:*

(a) $\mathrm{Res}_B^G \rho$ *is an irreducible B-representation;*
(b) $\rho(b_1 w b_2) = \rho(b_1)\rho(w)\rho(b_2)$ *for all $b_1, b_2 \in B$;*
(c) $\rho(wdw) = \rho(w)\rho(d)\rho(w)$ *for all $d \in D$;*
(d) $\rho(w b_0 w) = \rho(w)\rho(b_0)\rho(w)$.

Then (ρ, V) is an irreducible G-representation.

Proof. We show that $\rho(g_1 g_2) = \rho(g_1)\rho(g_2)$ for all $g_1, g_2 \in G$. Note that, this gives, in particular, that $\rho(g) \in$ GL(V) for all $g \in G$. When $g_1, g_2 \in B$, this follows from the hypothesis (a) (which also implies that $\rho(1_G) = I_V$). By virtue of the Bruhat decomposition (cf. Lemma 14.2.4) we have the following remaining cases:

First case: $g_1 = b \in B$ and $g_2 = b_1 w b_2 \in BwB$. Then

$$\rho(g_1 g_2) = \rho(b b_1 w b_2)$$
$$\text{(by hypothesis (b))} \quad = \rho(b b_1)\rho(w)\rho(b_2)$$
$$\text{(by hypothesis (a))} \quad = \rho(b)\rho(b_1)\rho(w)\rho(b_2)$$
$$\text{(by hypothesis (b))} \quad = \rho(b)\rho(b_1 w b_2)$$
$$= \rho(g_1)\rho(g_2).$$

The case $g_1 \in BwB$ and $g_2 \in B$ can be treated in the same way.

Second case: $g_1 = b_1 w b_2 \in BwB$ and $g_2 = b_3 w b_4 \in BwB$. We must further distinguish two subcases:

First subcase: $b_2 b_3 = d \in D$. Then

$$\rho(g_1 g_2) = \rho(b_1 w d w b_4)$$
$$\text{(by (14.30))} \quad = \rho(b_1 \tilde{d} b_4)$$
$$\text{(by hypothesis (a))} \quad = \rho(b_1)\rho(\tilde{d})\rho(b_4)$$
$$\text{(by hypothesis (c))} \quad = \rho(b_1)\rho(w)\rho(d)\rho(w)\rho(b_4)$$
$$\text{(by hypothesis (a))} \quad = \rho(b_1)\rho(w)\rho(b_2)\rho(b_3)\rho(w)\rho(b_4)$$
$$\text{(by hypothesis (b))} \quad = \rho(g_1)\rho(g_2).$$

Second subcase: $b_2 b_3 \in B \setminus D$. By (14.29) there exist $d_1, d_2 \in D$ such that

$$b_2 b_3 = d_1 b_0 d_2. \tag{14.31}$$

Then

$$\rho(g_1g_2) = \rho(b_1wd_1b_0d_2dwb_4)$$
$$\text{(by (14.30))} \quad = \rho(b_1\tilde{d_1}wb_0w\tilde{d_2}b_4)$$
$$\text{(by the first case for } wb_0w \in BwB) \quad = \rho(b_1\tilde{d_1})\rho(wb_0w)\rho(\tilde{d_2}b_4)$$
$$\text{(by hypothesis (d))} \quad = \rho(b_1\tilde{d_1})\rho(w)\rho(b_0)\rho(w)\rho(\tilde{d_2}b_4)$$
$$\text{(by the first case)} \quad = \rho(b_1\tilde{d_1}w)\rho(b_0)\rho(w\tilde{d_2}b_4)$$
$$\text{(by (14.30))} \quad = \rho(b_1wd_1)\rho(b_0)\rho(d_2wb_4)$$
$$\text{(by the first case and hypothesis (a))} \quad = \rho(b_1w)\rho(d_1b_0d_2)\rho(wb_4)$$
$$\text{(by (14.31))} \quad = \rho(b_1w)\rho(b_2b_3)\rho(wb_4)$$
$$\text{(by the first case and hypothesis (a))} \quad = \rho(b_1wb_2)\rho(b_3wb_4).$$

This shows that ρ is a representation. Its G-irreducibility follows from B-irreducibility (hypothesis (a)). $\qquad\square$

We now fix $\chi \in \widehat{\mathbb{F}_q}$ and consider an indecomposable character $\nu \in \widehat{\mathbb{F}_{q^2}^*}$ (cf. Definition 7.2.1). Let $j = j_{\chi,\nu}$ be the associated generalized Kloostermann sum (cf. (7.16)). Set $V = L(\mathbb{F}_q^*)$. We define a map $\rho\colon G \to \text{End}(V)$ by setting, for all $f \in V$ and $y \in \mathbb{F}_q^*$,

$$[\rho(g)f](y) = \nu(\delta)\chi(\delta^{-1}\beta y^{-1})f(\delta\alpha^{-1}y) \tag{14.32}$$

if $g = \begin{pmatrix} \alpha & \beta \\ 0 & \delta \end{pmatrix} \in B$ and

$$[\rho(g)f](y) = -\sum_{x\in\mathbb{F}_q^*} \nu(-\gamma x)\chi(\alpha\gamma^{-1}y^{-1} + \gamma^{-1}\delta x^{-1})j(\gamma^{-2}y^{-1}x^{-1}\det(g))f(x) \tag{14.33}$$

if $g = \begin{pmatrix} \alpha & \beta \\ \gamma & \delta \end{pmatrix} \in G \setminus B \equiv BwB$ (that is, if $\gamma \neq 0$).

Remark 14.6.8 As noted by Terras [159, p. 372], the minus sign in the right hand side of (14.33) is essential for the definition of $\rho(g)$ for $g \in G \setminus B$. Note that Piatetski-Shapiro [123] defines an induced representation by a *right*-translation action, namely, given a K-representation (σ, V), he defines $(\rho, \text{Ind}_K^G V)$ by setting

$$\text{Ind}_K^G V = \{f\colon G \to V : f(kg) = \sigma(k)f(g) \text{ for all } k \in K \text{ and } g \in G\} \tag{14.34}$$

and

$$[\rho(g_1)f](g_2) = f(g_2g_1)$$

for all $f \in \mathrm{Ind}_K^G V$, and $g_1, g_2 \in G$ (compare with (11.1) and (11.2)). Moreover, if $k(y, x; g)$ is as in [123, p. 40], *our ρ is defined by*

$$[\rho(g)f](y) = \sum_{x \in \mathbb{F}_q^*} k(y^{-1}, x^{-1}; g)f(x)$$

for all $f \in \mathrm{Ind}_K^G V$, $g \in G$, and $y \in \mathbb{F}_q^*$.

Theorem 14.6.9 *The above defined map ρ is an irreducible unitary G-representation and* $\mathrm{Res}_{\mathrm{Aff}(\mathbb{F}_q)}^G \rho = \pi$ *(cf. Proposition 14.5.9).*

Proof. The proof is an application of Proposition 14.6.7.

First of all, we prove that

$$\mathrm{Res}_B^G \rho \sim \left(\mathrm{Res}_{\mathbb{F}_q^*}^{\mathbb{F}_{q^2}^*} \nu \right) \boxtimes \pi.$$

Indeed, using Theorem 14.4.1, we get

$$\left\{ \left[\left(\mathrm{Res}_{\mathbb{F}_q^*}^{\mathbb{F}_{q^2}^*} \nu \boxtimes \pi \right) \begin{pmatrix} \alpha & \beta \\ 0 & \delta \end{pmatrix} \right] f \right\} (y) = \nu(\delta) \left[\pi \begin{pmatrix} \alpha\delta^{-1} & \beta\delta^{-1} \\ 0 & 1 \end{pmatrix} f \right] (y)$$

$$\text{(by Proposition 12.1.4)} \quad = \nu(\delta)\chi(\beta\delta^{-1}y^{-1})f(\alpha^{-1}\delta y)$$

$$\text{(by (14.32))} \quad = [\rho(g)f](y),$$

for all $\begin{pmatrix} \alpha & \beta \\ 0 & \delta \end{pmatrix} \in B$, $f \in V$, and $y \in \mathbb{F}_q^*$. This shows that $\mathrm{Res}_B^G \rho$ is B-irreducible, and condition (a) in Proposition 14.6.7 is satisfied.

We also note that, for all $y \in \mathbb{F}_q^*$,

$$[\rho(w)f](y) = - \sum_{x \in \mathbb{F}_q^*} \nu(-x)j(-x^{-1}y^{-1})f(x). \tag{14.35}$$

Let now $b_1 = \begin{pmatrix} \alpha_1 & \beta_1 \\ 0 & \delta_1 \end{pmatrix}$, $b_2 = \begin{pmatrix} \alpha_2 & \beta_2 \\ 0 & \delta_2 \end{pmatrix} \in B$. Then

$$b_1 w b_2 = \begin{pmatrix} \beta_1\alpha_2 & \beta_1\beta_2 + \alpha_1\delta_2 \\ \delta_1\alpha_2 & \delta_1\beta_2 \end{pmatrix}$$

and $\det(b_1 w b_2) = -\alpha_1\alpha_2\delta_1\delta_2$ so that

$$[\rho(b_1 w b_2)f](y) = - \sum_{x \in \mathbb{F}_q^*} \nu(-\delta_1\alpha_2 x)\chi(\beta_1\delta_1^{-1}y^{-1} + \alpha_2^{-1}\beta_2 x^{-1}) \cdot$$

$$j(-\alpha_1\delta_2\alpha_2^{-1}\delta_1^{-1}x^{-1}y^{-1})f(y)$$

and

$$[\rho(b_1)\rho(w)\rho(b_2)f](y) = \nu(\delta_1)\chi(\delta_1^{-1}\beta_1 y^{-1})[\rho(w)\rho(b_2)f](\delta_1\alpha_1^{-1}y)$$

$$\text{(by (14.35))} \quad = -\nu(\delta_1)\chi(\delta_1^{-1}\beta_1 y^{-1})\sum_{x\in\mathbb{F}_q^*}\nu(-x)\cdot$$

$$j(-x^{-1}y^{-1}\delta_1^{-1}\alpha_1)[\rho(b_2)f](x)$$

$$= -\sum_{x\in\mathbb{F}_q^*}\nu(-x\delta_1\delta_2)\chi(\delta_1^{-1}\beta_1 y^{-1} + \delta_2^{-1}\beta_2 x^{-1})\cdot$$

$$j(-x^{-1}y^{-1}\delta_1^{-1}\alpha_1)f(\delta_2\alpha_2^{-1}x)$$

$$\text{(setting } z = \delta_2\alpha_2^{-1}x) \quad = -\sum_{z\in\mathbb{F}_q^*}\nu(-z\delta_1\alpha_2)\chi(\beta_1\delta_1^{-1}y^{-1} + \alpha_2^{-1}\beta_2 z^{-1})\cdot$$

$$j(-z^{-1}y^{-1}\alpha_1\delta_2\alpha_2^{-1}\delta_1^{-1})f(z).$$

This shows that $\rho(b_1 w b_2) = \rho(b_1)\rho(w)\rho(b_2)$, and we have proved condition (b) in Proposition 14.6.7.

We now consider $d = \begin{pmatrix} \alpha & 0 \\ 0 & \delta \end{pmatrix} \in D$ so that $wdw = \begin{pmatrix} \delta & 0 \\ 0 & \alpha \end{pmatrix}$ (cf. (14.30)). Then, by (14.35),

$$[\rho(w)\rho(d)\rho(w)f](y) = -\sum_{x\in\mathbb{F}_q^*}\nu(-x)j(-x^{-1}y^{-1})[\rho(d)\rho(w)f](x)$$

$$= -\sum_{x\in\mathbb{F}_q^*}\nu(-x\delta)j(-x^{-1}y^{-1})[\rho(w)f](\alpha^{-1}\delta x)$$

$$= \sum_{x,z\in\mathbb{F}_q^*}\nu(xz\delta)j(-x^{-1}y^{-1})j(-\alpha\delta^{-1}x^{-1}z^{-1})f(z)$$

$$\text{(set } t = -x^{-1}z^{-1}\alpha\delta^{-1}) \quad = \nu(-\alpha)\sum_{z\in\mathbb{F}_q^*}\left[\sum_{t\in\mathbb{F}_q^*}\nu(t^{-1})j(t)j(y^{-1}z\alpha^{-1}\delta t)\right]f(z)$$

$$\text{(by Corollary 7.3.6)} \quad = \sum_{z\in\mathbb{F}_q^*}\nu(\alpha)\delta_{1,y^{-1}z\alpha^{-1}\delta}f(z)$$

$$= \nu(\alpha)f(\alpha\delta^{-1}y)$$

$$\text{(by (14.32))} \quad = [\rho(wdw)f](y)$$

and condition (c) also is proved.

Finally, if $b_0 = \begin{pmatrix} -1 & -1 \\ 0 & 1 \end{pmatrix}$ is as in (14.28), then $w b_0 w = \begin{pmatrix} 1 & 0 \\ -1 & -1 \end{pmatrix}$ so that

$$[\rho(w b_0 w)f](y) = -\sum_{z \in \mathbb{F}_q^*} \nu(z) \chi(z^{-1} - y^{-1}) j(-z^{-1} y^{-1}) f(z)$$

while, using again (14.35) and (14.32),

$$[\rho(w)\rho(b_0)\rho(w)f](y) = -\sum_{x \in \mathbb{F}_q^*} \nu(-x) j(-x^{-1}y^{-1})[\rho(b_0)\rho(w)f](x)$$

$$= -\sum_{x \in \mathbb{F}_q^*} \nu(-x) j(-x^{-1}y^{-1}) \chi(-x^{-1})[\rho(w)f](-x)$$

$$= \sum_{x,z \in \mathbb{F}_q^*} \nu(xz) j(-x^{-1}y^{-1}) j(x^{-1}z^{-1}) \chi(-x^{-1}) f(z)$$

$$(\text{setting } w = -x^{-1}) = \sum_{z \in \mathbb{F}_q^*} \nu(-z) \left[\sum_{w \in \mathbb{F}_q^*} j(wy^{-1}) j(w(-z^{-1})) \nu(w^{-1}) \chi(w) \right] f(z)$$

$$= -\sum_{z \in \mathbb{F}_q^*} \nu(z) j(-y^{-1}z^{-1}) \chi(z^{-1} - y^{-1}) f(z),$$

where the last equality follows from Proposition 7.3.4. Thus condition (d) is proved as well.

We are only left to show that ρ is unitary. Let $f_1, f_2 \in L(\mathbb{F}_q^*)$. If $g = \begin{pmatrix} \alpha & \beta \\ 0 & \delta \end{pmatrix}$ then we have

$$\langle \rho(g)f_1, \rho(g)f_2 \rangle = \sum_{x \in \mathbb{F}_q^*} \nu(\delta) \chi(\delta^{-1}\beta x^{-1}) f_1(\delta\alpha^{-1}x) \overline{\nu(\delta) \chi(\delta^{-1}\beta x^{-1}) f_2(\delta\alpha^{-1}x)}$$

$$=_{(*)} \sum_{y \in \mathbb{F}_q^*} f_1(y) \overline{f_2(y)}$$

$$= \langle f_1, f_2 \rangle$$

where $=_{(*)}$ follows from the substitution $y = \delta\alpha^{-1}x$ and the fact that $|\nu(\cdot)| = |\chi(\cdot)| = 1$.

Similarly, if $g = \begin{pmatrix} \alpha & \beta \\ \gamma & \delta \end{pmatrix}$ with $\gamma \neq 0$, then

$$\langle \rho(g)f_1, \rho(g)f_2 \rangle = \sum_{y \in \mathbb{F}_q^*} \sum_{x,z \in \mathbb{F}_q^*} \nu(-\gamma x)\chi(\alpha\gamma^{-1}y^{-1} + \gamma^{-1}\delta x^{-1}) \cdot$$

$$j(\gamma^{-2}y^{-1}x^{-1}\det(g))f_1(x)\overline{\nu(-\gamma z)} \cdot$$

$$\overline{\chi(\alpha\gamma^{-1}y^{-1} + \gamma^{-1}\delta z^{-1})j(\gamma^{-2}y^{-1}z^{-1}\det(g))f_2(z)}$$

$$= \sum_{x,z \in \mathbb{F}_q^*} f_1(x)\overline{f_2(z)}\nu(xz^{-1})\chi[\gamma^{-1}\delta(x^{-1} - z^{-1})] \cdot$$

$$\sum_{y \in \mathbb{F}_q^*} j(\gamma^{-2}y^{-1}x^{-1}\det(g))\overline{j(\gamma^{-2}y^{-1}z^{-1}\det(g))}$$

(by Proposition 7.3.5) $= \sum_{x,z \in \mathbb{F}_q^*} f_1(x)\overline{f_2(z)}\nu(xz^{-1})\chi[\gamma^{-1}\delta(x^{-1} - z^{-1})]\delta_{x,z}$

$$= \langle f_1, f_2 \rangle. \qquad \square$$

In the following, we write ρ_ν (resp. j_ν) to emphasize the dependence of the representation ρ (resp. the generalized Kloosterman sum) from the indecomposable character ν.

Theorem 14.6.10 *Let μ and ν be indecomposable characters of $\mathbb{F}_{q^2}^*$. Then the following conditions are equivalent.*

(a) *the representations ρ_μ and ρ_ν are equivalent;*
(b) *$\mu = \nu$ or $\overline{\mu} = \nu$;*
(c) *$j_\mu = j_\nu$ and $\mu|_{\mathbb{F}_q^*} = \nu|_{\mathbb{F}_q^*}$.*

Proof. The implication (b) $\Rightarrow$ (c) follows immediately from the definitions, and the converse, namely (c) $\Rightarrow$ (b), is Theorem 7.3.7. The fact that (c) implies (a) is trivial. We are only left to prove (a) $\Rightarrow$ (c). We thus suppose that $\rho_\mu \sim \rho_\nu$. Then there exists an invertible operator $T : L(\mathbb{F}_q^*) \to L(\mathbb{F}_q^*)$ such that

$$T\rho_\mu(g) = \rho_\nu(g)T$$

for all $g \in G$. Since, taking into account Theorem 14.6.9,

$$\text{Res}^G_{\text{Aff}(\mathbb{F}_q)}\rho_\mu = \text{Res}^G_{\text{Aff}(\mathbb{F}_q)}\rho_\nu = \pi$$

and π is Aff($\mathbb{F}_q$)-irreducible, we deduce that $T = \lambda I_{L(\mathbb{F}_q^*)}$ for some $\lambda \in \mathbb{C} \setminus \{0\}$, so that

$$\rho_\mu(g) = \rho_\nu(g)$$

for all $g \in G$.

In particular, for all $x, \delta \in \mathbb{F}_q^*$ and $f \in L(\mathbb{F}_q^*)$ we have:

$$\rho_\mu \begin{pmatrix} 1 & 0 \\ 0 & \delta \end{pmatrix} = \rho_\nu \begin{pmatrix} 1 & 0 \\ 0 & \delta \end{pmatrix}$$

so that

$$\mu(\delta)f(\delta x) = \nu(\delta)f(\delta x)$$

and therefore

$$\mu(\delta) = \nu(\delta). \tag{14.36}$$

This shows that $\mu|_{\mathbb{F}_q^*} = \nu|_{\mathbb{F}_q^*}$. Similarly, from (14.35) and the equality $\rho_\mu(w) = \rho_\nu(w)$ we deduce

$$\sum_{x \in \mathbb{F}_q^*} \mu(-x)j_\mu(-x^{-1}y^{-1})f(x) = \sum_{x \in \mathbb{F}_q^*} \nu(-x)j_\nu(-x^{-1}y^{-1})f(x),$$

for all $y \in \mathbb{F}_q^*$, that implies (taking into account (14.36)) that

$$j_\mu(x) = j_\nu(x) \tag{14.37}$$

for all $x \in \mathbb{F}_q^*$. □

Corollary 14.6.11 *The set* $\{\rho_\nu : \nu$ *indecomposable character of* $\mathbb{F}_{q^2}^*\}$ *coincides with the set* Cusp *of all irreducible cuspidal representations of G.*

Proof. Let ν be an indecomposable character of $\mathbb{F}_{q^2}^*$. By Theorem 14.6.9, $\operatorname{Res}_{\mathrm{Aff}(\mathbb{F}_q)}^G \rho_\nu = \pi$ (and ρ_ν is irreducible) so that, by virtue of Corollary 14.6.5, $\rho_\nu \in \operatorname{Cusp}$ (alternatively, keeping in mind $\dim \rho_\nu = q - 1$, to show that ρ_ν is cuspidal one may refer to the discussion in the second proof of Theorem 14.6.3). By Remark 14.6.4 (cf. also the second proof of Theorem 14.6.3), there are exactly $\frac{q(q-1)}{2}$ pairwise nonequivalent irreducible cuspidal representations. On the other hand, the number of indecomposable characters is $q(q - 1)$: thus, the ρ_νs exhaust Cusp (and, in fact, since $\rho_\nu = \rho_{\bar\nu}$, each cuspidal representation is listed twice). □

14.7 Whittaker models and Bessel functions

In this section, we expose Piatetsky-Schapiro's theory of Whittaker models and Bessel functions. Our approach, however, is based on our theory of multiplicity-free triples (see Chapter 13): this way, we clarify many intricate points and simplify calculations.

Fix a nontrivial character $(\chi, \mathbb{C}) \in \widehat{U} \equiv \widehat{\mathbb{F}_q}$. By Theorem 14.6.3, the induced representation $(\mathrm{Ind}_U^G \chi, \mathrm{Ind}_U^G \mathbb{C})$ is multiplicity free and contains all the irreducible representations of G of dimension greater than 1. Let (ρ, V) be an arbitrary irreducible G-representation with $\dim V > 1$, so that, by the above, $\dim \mathrm{Hom}_G(\rho, \mathrm{Ind}_U^G \chi) = 1$. We fix an operator $T^\rho \in \mathrm{Hom}_G(\rho, \mathrm{Ind}_U^G \chi)$, which is also an isometry (so that, T^ρ is defined up to a complex constant of modulus 1). The subspace $T^\rho V \leq \mathrm{Ind}_U^G \mathbb{C}$ is called the *Whittaker model* of ρ. Note that it does not depend on T^ρ and, for all $v \in V$, the function $T^\rho v : G \to \mathbb{C}$ satisfies

$$[T^\rho v](gu) = \overline{\chi(u)}[T^\rho v](g) \tag{14.38}$$

for all $g \in G$, $v \in V$, and $u \in U$ (by definition of $\mathrm{Ind}_U^G \chi$), and

$$[T^\rho v](h^{-1}g) = [T^\rho \rho(h)v](g) \tag{14.39}$$

for all $g, h \in G$, $v \in V$ (because T^ρ is an intertwiner and, again, by definition of $\mathrm{Ind}_U^G \chi$). Finally, since T^ρ is an isometry we have

$$\|T^\rho v\|_{\mathrm{Ind}_U^G \mathbb{C}} = \|v\|_V.$$

In particular, $T^\rho v = 0 \Leftrightarrow v = 0$.

Proposition 14.7.1 *Let (ρ, V) be an irreducible G-representation satisfying $\dim V > 1$. Then*

(i)

$$\mathrm{Res}_{\mathrm{Aff}(\mathbb{F}_q)}^G V \sim J(V) \oplus V_\pi$$

where $J(V)$ is the Jacquet module (see Section 14.5) and (π, V_π) is the unique $q - 1$ dimensional irreducible representation of $\mathrm{Aff}(\mathbb{F}_q)$.

(ii) *Let $v \in J(V)$ then*

$$\rho \begin{pmatrix} 1 & \beta \\ 0 & 1 \end{pmatrix} v = v$$

for all $\begin{pmatrix} 1 & \beta \\ 0 & 1 \end{pmatrix} \in U.$

(iii) $\dim V > \dim J(V).$

Proof.

(i) It is an immediate consequence of the following facts: (ρ, V) is contained in $\mathrm{Ind}_U^G \chi \sim \mathrm{Ind}_{\mathrm{Aff}(\mathbb{F}_q)}^G \pi$ so that $\mathrm{Res}_{\mathrm{Aff}(\mathbb{F}_q)}^G V$ contains V_π with multiplicity one. If (ρ, V) is cuspidal, then $\mathrm{Res}_{\mathrm{Aff}(\mathbb{F}_q)}^G \rho \sim \pi$ (cf. Corollary 14.6.5) and $J(V) = 0$ (by definition). If (ρ, V) is parabolic, we may invoke Proposition 14.5.9.

(ii) If $J(V)$ is nontrivial, then by Theorem 12.1.3 and Proposition 14.5.9 we have

$$\text{Res}^G_{\text{Aff}(\mathbb{F}_q)}\left[\rho|_{J(V)}\right] = \begin{cases} \text{either} & \Psi_1 \oplus \Psi_2 \\ \text{or} & \Psi \end{cases}$$

and Ψ is trivial on U.

(iii) This follows immediately from (i). □

The following is an elementary but useful identity.

Lemma 14.7.2 *Let* (ρ, V) *be an irreducible G-representation with* $\dim V > 1$. *Let also* $v \in V$, $\alpha \in \mathbb{F}_q^*$ *and* $\beta \in \mathbb{F}_q$. *Then we have:*

$$[T^\rho v]\begin{pmatrix} \alpha & \beta \\ 0 & 1 \end{pmatrix} = \overline{\chi(\alpha^{-1}\beta)}[T^\rho v]\begin{pmatrix} \alpha & 0 \\ 0 & 1 \end{pmatrix}.$$

Proof.

$$[T^\rho v]\begin{pmatrix} \alpha & \beta \\ 0 & 1 \end{pmatrix} = [T^\rho v]\left[\begin{pmatrix} \alpha & 0 \\ 0 & 1 \end{pmatrix}\begin{pmatrix} 1 & \alpha^{-1}\beta \\ 0 & 1 \end{pmatrix}\right]$$

$$(\text{by } (14.38)) \quad = \overline{\chi(\alpha^{-1}\beta)}[T^\rho v]\begin{pmatrix} \alpha & 0 \\ 0 & 1 \end{pmatrix}. \quad \square$$

Proposition 14.7.3 *Let* (ρ, V) *be an irreducible G-representation with* $\dim V > 1$ *and define a linear map* $R\colon V \to L(\mathbb{F}_q^*)$ *by setting*

$$[Rv](x) = [T^\rho v]\begin{pmatrix} x & 0 \\ 0 & 1 \end{pmatrix}$$

for all $v \in V$, $x \in \mathbb{F}_q^*$. *Then R is a surjective A-homomorphism (cf. (12.2)) and its kernel is exactly* $J(V)$.

Proof. Suppose that $v \in J(V)$. Then, for $\alpha \in \mathbb{F}_q^*$, $\beta \in \mathbb{F}_q$ we have

$$[T^\rho v]\begin{pmatrix} \alpha & \beta \\ 0 & 1 \end{pmatrix} = [T^\rho v]\left[\begin{pmatrix} \alpha & -\beta \\ 0 & 1 \end{pmatrix}^{-1}\begin{pmatrix} \alpha & 0 \\ 0 & 1 \end{pmatrix}\right]$$

$$(\text{by } (14.39)) \quad = \left[T^\rho \rho\begin{pmatrix} 1 & -\beta \\ 0 & 1 \end{pmatrix}v\right]\begin{pmatrix} \alpha & 0 \\ 0 & 1 \end{pmatrix}$$

$$(\text{by Proposition } 14.7.1.(\text{ii})) \quad = [T^\rho v]\begin{pmatrix} \alpha & 0 \\ 0 & 1 \end{pmatrix}.$$

Then, using Lemma 14.7.2, we deduce that

$$[T^\rho v]\begin{pmatrix} \alpha & 0 \\ 0 & 1 \end{pmatrix} = [T^\rho v]\begin{pmatrix} \alpha & \beta \\ 0 & 1 \end{pmatrix} = \overline{\chi(\alpha^{-1}\beta)}[T^\rho v]\begin{pmatrix} \alpha & 0 \\ 0 & 1 \end{pmatrix}$$

for all $\beta \in \mathbb{F}_q$, and this implies that $[T^\rho v]\begin{pmatrix} \alpha & 0 \\ 0 & 1 \end{pmatrix} \equiv [Rv](\alpha) = 0$ for all $\alpha \in \mathbb{F}_q^*$ (since χ is nontrivial). That is, $v \in \text{Ker}R$, showing that $J(V) \subset \text{Ker}(R)$.

Let us prove that $\text{Ker}R$ is $\text{Aff}(\mathbb{F}_q)$-invariant. If $\alpha, \gamma \in \mathbb{F}_q^*$, $\beta \in \mathbb{F}_q$ and $v \in \text{Ker}R$ then, taking into account (14.39), we have

$$\left[T^\rho \rho \begin{pmatrix} \gamma & \beta \\ 0 & 1 \end{pmatrix} v\right]\begin{pmatrix} \alpha & 0 \\ 0 & 1 \end{pmatrix} = [T^\rho v]\begin{pmatrix} \gamma^{-1}\alpha & -\gamma^{-1}\beta \\ 0 & 1 \end{pmatrix}$$

$$\text{(by Lemma 14.7.2)} \quad = \overline{\chi(-\alpha^{-1}\beta)}[T^\rho v]\begin{pmatrix} \gamma^{-1}\alpha & 0 \\ 0 & 1 \end{pmatrix}$$

$$(v \in \text{Ker}R) \quad = 0.$$

Then, by Proposition 14.7.1.(i), the kernel of R must equal either $J(V)$ or $J(V) \oplus V_\pi = V$. Let us show that the second possibility cannot occur. Indeed, $\text{Ker}(R) = V$ implies $[T^\rho v](1_G) = 0$ for all $v \in V$. From (14.39) we then deduce that $[T^\rho v](g) = [T^\rho \rho(g^{-1})v](1_G) = 0$ for all $v \in V$ and $g \in G$, contradicting the fact that T^ρ is an isometry. The fact that R commutes with the A-representations on V and $L(\mathbb{F}_q^*)$ is obvious. $\qquad\Box$

Now consider again an irreducible G-representation (ρ, V) with $\dim V > 1$. Since it is contained in $\text{Ind}_U^G \chi$ with multiplicity one, by Frobenius reciprocity $\text{Res}_U^G \rho$ contains χ with multiplicity one. That is, there exists $v_0 \in V$, $\|v_0\| = 1$ such that

$$\rho(u)v_0 = \chi(u)v_0 \tag{14.40}$$

for all $u \in U$. Moreover, if $v \in V$ satisfies $\rho(u)v = \chi(u)v$ for all $u \in U$, then v must be a multiple of v_0. Clearly, v_0 is defined up to a complex multiple of modulus one; Piatetski-Shapiro called it the *Bessel vector* associated with the representation (ρ, V) (and the character $\chi \in \widehat{U}$).

We can now apply our theory of multiplicity-free triples developed in Chapter 13. By (13.31), T^ρ may be expressed by means of

$$[T^\rho v](g) = \sqrt{\frac{d_\rho}{|G/U|}} \langle v, \rho(g)v_0 \rangle. \tag{14.41}$$

The *Bessel (or spherical) function* associated with ρ (and χ) is defined by setting

$$\varphi^\rho(g) = \langle v_0, \rho(g)v_0 \rangle \equiv \sqrt{\frac{|G/U|}{d_\rho}}[T^\rho v_0](g) \tag{14.42}$$

for all $g \in G$, see (13.32). Clearly $\varphi^\rho(1_G) = 1$.

Proposition 14.7.4 *The Bessel function* φ^ρ *satisfies*

$$\varphi^\rho \begin{pmatrix} \alpha & 0 \\ 0 & 1 \end{pmatrix} = 0$$

for all $\alpha \in \mathbb{F}_q^* \setminus \{1\}$.

Proof. On the one hand, for all $\alpha \in \mathbb{F}_q^*$, $\beta \in \mathbb{F}_q$, we have

$$\varphi^\rho \begin{pmatrix} \alpha & \beta \\ 0 & 1 \end{pmatrix} = \left\langle v_0, \rho \left[\begin{pmatrix} 1 & -\beta \\ 0 & 1 \end{pmatrix}^{-1} \begin{pmatrix} \alpha & 0 \\ 0 & 1 \end{pmatrix} \right] v_0 \right\rangle$$

$$= \left\langle \rho \begin{pmatrix} 1 & -\beta \\ 0 & 1 \end{pmatrix} v_0, \rho \begin{pmatrix} \alpha & 0 \\ 0 & 1 \end{pmatrix} v_0 \right\rangle$$

$$\text{(by (14.40))} \quad = \overline{\chi(\beta)} \varphi^\rho \begin{pmatrix} \alpha & 0 \\ 0 & 1 \end{pmatrix}.$$

On the other hand

$$\varphi^\rho \begin{pmatrix} \alpha & \beta \\ 0 & 1 \end{pmatrix} = \sqrt{\frac{|G/U|}{d_\rho}} [T^\rho v_0] \begin{pmatrix} \alpha & \beta \\ 0 & 1 \end{pmatrix}$$

$$\text{(by Lemma 14.7.2)} \quad = \overline{\chi(\alpha^{-1}\beta)} \sqrt{\frac{|G/U|}{d_\rho}} [T^\rho v_0] \begin{pmatrix} \alpha & 0 \\ 0 & 1 \end{pmatrix}$$

$$= \overline{\chi(\alpha^{-1}\beta)} \varphi^\rho \begin{pmatrix} \alpha & 0 \\ 0 & 1 \end{pmatrix}.$$

If $\alpha \neq 1$, letting β vary in $\mathbb{F}_q$, we deduce that $\varphi^\rho \begin{pmatrix} \alpha & 0 \\ 0 & 1 \end{pmatrix} = 0$. $\qquad \square$

First of all, we determine the Bessel vectors and Bessel functions associated with parabolic representations. These representations (see Section 14.5) are obtained as induced representations: if $\mu = \chi_{\psi_1, \psi_2}$ (with $\psi_1 \neq \psi_2$ or $\psi_1 = \psi_2$) then the representation space of $\mathrm{Ind}_B^G \mu$ is

$$V = \{f \colon G \to \mathbb{C} \colon f(gb) = \overline{\mu(b)} f(g), \text{ for all } g \in G, b \in B\}. \quad (14.43)$$

Now, if $\psi_1 \neq \psi_2$, then it is irreducible, while if $\psi_1 = \psi_2 = \psi$, we have (see Theorem 14.5.6.(ii)) $\mathrm{Ind}_B^G \mu = \widehat{\chi}_{\psi, \psi} = \widehat{\chi}_\psi^0 \oplus \widehat{\chi}_\psi^1$, where $\widehat{\chi}_\psi^0$ is one-dimensional and $\widehat{\chi}_\psi^1$ is (irreducible and) q-dimensional. Since $\mathrm{Ind}_U^G \chi$ does not contain one-dimensional G-representations (by Theorem 14.6.3), for every $T \in \mathrm{Hom}_G(\widehat{\chi}_{\psi, \psi}, \mathrm{Ind}_U^G \chi)$ we have $V_{\widehat{\chi}_\psi^0} \subseteq \mathrm{Ker} T$.

Proposition 14.7.5 *With the notation above and keeping in mind the Bruhat decomposition (cf. Lemma 14.2.4), the Bessel vector $f_0 \in V$ is given by*

$$\begin{cases} f_0(b) = 0 & \text{for all } b \in B \\ f_0(uwb) = \frac{1}{\sqrt{q}} \overline{\mu(b)\chi(u)} & \text{for all } b \in B, u \in U. \end{cases} \tag{14.44}$$

Proof. Let f_0 be a function satisfying (14.44). It is a straightforward computation to check that f_0 belongs to V (cf. (14.43)). Moreover, for all $u, u' \in U$ and $b \in B$, we have

$$f_0(u^{-1}b) = 0 = \chi(u)f_0(b)$$

and

$$f_0(u^{-1}u'wb) = \frac{1}{\sqrt{q}} \chi(u)\overline{\mu(b)\chi(u')} = \chi(u)f_0(u'wb)$$

that is, f_0 belongs to the χ-component of $\operatorname{Res}_U^G \operatorname{Ind}_B^G \mu$.

In the case $\psi_1 = \psi_2 = \psi$, the one-dimensional representation $\widehat{\chi}_\psi^0$ cannot contain a χ-component, since $\chi \in \widehat{U}$ is non-trivial, while $\operatorname{Res}_U^G \widehat{\chi}_\psi^0$ is trivial by (14.18) since $\det(u) = 1$ for all $u \in U$. This can be alternatively deduced by using Frobenius reciprocity and recalling that $\widehat{\chi}_\psi^0$ is not contained in $\operatorname{Ind}_U^G \chi$ (cf. Theorem 14.6.3).

Finally, by (11.4) and using the Bruhat decomposition, we have

$$\langle f_0, f_0 \rangle = \frac{1}{|B|} \sum_{g \in G} |f_0(g)|^2$$

$$\text{(by (14.44))} \quad = \frac{1}{|B|} \sum_{g \in UwB} |f_0(g)|^2$$

$$= \frac{1}{|B|} \sum_{u \in U} \sum_{b \in B} |f_0(uwb)|^2$$

$$\text{(by (14.44) and } |U| = q) \quad = \frac{1}{|B| \cdot |U|} \sum_{u \in U} \sum_{b \in B} |\mu(b)| \cdot |\chi(u)|$$

$$= \frac{1}{|B|} \sum_{b \in B} |\mu(b)| \cdot \frac{1}{|U|} \sum_{u \in U} |\chi(u)|$$

$$= 1. \qquad \square$$

Corollary 14.7.6 *Let $\rho = \widehat{\chi}_{\psi_1, \psi_2}$ be a parabolic representation. Then, with the same notation as in Proposition 14.7.5, we have*

$$[T^\rho f](g) = \sqrt{\frac{d_\rho}{|G|}} \sum_{u \in U} f(guw)\chi(u)$$

for all $f \in V$ (cf. (14.43)) and $g \in G$.

Proof. Let $f \in V$ and $g \in G$. By (14.41) we have

$$[T^\rho f](g) = \sqrt{\frac{d_\rho}{|G/U|}} \langle f, \rho(g)f_0 \rangle \mathrm{Ind}_B^G \mu$$

$$= \sqrt{\frac{d_\rho}{|G/U|}} \frac{1}{|B|} \sum_{h \in G} f(h)\overline{f_0(g^{-1}h)}$$

$$(\text{setting } h = gt) \quad = \sqrt{\frac{d_\rho}{|G/U|}} \frac{1}{|B|} \sum_{t \in G} f(gt)\overline{f_0(t)}$$

$$(\text{by Proposition 14.7.5}) \quad = \sqrt{\frac{d_\rho}{|G|}} \frac{1}{|B|} \sum_{u \in U} \sum_{b \in B} f(guwb)\mu(b)\chi(u)$$

$$(\text{by (14.43)}) \quad = \sqrt{\frac{d_\rho}{|G|}} \sum_{u \in U} f(guw)\chi(u). \qquad \square$$

Corollary 14.7.7 *With the same notation as in Corollary 14.7.6, the spherical function associated with ρ is given by*

$$\varphi^\rho(g) = \frac{1}{\sqrt{q}} \sum_{u \in U} f_0(guw)\chi(u)$$

for all $g \in G$.

Proof. Set $f = f_0$ in Corollary 14.7.6 and use (14.42). $\qquad \square$

It is interesting to analyze a special value of φ^ρ.

Proposition 14.7.8 *With the same notation as in Corollary 14.7.7, we have*

$$\varphi^\rho \begin{pmatrix} 0 & \alpha \\ 1 & 0 \end{pmatrix} = \frac{1}{q} \sum_{\substack{x,y \in \mathbb{F}_q^*: \\ xy=-\alpha}} \overline{\psi_1(x)\psi_2(y)}\chi(x+y)$$

for all $\alpha \in \mathbb{F}_q^*$.

Proof. First of all, note that, for $x \neq 0$, the Bruhat decomposition yields

$$\begin{pmatrix} \alpha & 0 \\ x & 1 \end{pmatrix} = \begin{pmatrix} 1 & \alpha x^{-1} \\ 0 & 1 \end{pmatrix} \begin{pmatrix} 0 & 1 \\ 1 & 0 \end{pmatrix} \begin{pmatrix} x & 1 \\ 0 & -\alpha x^{-1} \end{pmatrix}$$

so that by Proposition 14.7.5

$$f_0 \begin{pmatrix} \alpha & 0 \\ x & 1 \end{pmatrix} = \frac{1}{\sqrt{q}} \overline{\mu \begin{pmatrix} x & 1 \\ 0 & -\alpha x^{-1} \end{pmatrix} \chi \begin{pmatrix} 1 & \alpha x^{-1} \\ 0 & 1 \end{pmatrix}}$$

$$= \frac{1}{\sqrt{q}} \overline{\psi_1(x)} \overline{\psi_2(-\alpha x^{-1})} \chi(\alpha x^{-1}). \tag{14.45}$$

From Corollary 14.7.7, the identity $\begin{pmatrix} 0 & \alpha \\ 1 & 0 \end{pmatrix} \begin{pmatrix} 1 & x \\ 0 & 1 \end{pmatrix} \begin{pmatrix} 0 & 1 \\ 1 & 0 \end{pmatrix} = \begin{pmatrix} \alpha & 0 \\ x & 1 \end{pmatrix}$,

and $f_0 \begin{pmatrix} \alpha & 0 \\ 0 & 1 \end{pmatrix} = 0$, we then deduce that

$$\varphi^\rho \begin{pmatrix} 0 & \alpha \\ 1 & 0 \end{pmatrix} = \frac{1}{\sqrt{q}} \sum_{x \in \mathbb{F}_q^*} f_0 \begin{pmatrix} \alpha & 0 \\ x & 1 \end{pmatrix} \chi(x)$$

$$\text{(by (14.45))} = \frac{1}{q} \sum_{x \in \mathbb{F}_q^*} \overline{\psi_1(x)} \overline{\psi_2(-\alpha x^{-1})} \chi(x - \alpha x^{-1})$$

$$(y = -\alpha x^{-1}) = \frac{1}{q} \sum_{\substack{x,y \in \mathbb{F}_q^*: \\ xy = -\alpha}} \overline{\psi_1(x)} \overline{\psi_2(y)} \chi(x + y). \qquad \square$$

We now examine the Bessel vector and the Bessel function for a cuspidal representation (ρ, V) (cf. Definition 14.6.2). Let $\{f_x : x \in \mathbb{F}_q^*\}$ be the orthonormal basis of $V = L(\mathbb{F}_q^*)$, where

$$f_x(y) = \delta_{x,y} = \begin{cases} 1 & \text{if } y = x \\ 0 & \text{if } y \neq x \end{cases} \tag{14.46}$$

for all $x, y \in \mathbb{F}_q^*$.

Proposition 14.7.9

(i) f_1 is the Bessel vector for ρ.

(ii) The associated intertwining operator is given by:

$$[T^\rho f](g) = \frac{1}{\sqrt{q^2 - 1}} \overline{\nu(\delta)} \overline{\chi(\beta \alpha^{-1})} f(\alpha \delta^{-1})$$

if $g = \begin{pmatrix} \alpha & \beta \\ 0 & \delta \end{pmatrix} \in B$ and by

$$[T^\rho f](g) = -\frac{1}{\sqrt{q^2 - 1}} \sum_{x \in \mathbb{F}_q^*} \nu[\gamma x \det(g)^{-1}] \overline{\chi(\delta \gamma^{-1} + \gamma^{-1} \alpha x^{-1})}$$

$$\cdot j(\gamma^{-2} x^{-1} \det(g)) f(x)$$

$$if \ g = \begin{pmatrix} \alpha & \beta \\ \gamma & \delta \end{pmatrix} \in G \setminus B, \ for \ all \ f \in V.$$

(iii) *The spherical function of ρ is given by:*

$$\varphi^\rho(g) = \overline{\nu(\delta)\chi(\beta\alpha^{-1})}\delta_{\alpha,\delta}$$

$$if \ g = \begin{pmatrix} \alpha & \beta \\ 0 & \delta \end{pmatrix} \in B \ and$$

$$\varphi^\rho(g) = -\nu[\gamma \det(g)^{-1}]\overline{\chi(\delta\gamma^{-1} + \gamma^{-1}\alpha)}j(\gamma^{-2}\det(g))$$

$$if \ g = \begin{pmatrix} \alpha & \beta \\ \gamma & \delta \end{pmatrix} \in G \setminus B.$$

Proof. Let $f \in V$.

(i) From (14.32) we have

$$[\rho \begin{pmatrix} 1 & \beta \\ 0 & 1 \end{pmatrix} f](x) = \chi(\beta x^{-1})f(x)$$

for all $x \in \mathbb{F}_q^*$, so that f is a Bessel vector if and only if

$$\chi(\beta x^{-1})f(x) = \chi(\beta)f(x)$$

for all $x \in \mathbb{F}_q^*$ and $\beta \in \mathbb{F}_q$. Since χ is nontrivial, this forces $f = \lambda f_1$ for some $\lambda \in \mathbb{C}$. In particular, f_1 is a Bessel vector. Note that we have actually reproved that $\operatorname{Res}_U^G \rho$ contains χ with multiplicity one and therefore that ρ is contained in $\operatorname{Ind}_U^G \chi$ with multiplicity one.

(ii) Note that, by (14.41),

$$[T^\rho f](g) = \sqrt{\frac{q-1}{|G/U|}} \langle f, \rho(g)f_1 \rangle_V$$

$$(\text{by Proposition 14.3.1}) = \frac{1}{\sqrt{q^2-1}} \langle \rho(g^{-1})f, f_1 \rangle_V$$

$$= \frac{1}{\sqrt{q^2-1}} [\rho(g^{-1})f](1),$$

and that

$$\begin{pmatrix} \alpha & \beta \\ \gamma & \delta \end{pmatrix}^{-1} = \begin{pmatrix} \delta \det(g)^{-1} & -\beta \det(g)^{-1} \\ -\gamma \det(g)^{-1} & \alpha \det(g)^{-1} \end{pmatrix}$$

in particular,

$$\begin{pmatrix} \alpha & \beta \\ 0 & \delta \end{pmatrix}^{-1} = \begin{pmatrix} \alpha^{-1} & -\beta\alpha^{-1}\delta^{-1} \\ 0 & \delta^{-1} \end{pmatrix}.$$

Then it suffices to apply (14.32) and (14.33), respectively (and $\det(g^{-1}) = (\det g)^{-1}$).

(iii) It is an immediate consequence of (14.41), (ii), and the definition of f_1: indeed, $\varphi^\rho(g) = [\rho(g^{-1})f_1](1)$ for all $g \in G$. $\square$

Corollary 14.7.10 *Let* (ρ, V) *be a cuspidal representation,* $f \in V$ *and* $x \in \mathbb{F}_q^*$. *Then*

$$[T^\rho f]\begin{pmatrix} x & 0 \\ 0 & 1 \end{pmatrix} = \frac{1}{\sqrt{q^2-1}} f(x) \tag{14.47}$$

and

$$\varphi^\rho \begin{pmatrix} x & 0 \\ 0 & 1 \end{pmatrix} = f_1(x). \tag{14.48}$$

Moreover, for all $\beta, \gamma \in \mathbb{F}_q^*$,

$$\varphi^\rho \begin{pmatrix} 0 & \beta \\ \gamma & 0 \end{pmatrix} = -\overline{\nu(-\beta)}j(-\beta\gamma^{-1}). \tag{14.49}$$

Proof. (14.47) is immediate after Proposition 14.7.9.(ii). (14.48) follows from Proposition 14.7.9.(iii) (or Proposition 14.7.4) and the definition of f_1. Finally, (14.49) is just a particular case of Proposition 14.7.9.(iii). $\square$

Remark 14.7.11 With $\beta = -1$ and γ^{-1} in place of γ, (14.49) yields

$$j(\gamma) = -\varphi^\rho \begin{pmatrix} 0 & -1 \\ \gamma^{-1} & 0 \end{pmatrix} = -\overline{\varphi^\rho \begin{pmatrix} 0 & \gamma \\ -1 & 0 \end{pmatrix}},$$

where the last equality follows from $\begin{pmatrix} 0 & -1 \\ \gamma^{-1} & 0 \end{pmatrix}^{-1} = \begin{pmatrix} 0 & \gamma \\ -1 & 0 \end{pmatrix}$ and $\varphi^\rho(g^{-1}) = \overline{\varphi^\rho(g)}$. Analogously, setting $\gamma = -1$ we get

$$j(\beta) = -\nu(-\beta)\varphi^\rho \begin{pmatrix} 0 & \beta \\ -1 & 0 \end{pmatrix}.$$

Remark 14.7.12 With Piatetski-Shapiro's definition of an induced representation (cf. (14.34)), the intertwining operator T^ρ in (14.41) and the associated spherical function in (14.42) become

$$[T^\rho v](g) = \sqrt{\frac{d_\rho}{|G/U|}} \langle \rho(g)v, v_0 \rangle$$

and

$$\varphi^\rho(g) = \langle \rho(g)v_0, v_0 \rangle,$$

for all $v \in V$ and $g \in G$, respectively. Therefore, our spherical functions are the *conjugate* of the Bessel functions J_ρ in [123]: indeed, one has

$$J_\rho \begin{pmatrix} 0 & x \\ -1 & 0 \end{pmatrix} = -j(x)$$

for all $x \in \mathbb{F}_q^*$.

For the last result of this section, we identify the subgroup

$$A = \left\{ \begin{pmatrix} a & 0 \\ 0 & 1 \end{pmatrix} : a \in \mathbb{F}_q^* \right\} \subset \mathrm{Aff}(\mathbb{F}_q)$$

with $\mathbb{F}_q^*$ via the isomorphism $\begin{pmatrix} a & 0 \\ 0 & 1 \end{pmatrix} \mapsto a$.

Proposition 14.7.13 *Let* (ρ, V) *be a cuspidal representation of G. Then*

$$[T^\rho f](g) = \sum_{a \in A} [T^\rho f](a) \varphi^\rho(a^{-1}g) \tag{14.50}$$

and

$$[\rho(g)f](a) = \sum_{a_1 \in A} f(a_1) \varphi^\rho(a_1^{-1}g^{-1}a) \tag{14.51}$$

for all $f \in V$, $g \in G$, *and* $a \in A$.

Proof. (14.50) is an immediate consequence of (14.47) and the explicit expressions in Proposition 14.7.9.(ii) and (iii).

We now prove (14.51). Let $g \in G$ and $a \in A$. Then, by (14.47),

$$[\rho(g)f](a) = \sqrt{q^2 - 1}[T^\rho \rho(g)f](a)$$

$$\text{(by (14.39))} \quad = \sqrt{q^2 - 1}[T^\rho f](g^{-1}a)$$

$$\text{(by (14.50))} \quad = \sqrt{q^2 - 1} \sum_{a_1 \in A} [T^\rho f](a_1) \varphi^\rho(a_1^{-1}g^{-1}a)$$

$$\text{(by (14.47))} \quad = \sum_{a_1 \in A} f(a_1) \varphi^\rho(a_1^{-1}g^{-1}a). \qquad \square$$

For another approach, we refer to [86].

14.8 Gamma coefficients

Following Piatetski-Schapiro [123], we introduce another set of functions, connected with the representation theory of GL(2, $\mathbb{F}_q$) that may be expressed in terms of Gauss sums (cf. Section 7.4). We recall (see Section 10.5) that if (ρ, V)

is a representation of a finite group G, then, denoting by V' the dual space of V, the associated adjoint representation is the G-representation (ρ', V') defined by setting

$$[\rho'(g)\varphi](v) = \varphi[\rho(g^{-1})v]$$

for all $g \in G$, $v \in V$ and $\varphi \in V'$. Moreover, the associated character is given by $\chi^{\rho'}(g) = \chi^\rho(g^{-1}) = \overline{\chi^\rho(g)}$, for all $g \in G$.

Suppose now that (ρ, V) is an irreducible representation of $G = \mathrm{GL}(2, \mathbb{F}_q)$ with $\dim V > 1$. We say that $\omega \in \widehat{\mathbb{F}_q^*}$ is an *exceptional character* for ρ if ρ is parabolic and

$$\rho = \widehat{\chi}_{\psi_1,\psi_2} \text{ with } \psi_1 = \overline{\omega} = \omega^{-1} \text{ or } \psi_2 = \overline{\omega} = \omega^{-1}$$

or

$$\rho = \widehat{\chi}_\psi^1 \text{ with } \psi = \overline{\omega} = \omega^{-1}.$$

By Proposition 14.5.9, ω is exceptional for (ρ, V) if and only if $\overline{\omega}$ is contained in $\mathrm{Res}_A^G \rho|_{J(V)}$, that is, ω is contained in $\left(\mathrm{Res}_A^G \rho\right)'|_{J(V')}$.

Proposition 14.8.1 *Let* $\omega \in \widehat{\mathbb{F}_q^*}$ *and suppose that it is not exceptional for* ρ. *Then* ω *is contained in* $\left(\mathrm{Res}_A^G \rho\right)'$ *with multiplicity one.*

Proof. If $\omega \in \widehat{\mathbb{F}_q^*}$ is not exceptional, then $\overline{\omega}$ it is not contained in $\mathrm{Res}_A^G \rho|_{J(V)}$ and, by Corollary 12.1.5, it is contained in $\mathrm{Res}_A^G \rho|_{V_\pi}$ with multiplicity one. By Proposition 14.7.1.(i) it is contained in $\mathrm{Res}_A^G \rho$ with multiplicity one. From the discussion above we deduce that ω is contained in $\left(\mathrm{Res}_A^G \rho\right)'$ with multiplicity one. $\square$

Lemma 14.8.2 (Definition and existence of $\Gamma_\rho(\omega)$) *Let* $\omega \in \widehat{\mathbb{F}_q^*}$ *and suppose that it is nonexceptional for* (ρ, V). *Then there exists* $\Gamma_\rho(\omega) = \Gamma_{\rho,\chi}(\omega) \in \mathbb{C}$ *such that*

$$\Gamma_\rho(\omega) \sum_{x \in \mathbb{F}_q^*} [T^\rho v] \begin{pmatrix} x & 0 \\ 0 & 1 \end{pmatrix} \overline{\omega(x)} = \sum_{x \in \mathbb{F}_q^*} [T^\rho v] \begin{pmatrix} 0 & x \\ 1 & 0 \end{pmatrix} \overline{\omega(x)}$$

for all $v \in V$.

Proof. Define φ and ψ in V' by setting

$$\varphi(v) = \sum_{x \in \mathbb{F}_q^*} [T^\rho v] \begin{pmatrix} x & 0 \\ 0 & 1 \end{pmatrix} \overline{\omega(x)}$$

and

$$\psi(v) = \sum_{x \in \mathbb{F}_q^*} [T^\rho v] \begin{pmatrix} 0 & x \\ 1 & 0 \end{pmatrix} \overline{\omega(x)}$$

for all $v \in V$. Then

$$\varphi \left[\rho \begin{pmatrix} \alpha & 0 \\ 0 & 1 \end{pmatrix} v \right] = \sum_{x \in \mathbb{F}_q^*} \left[T^\rho \rho \begin{pmatrix} \alpha & 0 \\ 0 & 1 \end{pmatrix} v \right] \begin{pmatrix} x & 0 \\ 0 & 1 \end{pmatrix} \overline{\omega(x)}$$

$$\text{(by (14.39))} = \sum_{x \in \mathbb{F}_q^*} [T^\rho v] \begin{pmatrix} x\alpha^{-1} & 0 \\ 0 & 1 \end{pmatrix} \overline{\omega(x)}$$

$$\text{(setting } x = y\alpha) = \overline{\omega(\alpha)} \sum_{y \in \mathbb{F}_q^*} [T^\rho v] \begin{pmatrix} y & 0 \\ 0 & 1 \end{pmatrix} \overline{\omega(y)}$$

$$= \overline{\omega(\alpha)} \varphi(v),$$

so that, for $\alpha \in A$,

$$[\rho'(\alpha)\varphi](v) = \varphi[\rho(\alpha^{-1})v] = \omega(\alpha)\varphi(v)$$

for all $v \in V$, that is, $\rho'(\alpha)\varphi = \omega(\alpha)\varphi$.

Similarly,

$$\psi \left[\rho \begin{pmatrix} \alpha & 0 \\ 0 & 1 \end{pmatrix} v \right] = \overline{\omega(\alpha)} \psi(v),$$

so that we also have $\rho'(\alpha)\psi = \omega(\alpha)\psi$, for $\alpha \in A$, and, by Proposition 14.8.1, there exists $\Gamma_\rho(\omega) \in \mathbb{C}$ such that $\psi = \Gamma_\rho(\omega)\varphi$. $\qquad\square$

Corollary 14.8.3 $\Gamma_\rho(\omega)$ *may be expressed in terms of the Bessel function* φ^ρ *(see (14.42)):*

$$\Gamma_\rho(\omega) = \sum_{x \in \mathbb{F}_q^*} \varphi^\rho \begin{pmatrix} 0 & x \\ 1 & 0 \end{pmatrix} \overline{\omega(x)}. \tag{14.52}$$

Proof. If v_0 is a Bessel vector, then Lemma 14.8.2 with $v = v_0$ implies (recall that $\varphi^\rho \begin{pmatrix} x & 0 \\ 0 & 1 \end{pmatrix} = \sqrt{\frac{|G/U|}{d_\rho}} [T^\rho v_0] \begin{pmatrix} x & 0 \\ 0 & 1 \end{pmatrix} = 0$ for $x \neq 1$, see Proposition 14.7.4, and $\varphi^\rho(1_G) = 1$)

$$\Gamma_\rho(\omega)[T^\rho v_0] \begin{pmatrix} 1 & 0 \\ 0 & 1 \end{pmatrix} = \sum_{x \in \mathbb{F}_q^*} [T^\rho v_0] \begin{pmatrix} 0 & x \\ 1 & 0 \end{pmatrix} \overline{\omega(x)}$$

which in turn yields the desired identity. $\qquad\square$

We can use (14.52) to define $\Gamma_\rho(\omega)$ also for exceptional characters and cuspidal representations.

Definition 14.8.4 Let ρ be an irreducible G-representation with $\dim\rho > 1$. Then the complex-valued function $\Gamma_\rho(\cdot)$, defined by means of (14.52), is called the *Gamma coefficient* associated with ρ (and the fixed character $\chi \in \widehat{U}$).

We recall (see Definition 7.4.1) that for $\chi \in \widehat{\mathbb{F}_q}$ and $\psi \in \widehat{\mathbb{F}_q^*}$, the associated Gauss sum is defined as

$$g(\psi, \chi) = \sum_{x \in \mathbb{F}_q} \chi(x)\psi(x)$$

where we have set $\psi(0) = \begin{cases} 0 & \text{if } \psi \neq \mathbf{1} \\ 1 & \text{if } \psi = \mathbf{1}. \end{cases}$

Proposition 14.8.5 *Suppose that ρ is parabolic. Then, with the same notation as in Theorem 14.5.6, and the beginning of this section, we have*

$$\Gamma_\rho(\omega) = \frac{\omega(-1)}{q} g(\overline{\psi_1\omega}, \chi)g(\overline{\psi_2\omega}, \chi).$$

In particular, $|\Gamma_\rho(\omega)| = 1$.

Proof. By Proposition 14.7.8 and Corollary 14.8.3 we have:

$$\Gamma_\rho(\omega) = \frac{1}{q} \sum_{x \in \mathbb{F}_q^*} \sum_{\substack{r,s \in \mathbb{F}_q^*: \\ rs = -x}} \overline{\psi_1(r)\psi_2(s)}\chi(r+s)\overline{\omega(-rs)}$$

$$= \frac{\omega(-1)}{q} \sum_{x \in \mathbb{F}_q^*} \sum_{\substack{r,s \in \mathbb{F}_q^*: \\ rs = -x}} (\overline{\psi_1(r)\omega(r)}\chi(r))(\overline{\psi_2(s)\omega(s)}\chi(s))$$

$$= \frac{\omega(-1)}{q} \sum_{r \in \mathbb{F}_q^*} \overline{\psi_1(r)\omega(r)}\chi(r) \sum_{s \in \mathbb{F}_q^*} \overline{\psi_2(s)\omega(s)}\chi(s)$$

$$= \frac{\omega(-1)}{q} g(\overline{\psi_1\omega}, \chi)g(\overline{\psi_2\omega}, \chi).$$

Just note that $\overline{\psi_1\omega}, \overline{\psi_2\omega} \neq \mathbf{1}$, because ω is not exceptional for ρ so that the sum $\sum_{r \in \mathbb{F}_q^*}$ is in fact the sum $\sum_{r \in \mathbb{F}_q}$ (and, similarly, for the sums in s).
Since $|g(\psi, \chi)| = \sqrt{q}$ (cf. Theorem 7.4.3.(vii)), we get $|\Gamma_\rho(\omega)| = 1$. $\square$

Remark 14.8.6 If we use a different character in place of χ, say $\tilde{\chi}$, we get a different value of $\Gamma_\rho(\omega)$. Since there exists $\alpha \in \mathbb{F}_q^*$ such that $\tilde{\chi}(x) = \chi(\alpha x)$ for all $x \in \mathbb{F}_q$ (cf. Proposition 7.1.1), we deduce that, for ρ parabolic, the Gamma coefficient with respect to $\tilde{\chi}$ is

$$\Gamma_{\rho,\tilde{\chi}}(\omega) = \omega(\alpha)^2 \psi_1(\alpha)\psi_2(\alpha)\Gamma_{\rho,\chi}(\omega).$$

Proposition 14.8.7 *Suppose that ρ is the cuspidal representation associated with the indecomposable character $\nu \in \widehat{\mathbb{F}^*_{q^2}}$. Then, denoting simply by Tr and N the trace and the norm of the extension $\mathbb{F}_{q^2}/\mathbb{F}_q$ (see Section 6.7), we have*

$$\Gamma_\rho(\omega) = -\frac{\omega(-1)}{q} \sum_{t \in \mathbb{F}^*_{q^2}} \overline{\nu(t)\omega(t\bar{t})}\chi(t + \bar{t})$$

$$= -\frac{\omega(-1)}{q} g(\nu^{-1}(\omega \circ \mathrm{Tr})^{-1}, \chi \circ \mathrm{N})$$

for every $\omega \in \widehat{\mathbb{F}^*_q}$. In particular, $|\Gamma_\rho(\omega)| = 1$.

Proof. By Definition 14.8.4 we have

$$\Gamma_\rho(\omega) = \sum_{x \in \mathbb{F}^*_q} \varphi^\rho \begin{pmatrix} 0 & x \\ 1 & 0 \end{pmatrix} \overline{\omega(x)}$$

$$(\text{by } (14.49)) = -\sum_{x \in \mathbb{F}^*_q} \overline{\nu(-x)}j(-x)\overline{\omega(x)}$$

$$(\text{by } (7.16)) = -\frac{1}{q}\sum_{x \in \mathbb{F}^*_q} \overline{\nu(-x)\omega(x)} \sum_{\substack{t \in \mathbb{F}^*_{q^2}: \\ t\bar{t}=-x}} \chi(t+\bar{t})\nu(t)$$

$$= -\frac{1}{q}\sum_{x \in \mathbb{F}^*_q} \overline{\nu(x)\omega(-x)} \sum_{\substack{t \in \mathbb{F}^*_{q^2}: \\ t\bar{t}=x}} \chi(t+\bar{t})\nu(t)$$

$$(\text{Hilbert Satz } 90) = -\frac{1}{q}\sum_{t \in \mathbb{F}^*_{q^2}} \overline{\nu(t\bar{t})\omega(-t\bar{t})}\chi(t+\bar{t})\nu(t)$$

$$= -\frac{\omega(-1)}{q}\sum_{t \in \mathbb{F}^*_{q^2}} \overline{\nu(\bar{t})\omega(t\bar{t})}\chi(t+\bar{t})$$

$$= -\frac{\omega(-1)}{q}\sum_{t \in \mathbb{F}^*_{q^2}} \overline{\nu(t)\omega(t\bar{t})}\chi(t+\bar{t})$$

$$= -\frac{\omega(-1)}{q}g(\nu^{-1}(\omega \circ \mathrm{Tr})^{-1}, \chi \circ \mathrm{N}).$$

Since $|g(\cdot, \cdot)| = \sqrt{|\mathbb{F}_{q^2}|} = q$ (cf. Theorem 7.4.3.(vii)), we also have $|\Gamma_\rho(\omega)| = 1$. $\qquad\qquad\square$

Remark 14.8.8 As in Remark 14.8.6, if $\tilde{\chi}$ is another character of $\mathbb{F}_q$ and $\tilde{\chi}(x) = \chi(\alpha x)$, then, for a cuspidal representation ρ we have

$$\Gamma_{\rho,\tilde{\chi}}(\omega) = \nu(\alpha)\omega(\alpha)^2 \Gamma_{\rho,\chi}(\omega).$$

14.9 Character theory of GL(2, $\mathbb{F}_q$)

In this section we compute the characters of all irreducible representations of G as well as the *Gelfand-Graev character* ξ of $\mathrm{Ind}_U^G \chi$, where χ is, as usual, a fixed nontrivial character of U.

Proposition 14.9.1 *Let ξ denote the character of* $\mathrm{Ind}_U^G \chi$. *Then*

$$
\xi(g) = \begin{cases}
(q-1)^2(q+1) & \text{if } g = 1_G \\[2mm]
1-q & \text{if } g \text{ is conjugate to } \begin{pmatrix} 1 & 1 \\ 0 & 1 \end{pmatrix} \\[2mm]
0 & \text{otherwise,}
\end{cases}
$$

for all $g \in G$.

Proof. First of all, note that $D \coprod DU w$ is a set of representatives for the left cosets of U in G:

$$G = \left(\coprod_{d \in D} dU \right) \coprod \left(\coprod_{\substack{d \in D, \\ u \in U}} duwU \right). \tag{14.53}$$

Indeed, one just needs to recall the Bruhat decomposition and to note that, for $g = \begin{pmatrix} \alpha & \beta \\ \gamma & \delta \end{pmatrix} \in G \setminus B$ (i.e. with $\gamma \neq 0$) we have

$$\begin{pmatrix} x & 0 \\ 0 & y \end{pmatrix} \begin{pmatrix} 1 & z \\ 0 & 1 \end{pmatrix} \begin{pmatrix} 0 & 1 \\ 1 & 0 \end{pmatrix} \begin{pmatrix} 1 & v \\ 0 & 1 \end{pmatrix} = \begin{pmatrix} xz & x+xzv \\ y & yv \end{pmatrix} = \begin{pmatrix} \alpha & \beta \\ \gamma & \delta \end{pmatrix}$$

if and only if $y = \gamma$, $v = \delta\gamma^{-1}$, $x = \beta - \alpha\delta\gamma^{-1} \equiv -\gamma^{-1}\det(g)$ and $z = -\alpha\gamma\det(g)^{-1}$. In other words, any $g \in G \setminus B$ may be written in a unique way in the form $g = duwu_1$, with $d \in D$ and $u, u_1 \in U$.

First of all we clearly have

$$\xi(1_G) = \dim\mathrm{Ind}_U^G \chi = \frac{|G|}{|U|} = (q^2-1)(q-1).$$

From Frobenius character formula (cf. (11.18)) it follows that

$$\xi(g) = \sum_{\substack{d \in D: \\ d^{-1}gd \in U}} \chi(d^{-1}gd) + \sum_{\substack{d \in D, u \in U: \\ (duw)^{-1}gduw \in U}} \chi(wu^{-1}d^{-1}gduw). \tag{14.54}$$

In particular, if g is not conjugated to an element of U, we have $\xi(g) = 0$. Recalling Theorem 14.3.2, we have that $U \setminus \{1_G\}$ is contained in the conjugacy class of $\begin{pmatrix} 1 & 1 \\ 0 & 1 \end{pmatrix}$. We deduce that $\xi(g) = 0$ if g is not conjugated to $\begin{pmatrix} 1 & 1 \\ 0 & 1 \end{pmatrix}$.

We are only left to the case when g is conjugated to $\begin{pmatrix} 1 & 1 \\ 0 & 1 \end{pmatrix}$. If $h = \begin{pmatrix} \alpha & \beta \\ \gamma & \delta \end{pmatrix} \in G$, and setting $\Delta = \det(h)$, we have

$$\begin{pmatrix} \alpha & \beta \\ \gamma & \delta \end{pmatrix}^{-1} \begin{pmatrix} 1 & 1 \\ 0 & 1 \end{pmatrix} \begin{pmatrix} \alpha & \beta \\ \gamma & \delta \end{pmatrix} = \begin{pmatrix} \alpha & \beta \\ \gamma & \delta \end{pmatrix}^{-1} \left[\begin{pmatrix} 1 & 0 \\ 0 & 1 \end{pmatrix} + \begin{pmatrix} 0 & 1 \\ 0 & 0 \end{pmatrix} \right] \begin{pmatrix} \alpha & \beta \\ \gamma & \delta \end{pmatrix}$$

$$= \begin{pmatrix} 1 & 0 \\ 0 & 1 \end{pmatrix} + \begin{pmatrix} \delta\Delta^{-1} & -\beta\Delta^{-1} \\ -\gamma\Delta^{-1} & \alpha\Delta^{-1} \end{pmatrix} \begin{pmatrix} 0 & 1 \\ 0 & 0 \end{pmatrix} \begin{pmatrix} \alpha & \beta \\ \gamma & \delta \end{pmatrix}$$

$$= \begin{pmatrix} 1 + \gamma\delta\Delta^{-1} & \delta^2\Delta^{-1} \\ -\gamma^2\Delta^{-1} & 1 - \gamma\delta\Delta^{-1} \end{pmatrix} \tag{14.55}$$

so that $h^{-1} \begin{pmatrix} 1 & 1 \\ 0 & 1 \end{pmatrix} h$ is not in U if $\gamma \neq 0$. Therefore, for the expression of $\xi \begin{pmatrix} 1 & 1 \\ 0 & 1 \end{pmatrix}$ in (14.54), only the first sum may be different from 0 (the second one vanishes since $(duw)^{-1} \begin{pmatrix} 1 & 1 \\ 0 & 1 \end{pmatrix} duw$ does not even belong to B).

Thus,

$$\xi \begin{pmatrix} 1 & 1 \\ 0 & 1 \end{pmatrix} = \sum_{x,y\in\mathbb{F}_q^*} \chi \left[\begin{pmatrix} x & 0 \\ 0 & y \end{pmatrix}^{-1} \begin{pmatrix} 1 & 1 \\ 0 & 1 \end{pmatrix} \begin{pmatrix} x & 0 \\ 0 & y \end{pmatrix} \right]$$

$$= \sum_{x,y\in\mathbb{F}_q^*} \chi \begin{pmatrix} 1 & x^{-1}y \\ 0 & 1 \end{pmatrix}$$

$$= \sum_{x,y\in\mathbb{F}_q^*} \chi(x^{-1}y)$$

$$= (q-1) \sum_{x\in\mathbb{F}_q^*} \chi(x)$$

$$= 1 - q,$$

where the last equality follows from the orthogonality relation

$$0 = \langle \chi, 1 \rangle = \sum_{x\in\mathbb{F}_q} \chi(x) = 1 + \sum_{x\in\mathbb{F}_q^*} \chi(x). \tag{14.56}$$

$\square$

In Table 14.2 (where in the first column there are the irreducible representations and in the first line the representatives of the conjugacy classes), we give the values of the characters of the higher dimensional representations of G on each conjugacy class, as well as the cardinality of the corresponding irreducible representations (here, $x, y \in \mathbb{F}_q^*$ and $z \in \mathbb{F}_{q^2} \backslash \mathbb{F}_q$).

Table 14.2. *The character table of* GL$(2, \mathbb{F}_q)$.

	$\begin{pmatrix} x & 0 \\ 0 & x \end{pmatrix}$	$\begin{pmatrix} x & 0 \\ 0 & y \end{pmatrix}_{y \neq x}$	$\begin{pmatrix} x & 1 \\ 0 & x \end{pmatrix}$	$\begin{pmatrix} 0 & -z\bar{z} \\ 1 & z+\bar{z} \end{pmatrix}$	$|\mathrm{irr}|$
$\widehat{\chi}_\psi^0$	$\psi(x^2)$	$\psi(xy)$	$\psi(x^2)$	$\psi(z\bar{z})$	$q-1$
$\widehat{\chi}_\psi^1$	$q\psi(x^2)$	$\psi(xy)$	0	$-\psi(z\bar{z})$	$q-1$
$\widehat{\chi}_{\psi_1,\psi_2}$	$(q+1)$ $\psi_1(x)\psi_2(x)$	$\psi_1(x)\psi_2(y)$ $+\psi_1(y)\psi_2(x)$	$\psi_1(x)\psi_2(x)$	0	$\frac{(q-1)(q-2)}{2}$
ρ_ν	$(q-1)\nu(x)$	0	$-\nu(x)$	$-\nu(z) - \nu(\bar{z})$	$\frac{q(q-1)}{2}$

In order to compute the characters of $\widehat{\chi}_\psi^1$ and $\widehat{\chi}_{\psi_1,\psi_2}$ we need the following remarks:

(a) $h^{-1} \begin{pmatrix} x & 1 \\ 0 & x \end{pmatrix} h \in B$ if and only if $h \in B$. The proof follows the same lines as in (14.55).

(b) An element $(uw)^{-1} duw$, with $u \in U$ and $d \in D \backslash Z$, belongs to B if and only if $u = 1_G$. Indeed, an element in Uw is of the form

$$uw = \begin{pmatrix} 1 & \beta \\ 0 & 1 \end{pmatrix} \begin{pmatrix} 0 & 1 \\ 1 & 0 \end{pmatrix} = \begin{pmatrix} \beta & 1 \\ 1 & 0 \end{pmatrix}$$

and its inverse is

$$\begin{pmatrix} 0 & 1 \\ 1 & 0 \end{pmatrix} \begin{pmatrix} 1 & -\beta \\ 0 & 1 \end{pmatrix} = \begin{pmatrix} 0 & 1 \\ 1 & -\beta \end{pmatrix}$$

so that if $d = \begin{pmatrix} x & 0 \\ 0 & y \end{pmatrix} \in D \backslash Z$ ($x, y \in \mathbb{F}_q^*$, $x \neq y$) then

$$(uw)^{-1} duw = \begin{pmatrix} 0 & 1 \\ 1 & -\beta \end{pmatrix} \begin{pmatrix} x & 0 \\ 0 & y \end{pmatrix} \begin{pmatrix} \beta & 1 \\ 1 & 0 \end{pmatrix}$$

$$= \begin{pmatrix} y & 0 \\ \beta(x-y) & x \end{pmatrix}.$$

(c) An element in $C \backslash Z$ is not conjugate to any element in B (see table in Theorem 14.3.2) because its eigenvalues (as a 2×2 matrix) are not in $\mathbb{F}_q$.

(d) $G = B \bigsqcup (\bigsqcup_{u \in U} uwB)$ is the decomposition into left B-cosets (cf. (14.53) and the Bruhat decomposition).

Proof of the character table. The first row follows from (14.18).

From (d) and Frobenius character formula, it follows that the character of $\widehat{\chi}_{\psi_1, \psi_2}$ evaluated at $g \in G$ equals

$$\sum_{\substack{u \in U: \\ wu^{-1}guw \in B}} \chi_{\psi_1, \psi_2}(wu^{-1}guw) + \chi_{\psi_1, \psi_2}(g)\mathbf{1}_B(g). \tag{14.57}$$

By (c), this is equal to 0 if $g \in C \setminus Z$. If $g = \begin{pmatrix} x & 0 \\ 0 & x \end{pmatrix} \in Z$, then it is equal to

$$(q+1)\chi_{\psi_1, \psi_2}\begin{pmatrix} x & 0 \\ 0 & x \end{pmatrix} = (q+1)\psi_1(x)\psi_2(x).$$

From (b), it follows that if $g = \begin{pmatrix} x & 0 \\ 0 & y \end{pmatrix} \in D \setminus Z$, then all terms but the one corresponding to $u = 1_G$ in the summation in (14.57) are equal to zero, so that (14.57) is equal to

$$\chi_{\psi_1, \psi_2}\begin{pmatrix} x & 0 \\ 0 & y \end{pmatrix} + \chi_{\psi_1, \psi_2}\begin{pmatrix} y & 0 \\ 0 & x \end{pmatrix} = \psi_1(x)\psi_2(y) + \psi_1(y)\psi_2(x).$$

From (a), it follows that if $g = \begin{pmatrix} x & 1 \\ 0 & x \end{pmatrix}$, then all terms in the summation (14.57) are equal to zero, so that $\chi_{\psi_1, \psi_2}\begin{pmatrix} x & 1 \\ 0 & x \end{pmatrix} = \psi_1(x)\psi_2(x)$.

The values of the character of $\widehat{\chi}_\psi^1$ may be found in the same way, setting $\psi_1 = \psi_2$ in the previous formulas and using the identities

$$\widehat{\chi}_{\psi, \psi} = \widehat{\chi}_\psi^0 + \widehat{\chi}_\psi^1 \quad \text{and} \quad \widehat{\chi}_\psi^0 = \psi(\det(g)).$$

In order to compute the character of a cuspidal representation, we use (14.51), which yields the matrix coefficients of ρ_v in terms of the spherical functions. Indeed, if $\{f_x : x \in \mathbb{F}_q^*\}$ is as (14.46), then the character of ρ_v has the following expression:

$$\sum_{x \in \mathbb{F}_q^*} \langle \rho_v(g)f_x, f_x \rangle = \sum_{x \in \mathbb{F}_q^*} [\rho_v(g)f_x](x)$$

$$(\text{by } (14.51) \text{ and } A \cong \mathbb{F}_q^*) = \sum_{a \in A} \varphi^{\rho_v}(a^{-1}g^{-1}a). \tag{14.58}$$

For $g = \begin{pmatrix} x & 0 \\ 0 & x \end{pmatrix}$, (14.58) is equal to

$$(q-1)\varphi^{\rho_\nu}(g^{-1}) = (q-1)\overline{\nu(x^{-1})} = (q-1)\nu(x)$$

where the first equality follows from Proposition 14.7.9.(iii). For $g = \begin{pmatrix} x & 1 \\ 0 & x \end{pmatrix}$
we have $g^{-1} = \begin{pmatrix} x^{-1} & -x^{-2} \\ 0 & x^{-1} \end{pmatrix}$ and

$$\begin{pmatrix} \alpha^{-1} & 0 \\ 0 & 1 \end{pmatrix}\begin{pmatrix} x^{-1} & -x^{-2} \\ 0 & x^{-1} \end{pmatrix}\begin{pmatrix} \alpha & 0 \\ 0 & 1 \end{pmatrix} = \begin{pmatrix} x^{-1} & -\alpha^{-1}x^{-2} \\ 0 & x^{-1} \end{pmatrix}$$

so that, in this case, (14.58) is equal to

$$\sum_{\alpha \in \mathbb{F}_q^*} \varphi^{\rho_\nu}\begin{pmatrix} x^{-1} & -\alpha^{-1}x^{-2} \\ 0 & x^{-1} \end{pmatrix} = \sum_{\alpha \in \mathbb{F}_q^*} \overline{\nu(x^{-1})\chi(-x \cdot \alpha^{-1}x^{-2})}$$

$$= \nu(x) \sum_{\alpha \in \mathbb{F}_q^*} \chi(\alpha^{-1}x^{-1})$$

$$= -\nu(x),$$

where the first equality follows from Proposition 14.7.9.(iii) and the last one
from (14.56).
For $g = \begin{pmatrix} x & 0 \\ 0 & y \end{pmatrix}$, with $x \neq y$, we have

$$\begin{pmatrix} \alpha^{-1} & 0 \\ 0 & 1 \end{pmatrix}\begin{pmatrix} x^{-1} & 0 \\ 0 & y^{-1} \end{pmatrix}\begin{pmatrix} \alpha & 0 \\ 0 & 1 \end{pmatrix} = \begin{pmatrix} x^{-1} & 0 \\ 0 & y^{-1} \end{pmatrix}$$

so that, in this case, (14.58) is equal to $(q-1)\varphi^{\rho_\nu}\begin{pmatrix} x^{-1} & 0 \\ 0 & y^{-1} \end{pmatrix}$ and this vanishes, by Proposition 14.7.9.(iii).
Finally, if $g = \begin{pmatrix} 0 & -z\bar{z} \\ 1 & z+\bar{z} \end{pmatrix}$, $z \in \mathbb{F}_{q^2} \setminus \mathbb{F}_q$, setting $\beta = -z\bar{z}, \delta = z + \bar{z}$ we have
$g^{-1} = \begin{pmatrix} -\beta^{-1}\delta & 1 \\ \beta^{-1} & 0 \end{pmatrix}$ and

$$\begin{pmatrix} \alpha^{-1} & 0 \\ 0 & 1 \end{pmatrix}\begin{pmatrix} -\beta^{-1}\delta & 1 \\ \beta^{-1} & 0 \end{pmatrix}\begin{pmatrix} \alpha & 0 \\ 0 & 1 \end{pmatrix} = \begin{pmatrix} -\beta^{-1}\delta & \alpha^{-1} \\ \alpha\beta^{-1} & 0 \end{pmatrix}$$

so that (14.58) is equal to (by Proposition 14.7.9.(iii))

$$-\sum_{\alpha\in\mathbb{F}_q^*} \nu(-\beta\cdot\alpha\beta^{-1})\overline{\chi(-\beta^{-1}\delta\cdot\alpha^{-1}\beta)}j[\alpha^{-2}\beta^2(-\beta^{-1})]$$

$$= -\sum_{\alpha\in\mathbb{F}_q^*} \nu(-\alpha)\chi(\alpha^{-1}\delta)j(-\alpha^{-2}\beta)$$

$$(\text{by } (7.16)) = -\frac{1}{q}\sum_{\alpha\in\mathbb{F}_q^*} \chi(\alpha^{-1}\delta) \sum_{\substack{x\bar{x}\in\mathbb{F}_{q^2}^*:\\ x\bar{x}=-\alpha^{-2}\beta}} \chi(x+\bar{x})\nu(-\alpha x)$$

$$(y = -\alpha x) = -\frac{1}{q}\sum_{\alpha\in\mathbb{F}_q^*} \chi(\alpha^{-1}\delta) \sum_{\substack{y\in\mathbb{F}_{q^2}^*:\\ y\bar{y}=-\beta}} \chi[-\alpha^{-1}(y+\bar{y})]\nu(y)$$

$$(\alpha^{-1}\mapsto\alpha) = -\frac{1}{q}\sum_{\substack{y\in\mathbb{F}_{q^2}^*:\\ y\bar{y}=-\beta}} \nu(y) \sum_{\alpha\in\mathbb{F}_q^*} \chi(\alpha[\delta-(y+\bar{y})])$$

$$= -\frac{1}{q}\sum_{\substack{y\in\mathbb{F}_{q^2}^*:\\ y\neq z,\bar{z}\\ y\bar{y}=-\beta}} \nu(y) \sum_{\alpha\in\mathbb{F}_q^*} \chi(\alpha[\delta-(y+\bar{y})])$$

$$-\frac{1}{q}\nu(z) \sum_{\alpha\in\mathbb{F}_q^*} \chi(\alpha[\delta-(z+\bar{z})])$$

$$-\frac{1}{q}\nu(\bar{z}) \sum_{\alpha\in\mathbb{F}_q^*} \chi(\alpha[\delta-(\bar{z}+z)])$$

$$(\delta = z+\bar{z}) =_* -\frac{1}{q}\sum_{\substack{y\in\mathbb{F}_{q^2}^*:\\ y\neq z,\bar{z}\\ y\bar{y}=-\beta}} \nu(y) \sum_{\gamma\in\mathbb{F}_q^*} \chi(\gamma) - \frac{q-1}{q}[\nu(z)+\nu(\bar{z})]$$

$$(\text{by } (14.56)) = -\frac{1}{q}[(q-1)[\nu(z)+\nu(\bar{z})] - \sum_{\substack{y\in\mathbb{F}_{q^2}^*:\\ y\neq z,\bar{z}\\ y\bar{y}=-\beta}} \nu(y)]$$

$$= -\frac{1}{q}[q[\nu(z)+\nu(\bar{z})] - \sum_{\substack{y\in\mathbb{F}_{q^2}^*:\\ y\bar{y}=-\beta}} \nu(y)]$$

$$(\text{by Proposition } 7.2.3) = -\nu(z)-\nu(\bar{z}),$$

where $=_*$ follows from the fact that, assuming $y\bar{y} = -\beta$, we have $\delta = y + \bar{y}$ if and only if $y = z$ or $y = \bar{z}$ (see Section 6.8) and, if $\delta \neq y + \bar{y}$, then we may set $\gamma = \alpha[\delta - (y + \bar{y})] \in \mathbb{F}_q^*$. $\square$

Proposition 14.9.2 *Let ρ_μ and ρ_ν be cuspidal representations associated with the indecomposable characters μ and ν, respectively. Suppose that*

- ρ_μ *and ρ_ν have the same central character;*
- $\Gamma_{\rho_\mu} = \Gamma_{\rho_\nu}$.

Then $\rho_\mu \sim \rho_\nu$.

Proof. From the character table of GL(2, $\mathbb{F}_q$) (cf. Table 14.2) we deduce that

$$\mu|_{\mathbb{F}_q^*} = \nu|_{\mathbb{F}_q^*}. \tag{14.59}$$

Moreover, Corollary 14.8.3 implies that

$$\sum_{x \in \mathbb{F}_q^*} \varphi^{\rho_\mu} \begin{pmatrix} 0 & x \\ 1 & 0 \end{pmatrix} \overline{\omega(x)} = \sum_{x \in \mathbb{F}_q^*} \varphi^{\rho_\nu} \begin{pmatrix} 0 & x \\ 1 & 0 \end{pmatrix} \overline{\omega(x)}$$

for all $\omega \in \widehat{\mathbb{F}_q^*}$, so that

$$\varphi^{\rho_\mu} \begin{pmatrix} 0 & x \\ 1 & 0 \end{pmatrix} = \varphi^{\rho_\nu} \begin{pmatrix} 0 & x \\ 1 & 0 \end{pmatrix}.$$

By using (14.49) and taking into account (14.59), we deduce that $j_{\rho_\mu} = j_{\rho_\nu}$. From Theorem 14.6.10, we finally deduce that $\rho_\mu \sim \rho_\nu$. $\square$

14.10 Induced representations from GL(2, $\mathbb{F}_q$) to GL(2, $\mathbb{F}_{q^m}$)

In this section we give a series of formulas for the decomposition of the induced representation $\mathrm{Ind}_{\mathrm{GL}(2,\mathbb{F}_q)}^{\mathrm{GL}(2,\mathbb{F}_{q^m})} \rho$ for every irreducible representation ρ of GL(2, $\mathbb{F}_q$). These formulas may be easily obtained from the character table of GL(2, $\mathbb{F}_q$) (see Table 14.2). The proofs are tedious calculations, but the results are very interesting. We limit ourselves to:

- give all the preliminary results and introduce a suitable notation in order to simplify the exposition;
- give all the formulas;
- prove one formula to indicate the method and leaving the remaining formulas as exercises;
- indicate an alternative proof for one formula that avoids the use of the character table, suggesting to the reader how to develop similar techniques.

We fix a prime power $q = p^n$ and an integer $m \geq 2$. We set $G = \text{GL}(2, \mathbb{F}_q)$ and $G_m = \text{GL}(2, \mathbb{F}_{q^m})$.

- We indicate by ψ, ψ_1, and ψ_2 characters of $\mathbb{F}_q^*$ and by $\widehat{\chi}_\psi^0$, $\widehat{\chi}_\psi^1$, and $\widehat{\chi}_{\psi_1, \psi_2}$ the associated parabolic representations of G.
- Similarly, ξ denotes a character of $\mathbb{F}_{q^m}^*$.
- Also, ν (respectively μ) denotes an indecomposable character of $\mathbb{F}_{q^2}^*$ (respectively $\mathbb{F}_{q^{2m}}^*$) and ρ_ν (respectively ρ_μ) the associated cuspidal representation of G (respectively G_m).
- By $\xi^\sharp$, $\nu^\sharp$, and $\mu^\sharp$ we denote the restriction of these characters to $\mathbb{F}_q^*$, that is, $\xi^\sharp = \text{Res}_{\mathbb{F}_q^*}^{\mathbb{F}_{q^m}^*} \xi$, and so on.
- By $\mu^\flat$ we denote the restriction of μ to $\mathbb{F}_{q^2}^*$, that is, $\mu^\flat = \text{Res}_{\mathbb{F}_{q^2}^*}^{\mathbb{F}_{q^{2m}}^*} \mu$. If m is even, so that $\mathbb{F}_{q^2} \subseteq \mathbb{F}_{q^m}$, then $\xi^\flat$ is the restriction of ξ to $\mathbb{F}_{q^2}^*$, that is, $\xi^\flat = \text{Res}_{\mathbb{F}_{q^2}^*}^{\mathbb{F}_{q^m}^*} \xi$.
- By $\overline{\nu}$, $\overline{\mu}$ (and $\overline{\xi}$, if m is even) we denote the conjugate character, as in Section 7.2, that is $\overline{\nu}(z) = \nu(\overline{z})$, for all $z \in \mathbb{F}_{q^2}^*$. Warning: recall that $\overline{\nu(z)}$ is the complex conjugate of $\nu(z)$.
- As in Section 7.5, we set $\Psi = \psi \circ N$, where $N \colon \mathbb{F}_{q^2}^* \to \mathbb{F}_q$ is the norm, that is $\Psi(z) = \psi(z\overline{z})$, for all $z \in \mathbb{F}_q^*$. Similarly, we set $\Xi = \xi^\sharp \circ N$, that is, $\Xi(z) = \xi(z\overline{z})$, for all $z \in \mathbb{F}_{q^2}^*$.

Clearly,

$$\begin{aligned}
\widehat{\mathbb{F}_{q^m}^*} &\to \widehat{\mathbb{F}_q^*} \\
\xi &\mapsto \xi^\sharp
\end{aligned} \tag{14.60}$$

is a surjective homomorphism of Abelian (indeed cyclic) groups and each ψ is the image of $\frac{q^m - 1}{q - 1}$ characters of $\mathbb{F}_{q^m}^*$.

Exercise 14.10.1 Consider the map (14.60) for $m = 2$, so that $\frac{q^m - 1}{q - 1} = q + 1$. Prove that

(1) if ψ is not a square, then it is the image of $q + 1$ indecomposable characters;
(2) if ψ is a square and q is odd, then ψ is the image of $q - 1$ indecomposable characters and 2 decomposable characters;
(3) if q is even, then each ψ is a square and the image of q indecomposable characters and 1 decomposable character.

Hint: Recall Proposition 6.4.4.

When restricting an irreducible representation from G_m to G we need the following remarks:

- if m is even, then the conjugacy class of G of type (b_3) represented by $\begin{pmatrix} 0 & -z\bar{z} \\ 1 & z+\bar{z} \end{pmatrix}$ is contained in a conjugacy class of type (b_2) in G_m (because G_m contains G_2), and it is represented by $\begin{pmatrix} z & 0 \\ 0 & \bar{z} \end{pmatrix}$;

- if m is odd, then $\begin{pmatrix} 0 & -z\bar{z} \\ 1 & z+\bar{z} \end{pmatrix}$ is of type (b_3) also in G_m.

Table 14.3. *The "character table" of the restrictions from G_m to G.*

	$\begin{pmatrix} x & 0 \\ 0 & x \end{pmatrix}$	$\begin{pmatrix} x & 0 \\ 0 & y \end{pmatrix}_{y \neq x}$	$\begin{pmatrix} x & 1 \\ 0 & x \end{pmatrix}$	$\begin{pmatrix} 0 & -z\bar{z} \\ 1 & z+\bar{z} \end{pmatrix}$	
$\mathrm{Res}_G^{G_m} \widehat{\chi}_\xi^0$	$\xi(x^2)$	$\xi(xy)$	$\xi(x^2)$	$\xi(z\bar{z})$	
$\mathrm{Res}_G^{G_m} \widehat{\chi}_\xi^1$	$q^m \xi(x^2)$	$\xi(xy)$	0	$\xi(z\bar{z})$ $-\xi(z\bar{z})$	m even m odd
$\mathrm{Res}_G^{G_m} \widehat{\chi}_{\xi_1,\xi_2}$	(q^m+1) $\xi_1(x)\xi_2(x)$	$\xi_1(x)\xi_2(y)$ $+\xi_1(y)\xi_2(x)$	$\xi_1(x)\xi_2(x)$	$\xi_1(z)\xi_2(\bar{z})$ $+\xi_1(\bar{z})\xi_2(z)$ 0	m even m odd
$\mathrm{Res}_G^{G_m} \rho_\mu$	$(q^m - 1)\mu(x)$	0	$-\mu(x)$	0 $-\mu(z) - \mu(\bar{z})$	m even m odd

We shall use a series of abbreviated notation:

- $\bigoplus_{\xi^\sharp = \psi}$ indicates the direct sum over all $\xi \in \widehat{\mathbb{F}_{q^m}^*}$ such that $\xi^\sharp = \psi$;
- $\bigoplus_{(\xi^\sharp)^2 = \nu^\sharp}$ indicates the direct sum over all $\xi \in \widehat{\mathbb{F}_{q^m}^*}$ such that $(\xi^\sharp)^2 = \nu^\sharp$, that is, $\xi(x^2) = \nu(x)$ for all $x \in \mathbb{F}_q^*$;
- $\bigoplus_{(\xi_1\xi_2)^\sharp = \nu^\sharp}$ indicates the direct sum over all pairs $\{\xi_1, \xi_2\}$ where $\xi_1, \xi_2 \in \widehat{\mathbb{F}_{q^m}^*}$, $\xi_1 \neq \xi_2$ such that $(\xi_1\xi_2)^\sharp = \nu^\sharp$: each unordered pair is counted once;
- $\ominus_{(\overline{\xi_1}\xi_2)^\flat = \nu}$ indicates that we subtract (from the previous sum) the sum over all pairs $\{\xi_1, \xi_2\}$ such that $(\overline{\xi_1}\xi_2)^\flat = \nu$, that is, $\xi_1(\bar{z})\xi_2(z) = \nu(z)$ for all $z \in \mathbb{F}_{q^2}^*$; note that $(\overline{\xi_1}\xi_2)^\flat = \nu$ implies $(\xi_1\xi_2)^\sharp = \nu^\sharp$, so that we subtract terms that are effectively present (in the previous sum).

Other notations will be clear from the context. Finally, we observe that $\mathrm{Res}_G^{G_m} \widehat{\chi}_\xi^0$ cannot contain $\widehat{\chi}_\psi^1$, $\widehat{\chi}_{\psi_1,\psi_2}$, nor ρ_ν, because it is one-dimensional.

Therefore, by Frobenius reciprocity, $\mathrm{Ind}_G^{G_m}\widehat{\chi}_\psi^1$, $\mathrm{Ind}_G^{G_m}\widehat{\chi}_{\psi_1,\psi_2}$, and $\mathrm{Ind}_G^{G_m}\rho_\nu$ do not contain one-dimensional representation of G_m (cf. Corollary 11.2.3).

We are now in position to give the desired decomposition formulas for the induced representations. For three cases we have to distinguish between the case where m is odd or even.

Suppose that m is odd. Then,

$$
\mathrm{Ind}_G^{G_m}\widehat{\chi}_\psi^0 = \frac{q^{m-1}-1}{q^2-1}\left[\left(\bigoplus_{(\xi^\sharp)^2=\psi^2}\widehat{\chi}_\xi^1\right)\oplus\left(\bigoplus_{(\xi_1\xi_2)^\sharp=\psi^2}\widehat{\chi}_{\xi_1,\xi_2}\right)\right.
$$
$$
\left.\oplus\left(\bigoplus_{\mu^\sharp=\psi^2}\rho_\mu\right)\right]\oplus\left(\bigoplus_{\xi^\sharp=\psi}\widehat{\chi}_\xi^0\right) \tag{14.61}
$$
$$
\oplus\left(\bigoplus_{\xi_1^\sharp=\xi_2^\sharp=\psi}\widehat{\chi}_{\xi_1,\xi_2}\right)\ominus\left(\bigoplus_{\mu^\flat=\Psi}\rho_\mu\right),
$$

$$
\mathrm{Ind}_G^{G_m}\widehat{\chi}_\psi^1 = \frac{q(q^{m-1}-1)}{q^2-1}\left[\left(\bigoplus_{(\xi^\sharp)^2=\psi^2}\widehat{\chi}_\xi^1\right)\oplus\left(\bigoplus_{(\xi_1\xi_2)^\sharp=\psi^2}\widehat{\chi}_{\xi_1,\xi_2}\right)\right.
$$
$$
\left.\oplus\left(\bigoplus_{\mu^\sharp=\psi^2}\rho_\mu\right)\right]\oplus\left(\bigoplus_{\xi^\sharp=\psi}\widehat{\chi}_\xi^1\right) \tag{14.62}
$$
$$
\oplus\left(\bigoplus_{\xi_1^\sharp=\xi_2^\sharp=\psi}\widehat{\chi}_{\xi_1,\xi_2}\right)\oplus\left(\bigoplus_{\mu^\flat=\Psi}\rho_\mu\right),
$$

and

$$
\mathrm{Ind}_G^{G_m}\rho_\nu = \frac{q^{m-1}-1}{q+1}\left[\left(\bigoplus_{(\xi^\sharp)^2=\nu^\sharp}\widehat{\chi}_\xi^1\right)\oplus\left(\bigoplus_{(\xi_1\xi_2)^\sharp=\nu^\sharp}\widehat{\chi}_{\xi_1,\xi_2}\right)\right.
$$
$$
\left.\oplus\left(\bigoplus_{\mu^\sharp=\nu^\sharp}\rho_\mu\right)\right]\oplus\left(\bigoplus_{\mu^\flat=\nu}\rho_\mu\right). \tag{14.63}
$$

Suppose now that m is even. Then,

$$
\mathrm{Ind}_G^{G_m} \widehat{\chi}_\psi^0 = \frac{q(q^{m-2}-1)}{q^2-1} \left[\left(\bigoplus_{(\xi^\sharp)^2=\psi^2} \widehat{\chi}_\xi^1 \right) \oplus \left(\bigoplus_{(\xi_1\xi_2)^\sharp=\psi^2} \widehat{\chi}_{\xi_1,\xi_2} \right) \right.
$$

$$
\left. \oplus \left(\bigoplus_{\mu^\sharp=\psi^2} \rho_\mu \right) \right] \oplus \left(\bigoplus_{\xi^\sharp=\psi} \widehat{\chi}_\xi^0 \right)
$$

$$
\oplus \left(\bigoplus_{\xi^\sharp=\psi} \widehat{\chi}_\xi^1 \right) \oplus \left(\bigoplus_{\xi_1^\sharp=\xi_2^\sharp=\psi} \widehat{\chi}_{\xi_1,\xi_2} \right)
$$

$$
\oplus \left(\bigoplus_{(\overline{\xi_1}\xi_2)^\rho=\Psi} \widehat{\chi}_{\xi_1,\xi_2} \right),
$$

$$
\mathrm{Ind}_G^{G_m} \widehat{\chi}_\psi^1 = \frac{q^m-1}{q^2-1} \left[\left(\bigoplus_{(\xi^\sharp)^2=\psi^2} \widehat{\chi}_\xi^1 \right) \oplus \left(\bigoplus_{(\xi_1\xi_2)^\sharp=\psi^2} \widehat{\chi}_{\xi_1,\xi_2} \right) \right.
$$

$$
\left. \oplus \left(\bigoplus_{\mu^\sharp=\psi^2} \rho_\mu \right) \right] \oplus \left(\bigoplus_{\xi_1^\sharp=\xi_2^\sharp=\psi} \widehat{\chi}_{\xi_1,\xi_2} \right)
$$

$$
\ominus \left(\bigoplus_{(\xi_1\overline{\xi_2})^\rho=\Psi} \widehat{\chi}_{\xi_1,\xi_2} \right),
$$

and

$$
\mathrm{Ind}_G^{G_m} \rho_\nu = \frac{q^{m-1}+1}{q+1} \left[\left(\bigoplus_{(\xi^\sharp)^2=\nu^\sharp} \widehat{\chi}_\xi^1 \right) \oplus \left(\bigoplus_{(\xi_1\xi_2)^\sharp=\nu^\sharp} \widehat{\chi}_{\xi_1,\xi_2} \right) \right.
$$

$$
\left. \oplus \left(\bigoplus_{\mu^\sharp=\nu^\sharp} \rho_\mu \right) \right] \ominus \left(\bigoplus_{(\overline{\xi_1}\xi_2)^\rho=\nu} \widehat{\chi}_{\xi_1,\xi_2} \right).
$$

Finally, the next formula does not depend on the parity of m:

$$\mathrm{Ind}_G^{G_m} \widehat{\chi}_{\psi_1,\psi_2} = \frac{q^{m-1}-1}{q-1} \left[\left(\bigoplus_{(\xi^\sharp)^2=\psi_1\psi_2} \widehat{\chi}_\xi^1 \right) \oplus \left(\bigoplus_{(\xi_1\xi_2)^\sharp=\psi_1\psi_2} \widehat{\chi}_{\xi_1,\xi_2} \right) \right.$$

$$\left. \oplus \left(\bigoplus_{\mu^\sharp=\psi_1\psi_2} \rho_\mu \right) \right] \oplus \left(\bigoplus_{\substack{\xi_1^\sharp=\psi_1 \\ \xi_2^\sharp=\psi_2}} \widehat{\chi}_{\xi_1,\xi_2} \right). \tag{14.64}$$

Exercise 14.10.2 Prove the seven last decomposition formulas; see Example 14.10.4.

Exercise 14.10.3 Prove that $\mathrm{Ind}_G^{G_2} \rho_\nu$ decomposes without multiplicity, write down the decomposition (it is just (14.63) for $m = 2$), and check that the dimension of the left hand side equals the sum of the dimensions of the irreducible representations in the right hand side.

Example 14.10.4 We show how to derive the seven decomposition formulas above. We just compute the multiplicity of $\widehat{\chi}_\psi^0$ in $\mathrm{Res}_G^{G_m} \rho_\mu$ for m odd. Let χ^μ denote the character of $\mathrm{Res}_G^{G_m} \rho_\mu$. From Table 14.1, Table 14.2, and Table 14.3 we get

$$\langle \widehat{\chi}_\psi^0, \xi^\mu \rangle = (q^m - 1) \sum_{x \in \mathbb{F}_q^*} \psi(x^2)\overline{\mu(x)} - (q^2 - 1) \sum_{x \in \mathbb{F}_q^*} \psi(x^2)\overline{\mu(x)}$$

$$- \frac{q^2-q}{2} \sum_{z \in \mathbb{F}_{q^2} \setminus \mathbb{F}_q} \psi(z\bar{z})[\overline{\mu(z)} + \overline{\mu(\bar{z})}]$$

$$=_{(*)} (q^m - q^2) \sum_{x \in \mathbb{F}_q^*} \psi(x^2)\overline{\mu(x)} - (q^2 - q) \sum_{z \in \mathbb{F}_{q^2}^*} \psi(z\bar{z})\overline{\mu(z)}$$

$$+ (q^2 - q) \sum_{x \in \mathbb{F}_q^*} \psi(x^2)\overline{\mu(x)}$$

$$= (q^m - q) \sum_{x \in \mathbb{F}_q^*} \psi^2(x)\overline{\mu(x)} + (q^2 - q) \sum_{z \in \mathbb{F}_{q^2}^*} \Psi(z)\overline{\mu(z)}$$

$$=_{(**)} q(q^2 - 1)(q - 1) \left[\frac{q^{m-1}-1}{q^2-1} \delta_{\psi^2,\mu^\sharp} - \delta_{\Psi,\mu^\flat} \right]$$

where $=_{(*)}$ follows from $\mathbb{F}_{q^2} \setminus \mathbb{F}_q = \mathbb{F}_{q^2}^* \setminus \mathbb{F}_q^*$ and $=_{(**)}$ follows from Proposition 2.3.5 and Theorem 6.7.2. That is, since $|G| = q(q^2 - 1)(q - 1)$, by Proposition 10.2.18, the multiplicity of $\widehat{\chi}_\psi^0$ in $\mathrm{Res}_G^{G_m} \rho_\mu$ is equal to $\frac{q^{m-1}-1}{q^2-1}$ if $\psi^2 = \mu^\sharp$

and $\Psi \neq \mu^\flat$, while it is equal to $\frac{q^{m-1}-1}{q^2-1} - 1$ if $\psi^2 = \mu^\sharp$ and $\Psi = \mu^\flat$ (note that $\Psi = \mu^\flat \Rightarrow \psi^2 = \mu^\sharp$). By Frobenius reciprocity, these are also the multiplicities of ρ_μ in $\mathrm{Ind}_G^{G_m} \widehat{\chi}_\psi^0$. This leads to the terms

$$\left(\frac{q^{m-1}-1}{q^2-1} \bigoplus_{\mu^\sharp = \psi^2} \rho_\mu \right) \ominus \left(\bigoplus_{\mu^\flat = \Psi} \rho_\mu \right)$$

in (14.61).

Exercise 14.10.5

(1) Recalling the notation in Section 14.4 (so that, in particular, Ψ is not $\psi \circ N$), prove that
- $\mathrm{Res}_B^G \widehat{\chi}_\psi^1 = [\pi \boxtimes \Psi^2] \oplus \chi_{\psi,\psi}$;
- $\mathrm{Res}_B^G \widehat{\chi}_{\psi_1,\psi_2}^1 = [\pi \boxtimes \Psi_1 \Psi_2] \oplus 2\chi_{\psi_1,\psi_2}$;
- $\mathrm{Res}_B^G \rho_\nu = \pi \boxtimes \nu^\sharp$.

  Hint. Use the decomposition $B = \mathrm{Aff}(\mathbb{F}_q) \times Z$ and compute Res_Z^G by means of the character table of G.

(2) Deduce that

$$\mathrm{Ind}_B^G[\pi \boxtimes \Psi] = \left(\bigoplus_{\psi_1^2 = \psi} \widehat{\chi}_{\psi_1}^1 \right) \oplus \left(\bigoplus_{\psi_1 \psi_2 = \psi} \widehat{\chi}_{\psi_1,\psi_2} \right) \oplus \left(\bigoplus_{\nu^\sharp = \psi} \rho_\nu \right)$$

(clearly, the first term is absent if ψ is not a square).

Exercise 14.10.6 Denote by B_m the Borel subgroup of G_m and, for $\xi_1, \xi_2 \in \widehat{\mathbb{F}_{q^m}^*}$, denote by $\Xi_1 \boxtimes \Xi_2$ the corresponding representation of B_m. From Exercise 12.1.9, Exercise 11.1.10, and the decomposition $B = \mathrm{Aff}(\mathbb{F}_q) \times Z$, deduce that

$$\mathrm{Ind}_B^{B_m}[\Xi_1 \boxtimes \Xi_2] = \frac{q^{m-1}-1}{q-1} \left[\bigoplus_{\xi_2^\sharp = \psi_2} (\pi_{q^m} \boxtimes \Xi_2) \right] \oplus \left[\bigoplus_{\substack{\xi_1^\sharp = \psi_1 \\ \xi_2^\sharp = \psi_2}} (\Psi_1 \boxtimes \Psi_2) \right].$$

Exercise 14.10.7

(1) Use Exercise 14.10.6, the definition of $\widehat{\chi}_{\psi_1,\psi_2}$, and transitivity of induction, to give another proof of (14.64).

   Hint. Recall that $\chi_{\psi_1,\psi_2} = \Psi_1 \boxtimes (\Psi_1 \Psi_2)$.

(2) For the remaining six decomposition formulas for $\mathrm{Ind}_G^{G_m}$, try to find alternative proofs that avoid the character tables but make use of the theory of induced representations.

14.11 Decomposition of tensor products

In this section we give a complete series of formulas for the decomposition of the tensor products of irreducible representations of GL$(2, \mathbb{F}_q)$. In general, this is a very difficult problem: for instance, for the symmetric group (cf. Section 2.9 of the monograph by James and Kerber [82]) no complete solution is known; nowadays it constitutes an active area of research (see [162] for a recent contribution and a reference to the current literature). See also our recent papers [35, 36] for a suitable harmonic analysis of tensor products of irreducible representations. The style is the same as in the previous section and we keep the same notation therein. In addition, we also write

- $\bigoplus_{\nu^\sharp = (\psi_1 \psi_2)^2}$ to denote the direct sum over all indecomposable characters $\nu \in \mathbb{F}^*_{q^2}$ such that $\nu^\sharp = (\psi_1 \psi_2)^2$;
- $\bigoplus_{\psi_3 \psi_4 = \psi_1^2 \nu_1^\sharp}$ for the direct sum over all unordered pairs $\{\psi_3, \psi_4\} \subset \widehat{\mathbb{F}^*_q}$, with $\psi_3 \neq \psi_4$ and such that $\psi_3 \psi_4 = \psi_1^2 \nu_1^\sharp$, and so on.

The formulas below are given without proof; they may be proved by means of the character table of GL$(2, \mathbb{F}_q)$ (see Table 14.2) and the table of conjugacy classes (see Table 14.1). At the end, we give an example of such computations.

We have the following trivial identities:

$$\widehat{\chi}^0_\psi \otimes \widehat{\chi}^0_{\psi_0} = \widehat{\chi}^0_{\psi \psi_0} \qquad\qquad \widehat{\chi}^0_{\psi_0} \otimes \widehat{\chi}^1_\psi = \widehat{\chi}^1_{\psi \psi_0}$$

$$\widehat{\chi}^0_{\psi_0} \otimes \widehat{\chi}_{\psi_1, \psi_2} = \widehat{\chi}_{\psi_0 \psi_1, \psi_0 \psi_2} \qquad \widehat{\chi}^0_\psi \otimes \rho_\nu = \rho_{\psi \nu}.$$

Moreover,

$$\widehat{\chi}^1_{\psi_1} \otimes \widehat{\chi}^1_{\psi_2} = \widehat{\chi}^0_{\psi_1 \psi_2} \oplus \widehat{\chi}^1_{\psi_1 \psi_2} \oplus \widehat{\chi}^1_{-\psi_1 \psi_2}$$

$$\oplus \left(\bigoplus_{\psi_3 \psi_4 = (\psi_1 \psi_2)^2} \widehat{\chi}_{\psi_3, \psi_4} \right) \oplus \left(\bigoplus_{\nu^\sharp = (\psi_1 \psi_2)^2} \rho_\nu \right),$$

where the third term appears only if q is odd.

$$\widehat{\chi}^1_{\psi_1} \otimes \widehat{\chi}_{\psi_2, \psi_3} = \left(\bigoplus_{\psi^2 = \psi_1^2 \psi_2 \psi_3} \widehat{\chi}^1_\psi \right) \oplus \left(\bigoplus_{\psi_4 \psi_5 = \psi_1^2 \psi_2 \psi_3} \widehat{\chi}_{\psi_4, \psi_5} \right)$$

$$\oplus \widehat{\chi}_{\psi_1 \psi_2, \psi_1 \psi_3} \oplus \left(\bigoplus_{\nu^\sharp = \psi_1^2 \psi_2 \psi_3} \rho_\nu \right).$$

$$\widehat{\chi}_{\psi_1}^1 \otimes \rho_\nu = \left(\bigoplus_{\psi^2 = \psi_1^2 \nu^\sharp} \widehat{\chi}_\psi^1 \right) \oplus \left(\bigoplus_{\psi_2 \psi_3 = \psi_1^2 \nu^\sharp} \widehat{\chi}_{\psi_2,\psi_3} \right)$$

$$\oplus \left(\bigoplus_{\substack{\nu_1^\sharp = \psi_1^2 \nu^\sharp \\ \nu_1 \neq \Psi_1 \nu, \Psi_1 \overline{\nu}}} \rho_{\nu_1} \right).$$

$$\widehat{\chi}_{\psi_1,\psi_2} \otimes \widehat{\chi}_{\psi_3,\psi_4} = \left(\delta_{\psi_1\psi_3,\psi_2\psi_4} \widehat{\chi}_{\psi_1\psi_3}^0 \right) \oplus \left(\delta_{\psi_1\psi_4,\psi_2\psi_3} \widehat{\chi}_{\psi_1\psi_4}^0 \right)$$

$$\oplus \left(\bigoplus_{\psi^2 = \psi_1\psi_2\psi_3\psi_4} \widehat{\chi}_\psi^1 \right) \oplus \left(\delta_{\psi_1\psi_3,\psi_2\psi_4} \widehat{\chi}_{\psi_1\psi_3}^1 \right)$$

$$\oplus \left(\delta_{\psi_1\psi_4,\psi_2\psi_3} \widehat{\chi}_{\psi_1\psi_4}^1 \right) \oplus \left(\bigoplus_{\psi_5\psi_6 = \psi_1\psi_2\psi_3\psi_4} \widehat{\chi}_{\psi_5,\psi_6} \right)$$

$$\oplus \left(\bigoplus_{\nu^\sharp = \psi_1\psi_2\psi_3\psi_4} \rho_\nu \right) \oplus \widehat{\chi}_{\psi_1\psi_3,\psi_2\psi_4} \oplus \widehat{\chi}_{\psi_1\psi_4,\psi_2\psi_3},$$

where the last but one (respectively, last) term appears only if $\psi_1\psi_3 \neq \psi_2\psi_4$ (respectively, $\psi_1\psi_4 \neq \psi_2\psi_3$).

$$\widehat{\chi}_{\psi_1,\psi_2} \otimes \rho_{\nu_1} = \left(\bigoplus_{\psi_3\psi_4 = \psi_1\psi_2\nu_1^\sharp} \widehat{\chi}_{\psi_3,\psi_4} \right)$$

$$\oplus \left(\bigoplus_{\nu^\sharp = \psi_1\psi_2\nu_1^\sharp} \rho_\nu \right) \oplus \left(\bigoplus_{\psi^2 = \psi_1\psi_2\nu_1^\sharp} \widehat{\chi}_\psi^1 \right),$$

where the last term appears only if $\psi_1\psi_2\nu_1^\sharp$ is a square.

Finally,

$$\rho_{\nu_1} \otimes \rho_{\nu_2} = \left(\delta_{\nu_1,\nu_2} + \delta_{\nu_1,\overline{\nu_2}} \right) \widehat{\chi}_{\nu_1^\sharp}^0 \oplus \left(\bigoplus_{\substack{\psi^2 = (\nu_1\nu_2)^\sharp \\ \Psi \neq \nu_1\nu_2, \overline{\nu_1}\nu_2}} \widehat{\chi}_\psi^1 \right)$$

$$\quad (14.65)$$

$$\oplus \left(\bigoplus_{\psi_1\psi_2 = (\nu_1\nu_2)^\sharp} \widehat{\chi}_{\psi_1,\psi_2} \right) \oplus \left(\bigoplus_{\substack{\nu^\sharp = (\nu_1\nu_2)^\sharp \\ \nu \neq \nu_1\nu_2, \overline{\nu_1}\nu_2, \nu_1\overline{\nu_2}, \overline{\nu_1}\nu_2}} \rho_\nu \right),$$

where the second term appears only if $(\nu_1\nu_2)^\sharp$ is a square and $\nu_1 \neq \nu_2, \overline{\nu_2}$.

Exercise 14.11.1 Prove the above decomposition formulas (cf. Example below).

Example 14.11.2 We show how to compute the multiplicity of ρ_ν in $\rho_{\nu_1} \otimes \rho_{\nu_2}$. Denoting by χ^ν, χ^{ν_1}, and χ^{ν_2} the characters of ρ_ν, ρ_{ν_1}, and ρ_{ν_2}, respectively, and recalling that, by (10.63), the character of $\rho_{\nu_1} \otimes \rho_{\nu_2}$ is $\chi^{\nu_1}\chi^{\nu_2}$, we have

$$\langle \chi^{\nu_1}\chi^{\nu_2}, \chi^\nu \rangle = (q-1)^3 \sum_{x \in \mathbb{F}_q^*} \nu_1(x)\nu_2(x)\overline{\nu(x)} - (q^2-1)\sum_{x \in \mathbb{F}_q^*} \nu_1(x)\nu_2(x)\overline{\nu(x)}$$

$$- \frac{q^2-q}{2} \sum_{z \in \mathbb{F}_{q^2}\backslash\mathbb{F}_q} [\nu_1(z) + \nu_1(\overline{z})] \cdot [\nu_2(z) + \nu_2(\overline{z})] \cdot \left[\overline{\nu(z)} + \overline{\nu(\overline{z})}\right]$$

$$= \left[(q-1)^3(q-1) - (q^2-1)(q-1)\right]\delta_{(\nu_1\nu_2)^\sharp, \nu^\sharp}$$

$$+ 4(q^2-q)(q-1)\sum_{x \in \mathbb{F}_q^*} \nu_1(x)\nu_2(x)\overline{\nu(x)}$$

$$- (q^2-q)\sum_{z \in \mathbb{F}_{q^2}^*} \left[\nu_1(z)\nu_2(z)\overline{\nu(z)} + \nu_1(\overline{z})\nu_2(z)\overline{\nu(z)}\right.$$

$$\left. + \nu_1(\overline{z})\nu_2(\overline{z})\overline{\nu(z)} + \nu_1(z)\nu_2(\overline{z})\overline{\nu(z)}\right]$$

$$= |G| \left[\delta_{(\nu_1\nu_2)^\sharp, \nu^\sharp} - (\delta_{\nu_1\nu_2, \nu} + \delta_{\overline{\nu_1}\nu_2, \nu} + \delta_{\overline{\nu_1\nu_2}, \nu} + \delta_{\nu_1\overline{\nu_2}, \nu})\right],$$

and this explains the last term in (14.65).

Appendix
Chebyshëv polynomials

In this appendix we define and study in detail the notions of a Chebyshëv set, Chebyshëv polynomials of the first and second kind, and some modified versions of the latter. These play a crucial role in the spectral analysis of the DFT in Section 4.2 as well as in the proof of the Alon-Boppana-Serre theorem (Theorem 9.2.6). Our main sources are [104] and the monographs by Briggs and Henson [22] and by Davidoff, Sarnak, and Valette [49].

Definition A.1 Let $I \subseteq \mathbb{R}$ be an interval. We say that the real valued functions $\phi_1, \phi_2, \ldots, \phi_n$ defined on I form a *Chebyshëv set* on I if, for all choices of $a_1, a_2, \ldots, a_n \in \mathbb{R}$, the function $\sum_{j=1}^{n} a_j \phi_j$ has at most $n-1$ distinct zeroes in I.

Proposition A.2 *Let* $\{\phi_1, \phi_2, \ldots, \phi_n\}$ *be a Chebyshëv set on the interval I. Then*

(i) *if $t_1, t_2, \ldots, t_n \in I$ are distinct, then the vectors*
$$\mathbf{z}_k = (\phi_k(t_1), \phi_k(t_2), \ldots, \phi_k(t_n)),$$
$k = 1, 2, \ldots, n$ *are $\mathbb{R}$-linearly independent in $\mathbb{R}^n$;*

(ii) *if $t_1, t_2, \ldots, t_{n+1} \in I$ are distinct and $s_1, s_2, \ldots, s_{n+1}$ are real numbers that alternate in sign (i.e. $s_j s_{j+1} < 0$ for $j = 1, 2, \ldots, n$), then the vectors $\mathbf{w}_k = (\phi_k(t_1), \phi_k(t_2), \ldots, \phi_k(t_{n+1})) \, k = 1, 2, \ldots, n$ and $\mathbf{w}_{n+1} = (s_1, s_2, \ldots, s_{n+1})$ are $\mathbb{R}$-linearly independent in $\mathbb{R}^{n+1}$.*

Proof.

(i) The linear relation $\sum_{j=1}^{n} a_j \mathbf{z}_j = 0$ yields $\sum_{j=1}^{n} a_j \phi_j(t_k) = 0$, for $k = 1, 2, \ldots, n$, which forces, by definition of a Chebyshëv set, $a_j = 0$ for all $j = 1, 2, \ldots, n$.

(ii) Suppose that there exist $a_1, a_2, \ldots, a_{n+1} \in \mathbb{R}$ such that $\sum_{j=1}^{n+1} a_j \mathbf{w}_j = 0$. This is equivalent to saying

$$a_1 \phi_1(t_k) + a_2 \phi_2(t_k) + \cdots + a_n \phi_n(t_k) = -a_{n+1} s_k$$

for all $k = 1, 2, \ldots, n+1$. If $a_{n+1} = 0$ we can argue as in (i). Otherwise we deduce that $\sum_{j=1}^{n} a_j \phi_j$ alternates the sign at the points $t_1, t_2, \ldots, t_{n+1}$. We may suppose that $t_1 < t_2 < \cdots < t_{n+1}$ and conclude, by virtue of the intermediate value theorem, that there exist $\tilde{t}_k \in (t_k, t_{k+1})$ such that $\sum_{j=1}^{n} a_j \phi_j(\tilde{t}_k) = 0$ for $k = 1, 2, \ldots, n$. By definition of a Chebyshëv set, we get the $a_j = 0$ for all $j = 1, 2, \ldots, n$ and thus also $a_{n+1} = 0$. $\qquad\square$

Proposition A.3

(i) *The functions* $1, \cos \theta, \cos 2\theta, \ldots, \cos n\theta$ *constitute a Chebyshëv set in* $[0, \pi]$.

(ii) *The functions* $\sin \theta, \sin 2\theta, \ldots, \sin n\theta$ *constitute a Chebyshëv set in* $(0, \pi)$.

Proof.

(i) First of all, note that $\cos k\theta$ may be written as a polynomial of degree k in $\cos \theta$. Indeed, De Moivre's formula yields

$$\cos k\theta + i \sin k\theta = (\cos \theta + i \sin \theta)^k = \sum_{h=0}^{k} \binom{k}{h} (\cos \theta)^{k-h} i^h (\sin \theta)^h$$

$$(\text{A.1})$$

so that (since i^h is real if and only if h is even)

$$\cos k\theta = \sum_{h=0}^{[k/2]} \binom{k}{2h} (-1)^h (\cos \theta)^{k-2h} (\sin \theta)^{2h}$$

and, using the identity $\sin^2 \theta = 1 - \cos^2 \theta$, we get the desired expression. Therefore, a function of the form $\phi(\theta) = a_0 + a_1 \cos \theta + \cdots + a_n \cos n\theta$ can be written in the form $\phi(\theta) = P(\cos \theta)$ where P is a real polynomial of degree $\leq n$. Since P has at most n roots in $[-1, 1]$ and the map $\theta \mapsto \cos \theta$ is a bijection between $[0, \pi]$ and $[-1, 1]$, we deduce that $\phi(\theta)$ has at most n roots in $[0, \pi]$.

(ii) From (A.1) we also deduce that

$$\sin k\theta = \sum_{h=0}^{[(k-1)/2]} \binom{k}{2h+1} (-1)^h (\cos \theta)^{k-2h-1} (\sin \theta)^{2h+1}$$

that yields an expression of $\frac{\sin k\theta}{\sin \theta}$ as a polynomial of degree $k - 1$ in $\cos \theta$. Then, for $0 < \theta < \pi$, we have that $\psi(\theta) = b_1 \sin \theta + b_2 \sin 2\theta + \cdots + b_n \sin n\theta$ can be written in the form

$$\psi(\theta) = \sin \theta \left(b_1 + b_2 \frac{\sin 2\theta}{\sin \theta} + \cdots + b_n \frac{\sin n\theta}{\sin \theta} \right) = \sin \theta P(\cos \theta)$$

where P is a polynomial of degree $\leq n - 1$. Then we may conclude as in (i). $\qquad \square$

In the proof of Proposition A.3 we have shown the existence of polynomials $T_n \in \mathbb{R}[x]$ and $U_n \in \mathbb{R}[x]$ of degree n such that

$$\cos n\theta = T_n(\cos \theta) \text{ and } \frac{\sin(n+1)\theta}{\sin \theta} = U_n(\cos \theta).$$

The T_n's are called the *Chebyshëv polynomials of the first kind*. As we shall see (cf. Lemma A.3) the U_n's are the so-called *Chebyshëv polynomials of the second kind*.

Exercise A.4 Show that the Chebyshëv polynomials of the first kind are expressed as

$$T_n(x) = \sum_{k=0}^{[n/2]} \binom{n}{2k} (x^2 - 1)^k x^{n-2k}$$

and satisfy:

(1) the recurrence relation

$$T_{n+1}(x) = 2xT_n(x) - T_{n-1}(x) \text{ for } n \geq 1 \text{ with } T_0(x) = 1, \ T_1(x) = x;$$

(2) the differential equation

$$(1 - x^2)y'' - xy' + n^2 y = 0;$$

(3) the orthogonality relations

$$\int_{-1}^{1} T_n(x) T_m(x) \frac{dx}{\sqrt{1 - x^2}} = \begin{cases} 0 & \text{if } n \neq m \\ \pi & \text{if } n = m = 0 \\ \pi/2 & \text{if } n = m \neq 0; \end{cases}$$

(4) the multiplicative property: $T_m T_n = \frac{1}{2}(T_{n+m} + T_{|m-n|})$;

(5) the semigroup property: $T_m(T_n(x)) = T_{mn}(x)$;

(6) the discrete orthogonality relations

$$\frac{1}{2}T_0\left(\cos\frac{j\pi}{n}\right)T_0\left(\cos\frac{k\pi}{n}\right) + \sum_{r=1}^{n-1}T_r\left(\cos\frac{j\pi}{n}\right)T_r\left(\cos\frac{k\pi}{n}\right)$$

$$+ \frac{1}{2}T_n\left(\cos\frac{j\pi}{n}\right)T_n\left(\cos\frac{k\pi}{n}\right) = \begin{cases} 0 & \text{if } j \neq k \\ n/2 & \text{if } j = k \neq 0, n \\ n & \text{if } j = k = 0, n; \end{cases}$$

(7) the dual discrete orthogonality relations:

$$\frac{1}{2}T_j(1)T_k(1) + \sum_{r=1}^{n-1}T_j\left(\cos\frac{\pi r}{n}\right)T_k\left(\cos\frac{\pi r}{n}\right)$$

$$+ \frac{1}{2}T_j(-1)T_k(-1) = \begin{cases} 0 & \text{if } j \neq k \\ n/2 & \text{if } j = k \neq 0, n \\ n & \text{if } j = k = 0, n; \end{cases}$$

(8) the associated generating function is:

$$\sum_{n=0}^{\infty}T_n(x)t^n = \frac{1-tx}{1-2tx+t^2}.$$

Exercise A.5 Let $X_n = \{0, 1, \ldots, n\}$ and $\widetilde{X}_n = \{\cos\frac{j\pi}{n} : j = 0, 1, \ldots, n\}$. The map $\mathfrak{F} : L(\widetilde{X}_n) \to L(X_n)$, defined by setting

$$[\mathfrak{F}f](k) = \frac{1}{n}f(1)T_k(1) + \frac{2}{n}\sum_{j=1}^{n-1}f\left(\cos\frac{j\pi}{n}\right)T_k\left(\cos\frac{j\pi}{n}\right) + \frac{1}{n}f(-1)T_k(-1)$$

for all $f \in L(\widetilde{X}_n)$ and $k \in X_n$, is called the *Discrete Chebyshëv Transform* (see the monograph [22] by Briggs and Henson for more on this). Show that the following inversion formula holds:

$$f\left(\cos\frac{j\pi}{n}\right) = \frac{1}{2}[\mathfrak{F}f](0)T_0\left(\cos\frac{j\pi}{n}\right) + \sum_{k=1}^{n-1}[\mathfrak{F}f](k)T_k\left(\cos\frac{j\pi}{n}\right)$$

$$+ \frac{1}{2}[\mathfrak{F}f](n)T_n\left(\cos\frac{j\pi}{n}\right),$$

for all $f \in L(\widetilde{X}_n)$ and $j = 0, 1, \ldots, n$. Moreover, for n even, analyze the relations between the Discrete Chebyshëv Transform and the Discrete Fourier Transform of an even function (see Exercise 4.1.7 and Exercise 4.1.8).

Definition A.6 The *Chebyshëv polynomials of the second kind* are the polynomials $U_m(x)$, $m \in \mathbb{N}$, defined by means of the initial positions $U_0(x) = 1$ and $U_1(x) = 2x$ and the recurrence relation

$$U_{m+1}(x) = 2xU_m(x) - U_{m-1}(x) \tag{A.2}$$

for all $m \geq 1$.

Note that $\deg U_m(x) = m$ and the leading coefficient of $U_m(x)$ is 2^m, for all $m \in \mathbb{N}$.

Exercise A.7 Show that the Chebyshëv polynomials of the second kind are expressed as

$$U_n(x) = \sum_{k=0}^{[n/2]} \binom{n+1}{2k+1} (x^2 - 1)^k x^{n-2k}$$

and satisfy:

(1) the differential equation

$$(1 - x^2)y'' - 3xy' + n(n+2)y = 0;$$

(2) the orthogonality relations

$$\int_{-1}^{1} U_n(x)U_m(x)\sqrt{1 - x^2}dx = \frac{\pi}{2}\delta_{n,m};$$

(3) the associated generating function is:

$$\sum_{n=0}^{\infty} U_n(x)t^n = \frac{1}{1 - 2tx + t^2};$$

(4) finally prove that $T'_{n+1}(x) = (n+1)U_n(x)$.

Lemma A.8

$$U_m(\cos\theta) = \frac{\sin(m+1)\theta}{\sin\theta} \tag{A.3}$$

for all $m \in \mathbb{N}$ and $\theta \in \mathbb{R} \setminus \pi\mathbb{Z}$.

Note that we may interpret $\dfrac{\sin(m+1)k\pi}{\sin k\pi}$, $k \in \mathbb{Z}$, as the limit of $\dfrac{\sin(m+1)\theta}{\sin\theta}$ for $\theta \to k\pi$ that may be evaluated by means of L'Hôpital's rule, so that (A.3) becomes

$$U_m(\cos k\pi) \equiv U_m((-1)^k) = (-1)^{km}(m+1).$$

Proof. We prove it by induction on m. Clearly,

$$U_0(\cos\theta) = 1 = \frac{\sin(0+1)\theta}{\sin\theta}$$

and

$$U_1(\cos\theta) = 2\cos\theta = \frac{\sin 2\theta}{\sin\theta} = \frac{\sin(1+1)\theta}{\sin\theta},$$

showing the base of induction. Moreover,

$$\sin(m+2)\theta = \sin m\theta \cos 2\theta + \sin 2\theta \cos m\theta$$

$$(\cos 2\theta = 2\cos^2\theta - 1) = 2\cos^2\theta \sin m\theta - \sin m\theta + 2\sin\theta\cos\theta\cos m\theta$$

$$= 2\cos\theta(\cos\theta\sin m\theta + \sin\theta\cos m\theta) - \sin m\theta$$

$$= 2\cos\theta\sin(m+1)\theta - \sin m\theta$$

and therefore, assuming that (A.3) holds both for m and $m-1$, we have:

$$\frac{\sin(m+2)\theta}{\sin\theta} = 2\cos\theta\frac{\sin(m+1)\theta}{\sin\theta} - \frac{\sin m\theta}{\sin\theta}$$

$$\text{(by inductive hypothesis)} = 2\cos\theta U_m(\cos\theta) - U_{m-1}(\cos\theta)$$

$$\text{(by (A.2))} = U_{m+1}(\cos\theta). \qquad \square$$

We now define a first set of modified Chebyshëv polynomials of the second kind. Let us fix, once and for all, a positive integer k, and define $P_m \in \mathbb{R}[x]$, $m \in \mathbb{N}$, by setting

$$P_m(x) = (k-1)^{\frac{m}{2}}U_m\left(\frac{x}{2\sqrt{k-1}}\right). \tag{A.4}$$

Lemma A.9 *We have $P_0(x) = 1$, $P_1(x) = x$ and, for all $m \geq 1$,*

$$P_{m+1}(x) = xP_m(x) - (k-1)P_{m-1}(x).$$

Proof.

$$xP_m(x) - (k-1)P_{m-1}(x) = x(k-1)^{\frac{m}{2}}U_m\left(\frac{x}{2\sqrt{k-1}}\right)$$

$$- (k-1)^{\frac{m+1}{2}}U_{m-1}\left(\frac{x}{2\sqrt{k-1}}\right)$$

$$= (k-1)^{\frac{m+1}{2}}\left[2\frac{x}{2\sqrt{k-1}}U_m\left(\frac{x}{2\sqrt{k-1}}\right)\right.$$

$$\left. -U_{m-1}\left(\frac{x}{2\sqrt{k-1}}\right)\right]$$

$$(\text{by (A.2)}) = (k-1)^{\frac{m+1}{2}}U_{m+1}\left(\frac{x}{2\sqrt{k-1}}\right)$$

$$= P_{m+1}(x). \qquad \square$$

Another modified version of the U_m's is provided by the polynomials $X_m \in \mathbb{R}[x]$, $m \in \mathbb{N}$, defined by setting

$$X_m(x) = U_m\left(\frac{x}{2}\right). \tag{A.5}$$

Lemma A.10 *The following properties hold for the polynomials X_m, $m \in \mathbb{N}$:*

(i) $X_m(2\cos\theta) = \dfrac{\sin(m+1)\theta}{\sin\theta}$.

(ii) $X_{m+1}(x) = xX_m(x) - X_{m-1}(x)$.

(iii) *The roots of X_m are $A_h = 2\cos\dfrac{h\pi}{m+1}$ for $h = 1, 2, \ldots, m$.*

Proof.

(i) follows immediately from Lemma (A.8), and (ii) is obvious. Since $\deg X_m = m$, the polynomial X_m has at most m roots. But by (i) we have

$$X_m(2\cos\theta) = 0 \Leftrightarrow \sin(m+1)\theta = 0 \text{ and } \sin\theta \neq 0$$

$$\Leftrightarrow (m+1)\theta = h\pi \text{ with } h \in \mathbb{Z} \text{ and } (m+1)\nmid h,$$

so that the A_h's as in the statement are precisely the m distinct roots of X_m. $\qquad\square$

Comparing (A.4) and (A.5), we deduce that

$$P_m(x) = (k-1)^{m/2}X_m(\frac{x}{\sqrt{k-1}}) \tag{A.6}$$

for all $m \in \mathbb{N}$.

Now we give deeper and more difficult properties of the polynomials X_m's.

Lemma A.11

(i) *For $0 \leq \ell \leq h$ we have:*

$$X_\ell X_h = \sum_{m=0}^{\ell} X_{\ell+h-2m}.$$

(ii) *For $m \in \mathbb{N}$*

$$\frac{X_m(x)}{x - \alpha_m} = \sum_{j=0}^{m-1} X_{m-1-j}(\alpha_m) X_j(x),$$

where $\alpha_m = 2\cos\frac{\pi}{m+1}$.

Proof.

(i) The proof is by induction of ℓ. For $\ell = 0$ it is trivial ($X_0 = 1$), while for $\ell = 1$ we have $X_1(x) = x$ and, by virtue of Lemma A.10.(ii),

$$X_1 X_h = x X_h = X_{h+1} + X_{h-1}.$$

The inductive step is the following: for $2 \leq \ell \leq h$ we have, taking into account Lemma A.10.(ii),

$$X_\ell X_h = x X_{\ell-1} X_h - X_{\ell-2} X_h$$

$$\text{(by inductive hypothesis)} = x \sum_{m=0}^{\ell-1} X_{\ell-1+h-2m} - \sum_{m=0}^{\ell-2} X_{\ell-2+h-2m}$$

$$= \sum_{m=0}^{\ell-2} (x X_{\ell+h-2m-1} - X_{\ell+h-2m-2}) + x X_{h-\ell+1}$$

$$\text{(by Lemma A.10.(ii))} = \sum_{m=0}^{\ell-2} X_{\ell+h-2m} + X_{h-\ell} + X_{h-\ell+2}$$

$$= \sum_{m=0}^{\ell} X_{\ell+h-2m}.$$

(ii) First of all, note that Lemma A.10.(ii) may be rewritten as

$$x X_j = X_{j-1} + X_{j+1}. \tag{A.7}$$

Moreover,

$$X_0(\alpha_m) = 1 \tag{A.8}$$

$$X_1(\alpha_m) - \alpha_m X_0(\alpha_m) = \alpha_m - \alpha_m = 0 \tag{A.9}$$

and, for $m \geq 2$:

$$X_{m-2}(\alpha_m) - \alpha_m X_{m-1}(\alpha_m) = -X_m(\alpha_m) = 0 \qquad (A.10)$$

where the first (resp. second) equality follows from Lemma A.10.(ii) (resp. (iii)). Therefore,

$$(x - \alpha_m) \sum_{j=0}^{m-1} X_{m-1-j}(\alpha_m) X_j(x) = x X_{m-1}(\alpha_m)$$

$$+ \sum_{j=1}^{m-1} X_{m-1-j}(\alpha_m) x X_j(x)$$

$$- \sum_{j=0}^{m-1} X_{m-j-1}(\alpha_m) \alpha_m X_j(x)$$

(by (A.7) and $X_1(x) = x$) $= X_1(x) X_{m-1}(\alpha_m)$

$$+ \sum_{j=1}^{m-1} X_{m-j-1}(\alpha_m) \left[X_{j+1}(x) + X_{j-1}(x) \right]$$

$$- \sum_{j=0}^{m-1} X_{m-j-1}(\alpha_m) \alpha_m X_j(x)$$

(by rearranging) $= X_0(x) \left[X_{m-2}(\alpha_m) - X_{m-1}(\alpha_m) \alpha_m \right]$

$$+ \sum_{j=1}^{m-2} X_j(x) \left[X_{m-j}(\alpha_m) \right.$$

$$\left. + X_{m-j-2}(\alpha_m) - \alpha_m X_{m-j-1}(\alpha_m) \right]$$

$$+ \left[\alpha_m - \alpha_m X_0(\alpha_m) \right] X_{m-1}(x)$$

$$+ X_0(\alpha_m) X_m(x)$$

$$= X_m(x)$$

where the last equality follows from (A.8), (A.9), (A.10) and Lemma A.10.(ii) applied to the main sum. □

We now define a further family of polynomials:

$$Y_m(x) = \frac{X_m^2(x)}{x - \alpha_m}. \qquad (A.11)$$

Since $X_m(x)$ is divisible by $x - \alpha_m$, we deduce that Y_m is indeed a polynomial of degree $2m - 1$.

Lemma A.12

$$Y_m(x) = \sum_{i=1}^{2m-1} y_i X_i(x)$$

where the coefficients $y_i \in \mathbb{R}$ are given by the rule

$$y_i = \sum_{\ell} X_{\ell}(\alpha_m), \qquad (A.12)$$

the sum running over all ℓ satisfying the following conditions:

(1) $0 \le \ell \le \min\{i - 1, 2m - 1 - i\}$;
(2) $2m - 1 - i - \ell$ *is even.*

Proof. We have

$$Y_m(x) = \frac{X_m^2(x)}{x - \alpha_m}$$

$$\text{(by Lemma A.11.(ii))} = X_m(x) \sum_{j=0}^{m-1} X_{m-j-1}(\alpha_m) X_j(x) \qquad (A.13)$$

$$\text{(by Lemma A.11.(i))} = \sum_{j=0}^{m-1} X_{m-j-1}(\alpha_m) \sum_{h=0}^{j} X_{m+j-2h}(x).$$

In the above sums the summation indices j and h satisfy $0 \le j \le m - 1$ and $-2j \le -2h \le 0$. Thus, if we set $i = m + j - 2h$ we have

$$1 \le m - j \le i = m + j - 2h \le m + j \le 2m - 1$$

so that

$$Y_m(x) = \sum_{i=1}^{2m-1} y_i X_i(x), \qquad (A.14)$$

where $y_i = \sum_{\ell} X_{\ell}(\alpha_m)$ with $\ell = m - j - 1$. It remains to determine the range of ℓ in terms of the new summation index i. Since $1 \le i \le 2m - 1$ and $0 \le \ell \le m - 1$, then the product $X_{\ell}(\alpha_m) X_i(x)$ appears in (A.13) (and therefore in (A.14)) if and only if, recalling that $j = m - 1 - \ell$, there exists $0 \le h \le j$ such that $i = m + j - 2h$. Since $i + \ell = 2m - 1 - 2h$ then $2m - 1 - i - \ell$ must be even $(= 2h)$, thus showing (2), and the condition $0 \le h \le j$ is equivalent to

$$0 \le \frac{2m - 1 - i - \ell}{2} (\equiv \frac{m + j - i}{2} \equiv h) \le m - 1 - \ell (\equiv j)$$

that is,

$$0 \leq 2m - 1 - i - \ell \leq 2m - 2 - 2\ell.$$

This is equivalent to (1). □

Proposition A.13 *The coefficients y_i's in Lemma A.12 are all positive, that is, Y_m is a positive linear combination of the X_i's, $1 \leq i \leq 2m - 1$.*

Proof. By taking the arithmetical mean of the terms appearing in the upper bound for the index ℓ in (A.12), we have

$$\min\{i - 1, 2m - i - 1\} \leq \frac{(i - 1) + (2m - i - 1)}{2} = m - 1$$

so that $\ell \leq m - 1$. Since $2\cos\frac{\pi}{\ell+1} < \alpha_m = 2\cos\frac{\pi}{m+1}$, $\lim_{x\to+\infty} X_\ell(x) = +\infty$, and $2\cos\frac{\pi}{\ell+1}$ is the largest root of X_ℓ (by Lemma A.10.(iii)), we conclude that $X_\ell(\alpha_m) > 0$ for $\ell = 0, 1, \ldots, m - 1$. As a consequence, (A.12) ensures that $y_i > 0$ for $i = 1, 2, \ldots, 2m - 1$. □

Corollary A.14 *For every $\varepsilon \in (0, 1)$ there exists a polynomial $Z_\varepsilon \in \mathbb{R}[x]$ such that*

(i) $Z_\varepsilon(x) = \sum_{j\geq 0} z_{\varepsilon,j} X_j(x)$ *with* $z_{\varepsilon,j} \geq 0$;
(ii) $Z_\varepsilon(x) \leq -1$ *for* $x \leq 2 - \varepsilon$;
(iii) $Z_\varepsilon > 0$ *for* $x > 2$.

Proof. We look for Z_ε of the form

$$Z_\varepsilon = zY_m + z'Y_{m'} \tag{A.15}$$

for suitable $m, m' \in \mathbb{N}$ and $z, z' > 0$. With this choice of the form of Z_ε, condition (i) follows from Proposition A.13. Similarly, (iii) follows from the definition of Y_m (see (A.11)) and the fact that $Y_m(x) > 0$ for $x > \alpha_m$ and, by definition, one always has $\alpha_m < 2$.

Now, if we choose m, m' in such a way that $\alpha_m, \alpha_{m'} > 2 - \varepsilon$, then, arguing as above, from (A.11) we deduce that the corresponding Z_ε in (A.15) satisfies $Z_\varepsilon(x) \leq 0$ for $x \leq 2 - \varepsilon$. If, in addition, m and m' are chosen in such a way that the numbers (cf. Lemma A.10.(iii)) $2\cos\frac{j\pi}{m+1}$, $j = 1, 2, \ldots, m$ (the roots of Y_m) and $2\cos\frac{h\pi}{m'+1}$, $h = 1, 2, \ldots, m'$ (the roots of $Y'_{m'}$) are all distinct (for instance, it suffices to take $m' = m + 1$: see Exercise A.15) then we have

$$Z_\varepsilon(x) < 0 \text{ for } x \leq 2 - \varepsilon. \tag{A.16}$$

Since $\lim_{x \to -\infty} Z_\varepsilon(x) = -\infty$ we deduce that $M = \max_{(-\infty, 2-\varepsilon]} Z_\varepsilon(x)$ is negative. Thus from (A.16) we get (ii) by replacing z and z' by $\frac{z}{-M}$ and $\frac{z'}{-M}$, respectively. $\square$

Exercise A.15 Show that, for $1 \le j \le m$ and $1 \le h \le m + 1$, we have $\frac{j}{m+1} \ne \frac{h}{m+2}$.

Hint: Write the equation $\frac{j}{m+1} = \frac{h}{m+2}$ in the form $\frac{j}{h} = 1 - \frac{1}{m+2}$.

Bibliography

[1] A. Abdollahi and A. Loghman, On one-factorizations of replacement products, *Filomat* **27** (2013), no. 1, 57–63.

[2] R.C. Agarwal and J.W. Cooley, New algorithms for digital convolution, *IEEE Trans. Acoust. Speech, Signal Processing*, ASS-**25**, 392–410.

[3] L.V. Ahlfors, *Complex analysis. An introduction to the theory of analytic functions of one complex variable.* Third edition. International Series in Pure and Applied Mathematics. McGraw-Hill Book Co., New York, 1978.

[4] S. Ahmad, Cycle structure of automorphisms of finite cyclic groups, *J. Combinatorial Theory* **6** (1969), 370–374.

[5] M. Aigner and G.M. Ziegler, *Proofs from The Book.* Fifth edition. Springer-Verlag, Berlin, 2014.

[6] M. Ajtai, J. Komlós, and E. Szemerédi, An $O(n \log n)$ sorting network. Proceedings of the 15th Annual ACM Symposium on Theory of Computing, pp. 1–9, (1983).

[7] N. Alon, Eigenvalues and expanders, *Combinatorica* **6** (1986), 83–96.

[8] N. Alon, A. Lubotzky, and A. Wigderson, Semi-direct product in groups and zig-zag product in graphs: connections and applications (extended abstract). 42nd IEEE Symposium on Foundations of Computer Science (Las Vegas, NV, 2001), 630–637, IEEE Computer Soc., Los Alamitos, CA, 2001.

[9] N. Alon and V.D. Milman, λ_1, isoperimetric inequalities for graphs, and superconcentrators, *J. Combin. Theory Ser. B* **38** (1985), no. 1, 73–88.

[10] N. Alon, O. Schwartz, and A. Shapira, An elementary construction of constant-degree expanders, *Combin. Probab. Comput.* **17** (2008), no. 3, 319–327.

[11] N. Alon and J.H. Spencer, *The probabilistic method.* Third edition. With an appendix on the life and work of Paul Erdös. Wiley-Interscience Series in Discrete Mathematics and Optimization. John Wiley & Sons, Inc., Hoboken, NJ, 2008.

[12] J.L. Alperin and R.B. Bell, *Groups and representations.* Graduate Texts in Mathematics, 162. Springer-Verlag, New York, 1995.

[13] T.M. Apostol, *Introduction to analytic number theory.* Undergraduate Texts in Mathematics. Springer-Verlag, New York-Heidelberg, 1976.

[14] L. Auslander, E. Feig, and S. Winograd, New algorithms for the multidimensional discrete Fourier transform, *IEEE Trans. Acoust. Speech, Signal, Proc.* ASSP-**31** (2) (1984), no. 1, 388–403.

[15] L. Auslander and R. Tolimieri, Is computing with the finite Fourier transform pure or applied mathematics? *Bull. Amer. Math. Soc. (N.S.)* **1** (1979), no. 6, 847–897.

[16] J. Ax, Zeroes of polynomials over finite fields, *Amer. J. Math.* **86** (1964), 255–261.

[17] L. Bartholdi and W. Woess, Spectral computations on lamplighter groups and Diestel-Leader graphs, *J. Fourier Anal. Appl.* **11** (2005), no. 2, 175–202.

[18] R. Beals, On orders of subgroups in Abelian groups: an elementary solution of an exercise of Herstein, *Amer. Math. Monthly* **116** (2009), no. 10, 923–926.

[19] M.B. Bekka, P. de la Harpe, and A. Valette, *Kazhdan's property (T)*. New Mathematical Monographs, 11. Cambridge University Press, 2008.

[20] B.C. Berndt, R.J. Evans, and K.S. Williams, *Gauss and Jacobi sums*. Canadian Mathematical Society Series of Monographs and Advanced Texts. A Wiley-Interscience Publication. John Wiley & Sons, Inc., New York, 1998.

[21] K.P. Bogart, An obvious proof of Burnside's lemma, *Amer. Math. Monthly* **98** (1991), no. 10, 927–928.

[22] W.L. Briggs and V.E. Henson, *The DFT. An owner's manual for the discrete Fourier transform*. Society for Industrial and Applied Mathematics (SIAM), Philadelphia, PA, 1995.

[23] D. Bump, *Lie groups*. Graduate Texts in Mathematics, 225. Springer-Verlag, New York, 2004.

[24] D. Bump, P. Diaconis, A. Hicks, L. Miclo, and H. Widom, An Exercise(?) in Fourier Analysis on the Heisenberg Group, *Ann. Fac. Sci. Toulouse Math.* (6) **26** (2017), no. 2, 263–288.

[25] D. Bump and D. Ginzburg, Generalized Frobenius-Schur numbers, *J. Algebra* **278** (2004), no. 1, 294–313.

[26] P. Buser, Über eine Ungleichung von Cheeger, *Math. Z.* **158** (1978), no. 3, 245–252.

[27] P. Buser, A note on the isoperimetric constant, *Ann. Sci. École Norm. Sup.* (4) **15** (1982), no. 2, 213–230.

[28] T. Ceccherini-Silberstein, F. Scarabotti, and F. Tolli, Trees, wreath products and finite Gelfand pairs, *Adv. Math.* **206** (2006), no. 2, 503–537.

[29] T. Ceccherini-Silberstein, F. Scarabotti, and F. Tolli, *Harmonic analysis on finite groups: representation theory, Gelfand pairs and Markov chains*. Cambridge Studies in Advanced Mathematics 108, Cambridge University Press, 2008.

[30] T. Ceccherini-Silberstein, A. Machì, F. Scarabotti, and F. Tolli, Induced representation and Mackey theory, *J. Math. Sci.* (New York) **156** (2009), no. 1, 11–28.

[31] T. Ceccherini-Silberstein, F. Scarabotti, and F. Tolli, Clifford theory and applications, *J. Math. Sci.* (New York) **156** (2009), no. 1, 29–43.

[32] T. Ceccherini-Silberstein, F. Scarabotti, and F. Tolli, Representation theory of wreath products of finite groups, *J. Math. Sci.* (New York) **156** (2009), no. 1, 44–55.

[33] T. Ceccherini-Silberstein, F. Scarabotti, and F. Tolli, *Representation theory of the symmetric groups: the Okounkov-Vershik approach, character formulas, and partition algebras*. Cambridge Studies in Advanced Mathematics 121, Cambridge University Press, 2010.

[34] T. Ceccherini-Silberstein, F. Scarabotti, and F. Tolli, *Representation theory and harmonic analysis of wreath products of finite groups*. London Mathematical Society Lecture Note Series 410, Cambridge University Press, 2014.

[35] T. Ceccherini-Silberstein, F. Scarabotti, and F. Tolli, Mackey's theory of τ-conjugate representations for finite groups, *Jpn. J. Math.* **10** (2015), no. 1, 43–96.

[36] T. Ceccherini-Silberstein, F. Scarabotti, and F. Tolli, Mackey's criterion for subgroup restriction of Kronecker products and harmonic analysis on Clifford groups, *Tohoku Math. J.* (2) **67** (2015), no. 4, 553–571.

[37] T. Ceccherini-Silberstein, F. Scarabotti, and F. Tolli, Harmonic analysis and spherical functions for multiplicity-free induced representations. In preparation.

[38] J. Cheeger, A lower bound for the smallest eigenvalue of the Laplacian. In Problems in analysis (Papers dedicated to Salomon Bochner, 1969), pp. 195–199. Princeton Univ. Press, 1970.

[39] C. Chevalley, Démonstration d'une hypothèse de M. Artin, *Abh. Math. Sem. Univ. Hamburg* **11** (1935), no. 1, 73–75.

[40] P. Chiu, Cubic Ramanujan graphs, *Combinatorica* **12** (1992), no. 3, 275–285.

[41] J.W. Cooley and J.W. Tukey, An algorithm for the machine calculation of complex Fourier series, *Math. Comp.* **19** (1965), 297–301.

[42] Ch.W. Curtis and I. Reiner, *Representation theory of finite groups and associative algebras.* Reprint of the 1962 original. Wiley Classics Library. A Wiley-Interscience Publication. John Wiley & Sons, Inc., New York, 1988.

[43] Ch.W. Curtis and I. Reiner, *Methods of representation theory. With applications to finite groups and orders.* Voll. I and II. Pure and Applied Mathematics (New York). A Wiley-Interscience Publication. John Wiley & Sons, Inc., New York, 1981 and 1987.

[44] D. D'Angeli and A. Donno, Crested products of Markov chains, *Ann. Appl. Probab.* **19** (2009), no. 1, 414–453.

[45] D. D'Angeli and A. Donno, Wreath product of matrices, *Linear Algebra Appl.* **513** (2017), 276–303.

[46] D. D'Angeli and A. Donno, Shuffling matrices, Kronecker product and Discrete Fourier Transform, *Discrete Appl. Math.* **233** (2017), 1–18.

[47] H. Davenport, *The higher arithmetic. An introduction to the theory of numbers.* Eighth edition. With editing and additional material by James H. Davenport. Cambridge University Press, Cambridge, 2008.

[48] H. Davenport and H. Hasse, Die Nullstellen der Kongruenzzetafunktionen in gewissen zyklischen Fällen, *J. Reine Angew. Math.* **172** (1935), 151–182.

[49] G. Davidoff, P. Sarnak, and A. Valette, *Elementary number theory, group theory, and Ramanujan graphs.* London Mathematical Society Student Texts, 55. Cambridge University Press, Cambridge, 2003.

[50] M. Davio, Kronecker products and shuffle algebra, *IEEE Trans. Comput.* **30** (1981), no. 2, 116–125.

[51] Ph.J. Davis, *Circulant matrices*, Pure and Applied Mathematics. John Wiley & Sons, New York-Chichester-Brisbane, 1979.

[52] P. Deligne, La conjecture de Weil. I, *Inst. Hautes Études Sci. Publ. Math.* **43** (1974), 273–307.

[53] P. Diaconis, *Group representations in probability and statistics.* Institute of Mathematical Statistics Lecture Notes—Monograph Series, 11. Institute of Mathematical Statistics, Hayward, CA, 1988.

[54] P. Diaconis and D. Rockmore, Efficient computation of the Fourier transform on finite groups, *J. Amer. Math. Soc.* **3** (1990), no. 2, 297–332.

[55] B.W. Dickinson and K. Steiglitz, Eigenvectors and functions of the discrete Fourier transform, *IEEE Trans. Acoust. Speech Signal Process.* **30** (1982), no. 1, 25–31.

[56] I. Dinur, The PCP theorem by gap amplification, *Journal of the ACM*, **54** (2007) No. 3, Art. 12, 44 p.

[57] J. Dodziuk, Difference equations, isoperimetric inequality and transience of certain random walks, *Trans. Amer. Math. Soc.* **284** (1984), no. 2, 787–794.

[58] A. Donno, Replacement and zig-zag products, Cayley graphs and Lamplighter random walk, *Int. J. Group Theory* **2** (2013), no. 1, 11–35.

[59] A. Donno, Generalized wreath products of graphs and groups, *Graphs Combin.* **31** (2015), no. 4, 915–926.

[60] P. Erdős, Über die Reihe $\sum \frac{1}{p}$, *Mathematica, Zutphen B* **7** (1938), 1–2.

[61] W. Feller, *An introduction to probability theory and its applications.* Vol. II. Second edition John Wiley & Sons, Inc., New York-London-Sydney 1971.

[62] G.B. Folland, *Harmonic analysis in phase space.* Annals of Mathematics Studies, 122. Princeton University Press, 1989.

[63] W. Fulton and J. Harris, *Representation Theory. A first course.* Springer-Verlag, New York, 1991.

[64] O. Gabber and Z. Galil, Explicit constructions of linear-sized superconcentrators. Special issued dedicated to Michael Machtey. *J. Comput. System Sci.* **22** (1981), no. 3, 407–420.

[65] C. Godsil and G. Royle, *Algebraic graph theory.* Graduate Texts in Mathematics, 207. Springer-Verlag, New York, 2001.

[66] I.J. Good, The interaction algorithm and practical Fourier analysis, *J. Roy. Statist. Soc. Ser. B* **20** (1958), 361–372.

[67] B. Green and T. Tao, The primes contain arbitrarily long arithmetic progressions, *Ann. of Math.* **167** (2008), no. 2, 481–547.

[68] R.E. Greenwood and A.M. Gleason, Combinatorial relations and chromatic graphs, *Canad. J. Math.* **7** (1955), 1–7.

[69] R.I. Grigorchuk, P.-H. Leemann, and T. Nagnibeda, Lamplighter groups, de Brujin graphs, spider-web graphs and their spectra, *J. Phys. A* **49** (2016), no. 20, 205004, 35 p.

[70] R.I. Grigorchuk and A. Żuk, The lamplighter group as a group generated by a 2-state automaton, and its spectrum, *Geom. Dedicata* **87** (2001), no. 1–3, 209–244.

[71] I.N. Herstein, *Topics in algebra.* Second edition. Xerox College Publishing, Lexington, Mass.-Toronto, Ont., 1975.

[72] Ch.J. Hillar and D.L. Rhea, Automorphisms of finite Abelian groups, *Amer. Math. Monthly* **114** (2007), no. 10, 917–923.

[73] D.A. Holton and J. Sheehan, *The Petersen graph.* Australian Mathematical Society Lecture Series, 7. Cambridge University Press, Cambridge, 1993.

[74] Sh. Hoory, N. Linial, and A. Wigderson, Expander graphs and their applications, *Bull. Amer. Math. Soc. (N.S.)* **43** (2006), no. 4, 439–561.

[75] R.A. Horn and R.Ch. Johnson, *Matrix analysis.* Second edition. Cambridge University Press, Cambridge, 2013.

[76] R. Howe, On the role of the Heisenberg group in harmonic analysis, *Bull. Amer. Math. Soc. (N.S.)* **3** (1980), no. 2, 821–843.

[77] L.K. Hua and H.S. Vandiver, Characters over certain types of rings with applications to the theory of equations in a finite field, *Proc. Nat. Acad. Sci. U.S.A.* **35** (1949), 94–99.

[78] B. Huppert, *Character theory of finite groups.* De Gruyter Expositions in Mathematics, **25**, Walter de Gruyter, 1998.

[79] K. Ireland and M. Rosen, *A classical introduction to modern number theory.* Second edition. Graduate Texts in Mathematics, 84. Springer-Verlag, New York, 1990.

[80] I.M. Isaacs, *Character theory of finite groups.* Corrected reprint of the 1976 original [Academic Press, New York]. Dover Publications, Inc., New York, 1994.

[81] H. Iwaniec and E. Kowalski, *Analytic number theory.* American Mathematical Society Colloquium Publications, 53. American Mathematical Society, Providence, RI, 2004.

[82] G.D. James and A. Kerber, *The representation theory of the symmetric group.* Encyclopedia of Mathematics and its Applications, **16**, Addison-Wesley, Reading, MA, 1981.

[83] S. Jimbo and A. Maruoka, Expanders obtained from affine transformations, *Combinatorica* **7** (1987), no. 4, 343–355.

[84] S. Karlin and H.M. Taylor, *An introduction to stochastic modeling.* Third edition. Academic Press, Inc., San Diego, CA, 1998.

[85] Y. Katznelson, *An introduction to harmonic analysis.* Third edition. Cambridge Mathematical Library. Cambridge University Press, Cambridge, 2004.

[86] S.P. Khekalo, The Bessel function over finite fields, *Integral Transforms Spec. Funct.* **16** (2005), no. 3, 241–253.

[87] A.W. Knapp, *Basic algebra.* Cornerstones. Birkhäuser Boston, Inc., Boston, MA, 2006.

[88] A.W. Knapp, *Advanced algebra.* Cornerstones. Birkhäuser Boston, Inc., Boston, MA, 2007.

[89] A. Kurosh, *Higher algebra*. Translated from the Russian by George Yankovsky. Reprint of the 1972 translation. "Mir," Moscow, 1988.

[90] H. Kurzweil and B. Stellmacher, *The theory of finite groups. An introduction*. Translated from the 1998 German original. Universitext. Springer-Verlag, New York, 2004.

[91] P. Lancaster and M. Tismenetsky, *The theory of matrices*. Second edition. Computer Science and Applied Mathematics. Academic Press, Inc., Orlando, FL, 1985.

[92] S. Lang, $SL_2(R)$. Reprint of the 1975 edition. Graduate Texts in Mathematics, 105. Springer-Verlag, New York, 1985.

[93] S. Lang, *Algebra*. Revised third edition. Graduate Texts in Mathematics, 211. Springer-Verlag, New York, 2002.

[94] F. Lehner, M. Neuhauser, and W. Woess, On the spectrum of lamplighter groups and percolation clusters, *Math. Ann.* **342** (2008), no. 1, 69–89.

[95] W.C.W. Li, *Number theory with applications*. Series on University Mathematics, 7. World Scientific Publishing Co., Inc., River Edge, NJ, 1996.

[96] R. Lidl and H. Niederreiter, *Finite fields*. With a foreword by P.M. Cohn. Second edition. Encyclopedia of Mathematics and its Applications, 20. Cambridge University Press, Cambridge, 1997.

[97] J.H. van Lint and R.M. Wilson, *A course in combinatorics*. Second edition. Cambridge University Press, Cambridge, 2001.

[98] L.H. Loomis, *An introduction to abstract harmonic analysis*. D. Van Nostrand Company, Inc., Toronto-New York-London, 1953.

[99] A. Lubotzky, *Discrete groups, expanding graphs and invariant measures*. With an appendix by Jonathan D. Rogawski. Progress in Mathematics, 125. BirkhÄuser Verlag, Basel, 1994.

[100] A. Lubotzky, Expander graphs in pure and applied mathematics, *Bull. Amer. Math. Soc. (N.S.)* **49** (2012), no. 1, 113–162.

[101] A. Lubotzky, R. Phillips, and P. Sarnak, Ramanujan graphs, *Combinatorica* **8** (1988), no. 3, 261–277.

[102] A. Machì, *Teoria dei gruppi*. Milano, Feltrinelli, 1974.

[103] A. Machì, *Groups. An introduction to ideas and methods of the theory of groups*. Unitext, 58. Springer, Milan, 2012.

[104] J.H. MacClellan and T.W. Parks, Eigenvalue and eigenvector decomposition of the discrete Fourier transform, *IEEE Trans. Audio Electroacoust.* AU-**20** (1972), no. 1, 66–74.

[105] I.G. Macdonald, *Symmetric functions and Hall polynomials*. Second edition. With contributions by A. Zelevinsky. Oxford Mathematical Monographs. Oxford Science Publications. The Clarendon Press, Oxford University Press, New York, 1995.

[106] J.H. MacKay, Another proof of Cauchy's group theorem, *Amer. Math. Monthly* **66** (1959), 119.

[107] G.W. Mackey, Unitary representations of group extensions. I, *Acta Math.* **99** (1958), 265–311.

[108] G.W. Mackey, *Unitary group representations in physics, probability, and number theory*, Second edition. Advanced Book Classics. Addison-Wesley Publishing Company, Advanced Book Program, Redwood City, CA, 1989.

[109] A.W. Marcus, D.A. Spielman, and N. Srivastava, Interlacing families I: Bipartite Ramanujan graphs of all degrees, *Ann. of Math.* (2) **182** (2015), no. 1, 307–325.

[110] A.W. Marcus, D.A. Spielman, and N. Srivastava, Interlacing families IV: Bipartite Ramanujan graphs of all sizes. 2015 IEEE 56th Annual Symposium on Foundations of Computer Science–FOCS 2015, 1358–1377, IEEE Computer Soc., Los Alamitos, CA, 2015.

[111] G.A. Margulis, Explicit constructions of expanders, *Problemy Peredachi Informatsii* **9** (1973), no. 4, 71–80.

[112] G.A. Margulis, Explicit group-theoretic constructions of combinatorial schemes and their applications in the construction of expanders and concentrators, *Problemy Peredachi Informatsii* **24** (1988), no. 1, 51–60.

[113] S. Mac Lane and G. Birkhoff, *Algebra*. Third edition. Chelsea Publishing Co., New York, 1988.

[114] A.I. Mal'cev, *Foundations of linear algebra*. Translated from the Russian by Thomas Craig Brown. San Francisco, Calif.-London, 1963.

[115] A.I. Markushevich, *The theory of analytic functions: a brief course*. Translated from the Russian by Eugene Yankovsky. "Mir," Moscow, 1983.

[116] M. Morgenstern, Ramanujan diagrams, *SIAM J. Discrete Math.* **7** (1994), no. 4, 560–570.

[117] T. Nagell, *Introduction to number theory*. Second edition. Chelsea Publishing Co., New York 1964.

[118] M. Nathanson, *Elementary methods in number theory*. Graduate Texts in Mathematics, Vol. 195, Springer-Verlag, New York, 2000.

[119] M.A. Naimark and A.I. Stern, *Theory of group representations*. Springer-Verlag, New York, 1982.

[120] G. Navarro, On the fundamental theorem of finite abelian groups, *Amer. Math. Monthly* **110** (2003), no. 2, 153–154.

[121] P.M. Neumann, A lemma that is not Burnside's, *Math. Sci.* **4** (1979), no. 2, 133–141.

[122] A. Nilli, Tight estimates for eigenvalues of regular graphs, *Electron. J. Combin.* **11** (2004), no. 1, Note 9, 4 p.

[123] I. Piatetski-Shapiro, *Complex representations of* GL(2, *K*) *for finite fields K*. Contemporary Mathematics, 16. American Mathematical Society, Providence, R.I., 1983.

[124] C. Procesi, *Lie groups. An approach through invariants and representations*. Universitext. Springer, New York, 2007.

[125] Ch.M. Rader, Discrete Fourier transforms when the number of data samples is prime, *Proc. IEEE* **56** (1968), 1107–1108.

[126] O. Reingold, Undirected connectivity in log-space, *Journal of the ACM*, **55** (2008), no. 4, Art. 17, 24 p.

[127] O. Reingold, L. Trevisan, and S. Vadhan, Pseudorandom walks on regular digraphs and the RL vs. L problem. STOC'06: Proceedings of the 38th Annual ACM Symposium on Theory of Computing, 457–466, ACM, New York, 2006.

[128] O. Reingold, S. Vadhan, and A. Wigderson, Entropy waves, the zig-zag graph product, and new constant-degree expanders, *Ann. of Math.* (2) **155** (2002), no. 1, 157–187.

[129] D.J.S. Robinson, *A course in the theory of groups*. Second edition. Graduate Texts in Mathematics, 80. Springer-Verlag, New York, 1996.

[130] D.J. Rose, Matrix identities of the fast Fourier transform, *Linear Algebra Appl.* **29** (1980), 423–443.

[131] K.F. Roth, On certain sets of integers, *J. London Math. Soc.* **28** (1953), 104–109.

[132] J.J. Rotman, *An introduction to the theory of groups*. Fourth edition. Graduate Texts in Mathematics, 148. Springer-Verlag, New York, 1995.

[133] W. Rudin, *Real and complex analysis*. Third edition. McGraw-Hill Book Co., New York, 1987.

[134] W. Rudin, *Fourier analysis on groups*. Reprint of the 1962 original. Wiley Classics Library. A Wiley-Interscience Publication. John Wiley & Sons, Inc., New York, 1990.

[135] P. Sarnak, *Some applications of modular forms*. Cambridge Tracts in Mathematics, 99, Cambridge University Press, Cambridge, 1990.

[136] F. Scarabotti and F. Tolli, Harmonic analysis of finite lamplighter random walks, *J. Dyn. Control Syst.* **14** (2008), no. 2, 251–282.

[137] F. Scarabotti and F. Tolli, Harmonic analysis on a finite homogeneous space, *Proc. Lond. Math. Soc.* (3) **100** (2010), no. 2, 348–376.

[138] F. Scarabotti and F. Tolli, Harmonic analysis on a finite homogeneous space II: the Gelfand-Tsetlin decomposition, *Forum Math.* **22** (2010), no. 5, 879–911.

[139] F. Scarabotti and F. Tolli, Hecke algebras and harmonic analysis on finite groups, *Rend. Mat. Appl.* (7) **33** (2013), no. 1–2, 27–51.

[140] F. Scarabotti and F. Tolli, Induced representations and harmonic analysis on finite groups, *Monatsh. Math.* **181** (2016), no. 4, 937–965.

[141] M. Scafati and G. Tallini, *Geometria di Galois e teoria dei codici.* Ed. CISU 1995.

[142] J. Schulte, Harmonic analysis on finite Heisenberg groups, *European J. Combin.* **25** (2004), no. 3, 327–338.

[143] A. Selberg, An elementary proof of Dirichlet's theorem about primes in an arithmetic progression, *Ann. of Math.* (2) **50** (1949), 297–304.

[144] J.P. Serre, *A course in arithmetic.* Translated from the French. Graduate Texts in Mathematics, No. 7. Springer-Verlag, New York-Heidelberg, 1973.

[145] J.P. Serre, *Linear representations of finite groups.* Graduate Texts in Mathematics, Vol. 42. Springer-Verlag, New York-Heidelberg, 1977.

[146] J.P. Serre, *Répartition asymptotique des valeurs propres de l'opérateur de Hecke* T_p, *J. Amer. Math. Soc.* **10** (1997), no. 1, 75–102.

[147] R. Shaw, *Linear algebra and group representations. Vol. II. Multilinear algebra and group representations.* Academic Press, Inc. [Harcourt Brace Jovanovich, Publishers], London-New York, 1983.

[148] B. Simon, *Representations of finite and compact groups.* American Math. Soc., 1996.

[149] R.P. Stanley, *Enumerative combinatorics, Vol.1.* Cambridge University Press, 1997.

[150] E.M. Stein and R. Shakarchi, *Fourier analysis. An introduction.* Princeton Lectures in Analysis, 1. Princeton University Press, 2003.

[151] E.M. Stein and R. Shakarchi, *Complex analysis.* Princeton Lectures in Analysis, 2. Princeton University Press, 2003.

[152] J.R. Stembridge, On Schur's Q-functions and the primitive idempotents of a commutative Hecke algebra, *J. Algebraic Combin.* **1** (1992), no. 1, 71–95.

[153] C. Stephanos, Sur une extension du calcul des substitutions linéaires, *Journal de Mathmatiques Pures et Appliquées* V, **6** (1900), 73–128.

[154] S. Sternberg, *Group theory and physics.* Cambridge University Press, Cambridge, 1994.

[155] E. Szemerédi, On sets of integers containing no four elements in arithmetic progression, *Acta Math. Acad. Sci. Hungar.* **20** (1969), 89–104.

[156] E. Szemerédi, On sets of integers containing no k elements in arithmetic progression, *Acta Arith.* **27** (1975), 199–245.

[157] T. Tao, An uncertainty principle for cyclic groups of prime order, *Math. Res. Lett.* **12** (2005), no. 1, 121–127.

[158] T. Tao, The ergodic and combinatorial approaches to Szemerédi's theorem. Additive combinatorics, 145–193, CRM Proc. Lecture Notes, 43, Amer. Math. Soc., Providence, RI, 2007.

[159] A. Terras, *Fourier analysis on finite groups and applications.* London Mathematical Society Student Texts, 43. Cambridge University Press, Cambridge, 1999.

[160] R. Tolimieri, M. An, and C. Lu, *Mathematics of multidimensional Fourier transform algorithms.* Second edition. Signal Processing and Digital Filtering. Springer-Verlag, New York, 1997.

[161] A. Valette, Graphes de Ramanujan et applications. Séminaire Bourbaki, Vol. 1996/97. *Astérisque* No. **245** (1997), Exp. No. 829, 4, 247–276.

[162] E. Vallejo, A diagrammatic approach to Kronecker squares, *J. Combin. Theory Ser. A* **127** (2014), 243–285.

[163] Ch.F. Van Loan, *Computational frameworks for the fast Fourier transform.* Frontiers in Applied Mathematics, 10. Society for Industrial and Applied Mathematics (SIAM), Philadelphia, PA, 1992.

[164] E. Warning, Bemerkung zur vorstehenden Arbeit, *Abh. Math. Sem. Univ. Hamburg* **11** (1935), no. 1, 76–83.

[165] A. Weil, Numbers of solutions of equations in finite fields, *Bull. Amer. Math. Soc.* **55** (1949), 497–508.

[166] H. Weyl, *Algebraic theory of numbers.* Annals of Mathematics Studies, no. 1. Princeton University Press, 1940.

[167] H. Wielandt, *Finite permutation groups.* Academic Press, New York-London, 1964.

[168] S. Winograd, On computing the discrete Fourier transform, *Math. Comp.* **32** (1978), no. 141, 175–199.

[169] E.M. Wright, Burnside's lemma: a historical note, *J. Combin. Theory Ser. B* **30** (1981), no. 1, 89–90.

[170] G. Zappa, *Fondamenti di teoria dei gruppi, Vol. I.* Consiglio Nazionale delle Ricerche Monografie Matematiche, 13 Edizioni Cremonese, Rome, 1965.

Index

2-regular segment, 252
GL(2, $\mathbb{F}$), 482
  center of —, 486
  conjugacy classes of —, 482
  spherical function for —, 515
GL(2, $\mathbb{F}_q$), 488
  Borel subgroup of —, 488
  character table for —, 529
  conjugacy classes of —, 489
  cuspidal representation of —, 502
  decomposition of tensor products of
    representations of —, 540
  Gelfand-Graev character for —, 527
  induced representations from — to
    GL(2, $\mathbb{F}_{q^m}$), 533
  one-dimensional representations of —, 498
  order of —, 488
  parabolic induction for —, 494
  representation theory of the Borel subgroup
    of —, 492
  Whittaker model for —, 513
p-group, 22
p-primary group, 22

Abel formula of summation by parts, 77
Abelian
  — algebra, 55, 362
  automorphism of a finite — group, 27
  Cauchy theorem for — groups, 20
  character of an — group, 50
  convolution on the group algebra of an —
    group, 54
  dual of an — group, 50
  endomorphism of a finite — group, 26
  Fourier transform on an — group, 53
  invariant factors decomposition of a finite
    — group, 18
  primary component of an — group, 22

primary decomposition of a finite — group,
  21, 22
action
  — of a finite group on a finite set, 372
  diagonal —, 377
  doubly transitive —, 379
  transitive —, 372
adapted basis, 397
additive character of $\mathbb{F}_q$, 197
  principal —, 198
adjacency
  — matrix, 238
  — operator, 238
adjacent vertex, 236
adjoint
  — in $L(G)$, 363
  — operator, 345
  — representation, 380
adjugate matrix, 36
affine group
  — over $\mathbb{F}_q$, 374, 426
  — over $\mathbb{Z}/n\mathbb{Z}$, 432
  — over a field, 487
algebra, 55, 361
  *- —, 362
  — *-anti-homomorphism, 363
  — *-anti-isomorphism, 363
  — *-homomorphism, 362
  — *-isomorphism, 363
  Abelian —, 55, 362
  anti-automorphism of an —, 467
  center of an —, 362
  commutative —, 55, 362
  convolution on the group — of an Abelian
    group, 54
  group —, 363
  Hecke —, 462
  involutive —, 362

algebra (*cont.*)
  involutive anti-automorphism of an —, 467
  sub- —, 361
  unital —, 55, 362
algebraic
  — element, 171
  — extension, 173
  — number, 65
algorithm
  Cooley-Tukey —, 129, 161, 162
  decimation in frequency of the
    Cooley-Tukey —, 162
  decimation in time form of the
    Cooley-Tukey —, 162
  Diaconis and Rockmore —, 397
  parallel form of the Cooley-Tukey —, 162
  Rader —, 159
  Rader-Winograd —, 158
  vector form of the Cooley-Tukey —, 162
Alon-Boppana theorem, 299
Alon-Boppana-Serre theorem, 298
  Nilli's proof, 305
Alon-Milman theorem, 287
Alon-Schwartz-Shapira theorem, 320
ambivalent group, 468
anti-automorphism of a group, 466
  involutive —, 467
anti-automorphism of an algebra, 467
  involutive —, 467
Auslander-Feigh-Winograd theorem, 230
automorphism of a finite Abelian group, 27

Bézout identity, 4
  generalized —, 5
Bessel
  — function for GL(2, $\mathbb{F}_q$), 515
  — vector, 515
bicolorable graph, 246
bidual of a group, 52
bipartite graph, 245
  complete —, 247
  partite sets of a —, 245
block diagonal power of a matrix, 148
Borel subgroup
  — of GL(2, $\mathbb{F}$), 486
  — of GL(2, $\mathbb{F}_q$), 488
  representation theory of the — of
    GL(2, $\mathbb{F}_q$), 492
boundary of a set of vertices in a graph, 284
Bruhat decomposition, 487
Bump-Ginzburg criterion, 468
Burnside lemma, 376

canonical form of a matrix, 482
  Jordan —, 485
  rational —, 485
Cartan subgroup, 488

Cartesian product of graphs, 258
Cauchy
  — theorem for (not necessarily Abelian)
    groups, 21
  — theorem for Abelian groups, 20
Cayley graph, 280
Cayley-Hamilton Theorem, 484
center
  — of GL(2, $\mathbb{F}$), 486
  — of a group, 431
  — of an algebra, 362
central function, 364
centralizer subgroup, 487
character
  — of $\mathbb{Z}_n$, 49
  — of a representation, 355
  — of an Abelian group, 50
  — table for GL(2, $\mathbb{F}_q$), 529
  additive — of $\mathbb{F}_q$, 197
  conjugate —, 202
  decomposable —, 201
  Dirichlet —, 84
  dual orthogonality relations for —s of $\mathbb{Z}_n$,
    50
  dual orthogonality relations for —s of a
    group, 370
  dual orthogonality relations for —s of an
    Abelian group, 52
  exceptional —, 523
  fixed point — formula, 375
  Fourier transform of a —, 382
  Frobenius — formula, 405
  Gelfand-Graev —, 527
  indecomposable —, 201
  multiplicative — of $\mathbb{F}_q$, 199
  multiplicative — of $\mathbb{Z}/m\mathbb{Z}$, 84
  permutation —, 375
  principal Dirichlet —, 85
  real Dirichlet —, 87
characteristic
  — function, 47
  — of a field, 171
  — polynomial of $\mathcal{F}$, 116
  — polynomial of $\mathcal{F}^2$, 103
  — polynomial of a matrix, 484
  — subgroup, 431
Chebotarëv theorem, 68
Chebyshëv polynomials of the second kind,
    547
  modified —, 548, 549
Cheeger constant, 284
Chevalley theorem, 227
Chinese remainder
  — map, 138
  — theorem, 9, 13
circulant matrix, 58
  elementary permutation —, 147

class function, 364
Clebsch graph, 243
closed path, 237
coefficient
  (matrix) — of a representation, 351
  Gamma —, 525
coloring
  — of a graph, 243
  — of an edge, 243
combinatorial Laplacian, 286
commutant
  — of one representation, 349, 390
  — of two representations, 349
commutative algebra, 55, 362
companion matrix of a monic polynomial,
  191
complement of a graph, 242
complete graph, 247, 271, 292
  lamplighter on the —, 270
composite bijection permutation, 136
composition of paths, 237
congruence permutation
  elementary —, 135
  product —, 135
conjugate
  — character, 202
  — of an element in $\mathbb{F}_{q^2}$, 196
  — representation, 380
conjugation homomorphism, 280
connected
  — components of a graph, 237
  — graph, 237
convolution, 363
  — formula for the spherical Fourier
    transform, 478
  — operator, 365
  — on $L(A)$, 56
  — on the group algebra of an Abelian
    group, 54
Cooley-Tukey algorithm, 129, 161, 162
  decimation in frequency of the —, 162
  decimation in time form of the —, 162
  parallel form of the —, 162
  vector form of the —, 162
core matrix, 159, 228
Courant-Fischer min-max formula, 304
Curtis and Fossum basis, 464
cuspidal representation, 502
cycle
  — in a graph, 237
  discrete — graph, 250
cyclic group, 48
  endomorphism of a finite —, 30

decomposable character, 201
decomposition
  — of a representation, 344

invariant factors — of a finite Abelian
  group, 18
  primary — of a finite Abelian group, 21, 22
degree
  — of a field extension, 171
  — of a polynomial, 168
  — of a representation, 344
  — of a regular graph, 236
  — of a vertex, 236
derived subgroup, 431, 486
Diaconis and Rockmore, 397, 398
diagonal
  action, 377
  — matrix of twiddle factors, 153
  — operator, 361
  block — power of a matrix, 148
diameter of a finite graph, 237
differential operator, 68
dihedral group, 359
dimension of a representation, 344
Dirac function, 46
direct sum of representations, 344
directed graph, 236
Dirichlet
  — L-function, 89
  — character, 84
  — double summation method, 87
  — form, 286
  — formula, 89
  — series, 77
  — theorem $L(1, \chi) \neq 0$, 95
  — theorem on primes in arithmetic
    progressions, 99
  principal — character, 85
  real — character, 87
discrete
  — circle, 250
  — cycle graph, 250
  — Fourier transform (DFT), 53, 59
  — Fourier transform (DFT) revisited, 443,
    445
  Gauss-Schur theorem on the trace of the —
    Fourier transform (DFT), 116
distance
  geodesic — in a graph, 237
  Hamming —, 248, 264
Dodziuk theorem, 288
domain
  integral —, 167
  principal ideal —, 168
  unique factorization — (UFD), 169
doubly transitive action, 379
dual
  — group of $\mathbb{F}_q$, 197
  — group of $\mathbb{F}_q^*$, 199
  — group of an Abelian group, 50
  — of a finite dimensional vector space, 380

dual (*cont.*)
— of a finite group, 347
— orthogonality relations for characters of $\mathbb{Z}_n$, 50
— orthogonality relations for characters of a group, 370
— orthogonality relations for characters of an Abelian group, 52

edge
— coloring, 243
— of a graph, 235
multiple —, 235
oriented — of a graph, 236
eigenidentities, 152
tensor form of the —, 153
Eisenstein criterion, 63
element
algebraic —, 171
primitive — of a Galois field, 176
elementary congruence permutation, 135
endomorphism
— of a finite Abelian group, 26
— of a finite cyclic group, 30
equivalent representations, 344
Erdős' proof of Euler theorem
$\sum_{p\ prime} \frac{1}{p} = +\infty$, 98
Euclid's proof of the infinitude of primes, 6
Euclidean algorithm, 5
Euler
— identity, 31
— product formula, 82
— theorem $\sum_{p\ prime} \frac{1}{p} = +\infty$, 97
— theorem $\sum_{p\ prime} \frac{1}{p} = +\infty$ (Erdős' proof), 98
— totient function, 7
Euler-Mascheroni constant, 82
exceptional character, 523
expander, 309, 310
— via zig-zag products, 338
Margulis —, 319
exponential set, 257
extension, 171
algebraic —, 173
degree of a field —, 171
finite —, 171
Galois group of an —, 174
infinite —, 171
norm of a field —, 187
quadratic —, 173
trace of a field —, 187

faithful representation, 344
fast Fourier transform (FFT), 129
— over a noncommutative group, 397
— revisited, 447, 455

algorithmic aspects of the —, 161
matrix form of the —, 151
Fermat
— identity, 31
— little theorem, 9
field, 168
— extension, 171
Galois —, 178, 181
primitive element of a Galois —, 176
splitting — of a polynomial, 174
sub—, 171
finite
— extension, 171
— graph, 236
fixed point character formula, 375
formula
Abel — of summation by parts, 77
Courant-Fischer min-max —, 304
Dirichlet —, 89
Euler product —, 82
Frobenius character —, 405
Gauss —, 116
Mackey — for invariants, 414, 417
Parseval — for $\mathbb{Z}_n^2$, 312
Parseval — for an Abelian group, 54
Plancherel — for $\mathbb{Z}_n^2$, 312
Plancherel — for a finite group, 371
Plancherel — for an Abelian group, 54
Plancherel — for the spherical Fourier transform, 478
Poisson summation —s, 60
Fourier
— transform, 367
— coefficient, 53
— inversion formula, 368
— inversion formula for an Abelian group, 53
— transform of a character, 382
— transform on an Abelian group, 53
— matrix of $\mathbb{F}_q$, 227
convolution formula for the spherical — transform, 478
discrete — transform (DFT), 53, 59
discrete — transform (DFT) revisited, 443, 445
fast — transform (FFT), 129
fast — transform (FFT) revisited, 447, 455
Gauss-Schur theorem on the trace of the discrete — transform (DFT), 116
inverse — transform, 370
inversion formula for the spherical — transform, 477
normalized — transform, 53
Plancherel formula for the spherical — transform, 478
spherical — transform, 477

Frobenius
— automorphism, 177
— character formula, 405
— reciprocity law, 409
— reciprocity law (other side), 411
— reciprocity law for one-dimensional
  representations, 412
function
  Bessel — for GL(2, $\mathbb{F}_q$), 515
  central —, 364
  characteristic —, 47
  class —, 364
  Dirac —, 46
  Dirichlet L—, 89
  Euler totient —, 7
  inflation of a —, 59
  Riemann zeta —, 82
  spherical —, 469
  spherical — for GL(2, $\mathbb{F}_q$), 515
fundamental theorem of arithmetic, 5

Galois
— field, 178, 181
— group of an extension, 174
Gamma coefficient, 525
Gauss
— formula, 116
— law of quadratic reciprocity, 127
— law of quadratic reciprocity (second
  proof), 183
— sum, 126, 210
— theorem on cyclicity of $\mathcal{U}(\mathbb{Z}/n\mathbb{Z})$, 35
— totient function theorem, 8
— lemma, 64
Gauss-Schur theorem on the trace of the DFT,
  116
Gelfand pair, 466
  symmetric —, 468
  weakly symmetric —, 468
Gelfand-Graev character, 527
general radix identity, 154
generalized quaternion group, 360
generalized Winograd's method, 157
geodesic distance in a graph, 237
Good's method, 158
graph
  d-edge-colorable —, 275
  — edge, 235, 236
  — isomorphism, 237
  — multiple edge, 235
  — vertex, 235, 236
  primitive —, 242
  bicolorable —, 246
  bipartite —, 245
  boundary of a set of vertices in a —, 284
  Cartesian product of —s, 258
  Cayley —, 280

Cheeger constant of a —, 284
Clebsch —, 243
complement of a —, 242
complete —, 247, 292
complete bipartite —, 247
connected —, 237
connected component of a —, 237
degree of a regular —, 236
diameter of a finite —, 237
directed —, 236
directed — isomorphism, 237
discrete cycle —, 250
expander —, 309, 310
finite —, 236
geodesic distance in a —, 237
Hamming —, 264
isoperimetric constant of a —, 284
lamplighter —, 268
lexicographic product of —s, 260
Margulis —, 319
non-oriented square of a —, 338
Paley —, 308
partite sets of a bipartite —, 245
Petersen —, 243
Ramanujan —, 307
regular —, 236
replacement product of —s, 275
simple —, 235
spectral gap of a —, 292
spectrum of a —, 238
strongly regular —, 241
subgraph of a —, 236
tensor product of —s, 259
triangular —, 242
undirected —, 235
wreath product of —s, 267
zig-zag product of —s, 277
greatest common divisor, 5
Green-Tao theorem, 100
group
  p- —, 22
  p-primary —, 22
  — GL(2, $\mathbb{F}_q$), 488
  — GL(h, $\mathbb{F}_p$), 40
  — algebra, 363
  — of units of a unital ring, 28
  affine — over $\mathbb{F}_q$, 374, 426
  affine — over $\mathbb{Z}/n\mathbb{Z}$, 432
  affine — over a field, 487
  ambivalent —, 468
  anti-automorphism of a —, 466
  bidual of a —, 52
  center of a —, 431
  characteristic subgroup of a —, 431
  cyclic —, 48
  derived subgroup of a —, 431
  dihedral —, 359

group (*cont.*)
dual — of an Abelian group, 50
Galois — of an extension, 174
generalized quaternion —, 360
Heisenberg — over $\mathbb{F}_q$, 457
Heisenberg — over $\mathbb{Z}/n\mathbb{Z}$, 437
inertia —, 421
involutive anti-automorphism of a —, 467
solvable —, 486
symmetric —, 349

Hamming
— distance, 248, 264
— graph, 264
Hankel matrix, 159
Hasse-Davenport identity, 216
Hecke
— algebra, 462
— operator, 295
— relations, 296
commutative — algebra, 466
Curtis and Fossum basis of a — algebra, 464
multiplicative linear functional on a —
algebra, 472
structure constants of a — algebra, 465
Heisenberg
— group over $\mathbb{F}_q$, 457
— group over $\mathbb{Z}/n\mathbb{Z}$, 437
Hilbert Satz 90, 187, 188
Hilbert-Schmidt inner product, 395
homogenous space, 372
homomorphism
conjugation —, 280
Hua-Vandiver-Weil theorem
— (homogeneous case), 224
— (non-homogeneous case), 225
hypercube, 248
weight of a vertex of the —, 249

ideal
— of a commutative ring, 167
maximal —, 170
principal —, 168
principal — domain, 168
idempotent, 390
identity
Bézout —, 4
eigen—, 152
Euler —, 31
Fermat —, 31
general radix —, 154
generalized Bézout —, 5
Hasse-Davenport —, 216
permutational reverse radix —, 138
reverse radix —, 149
similarity —, 158
twiddle free —, 157

twiddle —, 155
indecomposable character, 201
induced representation, 399
— and direct sums, 408
— and tensor products, 406
— from GL(2, $\mathbb{F}_q$) to GL(2, $\mathbb{F}_{q^m}$), 533
— of a one-dimensional representation,
403
character of an —, 404, 405
matrix coefficients of an —, 404
transitivity of —, 401
inertia group, 421
infinite
— extension, 171
— product, 76
converging — product, 76
diverging — product, 76
inflation
— of a function, 59
— of a representation, 421, 495
initial vertex of an oriented edge, 236
inner product, 345
integral domain, 167
intertwiner, 349
invariant
— factors decomposition of a finite Abelian
group, 18
— operator, 56
— subspace, 344
— vector, 344
subspace of — vectors, 344
inverse path, 237
inversion formula
— for the spherical Fourier transform,
477
Fourier — for an Abelian group, 53
invertible element in a commutative ring, 168
involutive
— algebra, 362
— anti-automorphism of a group, 467
— anti-automorphism of an algebra, 467
irreducible
— element in an integral domain, 169
— polynomial, 169
— representation, 344
isomorphism
— of directed graphs, 237
— of graphs, 237
isoperimetric
— constant, 284
Alon-Milman — inequality, 287
Alon-Schwartz-Shapira — inequality, 320
Dodziuk — inequality, 288
Reingold-Vadhan-Wigderson — inequality,
333
isotypic component, 357
— of $L(G)$, 384

Jacobi sum, 217, 219
Jacquet module of a representation, 496
Jordan canonical form, 485
kernel
 — of a convolution operator, 365
 — of a convolution operator on an Abelian
  group, 56
 — of a representation, 344
Kloosterman sum, 210
 generalized —, 203
 orthogonality relations for generalized —s,
  206
Kronecker
 — product, 142
 — sum of linear operators, 253
 factorizations of — products, 151
 similarity of — products by stride
  permutations, 144

lamplighter
 — graph, 268
 — on the complete graph, 270, 271
Laplacian
 combinatorial —, 286
left regular representation, 348
Legendre symbol, 120
 — on $\mathbb{F}_q$, 307
lemma
 Burnside —, 376
 converse to Schur —, 351
 Gauss —, 64
 Mackey —, 419
 Schur —, 350
 Wielandt —, 378
length of a path, 237
lexicographic product of graphs, 260
little group method, 423
loop in a graph, 235

Mackey
 — formula for invariants, 414, 417
 — intertwining number theorem, 417
 — irreducibility criterion, 417
 — lemma, 419
 — tensor product theorem, 421
 — theory, 413
Mackey-Wigner little group method, 421, 423
Margulis graph, 319
matrix
 — form of the FFT, 151
 — factorization of composite bijection
  permutations, 149
 adjacency —, 238
 adjugate —, 36
 block diagonal power of a —, 148
 canonical form of a —, 482

circulant —, 58
companion — of a monic polynomial, 191
core —, 159, 228
diagonal — of twiddle factors, 153
elementary circulant permutation —, 147
Fourier — of $\mathbb{F}_q$, 227
Hankel —, 159
permutation —, 140
skew circulant —, 228
unipotent —, 486
unitary —, 345
maximal ideal, 170
minimal
 — central projection, 395
 — polynomial, 65, 172, 484
modified replacement product, 282
monic polynomial, 168
multiple edge, 235
multiplicative character
 — of $\mathbb{F}_q$, 199
 — of $\mathbb{Z}/m\mathbb{Z}$, 84
 order of a — of $\mathbb{F}_q$, 200
 principal — of $\mathbb{F}_q$, 201
multiplicative linear functional, 472
multiplicity
 — of a representation, 357
multiplicity-free
 — representation, 394
 — triple, 466
 Bump-Ginzburg criterion for a — triple, 468
 spherical function associated with a —
  triple, 469
multiplier operator, 444

neighborhood of a vertex, 236
non-backtraking path, 295
non-oriented square, 338
norm of a field extension, 187

operator
 (monomial) differential —, 68
 adjacency —, 238
 adjoint —, 345
 convolution —, 365
 convolution — on $L(A)$, 56
 convolution kernel — on an Abelian group,
  56
 diagonal —, 361
 invariant —, 56
 multiplier —, 444
 polar decomposition of an —, 346
 translation —, 55, 444
 unitary —, 345
orbit of a point, 372
order
 — of a multiplicative character of $\mathbb{F}_q$, 200
 — of a differential operator, 68

order (*cont.*)
— of a finite cyclic group, 7, 48
— of a finite field, 176
orientation of a graph, 236
orthogonality relations
— for characters, 356
— for characters of $\mathbb{Z}_n$, 49
— for characters of an Abelian group, 51
— for generalized Kloosterman sums, 206
— for matrix coefficients, 354
— for spherical functions, 476, 480
— on $\mathbb{F}_q^*$, 201
— on $\widehat{\mathbb{F}_q}$, 198

Paley graph, 308
parabolic induction, 494
Parseval formula
— for $\mathbb{Z}_n^2$, 312
— for an Abelian group, 54
partial stride permutation, 134
partite set, 245
path
— composition, 237
— in a graph, 237
closed —, 237
initial vertex of a —, 237
inverse —, 237
length of a —, 237
non-backtraking —, 295
terminal vertex of a —, 237
trivial —, 237
permutation
— character, 375
— matrix, 140
— representation, 373
— representation of S_n, 374
composite bijection —, 136
elementary circulant — matrix, 147
elementary congruence —, 135
matrix factorization of composite bijection
    —, 149
partial stride —, 134
product congruence —, 135
shuffle —, 132
stride —, 132
permutational reverse radix identity, 138
Peter-Weyl theorem, 357
Petersen graph, 243
Plancherel formula, 371
— for $\mathbb{Z}_n^2$, 312
— for a finite group, 371
— for an Abelian group, 54
— for the spherical Fourier transform,
    478
Poisson summation formulas, 60
polar decomposition of a linear operator,
    346

polynomial
characteristic — of $\mathcal{F}$, 116
characteristic — of $\mathcal{F}^2$, 103
characteristic — of a matrix, 484
companion matrix of a monic —, 191
degree of a —, 168
irreducible —, 169
minimal —, 65, 172, 484
monic —, 168
primitive —, 64
root of a —, 65
splitting field of a —, 174
Pontrjagin duality, 53
primary
— component of an Abelian group, 22
— decomposition of a finite Abelian group,
    21, 22
primitive
— element of a Galois field, 176
— graph, 242
— polynomial, 64
— root, 35
principal
— Dirichlet character, 85
— additive character of $\mathbb{F}_q$, 198
— ideal, 168
— ideal domain, 168
— multiplicative character of $\mathbb{F}_q$, 201
product
— congruence permutation, 135
Cartesian — of graphs, 258
converging infinite —, 76
diverging infinite —, 76
infinite —, 76
inner —, 345
internal tensor — of representations, 387
Kronecker —, 142
lexicographic — of graphs, 260
outer tensor — of representations, 386
replacement — of graphs, 275
tensor — of functions, 253
tensor — of linear operators, 253
tensor — of subspaces, 253
tensor — of two spaces, 384
wreath — of graphs, 267
zig-zag — of graphs, 277
projection, 390
minimal central —, 395
orthogonal —, 390

quadratic
— extension, 173
— nonresidue, 117
— residue, 117
Gauss law of — reciprocity, 127
Gauss law of — reciprocity (second proof),
    183

Rader
— Winograd algorithm, 158
— algorithm, 159
radix identity
general —, 154
permutational reverse —, 138
reverse —, 149
Ramanujan graph, 307
rational canonical form, 485
regular
2 — segment, 252
— graph, 236
strongly — graph, 241
Reingold-Vadhan-Wigderson theorem, 333
replacement product of graphs, 275
modified —, 282
representation, 343
(matrix) coefficient of a —, 351
adjoint —, 380
character of a —, 355
commutant of a —, 349
conjugate —, 380
cuspidal —, 502
decomposition of a —, 344
decomposition of tensor products of —s of
GL(2, $\mathbb{F}_q$), 540
degree of a —, 344
dimension of a —, 344
direct sum of —s, 344
equivalence of —s, 344
faithful —, 344
induced —, 399
induced — from GL(2, $\mathbb{F}_q$) to GL(2, $\mathbb{F}_{q^m}$),
533
inflation of a —, 421, 495
irreducible —, 344
isotypic component of a —, 357
Jacquet module of a —, 496
kernel of a —, 344
left regular —, 348
multiplicity of a —, 357
multiplicity-free —, 394
permutation —, 373
permutation — of S_n, 374
restriction of a — to a subgroup, 344
restriction of a — to an invariant subspace,
344
right regular —, 348
sign —, 349
spherical —, 474
sub- —, 344
unitary —, 345
restriction
— of a representation to a subgroup, 344
— of a representation to an invariant
subspace, 344
reverse radix identity, 149

Riemann zeta function, 82
elementary asymptotics for the —, 82
Euler product formula for the —, 82
right regular representation, 348
root
— of a polynomial, 65
primitive —, 35
rotation map, 273
Ruritanian map, 138

Schur
— lemma, 350
— theorem on the DFT, 115
converse to — lemma, 351
self-adjoint
— element in a ∗-algebra, 362
— projection, 390
semidirect product
— with an Abelian group, 424
external —, 281
internal —, 280
sequence
strictly multiplicative —, 79
shuffle permutation, 132
sign representation, 349
similarity identity, 158
simple
— tensor, 384
— graph, 235
solvable group, 486
spectral gap of a graph, 292
spectrum of a graph, 238
spherical
— Fourier transform, 477
— function associated with a — triple, 469
— function for GL(2, $\mathbb{F}_q$), 515
— representation, 474
convolution formula for the — Fourier
transform, 478
inversion formula for the — Fourier
transform, 477
orthogonality relations for — functions,
476, 480
Plancherel formula for the — Fourier
transform, 478
splitting field, 174
existence and uniqueness, 174
stabilizer of a point, 372
strictly multiplicative sequence, 79
stride permutation, 132
partial —, 134
strongly regular graph, 241
structure constants of an Hecke algebra, 465
sub-representation, 344
subalgebra, 361
subfield, 171
subgraph, 236

symmetric Gelfand pair, 468
symmetric group, 349

Tao's uncertainty principle for cyclic groups,
    62
tensor
    — form of the eigenidentities, 153
    — product of functions, 253
    — product of graphs, 259
    — product of linear operators, 253
    — product of subspaces, 253
    — product of two spaces, 384
    — product and induced representations, 406
    decomposition of — products of
        representations of GL(2, $\mathbb{F}_q$), 540
    internal — product of representations, 387
    outer — product of representations, 386
    simple —, 384
terminal vertex of an oriented edge, 236
theorem
    Alon-Boppana —, 299
    Alon-Boppana-Serre —, 298
    Alon-Boppana-Serre — (Nilli's proof), 305
    Alon-Milman —, 287
    Alon-Schwartz-Shapira —, 320
    Auslander-Feigh-Winograd —, 230
    Cauchy — for (not necessarily Abelian)
        groups, 21
    Cauchy — for Abelian groups, 20
    Cayley-Hamilton —, 484
    Chebotarëv —, 68
    Chevalley —, 227
    Chinese remainder —, 9, 13
    Dirichlet — $L(1, \chi) \neq 0$, 95
    Dirichlet — on primes in arithmetic
        progressions, 99
    Dodziuk —, 288
    Euler — $\sum_{p\ prime} \frac{1}{p} = +\infty$, 97
    Euler — $\sum_{p\ prime} \frac{1}{p} = +\infty$ (Erdős' proof),
        98
    Fermat little —, 9
    fundamental — of arithmetic, 5
    Gauss — on cyclicity of $\mathcal{U}(\mathbb{Z}/n\mathbb{Z})$, 35
    Gauss totient function —, 8
    Gauss-Schur — on the trace of the DFT, 116
    Green-Tao —, 100
    Hasse-Davenport —, 216
    Hilbert Satz 90, 187, 188
    Hua-Vandiver-Weil — (homogeneous case),
        224
    Hua-Vandiver-Weil — (non-homogeneous
        case), 225
    Mackey intertwining number —, 417
    Mackey tensor product —, 421
    Mackey-Wigner little group method —,
        423

Peter-Weyl —, 357
Reingold-Vadhan-Wigderson —, 333
Schur — on the DFT, 115
Tao's uncertainty principle – for cyclic
    groups, 159
Warning —, 227
trace
    — of a field extension, 187
    — of a linear operator, 353
    Gauss-Schur theorem on the — of the DFT,
        116
    Hasse-Davenport identity, 216
transitive
    — action, 372
    doubly — action, 379
translation operator, 55, 444
triangular graph, 242
trivial path, 237
twiddle
    — free identity, 157
    — identity, 155
    diagonal matrix of — factors, 153

uncertainty principle
    — for Abelian groups, 61
    Tao's — for cyclic groups, 62
undirected graph, 235
unipotent matrices subgroup, 486
unipotent matrix, 486
unique factorization domain (UFD), 169
unit, 55
    — in a commutative ring, 168
    — in an algebra, 362
unital algebra, 55
unitary
    — matrix, 345
    — operator, 345
    — representation, 345

vector
    Bessel —, 515
    invariant —, 344
vertex
    — of a graph, 235, 236
    —neighbor, 236
    adjacent —, 236
    degree of a —, 236
    initial — of a path, 237
    initial — of an oriented edge, 236
    terminal — of a path, 237
    terminal — of an oriented edge, 236

Warning theorem, 227
weight of a vertex of the hypercube, 249
Weil-Berezin map, 447
Whittaker model, 513
Wielandt lemma, 378

Winograd
— method, 157
— similarity, 158
generalized — method, 157

Rader — algorithm, 158
wreath product of graphs, 267

zig-zag product of graphs, 277

Printed in the United States
by Baker & Taylor Publisher Services